W9-AXL-826

Principles of Combustion

Principles of Combustion

Kenneth Kuan-yun Kuo
Distinguished Alumni Professor
Department of Mechanical Engineering
The Pennsylvania State University
University Park, Pennsylvania

A WILEY-INTERSCIENCE PUBLICATION
JOHN WILEY & SONS
New York · Chichester · Brisbane · Toronto · Singapore

Library of Congress Cataloging in Publication Data:

Kuo, Kenneth K.
Principles of combustion.

"A Wiley-Interscience publication."
Includes bibliographies and index.
1. Combustion. I. Title.
QD516.K86 1986 541.3'61 85-22627
ISBN 0-471-09852-3

Printed in the United States of America

10 9 8

To my parents—General and Mrs. Wei-li Kuo—for their love and guidance

To my wife—Olivia Jeon-lin Kuo—for her love, understanding, and encouragement

To my Ph.D. thesis advisor at Princeton University—Professor Martin Summerfield—for inspiring my interest in combustion

Preface

Today's combustion engineers and scientists are often confronted with complex phenomena which depend upon interrelated processes of fluid mechanics, heat and mass transfer, chemical kinetics, thermodynamics, and turbulence. Understanding of the fundamental concepts of these coupled processes will provide engineers and scientists with the technical background and training required to solve various combustion problems. This book is devoted to the fundamentals of chemically reacting flow systems with application to power production, jet and rocket propulsion, fire prevention and safety, pollution control, material processing industries, and so on.

As is evident from the above applications, the subject of combustion is wide-ranging and important. Although many idealized problems of combustion have been attacked and solved in simplified forms in the past, there are still many practical problems that defy exact solution today. Real life is indeed much more complicated than the idealized situations considered in the past. The investigation of real-life problems in the combustion field has really just begun. The techniques for formulating and solving coupled problems certainly will and should receive increasingly greater attention in the years ahead. The need for instructional material which stresses interrelations of heat, mass, and momentum transfer in chemical reacting systems is therefore quite obvious and indeed crucial.

Only a few books have been published on combustion, and of them, many do not include recent major developments. Other books are specialized and often too advanced for a general combustion course. Having taught a graduate level combustion course for the last several years, I am actually aware of the need for a book which gives a comprehensive and modern treatment of combustion. To do justice to the subject, I had to collect material from many sources to form my lecture notes. This book has been developed through a continuous updating and modification of these lecture notes.

Another problem with existing textbooks on combustion is the paucity of examples to demonstrate the application of theory. In this book—especially in the first few chapters, which deal with the fundamental aspects of combustion—many examples have been presented to help readers assimilate the important concepts of combustion.

In keeping with recent practice, metric units have been employed throughout the book.

The level of this book is suitable for either senior or graduate courses in combustion. Full coverage of the material will generally require six semester credits of course work. It is suggested that Chapters 1 through 6 be covered in the first semester and Chapters 7 through 10 in the second.

The emphasis of this book is on the theoretical modeling of combustion, though experimental techniques are described wherever relevant. Particular attention has been paid to theoretical formulations of combustion problems for solution with the aid of digital computers. It should be noted that rapid progress in the combustion field precludes any possibility of including all theoretical models for exhaustive treatment.

The book has been organized into 10 chapters with exercises and/or projects at the end of each chapter. Many chapters are provided with a list of the new symbols introduced in them. A short, perspicuous introduction to combustion modeling is provided at the beginning of this book to give the reader an overview of the need, basic procedures, and application of combustion modeling. Chapters 1, 2, and 3 are essential background material for the study of combustion, dealing with chemical thermodynamics, chemical kinetics, and conservation equations for multicomponent reacting flows. Chapter 4 deals with deflagration and detonation waves in premixed gases as well as the deflagration-to-detonation transition. The next chapter deals with premixed laminar flames. Chapter 6 is a treatment of laminar diffusion flame jets and the combustion of a single liquid droplet. Turbulent flames and their modeling techniques are discussed in Chapter 7. Spray combustion of fuel droplets and burning of solid particles in convective streams are treated in Chapter 8. Chapter 9 deals with various types of chemically reacting boundary-layer flows. The final chapter covers the topic of ignition, which is intimately related to combustion.

Those who are familiar with my research in the field of solid propellant combustion may be surprised to see a limited coverage of it in this book. I must mention that the recent developments in this field were summarized by many specialists in the AIAA Progress Series, Volume 90, entitled *Fundamentals of Solid Propellant Combustion*, edited by Professor Martin Summerfield and me, published in December 1984 by the AIAA.

I would like to give special thanks to my colleague Professor Gerard M. Faeth, not only for his encouragement for writing this book, but also for his valuable suggestions for improving this book. He generously provided much useful information on spray combustion, which forms a major portion of Chapter 8; to the field of spray combustion, he and his coworkers have made many significant contributions. I must also thank Dr. Kevin White and Mr. Leland A. Watermeier of the U.S. Army Ballistic Research Laboratory (BRL) for organizing a combustion course at BRL during my sabbatical leave. Through the teaching of that course, I have obtained valuable feedback from many combustion scientists and engineers. I would also like to thank Dr. Joseph M. Heimerl of BRL for his valuable suggestions and comments. All the graduate students in my combustion course have contributed immensely toward the improvement of my class notes. Through their term papers and project assignments, I have been able to incorporate many up-to-date theoretical models and techniques. The painstaking efforts of Dr. Louis K. Chang in drawing most of the illustrations deserve special thanks. The help of Mrs. Mary Jane Coleman in typing the manuscript is also highly appreciated. Last, but by no means least, I would like to thank my wife, Olivia, for her understanding, love, and patience. I would also like to thank my daughters, Phyllis and Angela, who gave me love and happiness during my undertaking of this time-consuming but enjoyable task.

KENNETH K. KUO

University Park, Pennsylvania
March 1986

Contents

CHAPTER 6. GASEOUS DIFFUSION FLAMES AND COMBUSTION OF A SINGLE LIQUID FUEL DROPLET 347

CHAPTER 7. TURBULENT FLAMES 402

Principles of Combustion

Introduction

A short introduction is given here to provide readers with an overall picture for the theoretical modeling of combustion problems. Another objective of this introduction is to help readers link the material covered in various chapters. The material is presented here by summarizing important points in an easy-to-read format. The introduction includes: (1) general background required for modeling combustion problems, (2) objectives of combustion modeling, (3) applications of combustion modeling, (4) classification of combustion modeling, (5) essential components of a typical theoretical model, (6) governing equations of combustion models, (7) some common assumptions made in combustion models, and (8) general procedures in the development of a theoretical model. This short summary should enable readers to appreciate the effort required in theoretical-model formulation and validation.

BACKGROUND REQUIRED FOR MODELING COMBUSTION PROBLEMS

The science of combustion involves complex interactions between many constituent disciplines, including:

- Thermodynamics
- Chemical kinetics
- Fluid mechanics
- Heat and mass transfer
- Turbulence
- Materials structure and behavior

The theoretical formulation and solution of combustion problems requires:

- ☐ Mathematics
- ☐ Numerical methods

Model validation by comparison with experimental data requires:

- ☐ Design of test apparatus for combustion research
- ☐ Instrumentation and data acquisition
- ☐ Data analysis and correlation

OBJECTIVES OF COMBUSTION MODELING

- ☐ To simulate combustion processes and to develop predictive capability for combustion behavior under various conditions.
- ☐ To help in interpreting and understanding observed combustion phenomena.
- ☐ To substitute for difficult or expensive experiments.
- ☐ To guide the design of combustion experiments.
- ☐ To help establish the influence of individual parameters in combustion processes by conducting parametric studies.

APPLICATIONS OF COMBUSTION MODELING

- ☐ Power production
  - Coal combustion in power stations
  - Liquid fuels for automobiles, aircrafts, ships, and so on
  - Natural gases for gas-turbine engines, and so on
  - Solid and liquid propellants in rocket motors
- ☐ Process industry for production of materials such as steel, glass, ceramics, cement, refined fuels, carbon black
- ☐ Fire prevention and safety
- ☐ Household and industrial heating
- ☐ Combustion effects on the environment
  - Formation of pollutants such as NO_x, SO_x, CO
  - Formation of particulates such as soot, coke
  - Methods for composition and temperature control of exhaust

The classification of combustion problems is based upon the time and spatial dependence, initial reactant mixing condition, flow condition, phases of reactants, sites of reactions, rates of reactions, natural or forced convection, degree of compressibility of the flow, and speed of the combustion wave. The accompanying table summarizes the various classifications of combustion

Classification of combustion problems.

Condition of Combustion	Classification
Time dependence	Steady, unsteady
Spatial dependence	1*D*, 2*D*, 3*D*
Initial reactant mixing condition	Premixed, nonpremixed (diffusion)
Flow condition	Laminar, turbulent
Phases of reactants	Single-phase, multiphase
Sites of reactions	Homogeneous, heterogeneous
Rate of reaction	Equilibrium chemistry (infinite rate), finite rate
Convective condition	Natural convection, forced convection
Compressibility effects	Incompressible, compressible
Speed of combustion waves	Deflagration (subsonic wave), detonation (supersonic wave)

problems. This table is particularly useful to beginners in becoming familiar with the terminology used in chemically reacting flows.

The figure illustrates the essential elements required to form a theoretical model. It also describes the relationship between various components of the model. It can be seen clearly that the governing equations are coupled to all other branches of the model. Depending upon the complexity of the problem, there could also exist direct coupling relationships between various branches of the model.

In order to verify the predictability of a model, it is often necessary to have a set of input data and/or empirical correlations. These required input parameters and information are sometimes difficult to obtain.

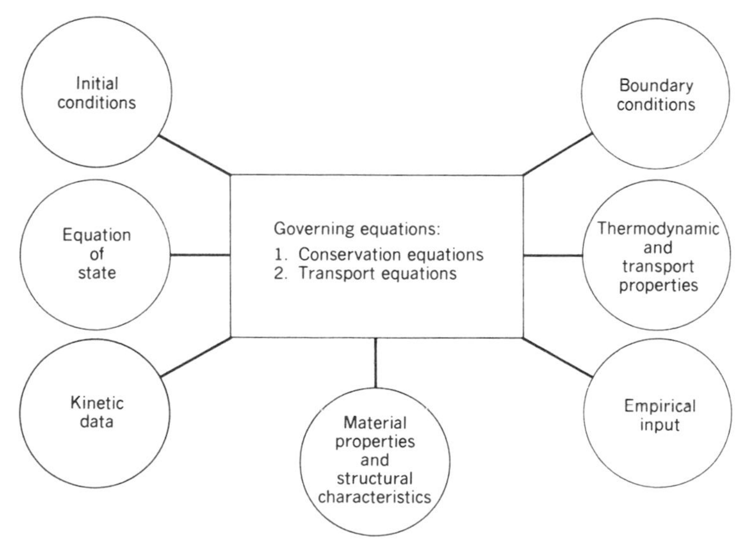

Components of a theoretical model.

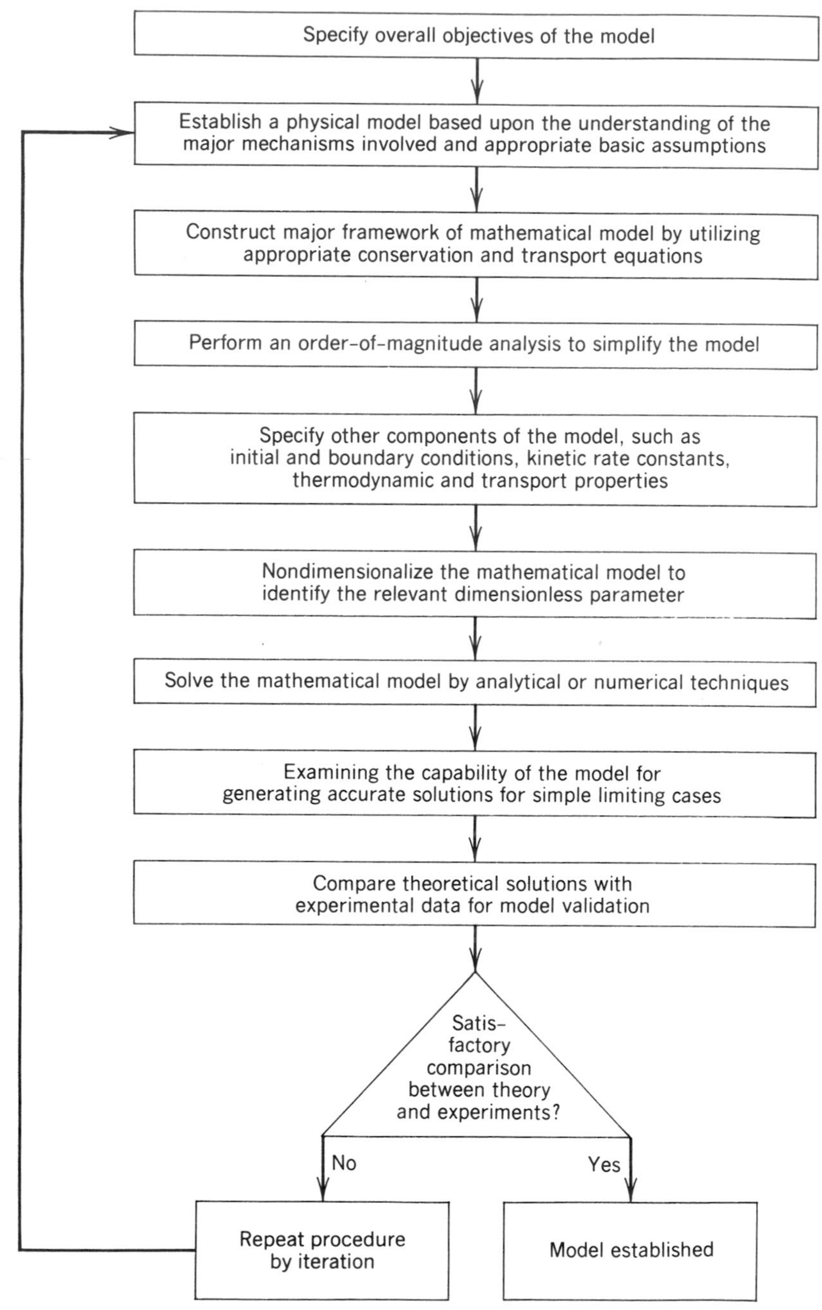

Procedures in the development of a theoretical model.

GOVERNING EQUATIONS FOR COMBUSTION MODELING

Conservation Equations:

- □ Conservation of mass (continuity equation)
- □ Conservation of molecular species (or conservation of atomic species)
- □ Conservation of momentum (for each independent spatial direction)
- □ Conservation of energy
- □ Conservation of angular momentum

Transport Equations:

These equations are usually required for turbulent combustion problems; typical examples are:

- □ Transport of turbulent kinetic energy
- □ Transport of turbulent dissipation rate
- □ Transport of turbulent Reynolds stresses
- □ Transport of probability density function
- □ Transport of moments such as $\overline{u''Y_i''}$, $\overline{T''^2}$, $\overline{Y_i''^2}$, ...

SOME COMMON ASSUMPTIONS MADE IN COMBUSTION MODELS

- □ Reacting fluid can be treated as a continuum
- □ Infinitely fast chemistry
- □ Simple, one-step, forward irreversible reaction
- □ Ideal gas law
- □ Lewis, Schmidt, and Prandtl numbers equal to one
- □ Equal mass diffusivities of all species
- □ Fick's law of diffusion is valid
- □ Constant specific heats of the gas phase
- □ Reacting solid surfaces are energetically homogeneous
- □ Uniform pressure for low-speed combustion situations
- □ Dufour and Soret effects are negligible
- □ Bulk viscosity is negligibly small
- □ Negligible combustion-generated turbulence

It is important to note that with the advancement of computer and numerical techniques, many of these assumptions can now be avoided.

The usual procedures followed in the development of a theoretical model are presented in the accompanying flow chart, which is self-explanatory.

CHAPTER

1 Review of Chemical Thermodynamics

ADDITIONAL SYMBOLS

Symbol	Description	Dimension
a	Affinity	—
C	Molar concentration	N/L^3
C_p	Constant-pressure specific heat	$Q/(MT)$
C_v	Constant-volume specific heat	$Q/(MT)$
e	Specific internal energy	Q/M
e_t	Total stored energy per unit mass	Q/M
E	Stored energy	Q
f	Fugacity or mixture fraction	F/L^2 or —
F/O	Fuel–oxidant ratio	—
G	Gibbs free energy	Q
$\mathcal{G}$	Molar specific Gibbs free energy	Q/N
g	Gravitational acceleration	L/t^2
g_c	Factor of proportionality	ML/Ft^2
H	Enthalpy	Q
$\mathcal{H}$	Molar specific enthalpy	Q/N
K_C	Equilibrium constant based upon species concentration	$(N/L^3)^{\Sigma(\nu''_{i,e}-\nu'_{i,e})}$
K_n	Equilibrium constant based upon moles of species	$N^{\Sigma(\nu''_{i,e}-\nu'_{i,e})}$
K_p	Partial-pressure equilibrium constant	$(F/L^2)^{\Sigma(\nu''_{i,e}-\nu'_{i,e})}$

K_X	Mole-fraction equilibrium constant	—
K_Y	Mass-fraction equilibrium constant	—
KE	Kinetic energy	Q
m	Order of chemical reaction	—
M	Total mass	M
M_i	Chemical symbol for species i	—
m_i	Mass of species i	M
n_i	Number of moles of species i	N
p	Total system pressure	F/L^2
p_i	Partial pressure of species i	F/L^2
PE	Potential energy	Q
$\hat{Q}$	Heat	Q
R_u	Universal gas constant	$Q/(NT)$
S	Entropy	Q/T
s	Molar specific entropy	$Q/(NT)$
T	Temperature	T
t	Time	t
u_i	Velocity in the x_i-direction	L/t
U	Internal energy	Q
V	Volume	L^3
$\mathbf{v}$	Resultant velocity	L/t
W_i	Molecular weight of species i	M/N
$\hat{W}$	Work	Q
x_i	Space coordinate in the ith direction	L
X	Mole fraction	—
Y	Mass fraction	—
z	Elevation	L
Z	Compressibility factor, $pv = ZRT$	—
$\Delta\mathscr{H}^\circ_{f,i}$	Standard heat of formation of species i	Q/N
ΔH_r	Heat of reaction	Q
$\Delta\mathscr{H}_v$	Heat of evaporization	Q/N
Γ	Proportionality constant f_i/p_i	—
ε	Reaction progress variable	—
$\gamma^\ddagger$	Parameter defined in Eq. (1-125)	—
μ	Chemical potential	Q/N
ν'_i	Stoichiometric coefficient for species i appearing as a reactant	N
ν''_i	Stoichiometric coefficient for species i appearing as a product	N

ρ	Density	M/L^3
ϕ	Equivalence ratio	—

Superscript

$\circ$	Standard state condition

Subscripts

g	Gas phase
i	Index for ith species; initial condition
l	Liquid phase
p	Constant-pressure process
s	Solid phase
v	Constant-volume process

Dimension Symbols

F	Force
L	Length
M	Mass
N	Mole
Q	Heat
t	Time
T	Temperature

In this chapter, we briefly introduce and discuss some of the fundamental concepts of chemical thermodynamics, and their application in solving those types of combustion problems in which only the initial and final thermodynamic states of a system are concerned. The application of a well-established chemical-equilibrium calculation program (CEC76),[1,2] developed by the NASA Lewis Center to solve for the final equilibrium product composition and adiabatic flame temperature of a chemical system, is also discussed. Although many examples are given in this chapter to illustrate the use of chemical thermodynamic principles, the scope is too limited to provide an extensive discussion of every aspect of chemical thermodynamics. The reader is referred to the bibliography at the end of this chapter[3–10,25] for more complete works on chemical thermodynamics.

To set the scope of the chemical thermodynamics, let us first make a distinction between heat-transfer theory and thermodynamics. While heat-transfer theory deals with energy transfer processes within a system or between system and surrounding due to a temperature gradient, thermodynamics deals with systems in equilibrium and therefore is applicable to phenomena involving flow and irreversible chemical reactions only when departures from equilibrium are small.

It is also important to note that the approach used in chemical thermodynamics is not a microscopic approach. Chemical thermodynamics treats matter in bulk, that is, with no regard for its detailed molecular structure. Unlike kinetic theory, which can provide a certain amount of information about the rates of chemical processes, chemical thermodynamics can only give the final, equilibrium conditions.

It is useful to define "thermodynamic equilibrium" by considering the following three distinct kinds of equilibrium:

- *Mechanical equilibrium*—exists when there are no unbalanced forces in the interior of a system or between a system and its surroundings.
- *Thermal equilibrium*—exists when all parts of a system are at the same temperature, which is the same as that of the surroundings.
- *Chemical equilibrium*—exists when a system has no tendency to undergo a spontaneous change in chemical composition, no matter how slow.

When all three kinds of equilibrium are satisfied, the system is said to be in a state of *thermodynamic equilibrium*. In that case the analysis becomes simpler, since the state variables do not change with respect to time. Then the state of complete equilibrium can be described in terms of macroscopic coordinates. It is found that the thermodynamic coordinates which are independent and appropriate for combustion studies are the pressure p, the volume V, and the total number of moles of a chemical species in a given phase n_i.

The approach to be taken here will be that of irreversible (nonequilibrium) thermodynamics rather than classical. It is useful to understand the major difference between the two and the reason for using the nonequilibrium approach in combustion studies. First of all, classical thermodynamics can make predictions only about states of thermodynamic equilibrium; it can tell us nothing about the rates at which processes take place. When a process is followed by means of classical thermodynamics, it must be regarded as consisting of a succession of states of thermodynamic equilibrium, that is, as taking place infinitely slowly. Such processes are necessarily reversible. A *reversible* process is one which performed in such a way that the system and its surroundings can both return to their initial states. Such a process must be carried out very slowly so that the system remains in equilibrium throughout. In the real world we must deal with irreversible processes that are a succession of nonequilibrium states. To do this in any detail, the notions of classical thermodynamics must be supplemented.

For systems in mechanical or thermal nonequilibrium, the usual procedure is to divide the system into a large number of subsystems that are of infinitesimal size relative to the original system, but still of macroscopic size relative to the molecular structure of the medium. By assuming that each subsystem is in local equilibrium internally (but not necessarily in equilibrium with its surrounding subsystems), we can apply equilibrium thermodynamics and the concept of state variables to the subsystems. We can thus construct, by integration, a picture of the behavior of the entire nonequilibrium system. This is normally done with such success in fluid dynamics and heat transfer that it is scarcely given a second thought.

When we come to the description of a system in chemical nonequilibrium, the procedure is somewhat different. We shall assume that the system is in mechanical and thermal equilibrium and that it is homogeneous in space. We assume the system to have a definite volume V; since it is in mechanical and thermal equilibrium, we can assign to it a definite pressure p. Furthermore, since the system is homogeneous, its composition can be specified by giving the number of moles,† n_i, of each of the constituent chemical species, M_i. Thus, the thermodynamic state of a system of N chemical species can be completely stated by specifying the values of $p, V, n_1, n_2, \ldots, n_N$; the pressure, p, is an intensive property, while V and n_i $(i = 1, 2, \ldots, N)$ are extensive properties. Intensive and extensive properties in the conventional manner are defined as follows:

An *intensive* property is one that is unchanged when the size of the system is increased by adding to it any number of systems that are identical to the original system. Some intensive properties are density, pressure, temperature, specific internal energy, specific entropy, chemical potential, and so on.

An *extensive* property is one that increases in proportion to the size of the system in such a process. Some extensive properties are volume, mass, total stored energy, total enthalpy, Gibbs free energy, and so on.

An intensive property can be formed by dividing an extensive property by another extensive property.

1 STATEMENT OF THERMODYNAMIC LAWS

When discussing the laws of thermodynamics, it is useful to classify systems according to the exchanges of energy (heat and work) and mass that can take place across the system boundary:

a. *Isolated* systems exchange neither energy nor mass with their surroundings.
b. *Closed* systems exchange energy but no mass with their surroundings.
c. *Open* systems exchange both energy and mass with their surroundings.

† The unit g-mole is commonly used for n_i; one g-mole designates the quantity of substance M_i whose mass in grams is equal to the molecular weight of the substance.

A. The zeroth law of thermodynamics states that there exists an intensive variable, the temperature

$$T = T(p, V, n_i), \tag{1-1}$$

the value of which is the same for all systems in equilibrium with each other. In other words, when two bodies have the same temperature as a third body, they have the same temperature as each other, and so will be in equilibrium if placed in thermal contact. This seems very obvious to us because we are so familiar with the experimental result. Though formulated after the other laws, it precedes them logically, and so has been called the zeroth law. This law suggests the need for a standard scale for temperature measurements. Equation (1-1) is called the equation of state.

B. The first law of thermodynamics (conservation of energy) states that there exists an extensive function called the stored energy E, composed of the following:

1. Internal energy U
2. Kinetic energy KE
3. Potential energy PE

Thus, we may write

$$E = U + \mathrm{KE} + \mathrm{PE} \tag{1-2}$$

where

$$E = E(p, V, n_i) \tag{1-3}$$

The function E has the property that, for a closed system (no mass exchange with its surroundings), the heat added to the system in an infinitesimal process is

$$\delta\hat{Q} = dE + \delta\hat{W} \tag{1-4}$$

where the caret over a variable indicates that the variable is not a thermodynamic property, and δ indicates an inexact differential, since $\hat{Q}$ and $\hat{W}$ are path-dependent functions. Also note that $\delta\hat{W}$ is the work done by the system, which, for a system in chemical equilibrium, equals $p\,dV$. The work terms in the energy conservation equation in a flow system are derived in Chapter 3.

C. The second law of thermodynamics states that there exists an absolute scale for the temperature and an extensive function called the entropy

$$S = S(p, V, n_i) \tag{1-5}$$

Thus, for an infinitesimal process in a closed system

$$T\,dS \geq \delta\hat{Q} \tag{1-6}$$

where the equality is valid for reversible processes and the inequality valid for natural (irreversible) processes.

D. The third law of thermodynamics, according to W. H. Nernst and M. Planck, states that the entropy of a perfect crystal is zero at the absolute zero of temperature. This is used as the base or the reference value for evaluating entropies of various substances. As stated clearly by Van Wylen and Sonntag,[3] from a statistical point of view this means that the crystal structure is such that it has the maximum degree of order. It also follows that a substance that does not have a perfect crystalline structure at absolute zero, but instead has a degree of randomness (such as a solid solution or a glassy solid), has a finite value of entropy at absolute zero. The experimental evidence on which the third law rests is primarily data on chemical reactions at low temperatures and measurements of the heat capacity at temperatures approaching absolute zero.

2 EQUATION OF STATE

In general, for a closed system of known material at a volume V and temperature T, there will be one set of values of n_i's for which the system is in chemical equilibrium. Then,

$$n_i^* = n_i^*(V, T) \tag{1-7}$$

where the values n_i^* are the equilibrium values. The equation of state for a system in equilibrium becomes

$$p = p(V, T, n_1^*, n_2^*, \ldots, n_N^*) \tag{1-8}$$

From Dalton's law of partial pressures, we know that for a mixture of thermally perfect gases in thermodynamic equilibrium,

$$p = \frac{1}{V} \sum_{i=1}^{N} n_i^* R_u T \tag{1-9}$$

The pressure of the system in chemical nonequilibrium can be represented by simply removing the asterisk:

$$p = \frac{1}{V} \sum_{i=1}^{N} n_i R_u T \tag{1-10}$$

3 CONSERVATION OF MASS

In a closed system, the total mass of the contents cannot change. If chemical nonequilibrium exists, however, the amounts of the individual species will vary.

A single arbitrary chemical reaction may be written in the form

$$\sum_{i=1}^{N} \nu_i' M_i \to \sum_{i=1}^{N} \nu_i'' M_i \tag{1-11}$$

where ν_i' is the stoichiometric coefficient for species i appearing as a reactant, ν_i'' is the coefficient for species i appearing as a product, and M_i represents the chemical symbol for species i. Species which are not reactants have $\nu_i' = 0$, while those that do not appear as products have $\nu_i'' = 0$. It is purely a matter of choice which are taken as reactants and which as products, so long as we are consistent once the choice has been made.

Equation (1-11) implies that when $\nu_i'' - \nu_i'$ moles of M_i are formed, $\nu_j' - \nu_j''$ moles of M_j disappear due to the chemical reaction (note that $j \neq i$). This equation indicates a relationship between the change in the number of moles of each species.

Example 1.1

$$CO + \tfrac{1}{2}O_2 \to CO_2$$

Let

$$M_1 = CO$$

$$M_2 = O_2$$

$$M_3 = CO_2$$

Then

$$\nu_1' = 1, \qquad \nu_1'' = 0$$

$$\nu_2' = \tfrac{1}{2}, \qquad \nu_2'' = 0$$

$$\nu_3' = 0, \qquad \nu_3'' = 1$$

When

$\nu_3'' - \nu_3' = 1$ moles of CO_2 is formed, $\Delta n_3 = 1$;
$\nu_1' - \nu_1'' = 1$ moles of CO disappear, $\Delta n_1 = -1$;
$\nu_2' - \nu_2'' = \frac{1}{2}$ moles of O_2 disappear, $\Delta n_2 = -\frac{1}{2}$.

Then,

$$\frac{\Delta n_1}{\nu_1'' - \nu_1'} = \frac{\Delta n_2}{\nu_2'' - \nu_2'} = \frac{\Delta n_3}{\nu_3'' - \nu_3'}$$

For an infinitesimal change, it is convenient to introduce a dimensionless single-reaction progress variable ε, so that

$$dn_i = (\nu_i'' - \nu_i')\, d\varepsilon, \qquad i = 1, 2, \ldots, N \tag{1-12}$$

If $n_{i,r}$ denotes the number of moles of the various species at the same initial or reference condition at which ε is zero, the above equation can be integrated to obtain

$$n_i - n_{i,r} = (\nu_i'' - \nu_i')\varepsilon, \qquad i = 1, 2, \ldots, N \tag{1-13}$$

It follows from Eq. (1-13) that for a closed system in which a single reaction occurs, the n_i's in the thermodynamic state relations can be replaced by the quantities $n_{i,r}$ and the degree of reaction ε. For a system in which the composition at some reference condition is known, the thermochemical state of the system can be specified by

$$p = p(V, T, \varepsilon) \tag{1-14}$$

where the quantity ε can itself be regarded as a state variable. Chemical equilibrium for a given V and T will correspond to certain equilibrium values of n_i^*, n_i, and hence to a specific value ε^* of ε.

If m_i is the mass of the ith species and W_i the molecular weight of that species, then from Eq. (1-12)

$$dm_i = (\nu_i'' - \nu_i')W_i\, d\varepsilon, \qquad i = 1, 2, \ldots, N \tag{1-15}$$

Since the total mass of a closed system is constant,

$$M = \sum_{i=1}^{N} m_i = \text{constant} \tag{1-16}$$

we therefore have

$$\sum_{i=1}^{N} dm_i = 0 \tag{1-17}$$

Substituting Eq. (1-15) into Eq. (1-17), we have

$$\sum_{i=1}^{N} \left[(\nu_i'' - \nu_i')W_i\right] d\varepsilon = 0 \tag{1-18}$$

If the degree of the reaction is not zero (i.e., $d\varepsilon \neq 0$), then

$$\sum_{i=1}^{N} (\nu_i'' - \nu_i')W_i = 0 \tag{1-19}$$

which is known as the stoichiometric equation. If Eq. (1-12) is differentiated with respect to time, we obtain

$$\frac{dn_i}{dt} = (\nu_i'' - \nu_i')\frac{d\varepsilon}{dt} \tag{1-20}$$

which is the rate equation.

4 THE FIRST LAW OF THERMODYNAMICS; CONSERVATION OF ENERGY

The first law of thermodynamics states that during any cycle a system undergoes, the cyclic integral of the heat is proportional to the cyclic integral of work, i.e.,

$$\oint \delta\hat{Q} = \oint \delta\hat{W} \tag{1-21}$$

where $\oint \delta\hat{Q}$ is the cyclic integral of the heat transfer or the net heat transfer during the cycle, and $\oint \delta\hat{W}$ is the cyclic integral of the work or the net work during the cycle. The first law also implies the existence of a function of state called the energy E of the system, and relates the change of this function to the flow of energy from the surroundings.

The existence of the thermodynamic property E can be demonstrated as follows: By Eq. (1-21), we can write the cyclic integral of heat and work by taking paths a and b (Fig. 1.1). Then we have

$$\int_{1a}^{2a} \delta\hat{Q} + \int_{2b}^{1b} \delta\hat{Q} = \int_{1a}^{2a} \delta\hat{W} + \int_{2b}^{1b} \delta\hat{W} \tag{1-22a}$$

We now consider another cycle, the system changing from state 1 to state 2 by process a, as before, and returning to state 1 by process c. Then,

$$\int_{1a}^{2a} \delta\hat{Q} + \int_{2c}^{1c} \delta\hat{Q} = \int_{1a}^{2a} \delta\hat{W} + \int_{2c}^{1c} \delta\hat{W} \tag{1-22b}$$

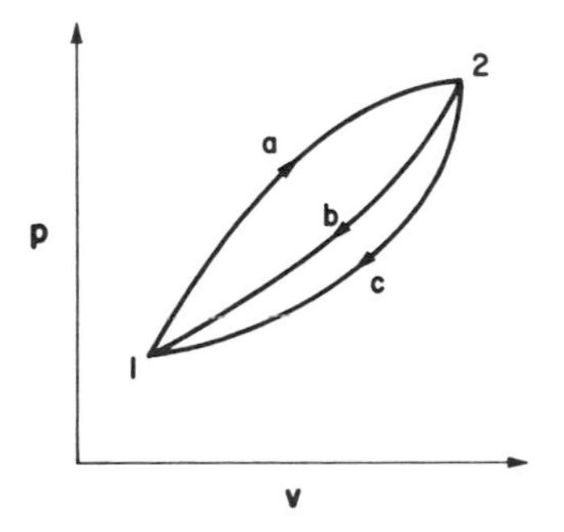

Figure 1.1 Demonstration of the existence of the thermodynamic property E.

After subtracting the second of these equations from the first and rearranging them, we have

$$\int_{2b}^{1b} (\delta\hat{Q} - \delta\hat{W}) = \int_{2c}^{1c} (\delta\hat{Q} - \delta\hat{W}) \tag{1-23}$$

Since b and c represent arbitrary processes between states 1 and 2, $\delta\hat{Q} - \delta\hat{W}$ is the same for all processes between 1 and 2. Therefore $\delta\hat{Q} - \delta\hat{W}$ depends only on the initial and final state and not on the path followed between the two states. We may conclude that $\delta\hat{Q} - \delta\hat{W}$ is a point function, and therefore is the exact differential of the system. This property is the energy of the system and is represented by the symbol E. We have

$$dE = \delta\hat{Q} - \delta\hat{W} \tag{1-24}$$

where the sign preceding $\delta\hat{W}$ is negative because $\delta\hat{W}$ is the work done *by* the system.

The physical significance of the property E is that it represents all the energy of the system in the given state. This energy can be present in many forms, including thermal, kinetic, potential (with respect to the chosen coordinate frame), energy associated with the motion and position of molecules, energy associated with atomic structure, chemical energy (e.g., in a storage battery), electrostatic (e.g., in a charged capacitor), and so on.

In the study of thermodynamics, it is convenient to consider the bulk kinetic and potential energy separately, and then to lump all other forms of energy of the system in a single property, which we shall call *internal energy* and give the symbol U. Therefore,

$$\begin{aligned} E &= \text{internal energy} + \text{kinetic energy} + \text{potential energy} \\ &= U + \text{KE} + \text{PE} \end{aligned}$$

where

$$\text{KE} = \frac{1}{2g_c} m|\mathbf{v}|^2$$

$$\text{PE} = \frac{mgz}{g_c} \tag{1-25}$$

In differential form $dE = dU + d(\text{KE}) + d(\text{PE})$ which from Eqs. (1-24) and (1-25) may be written

$$\delta\hat{Q} = dU + d\left(\frac{1}{2g_c} m|\mathbf{v}|^2\right) + d\left(\frac{mgz}{g_c}\right) + \delta\hat{W} \tag{1-26}$$

Assuming g is a constant, the above equation may be integrated between states 1 and 2 to give

$$_1\hat{Q}_2 = (U_2 - U_1) + \frac{m\left(|\mathbf{v}_2|^2 - |\mathbf{v}_1|^2\right)}{2g_c} + \frac{mg(z_2 - z_1)}{g_c} + {}_1\hat{W}_2 \qquad (1\text{-}27)$$

Here the internal energy U is an extensive property like the kinetic and potential energies, since all depend upon the mass of the system.

The work term includes three forms of work. *Shaft work* $\hat{W}_s$ is work done to produce an effect external to the system, that is, work which can be made to turn a shaft or raise a weight. *Flow work* is work performed to overcome pressure effects at any point on the boundary where mass flow occurs. The flow work rate can be written as

$$p\frac{dV}{dt} = \frac{p}{\rho}\left(\frac{\rho\, dV}{dt}\right) = \frac{p}{\rho}\dot{m} \qquad (1\text{-}28)$$

Viscous work $\hat{W}_\mu$ is work done to overcome fluid friction effects on the boundary where mass flow occurs. The work rate can be written as

$$\frac{\delta\hat{W}}{\delta t} = \frac{\delta\hat{W}_s}{dt} + \frac{\delta\hat{W}_\mu}{dt} + \int\frac{p}{\rho}\, d\dot{m}_{\text{out}} - \int\frac{p}{\rho}\, d\dot{m}_{\text{in}} \qquad (1\text{-}29)$$

For a closed system undergoing an infinitesimal reversible process, the law is usually expressed in classical (i.e., equilibrium) thermodynamics, by the equation

$$dU = \delta\hat{Q} - p\, dV \qquad (1\text{-}30)$$

where we have assumed no viscous or shaft work, and no kinetic or potential energy change. $\delta\hat{Q}$ represents the amount of heat the system receives from its surroundings and $p\,dV$ represents the flow work done by the system. Here, p and U are state functions related to the other variable V and T by state equations of the form

$$p = p(V, T), \qquad U = U(V, T) \qquad (1\text{-}31)$$

The law of conservation of energy can be applied to the study of chemical nonequilibrium. The only necessary adjustment is to redefine the state functions p and U as

$$p = p(V, T, n_1, n_2, \ldots, n_N), \qquad U = U(V, T, n_1, n_2, \ldots, n_N) \quad (1\text{-}32)$$

When the system is in equilibrium, the n_i reduce to $n_i^*(T, V)$ and the state equations (1-32) reduce to the equilibrium relations (1-31). We can thus consider complete thermodynamic equilibrium as a special, limiting case of chemical nonequilibrium.

In nonequilibrium thermodynamics, as in equilibrium thermodynamics, it is useful to define the enthalpy by the equation

$$H = U + pV \tag{1-33}$$

For chemical nonequilibrium, H is also given by a state relation of the form

$$H = H(V, T, n_1, n_2, \ldots, n_N) \tag{1-34}$$

or

$$H = H(V, T, \varepsilon) \tag{1-35}$$

Example 1.2

Consider the following reaction occurring at a constant volume V:

$$H_{2(g)} + \tfrac{1}{2}O_{2(g)} \rightarrow H_2O_{(g)} - 57.5 \text{ kcal} \qquad \text{at } 298.16 \text{ K}$$

For a constant volume process, the first law becomes

$$dU = \delta\hat{Q} \tag{1-36}$$

Since the internal energy is a point function and independent of the path, the energy change equals the heat change for the chemical process and

$$\Delta U = (\Delta\hat{Q})_v = -57.5 \text{ kcal/mole}$$

Consider the same reaction occurring under constant pressure. The first law for this case is

$$dU = \delta\hat{Q} - p\,dV \tag{1-37}$$

which integrates to

$$\Delta U = (\Delta\hat{Q})_p - p(\Delta V)_p \tag{1-38}$$

If the reaction is carried out isothermally at T and a constant pressure p, then per mole of H_2O we have

$$\Delta U = (\Delta\hat{Q})_p - RT\Delta n$$

where $\Delta n = (n_{\text{products}} - \text{n}_{\text{reactants}})_{\text{ideal gas}} = -\frac{1}{2}$; therefore

$$(\Delta\hat{Q})_p = \Delta U - \tfrac{1}{2}RT = -57.798 \text{ kcal}$$

This demonstrates that 0.298 kcal more heat is evolved when the reaction is carried out at a constant pressure than at a constant volume. The additional heat, $RT\Delta n$, is a result of the work of the surroundings on the gas in maintaining it at a constant pressure p during the reaction, while its volume is decreasing. This shows that the heat evolved in a chemical reaction depends upon the physical conditions (path-dependent) under which the reaction occurs.

5 THE SECOND LAW OF THERMODYNAMICS

5.1 Equilibrium Thermodynamics

The second law of thermodynamics postulates the existence of a state function called the entropy S, and defines the basic properties of this function. For a closed system that undergoes a change from one state of thermodynamic equilibrium 1 to another state 2, the change in entropy is given by

$$S_2 - S_1 = \int_1^2 \left(\frac{\delta \hat{Q}}{T} \right)_{\text{rev}} \tag{1-39}$$

where rev implies any reversible path between 1 and 2, $\delta \hat{Q}$ is the heat received from or added to the system, and T is the corresponding absolute temperature. The important point to note here is that since the change in the entropy of a substance is path-independent, it is also the same for all processes, both reversible and irreversible. The equation given here enables us to find the change in entropy only along a reversible path, but once evaluated, the magnitude of change is that for all processes between these two states. We may also note here that T takes the role of an integrating factor in that it has converted the inexact differential $\delta \hat{Q}$ to the exact differential $(\delta \hat{Q}/T)_{\text{rev}}$.

If the same system undergoes an irreversible or real process between the same two equilibrium end states 1 and 2, we have

$$S_2 - S_1 > \int_1^2 \left(\frac{\delta \hat{Q}}{T} \right) \tag{1-40}$$

where $\delta \hat{Q}$ is the heat added to the system in the particular process.

The above result is arrived at by the following considerations. Consider a system which undergoes two cycles between states 1 and 2: one cycle consists of two reversible processes a and b, and the other cycle consists of path a and an irreversible process c. (See Fig. 1.2.) For the reversible cycle,

$$\oint \frac{\delta \hat{Q}}{T} = \int_{1a}^{2a} \frac{\delta \hat{Q}}{T} + \int_{2b}^{1b} \frac{\delta \hat{Q}}{T} = 0 \tag{1-41}$$

For the cycle consisting of the reversible process a and the irreversible process

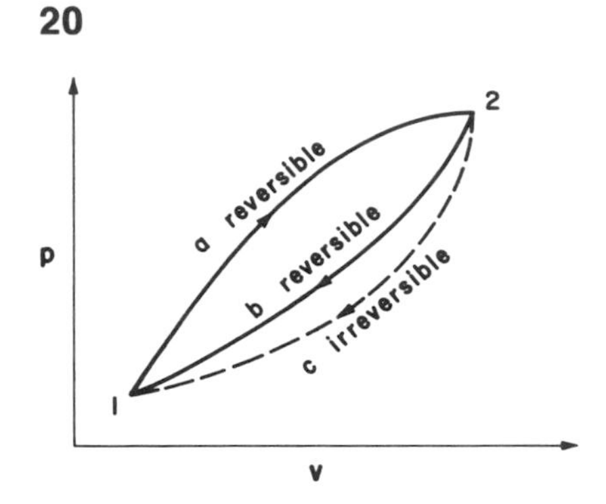

Figure 1.2 Two reversible cycles demonstrating the fact that entropy is a property of a substance.

c, we have

$$\oint \frac{\delta \hat{Q}}{T} = \int_{1a}^{2a} \frac{\delta \hat{Q}}{T} + \int_{2c}^{1c} \frac{\delta \hat{Q}}{T} < 0 \tag{1-42}$$

which is a statement of the inequality† of Clausius.[3] Subtracting the second equation from the first gives

$$\int_{2b}^{1b} \frac{\delta \hat{Q}}{T} > \int_{2c}^{1c} \frac{\delta \hat{Q}}{T} \tag{1-43}$$

Since entropy is a thermodynamic property and b is a reversible process,

$$\int_{2b}^{1b} \frac{\delta \hat{Q}}{T} = \int_{2b}^{1b} dS = \int_{2c}^{1c} dS \tag{1-44}$$

Therefore,

$$\int_{2b}^{1b} dS > \int_{2c}^{1c} \frac{\delta \hat{Q}}{T} \tag{1-45}$$

or, in general,

$$dS \geq \frac{\delta \hat{Q}}{T} \tag{1-46}$$

which when integrated between states 1 and 2 gives the desired result

$$S_2 - S_1 > \int_1^2 \left(\frac{\delta \hat{Q}}{T} \right) \tag{1-47}$$

† The inequality of Clausius is a corollary of the second law of thermodynamics. It has been demonstrated to be valid for all possible cycles.

In both cases considered, the temperature T is the absolute temperature of the reservoir that supplies that heat $\delta\hat{Q}$. Since the temperature of the reservoir and the temperature of the system are equal, in a reversible process, T in Eq. (1-39) is also the temperature of the system. For irreversible processes, however, the temperature of the system is undefined within the context of classical thermodynamics.

5.2 Nonequilibrium Thermodynamics

Given the existence of a variable of state called the entropy S, the change in entropy, dS, in any closed system undergoing any process can be split into two parts:

$$dS = d_eS + d_iS \tag{1-48}$$

where d_eS is the change in entropy resulting from interaction between the system and its surroundings (i.e., heat transfer to or from the system), and d_iS is the change in entropy resulting from a process taking place within the system (e.g., chemical reaction, constant-pressure mixing of separated gases). These differential quantities may also be thought of respectively as the flow of entropy into the system from the surroundings, and the production of entropy by irreversible processes within the system.

The entropy change d_iS is never negative:

$$d_iS = 0 \qquad \text{(reversible process)} \tag{1-49}$$

$$d_iS > 0 \qquad \text{(irreversible process)} \tag{1-50}$$

For a *closed* system undergoing any process, reversible or irreversible, d_eS is given by

$$d_eS = \frac{\delta\hat{Q}}{T} \tag{1-51}$$

For a closed system undergoing an irreversible process, Eq. (1-48) becomes, after integration,

$$S_2 - S_1 = \int_1^2 \left(\frac{\delta\hat{Q}}{T}\right) + \int_1^2 d_iS \tag{1-52}$$

For an irreversible process $d_iS > 0$, and thus Eq. (1-52) includes the inequality of the classical statement. The latter relationship [Eq. (1-52)] is much more useful than the former [Eq. (1-47)] in that it replaces an inequality by an equality, and thus implies an ability to actually calculate d_iS.

For an *isolated* system, $\delta \hat{Q} = 0$ hence $d_e S = 0$ and Eq. (1-48) reduces to

$$dS = d_i S \geq 0 \tag{1-53}$$

This is equivalent to the familiar classical statement that the entropy of an isolated system can never decrease.

If mass is added to an *inert open reversible* system, the entropy associated with the additional mass can be considered to be a part of the entropy flow into the system from the surroundings. We then have

$$dS = d_e S = \frac{\delta \hat{Q}}{T} + \sum_{j=1}^{N} s_j d_e n_j \tag{1-54}$$

where s_j is the entropy per mole of the jth species added, and $d_e n_j$ is the change in number of moles of the jth species flowing into the system. With mass addition, the first law becomes

$$dU = \delta \hat{Q} - p\, dV + \sum_{j=1}^{N} u_j d_e n_j \tag{1-55}$$

where u_j is the internal energy per mole associated with the jth species flowing into the system.

Let us define a new parameter μ_j in which

$$\mu_j \equiv u_j - T s_j \tag{1-56}$$

where μ_j represents the chemical potential which will later be shown to equal the change of free energy per mole. [The more fundamental definition of μ_j is given by Eq. (1-80), to be discussed extensively below.] Dividing Eq. (1-55) by T, we obtain

$$\frac{\delta \hat{Q}}{T} = \frac{dU}{T} + \frac{p\, dV}{T} - \frac{1}{T} \sum_{j=1}^{N} u_j d_e n_j \tag{1-57}$$

Substituting Eqs. (1-57) and (1-56) into Eq. (1-54) gives

$$dS = d_e S = \frac{1}{T} dU + \frac{p}{T} dV - \frac{1}{T} \sum_{j=1}^{N} \mu_j d_e n_j \tag{1-58}$$

for an *inert open reversible* system.

Equation (1-58) is based upon the assumption that all the species involved were chemically inert; hence there exist no irreversible processes resulting from chemical reactions within the system. In case chemical reactions occur, the mole numbers may change both by mass addition from outside the system and by chemical reaction inside. Under these conditions an infinitesimal process involving change in p, T, and the n_j's will, in general, be irreversible, since chemical equilibrium may not exist. When $d_i n_j$ (note that the subscript on the d indicates an internal differential change, while the subscript j on the n indicates the jth component) represents the change in the number of moles of the jth species due to chemical reaction,

$$dn_j = d_i n_j + d_e n_j \tag{1-59}$$

If we consider each species j to occupy a separate subsystem of the total system, each subsystem can be regarded as an inert open system. This means that for each species

$$dS_j = \frac{1}{T}\, dU_j + \frac{p_j}{T}\, dV - \frac{1}{T}\mu_j\, dn_j \tag{1-60}$$

where p_j is the partial pressure due to species j, and U_j is the internal energy associated with species j. Summing the preceding equation over all species (i.e., over all subsystems), and noting that

$$p = \sum_{j=1}^{N} p_j \tag{1-61}$$

$$U = \sum_{j=1}^{N} U_j \tag{1-62}$$

$$S = \sum_{j=1}^{N} S_j \tag{1-63}$$

we obtain for the total system

$$\boxed{dS = \frac{1}{T}\, dU + \frac{p}{T}\, dV - \frac{1}{T}\sum_{j=1}^{N} \mu_j\, dn_j} \tag{1-64}$$

which is the fundamental equation of chemical thermodynamics for *open irreversible chemical processes*. This is identical to Eq. (1-58) with the subscript e dropped from the d. Equation (1-64) is the result obtained by combining the first and second laws to describe an open system with chemical reactions.

From the equation of state, we have

$$S = S(U, V, n_1, n_2, \ldots, n_N) \tag{1-65}$$

or in differential form

$$dS = \left(\frac{\partial S}{\partial U}\right)_{V, n_j} dU + \left(\frac{\partial S}{\partial V}\right)_{U, n_j} dV + \sum_{j=1}^{N} \left(\frac{\partial S}{\partial n_j}\right)_{U, V, n'_j} dn_j \quad (1\text{-}66)$$

where the symbol n'_j in the last term indicates that all mole numbers except n_j itself are held constant when the derivative is taken. By comparing Eqs. (1-64) and (1-66) we have

$$\left(\frac{\partial S}{\partial U}\right)_{V, n_j} = \frac{1}{T} \quad (1\text{-}67)$$

$$\left(\frac{\partial S}{\partial V}\right)_{U, n_j} = \frac{p}{T} \quad (1\text{-}68)$$

$$-T\left(\frac{\partial S}{\partial n_j}\right)_{U, V, n'_j} = \mu_j \quad (1\text{-}69)$$

Since

$$H = U + pV \quad (1\text{-}70)$$

we have

$$dU = dH - p\,dV - V\,dP \quad (1\text{-}71)$$

Substituting Eq. (1-71) into Eq. (1-64), we obtain

$$dS = \frac{1}{T} dH - \frac{V}{T} dp - \frac{1}{T} \sum_{j=1}^{N} \mu_j \, dn_j \quad (1\text{-}72)$$

Similarly we can obtain

$$\mu_j = -T\left(\frac{\partial S}{\partial n_j}\right)_{H, p, n'_j} \quad (1\text{-}73)$$

Combining Eqs. (1-67), (1-68), (1-69) with Eq. (1-73) gives

$$\left(\frac{\partial S}{\partial n_j}\right)_{H, p, n'_j} = \left(\frac{\partial S}{\partial n_j}\right)_{U, V, n'_j} \quad (1\text{-}74)$$

In the following, we shall develop the relationship between μ_j and the change of Gibbs free energy. The *Gibbs free energy G*, an extensive property, is defined by the relation

$$G \equiv H - TS = U + pV - TS \tag{1-75}$$

where G, like H, may be considered as a secondary state variable. It should be noted that the entire content of thermodynamics has already been embodied in the properties of p, V, T, U, and S, and that the development of chemical thermodynamics could be carried out with three primary variables alone. It is purely a matter of convenience that we introduce the additional function G.

In differential form

$$dG = dH - T\,dS - S\,dT \tag{1-76}$$

which upon substitution into Eq. (1-72) gives

$$dG = V\,dp - S\,dT + \sum_{j=1}^{N} \mu_j\,dn_j \tag{1-77}$$

We can therefore write

$$V = \left(\frac{\partial G}{\partial p}\right)_{T,\,n_j} \tag{1-78}$$

$$S = -\left(\frac{\partial G}{\partial T}\right)_{p,\,n_j} \tag{1-79}$$

$$\mu_j = \left(\frac{\partial G}{\partial n_j}\right)_{p,\,T,\,n_j'} = -T\left(\frac{\partial S}{\partial n_j}\right)_{H,\,p,\,n_j'} \tag{1-80}$$

where Eq. (1-80) may be taken as the definition of μ_i, which is known as the chemical potential and plays an important role in chemical thermodynamics. The chemical potential, an intensive variable, is in general a function of the state of the system, as given by p, T, and the n_j's. Even though a species is not present in a system, its chemical potential nevertheless need not be zero. There is always the possibility of introducing it into the system, in which case the value of G will be altered, and the value of the corresponding μ_j must therefore be different from zero.

From the first law of thermodynamics we know that

$$dU = \delta\hat{Q} - p\,dV + d_eU \tag{1-81}$$

where d_eU is the internal energy carried in by mass addition. Substituting this into Eq. (1-64), we obtain

$$dS = d_eS + d_iS = \frac{1}{T}\overbrace{(\delta\hat{Q} - p\,dV + d_eU)}^{dU} + \frac{p}{T}\,dV$$

$$\overbrace{-\frac{1}{T}\sum_{j=1}^{N}\mu_j d_e n_j - \frac{1}{T}\sum_{j=1}^{N}\mu_j d_i n_j}^{-\frac{1}{T}\sum_{j=1}^{N}\mu_j\,dn_j}$$

$$= \frac{1}{T}\left(\delta\hat{Q} + d_eU - \sum_{j=1}^{N}\mu_j d_e n_j\right) - \frac{1}{T}\sum_{j=1}^{N}\mu_j d_i n_j \tag{1-82}$$

where the two terms on the right-hand side correspond to the terms d_eS and d_iS, respectively. Therefore,

$$d_iS = -\frac{1}{T}\sum_{j=1}^{N}\mu_j d_i n_j \tag{1-83}$$

which upon substitution of Eq. (1-73) becomes

$$d_iS = \sum_{j=1}^{N}\left(\frac{\partial S}{\partial n_j}\right)_{H,\,p,\,n_j'} d_i n_j \tag{1-84}$$

If we restrict ourselves to a single reaction, the changes $dn_j = d_i n_j$ for the closed system are given in terms of the change $d\varepsilon$ in the degree of reaction advancement, which we now write as

$$d_i n_j = (\nu_j'' - \nu_j')\,d\varepsilon \tag{1-85}$$

Substitution into Eq. (1-83) gives

$$d_iS = -\frac{1}{T}\left(\sum_{j=1}^{N}(\nu_j'' - \nu_j')\mu_j\right)d\varepsilon \tag{1-86}$$

This expression is required by the second law to be either positive, corresponding to an irreversible process, or zero, corresponding to a reversible process. Saying that a process is reversible in the present context is, however, equivalent to saying that the system is in chemical equilibrium at all times, or that the process is one of an infinitely slow succession of states of chemical equilibrium. The condition for chemical equilibrium is therefore that d_iS be zero at all times, which requires

$$\sum_{j=1}^{N} (\nu_j'' - \nu_j')\mu_j^* = 0 \tag{1-87}$$

where μ_j^* is the value of the chemical potential at the equilibrium state. Therefore, this is the equation of reaction equilibrium. It may be regarded as a universally valid formulation of the law of mass action, since it is not restricted to gases. The quantity $-\Sigma_{j=1}^{N}(\nu_j'' - \nu_j')\mu_j$ is known in chemical literature as the *affinity* of the chemical reaction, and is commonly denoted by the symbol a:

$$\text{chemical affinity} = a \equiv -\sum_{j=1}^{N} (\nu_j'' - \nu_j')\mu_j \tag{1-88}$$

The affinity can play a central role in chemical thermodynamics as elaborated by De Donder and his school.[14]

Let us now summarize some important points and physical interpretations about the chemical potential μ_j as follows:

A. μ_j is often referred to as the partial molar Gibbs function $\mu_j = (\partial G/\partial n_j)_{p,T,n_i'}$; it represents the change in Gibbs free energy as an infinitesimal amount of species j is added to the system while holding pressure, temperature, and the amount of other species constant.

B. It is an intensive property and has units of energy/mole.

C. Equation (1-77) can be integrated in a "process" in which the size of the system is increased by adding systems with the same intensive properties, all intensive properties remain constant, and all extensive properties increase proportionally. Hence, $dT = 0$, $dP = 0$, and $d\mu_i = 0$ in such a process, showing that Eq. (1-77) can readily be integrated from $(G = 0,\ n_i = 0)$ to (G, n_i), yielding

$$G = \sum_{i=1}^{N} \mu_i n_i \tag{1-89}$$

since

$$dG = \sum_{i=1}^{N} \mu_i\, dn_i + \sum_{i=1}^{N} n_i\, d\mu_i$$

$$= \sum_{i=1}^{N} d(\mu_i n_i) = d\left[\sum_{i=1}^{N} \mu_i n_i\right] \tag{1-90}$$

Therefore, μ_j may be regarded as the contribution of 1 mole of that constituent to the total value of G of the system. It might be imagined, in view of the above interpretation, that μ_j was equal to the value of G_j for 1 mole of the constituent j in the pure state. However, this is only true in certain limited circumstances.[7] In general, μ_j in a solution is not equal to G_j for the pure substance, and further, the value of μ_j varies as the composition of the system changes.

D. For a closed system at constant temperature and constant pressure, it can be shown from Eqs. (1-77) and (1-90) that

$$\sum_{j=1}^{N} n_j \, d\mu_j = -s\,dT + V\,dp$$

$$= 0 \qquad \text{for constant } T \text{ and } p \tag{1-91}$$

This relation is called the Gibbs–Duhem equation and has many applications, especially in connection with the study of liquid–vapor equilibria.

E. If a system at a given temperature and pressure is in chemical equilibrium, we have

$$\sum \mu_j \, dn_j = 0 \tag{1-92}$$

where the summation includes all the $\mu\, dn$ terms for *all the phases* constituting the system. This relation forms the basis of the well-known "phase rule."

F. When a system consists of a number of components having several phases in complete equilibrium under a given temperature and pressure condition, the chemical potential of each component is the same in all the phases, i.e.,

$$\mu_{j(g)} = \mu_{j(l)} = \mu_{j(s)} \qquad \text{for all } j \tag{1-93}$$

This is shown in Ref. 7 and Example 1.3b.

G. If the phases of a system are not in equilibrium, the chemical potentials of the components will not be the same in each phase. There will then be a tendency for component j to pass spontaneously from a phase in which its chemical potential μ_j is higher to that in which μ_j is lower. In other words, matter tends to move from a region of higher to one of lower chemical potential. This is also why μ_j is given the name of chemical potential. It is a measure of the driving force tending to cause a chemical reaction to take place.

*Example 1.3*a

Consider the following chemical equilibrium reaction to be established in any homogeneous mixture in a closed system:

$$\nu'_B B \rightleftharpoons \nu''_R R + \nu''_L L$$

Show that

$$\nu'_B \mu^*_B = \nu''_R \mu^*_R + \nu''_L \mu^*_L$$

Solution:

According to Eq. (1-92), the equilibrium condition is

$$\mu^*_B \, dn_B + \mu^*_R \, dn_R + \mu^*_L \, dn_L = 0$$

Also the variations in the numbers of molecules are not arbitrary, but are governed by the equations

$$dn_R = \frac{\nu''_R}{\nu'_B}(-dn_B) \quad \text{and} \quad dn_L = \frac{\nu''_L}{\nu'_B}(-dn_B)$$

which express that the numbers of product molecules formed is equal to the number of reactants disappearing, multiplied by appropriate coefficients. Upon substituting these equations into the first one, we find that

$$\nu'_B \mu^*_B = \nu''_R \mu^*_R + \nu''_L \mu^*_L$$

In general, the condition of homogeneous equilibrium is that the sum of chemical potentials of the products equals the sum of the chemical potentials of the reactants:

$$\sum \nu'_i \mu^*_{i(\text{reactants})} = \sum \nu''_j \mu^*_{j(\text{products})}$$

***Example 1.3*b**

Consider the following physical equilibrium to be established between two phases of a chemically identical substance:

$$B(\text{phase } 1) \rightleftharpoons B(\text{phase } 2)$$

Show that $\mu^*_1 = \mu^*_2$.

Solution:

The condition of equilibrium requires that

$$\mu^*_1 \, dn_1 + \mu^*_2 \, dn_2 = 0$$

Since the total mass is constant, we have

$$dn_1 + dn_2 = 0$$

In order to satisfy both equations under any arbitrary amount of variation of dn_1, we must have

$$\mu_1^* = \mu_2^*$$

6 CRITERIA FOR EQUILIBRIUM

Criteria for chemical equilibrium depend upon the condition at which certain thermodynamic properties (or property) are kept constant. For convenience in constant volume processes, we introduce another secondary thermodynamic function, called the *Helmholtz free energy* A, defined by

$$A \equiv U - TS \tag{1-94}$$

Physically, A represents the available useful work other than pressure–volume work, at constant temperature and volume. This can be seen by following Rossini's approach[15,9] in extending the formulation of the first law by writing

$$dU = \delta\hat{Q} - p\,dV + d\hat{\xi} \tag{1-95}$$

where $d\hat{\xi}$ designates the work other than pressure–volume work. For work done on the system $d\hat{\xi}$ is positive. For work done by the system $d\hat{\xi}$ is negative. Combining Eq. (1-95) with the second law of thermodynamics, we have

$$d\hat{\xi} = dU - T\,dS + p\,dV \tag{1-96}$$

and

$$d\hat{\xi} = dA + S\,dT + p\,dV \tag{1-97}$$

It is quite obvious from Eq. (1-97) that

$$d\hat{\xi} = dA \qquad \text{at constant } T \text{ and } V \tag{1-98}$$

Hence, dA represents the available useful work, other than pressure–volume work, at constant T and V. In particular, if dA is negative, then $d\hat{\xi}$ is negative and the system under consideration will do useful work. On the other hand, if $d\hat{\xi}$ is positive for any given process, then work must be done on the system. Finally, if $d\hat{\xi} = 0$, no useful work is done on the system or by the system and the system is at equilibrium.

By the same procedure described above, one can easily obtain the following equation written in terms of Gibbs free energy G:

$$d\hat{\xi} = dG + S\,dT - V\,dp \tag{1-99}$$

Hence

$$d\hat{\xi} = dG \qquad \text{at constant } T \text{ and } p \tag{1-100}$$

TABLE 1.1 General Equilibrium Criteria for Closed Thermodynamic System[15]

Variables Held Constant	Criteria for Thermodynamic Equilibrium in a Closed System
p	$dH - T\,dS = dG + S\,dT = 0$
V	$dU - T\,dS = dA + S\,dT = 0$
T	$d(U - TS) + p\,dV = dG - V\,dp = dA + p\,dV = 0$
S	$dU + p\,dV = dH - V\,dp = 0$
p, T	$dG = 0$
V, T	$dA = 0$
p, S	$dH = 0$
V, S	$dU = 0$
S, U or A, T	$dV = 0$
A, V or G, p	$dT = 0$
U, V or H, p	$dS = 0$
G, T or H, S	$dp = 0$

The general equilibrium criterion $d\hat{\xi} = 0$ leads to the results summarized in Table 1.1.

For open systems, Eq. (1-64) can be rewritten to show the change of A for an open irreversible chemical reaction process as

$$dA = -S\,dT - p\,dV + \sum_{j=1}^{N} \mu_j\,dn_j \tag{1-101}$$

Equations (1-64) and (1-72) can be rearranged to give

$$dU = T\,dS - p\,dV + \sum_{j=1}^{N} \mu_j\,dn_j \tag{1-102}$$

$$dH = T\,dS + V\,dp + \sum_{j=1}^{N} \mu_j\,dn_j \tag{1-103}$$

These equations together with Eq. (1-77) are very useful in chemical-equilibrium studies as well as in thermochemical calculations.

7 CONSERVATION OF ATOMIC SPECIES

For a generalized single reaction,

$$\sum_{i=1}^{N} \nu_i' M_i \rightarrow \sum_{i=1}^{N} \nu_i'' M_i \tag{1-104}$$

Once the values of ν_i' are assigned, there is a constraint on the values of ν_i''; they cannot be arbitrary. This constraint is due to the law of conservation of atomic species.

For each atomic species A, we define a as the number of that particular atomic species present in a particular molecular species M_i. We can then write its formula in general as

$$M_i = \left(A_{1\,a_1} A_{2\,a_2} A_{3\,a_3} \cdots\right)_i \tag{1-105}$$

Thus, for example,

$$M_{H_2O} = (H_2O_1)$$

$$M_{C_3H_8} = (C_3H_8) \qquad \text{(propane)}$$

$$M_{NH_4ClO_4} = (N_1H_4Cl_1O_4) \qquad \text{(ammonium perchlorate)}$$

The generalized single reaction can then be written

$$\sum_{i=1}^{N} \nu_i'\left(A_{1\,a_1} A_{2\,a_2} A_{3\,a_3} \cdots\right)_i = \sum_{i=1}^{N} \nu_i''\left(A_{1\,a_1} A_{2\,a_2} A_{3\,a_3} \cdots\right)_i \tag{1-106}$$

The total number of any given atomic species present in the reaction is conserved. Defining $[A_1]$ as the total number of atomic species A_1 present in the reaction,

$$[A_1] = \sum_{i=1}^{N} (\nu_i' a_{1i}) = \sum_{i=1}^{N} (\nu_i'' a_{1i}) = \text{constant} \tag{1-107}$$

It is important to note here that if, for example, we have five atomic species in a combustion system, five equations like Eq. (1-107) can be constructed in order to solve for the numbers of moles of products. There may, however, be more than five different molecular species which can be constructed from five different atomic species.

Example 1.4

Consider the reaction

$$2H_2 + O_2 \rightarrow 2H_2O$$

or, in a more helpful form,

$$2H_2 + 1O_2 + 0H_2O \rightarrow 0H_2 + 0O_2 + 2H_2O$$

Then the total number of hydrogen atoms is

$$[\mathrm{H}] = (2 \times 2) + (1 \times 0) + (0 \times 2)$$
$$= (0 \times 2) + (0 \times 0) + (2 \times 2) = 4$$

Similarly, for the oxygen atoms,

$$[\mathrm{O}] = (2 \times 0) + (1 \times 2) + (0 \times 1)$$
$$= (0 \times 0) + (0 \times 2) + (2 \times 1) = 2$$

Given the definition of Avagadro's number as representing the number of molecules per mole (6.02×10^{23}/mole), the molecular weight W is the weight of 6.02×10^{23} molecules and has the units of g/mole. In a gas mixture the number of moles n_i of gas i is equal to the weight of the gas divided by its molecular weight W_i:

$$n_i = \frac{m_i g / g_c}{W_i} \tag{1-108}$$

Then, for the reaction

$$2H_2 + O_2 \rightarrow 2H_2O$$

we have

$$\sum_{i=1}^{N} \left(\frac{m_i g}{g_c}\right)_{\text{reactant}} = \sum_{i=1}^{N} \left(\frac{m_i g}{g_c}\right)_{\text{product}} = n_{H_2} W_{H_2} + n_{O_2} W_{O_2} = n_{H_2O} W_{H_2O}$$
$$= (2 \times 2) + (1 \times 32) = 2 \times 18 = 36$$

In general, the mass balance equation for any generalized single reaction given in the beginning of this section can be written as

$$\sum_{i=1}^{N} \nu_i' W_i = \sum_{i=1}^{N} \nu_i'' W_i \tag{1-109}$$

8 CONVENTION FOR REACTANT-FRACTION SPECIFICATION

The weight of the reactant is generally specified by means of fractions and ratios as follows:

A. *Mass Fraction Y.* The mass fraction of the ith species is defined by

$$Y_i \equiv \frac{m_i}{\sum_{i=1}^{N} m_i} \qquad \text{where obviously} \quad \sum_{i=1}^{N} Y_i = 1 \tag{1-110}$$

for N different species in a given system.

B. *Mole Fraction X*. The mole fraction of the ith species is defined by

$$X_i \equiv \frac{n_i}{\sum_{i=1}^{N} n_i} \qquad \text{where obviously} \quad \sum_{i=1}^{N} X_i = 1 \tag{1-111}$$

for N different species in a given system. It is convenient to use Dalton's law to calculate partial pressures from mole fractions:

$$p_i V = n_i R_u T \tag{1-112}$$

Then,

$$p = \sum_{i=1}^{N} p_i = \frac{R_u T}{V} \sum_{i=1}^{N} n_i \tag{1-113}$$

Combining equations (1-112) and (1-113), we have

$$\frac{p_i}{p} = \frac{n_i}{\sum_{i=1}^{N} n_i} = X_i \tag{1-114}$$

and

$$p_i = X_i p \tag{1-115}$$

C. *Fuel–Oxidant Ratio F/O*:

$$F/O \equiv \frac{\text{mass of fuel}}{\text{mass of oxidant}} \tag{1-116}$$

For the reaction $2H_2 + O_2 \rightarrow$ product

$$F/O = \frac{2 \times 2.016}{1 \times 32} \approx \frac{1}{8}$$

D. *Equivalence Ratio* ϕ. The equivalence ratio may be defined as the ratio of the actual fuel–oxidant ratio to the ratio $(F/O)_{st}$ for a stoichiometric process—that is, one in which all products are in their most stable form. For example, $CH_4 + 2O_2 \rightarrow CO_2 + 2H_2O$ is a stoichiometric process because the products are in their most stable form; however, $CH_4 + 1\frac{1}{2}O_2 \rightarrow CO + 2H_2O$ is not a stoichiometric process, since the product CO is not stable but can still react with O_2 to form CO_2, which is the stable product.

A stoichiometric reaction is defined as a unique reaction in which all the reactants are consumed. It is the most economic reaction. Then the definition of the equivalence ratio is

$$\phi \equiv \frac{(F/O)}{(F/O)_{st}} \tag{1-117}$$

For fuel-lean conditions, we have $0 < \phi < 1$,
for stoichiometric conditions, we have $\phi = 1$,
and for fuel-rich conditions, we have $1 < \phi < \infty$.

E. *Parameter* $\gamma^{\ddagger}$. The parameter

$$\gamma^{\ddagger} \equiv \frac{(F/O)}{(F/O) + (F/O)_{st}} \tag{1-118}$$

serves basically the same purpose as the equivalence ratio ϕ, but unlike ϕ, $\gamma^{\ddagger}$ is bounded on both sides by finite values.

For fuel-lean conditions, we have $0 < \gamma^{\ddagger} < 0.5$,
for stoichiometric conditions, we have $\gamma^{\ddagger} = 0.5$,
and for fuel-rich conditions, we have $0.5 < \gamma^{\ddagger} < 1.0$.

F. *Mixture Fraction* f. Suppose we have a mixture (M) stream flowing at the rate of 1 kg/s which is made of two components: fuel (F) at the rate of f kg/s and air (A) at the rate of $(1 - f)$ kg/s. Then any extensive property ζ of the mixture resulting from this two-stream mixing process (shown in the sketch below) can be written as

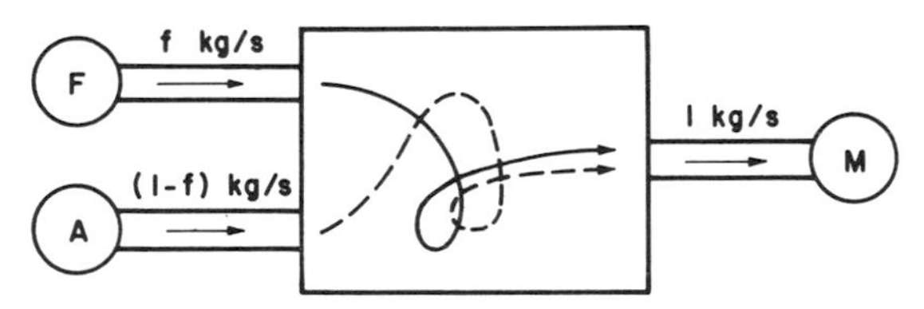

Mixing and combustion chamber for steady fuel and air streams.

$$f\zeta_F + (1 - f)\zeta_A = \zeta_M \tag{1-119}$$

which leads to

$$f = \frac{\zeta_M - \zeta_A}{\zeta_F - \zeta_A} \tag{1-120}$$

Any extensive property of a fluid which is free from sources and sinks and obeys the above relation is called a *conserved property*.[19]

It is easy to prove that the following properties are conserved properties in this sense:

$$Y_{inert}$$

—the mass fraction of a chemically inert mixture component;

$$Y_F - (F/O)_{\rm st} Y_O$$

—a "composite" mass fraction made of Y_F and Y_O;

$$Y_F + \frac{(F/O)_{\rm st}}{1 + (F/O)_{\rm st}} Y_P$$

—another "composite" mass fraction;

$$\left(\frac{F}{O}\right)_{\rm st} Y_O + \frac{(F/O)_{\rm st}}{1 + (F/O)_{\rm st}} Y_P$$

—another "composite" mass fraction,
where Y_P is the mass fraction of combustion product. In terms of mass, the global reaction can be written as

$$\{(F/O)_{\rm st} \text{ kg of } F\} + \{1 \text{ kg of } O\} \to \{[1 + (F/O)_{\rm st}] \text{ kg of } P\}$$

It should be noted that any linear combination of conserved properties,

$$a_0 + a_1\zeta_1 + a_2\zeta_2 + \cdots + a_n\zeta_n$$

is also a conserved property, where $a_0, a_1, \ldots,$ and a_n are constants.

Using the first composite mass fraction for a two-stream mixing process, the fuel and oxidant mass fractions are linked with f as follows:

$$f = \frac{[Y_F - (F/O)_{\rm st} Y_O]_M - [Y_F - (F/O)_{\rm st} Y_O]_A}{[Y_F - (F/O)_{\rm st} Y_O]_F - [Y_F - (F/O)_{\rm st} Y_O]_A} \tag{1-121}$$

If the F stream contains only fuel and the A stream contains oxidizer but no fuel, then we have

$$[Y_F]_A = 0, \qquad [Y_F]_F = 1, \qquad [Y_O]_F = 0$$

$$f = \frac{[Y_F - (F/O)_{\rm st} Y_O]_M + (F/O)_{\rm st} Y_{O,A}}{1 + (F/O)_{\rm st} Y_{O,A}} \tag{1-122}$$

If chemical reaction is complete within the mixing chamber, either fuel or oxidant will have zero concentration in the M state. Therefore, the stoichiometric value of f is

$$f_{st} = \frac{(F/O)_{\text{st}} Y_{O,A}}{1 + (F/O)_{\text{st}} Y_{O,A}} \tag{1-123}$$

and

$$\text{if} \quad f < f_{\text{st}}, \qquad f = \frac{-(F/O)_{\text{st}} Y_{O,M} + (F/O)_{\text{st}} Y_{O,A}}{1 + (F/O)_{\text{st}} Y_{O,A}}$$

$$\text{if} \quad f > f_{\text{st}}, \qquad f = \frac{Y_{F,M} + (F/O)_{\text{st}} Y_{OA}}{1 + (F/O)_{\text{st}} Y_{O,A}} \tag{1-124}$$

The equation (1-124) can be rearranged to give the following values after combustion:

for $f < f_{\text{st}}$:

$$Y_{F,M} = 0$$

$$Y_{O,M} = Y_{O,A} \frac{f_{\text{st}} - f}{f_{\text{st}}} \tag{1-125a}$$

for $f > f_{\text{st}}$:

$$Y_{O,M} = 0$$

$$Y_{F,M} = \frac{f - f_{\text{st}}}{1 - f_{\text{st}}} \tag{1-125b}$$

for any f:

$$Y_{D,M} = Y_{D,A}(1 - f)$$

$$Y_{P,M} = 1 - Y_{D,M} - Y_{O,M} - Y_{F,M} \tag{1-125c}$$

where $Y_{D,M}$ represents the mass fraction of diluent in the M state. The relationship between Y_O, Y_F, Y_P, Y_D, and f on the burned plane can be expressed by the accompanying graph, which consists *entirely of straight lines*. It should be noted that this graph is not the most general form: the F stream could have some diluent, and both streams might be contaminated with product. Also, the mixture fraction should be distinguished quite clearly from the mass fraction of fuel, Y_F, and the fuel/oxidant ratio.

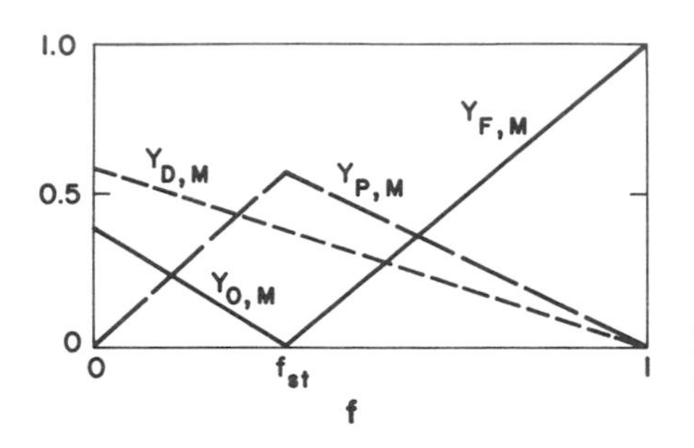

Mass fractions of various species as functions of mixture fraction on the totally burned plane.

9 STANDARD HEATS OF FORMATION

The standard heat of formation of a substance, $\Delta \mathscr{H}_f^\circ$ (kcal/mole), is defined as the heat evolved when one mole of the substance is formed from its elements in their respective standard states at 298.15 K and 1 atmosphere, that is, the standard temperature and pressure, respectively. The standard heat of formation can also be stated as the enthalpy ΔH_f° (kcal) of the substance in its standard state, in reference to its elements in their standard state at the same temperature. The subscript f indicates formation of the compound from elements, and the index ° refers to all products and reactants in their standard states.

The *standard state*, as used in reference to the elements, is the reference state of the aggregate. For gases the reference state is the ideal gaseous state at one atmosphere pressure and at a given temperature. The ideal gaseous state is that in which isolated molecules have no interaction and obey the equation of the state of a perfect gas. For pure liquids and solids the reference state is the real state of the substance at one atmosphere pressure and a given temperature. By convention, each element in its standard state is assigned an enthalpy of zero, or has a standard heat of formation of zero. The standard states of some elements are as follows: $H_{2(g)}$, $O_{2(g)}$, $N_{2(g)}$, $Hg_{(l)}$, $C_{(s,\text{graphite})}$.

Table 1.2 contains the standard heats of formation of various compounds at 298.15 K or 298.16 K. More complete information can be found either from *JANAF Thermochemical Tables*[16] or the *Handbook of Chemistry and Physics*.[17]

Example 1.5

As an example of the standard heat of formation, reference may be made to the reaction

$$C_{(s)} + O_{2(g)} \rightarrow CO_{2(g)} \underbrace{-94.054 \text{ kcal/mole}}_{(\Delta \mathscr{H}_f^\circ)_{CO_2} \text{ at } 298 \text{ K}}$$

When heat is evolved in formation of the compound, then the $\Delta \mathscr{H}_f^\circ$ of that compound is a negative quantity, since heat must be taken away from the system in order to maintain an isothermal reaction process.

TABLE 1.2 **Standard Heats of Formation of Selected Substances at 298.15 K**[a]

Substance	$\Delta\mathcal{H}_f^\circ$ (kcal/mole)	Substance	$\Delta\mathcal{H}_f^\circ$ (kcal/mole)
$B_{(c)}$	0.00	$F_{2(g)}$	0.00
$B_{(g)}$	132.8	$F_2O_{(g)}$	5.86
$B_{2(g)}$	195.0	$HF_{(g)}$	−65.4
$B_2H_{6(g)}$ diborane	9.8	$H_{(g)}$	52.100
$B_5H_{9(l)}$ pentaborane	10.24	$H_{2(g)}$	0.00
$BO_{(g)}$	−5.3	$OH_{(g)}$	9.432
$B_2O_{3(c)}$	−302.0	$H_2O_{(g)}$	−57.798
$BF_{3(g)}$	−271.42	$H_2O_{(l)}$	−68.32
$Br_{(g)}$	26.74	$H_2O_{2(g)}$	−31.83
$Br_{2(g)}$	7.34	$H_2O_{2(l)}$	−44.84
$HBr_{(g)}$	−8.71	$I_{(g)}$	25.633
$C_{(g)}$	170.89	$I_{2(g)}$	14.924
$C_{(c,\text{diamond})}$	0.45	$I_{2(c)}$	0.000
$C_{(c,\text{graphite})}$	0.00	$HI_{(g)}$	6.30
$CO_{(g)}$	−26.42	$Li_{(g)}$	38.41
$CO_{2(g)}$	−94.054	$Li_{(c)}$	0
$CH_{4(g)}$	−17.895	$Li_2O_{(c)}$	−143.1
$C_2H_{6(g)}$	−20.236	$Li_2O_{2(c)}$	−151.7
$C_3H_{8(g)}$	−24.82	$LiH_{(g)}$	30.7
$C_4H_{10(g)}$ *n*-butane	−29.812	$LiH_{(c)}$	−21.61
$C_4H_{10(g)}$ isobutane	−31.452	$LiOH_{(c)}$	−116.45
$C_5H_{12(g)}$ *n*-pentane	−35.0	$LiOH \cdot H_2O_{(c)}$	−188.77
$C_6H_{6(g)}$ benzene	19.82	LiF	−146.3
$C_7H_{8(g)}$ toluene	11.950	$N_{(g)}$	113.0
$C_8H_{10(g)}$ ethylbenzene	7.120	$NH_{3(g)}$	−10.97
$C_8H_{10(g)}$ *o*-xylene	4.540	$N_{2(g)}$	0.00
$C_8H_{10(g)}$ *m*-xylene	4.120	$NO_{(g)}$	21.58
$C_8H_{10(g)}$ *p*-xylene	4.290	$NO_{2(g)}$	7.91
$C_6H_5NH_{2(l)}$ aniline	−4.451	$NO_{3(g)}$	17.0
$CH_2O_{(g)}$ formaldehyde	−27.7	$N_2O_{(g)}$	19.61
$CH_3OH_{(g)}$	−48.08	$N_2O_{3(g)}$	19.80
$CH_3OH_{(l)}$	−57.02	$N_2O_{4(g)}$	2.17
$CF_{4(g)}$	−223.04	$N_2O_{5(g)}$	2.7
$CN_{4(c)}$ cyanogen azide	92.6	$N_2O_{5(c)}$	−10.0
$HCN_{(g)}$	32.3	$N_2H_{4(l)}$	12.10
$C(NO_2)_{4(l)}$	8.8	$N_2H_4 \cdot H_2O_{(l)}$	−57.95
$CH_5N_{(g)}$ methylamine	−6.7	$Na_{(g)}$	25.755
$CH_2N_{2(c)}$ cyanamide	9.2	$Na_{(c)}$	0
$CH_{3(g)}$	34.82	$Na_{2(g)}$	32.87
$CH_5N_{3(c)}$ guanidine	−17.0	$NaO_{2(c)}$	−61.9
$CH_3ON_{(c)}$ formamide	−61.6	$Na_2O_{(c)}$	−99.90
$CH_3O_2N_{(l)}$ nitromethane	−21.28	$Na_2O_{2(c)}$	−122.66
$CH_4ON_{2(c)}$ urea	−79.634	$NaH_{(g)}$	29.88
$CH_5O_4N_{3(c)}$ urea nitrate	−114.8	$NaH_{(c)}$	−13.7
$CH_2O_2N_{4(c)}$	22.14	$NaOH_{(c,\text{II})}$	−101.99

TABLE 1.2 (Continued)

Substance	$\Delta\mathscr{H}_f^\circ$ (kcal/mole)	Substance	$\Delta\mathscr{H}_f^\circ$ (kcal/mole)
$CH_6O_3N_{4(c)}$ guanidine nitrate	91.4	$NaF_{(c)}$	−136.0
$C_2H_{2(g)}$	54.19	$HNO_{3(l)}$	−41.40
$C_2H_{4(g)}$	12.54	$HNO_3 \cdot H_2O_{(l)}$	−112.96
$C_2H_2O_{(g)}$ ketene	−14.6	$NH_2OH_{(c)}$	−25.5
$C_2H_4O_{(g)}$ ethylene oxide	−12.58	$NH_4NO_{3(c)}$	−87.27
$C_2N_{2(g)}$	73.87	$NH_2OH \cdot HNO_{3(c)}$	−86.3
$C_2H_3N_{(g)}$ acetonitrile	21.0	$NF_{3(g)}$	−31.43
$C_2H_3N_{(g)}$ methyl isocyanide	35.9	$NH_4Cl_{(c)}$	−75.18
$C_2H_5O_2N_{(l)}$ nitroethane	−30.0	$NH_4ClO_{4(c)}$	−70.69
$C_2H_7N_{(g)}$ethylamine	−11.6	$O_{(g)}$	59.559
$C_2H_5O_2N_{(g)}$ ethyl nitrite	−24.8	$O_{2(g)}$	0.00
$C_2H_5O_3N_{(l)}$ ethyl nitrate	−44.3	$O_{3(g)}$	34.2
		$P_{(g)}$	79.8
$C_2H_4O_6N_{2(l)}$ glycol dinitrate	58	$P_{(cIII,white)}$	0.00
$Cl_{(g)}$	28.922	$PH_{3(g)}$	5.51
$Cl_{2(g)}$	0	$S_{(g)}$	66.68
$HCl_{(g)}$	−22.063	$SO_{2(g)}$	−70.947
$Cl_2O_{7(g)}$	63.4	$SO_{3(g)}$	−94.59
$ClI_{(g)}$	4.184	$H_2S_{(g)}$	−4.88
$F_{(g)}$	18.86		

[a]Constructed from data in Refs. 13 and 16.

Consider the reaction

$$H_{2(g)} \rightarrow 2H_{(g)} + 104.2 \text{ kcal}$$

which may be more compactly written as

$$H_{2(g)} \rightarrow 2H_{(g)} + 2\,\Delta\mathscr{H}^\circ_{f,H}$$

[note: $(\Delta\mathscr{H}_f^\circ)_{H,298\text{ K}} = 52.1$ kcal/g-mole]. When heat is absorbed in formation of the compound, then the $\Delta\mathscr{H}_f^\circ$ of that compound is a positive quantity, since heat must be added to the system. We may also note here that a substance with a larger positive standard heat of formation $\Delta\mathscr{H}_f^\circ$ tends to be a more chemically active species.

Example 1.6

Express the heat of formation of 1 mole of a given compound at 1 atmosphere pressure and an arbitrary temperature T, in terms of the heats of formation of the compound at the standard state and the enthalpy changes of all the constituents involved in the chemical reaction to form such a compound.

Solution:

The accompanying figure shows that path C can be replaced by the sum of paths A and B:

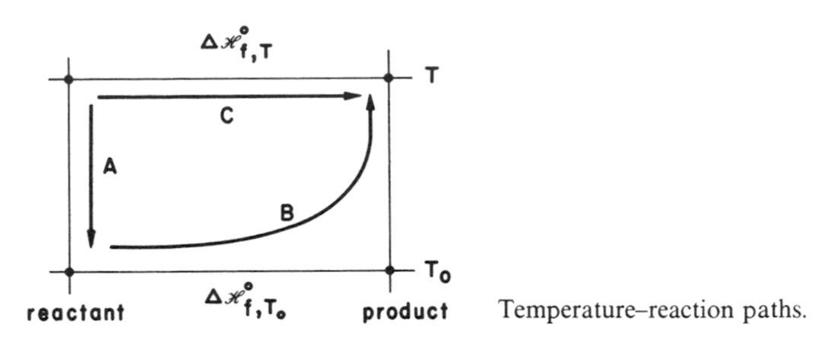

Temperature–reaction paths.

$$C = B + A$$

or

$$\Delta \mathcal{H}_{f,T}^{\circ} = 1\left[\left(\mathcal{H}_T^{\circ} - \mathcal{H}_{T_0}^{\circ}\right) + \Delta \mathcal{H}_{f,T_0}^{\circ}\right]_{\text{compound}} - \sum_{j \text{ elements}} \nu_j'\left(\mathcal{H}_T^{\circ} - \mathcal{H}_{T_0}^{\circ}\right)_j$$

where the number 1 stands for 1 mole of compound formed.

10 THERMOCHEMICAL LAWS

A. L. Lavoisier and P. S. Laplace (1780) are credited with the law which states: *The quantity of heat which must be supplied to decompose a compound into its elements is equal to the heat evolved when the compound is formed from its elements.* A more general form of the law states that *the heat change accompanying a chemical reaction in one direction is exactly equal in magnitude, but opposite in sign, to that associated with the same reaction in the reverse direction.*

Example 1.7

We have

$$CH_{4(g)} + 2O_{2(g)} \rightarrow CO_{2(g)} + 2H_2O_{(1)} - 212.8 \text{ kcal}$$

and

$$CO_{2(g)} + 2H_2O_{(1)} \rightarrow CH_{4(g)} + 2O_{2(g)} + 212.8 \text{ kcal}$$

where both reactions occur at 298.16 K.

In 1840 G. H. Hess developed empirically what is known as the law of constant heat summation. This law states that *the resultant heat change, at constant pressure or constant volume, in a given chemical reaction is the same whether it takes place in one or in several stages*. This means that the net heat of reaction depends only on the initial and final states. One result of Hess's law is that since ΔH for a constant-pressure process or ΔU for a constant volume process is path-independant, thermochemical equations can be added and subtracted just like algebraic equations. The following example demonstrates the application of this powerful law.

Example 1.8

The standard heat of formation of carbon dioxide is -94.05 kcal/g-mole. Consider the following method of determining this value:

$$CO_{(g)} + \tfrac{1}{2}O_{2(g)} \overset{298.16\text{ K}}{\rightarrow} CO_{2(g)} - 67.63\text{ kcal/g-mole}$$

where $\Delta\mathscr{H}$ is not the standard heat of formation, because the reactant CO is not an element in its standard state. In the reaction

$$C_{(s)} + \tfrac{1}{2}O_{2(g)} \overset{298.16\text{ K}}{\rightarrow} CO_{(g)} - 26.42\text{ kcal/g-mole}$$

$\Delta\mathscr{H}$ *is* the standard heat of formation of CO, $(\Delta\mathscr{H}_f^\circ)_{CO,298.16\text{ K}}$, because both C and O_2 are elements in their standard states. Summing the above two thermochemical equations gives

$$CO_{(g)} + \tfrac{1}{2}O_{2(g)} + C_{(s)} + \tfrac{1}{2}O_{2(g)} \overset{298.16\text{ K}}{\rightarrow} CO_{2(g)} + CO_{(g)} - 94.05\text{ kcal/g-mole}$$

cancelling molecular species which appear on both sides

$$C_{(s)} + O_{2(g)} \overset{298.16\text{ K}}{\rightarrow} CO_{2(g)} - 94.05\text{ kcal/g-mole}$$

which is the correct thermochemical equation for the formation of carbon dioxide. Thus the heat evolved is the standard heat of formation of CO_2, $(\Delta\mathscr{H}_f^\circ)_{CO_2,298.16\text{ K}}$.

Since Hess's law applies to both constant-pressure ($\Delta\mathscr{H}$) and constant-volume (ΔU) processes, we note that at absolute zero

$$\Delta\mathscr{H}_{f,T}^\circ = \Delta\mathscr{H}_{f,0}^\circ = \Delta U_{f,0}^\circ + (\Delta n) R_u \cancelto{0}{T} = \Delta U_{f,0}^\circ$$

11 BOND ENERGIES AND HEATS OF FORMATION

It is frequently possible by using tables of bond and resonance energies to estimate the standard heats of formation for compounds which have never been synthesized or are unsuitable for burning in a calorimeter. The bond

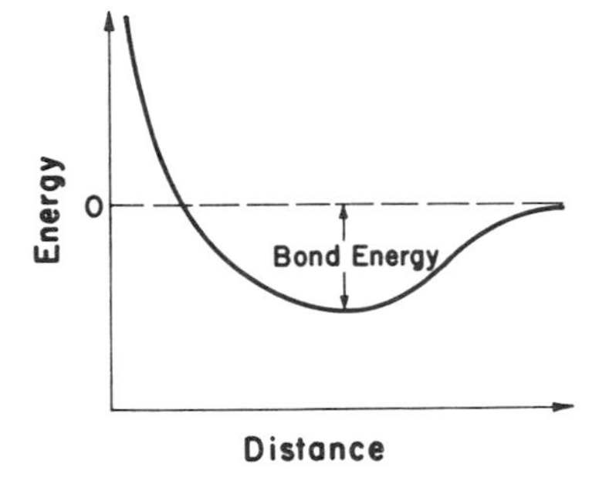

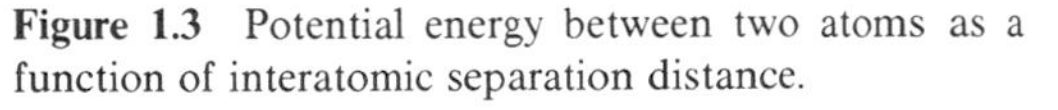
Figure 1.3 Potential energy between two atoms as a function of interatomic separation distance.

energy is defined as the contribution a particular bond makes to the energy required to dissociate the molecule into atoms. For this purpose one may use an average amount of energy, per mole, required to break that kind of bond in a molecule, since the energy needed to break a particular type of bond between two atoms is approximately the same, regardless of the molecule in which the bond occurs.

In general, the energy required to pull two atoms apart is a function of the distance between the two atoms, as shown in Fig. 1.3; the bond energy is the value at the potential minimum, relative to the energy level at infinity. Some bond energies obtained from Refs. 9, 12, and 13 are tabulated in Table 1.3. These values are given at 25°C rather than at 0 K.

In addition to the bond energy, one should take into account the possibility of resonance in a molecule in order to estimate its heat of formation. For example, the benzene molecule C_6H_6 can resonate between the following five structures:[9]

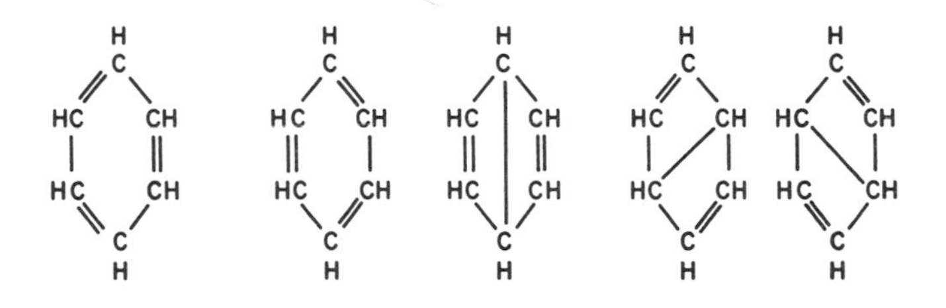

As a result of resonance, $\Delta\mathscr{H}^\circ_{f,C_6H_6}$ is much larger than the sum of three C=C, three C–C, and six C–H bond energies. The additional energy is termed the resonance energy; it must be considered in addition to the bond energies to calculate the actual heat of formation of a given chemical compound. The resonance energy for several compounds is also listed in Table 1.3.

In that table, the bond energies are based on data for substances in the gaseous state and, therefore, should be used for reactions involving gases only. Note also that the bond energy is the negative of the energy required to form the bond.

TABLE 1.3 Some Bond and Resonance Energies

Bond	Energy (kcal/mole)	Bond	Energy (kcal/mole)
H—H	104.18	C—O	86
C—C	85.5	C—Cl	66.5
Cl—Cl	57.87	C—Br	95.6
Br—Br	46.08	C—I	64.2
I—I	36.06	C—F	107
C—H	98.1	N≡N	225.5
N—H	86	C=C	143
O—H	109	C≡C	194.3
H—Cl	103.1	C=O[a]	167
H—Br	87.4	C≡N	210.6
H—I	71.4	O—O	33.1
C—N	174	$(O{=}O)_{O_2}$	118.86

Compound	Resonance Energy (kcal/mole)
Benzene, C_6H_6	48.9
-COOH group	28
CO_2	33
Naphthalene, $C_{10}H_8$	88.0
Aniline, $C_6H_5NH_2$	69.6
Furfural, C_4H_3OCHO	30.1

[a] In formaldehyde.

The following examples serve to demonstrate the use of bond and resonance energies to determine heats of formation of certain compounds, and heats of combustion of certain reactions.

Example 1.9

Determine the heat of formation of ethanol, C_2H_6O.

Solution:

The reference structure of C_2H_6O is

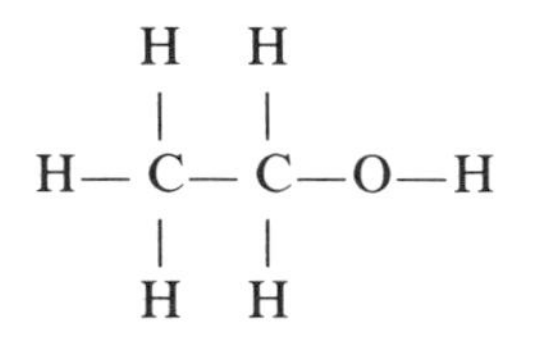

The energy required to form one mole of C_2H_6O is

$$5(C{-}H) + (C{-}C) + (C{-}O) + (O{-}H)$$

$$= 5(-98.1) + (-85.5) + (-86) + (-109)$$

$$= -771 \text{ kcal/mole}$$

Therefore, we may write the thermochemical equation as

$$6H_{(g)} + O_{(g)} + 2C_{(g)} \rightarrow C_2H_6O_{(g)} - 771 \text{ kcal/g-mole}$$

In order to get the heat of formation, we must utilize Hess's law and the thermochemical equations

$$3H_{2(g)} \rightarrow 6H(g) + 312.54 \text{ kcal}$$

$$\tfrac{1}{2}O_{2(g)} \rightarrow O_{(g)} + 59.16 \text{ kcal}$$

$$2C_{(s)} \rightarrow 2C_{(g)} + 343.4 \text{ kcal}$$

to obtain

$$3H_{2(g)} + \tfrac{1}{2}O_{2(g)} + 2C_{(s)}$$

$$\rightarrow C_2H_6O_{(g)} + (-771 + 715.10) \text{ kcal/g-mole of } C_2H_6O$$

Therefore, the estimated value for $\Delta \mathscr{H}^{\circ}_{f,T\ C_2H_6O}$ is -55.90 kcal/g-mole. The actual value of heat of formation is -53.3 kcal/g-mole; the difference is attributed to the fact that the estimated value is based upon average bond energies.

12 HEATS OF REACTION

The science of thermochemistry is concerned with the heat changes associated with chemical reaction; in other words, it deals essentially with the conversion of chemical energy into heat energy, and vice versa.

The heat change associated with a chemical reaction, like that for any other process, is in general an *indefinite* quantity, depending on the path taken. However, if the process is carried out at a constant pressure or constant volume, the heat change has a definite value, determined only by the initial and final states of the system. It is for this reason that heat changes of chemical reactions are measured under constant-pressure or constant-volume conditions.

There are a number of different ways in which a heat of reaction may be defined. One of the most general definitions is stated below.

If a closed system containing a given number of moles n_i of N different species at a given T and p is caused to undergo an isobaric process in which the values n_i are changed to prescribed final values and in which the initial and final values of T are the same, then the heat liberated by the system is the heat of reaction for this process.

When a system changes from one state to another, it may lose or gain energy in the form of heat and work. If, in a change from state A to state B, the energy content of the system is increased by ΔE, the work being done by the system being $\delta\hat{W}$ and the heat absorbed by the system $\delta\hat{Q}$, then by the first law of thermodynamics

$$\Delta E = \delta\hat{Q} - \delta\hat{W} \tag{1-126}$$

For chemical reactions under zero flow velocity and with no change in potential energy,

$$\Delta E = \Delta U + \cancelto{0}{\Delta \mathrm{PE}} + \cancelto{0}{\Delta \mathrm{KE}}$$

and the first law becomes

$$\Delta U = \delta\hat{Q} - \delta\hat{W} \tag{1-127}$$

If, in addition, the reaction occurs at a constant pressure, the first law is written as

$$\hat{Q}_p = \int(\delta\hat{Q})_p = \Delta U + p\,\Delta V \tag{1-128}$$

For a change from state A to state B,

$$\begin{aligned}\hat{Q}_p &= (U_B - U_A) + p(V_B - V_A)\\ &= (U_B + pV_B) - (U_A + pV_A)\\ &= H_B - H_A = \Delta H\end{aligned} \tag{1-129}$$

Therefore, for a constant-pressure, nonflow reaction,

$$\hat{Q}_p = \Delta H \tag{1-130}$$

Similarly, for a constant-volume, nonflow reaction, no external work is done

($\hat{W} = 0$), and the first law is

$$\hat{Q}_v = \Delta U \tag{1-131}$$

For a reacting flow, if there is no change in PE or KE, and no work other than that required for flow, then the net change in enthalpy is equal to the heat of reaction, i.e.,

$$\hat{Q} = \Delta H$$

It is a common practice to deal with moles in flow systems and with mass in closed systems.

The notation for the heat content or enthalpy of a substance at standard state is H_T°, where, as stated previously, the index $^\circ$ specifies standard state and the subscript T gives the temperature in kelvin. Therefore, H_0° represents the enthalpy of a substance in its standard state at 0 K.

The ideal or perfect-gas equation of state is

$$pV = mRT = nR_uT \tag{1-132}$$

which inserted into

$$H = U + pV$$

gives

$$H_T^\circ = U_T^\circ + (pV)^\circ = U_T^\circ + mRT = U_T^\circ + nR_uT \tag{1-133}$$

and at $T = 0$

$$H_0^\circ = U_0^\circ + mR(0) = U_0^\circ \tag{1-134}$$

Subtracting Eq. (1-134) from Eq. (1-133) gives a convenient way to calculate H_T° from U_T°, or vice versa:

$$H_T^\circ - H_0^\circ = (U_T^\circ - U_0^\circ) + mRT \tag{1-135}$$

From Eq. (1-129) we know that for a constant-pressure nonflow reaction,

$$\Delta H = \Delta U + p\,\Delta V \tag{1-136}$$

If V is the volume of one mole of any ideal gas at the constant temperature and pressure, then the change of pV is equal to $p\,\Delta V$. Using Eq. (1-132) for an ideal gas, one has

$$p\,\Delta V = (\Delta n)R_uT \tag{1-137}$$

Substituting Eq. (1-137) into Eq. (1-136), we have

$$\Delta H = \Delta U + (\Delta n)R_uT \tag{1-138}$$

where

$$\Delta n = \left[\left(\sum_{i=1}^{N} n_i\right)_{\text{products}} - \left(\sum_{i=1}^{N} n_i\right)_{\text{reactants}}\right]_{\text{gaseous species}} \tag{1-139}$$

From Eq. (1-135) the value of the heat of reaction at constant pressure can be calculated from the value of the heat of reaction at constant volume.

Consider the generalized reaction

$$\sum_{i=1}^{N} \nu_i' M_i \rightarrow \sum_{i=1}^{N} \nu_i'' M_i$$

The heat of reaction at standard state, $\Delta H_{r,T_0}$, at T_0 is

$$\Delta H_{r,T_0} = \sum_{i=1}^{N} \nu_i'' \Delta \mathscr{H}_{f,M_i}^{\circ} - \sum_{i=1}^{N} \nu_i' \Delta \mathscr{H}_{f,M_i}^{\circ} \tag{1-140}$$

In many cases, we would like to determine the heat of reaction at T_2. If the heat of reaction at T_1 is known, we then consider the two equilibrium states as indicated in Fig. 1.4:

$$(\text{heat change})_{\text{path } A} = (\text{heat change})_{\text{path } B}$$

$$\Delta H_{\text{reactants}} + \Delta H_{r,T_2} = \Delta H_{r,T_1} + \Delta H_{\text{products}}$$

$$\Delta H_{\text{reactants}} = \sum_{i=1}^{N} \nu_i' \int_{T_1}^{T_2} C_{p,M_i}\, dT$$

$$\Delta H_{\text{products}} = \sum_{i=1}^{N} \nu_i'' \int_{T_1}^{T_2} C_{p,M_i}\, dT$$

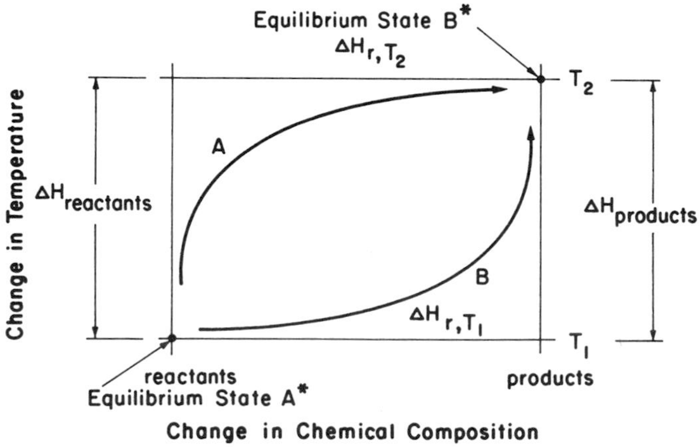

Figure 1.4 Two temperature–reaction paths, showing the relationship between heats of reaction at two different temperatures.

Therefore

$$\Delta H_{r,T_2} = \Delta H_{r,T_1} + \sum_{i=1}^{N} \nu_i'' \int_{T_1}^{T_2} C_{p,M_i}\, dT - \sum^{N} \nu_i' \int_{T_1}^{T_2} C_{p,M_i}\, dT \tag{1-141}$$

or

$$\Delta H_{r,T_2} = \Delta H_{r,T_1} + \left[H_{T_2} - H_{T_1}\right]_{\text{all products}} - \left[H_{T_2} - H_{T_1}\right]_{\text{all reactants}} \tag{1-142}$$

If we take the condition to be 1 atmosphere pressure and $T_1 = T_0 = 298.16$ K, then

$$\Delta H_{r,T} = \Delta H_{r,T_0} + \left[H_T - H_{T_0}\right]_{\text{products}} - \left[H_T - H_{T_0}\right]_{\text{reactants}} \tag{1-143}$$

Substituting in Eqs. (1-140) and (1-141) and changing to the appropriate integration limits gives

$$\Delta H_{r,T} = \left[\sum_{i=1}^{N} \nu_i'' \,\Delta\mathscr{H}^{\circ}_{f,M_i} - \sum_{i=1}^{N} \nu_i' \,\Delta\mathscr{H}^{\circ}_{f,M_i}\right] + \sum_{i=1}^{N} \nu_i'' \int_{T_0}^{T} C_{p,M_i}\, dT - \sum_{i=1}^{N} \nu_i' \int_{T_0}^{T} C_{p,M_i}\, dT \tag{1-144}$$

Rearranging,

$$\Delta H_{r,T} = \sum_{i=1}^{N} \nu_i'' \left(\Delta\mathscr{H}^{\circ}_{f,M_i} + \int_{T_0}^{T} C_{p,M_i}\, dT\right) - \sum_{i=1}^{N} \nu_i' \left(\Delta\mathscr{H}^{\circ}_{f,M_i} + \int_{T_0}^{T} C_{p,M_i}\, dT\right) \tag{1-145}$$

The above procedure may be clarified by considering the hypothetical reaction

$$A_{T_1} + B_{T_1} \rightarrow C_{T_1} + D_{T_1} + \Delta H_{r,T_1}$$

If the reaction takes place at T_1, after which the products are heated to T_2, the total heat evolved is

$$\Delta H = \Delta H_{r,T_1} + \int_{T_1}^{T_2} \left(C_{p,C} + C_{p,D}\right) dT$$

If the final state is reached by first heating the reactants to T_2, and then the reaction is again carried out isothermally, the total heat evolved is

$$\Delta H = \Delta H_{r,T_2} + \int_{T_1}^{T_2} \left(C_{p,A} + C_{p,B}\right) dT$$

The heat evolved by the above reaction processes is equal because of Hess's law, and also because the heat evolved in a constant-pressure process is path-independent. Therefore,

$$\Delta H_{r,T_2} + (C_{p,A} + C_{p,B})\,\Delta T = \Delta H_{r,T_1} + (C_{p,C} + C_{p,D})\,\Delta T \quad (1\text{-}146)$$

or

$$\Delta H_{r,T_2} - \Delta H_{r,T_1} = \int_{T_1}^{T_2} \sum_{i=1}^{N} (\nu_i'' C_{p,i})\, dT - \int_{T_1}^{T_2} \sum_{i=1}^{N} (\nu_i' C_{p,i})\, dT \quad (1\text{-}147)$$

which for $T_1 = 298$ K and $\Delta H_r = \Delta H_{r,T_0}$ gives the equation (1-144). The constant-pressure specific heat C_p is a function of T only, and is usually represented by a power series in T:

$$C_p = \alpha + \beta T + \gamma T^2 + \cdots \quad (1\text{-}148)$$

Values of the molar heat capacity (constant-pressure specific heat) C_p (cal/mole K) are presented in Tables 1.4a through 1.4d for a large number of substances over a temperature range of 298.16 to 5000 K. These values are useful in conjunction with equations which are expressed in terms of C_p. These tables, compiled by Penner,[9] are taken from more extensive sources, which are identified in the footnotes.

Expressions for C_p in terms of temperature are given in polynomial forms for 238 different gases by Andrews and Biblarz.[20] These expressions, based on least-squares polynomial approximation, are very convenient for numerical computations.

Values of the enthalpy differences $(\mathscr{H}_T^\circ - \mathscr{H}_0^\circ)$ have been tabulated for a number of substances, with the same temperature range as that of the previous tables. These values are useful in conjunction with equations which are expressed in terms of $\mathscr{H}_T^\circ - \mathscr{H}_0^\circ$. The values in Tables 1.5a through 1.5f were compiled by Penner;[9] more extensive data may be obtained from the original sources mentioned in the footnotes.

TABLE 1.4a Molar Heat Capacities C_p (cal / mole K)[a]

T (K)	$O_{2(g)}$	$H_{2(g)}$	$N_{2(g)}$	$O_{(g)}$	$H_{(g)}$	$N_{(g)}$	$C_{(c,\text{graphite})}$	$C_{(c,\text{diamond})}$
298.16	7.017	6.892	6.960	5.2364	4.9680	4.9680	2.066	1.449
300	7.019	6.895	6.961	5.2338	4.9680	4.9680	2.083	1.466
400	7.194	6.974	6.991	5.1341	4.9680	4.9680	2.851	2.38
500	7.429	6.993	7.070	5.0802	4.9680	4.9680	3.496	3.14
600	7.670	7.008	7.197	5.0486	4.9680	4.9680	4.03	3.79
700	7.885	7.035	7.351	5.0284	4.9680	4.9680	4.43	4.29
800	8.064	7.078	7.512	5.0150	4.9680	4.9680	4.75	4.66
900	8.212	7.139	7.671	5.0055	4.9680	4.9680	4.98	4.90
1000	8.335	7.217	7.816	4.9988	4.9680	4.9680	5.14	5.03
1100	8.449	7.308	7.947	4.9936	4.9680	4.9680	5.27	5.10
1200	8.530	7.404	8.063	4.9894	4.9680	4.9680	5.42	5.16
1300	8.608	7.505	8.165	4.9864	4.9680	4.9680	5.57	
1400	8.676	7.610	8.253	4.9838	4.9680	4.9680	5.67	
1500	8.739	7.713	8.330	4.9819	4.9680	4.9680	5.76	
1600	8.801	7.814	8.399	4.9805	4.9680	4.9680	5.83	
1700	8.859	7.911	8.459	4.9792	4.9680	4.9681	5.90	
1800	8.917	8.004	8.512	4.9784	4.9680	4.9683	5.95	
1900	8.974	8.092	8.560	4.9778	4.9680	4.9685	6.00	
2000	9.030	8.175	8.602	4.9776	4.9680	4.9690	6.05	
2100	9.085	8.254	8.640	4.9778	4.9680	4.9697	6.10	
2200	9.140	8.328	8.674	4.9784	4.9680	4.9708	6.14	
2300	9.195	8.398	8.705	4.9796	4.9680	4.9724	6.18	
2400	9.249	8.464	8.733	4.9812	4.9680	4.9746	6.22	
2500	9.302	8.526	8.759	4.9834	4.9680	4.9777	6.26	
2750	9.431	8.667	8.815	4.9917	4.9680	4.9900	6.34	
3000	9.552	8.791	8.861	5.0041	4.9680	5.0108	6.42	
3250	9.663	8.899	8.900	5.0207	4.9680	5.0426	6.50	
3500	9.763	8.993	8.934	5.0411	4.9680	5.0866	6.57	
3750	9.853	9.076	8.963	5.0649	4.9680	5.1437	6.64	
4000	9.933	9.151	8.989	5.0914	4.9680	5.2143	6.72	
4250	10.003	9.220	9.013	5.1199	4.9680	5.2977		
4500	10.063	9.282	9.035	5.1495	4.9680	5.3927		
4750	10.115	9.338	9.056	5.1799	4.9680	5.4977		
5000	10.157	9.389	9.076	5.2102	4.9680	5.6109		

[a] By permission, from NBS, *Tables of Selected Values of Chemical Thermodynamic Properties*, Series III, Volume I, March 1947 to June 1949 (after Penner[9]).

TABLE 1.4b Molar Heat Capacities C_p (cal / mole K)[a]

T (K)	$C_{(g)}$	$CO_{(g)}$	$NO_{(g)}$	$OH_{(g)}$	$H_2O_{(g)}$	$CO_{2(g)}$	$CH_{4(g)}$	$C_2H_{2(g)}$
298.16	4.9803	6.965	7.137	7.141	8.025	8.874	8.536	10.499
300	4.9801	6.965	7.134	7.139	8.026	8.894	8.552	10.532
400	4.9747	7.013	7.162	7.074	8.185	9.871	9.736	11.973
500	4.9723	7.120	7.289	7.048	8.415	10.662	11.133	12.967
600	4.9709	7.276	7.468	7.053	8.677	11.311	12.546	13.728
700	4.9701	7.451	7.657	7.087	8.959	11.849	13.88	14.366
800	4.9697	7.624	7.833	7.150	9.254	12.300	15.10	14.933
900	4.9693	7.787	7.990	7.234	9.559	12.678	16.21	15.449
1000	4.9691	7.932	8.126	7.333	9.861	12.995	17.21	15.922
1100	4.9691	8.058	8.243	7.440	10.145	13.26	18.09	16.353
1200	4.9697	8.167	8.342	7.551	10.413	13.49	18.88	16.744
1300	4.9705	8.265	8.426	7.663	10.668	13.68	19.57	17.099
1400	4.9725	8.349	8.498	7.772	10.909	13.85	20.18	17.418
1500	4.9747	8.419	8.560	7.875	11.134	13.99	20.71	17.704
1600	4.9783	8.481	8.614	7.973	11.34	14.1		
1700	7.9835	8.536	8.660	8.066	11.53	14.2		
1800	4.9899	8.585	8.702	8.152	11.71	14.3		
1900	4.9980	8.627	8.738	8.233	11.87	14.4		
2000	5.0075	8.665	8.771	8.308	12.01	14.5		
2100	5.0189	8.699	8.801	8.378	12.14	14.6		
2200	5.0316	8.730	8.828	8.828	12.26	14.6		
2300	5.0455	8.758	8.852	8.504	12.37	14.7		
2400	5.0607	8.784	8.874	8.561	12.47	14.8		
2500	5.0769	8.806	8.895	8.614	12.56	14.8		
2750	5.1208	(8.856)	8.941	8.733	12.8	14.9		
3000	5.1677	8.898	8.981	8.838	12.9	15.0		
3250	5.2150	8.933	9.017	8.931	13.1	15.1		
3500	5.2610	8.963	9.049	9.015	13.2	15.2		
3750	5.3043	8.990	9.079	9.092	13.2	15.3		
4000	5.3442	9.015	9.107	9.162	13.3	15.3		
4250	5.3800	9.038	9.133	9.228	13.4	15.4		
4500	5.4115	9.059	9.158	9.290	13.4	15.5		
4750	5.6375	9.078	9.183	9.350	13.5	15.5		
5000	5.9351	9.096	9.208	9.406	13.5	15.6		

[a] By permission, from NBS, *Tables of Selected Values of Chemical Thermodynamic Properties*, Series III, Volume I, March 1947 to June 1949 (after Penner[9]).

TABLE 1.4c Molar Heat Capacities C_p (cal / mole K)[a]

T (K)	$Cl_{2(g)}$	$Br_{2(g)}$	$I_{2(g)}$	$Cl_{(g)}$	$Br_{(g)}$	$I_{(g)}$	$HCl_{(g)}$	$HBr_{(g)}$	$HI_{(g)}$
298.16	8.11	8.60	8.81	5.2203	4.9680	4.9680	6.96	6.96	6.97
300	8.12	8.60	8.82	5.2237	4.9680	4.9680	6.96	6.96	6.97
400	8.44	8.77	8.90	5.3705	4.9683	4.9680	6.97	6.98	7.01
500	8.62	8.86	8.95	5.4363	4.9708	4.9680	7.00	7.04	7.11
600	8.74	8.91	8.98	5.4448	4.9793	4.9680	7.07	7.14	7.25
700	8.82	8.94	9.00	5.4232	4.9973	4.9680	7.17	7.27	7.42
800	8.88	8.97	9.02	5.3887	5.0258	4.9682	7.29	7.42	7.60
900	8.92	8.99	9.04	5.3506	5.0632	4.9688	7.42	7.58	7.77
1000	8.96	9.01	9.06	5.3133	5.1066	4.9700	7.56	7.72	7.92
1100	8.99	9.03	9.07	5.2788	5.1529	4.9726	7.69	7.86	8.06
1200	9.02	9.04	9.09	5.2477	5.1192	4.9770	7.81	7.99	8.18
1300	9.04	9.05	9.10	5.2201	5.2434	4.9836	7.93	8.10	8.29
1400	9.06	9.07	9.12	5.1958	5.2839	4.9925	8.04	8.20	8.38
1500	9.08	9.08	9.13	5.1745	5.3199	5.0039	8.14	8.30	8.46
1600				5.1557	5.3510	5.0178			
1700				5.1392	5.3771	5.0340			
1800				5.1246	5.3984	5.0521			
1900				5.1117	5.4152	5.0718			
2000				5.1002	5.4279	5.0928			
2100				5.0900	5.4369	5.1147			
2200				5.0809	5.4427	5.1371			
2300				5.0727	5.4458	5.1597			
2400				5.0654	5.4464	5.1822			
2500				5.0588	5.4450	5.2045			
2750				5.0449	5.4347	5.2571			
3000				5.0339	5.4178	5.3039			
3250				5.0251	5.3972	5.3437			
3500				5.0179	5.3748	5.3762			
3750				5.0120	5.3518	5.4016			
4000				5.0070	5.3292	5.4205			
4250				5.0028	5.3074	5.4337			
4500				4.9993	5.2867	5.4419			
4750				4.9964	5.2672	5.4458			
5000				4.9941	5.2490	5.1462			

[a] By permission, from NBS, *Tables of Selected Values of Chemical Thermodynamic Properties*, Series III, Volume I, March 1947 to June 1949 (after Penner[9]).

TABLE 1.4d Molar Heat Capacities C_p (cal / mole K)[a]

T (K)	$F_{2(g)}$	$F_{(g)}$	$HF_{(g)}$
100	6.957	5.068	6.961
200	7.097	5.403	6.959
298.16	7.487	5.436	6.960
300	7.495	5.435	6.960
400	7.895	5.361	6.961
500	8.200	5.282	6.973
600	8.420	5.220	6.987
700	8.581	5.171	7.015
800	8.702	5.134	7.063
900	8.796	5.108	7.129
1000	8.872	5.084	7.210
1100	8.934	5.067	7.304
1200	8.987	5.053	7.401
1300	9.033	5.042	7.503
1400	9.074	5.033	7.604
1500	9.111	5.025	7.703
1600	9.145	5.019	7.798
1700	9.177	5.014	7.886
1800	9.206	5.009	7.974
1900	9.230	5.005	8.054
2000	9.262	5.002	8.129
2100	9.287	4.999	8.199
2200	9.313	4.996	8.264
2300	9.337	4.994	8.326
2400	9.361	4.992	8.383
2500	9.384	4.990	8.436
2600	9.407	4.989	8.486
2700	9.426	4.987	8.532
2800	9.453	4.986	8.576
2900	9.474	4.985	8.617
3000	9.496	4.984	8.656
3200	9.539	4.982	8.727
3400	9.580	4.980	8.790
3600	9.622	4.979	8.847
3800	9.664	4.978	8.899
4000	9.705	4.977	8.946
4200	9.745	4.976	8.990
4400	9.786	4.975	9.028
4600	9.826	4.975	9.066
4800	9.866	4.974	9.102
5000	9.906	4.974	9.135

[a] From L. G. Cole, M. Farber, and G. W. Elverum, Jr., *J. Chem. Phys.*, Vol. 20, p. 586, 1952.

TABLE 1.5a Enthalpy Differences $\mathscr{H}_T^\circ - \mathscr{H}_0^\circ$ (cal / mole) for Various Substances[a]

T (K)	$O_{2(g)}$	$H_{2(g)}$	$N_{2(g)}$	$O_{(g)}$
298.16	2,069.8	2,023.8	2,072.3	1,607.4
300	2,082.7	2,036.5	2,085.1	1,617.0
400	2,792.4	2,731.0	2,782.4	2,134.9
500	3,524.2	3,429.5	3,485.0	2,645.4
600	4,279.2	4,128.6	4,198.0	3,151.7
700	5,057.4	4,831.5	4,925.3	3,655.5
800	5,852.1	5,537.4	5,668.6	4,157.6
900	6,669.6	6,248.0	6,428.0	4,658.7
1000	7,497.1	6,965.8	7,202.5	5,158.8
1100	8,335.2	7,692.0	7,991.5	5,658.4
1200	9,183.9	8,427.5	8,792.8	6,157.6
1300	10,041.0	9,173.2	9,604.7	6,656.4
1400	10,905.1	9,928.7	10,425.4	7,155.0
1500	11,776.4	10,694.2	11,253.6	7,653.3
1600	12,654	11,470.7	12,090.4	8,151.4
1700	13,537	12,257.2	12,933.6	8,649.4
1800	14,425	13,053.2	13,782.4	9,147.2
1900	15,320	13,858.2	14,636.5	9,645.0
2000	16,218	14,671.6	15,494.8	10,142.8
2100	17,123	15,493.0	16,357.1	10,640.5
2200	18,036	16,322.0	17,223.1	11,138.4
2300	18,952	17,158.2	18,092.3	11,636.4
2400	19,874	18,001.2	18,964.3	12,134.4
2500	20,800	18,850.5	19,839.0	12,632.5
2750	23,141	21,001.5	22,036.0	13,879.3
3000	25,515	23,185.8	24,245.4	15,128.7
3250	27,914	25,398	26,465	16,382
3500	30,342	27,635	28,697	17,639.3
3750	32,794	29,893	30,934	18,902.6
4000	35,264	32,172	33,176	20,172.0
4250	37,757	34,468	35,428	21,448.5
4500	40,271	36,780	37,688	22,732.2
4750	42,793	39,107	39,948	24,023.6
5000	45,320	41,449	42,220	25,322.0

[a] By permission, from NBS, *Tables of Selected Values of Chemical Thermodynamic Properties*, Series I, Volume I, March 1947 to June 1949 (after Penner[9]).

TABLE 1.5b Enthalpy Differences $\mathcal{H}_T^\circ - \mathcal{H}_0^\circ$ (cal / mole) for Various Substances[a]

T (K)	$H_{(g)}$	$N_{(g)}$	$C_{(c,\text{graphite})}$
298.16	1,481.2	1,481.2	251.56
300	1,490.4	1,490.4	255.31
400	1,987.2	1,987.2	502.6
500	2,484.0	2,484.0	820.8
600	2,980.8	2,980.8	1,198.1
700	3,477.6	3,477.6	1,622.0
800	3,974.4	3,974.4	2,081.7
900	4,471.2	4,471.2	2,569.4
1000	4,968.0	4,968.0	3,074.6
1100	5,464.7	5,464.7	3,596
1200	5,961.5	5,961.5	4,130
1300	6,458.3	6,458.3	4,680
1400	6,955.1	6,955.1	5,242
1500	7,451.9	7,451.9	5,814
1600	7,948.8	7,948.8	6,400
1700	8,445.6	8,455.6	6,987
1800	8,942.4	8,942.4	7,578
1900	9,439.2	9,439.2	8,170
2000	9,936.0	9,936.2	8,780
2100	10,432.8	10,433.0	9.387
2200	10,929.6	10,930.0	9,988
2300	11,426.4	11,427.3	10,600
2400	11,923.1	11,924.6	11,230
2500	12,420.0	12,422.3	11,850
2750	13,662.0	13,668.1	13,420
3000	14,904.0	14,917.8	15,030
3250	16,146.0	16,174.3	16,640
3500	17,388.0	17,440.2	18,270
3750	18,630.0	18,718.5	19,910
4000	19,872.0	20,013.2	21,600
4250	21,114.0	21,326.9	
4500	22,356.0	22,662.9	
4750	23,598.0	24,024.1	
5000	24,840.0	25,412.5	

[a] By permission, from NBS, *Tables of Selected Values of Chemical Thermodynamic Properties*, Series I, Volume I, March 1947 to June 1949 (after Penner[9]).

TABLE 1.5c Enthalpy Differences $\mathscr{H}_T^\circ - \mathscr{H}_0^\circ$ (cal / mole) for Various Substances[a]

T (K)	$C_{(g)}$	$OH_{(g)}$	$CO_{(g)}$	$NO_{(g)}$
298.16	1,558.9	2,106.2	2,072.63	2,194.2
300	1,568.1	2,122.5	2,085.45	2,206.8
400	2,065.8	2,829.6	2,783.8	2,920.8
500	2,563.1	3,535.0	3,490.0	3,644.0
600	3,060.3	4,240.8	4,209.5	4,381.2
700	3,557.3	4,946.9	4,945.8	5,136.6
800	4,054.3	5,658.4	5,699.8	5,909.6
900	4,551.3	6,377.4	6,470.6	6,700.5
1000	5,048.2	7,106.0	7,256.5	7,506.0
1100	5,545.1	7,844.1	8,056.2	8,323.7
1200	6,042.0	8,593.1	8,867.8	9,153.6
1300	6,539.1	9,354.8	9,689.9	9,991.8
1400	7,036.3	10,128	10,520.9	10,839
1500	7,533.4	10,910	11,358.8	11,694
1600	8,031.2	11,702	12,204	12,554
1700	8,529.2	12,504	13,055	13,416
1800	9,027.9	13,315	13,911	14,283
1900	9,527.2	14,134	14,722	15,154
2000	10,027.6	14,960	15,636	16,030
2100	10,528.8	15,794	16,505	16,907
2200	11,031.5	16,634	17,376	17,789
2300	11,535.2	17,480	18,251	18,674
2400	12,040.6	18,334	19,128	19,560
2500	12,547.5	19,193	20,007	20,450
2750	13,822.1	21,362	22,215	22,682
3000	15,108.0	23,559	24,434	24,924
3250	16,406.0	25,779	26,663	27,173
3500	17,715.6	28,025	28,900	29,428
3750	19,036.1	30,293	31,144	31,695
4000	20,367.2	32,576	33,395	33,972
4250	21,707.7	34,884	35,651	36,257
4500	23,057.1	37,206	37,913	38,552
4750	24,413.6	39,553	40,180	40,860
5000	25,776.0	41,910	42,452	43,180

TABLE 1.5c (Continued)

T (K)	$H_2O_{(g)}$	$CO_{2(g)}$	$CH_{4(g)}$	$C_2H_{2(g)}$
298.16	2,367.7	2,238.11	2,397	2,391.5
300	2,382.0	2,254.6	2,413	2,410.8
400	3,194.0	3,194.8	3,323	3,541.2
500	4,025.5	4,222.8	4,365	4,791.0
600	4,882.2	5,322.4	5,549	6,127
700	5,771.5	6,481.3	6,871	7,533
800	6,689.6	7,689.4	8,321	8,999
900	7,634.7	8,939.9	9,887	10,520
1000	8,608.0	10,222	11,560	12,090
1100	9,606.3	11,536	13,320	13,706
1200	10,630	12,872	15,170	15,362
1300	11,679	14,234	17,100	17,055
1400	12,753	15,611	19,090	18,782
1500	13,848	17,004	21,130	20,541
1600	14,966	18,400		
1700	16,106	19,820		
1800	17,264	21,260		
1900	18,440	22,690		
2000	19,630	24,140		
2100	20,834	25,600		
2200	22,053	27,060		
2300	23,283	28,520		
2400	24,521	30,000		
2500	25,770	31,480		
2750	28,930	35,200		
3000	32,160	38,940		
3250	35,393	42,710		
3500	38,675	46,520		
3750	42,000	50,330		
4000	45,360	54,160		
4250	48,705	58,010		
4500	52,065	61,880		
4750	55,433	65,740		
5000	58,850	69,650		

TABLE 1.5d Enthalpy Differences $\mathscr{H}_T^\circ - \mathscr{H}_0^\circ$ (cal / mole) for Various Substances[a]

T (K)	$Cl_{2(g)}$	$Br_{2(g)}$	$I_{2(g)}$	$Cl_{(g)}$	$Br_{(g)}$
298.16	2,193.9	2,324.5	2,417.8	1,499.1	1,481.3
300	2,208.9	2,340.3	2,433.9	1,508.7	1,490.4

TABLE 1.5d (Continued)

T (K)	$Cl_{2(g)}$	$Br_{2(g)}$	$I_{2(g)}$	$Cl_{(g)}$	$Br_{(g)}$
400	3,038.4	3,211.2	3,320.4	2,039.1	1,987.2
500	3,892.0	4,093.0	4,212.5	2,580.1	2,484.1
600	4,761.0	4,981.8	5,109.0	3,124.4	2,981.6
700	5,639.9	5,874.4	6,008.1	3,668.0	3,480.3
800	6,524.8	6,769.6	6,909.6	4,208.6	3,981.4
900	7,414.2	7,668.0	7,812.9	4,745.6	4,485.8
1000	8,309	8,568.0	8,718.0	5,278.8	4,994.2
1100	9,207	9,469.9	9,623.9	5,808.4	5,507.2
1200	10,108	10,373	10,532	6,334.7	6,024.7
1300	11,010	11,278	11,441	6,858.0	6,546.9
1400	11,914	12,184	12,354	7,378.8	7,073.4
1500	12,822	13,092	13,266	7,897.4	7,603.5
1600				8,413.8	8,137.1
1700				8,928.68	8,673.6
1800				9,441.7	9,212.4
1900				9,553.5	9,753.1
2000				10,464.0	10,295.4
2100				10,973.6	10.838.5
2200				11,482.2	11,382.6
2300				11,989.9	11,927.1
2400				12,496.8	12,471.6
2500				13,003.0	13,016.3
2750				14,265.9	14,376.5
3000				15,525.6	15,733.2
3250				16,783.0	17,084.9
3500				18,038.3	18,431.4
3750				19,291.9	19,772.5
4000				20,544.4	21,107.6
4250				21,795.7	22,437.0
4500				23,045.9	23,761.4
4750				24,295.3	25,080.5
5000				25,544.0	26,395.0

[a] By permission, from NBS, *Tables of Selected Values of Chemical Thermodynamic Properties*, Series I, Volume I, March 1947 to June 1949 (after Penner[9]).

TABLE 1.5e Enthalpy Differences $\mathscr{H}_T^\circ - \mathscr{H}_0^\circ$ (cal / mole) for Various Substances[a]

T (K)	$F_{2(g)}$	$F_{(g)}$	$HF_{(g)}$
100	694.99	498.5	675.7
200	1,395.8	1,024	1,372
298.16	2,110.5	1,558	2,055
300	2,124.4	1,568	2,068
400	2,894.5	2,108	2,765
500	3,700.1	2,640	3,462
600	4,531.7	3,165	4,160
700	5,382.2	3,685	4,859
800	6,243.9	4,200	5,563
900	7,121.8	4,712	6,273
1000	8,005.3	5,221	6,900
1100	8,895.6	5,729	7,715
1200	9,791.8	6,235	8,452
1300	10,693	6,739	9,198
1400	11,598	7,244	9,951
1500	12,507	7,746	10,720
1600	13,420	8,248	11,490
1700	14,336	8,750	12,280
1800	15,255	9,250	13,070
1900	15,177	9,753	13,870
2000	17,102	10,250	14,680
2100	18,030	10,750	15,470
2200	18,960	11,250	16,320
2300	19,892	11,750	17,150
2400	20,827	12,250	17,980
2500	21,765	12,750	18,830
2600	22,704	13,250	19,670
2700	23,646	13,750	20,520
2800	24,590	14,250	21,380
2900	25,536	14,740	22,240
3000	26,485	15,740	23,100
3200	28,388	16,240	24,840
3400	30,300	17,230	26,590
3600	32,220	18,230	28,350
3800	34,148	19,230	30,110
4000	36,086	20,220	31,920
4200	38,030	21,220	33,710
4400	39,984	22,210	35,510
4600	41,946	23,210	37,320
4800	43,914	24,200	39,140
5000	45,892	25,200	40,960

[a] From L. G. Cole, M. Farber, and G. W. Elverum, Jr., *J. Chem. Phys.*, Vol. **20**, p. 586, 1952.

TABLE 1.5f Enthalpy Differences $\mathscr{H}^\circ_T - \mathscr{H}^\circ_0$ (cal / mole) for Various Substances[a]

T (K)	$I_{(g)}$	$HCl_{(g)}$	$HBr_{(g)}$	$HI_{(g)}$
298.16	1,481.3	2,064.8	2,066.8	2,068.9
300	1,490.4	2,077.8	2,079.6	2,081.7
400	1,987.2	2,774.0	2,776.8	2,780.0
500	2,484.0	3,473.0	3,477.5	3,485.5
600	2,980.8	4,176.6	4,186.2	4,203.6
700	3,477.6	4,888.1	4,906.3	4,937.8
800	3,974.4	5,611.2	5,641.6	5,688.8
900	4,471.3	6,346.8	6,390.9	6,457.5
1000	4,968.2	7,095.0	7,156.0	7,242.0
1100	5,465.4	7,857.3	7,935.4	8,041.0
1200	5,962.8	8,632.8	8,727.6	8,852.4
1300	6,460.7	9,421.1	9,532.9	9,675.9
1400	6,959.5	10,220	10,347	10,510
1500	7,459.4	11,030	11,174	11,352
1600	7,960.5			
1700	8,463.1			
1800	8,967.2			
1900	9,473.4			
2000	9,981.6			
2100	10,492.0			
2200	11,004.6			
2300	11,519.5			
2400	12,036.5			
2500	12,556.0			
2750	13,863.8			
3000	15,183.9			
3250	16,515.2			
3500	17,855.2			
3750	19,202.6			
4000	20,555.6			
4250	21,912.6			
4500	23,272.2			
4750	24,633.0			
5000	25,994.5			

[a] From L. G. Cole, M. Farber, and G. W. Elverum, Jr., *J. Chem. Phys.*, Vol. **20**, p. 586, 1952.

Example 1.10

Evaluate the heat of reaction of ethane $C_2H_{6(g)}$.

Solution:

If the chemical reaction can be given by the following reaction, without any dissociation in the final product,

$$C_2H_{6(g)} + 3\tfrac{1}{2}O_{2(g)} \xrightarrow{298.16\text{ K}} 2CO_{2(g)} + 3H_2O_{(g)}$$

then to calculate ΔH_r at standard state conditions, we may use Eq. (1-140):

$$\Delta H^\circ_{r,T_0} = \sum_{i=1}^{N} \nu_i'' \Delta \mathscr{H}^\circ_{f,M_i} - \sum_{i=1}^{N} \nu_i' \Delta \mathscr{H}^\circ_{f,M_i}$$

$$= 0\,\Delta \mathscr{H}^\circ_{f,C_2H_6} + 0\,\Delta \mathscr{H}^\circ_{f,O_2} + 2\,\Delta \mathscr{H}^\circ_{f,CO_2} + 3\,\Delta \mathscr{H}^\circ_{f,H_2O_{(g)}}$$

$$-\left(1\,\Delta \mathscr{H}^\circ_{f,C_2H_6} + 3\tfrac{1}{2}\,\Delta \mathscr{H}^\circ_{f,O_2} + 0\,\Delta \mathscr{H}^\circ_{f,CO_2} + 0\,\Delta \mathscr{H}^\circ_{f,H_2O_{(g)}}\right)$$

$$= [-(2 \times 94) - (3 \times 57.8)] - [-(1 \times 20.24) + (3.5 \times 0)]$$

$$= -341.16 \text{ kcal}$$

When the heat of reaction is negative, heat is evolved and the process is called *exothermic*. When the heat of reaction is positive, heat must be absorbed by the system during the chemical reaction and the process is called *endothermic*.

The *heating value* of a fuel is a term commonly used in discussing combustion. The heating value is a positive number that is equal to the enthalpy of combustion but of opposite sign. There are many possible heating values for a fuel, depending on the phase of the water formed in the products (liquid or gas), the phase of the fuel (liquid or gas), and the conditions under which the combustion is carried out (constant pressure or constant volume).

The heat of combustion may be considered a special case of the heat of reaction. The *heat of combustion* of a substance is the heat liberated when a fuel (usually a hydrocarbon) reacts with oxygen to yield H_2O and CO_2.

Another method which is useful in determining an unknown heat of reaction is to use Hess's law to sum the appropriate reactions for which the heat of reaction is known.

Bond energies and resonance energies can be used for evaluating the heat of reaction or heat of combustion of a reaction, as illustrated in Example 1.11.

Example 1.11

Evaluate the heat of combustion of benzoic acid.

Solution:

The reaction is

$$C_6H_5COOH + 7\tfrac{1}{2}O_2 \rightarrow 7CO_2 + 3H_2O$$

In order to solve this problem, the reference structure of benzoic acid must be known:

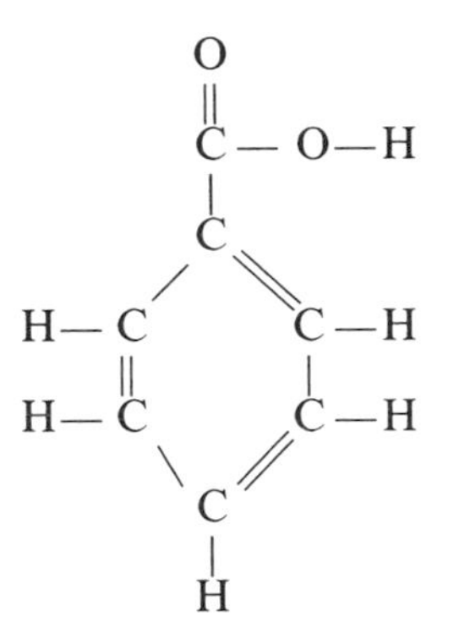

Benzoic acid

For the reactants, the bond energies are

$$4(\text{C—C}) + 3(\text{C=C}) + 5(\text{C—H}) + 1(\text{C=O})$$
$$+1(\text{C—O}) + 1(\text{O—H}) + 7\tfrac{1}{2}(\text{O=O})$$
$$+1(C_6H_6\text{, benzene-ring resonance}) + 1(\text{—COOH, carboxyl})$$

Inserting numerical values,

$$342 + 429 + 490.5 + 167 + 86 + 109 + 891.45 + 48.9 + 28 = 2591.9 \text{ kcal}$$

For the products, the bond energies are

$$14(\text{C=O}) + 6(\text{O—H}) + 7(CO_2 \text{ resonance}) = 2338 + 654 + 231 = 3223$$

recognizing that CO_2 is formed with two double bonds.[18] Thus, the heat of combustion of benzoic acid is

$$2591.9 - 3223 = -631.1 \text{ kcal.}$$

Example 1.12

Given the following heats of reaction at 298.16 K,

$$C_2H_{4(g)} + 3O_{2(g)} \rightarrow 2CO_{2(g)} + 2H_2O_{(l)} - 337.3 \text{ kcal} \tag{i}$$

$$H_{2(g)} + \tfrac{1}{2}O_{2(g)} \rightarrow H_2O_{(l)} - 68.3 \text{ kcal} \tag{ii}$$

$$C_2H_{6(g)} + 3\tfrac{1}{2}O_{2(g)} \rightarrow 2CO_{2(g)} + 3H_2O_{(l)} - 372.8 \text{ kcal} \tag{iii}$$

determine the heat of reaction for

$$C_2H_{4(g)} + H_{2(g)} \rightarrow C_2H_{6(g)} + \Delta H_r$$

Solution:

The result is obtained simply by adding (i) and (ii) and subtracting (iii), or by adding (i) and (ii), using Laplace's law on (iii), and adding the result to the sum of (i) and (ii):

$$C_2H_{4(g)} + \cancel{3O_{2(g)}} \rightarrow \cancel{2CO_{2(g)}} + \cancel{2H_2O_{(l)}} \quad -337.3 \frac{\text{kcal}}{\text{mole } C_2H_{4(g)}}$$

$$H_{2(g)} + \cancel{\tfrac{1}{2}O_{2(g)}} \rightarrow \cancel{H_2O_{(l)}} \quad -68.3 \frac{\text{kcal}}{\text{mole } H_{2(g)}}$$

$$+ \quad \cancel{2CO_{2(g)}} + \cancel{3H_2O_{(l)}} \rightarrow C_2H_{6(g)} + \cancel{3\tfrac{1}{2}O_{2(g)}} \quad +372.8 \frac{\text{kcal}}{\text{mole } C_2H_{6(g)}}$$

$$C_2H_{4(g)} + H_{2(g)} \rightarrow C_2H_{6(g)} \quad -32.8 \text{ kcal}$$

Heat of reactions may also be manipulated by Hess's law to arrive at heats of formation for substances. The heat of combustion of CH_4 is about ten times larger than the heat of formation of CH_4. Some tabulated heats of formation of compounds are deduced from heats of combustion, which can be measured experimentally in the laboratory. The heat of formation of CH_4 can be deduced from the heats of formation of CO_2 and H_2O and the heat of reaction of $CH_4 + 2O_2$.

Example 1.13

Deduce the heat of formation of CH_4 from those of CO_2 and H_2O, together with ΔH_r of the methane–oxygen reaction.

Solution:

$$C_{(s)} + O_{2(g)} \rightarrow CO_{2(g)} + \Delta\mathscr{H}^\circ_{f,\,CO_{2(g)}} \tag{a}$$

$$2H_{2(g)} + O_{2(g)} \rightarrow 2H_2O_{(g)} + 2\Delta\mathscr{H}^\circ_{f,\,H_2O_{(g)}} \tag{b}$$

$$CH_{4(g)} + 2O_{2(g)} \rightarrow CO_{2(g)} + 2H_2O_{(g)} + \Delta H_r \tag{c}$$

Using Laplace's law, the last equation becomes

$$CO_{2(g)} + 2H_2O_{(g)} \rightarrow CH_{4(g)} + 2O_{2(g)} - \Delta H_r \qquad \text{(d)}$$

Summing Eqs. (a), (b), and (d) gives

$$C_{(s)} + 2H_{2(g)} \rightarrow CH_{4(g)} + \left(\Delta\mathscr{H}^{\circ}_{f,\,CO_{2(g)}} + 2\,\Delta\mathscr{H}^{\circ}_{f,\,H_2O_{(g)}} - \Delta H_r\right)$$

Therefore,

$$\Delta\mathscr{H}^{\circ}_{f,\,CH_4} = \Delta\mathscr{H}^{\circ}_{f,\,CO_{2(g)}} + 2\,\Delta\mathscr{H}^{\circ}_{f,\,H_2O_{(g)}} - \Delta H_r$$

13 CALCULATION OF ADIABATIC FLAME TEMPERATURE

Consider a combustion process that takes place adiabatically and with no work or changes in kinetic or potential energy. For such a process the temperature of the products is referred to as the *adiabatic flame temperature*. This is the maximum temperature that can be achieved for the given reactants, because any heat transfer from the reacting substances and any incomplete combustion would tend to lower the temperature of the products. The adiabatic temperature T_f can be controlled with the amount of excess air that is used. In gas turbines, where the maximum permissible temperature is defined by metallurgical considerations in the turbine, close control of the temperature of the products is essential.

A few examples are given here to illustrate some of the steps used to calculate the adiabatic flame temperature.

Example 1.14

Calculate the adiabatic flame temperature of water vapor after the reaction of gaseous H_2 and O_2.

Solution:

This may be broken down as

$$2H_{2(g)} + O_{2(g)} \rightarrow 2H_2O_{(l)} + \left(2\,\Delta\mathscr{H}^{\circ}_{f,\,H_2O_{(l)}} = -136.6 \text{ kcal}\right)$$

$$2H_2O_{(l)} \rightarrow 2H_2O_{(g)} + \left(2\,\Delta\mathscr{H}_v = 21.0 \text{ kcal}\right)$$

($\Delta\mathscr{H}_v \equiv$ heat of vaporization). Adding these together,

$$2H_{2(g)} + O_{2(g)} \rightarrow 2H_2O_{(g)} - 115.6 \text{ kcal}$$

Therefore, the heat generated by the reaction is 115.6 kcal. If all of the heat is used to heat up the indicated product (adiabatic, no dissociation), then

$$115.6 = 2\int_{298}^{T_f} C_{p,\,H_2O}\,dT$$

$$57.8 = \mathscr{H}_{H_2O,\,T_f} - \mathscr{H}_{H_2O,\,298\,K}$$

Using Table 1.5, the approximate value of T_f is found to correspond to 57.8 kcal. The adiabatic flame temperature $T_f \approx 5000$ K balances the above equation. At this high temperature, dissociation usually occurs and the actual products are unknown, thus introducing additional complications. The adiabatic flame temperature for nondissociating products is normally called the adiabatic frozen-flame temperature.

An alternate means of evaluating ΔH_r at T_f follows. We take the reactants to be heated to T_f first and then reacted:

$$2H_{2(g),\,T_i} + O_{2(g),\,T_i} \xrightarrow{+\Delta H_{\text{reactant}}} 2H_{2(g),\,T_f} + O_{2(g),\,T_f} \rightarrow 2H_2O_{(g)} + \Delta H_{r,\,T_f}$$

where

$$\Delta H_{\text{reactant}} = 2\int_{T_i}^{T_f} C_{p,\,H_2}\,dT + \int_{T_i}^{T_f} C_{p,\,O_2}\,dT$$

Also, let us consider

$$2H_2O_{(g),\,T_i} \rightarrow 2H_2O_{(g),\,T_f} + \Delta H_{\text{prod}}$$

where

$$\Delta H_{\text{prod}} = 2\int_{298}^{T_f} C_{p,\,H_2O}\,dT$$

Therefore, from Eq. (1-143) we have

$$\Delta H_{r,\,T_f} + 2\int_{298}^{T_f} C_{p,\,H_2}\,dT + \int_{298}^{T_f} C_{p,\,O_2}\,dT = \Delta H_{r,\,298} + 2\int_{298}^{T_f} C_{p,\,H_2O}\,dT.$$

Rearranging the above equation, we have

$$\Delta H_{r,\,T_f} = \Delta H_{r,\,298,\,H_2O_{(g)}} + 2\int_{298}^{T_f} C_{p,\,H_2O}\,dT - 2\int_{298}^{T_f} C_{p,\,H_2}\,dT - \int_{298}^{T_f} C_{p,\,O_2}\,dT$$

which is of the form of Eq. (1-147).

For a fuel-rich reaction such as

$$10H_2 + O_2 \rightarrow 2H_2O + 8H_2$$

the heat of reaction is the same as that obtained from a stoichiometric burning.

The adiabatic frozen-flame temperature is lower, however, since the same amount of heat is used to heat additional material.

For a fuel-lean reaction such as

$$H_2 + 9O_2 \rightarrow H_2O + 8\tfrac{1}{2}O_2$$

the heat of reaction per mole of H_2 remains unchanged, but the flame temperature is lower than that of stoichiometric burning for the same reason as in the fuel-rich case.

To demonstrate the power of the adiabatic assumption in calculating the adiabatic frozen-flame temperature, consider a more detailed example of the combustion of liquid H_2 with liquid O_2. Let T_b represent the boiling temperature of $H_{2(l)}$, T_b' the boiling temperature of $O_{2(l)}$, and T_i the initial temperature of both reactants. Then we have

$$\begin{array}{lcl}
2H_{2(l),T_i} & + & O_{2(l),T_i'} \\
\quad\downarrow\ \hat{Q} = 2\int_{T_i}^{T_b} C_{p,H_{2(l)}}\,dT & & \quad\downarrow\ \hat{Q} = \int_{T_i'}^{T_b'} C_{p,O_{2(l)}}\,dT \\
2H_{2(l),T_b} & + & O_{2(l),T_b'} \\
\quad\downarrow\ \hat{Q} = 2\,\Delta\mathscr{H}_{v,H_2} & & \quad\downarrow\ \hat{Q} = \Delta\mathscr{H}_{v,O_2} \\
2H_{2(g),T_b} & + & O_{2(g),T_b'} \\
\quad\downarrow\ \hat{Q} = 2\int_{T_b}^{298} C_{p,H_{2(g)}}\,dT & & \quad\downarrow\ \hat{Q} = \int_{T_b'}^{298} C_{p,O_{2(g)}}\,dT \\
2H_{2(g),298} & + & O_{2(g),298} \rightarrow \quad 2H_2O_{(g),298} + 2\,\Delta\mathscr{H}^\circ_{f,H_2O_{(g)}} \\
 & & \qquad\qquad\qquad \downarrow\ \hat{Q} = 2\int_{298}^{T_f} C_{p,H_2O_{(g)}}\,dT \\
 & & \qquad\qquad\qquad 2H_2O_{(g),T_f}
\end{array}$$

Since the system is assumed to be adiabatic,

$$\sum \hat{Q} = 0$$

and

$$\begin{aligned}
&2\int_{T_i}^{T_b} C_{p,H_{2(l)}}\,dT + \int_{T_i'}^{T_b'} C_{p,O_{2(l)}}\,dT + 2\,\Delta\mathscr{H}_{v,H_2} + \Delta\mathscr{H}_{v,O_2} \\
&\quad + 2\int_{T_b}^{298} C_{p,H_{2(g)}}\,dT + \int_{T_b'}^{298} C_{p,O_{2(g)}}\,dT \\
&\quad + 2\,\Delta\mathscr{H}^\circ_{f,H_2O_{(g)}} + 2\int_{298}^{T_f} C_{p,H_2O_{(g)}}\,dT = 0 \qquad (1\text{-}149)
\end{aligned}$$

which may also be written as

$$2\mathscr{H}_{H_2}|_{T_i}^{298} + \mathscr{H}_{O_2}|_{T_i'}^{298} + \Delta H_r^\circ + 2\mathscr{H}_{H_2O}|_{298}^{T_f} = 0 \tag{1-150}$$

The results of the flame-temperature calculation procedure can be generalized by noting that any chemical equation may be expressed as

$$\sum_{i=1}^{N} \nu_i' M_i \to \sum_{i=1}^{N} \nu_i'' M_i \tag{1-151}$$

When a change of state takes place in a process, the different phases are considered as different species. The enthalpy equation for either frozen or equilibrium process is

$$\sum_{i=1}^{N} \nu_i'' \Delta\mathscr{H}_{f,M_i}^\circ - \sum_{i=1}^{N} \nu_i' \Delta\mathscr{H}_{f,M_i}^\circ - \sum_{i=1}^{N} \nu_i'\left(\mathscr{H}_{M_i,T_i} - \mathscr{H}_{M_i,298}\right) + \sum_{i=1}^{N} \nu_i''\left(\mathscr{H}_{M_i,T_f} - \mathscr{H}_{M_i,298}\right) = \Delta H \tag{1-152}$$

whereas for an adiabatic process the heat addition from outside the system, ΔH, is zero. This allows the calculation of the adiabatic frozen-flame temperature (ΔH positive for heat addition).

Generally speaking, the reactants may not be at the standard-state temperature T_0, but rather at T_i, where T_i may be higher or lower than T_0. During the combustion, a part of the heat evolved is used to heat the product to a temperature T_2. The energy-balance equation then becomes

$$\boxed{\begin{aligned}\Delta H = & \sum_{\substack{j=1\\ \text{products}}}^{N} \nu_j''\left\{\left[\left(\mathscr{H}_{T_f}^\circ - \mathscr{H}_0^\circ\right) - \left(\mathscr{H}_{T_0}^\circ - \mathscr{H}_0^\circ\right)\right] + \left(\Delta\mathscr{H}_f^\circ\right)_{T_0}\right\}_j \\ & - \sum_{\substack{j=1\\ \text{reactants}}}^{N} \nu_j'\left\{\left[\left(\mathscr{H}_{T_i}^\circ - \mathscr{H}_0^\circ\right) - \left(\mathscr{H}_{T_0}^\circ - \mathscr{H}_0^\circ\right)\right] + \left(\Delta\mathscr{H}_f^\circ\right)_{T_0}\right\}_j\end{aligned}} \tag{1-153}$$

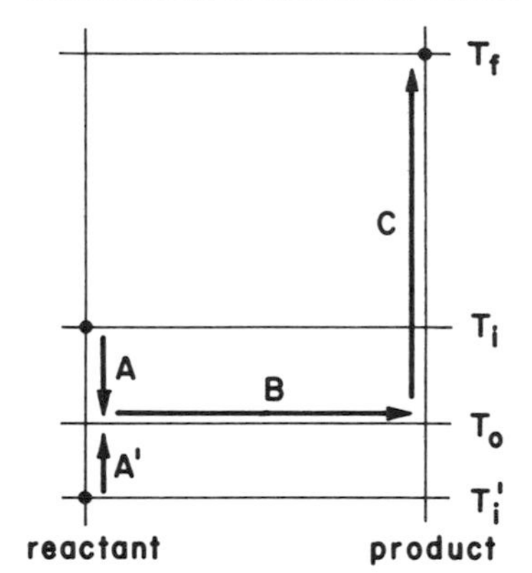

Figure 1.5 Temperature-reaction paths, showing the effect of initial reactant temperature.

If we let

$$B = \sum_{j=1}^{N} \nu_j''(\Delta\mathscr{H}_f^\circ)_{T_0,j} - \sum_{j=1}^{N} \nu_j'(\Delta\mathscr{H}_f^\circ)_{T_0,j}$$

$$A = -\sum_{j=1}^{N} \nu_j'\left[\left(\mathscr{H}_{T_i}^\circ - \mathscr{H}_0^\circ\right) - \left(\mathscr{H}_{T_0}^\circ - \mathscr{H}_0^\circ\right)\right]_j$$

$$C = \sum_{j=1}^{N} \nu_j''\left[\left(\mathscr{H}_{T_f}^\circ - \mathscr{H}_0^\circ\right) - \left(\mathscr{H}_{T_0}^\circ - \mathscr{H}_0^\circ\right)\right]_j$$

Figure 1.5 indicates that

$$\Delta H = A + B + C$$

If $T_i < T_0$, then

$$\Delta H = A' + B + C$$

The adiabatic flame temperature is of particular interest in connection with the combustion of gaseous hydrocarbons in oxygen or air. In such cases, it gives the "maximum flame temperature," the actual temperature being somewhat lower due to various disturbing factors. From the knowledge of heats of reaction and the variation of enthalpies of the reactants and products, it is possible to calculate the final temperature of the system.

Example 1.15

Assuming no dissociation of product species, determine the adiabatic flame temperature and product concentrations for the following chemical reaction at

$T_i = 298$ K:

$$CH_4 + 15\underbrace{\left(0.21O_2 + 0.79N_2\right)}_{\text{air}} \rightarrow n_{CO_2}CO_2 + n_{H_2O}H_2O_{(g)} + n_{N_2}N_2 + n_{O_2}O_2$$

Solution:

The four equations for conservation of atomic species are

$$\text{C:} \qquad 1 = n_{CO_2}$$

$$\text{O:} \qquad 15(0.21 \times 2) = 2n_{CO_2} + n_{H_2O} + 2n_{O_2}$$

$$\text{N:} \qquad 15(0.79 \times 2) = 2n_{N_2}$$

$$\text{H:} \qquad 4 = 2n_{H_2O}$$

These four equations may be solved simultaneously to yield values for n_{CO_2}, n_{H_2O}, n_{N_2}, and n_{O_2}. The adiabatic flame temperature T_f can then be calculated by letting $\Delta H = 0$ in the following equation:

$$\underbrace{\sum_{i=1}^{N} \nu_i'' \,\Delta\mathscr{H}^\circ_{f,M_i} - \sum_{i=1}^{N} \nu_i' \,\Delta\mathscr{H}^\circ_{f,M_i}}_{\Delta H_r} - \sum_{i=1}^{N} \nu_i'\left(\mathscr{H}_{T_i} - \mathscr{H}_{T_{298}}\right)_{M_i}$$

$$+ \sum_{i=1}^{N} \nu_i''\left(\mathscr{H}_{T_f} - \mathscr{H}_{298}\right)_{M_i} = \Delta H$$

Noting that $\mathscr{H}_{T_i} - \mathscr{H}_{298} = 0$ (since $T_i = T_{298}$) and that $\Delta\mathscr{H}_f^\circ$ of the elements in their standard state (O_2 and N_2) is 0, we have

$$\left[1\,\Delta\mathscr{H}^\circ_{f,CO_2} + 2\,\Delta\mathscr{H}^\circ_{f,H_2O_{(g)}}\right] - \left[1\,\Delta\mathscr{H}^\circ_{f,CH_{4(g)}}\right]$$

$$+\left\{1\left(\mathscr{H}_{T_f} - \mathscr{H}_{298}\right)_{CO_2} + 2\left(\mathscr{H}_{T_f} - \mathscr{H}_{298}\right)_{H_2O_{(g)}}\right.$$

$$\left. +11.85\left(\mathscr{H}_{T_f} - \mathscr{H}_{298}\right)_{N_2} + 1.15\left(\mathscr{H}_{T_f} - \mathscr{H}_{298}\right)_{O_2}\right\} = 0$$

$$\Delta H_r = [(-94{,}054) + 2(-57{,}798)] - [-17{,}895] = -191{,}755 \text{ cal}$$

Assume a flame temperature of 2000 K and using Table 1.5, the thermal energy change is

$$1(21901.89)_{CO_2} + 2(17262.3)_{H_2O} + 11.85(13422.5)_{N_2} + 1.15(14148.2)_{O_2}$$
$$= 231{,}753.55 \text{ cal}$$

Assume a flame temperature of 1700 K, the thermal-energy change is

$$1(17{,}582)_{CO_2} + 2(13{,}738)_{H_2O} + 11.85(10{,}861)_{N_2} + 1.15(11{,}467)_{O_2}$$
$$= 186{,}952.18 \text{ cal}$$

Interpolating, we arrive at $T_f = 1732$ K.

The above example shows that under the assumption of no product dissociation there are five equations to be solved for five unknowns. It is important to note that the above method of calculation is appropriate only if T_f is low ($T_f < 1200$ K). At higher temperatures the products dissociate and many other compounds are formed. The above chemical reaction could change into

$$CH_4 + n_{air}(0.21O_2 + 0.79N_2) \rightarrow n_{CO_2}CO_2 + n_{H_2O}H_2O + n_{N_2}N_2 + n_{O_2}O_2$$
$$+ n_{NO}NO + n_H H + n_{OH}OH + n_O O$$
$$+ n_N N + n_{CO}CO + n_{NO^+}NO^+$$
$$+ n_{e^-}e^- + \text{etc.}$$

Because we still have only four equations from the conservation of atomic species, there are more unknowns than equations, and the adiabatic flame temperature cannot be calculated as in the simplified reaction.

If the products are in a condition of chemical equilibrium, we can obtain more relations between the mole fractions of the product constituents at equilibrium.

$$N_2 + O_2 \rightleftharpoons 2NO$$
$$CO + \tfrac{1}{2}O_2 \rightleftharpoons CO_2$$
$$O_2 \rightleftharpoons 2O$$
$$N_2 \rightleftharpoons 2N$$
$$H_2O \rightleftharpoons \tfrac{1}{2}H_2 + OH$$
$$H_2O \rightleftharpoons H + OH$$
$$H_2 \rightleftharpoons 2H$$
$$NO \rightleftharpoons NO^+ + e^-$$

The equilibrium constants (to be discussed in the next section) for each individual reaction will give us further information on the determination of the adiabatic flame temperature. We note here that for air at high temperatures the reaction between O_2 and N_2 can produce many species. For example,

$$O_2 + 4N_2 \rightarrow n_{O_2}O_2 + n_{N_2}N_2 + n_O O + n_{NO}NO + n_N N + n_{NO^+}NO^+ + n_{e^-}e^-$$

14 EQUILIBRIUM CONSTANTS

From previous discussion, we know

$$G = H - TS = U + pV - TS \tag{1-154}$$

Then

$$dU = dH - p\,dV - V\,dp = T\,dS - p\,dV \tag{1-155}$$

and (considering that only $p\,dV$ work is done)

$$dG = dU + p\,dV + V\,dp - T\,dS - S\,dT \tag{1-156}$$

Then, substituting Eq. (1-155) into Eq. (1-156) gives

$$dG = V\,dp - S\,dT \tag{1-157}$$

Assuming the perfect-gas relation, for an isothermal process,

$$dG = V\,dp = \frac{nR_uT}{p}\,dp = nR_uT\,d(\ln p) \tag{1-158}$$

since

$$\frac{dp}{p} = d(\ln p)$$

Integrating from $p°$ to p Eq. (1-158) becomes

$$G - G° = nR_uT[\ln p - \ln p°] \tag{1-159}$$

For $p° = 1$ atm,

$$G = G° + nR_uT \ln p \tag{1-160}$$

Then for species i in the system,

$$G_i = G_i° + n_iR_uT \ln p_i \tag{1-161}$$

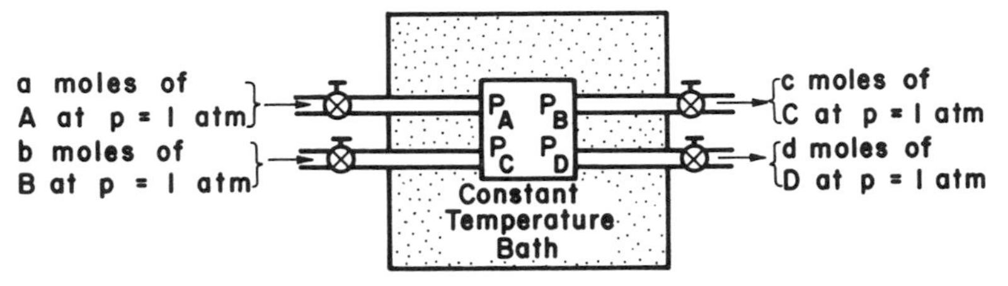

Figure 1.6 Van't Hoff equilibrium box.

Now consider the Van't Hoff equilibrium box in Fig. 1.6. From Eq. (1-161) we obtain

$$G_A = G_A^\circ + aR_uT \ln p_A$$

$$G_B = G_B^\circ + bR_uT \ln p_B$$

$$G_C = G_C^\circ + cR_uT \ln p_C$$

$$G_D = G_D^\circ + dR_uT \ln p_D$$

With reference to the equilibrium box, we assume a moles of A and b moles of B are pumped into the box, the reaction occurs isothermally and equilibrium is reached, and then c moles of C and d moles of D are pumped out. From these considerations we obtain

$$\begin{aligned}\Delta G &= G_{\text{product}} - G_{\text{reactant}} \\ &= (G_C + G_D) - (G_A + G_B) \\ &= \left[(G_C^\circ + G_D^\circ) - (G_A^\circ + G_B^\circ)\right] + R_uT \ln p_C^c p_D^d - R_uT \ln p_A^a p_B^b \\ &= \Delta G^\circ + R_uT \ln \frac{p_C^c p_D^d}{p_A^a p_B^b} \end{aligned} \tag{1-162}$$

Since we have assumed the system to be at equilibrium,

$$\Delta G = 0$$

and

$$-\Delta G^\circ = R_uT \ln \frac{p_C^c p_D^d}{p_A^a p_B^b} \tag{1-163}$$

We can define

$$K_p \equiv \frac{p_C^c p_D^d}{p_A^a p_B^b} \tag{1-164}$$

then

$$\boxed{\Delta G^\circ = -R_u T \ln K_p} \tag{1-165}$$

Equation (1-165) shows the relation between the standard-free-energy change and the equilibrium constant at any arbitrary pressure and temperature. The practical importance of the equilibrium constant K_p results from the fact that it is independent of total pressure and can therefore by listed as a unique function of temperature. K_p can be deduced from Eq. (1-165), if the value of ΔG° is known.

In order to gain some physical insight into the concept of the standard-free-energy change ΔG°, we note that the more negative ΔG° is, the larger K_p is and the more spontaneous the reaction. When the free energy of the reactant and product is the same, the reaction has no tendency to proceed in either direction; therefore, $\Delta G_{T,p} = 0$ (not ΔG°) for the equilibrium condition.

The equilibrium constant can be put into a more general form by considering the general chemical reaction

$$\sum_{i=1}^{N} \nu_i' M_i \rightleftharpoons \sum_{i=1}^{N} \nu_i'' M_i$$

where ν_i' and ν_i'' represent, respectively, the stoichiometric coefficients of reactants and products for the chemical species M_i, and N is the total number of chemical species involved. Then

$$\boxed{K_p = \prod_{i=1}^{N} (p_{i,e})^{(\nu_{i,s}'' - \nu_{i,s}')}} \tag{1-166}$$

where the subscript p means that the equilibrium constant is written in terms of partial pressures, and the subscript s identifies the coefficients as the stoichiometric coefficients for the individual chemical equilibrium reactions.

The same expression for the equilibrium constant can be obtained from thermodynamic considerations. Consider the hypothetical reaction between ideal gases

$$aA + bB \rightleftharpoons rR + sS$$

We shall now develop an expression for the free-energy change which occurs when the reactants at partial pressures p_A and p_B are converted into products at partial pressures p_R and p_S. From earlier discussion,

$$G(p,T) = H(T) - TS(p,T) \tag{1-167}$$

If we define a standard state $p_0 = 1$ atmosphere, then at the same temperature T we have

$$G^\circ(p_0, T) = H^\circ(T) - TS^\circ(p_0, T) \tag{1-168}$$

Subtracting Eq. (1-168) from (1-167) and noting that (since H is independent of pressure for a perfect gas) $H - H^\circ = 0$, we have

$$G - G^\circ = -T(S - S^\circ) \tag{1-169}$$

From the perfect-gas law, for an isothermal process

$$S - S^\circ = -nR_u \ln\left(\frac{p}{p_0}\right) \tag{1-170}$$

To see this, note that

$$\delta\hat{Q} = dE + \delta\hat{W}$$

For an isothermal process $dE = 0$ and

$$\delta\hat{Q} = \delta\hat{W} = p\,dV \tag{1-171}$$

But

$$d_eS = \frac{\delta\hat{Q}}{T} = \frac{p}{T}\,dV = \frac{nR_u}{V}\,dV = nR_u\,d\ln V = nR_u\,d\ln\frac{nR_uT}{p} \tag{1-172}$$

Since we have made the isothermal assumption,

$$d_eS = -nR_u\,d\ln p \tag{1-173}$$

and thus

$$S - S^\circ = -nR_u \ln\left(\frac{p}{p_0}\right)$$

which is (1-170).

Substituting Eq. (1-170) into Eq. (1-169) gives

$$G(T, p) = G^\circ + nR_uT \ln\left(\frac{p}{p_0}\right) \tag{1-174}$$

For the mixture, then,

$$G(T, p) = \sum_{i=1}^{N} n_i\left\{\mathscr{g}_i^\circ + R_uT \ln\left(\frac{p_i}{p_0}\right)\right\} \tag{1-175}$$

where $\mathcal{g}$ is the molar Gibbs free energy with units of (cal/g-mole) and p_i is the partial pressure as previously defined. At equilibrium

$$(dG)_{T,p} = 0 \tag{1-176}$$

If we take the differential of Eq. (1-175), we get

$$0 = \sum_{i=1}^{N} \mathcal{g}_i^\circ \, dn_i + R_u T \sum_{i=1}^{N} \left[\ln\left(\frac{p_i}{p_0}\right) \right] dn_i + R_u T \underbrace{\sum_{i=1}^{N} n_i \frac{dp_i}{p_i}}_{0} \tag{1-177}$$

The last term of Eq. (1-177) vanishes because

$$\sum_{i=1}^{N} n_i \frac{dp_i}{p_i} = \sum_{i=1}^{N} \left(\frac{n_{\text{total}}}{p} \right) dp_i = \frac{1}{p} \left(\sum_{i=1}^{N} n_i \right) \left(\sum_{i=1}^{N} dp_i \right)$$

$$= \frac{1}{p} \left(\sum_{i=1}^{N} n_i \right) d \left(\sum_{i=1}^{N} p_i \right) = 0 \tag{1-178}$$

since at equilibrium the total pressure is constant. The first term of Eq. (1-177) can be expressed as

$$\sum_{i=1}^{N} \mathcal{g}_i^\circ \, dn_i = \mathcal{g}_A^\circ \, dn_A + \mathcal{g}_B^\circ \, dn_B + \cdots + \mathcal{g}_R^\circ \, dn_R + \mathcal{g}_S^\circ \, dn_S + \cdots \tag{1-179}$$

where

$$dn_A = -a, \qquad dn_B = -b$$

$$dn_R = r, \qquad dn_S = s$$

Therefore,

$$\sum_{i=1}^{N} \mathcal{g}_i^\circ \, dn_i \equiv \Delta G^\circ = G^\circ_{\text{products}} - G^\circ_{\text{reactants}} \tag{1-180}$$

Eq. (1-177) then becomes

$$\Delta G^\circ = -R_u T \ln K_p \tag{1-181}$$

where

$$K_p = \frac{\left(\dfrac{p_R}{p_0} \right)^r \left(\dfrac{p_S}{p_0} \right)^s}{\left(\dfrac{p_A}{p_0} \right)^a \left(\dfrac{p_B}{p_0} \right)^b}, \qquad p_0 = 1, \tag{1-182}$$

and r, s, a, and b are the stoichiometric coefficients. We have thus arrived at the same result using two different methods.

In the preceding discussion the molar form of the free energy was introduced. We may also note that the molar standard free energy of formation $\Delta \mathscr{g}_f^\circ$ is tabulated for many compounds, allowing ΔG° to be calculated for any reaction as

$$\Delta G^\circ = \sum_{i=1}^{N} \nu_i'' \Delta \mathscr{g}_{fi}^\circ - \sum_{i=1}^{N} \nu_i' \Delta \mathscr{g}_{fi}^\circ \tag{1-183}$$

where ΔG° depends only on the nature of the reactants and products. For values of $\Delta \mathscr{g}_f^\circ$, see Table 1.6 or Ref. 4. It is important to note that the value of $\Delta \mathscr{g}_f^\circ$ for every *element* in its standard state is equal to zero.

Example 1.16

For the decomposition of gaseous hydrogen peroxide according to the reaction

$$H_2O_{2(g)} \rightleftharpoons H_2O_{(g)} + 0.5O_{2(g)}$$

What is the value of the equilibrium constant of this reaction at 298.15 K?

Solution:

The value of ΔG° for this equilibrium reaction can be calculated using Table 1.6 and Eq. (1-183) as shown below:

$$\Delta G^\circ = \sum_{i=1}^{N} \nu_i'' \Delta \mathscr{g}_{fi}^\circ - \sum_{i=1}^{N} \nu_i' \Delta \mathscr{g}_{fi}^\circ = \left[1\,\Delta \mathscr{g}_{f,H_2O_{(g)}}^\circ + \tfrac{1}{2}\,\Delta \mathscr{g}_{f,O_{2(g)}}^\circ + 0\,\Delta \mathscr{g}_{f,H_2O_{2(g)}}^\circ\right]$$

$$- \left[0\,\Delta \mathscr{g}_{f,H_2O_{(g)}}^\circ + 0\,\Delta \mathscr{g}_{f,O_{2(g)}}^\circ + 1\,\Delta \mathscr{g}_{f,H_2O_{2(g)}}^\circ\right]$$

$$= \left[1(-54.64) + \tfrac{1}{2}(0) + 0\right] - \left[0 + 0 + 1(-24.7)\right]$$

$$= -29.9 \text{ kcal}$$

From Eq. (1-181) ΔG° is related to K_p. However, 1 g-mole must be inserted in front of the right-hand side of Eq. (1-181) due to Eq. (1-161). After rearrangement, we have

$$K_p = \exp\left(-\frac{\Delta G^\circ}{R_u T}\right)$$

$$= \exp\left(\frac{29.9 \times 10^3 \text{ cal/g-mole}}{(1.987 \text{ cal/g-mole K})(298.16 \text{ K})}\right)$$

$$= 8.5 \times 10^{21}$$

TABLE 1.6 Free Energy of Formation[a] Δg_f° (kcal / mole) at 298 K

Gases		*Solids*	
H_2O	−54.64	AgCl	−26.22
H_2O_2	−24.7	AgBr	−22.39
O_3	39.06	AgI	−15.81
HCl	−22.77	BaO	−126.3
HBr	−12.72	$BaSO_4$	−350.2
HI	0.31	$BaCO_3$	−272.2
SO_2	−71.79	CaO	−144.1
SO_3	−88.52	$CaCO_3$	−269.8
H_2S	−7.89	$Ca(OH)_2$	−214.3
N_2O	24.9	SiO_2	−192.4
NO	20.72	Fe_2O_3	−177.1
NO_2	12.39	Al_2O_3	−376.8
NH_3	−3.97	CuO	−30.4
CO	−32.81	Cu_2O	−34.98
CO_2	−94.26	ZnO	−76.05
		Organic Compounds	
Gases			
Methane, CH_4	−12.14	Ethylene, C_2H_4	16.28
Ethane, C_2H_6	−7.86	Acetylene, C_2H_2	50.00
Propane, C_3H_8	−5.61	1-Butene, C_4H_8	17.09
n-Butane, C_4H_{10}	−3.75	*cis*-2-Butene, C_4H_8	15.74
Isobutane, C_4H_{10}	−4.3	*trans*-2-Butene, C_4H_8	15.05
n-Pentane, C_5H_{12}	−2.0	Isobutene, C_4H_8	13.88
Isopentane, C_5H_{12}	−3.5	1,3-Butadiene, C_4H_6	36.01
Neopentane, C_5H_{12}	−3.6	Methyl chloride, CH_3Cl	−14.0
Liquids			
Methanol, CH_3OH	−39.73	Benzene, C_6H_6	29.76
Ethanol, C_2H_5OH	−41.77	Chloroform, $CHCl_3$	−17.1
Acetic acid, CH_3COOH	−93.8	Carbon tetrachloride, CCl_4	−16.4
		Aqueous Ions	
H^+	0.0	OH^-	−37.59
Na^+	−62.59	Cl^-	−31.35
K^+	−67.47	Br^-	−24.57
Ag^+	18.13	I^-	−12.35
Ba^{2+}	−134.0	HS^-	3.01

TABLE 1.6 (Continued)

	Aqueous Ions		
Ca^{2+}	−132.48	S^{2-}	20.0
Cu^{2+}	15.53	SO_4^{2-}	−177.34
Zn^{2+}	−35.18	SO_3^{2-}	−126.2
	Gaseous Atoms		
H	48.57	I	16.77
F	14.2	C	160.84
Cl	25.19	N	81.47
Br	19.69	O	54.99

[a] From B. H. Mahan,[4] *Elementary Chemical Thermodynamics*, Benjamin, New York, 1964.

It is important to note that a large negative $\Delta G°$ corresponds to a large equilibrium constant. Large K_p means that once reaction starts the conversion of the reactants to products in their standard states will be quite complete at equilibrium with product concentrations much higher than those of reactants. Also, for any negative $\Delta G°$, K_p must be greater than unity.

Let us denote by $K_{p,i}$ the equilibrium constant for the formation of a certain molecular species i. Some examples are given below:

$$H_2 + \tfrac{1}{2}O_2 \rightleftharpoons H_2O, \qquad K_{p,H_2O} = \frac{p_{H_2O}}{p_{H_2}p_{O_2}^{1/2}}$$

$$\tfrac{1}{2}H_2 \rightleftharpoons H, \qquad K_{p,H} = \frac{p_H}{p_{H_2}^{1/2}}$$

$$\tfrac{1}{2}H_2 + \tfrac{1}{2}O_2 \rightleftharpoons OH, \qquad K_{p,OH} = \frac{p_{OH}}{p_{H_2}^{1/2}p_{O_2}^{1/2}}$$

We now consider a reaction involving all three species H_2O, H, and OH:

$$H_2O \rightleftharpoons H + OH$$

$$K_p = \frac{p_H p_{OH}}{p_{H_2O}}\frac{p_{H_2}p_{O_2}^{1/2}}{p_{H_2}p_{O_2}^{1/2}} = \frac{\dfrac{p_H}{p_{H_2}^{1/2}}\dfrac{p_{OH}}{p_{H_2}^{1/2}p_{O_2}^{1/2}}}{\dfrac{p_{H_2O}}{p_{H_2}p_{O_2}^{1/2}}} = \frac{K_{p,H}K_{p,OH}}{K_{p,H_2O}}$$

Therefore, K_p for the above reaction can be calculated from all the $K_{p,i}$ involved.

Consider the following general equilibrium chemical reaction with $\nu'_{i,s}$ and $\nu''_{i,s}$ representing the stoichiometric coefficients of reactants and products for the chemical species M_i; N is now the total number of chemical species involved:

$$\sum_{i=1}^{N} \nu''_{i,s} M_i \rightleftharpoons \sum_{i=1}^{N} \nu'_{i,s} M_i \tag{1-184}$$

The equilibrium constant K_p expressed in terms of ratios of equilibrium partial pressures can now be defined as

$$K_p = K_p(T) = \prod_{i=1}^{N} (p_{i,e})^{(\nu''_{i,s} - \nu'_{i,s})} \tag{1-185}$$

Other forms of the equilibrium constant are then

$$K_n \equiv \prod_{i=1}^{N} (n_{i,e})^{(\nu''_{i,s} - \nu'_{i,s})} \tag{1-186}$$

$$K_C \equiv \prod_{i=1}^{N} (C_{i,e})^{(\nu''_{i,s} - \nu'_{i,s})} \tag{1-187}$$

$$K_X \equiv \prod_{i=1}^{N} (X_{i,e})^{(\nu''_{i,s} - \nu'_{i,s})} \tag{1-188}$$

$$K_Y \equiv \prod_{i=1}^{N} (Y_{i,e})^{(\nu''_{i,s} - \nu'_{i,s})} \tag{1-189}$$

where for an ideal gas

$$C_i = (\text{concentration of species } i) = \frac{p_i}{R_u T}$$

$$n_i = (\text{actual number of moles of species } i) = \frac{p_i V}{R_u T} = \frac{p_i}{p} n_T$$

$$X_i = (\text{mole fraction of species } i) = \frac{p_i}{p} = \frac{n_i}{n_T}$$

$$Y_i = (\text{mass fraction of species } i) = \frac{W_i p_i}{\rho T R_u} = \frac{W_i C_i}{\rho} = \frac{\rho_i}{\rho}$$

$$n_T = (\text{total number of moles in the mixture}) = \sum_{i=1}^{N} n_i$$

$$p = (\text{total mixture pressure}) = \sum_{i=1}^{N} p_i$$

ρ = mixture density

If we define

$$\Delta n = \sum_{i=1}^{N} \nu_i'' - \sum_{i=1}^{N} \nu_i' \qquad (1\text{-}190)$$

the various equilibrium constants are then related according to

$$K_p = K_C(R_u T)^{\Delta n} = K_n\left(\frac{R_u T}{V}\right)^{\Delta n} = K_n\left(\frac{p}{n_T}\right)^{\Delta n} \qquad (1\text{-}191)$$

$$= K_X(p)^{\Delta n} = K_Y(R_u T\rho)^{\Delta n} \prod_{i=1}^{N} W_i^{(\nu_i' - \nu_i'')} \qquad (1\text{-}192)$$

When $\Delta n \neq 0$, K_p, unlike all the others, is a function of temperature only.

Consider the following reaction involving only ideal gases:

$$\nu_1' M_1 + \nu_2' M_2 \rightleftharpoons \nu_3'' M_3 + \nu_4'' M_4$$

Let $n°(M_1)$, $n°(M_2)$, $n°(M_3)$, and $n°(M_4)$ represent, respectively, the initial numbers of moles of M_1, M_2, M_3, and M_4 before equilibrium has been reached at a given temperature T. After equilibrium is reached, the following number of moles for each component will be present:

$$n_e(M_1) = n°(M_1) - \nu_1' X$$

$$n_e(M_2) = n°(M_2) - \nu_2' X$$

$$n_e(M_3) = n°(M_3) + \nu_3'' X$$

$$n_e(M_4) = n°(M_4) + \nu_4'' X$$

since for every $\nu_3'' X$ moles of M_3 formed, $\nu_4'' X$ moles of M_4 are formed, $\nu_1' X$ moles of M_1 disappear, and $\nu_2' X$ moles of M_2 disappear. From the definitions of K_n and K_p, it now follows that

$$K_n = \frac{[n°(M_3) + \nu_3'' X]^{\nu_3''}[n°(M_4) + \nu_4'' X]^{\nu_4''}}{[n°(M_i) - \nu_i' X]^{\nu_i'}[n°(M_2) - \nu_2' X]^{\nu_2'}}$$

and

$$K_p = K_n\left(\frac{p}{n_T}\right)^{(\nu_3'' + \nu_4'' - \nu_1' - \nu_2')}$$

where

$$n_T = n°(M_1) + n°(M_2) + n°(M_3) + n°(M_4) + X(\nu_3'' + \nu_4'' - \nu_1' - \nu_2')$$

It is evident that for given values of K_n or K_p, the value of X can be calculated readily for given initial concentrations of the various reactants.

Values of K_p presented in Tables 1.7a through 1.7d were compiled by Penner[9] from the sources indicated. The tabulated equilibrium constants are

TABLE 1.7a Equilibrium Constants with Respect to the Elements in Their Standard States[a]

T (K)	$K_{p,1}$	$K_{p,2}$	$K_{p,3}$	$K_{p,4}$	$K_{p,5}$
298.16	4.8978×10^{-41}	2.4831×10^{-36}	2.8340×10^{-7}	1.1143×10^{40}	1.9187×10^{-60}
300	9.0157×10^{-41}	4.2560×10^{-36}	3.1311×10^{-7}	6.1235×10^{39}	4.6559×10^{-60}
400	5.5335×10^{-30}	1.3459×10^{-26}	2.1419×10^{-5}	1.7418×10^{29}	1.8323×10^{-44}
500	1.7140×10^{-23}	6.9984×10^{-21}	2.6984×10^{-4}	7.6913×10^{22}	4.3351×10^{-35}
600	3.7239×10^{-19}	4.6452×10^{-17}	1.4555×10^{-3}	4.2954×10^{18}	7.8886×10^{-29}
700	4.7315×10^{-16}	2.5351×10^{-14}	4.8362×10^{-3}	3.8282×10^{15}	2.3768×10^{-24}
800	1.0162×10^{-13}	2.9040×10^{-12}	1.1869×10^{-2}	1.9454×10^{13}	5.4828×10^{-21}
900	6.6681×10^{-12}	1.1700×10^{-10}	2.3757×10^{-2}	3.1405×10^{11}	2.2856×10^{-18}
1000	1.9055×10^{-10}	2.2693×10^{-9}	4.1295×10^{-2}	1.1482×10^{10}	2.8708×10^{-16}
1100	2.9703×10^{-9}	2.5852×10^{-8}	6.4834×10^{-2}	7.6015×10^{8}	1.5066×10^{-14}
1200	2.9390×10^{-8}	1.9715×10^{-7}	9.4287×10^{-2}	7.8759×10^{7}	4.0926×10^{-13}
1300	2.0469×10^{-7}	1.1052×10^{-6}	1.2900×10^{-1}	1.1497×10^{7}	6.7143×10^{-12}
1400	1.0824×10^{-6}	4.8596×10^{-6}	1.6899×10^{-1}	2.2060×10^{6}	7.4131×10^{-11}
1500	4.5973×10^{-6}	1.7575×10^{-5}	2.1324×10^{-1}	5.2541×10^{5}	5.9402×10^{-10}
1600	1.6315×10^{-5}	5.4300×10^{-5}	2.6104×10^{-1}	1.4955×10^{5}	3.6787×10^{-9}
1700	4.9911×10^{-5}	1.4710×10^{-4}	3.1180×10^{-1}	4.9204×10^{4}	1.8420×10^{-8}
1800	1.3493×10^{-4}	3.5752×10^{-4}	3.6495×10^{-1}	1.8302×10^{4}	7.7215×10^{-8}
1900	3.2885×10^{-4}	7.9232×10^{-4}	4.2004×10^{-1}	7.5422×10^{3}	2.7861×10^{-7}
2000	7.3350×10^{-4}	1.6233×10^{-3}	4.7664×10^{-1}	3.3931×10^{3}	8.8491×10^{-7}
2100	1.5174×10^{-3}	3.1110×10^{-3}	5.3394×10^{-1}	1.6458×10^{3}	2.5194×10^{-6}
2200	2.9383×10^{-3}	5.6247×10^{-3}	5.9208×10^{-1}	8.5212×10^{2}	6.5283×10^{-6}
2300	5.3753×10^{-3}	9.6627×10^{-3}	6.5041×10^{-1}	4.6677×10^{2}	1.5585×10^{-5}
2400	9.3821×10^{-3}	1.5874×10^{-2}	7.0871×10^{-1}	2.6847×10^{2}	3.4610×10^{-5}
2500	1.5574×10^{-2}	2.5090×10^{-2}	7.6648×10^{-1}	1.6127×10^{2}	7.2161×10^{-5}
2750	4.7424×10^{-2}	6.8250×10^{-2}	9.0910×10^{-1}	5.3272×10^{1}	3.5917×10^{-4}
3000	1.2010×10^{-1}	1.5762×10^{-1}	1.0478	2.0999×10^{1}	1.3515×10^{-3}
3250	2.6381×10^{-1}	3.2048×10^{-1}	1.1788	9.5786	4.2678×10^{-3}
3500	5.1807×10^{-1}	5.8993×10^{-1}	1.3046	4.9295	1.1311×10^{-2}
3750	9.3022×10^{-1}	1.0000	1.4218	2.7498	2.6363×10^{-2}
4000	1.5528	1.5933	1.5315	1.6623	5.5361×10^{-2}
4250	2.4416	2.4010	1.6351	1.0568	1.0668×10^{-1}
4500	3.6521	3.4602	1.7322	7.0307×10^{-1}	1.9134×10^{-1}
4750	5.2349	4.8029	1.8229	4.9181×10^{-1}	3.2321×10^{-1}
5000	7.2395	6.4506	1.9067	3.5465×10^{-1}	5.1874×10^{-1}

[a] By permission, from NBS Circular 500, *Selected Values of Chemical Thermodynamic Properties*, 1 February 1952. The equilibrium constants are defined as follows: $K_{p,1} = p_{\mathrm{O}}/p_{\mathrm{O}_2}^{1/2}$, $K_{p,2} = p_{\mathrm{H}}/p_{\mathrm{H}_2}^{1/2}$, $K_{p,3} = p_{\mathrm{OH}}/p_{\mathrm{O}_2}^{1/2}p_{\mathrm{H}_2}^{1/2}$, $K_{p,4} = p_{\mathrm{H_2O}}/p_{\mathrm{H}_2}p_{\mathrm{O}_2}^{1/2}$, $K_{p,5} = p_{\mathrm{N}}/p_{\mathrm{N}_2}^{1/2}$. The partial pressures are those of the (ideal) gas, unless the contrary is indicated explicitly.

TABLE 1.7b Equilibrium Constants with Respect to the Elements in Their Standard States[a]

T (K)	$K_{p,6}$	$K_{p,7}$	$K_{p,8}$	$K_{p,9}$	$K_{p,10}$
298.16	6.5013×10^{-16}	3.1470×10^{-1}	1.2677×10^{-118}	1.0169×10^{24}	1.2331×10^{69}
300	8.1283×10^{-16}	3.1605×10^{-1}	7.4989×10^{-118}	8.4723×10^{23}	4.6559×10^{68}
400	6.9823×10^{-12}	3.7766×10^{-1}	1.4723×10^{-86}	1.3397×10^{19}	3.4356×10^{51}
500	1.6055×10^{-9}	4.1534×10^{-1}	9.2683×10^{-68}	1.7865×10^{16}	1.8113×10^{41}
600	6.0353×10^{-8}	4.4035×10^{-1}	3.2509×10^{-55}	2.1677×10^{14}	2.5177×10^{34}
700	8.0612×10^{-7}	4.5709×10^{-1}	2.9992×10^{-46}	9.2257×10^{12}	3.1842×10^{29}
800	5.6234×10^{-6}	4.6979×10^{-1}	1.5959×10^{-39}	8.5507×10^{11}	6.7143×10^{25}
900	2.5486×10^{-5}	4.7962×10^{-1}	2.7227×10^{-34}	1.3366×10^{11}	9.2470×10^{22}
1000	8.5487×10^{-5}	4.8742×10^{-1}	4.1687×10^{-30}	3.0061×10^{10}	4.7534×10^{20}
1100	2.3020×10^{-4}	4.9363×10^{-1}	1.1092×10^{-26}	8.8004×10^{9}	6.3387×10^{18}
1200	5.2590×10^{-4}	4.9785×10^{-1}	7.9068×10^{-24}	3.1499×10^{9}	1.7378×10^{17}
1300	1.0554×10^{-3}		2.0464×10^{-21}	1.3110×10^{9}	8.2414×10^{15}
1400	1.9178×10^{-3}		2.3988×10^{-19}	6.1660×10^{8}	6.0534×10^{14}
1500	3.2255×10^{-3}		1.4825×10^{-17}	3.1945×10^{8}	6.2951×10^{13}
1600	5.0781×10^{-3}		5.4828×10^{-16}	1.7960×10^{8}	8.6896×10^{12}
1700	7.5770×10^{-3}		1.3183×10^{-14}	1.0708×10^{8}	1.5031×10^{12}
1800	1.0817×10^{-2}		2.2803×10^{-13}	6.7422×10^{7}	3.1623×10^{11}
1900	1.4873×10^{-2}		2.7861×10^{-12}	4.4586×10^{7}	7.8524×10^{10}
2000	1.9815×10^{-2}		2.7102×10^{-11}	3.0641×10^{7}	2.2387×10^{10}
2100	2.5674×10^{-2}		2.1238×10^{-10}	2.1827×10^{7}	7.2028×10^{9}
2200	3.2516×10^{-2}		1.3671×10^{-9}	1.5389×10^{7}	2.5498×10^{9}
2300	4.0346×10^{-2}		7.5266×10^{-9}	1.1943×10^{7}	9.9426×10^{8}
2400	4.9125×10^{-2}		3.5818×10^{-8}	9.1348×10^{6}	4.1850×10^{8}
2500	5.8878×10^{-2}		1.5108×10^{-7}	7.1532×10^{6}	1.8915×10^{8}
2750	8.7466×10^{-2}		3.4380×10^{-6}	4.5426×10^{6}	3.3083×10^{7}
3000	1.2148×10^{-1}		4.6366×10^{-5}	2.6182×10^{6}	7.7108×10^{6}
3250	1.6040×10^{-1}		4.1812×10^{-4}	1.7527×10^{6}	2.2527×10^{6}
3500	2.0334×10^{-1}		2.7454×10^{-3}	1.1844×10^{6}	7.8109×10^{5}
3750	2.4939×10^{-1}		1.3948×10^{-2}	9.1496×10^{5}	3.1275×10^{5}
4000	2.9793×10^{-1}		5.7663×10^{-2}	6.9375×10^{5}	1.3880×10^{5}
4250	3.4839×10^{-1}				
4500	4.0085×10^{-1}				
4750	4.5188×10^{-1}				
5000	5.0405×10^{-1}				

[a] By permission, from NBS Circular 500, *Selected Values of Chemical Thermodynamic Properties*, 1 February 1952. The equilibrium constants are defined as follows: $K_{p,6} = p_{NO}/p_{N_2}^{1/2}p_{O_2}^{1/2}$, $\mathscr{K}_{p,7} = K_{p,7} = p_{C_{(c,diamond)}}/p_{C_{(c,graphite)}}$, $K_{p,8} = p_{C_{(g)}}$ and $\mathscr{K}_{p,8} = p_{C_{(g)}}/p_{C_{(c,graphite)}}$, $K_{p,9} = p_{CO}/p_{O_2}^{1/2}$ and $\mathscr{K}_{p,9} = p_{CO}/p_{C_{(c,graphite)}}p_{O_2}^{1/2}$, $K_{p,10} = p_{CO_2}/p_{O_2}$, $\mathscr{K}_{p,10} = p_{CO_2}/p_{C_{(c,graphite)}}p_{O_2}$.

TABLE 1.7c Equilibrium Constants with Respect to the Elements in their Standard States[a]

T (K)	$K_{p,11}$	$K_{p,12}$	$K_{p,14}$	$K_{p,15}$
298.16	7.9159×10^{8}	2.2439×10^{-37}	3.4277×10^{-19}	6.9343×10^{-15}
300	6.5826×10^{8}	3.9264×10^{-37}	4.6238×10^{-19}	8.8105×10^{-15}
400	3.0896×10^{5}	2.8774×10^{-27}	9.0991×10^{-14}	1.3970×10^{-10}
500	2.6749×10^{3}	2.3496×10^{-21}	1.4060×10^{-10}	4.7163×10^{-8}
600	9.9937×10^{1}	2.0184×10^{-17}	1.9151×10^{-8}	2.3030×10^{-6}
700	8.9578	1.2794×10^{-14}	6.4908×10^{-7}	3.7282×10^{-5}
800	1.4135	1.5922×10^{-12}	9.2003×10^{-6}	3.0227×10^{-4}
900	3.2501×10^{-1}	6.7608×10^{-11}	7.2812×10^{-5}	1.5453×10^{-3}
1000	9.8288×10^{-2}	1.3372×10^{-9}	3.8300×10^{-4}	5.7161×10^{-3}
1100	3.6771×10^{-2}	1.5283×10^{-8}	1.4949×10^{-3}	1.6719×10^{-2}
1200	1.6073×10^{-2}	1.1564×10^{-7}	4.6666×10^{-3}	4.0992×10^{-2}
1300	7.9177×10^{-3}	6.3738×10^{-7}	1.2249×10^{-2}	8.7740×10^{-2}
1400	4.3311×10^{-3}	2.7473×10^{-6}	2.8074×10^{-2}	1.6877×10^{-1}
1500	2.5586×10^{-3}	9.6962×10^{-6}	5.7717×10^{-2}	2.9813×10^{-1}

T (K)	$K_{p,16}$	$K_{p,18}$	$K_{p,19}$	$K_{p,20}$
298.16	2.5468×10^{-11}	4.8978×10^{16}	5.3629×10^{9}	2.9648×10^{1}
300	3.0620×10^{-11}	3.8994×10^{16}	3.4206×10^{9}	2.9174×10^{1}
400	5.9965×10^{-8}	3.6813×10^{12}	1.8923×10^{7}	1.7100×10^{1}
500	5.7332×10^{-6}	1.3964×10^{10}	8.2433×10^{5}	1.2050×10^{1}
600	1.2086×10^{-4}	3.3450×10^{8}	1.0044×10^{5}	9.4406
700	1.0725×10^{-3}	2.3073×10^{7}	2.2126×10^{4}	7.8524
800	5.5373×10^{-3}	3.0825×10^{6}	7.0746×10^{3}	6.7920
900	1.9911×10^{-2}	6.4062×10^{5}	2.8987×10^{3}	6.0534
1000	5.5578×10^{-2}	1.8159×10^{5}	1.4158×10^{3}	5.5208
1100	1.2897×10^{-1}	6.4670×10^{4}	7.8705×10^{2}	5.1050
1200	2.6062×10^{-1}	2.7296×10^{4}	4.8128×10^{2}	4.7863
1300	4.7315×10^{-1}	1.3131×10^{4}	3.1747×10^{2}	4.5186
1400	7.8977×10^{-1}	7.0146×10^{3}	2.2233×10^{2}	4.3152
1500	8.1111×10^{-1}	4.0654×10^{3}	1.6297×10^{2}	4.1400

[a] By permission, from NBS Circular 500, *Selected Values of Chemical Thermodynamic Properties*, 1 February 1952. The equilibrium constants are defined as follows: $K_{p,11} = p_{CH_4}/p^2_{H_2}$ and $\mathcal{K}_{p,11} = p_{CH_4}/p_{C_{(c,\ graphite)}}p^2_{H_2}$, $K_{p,12} = p_{C_2H_2}/p_{H_2}$ and $\mathcal{K}_{p,12} = p_{C_2H_2}/p^2_{C_{(c,\ graphite)}}p_{H_2}$, $K_{p,14} = p_{Cl}/p^{1/2}_{Cl_2}$, $K_{p,15} = p_{Br}/p^{1/2}_{Br_2}$, $K_{p,16} = p_I/p^{1/2}_{I_2}$, $K_{p,18} = p_{HCl}/p^{1/2}_{H_2}p^{1/2}_{Cl_2}$, $K_{p,19} = p_{HBr}/p^{1/2}_{H_2}p^{1/2}_{Br_2}$, $K_{p,20} = p_{HI}/p^{1/2}_{H_2}p^{1/2}_{I_2}$.

TABLE 1.7d Equilibrium Constants with Respect to the Elements in Their Standard States[a]

T (K)	$K_{p,13}$	$K_{p,17}$
298.16	1.998×10^{-21}	3.590×10^{-48}
300	2.908×10^{-21}	6.980×10^{-48}
400	1.319×10^{-14}	3.512×10^{-36}
500	1.399×10^{-10}	3.719×10^{-29}
600	7.019×10^{-7}	1.789×10^{-24}
700	6.095×10^{-6}	4.078×10^{-21}
800	1.767×10^{-4}	1.343×10^{-18}
900	2.439×10^{-3}	1.227×10^{-16}
1000	2.023×10^{-2}	4.558×10^{-15}
1200	4.874×10^{-1}	1.054×10^{-12}
1400	4.812	5.118×10^{-11}
1600	2.690×10^{1}	9.528×10^{-10}
1800	1.034×10^{2}	9.312×10^{-9}
2000	3.055×10^{2}	5.794×10^{-8}
2500	2.177×10^{3}	1.638×10^{-6}
3000	8.055×10^{3}	1.427×10^{-5}
3500	2.084×10^{4}	6.947×10^{-5}
4000	4.245×10^{4}	2.290×10^{-4}
4500	7.424×10^{4}	5.804×10^{-4}
5000	1.155×10^{5}	1.225×10^{-3}

[a] From L. G. Cole, M. Farber, and G. W. Elverum, Jr., *J. Chem. Phys.*, Vol. **20**, p. 586, 1952. The equilibrium constants are defined as follows: $K_{p,13} = p_{\mathrm{F}}^2/p_{\mathrm{F}_2}$, $K_{p,17} = p_{\mathrm{H}_2}^{1/2} p_{\mathrm{F}_2}^{1/2}/p_{\mathrm{HF}}$. Equilibrium constants involving the interhalogens ClF, BrF, IF, BrCl, ICl, and IBr are given by L. G. Cole and G. W. Elverum, Jr., *J. Chem. Phys.*, Vol. **20**, p. 1543, 1952.

written for the formation reaction to form one mole of substances from their elements, for example,

$$C_{(s)} + O_2 \rightleftarrows CO_2$$

When any elements are in the condensed phases, the equilibrium constants are written as κ_p, that is, $\kappa_p = p_{\mathrm{CO}_2}/(p_{\mathrm{O}_2} p_{\mathrm{C}_{(s)}})$ for the above equilibrium reaction. Replacing $p_{\mathrm{C}_{(s)}}$ by its vapor pressure and multiplying through by $P_{\mathrm{vp,C}}$, we have

$$K_p = \kappa_p P_{\mathrm{vp,C}} = \frac{p_{\mathrm{CO}_2}}{p_{\mathrm{O}_2}}$$

Values of κ_p defined either by the above equation or by Eq. (1-185) are tabulated in Tables 1.7a through 1.7d. K_p increases with increasing temperature for endothermic reactions and decreases for exothermic reactions.

Example 1.17

Consider the following chemical reaction:

$$CH_4 + 1.5O_2 \rightarrow CO + 2H_2O$$

In most cases, this reaction is not complete, and dissociation is usually very significant when the heat of combustion is high. Some of the products undergo a further reaction and form CO_2 and H_2; the equilibrium mixture is assumed to be

$$(1-\eta)CO + (2-\eta)H_2O + \eta CO_2 + \eta H_2$$

due to the reaction (called water-gas reaction)

$$CO + H_2O \rightleftarrows CO_2 + H_2$$

Write the necessary mathematical relationships required for solving the equilibrium composition and adiabatic flame temperature.

Solution:

For water–gas reaction, the equilibrium constant can be written as

$$K_p = \frac{p_{CO_2} p_{H_2}}{p_{CO} p_{H_2O}}$$

$$n_T = \sum_{i=1}^{N} n_i = (1-\eta) + (2-\eta) + \eta + \eta = 3$$

$$X_{CO} = \frac{1-\eta}{3}, \qquad X_{H_2O} = \frac{2-\eta}{3}, \qquad X_{CO_2} = X_{H_2} = \frac{\eta}{3}$$

$$p_{CO} = \frac{1-\eta}{3} p, \qquad p_{H_2O} = \frac{2-\eta}{3} p, \qquad p_{CO_2} = p_{H_2} = \frac{\eta}{3} p$$

Therefore,

$$K_p = \frac{\left(\frac{\eta}{3}\right)^2 p^2}{\left(\frac{1-\eta}{3}\right) p \left(\frac{2-\eta}{3}\right) p} = \frac{\eta^2}{(1-\eta)(2-\eta)}$$

$K_p(T_f)$ is obtained from suitable tables, η is determined, and the equilibrium composition is known. The adiabatic flame temperature can be determined through iteration, using the enthalpy balance equation. For the chemical reaction

$$CH_4 + 1.5O_2 \rightarrow (1-\eta)CO + (2-\eta)H_2O + \eta CO_2 + \eta H_2$$

the general enthalpy equation

$$\Delta H = \sum_{i=1}^{N} \nu_i''(\Delta \mathscr{H}_f^\circ)_i - \sum_{i=1}^{N} \nu_i'(\Delta \mathscr{H}_f^\circ)_i + \sum_{i=1}^{N} \nu_i''\mathscr{H}_i \Big|_{298}^{T_f} + \sum_{i=1}^{N} \nu_i'\mathscr{H}_i \Big|_{T_i}^{298} = 0$$

can be written as

$$\begin{aligned}
&(1-\eta)\,\Delta\mathscr{H}^\circ_{f,CO} + (2-\eta)\,\Delta\mathscr{H}^\circ_{f,H_2O} + \eta\,\Delta\mathscr{H}^\circ_{f,CO_2} - \Delta\mathscr{H}^\circ_{f,CH_4}\\
&\quad + \left[\mathscr{H}_{298} - \mathscr{H}_{T_i}\right]_{CH_4} + 1.5\left[\mathscr{H}_{298} - \mathscr{H}_{T_i}\right]_{O_2} + (1-\eta)\left[\mathscr{H}_{T_f} - \mathscr{H}_{298}\right]_{CO}\\
&\quad + (2-\eta)\left[\mathscr{H}_{T_f} - \mathscr{H}_{298}\right]_{H_2O} + \eta\left[\mathscr{H}_{T_f} - \mathscr{H}_{298}\right]_{CO_2} + \eta\left[\mathscr{H}_{T_f} - \mathscr{H}_{298}\right]_{H_2} = 0
\end{aligned}$$

Example 1.18

A mixture of 1 mole of N_2 and 0.5 mole of O_2 is heated to 4000 K at 1 atm pressure, resulting in an equilibrium mixture of N_2, O_2, and NO only. If the O_2 and N_2 were initially at 298.16 K and were heated steadily, how much heat was required to bring the final mixture to 4000 K on the basis of one initial mole of N_2?

Solution:

The reaction is written as

$$N_2 + 0.5O_2 \rightarrow a\,N_2 + b\,O_2 + c\,NO$$

The law of conservation of atomic species gives

$$N:\ 2 = 2a + c$$

$$O:\ 1 = 2b + c$$

$$\therefore\quad a = 0.5(2-c) = 1 - c/2$$

$$b = 0.5(1-c) = \tfrac{1}{2} - c/2$$

Letting $c = x$,

$$N_2 + 0.5O_2 \rightarrow (1 - x/2)N_2 + \left(\tfrac{1}{2} - x/2\right)O_2 + xNO$$

we are given $T_i = 298.16$ K and $T_f = 4000$ K, and from Table 1.7b for $0.5N_2 + 0.5O_2 \rightleftarrows NO$ we have $K_p = 0.29793 \approx 0.3$;

$$\therefore \quad K_p = 0.3 = \frac{p_{NO}}{p_{N_2}^{1/2} p_{O_2}^{1/2}} = \frac{x}{(1 - x/2)^{1/2}(\frac{1}{2} - x/2)^{1/2}} \left(\frac{p}{\Sigma n} \right)^{1-1/2-1/2}$$

and $0.09 = 4x^2/[(2 - x)(1 - x)]$, or $x^2 - 3x + 2 = 44.5x^2$; then $43.5x^2 + 3x - 2 = 0$, so that

$$x = \frac{-3 \pm \sqrt{9 + 8 \times 43.5}}{2 \times 43.5} = \frac{-3 \pm 18.89}{87}$$

$$= 0.1825 \text{ or } -0.2516 \text{ (no physical meaning)}$$

The final mixture is then

$$0.90875N_2 + 0.40875O_2 + 0.1825NO$$

Note (see accompanying graph):

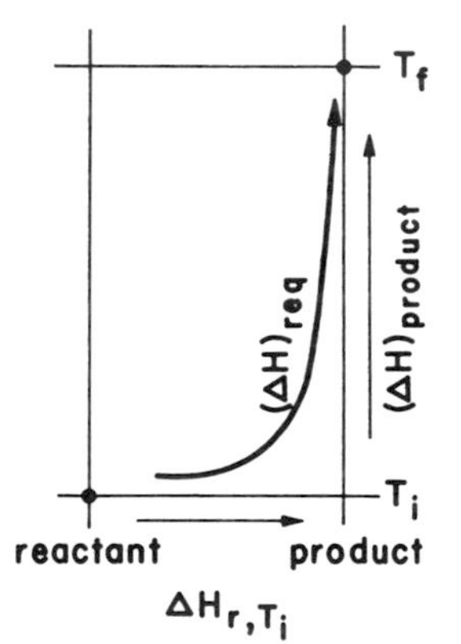

Temperature–reaction path showing $(\Delta H)_{\text{required}} = \Delta H_{r,T_i} + (\Delta H)_{\text{product}}$

$$\left(\Delta \mathscr{H}^{\circ}_{f,T_0} \right)_{NO} = +21.58 \text{ kcal/mole}$$

$$(\Delta H)_{\text{required}} = \underbrace{\left(\sum_{i=1}^{N} \nu_i''(\Delta \mathscr{H}_f^{\circ})_i - \sum_{i=1}^{N} \nu_i'(\Delta \mathscr{H}_f^{\circ})_i \right)}_{\Delta H_{r,T_i}} + (\Delta H)_{\text{product}}$$

$$= 0.1825 \times 21.58 + 0.90875(\mathscr{H}^{\circ}_{4000} - \mathscr{H}^{\circ}_{298.16})_{N_2}$$

$$+ 0.40875(\mathscr{H}^{\circ}_{4000} - \mathscr{H}^{\circ}_{298.16})_{O_2}$$

$$+ 0.1825(\mathscr{H}^{\circ}_{4000} - \mathscr{H}^{\circ}_{298.16})_{NO}$$

$$= 0.1825 \times 21.58 + 0.90875 \times 31.09 + 0.40875 \times 33.2$$

$$+ 0.1825 \times 31.7$$

$$= 51.53 \text{ kcal}$$

More complicated reactions (including dissociation) will be discussed following the introduction of the concepts of fugacity and activity.

15 FUGACITY

G. N. Lewis in 1901 introduced the concept of *fugacity*, which has proved of great value in representing the actual behavior of real gases, as distinct from the postulated behavior of ideal gases, especially at high pressures. For ideal gases at a constant temperature,

$$dG = nR_uT\,d(\ln p) \tag{1-193}$$

For a gas which does not behave ideally the above equation will not hold, but a function f, known as the fugacity, may be defined in such a manner that the relationship

$$dG = nR_uT\,d(\ln f) \tag{1-194}$$

is always satisfied, irrespective of whether the gas is ideal or not. Integrating Eq. (1-194),

$$G_2 - G_1 = nR_uT \ln \frac{f_2}{f_1} \tag{1-195}$$

In this respect, fugacity may be considered to be the corrected partial pressure:

$$\frac{f_i}{p_i} = \Gamma \text{ (proportionality constant)} \tag{1-196}$$

where Γ is a function of many parameters, such as the temperature and pressure. For an ideal gas

$$\frac{f}{p} = 1 \tag{1-197}$$

As the pressure of the real gas is decreased, however, the behavior approaches that for an ideal gas. Therefore, the gas at very low pressure is chosen as the reference state, and it is postulated that the ratio f/p then approaches unity. Thus

$$\lim_{p \to 0} \frac{f}{p} = 1 \tag{1-198}$$

where f has the units of pressure.

During an isothermal process at temperature T, the fugacity can be determined from the compressibility factor Z and the pressure p by the following relationship:

$$Z\,d(\ln p)_T = d(\ln f)_T \tag{1-199}$$

Integrating at constant temperature from $p = 0$ to a finite pressure, we have

$$\ln \frac{f}{p} = \int_0^{P_r} (Z - 1)\, d(\ln P_r)_T \tag{1-200}$$

At any temperature, the right-hand side of this equation can be integrated graphically, using the generalized compressibility chart to find Z at each reduced pressure P_r. The value of f is therefore determined.

16 MORE COMPLICATED DISSOCIATION IN THE COMBUSTION OF HYDROCARBONS

The frozen process in the combustion of methane, CH_4, is

$$CH_4 + 1.5O_2 \rightarrow CO + 2H_2O$$

In equilibrium, if the concentrations of C and O_2 are significant, the following three equilibrium reactions can be considered:

$$CO \rightleftharpoons C + \tfrac{1}{2}O_2, \qquad K_{p_1} = \frac{p_C p_{O_2}^{1/2}}{p_{CO}} \tag{1-201}$$

$$H_2O \rightleftharpoons H_2 + \tfrac{1}{2}O_2, \qquad K_{p_2} = \frac{p_{H_2} p_{O_2}^{1/2}}{p_{H_2O}} \tag{1-202}$$

$$CO_2 \rightleftharpoons C + O_2, \qquad K_{p_3} = \frac{p_C p_{O_2}}{p_{CO_2}} \tag{1-203}$$

The seven unknown quantities are the flame temperature and the partial pressures of the species $C, CO, CO_2, H_2, H_2O, O_2$.

The conservation of atomic species can be written for the three species present in the system:

$$\text{C:} \qquad 1 = n_C + n_{CO} + n_{CO_2}$$

$$\text{O:} \qquad 3 = 2n_{O_2} + n_{CO} + 2n_{CO_2} + n_{H_2O}$$

$$\text{H:} \qquad 4 = 2n_{H_2O} + 2n_{H_2}$$

where

$$\frac{n_i}{n_T} = \frac{p_i}{p} \tag{1-204}$$

n_T, although unknown, does not generally differ significantly from the stoichiometric value. n_T for $CH_4 + 1\frac{1}{2}O_2 \rightarrow CO + 2H_2O$ can be assumed to be 3. With p known, and substituting Eq. (1-204) into the atomic species conservation equations, we get

$$\frac{p}{n_T} = p_C + p_{CO} + p_{CO_2} \tag{1-205}$$

$$3\frac{p}{n_T} = 2p_{O_2} + p_{CO} + 2p_{CO_2} + p_{H_2O} \tag{1-206}$$

$$4\frac{p}{n_T} = 2p_{H_2O} + 2p_{H_2} \tag{1-207}$$

Assuming $T_f = T_{f(0)}$, we can find from tables the values for $K_{p_1}(T_{f(0)})$, $K_{p_2}(T_{f(0)})$, and $K_{p_3}(T_{f(0)})$. The values of p_C, p_{CO}, p_{CO_2}, p_{H_2O}, p_{H_2}, and p_{O_2} can be obtained by solving the six simultaneous equations (1-201), (1-202), (1-203), (1-205), (1-206), (1-207). n_C, n_{CO}, n_{CO_2}, n_{H_2O}, n_{H_2}, and n_{O_2} can be obtained from Eq. (1-204). The ν_i'''s are therefore known, and the enthalpy balance equation is

$$\sum_{i=1}^{N} \nu_i''\Delta\mathscr{H}^\circ_{f,M_i} - \sum_{i=1}^{N} \nu_i'\Delta\mathscr{H}^\circ_{f,M_i} + \sum_{i=1}^{N} \nu_i''\mathscr{H}_i\Bigg|_{298}^{T_f} + \sum_{i=1}^{N} \nu_i'\mathscr{H}_i\Bigg|_{T_i}^{298} = 0 \tag{1-208}$$

If Eq. (1-208) is not satisfied, we compute $n_T = \sum_{i=1}^{N} n_i$, reassume a value for T_f, and solve the six simultaneous equations again. This procedure is repeated until the enthalpy balance equation is satisfied. This is a general way of calculating the flame temperature and equilibrium compositions, and one which completely defines the whole system. One disadvantage of this method is that as the partial pressure, which is in the denominator of the K_p expression, of a particular species becomes very small, error can be introduced. In some cases, if the concentrations of C and O_2 are significant and only one dissociation equation

$$CO + H_2O \rightleftharpoons CO_2 + H_2$$

is assumed, the calculated flame temperature will be higher than the actual flame temperature and no indication of this will appear during the computation. Therefore, we should be careful to consider all the possible significant species in the equilibrium mixture; otherwise, the correct flame temperature will not be obtained.

If a certain substance in an equilibrium mixture has two phases, an additional unknown is introduced. If liquid water and water vapor coexist in one system, it is necessary to break $n_{H_2O_{total}}$ into $n_{H_2O_{(g)}}$ and $n_{H_2O_{(l)}}$; then

$$n_{H_2O_{total}} = n_{H_2O_{(g)}} + n_{H_2O_{(l)}}$$

and

$$f_{H_2O_{(g)}} \approx p_{H_2O_{(g)}} = \text{function of } T,$$

since the water vapor pressure is constant at a certain temperature. The equilibrium constant for phase equilibrium is known as a function of temperature:

$$K_p = \frac{p_{H_2O_{(g)}}}{p_{H_2O_{(l)}}}$$

For $H_2O \rightleftarrows \frac{1}{2}O_2 + H_2$, the equilibrium constant is

$$K_p = \frac{p_{O_2}^{1/2} p_{H_2}}{p_{H_2O_{(g)}}}$$

The vapor pressure in the denominator is a known quantity. Therefore, p_{O_2} can be expressed as a function of p_{H_2}, or $n_{O_2} = F(n_{H_2})$.

The new conservation equation of atomic species becomes

$$\text{O:} \qquad (\text{number of O}) = 2n_{O_2} + n_{H_2O_{total}} + \cdots$$

$$= 2F(n_{H_2}) + n_{H_2O_{total}} + \cdots$$

$$\text{H:} \qquad (\text{number of H}) = 2n_{H_2} + 2n_{H_2O_{total}}$$

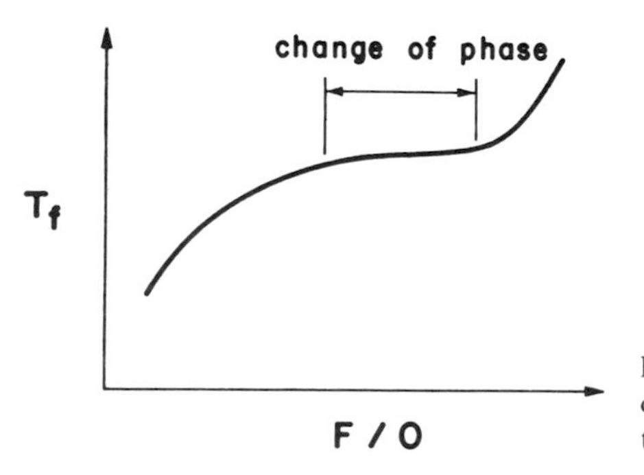

Figure 1.7 Exaggerated effect of phase change on the dependence of adiabatic flame temperature on fuel/oxidant ratio.

For $CO \rightleftharpoons C_{(s)} + \frac{1}{2}O_2$, $K_p = p_C p_{O_2}^{1/2}/p_{CO}$ should be considered; however, one can replace K_p by $K_p' = K_p/f_C = p_{O_2}^{1/2}/p_{CO}$, since p_C is negligible because the boiling temperature of C is very high.

In some cases, in order to insure that all species in the equilibrium mixture are considered, we must check to determine if the heat release is sufficient to change a given substance from one phase to another. In general, when a change in phase of certain species occurs, the adiabatic flame temperature versus F/O has a flat region. (See Fig. 1.7.) For some problems in which $H_2O_{(l)}$ is present in appreciable amounts, CO_2 may dissolve in the H_2O, resulting in the fugacity of the solution being different from that of pure $H_2O_{(l)}$.

17 THE CLAUSIUS – CLAPEYRON EQUATION: PHASE EQUILIBRIUM

Consider any system consisting of two phases, for example, liquid and vapor. As long as both phases are present, an appreciable transfer from one phase to the other will not disturb the equilibrium at constant temperature and pressure,

$$\Delta G = 0$$

Let G_A be the Gibbs free energy of the substance in one phase, and G_B that in the other phase. In phase change there is no work done other than that of expansion. Therefore,

$$dG_A = V_A\,dp - S_A\,dT \tag{1-209}$$

$$dG_B = V_B\,dp - S_B\,dT \tag{1-210}$$

where, because $\Delta G = 0$,

$$dG_A = dG_B \quad \text{and} \quad V_A\,dp - S_A\,dT = V_B\,dp - S_B\,dT \tag{1-211}$$

Therefore,

$$\frac{dp}{dT} = \frac{S_A - S_B}{V_A - V_B} = \frac{\Delta S}{\Delta V} \tag{1-212}$$

This is the Clausius–Clapeyron equation for phase equilibrium; it effectively relates the change in vapor pressure of a substance to a very small change in temperature. Since

$$\Delta S = \frac{\Delta H}{T},$$

and since $\Delta \mathscr{H}_v$ is the molar latent heat of the phase change for one mole of substance at temperature T,

$$\frac{dp}{dT} = \frac{\Delta \mathscr{H}_v}{T \Delta v} \tag{1-213}$$

where Δv is the difference in the molar volumes of the two phases. Equation (1-212) gives the variation of the equilibrium pressure with temperature for any two phases of a given substance.

For liquid–vapor equilibrium, Eq. (1-212) becomes

$$\boxed{\frac{dp}{dT} = \frac{\Delta \mathscr{H}_v}{T(v_g - v_l)}} \tag{1-214}$$

If the variation of vapor pressure with temperature is known, it is possible to calculate the heat of vaporization. If the temperature is not too near the critical point, then

$$v_g \gg v_l, \qquad \text{or} \quad v_g - v_l \simeq v_g$$

Equation (1-214) then becomes

$$\frac{dp}{dT} \approx \frac{\Delta \mathscr{H}_v}{T v_g} \tag{1-215}$$

Furthermore, in regions well below the critical point, the vapor pressure is relatively small, and the ideal-gas law may be assumed to be applicable, that is,

$pv_g = R_uT$ where v_g is the molar volume of the vapor. Thus

$$\frac{dp}{dT} = \frac{p\,\Delta\mathscr{H}_v}{R_uT^2}$$

Therefore, the Clausius–Clapeyron equation may be written as

$$\boxed{\frac{d(\ln p)}{dT} \approx \frac{\Delta\mathscr{H}_v}{R_uT^2}} \tag{1-216}$$

and the integrated form is

$$\boxed{\ln\frac{p_2}{p_1} \approx -\frac{\Delta\mathscr{H}_v}{R_u}\left(\frac{1}{T_2}-\frac{1}{T_1}\right)} \tag{1-217}$$

This equation implies that the vapor pressure p_2 at T_2 can be calculated if the vapor pressure p_1 at T_1 is known. However, it should be noted that when a change of state is involved between T_1 and T_2, the vapor-pressure-versus-temperature curve may have significant changes in slope, as shown by the accompanying figure.

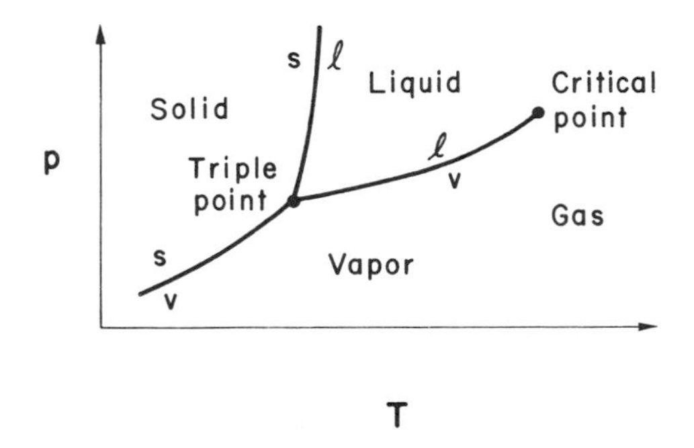

Pressure versus temperature of a substance, showing the abrupt change in slope at the triple point from solid-vapor interface to liquid–vapor interface

Table 1.8 gives thermodynamic data for phase changes of many substances.

TABLE 1.8 Thermodynamic Data for Phase Changes[a]

Substance	Process	Pressure (Torr)	Temperature (K)	$\Delta\mathcal{H}$ (kcal/mole)	ΔS (cal/mole K)	ΔC_p (cal/mole K)
C_2H_4O	$c \to l$	760	160.71	1.236	7.69	3.45
(ethylene oxide)	$l \to g$		283.72	6.101	21.50	−9.7
C_2H_6O	$c \to l$		131.66	1.180	8.96	6.8
(dimethylether)	$l \to g$		248.34	5.141	20.70	−10.6
C_2H_5OH	$c \to l$		158.6	1.200	7.57	5.70
(ethanol)	$l \to g$	760	351.7	9.22	26.22	
CH_4ON_2 (urea)	$c \to l$		405.8	3.60	8.9	
CH_3O_2N	$c \to l$		244.78	2.319	9.47	
(nitromethane)	$l \to g$	760	374.0			
CH_3O_2N (methyl nitrite)	$l \to g$	760	255	5.0	19.7	
CH_3O_3N (methyl nitrate)	$l \to g$	760	339.7	7.8	23.0	
CH_5O_2N	$c \to l$		18			
(nitroethane)	$l \to g$	17	293	9.1	31	
$C_2H_5O_2N$ (ethyl nitrite)	$l \to g$	760	290.1	6.64	22.9	
$C_2H_5O_3N$	$c \to l$		171			
(ethyl nitrate)	$l \to g$	760	361.9			
$C_2H_4O_6N_2$	$c \to l$		250.9	4.5	18	
(glycol dinitrate)	$l \to g$	19	378			
$C(NO_2)_4$	$c \to l$		286			
(tetranitromethane)	$l \to g$	760	398.9	9.2	23	
CO	$l \to g$	760	81.67	1.444	17.68	
CO_2	$l \to g$	760	194.68	6.031	30.98	
Cl_2	$c \to l$		172.18	1.531	8.892	2.75
	$l \to g$	760	239.11	4.878	20.4	−8.76
ClF	$l \to g$	760	172.9	5.34	30.88	
Cl_2O_7	$l \to g$	760	354.7	7.88	22.2	
F_2	$c \to l$		55.20	0.372	6.74	1.86
	$l \to g$	760	85.24	1.51	17.7	4.27
F_2O	$l \to g$	760	128.3	2.65	20.7	
H_2	$c \to l$	54.0	13.96	0.028	2.0	1.9
	$l \to g$	760	20.39	0.216	10.6	
HBr	$c \to l$		186.30	0.5751	3.087	1.64
	$l \to g$	760	206.44	4.210	20.39	−7.37
HCl	$c, I \to l$		158.97	0.4760	2.994	2.10
	$l \to g$	760	188.13	3.86	20.5	−7.14
HCN	$l \to g$	760	298.86	6.027	20.17	
HF	$c \to l$		190.09	1.094	15.756	2.55
	$l \to g$	760	293.1	1.8	6.1	10.9

TABLE 1.8(Continued)

Substance	Process	Pressure (Torr)	Temperature (K)	$\Delta\mathscr{H}$ (kcal/mole)	ΔS (cal/mole K)	ΔC_p (cal/mole K)
HI	$c \to l$					
	$c \to g$	0.31	298.16	14.88	49.91	
	$c \to l$		222.37	0.6863	3.086	1.10
HNO_3	$c \to l$		231.57	2.503	10.808	10.55
	$l \to g$	48	293	9.43	32.17	
H_2O	$c \to l$	760	273.16	1.4363	5.2581	8.911
	$l \to g$	4.58	273.16	10.767	39.416	−10.184
	$l \to g$	23.75	298.16	10.514	35.263	9.971
	$l \to g$	760	373.16	9.717	26.040	10.021
H_2O_2	$c \to l$		271.2	2.52	9.29	
	$l \to g$	2.1	298.16	13.01	43.64	
I_2	$c \to l$		386.8	3.74	9.67	
IF_7	$c \to l$	760	276.6	7.37	26.64	
N_2	$c, I \to l$	94	63.15	0.172	2.709	
	$l \to g$	760	77.36	1.333	17.231	
NH_3	$c \to l$	45.57	195.42	1.351	6.9133	
	$l \to g$	760	239.76	5.581	23.277	
N_2H_4	$c \to l$		274.7			
	$l \to g$	764	386.7	10	25.9	
NH_4N_3	$c \to g$	760	407	15.1	37.11	
NH_4NO_3	$c, V \to$ c, IV	760	255	0.13	0.511	
	$c, IV \to$ c, III	760	305.3	0.38	1.23	
	c, III c, II	6.32×10^5	336.5	0.20	0.594	
	$c, II \to$ c, I	760	398.4	1.01	2.535	
	$c, I \to l$	760	442.8	1.3	12.94	
$N_2H_1 \cdot HNO_2$	$c, \to l$		316			
$N_2H_4NO_3$	$c, I \to l$		343.9			
$N_2H_4 \cdot H_2O$	$c \to l$		233			
	$l \to g$		118.5	391.7		
NH_2OH	$c \to l$		306.3			
	$l \to g$	22	331			
NO	$c \to l$	164.4	109.55	0.5495	5.016	6.0
	$l \to g$	760	121.42	3.292	27.113	11.8
N_2O	$c \to l$	658.9	182.34	1.563	8.5719	4.67
	$l \to g$	760	184.68	3.956	21.421	
N_2O_3	$c \to l$		162			
	$l \to g$	760	275	9.4	34.2	

TABLE 1.8(Continued)

Substance	Process	Pressure (Torr)	Temperature (K)	$\Delta\mathscr{H}$ (kcal/mole)	ΔS (cal/mole K)	ΔC_p (cal/mole K)
N_2O_4	$c \to l$	139.78	261.96	3.502	13.368	6.12
	$l \to g$	760	294.31	9.110	30.954	
N_2O_5	$c \to g$	760	305.6	13.6	44.50	
NO_2F	$c \to l$		107.2			
	$l \to g$	760	200.8	4.31	21.46	
NO_3F	$c \to l$		92			
	$l \to g$	103	193			
O_2	$c, \text{I} \to l$	1.1	54.39	0.1063	1.95	1.74
	$l \to g$	760	90.19	1.6299	18.07	−6.00
O_3	$l \to g$	760	162.65	2.59	15.92	
Br_2	$c \to l$	760	265.9	2.52	9.48	0.9
	$l \to g$	214	298.16	7.34	24.6	
C	$c \to g$	760	4620			
CH_4	$c \to l$	87.7	90.68	0.225	2.48	
	$l \to g$	760	111.67	1.955	17.51	
C_2H_2	$c \to l$	900	191.7	0.9	5	
(ethyne,	$l \to g$	900	191.7	4.2	22	
acetylene)	$l \to g$	760	189.2	5.1	27	
C_2H_4	$c \to l$	0.9	103.97	0.8008	7.702	
(ethene, ethylene)	$l \to g$	760	169.45	3.237	19.10	
C_2H_6	$c \to l$	0.006	89.89	0.6829	7.597	2.2
(ethane)	$l \to g$	760	184.53	3.517	19.06	−11.5
CHF_3	$c \to l$		113			
(trifluoro-methane)	$l \to g$	760	189.0	4.4	2.3	
CH_5N	$c \to l$		179.70	1.466	8.16	
(methylamine)	$l \to g$	760	266.84	6.17	23.1	
C_2H_7N	$c \to l$		180.97	1.420	7.85	9.81
(dimethylamine)	$l \to g$	760	280.0	6.33	22.6	17.1
$C_2H_8N_2$ (2-dimethyl hydrazine)	$l \to g$	760	354			
CH_2O	$c \to l$		154.9			
(formaldehyde)	$l \to g$	760	253.9	5.85	23.0	
CH_4O	$c, \text{I} \to l$		175.26	0.757	4.32	4.2
(methanol)	$l \to g$	760	337.9	8.43	24.95	
CH_4O_2 (methyl hydrogen peroxide)	$l \to g$	34	298	7.9	26.5	

[a] By permission, from NBS Circular 500, *Selected Values of Chemical Thermodynamic Properties*, 1 February 1952.

18 CALCULATION OF COMPLEX EQUILIBRIUM WITH THE NASA – LEWIS COMPUTER PROGRAM

Many processes used today involve complex chemical mixtures, frequently at high temperatures. Some of these mixtures result from combustion processes such as occur in automobiles, aircraft, and rockets. Others occur in processing equipment in the chemical, petroleum, and natural-gas industries. Research equipment, such as shock tubes, also involves high-temperature gas mixtures.

The need frequently arises for the thermodynamic and transport properties of these mixtures, particularly for use in heat- and mass-transfer calculations. Usually, the temperature of the gases is too high for the properties to be measured directly. Consequently, the properties are calculated.

Numerous computer programs have been written to calculate both thermodynamic properties and transport properties. However, recently Svehla and McBride[2] at the NASA–Lewis Research Center have extended the widely used CEC71 program[1] (short for Chemical Equilibrium Calculations, 1971) to include transport properties calculations for any general chemically reacting system. The computer program was called TRAN72 (short for Transport Properties, 1972) and sometimes called CEC72. The latest revisions of the TRAN76 program were described by Gordon and McBride in 1976.[21] The purpose of this section is to highlight the capabilities, theory, and required input data for the multipurpose TRAN76 computer program. For a more comprehensive and in-depth review, the reader is referred to Refs. 1, 2, 21, and 22.

18.1 Assumptions and Capabilities

The program is designed to provide both thermodynamic and transport properties for a wide range of scientific and engineering applications and for a range of independent variables. Thermodynamic data for more than 400 chemical species (including gases, solids, and ions) are provided with the program for a temperature range of 300 to 5000 K. Transport and relaxation data are provided for many common species; transport data are also provided for interactions between unlike species. In contrast with the thermodynamic data, the temperature range of the transport data is not the same for all interactions.

The range of applicability of the thermodynamic calculations is approximately described by the limits of applicability of the ideal-gas law. The lower limit for temperature in the transport calculations occurs when ternary and higher order molecular collisions become important. This also defines the upper pressure limit for the transport property calculations. The upper limit for temperature occurs when ionization becomes appreciable. Incipient ionization can be included in the calculations. But for increasing ionization, higher approximations are needed in the transport calculations. The lower limit is given by the onset of free-molecular-flow regime, which occurs when the mean free path is of the same order of magnitude as the dimensions of the container.

Under these conditions the equations for the transport properties are no longer applicable.

Thermodynamic data contained in the TRAN76 data file is obtained from the JANAF Thermochemical Tables[16] and from data generated at NASA–Lewis Research Center. Heats of transition are included in the thermodynamic data, and the program automatically checks for the condensation of species. Thus, without any special instructions, the proper concentrations of gaseous, liquid, and solid phases of all species included in the THERMO library are calculated.

The usual equations for the conservation of mass, momentum, and energy are used along with the ideal-gas assumption. Composition and properties are calculated for equilibrium conditions and, in some situations, for frozen conditions (sometimes called nonreacting). The effects of chemical kinetics—that is, of finite reaction rates—are not included.

The program is capable of calculating several types of problems:

1. Equilibrium compositions for assigned thermodynamic states
2. Theoretical rocket performance
3. Shock-tube parameter calculations
4. Chapman–Jouguet detonations

In the first type of problem it can calculate the equilibrium composition of a mixture for assigned thermodynamic states. The thermodynamic states are assigned by specifying two thermodynamic state functions (code names are given in parenthesis):

1. Temperature and pressure (TP)
2. Enthalpy and pressure (HP)
3. Entropy and pressure (SP)
4. Temperature and volume or density (TV)
5. Internal energy and volume or density (UV)
6. Entropy and volume or density (SV)

As mentioned, the ideal-gas law is taken for the equation of state for the mixture. The analysis assumes that interactions between phases are negligible. In the event that condensed species are present, it is assumed that they occupy negligible volume and exert negligible pressure compared to the gaseous species.

18.2 Equations Describing Chemical Equilibrium

Prior to 1958 all equilibrium computations were carried out using the equilibrium constant formulation of the governing equations. In 1958, White, Johnson, and Dantzig[23] suggested that equilibrium compositions be calculated by free-energy minimization. Their procedure soon captured the fancy of some of the researchers making thermodynamic calculations, and the world of equilibrium computations was then divided into two camps, the free-energy minimizers and the traditional equilibrium-constant formulators. Most present-day multipurpose chemical-equilibrium computer programs use the free-energy minimization procedure; however, in some special-purpose computer programs, the equilibrium-constant formulation technique is used. The two formulations reduce to the same number of nonlinear iterative equations.

There are several disadvantages of the equilibrium-constant method which limit the versatility of a multipurpose program. Briefly, these disadvantages are more bookkeeping, numerical difficulties with the use of components, more difficulty in testing for the presence of condensed species, and more difficulties in extending the method to include the effects of nonideal equations of state. For these reasons, the free-energy minimization formulation was used to develop CEC76 (TRAN76). A very brief discussion of the development of the nonlinear iterative equations is presented below. For a more complete thermodynamic and mathematical treatise, the reader is referred to Ref. 22.

18.2.1 Thermodynamic Equations

As discussed earlier, it is assumed that all gases are ideal and that interactions among phases may be neglected. The equation of state for the mixture is

$$pV = nR_uT \tag{1-218}$$

or

$$\frac{p}{\rho} = RT \tag{1-218a}$$

wherein the units for the variables are those of the International System (SI). Equation (1-218) is assumed to be correct even when small amounts of condensed species (up to several percent by weight) are present. The volume V and number of moles n refer to the gases only, while the mass in the system is for the entire mixture including condensed species. The word "mixture" is used to designate the burned (reached) mixture of species at equilibrium, to distinguish from the mixture of unburned reactants, which is referred to as "total reactants." The molecular weight W of the mixture (including condensed species) is then defined to be

$$W = \frac{\sum_{j=1}^{N} n_j^* W_j}{\sum_{j=1}^{N_g} n_j^*} = \frac{1}{\sum_{j=1}^{N_g} n_j^*} = \frac{1}{n^*} \tag{1-219}$$

where n_j^* is the number of kilogram-moles of species j per kilogram of mixture, and W_j is the molecular weight of species j. Among the N possible species which may be considered, gases are indexed from 1 to N_g and condensed species from $N_g + 1$ to N.

18.2.2 Minimization of Gibbs Free Energy

The condition for equilibrium may be stated in terms of any of several thermodynamic functions, for example, the minimization of the Gibbs free energy or Helmholtz free energy, or the maximization of the entropy. If one wishes to use temperature and pressure to characterize a thermodynamic state, the Gibbs free energy is most easily minimized, inasmuch as temperature and pressure are its natural variables. Similarly, the Helmholtz free energy is most easily minimized if the thermodynamic state is characterized by temperature and volume (or density). In the following equations, the nonlinear iterative relations for minimization of the Gibbs free energy is presented. A parallel development for minimization of the Helmholtz free energy can be found in Ref. 1.

For a mixture of N species, the Gibbs free energy ($\mathcal{g}$) per kilogram of mixture is given by

$$\mathcal{g} = \sum_{j=1}^{N} \mu_j n_j^* \tag{1-220}$$

where the chemical potential per kilogram-mole of species j is defined to be

$$\mu_j = \left[\frac{\partial \mathcal{g}}{\partial n_j^*}\right]_{T,\,P,\,n_j^{*\prime}} \tag{1-221}$$

The condition for chemical equilibrium is the minimization of free energy. This minimization is usually subject to certain constraints, such as the following mass-balance constraints:

$$\sum_{j=1}^{n} a_{ij} n_j^* - b_i^\circ = 0, \qquad i = 1, \ldots, l \tag{1-222}$$

or

$$b_i - b_i^\circ = 0, \qquad i = 1, \ldots, l \tag{1-222a}$$

where the coefficient a_{ij} is the number of kilogram-atoms of element i per kilogram-mole of species j, b_i° is the assigned number of kilogram-atoms of element i per kilogram of total reactants (fuel and oxidant), and

$$b_i = \sum_{j=1}^{n} a_{ij} n_j^*, \qquad i = 1, \ldots, l \tag{1-223}$$

is the number of kilogram-atoms of element i per kilogram of mixture.

To minimize G, we shall follow the well-known Lagrange method[24] developed by the great eighteenth-century mathematician J. L. Lagrange. Let us define

$$\tilde{G} \equiv \mathcal{g} + \sum_{i=1}^{l} \lambda_i (b_i - b_i^\circ) \tag{1-224}$$

where λ_i are Lagrange multipliers that incorporate the l constraints in Eq. (1-222a). Then the condition for equilibrium becomes

$$\delta\tilde{G} = \sum_{j=1}^{N} \left(\mu_j + \sum_{i=1}^{l} \lambda_i a_{ij} \right) \delta n_j^* + \sum_{i=1}^{l} (b_i - b_i^\circ)\, \delta\lambda_i = 0 \tag{1-225}$$

Treating the variations δn_j^* and $\delta\lambda_i$ as independent gives

$$\mu_j + \sum_{i=1}^{l} \lambda_i a_{ij} = 0, \qquad j = 1, \ldots, N \tag{1-226}$$

and also the mass-balance equation (1-222a).

Based on the assumptions of an ideal gas the chemical potential may be written

$$\mu_j = \begin{cases} \mu_j^\circ + R_u T \ln\left(\dfrac{n_j^*}{n^*} \right) + R_u T \ln p(\text{atm}) & (j = 1, \ldots, N_g) \\ u_j^\circ & (j = N_g + 1, \ldots, N) \end{cases} \tag{1-227}$$

where μ_j° for gases ($j = 1$ to N_g) and for condensed phases ($j > N_g$) is the chemical potential in the standard state. The numerical values of μ_j° are generally found in JANAF Thermochemical Tables.[16] The pressure p(atm) in Eq. (1-227) must be in atmospheres. Equations (1-226) and (1-222a) permit the determination of equilibrium compositions for thermodynamic states specified by an assigned temperature T_0 and pressure p_0.

The equations required to obtain equilibrium composition are not all linear in the composition variables, and therefore an iteration procedure is generally required. Detailed iteration procedures are described in Ref. 1. Briefly speaking, the TRAN76 program follows a widely used steepest-descent Newton–Raphson method to solve for corrections to initial estimates of compositions n_j^*, Lagrange multipliers λ_i, mole number n, and (when required) temperature T. This method involves a Taylor series expansion of the appropriate equations with all terms containing derivatives higher than the first omitted. The correction variables used are $\Delta \ln n_j^*$ ($j = 1, \ldots, N_g$), Δn_j^* ($j = N_g + 1, \ldots, N$), $\Delta \ln n$, $\pi_i = -\lambda_i/(R_u T)$, and $\Delta \ln T$. As pointed out in Ref. 22, there is no restriction in starting each iteration with the estimate for the Lagrangian multipliers equal to zero, inasmuch as they appear linearly in Eq. (1-226).

For chemical systems containing many species, it would be necessary to solve a large number of simultaneous equations. They can be reduced quite simply to a much smaller number by algebraic substitution, eliminating

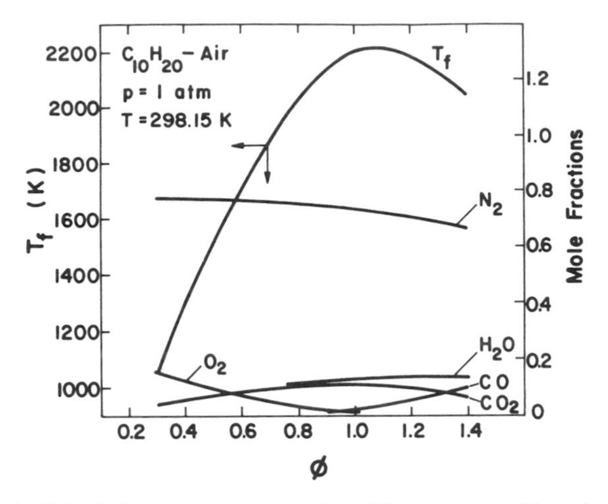

Figure 1.8 Calculated adiabatic flame temperature and equilibrium composition of products from combustion of kerosene in air.

$\Delta \ln n_j^*$ terms for gaseous species. The resulting equations are solved through continuous iteration until the corrections become smaller than the specified tolerance. Convergence of the Newton–Raphson iterative equations has always been obtained in less than the 35 iterations permitted by the program.[1] For most problems, a typical number of iterations is 8 to 12.

A typical output for the combustion of kerosene ($C_{10}H_{20}$) in air at initial temperature of 298.15 K and at constant pressure of 1 atm is shown in Fig. 1.8. The calculated equilibrium compositions and the flame temperature of the combustion product are plotted against equivalence ratio. As we can see from this figure, the adiabatic flame temperature reaches a peak very close to the stoichiometric condition ($\phi = 1.0$) on the slightly fuel-rich side. This is due to the fact that when the system is slightly underoxidized, the specific heat of the products is reduced and thus the flame temperature is increased.

REFERENCES

1. S. Gordon and B. J. McBride, *Computer Program for Calculation of Complex Chemical Equilibrium Compositions, Rocket Performance, Incident and Reflected Shocks, and Chapman-Jouguet Detonations*, NASA SP-273, 1971; Interim Revision, March 1976.
2. R. A. Svehla and B. J. McBride, *Fortran IV Computer Program for Calculation of Thermodynamic and Transport Properties of Complex Chemical Systems*, NASA TN D-7056, January 1973.
3. G. J. Van Wylen and R. E. Sonntag, *Fundamentals of Classical Thermodynamics*, John Wiley & Sons, New York, 1973.
4. B. H. Mahan, *Elementary Chemical Thermodynamics*, W. A. Benjamin, New York, 1964.
5. G. N. Lewis and M. Randall, *Thermodynamics*, McGraw-Hill Book Company, New York, 1961.
6. C. E. Mortimer, *Chemistry: A Conceptual Approach*, D. van Nostrand Company, Princeton, N.J., 1971.

7. S. Glasstone, *Thermodynamics for Chemists*, D. van Nostrand Company, Princeton, N.J., 1964.
8. R. A. Strehlow, *Fundamentals of Combustion*, International Textbook Co., Scranton, Pa., 1968.
9. S. S. Penner, *Chemistry Problems in Jet Propulsion*, Pergamon Press, New York, 1957.
10. F. A. Williams, *Combustion Theory*, Addison-Wesley Publishing Company, Reading, Mass., 1965.
11. I. Glassman, *Combustion*, Academic Press, New York, 1977.
12. A. M. Kanury, *Introduction to Combustion Phenomena*, Gordon and Breach, New York, September 1975.
13. R. C. Weast, *Handbook of Chemistry and Physics*, 49th ed., The Chemical Rubber Co., Cleveland, Ohio, 1968.
14. I. Prigogine and R. Defay, *Chemical Thermodynamics*, Longmans, Green, New York, 1954.
15. F. D. Rossini, *Chemical Thermodynamics*, Chapter 16, John Wiley & Sons, New York, 1950.
16. Anon., *JANAF Thermochemical Tables*, 2nd ed., U.S. Standard Reference Data System NSRDS-NBS 37, June 1971.
17. Anon., *Handbook of Chemistry and Physics*, 1983–84 ed., The Chemical Rubber Co., Cleveland, Ohio, 1983.
18. D. K. Sebera, *Electronic Structure and Chemical Bonding*, Chapter 7, Blaisdell Publishing Co., New York, 1964.
19. D. B. Spalding, *Combustion and Mass Transfer*, Chapter 6, Pergamon Press, New York, 1979.
20. J. R. Andrews and O. Biblarz, *Temperature Dependence of Gas Properties in Polynomial Form*, Naval Postgraduate School, Report NPS67-81-001, January 1981.
21. S. Gordon and B. J. McBride, *Computer Program for Calculation of Complex Chemical Equilibrium Compositions, Rocket Performance, Incident and Reflected Shocks, and Chapman–Jouguet Detonations*, NASA SP-273 Interim Revision N78-17724, March 1976.
22. F. J. Zeleznik and S. Gordon, "Calculation of Complex Equilibria," *Ind. Eng. Chem.*, Vol. 60, No. 6, pp. 27–57, June 1968.
23. W. B. White, S. M. Johnson, and G. B. Dantzig, *J. Chem. Phys.*, Vol. 28, p. 751, 1958.
24. A. E. Taylor, *Advanced Calculus*, Chapter 6, Ginn and Co., Lexington, Mass., 1955.
25. W. R. Smith and R. W. Missen, *Chemical Reaction Equilibrium Analysis: Theory and Algorithms*, John Wiley & Sons, New York, 1982.

HOMEWORK

1. Show that the total stored energy E is a thermodynamic property.

2. Show that the rate equation of the first law for a control volume (c.v.) can be expressed as:

$$\dot{Q}_{\text{c.v.}} + \sum \dot{m}_i \left(h_i + \frac{|\mathbf{v}_i|^2}{2g_c} + z_i \frac{g}{g_c} \right)$$

$$= \frac{dE_{\text{c.v.}}}{dt} + \sum \dot{m}_e \left(h_e + \frac{|\mathbf{v}_e|^2}{2g_c} + z_e \frac{g}{g_c} \right) + \dot{W}_{\text{c.v.}}$$

where h represents specific enthalpy ($h = u + p/\rho$) and subscripts i and e designate the inlet and exit conditions, respectively. Give the physical meaning of each term.

3. Air flows steadily through an air heater at the rate of 125 g/sec. It enters at 290 K and 1.2 atm with a velocity of 15 m/sec, and leaves at 325 K and 1 atm with a velocity of 18 m/sec. There is no shaft work done on or by the air, and the centerlines of the inlet and discharge ducts are in the same horizontal plane. The heating is accomplished with steam coils. What is the rate of heat input to the air in J/s? ($C_{p,\,\mathrm{air}} = 0.24$ cal/g K.)

   ANSWER: $\dot{Q}_{in} = 4401.3$ J/sec

4. Consider a general chemical equilibrium reaction

$$\sum \nu_i' M_i \rightleftharpoons \sum \nu_i'' M_i$$

   established in a heterogeneous mixture containing N chemical components. Some of the components have multiple phases in phase equilibrium. Show that

$$\mu_{j(g)} = \mu_{j(l)} = \mu_{j(s)} \qquad \text{for any component } j$$

5. Suppose gaseous ozone, $O_{3(g)}$, is formed from molecular oxygen, $O_{2(g)}$, by the following reaction:

$$\tfrac{3}{2}O_{2(g)} \rightleftharpoons O_{3(g)}$$

   What is the value of standard-free-energy change, $\Delta G°$, of this reaction? What is the equilibrium constant of this reaction at 298 K? If the value of K_p is very small, what does that imply?

   ANSWER: $\Delta G° = 39.06$ kcal, $K_p = 2.261 \times 10^{-29}$.

6. A combustible mixture of air and carbon monoxide which is 10% rich [air/fuel = $\frac{100}{110}$(air/fuel)$_{st}$] is compressed to a pressure of 8.28 bar and a temperature of 555 K. The mixture is ignited, and combustion occurs adiabatically at constant volume. When the maximum temperature is attained, analysis shows 0.228 moles of CO present for each mole of CO supplied. Show that the maximum temperature reached is 2950 K. If the pressure at this temperature is now doubled, calculate the amount of CO present.

   ANSWER: $n_{\mathrm{CO}} = 0.194$ when the pressure is doubled.

7. Using bond energies, determine the heat of formation of gaseous normal butane (C_4H_{10}).

   ANSWER: $\Delta \mathscr{H}^{\circ}_{f,\mathrm{C_4H_{10}}} = -32.94$ kcal/mole.

8. Methane supplied at 1 atm, 25°C is burned adiabatically in a steady-flow burner with the stoichiometric amount of air supplied at the same conditions. Assuming the reaction equation is

$$\mathrm{CH_4 + 2O_2 + 7.54N_2 \rightarrow \mathit{a}CO_2 + (1 - \mathit{a})CO + \mathit{b}H_2O}$$
$$\mathrm{+ (2 - \mathit{b})H_2 + (1.5 - 0.5\mathit{a} - 0.5\mathit{b})O_2 + 7.54N_2}$$

determine the temperature of the products.

ANSWER: $T_f = 2244.3$ K.

9. Show that the fugacity of a real gas is related to the pressure p and compressibility factor Z by

$$\ln \frac{f}{p} = \int_0^{P_r} (Z - 1)\, \mathrm{d}(\ln P_r)_T$$

where P_r is the reduced pressure ($P_r = p/p_{\text{critical}}$). Discuss the use of this equation. Hint: Integrate Eq. (1-199) from $p = 0$ to a finite pressure.

10. Consider the combustion of kerosene ($C_{10}H_{20}$) in air at initial temperature 298.16 K and at constant pressure 1 atm. Use the NASA–Lewis computer program (TRAN76 or CEC76) to carry out the thermochemical calculations for the equilibrium compositions and flame temperature for equivalence ratios of 0.3 to 1.4 ($\Delta\phi = 0.1$ increment). Plot T_f and the mole fractions of CO_2, H_2O, N_2, O_2, CO, NO, OH, O, H, H_2 and NO_2 versus ϕ. ($\Delta \mathscr{H}^\circ_{f,C_{10}H_{20}} = -59$ kcal/mole.)

ANSWER: For $\phi = 1$, $T_f = 2280$ K,

$$X_{CO} = 0.01455$$
$$X_{CO_2} = 0.1160$$
$$X_{H_2O} = 0.12584$$
$$X_{N_2} = 0.72753$$
$$X_{O_2} = 0.00653$$

11. Methane is burned with 80% of stoichiometric air in a steady-flow process at 1 atm. Methane and air are both supplied at 298 K, and the products leave at 1666 K. Assuming that no CH_4, OH, NO, or free oxygen appears in the products, determine the amount of heat transferred per kilogram of methane.

ANSWER: $Q = -2291$ kcal/(kg CH_4).

12. Consider the combustion of a homogeneous propellant in a close bomb. The ingredients of this propellant together with their chemical formulas, weight fractions, and heats of formation are listed below:

Ingredients	Chemical Formula	Heats of Formation (cal/g-mole)	Weight Fraction
NC nitrocellulose (13.25% N)	$C_6H_{7.329}O_5(NO_2)_{2.671}$	−163810	0.75
NG, nitroglycerin	$C_3H_5O_3(NO_2)_3$	−88600	0.15
DBP, dibutylphthalate	$C_{16}H_{22}O_4$	−201400	0.09
DPA, diphenylamine	$C_{12}NH_{11}$	31070	0.01

Use the computer program TRAN76 (CEC76) to compute the adiabatic flame temperature, chamber pressure, and product concentrations for loading densities of 0.05, 0.25, and 0.4 g/cm^3.

ANSWER: For loading density = 0.25 g/cm^3,

$$T_f = 2773 \text{ K}, \qquad p = 2480 \text{ atm}$$

$$X_{CO} = 0.47251, \qquad X_{H_2O} = 0.17751$$

$$X_{H_2} = 0.17099, \qquad X_{N_2} = 0.10436$$

$$X_{CO_2} = 0.07276$$

13. Sketch the most general burned and unburned planes of mass fractions of F, O, D, and P versus mixture fraction f, defined by Eq. (1-121). Consider both a fuel-rich stream and an oxidizer-rich stream, contaminated by the same kind of products and inert diluents.

14. The vapor pressure of liquid chlorine, in centimeters of mercury, can be represented by the expression

$$\log_{10} p = -\frac{1414.8}{T} + 9.91635 - 1.206 \times 10^{-2}T + 1.34 \times 10^{-5}T^2$$

The specific volume of chlorine gas at its boiling point is 269.1 cm^3/g, and that of the liquid is approximately 0.7 cm^3/g. Calculate the heat of vaporization of liquid chlorine in cal/g at its boiling point, 239.05 K. Note:

$$\frac{d \log_{10} p}{dT} = \frac{1}{2.303} \frac{d \ln p}{dt}$$

ANSWER: $\Delta \mathscr{H}_v = 68.3$ cal/g.

CHAPTER

2 Review of Chemical Kinetics

ADDITIONAL SYMBOLS

Symbol	Description	Dimension
A	Arrhenius factor	$(N/L^3)^{1-m}/t$
B	Collision-frequency factor	$(N/L^3)^{1-m}/(tT^\alpha)$
k	Specific reaction-rate constant	$(N/L^3)^{1-m}/t$
l	Length	L
N	Total number of species	—
$\underline{P}$	Steric factor	—
RR	Reaction rate	N/L^3t
r	Radius	L
z	Collision frequency	N/L^3t
σ	Diameter of a molecule or Collision radius	L
μ	Reduced mass	M

Subscripts

a	Activation
b	Backwards
e	Equilibrium
f	Forward
p	Pressure
t	Total

Some chemical reactions occur very rapidly and others occur very slowly. Most chemical reactions occur more rapidly as temperature increases. One objective of this chapter is to explain these observations. Chemical kinetics is the part of chemical science dealing with the quantitative study of the rates of chemical reactions and of the factors upon which they depend. It also deals with the interpretation of the empirical kinetic laws in terms of reaction mechanisms. The subject includes both the experimental study of reaction rates and the development of theories to explain experimental results and to predict the outcome of future experiments.

In terms of the physical state of the reacting chemical substances, chemical reactions can be classified into four different types: (1) gas-phase reactions, (2) liquid-phase reactions, (3) solid-phase reactions, and (4) heterogeneous reactions occurring at the interfaces of two substances of different phases, such as gas–solid interfaces. Most of the effort has been devoted to the first two types of reactions. Reactions at gas–solid interfaces have received some attention. Only a limited amount of work has been done on reactions in the solid phase.

In terms of the speed of reaction, chemical reactions can be divided into two categories:[2] (1) explosive and (2) nonexplosive. The study of explosive chemical reaction involves not only the determination of conditions under which chemical systems undergo very fast reaction, but also the examination of the reaction mechanism.

Although explosions are important, the subject of nonexplosive reaction is also of great interest. For instance, many pollutants are formed in reaction zones of rather steady reactions in various combustion systems.

Certain essential features of chemical kinetics which occur frequently in combustion phenomena will be reviewed in this chapter. For a more detailed understanding of these features, attention is directed to books on chemical kinetics, such as that by Laidler.[1]

1 RATES OF REACTIONS AND THEIR FUNCTIONAL DEPENDENCE

All chemical reactions, whether hydrolysis or combustion, take place at a definite rate, depending on the conditions of the system. Some important conditions[2] are (1) concentrations of the chemical compounds, (2) temperature, (3) pressure, (4) presence of a catalyst or inhibitor, and (5) radiative effects.

The rate of reaction may be expressed in terms of the concentration of any reactant as the rate of decrease of the concentration of that reactant (the rate of consumption of the reactant). It may also be expressed in terms of product concentration as the rate of increase of the product concentration. A conventional unit for reaction rate is moles/m^3 sec.

A one-step chemical reaction of arbitrary complexity can be represented by the following stoichiometric equation:[3]

$$\sum_{i=1}^{N} \nu_i' M_i \rightarrow \sum_{i=1}^{N} \nu_i'' M_i \tag{2-1}$$

where ν_i' are the stoichiometric coefficients of the reactants, ν_i'' the stoichiometric coefficients of the products, M the arbitrary specification of all chemical species, and N the total number of compounds involved. If a species represented by M_i does not occur as a reactant, then $\nu_i' = 0$; if the species does not occur as a product, then $\nu_i'' = 0$.

The meaning of Eq. (2-1) may be illustrated for the reaction between two hydrogen atoms in the presence of a third hydrogen atom to form a hydrogen molecule and a hydrogen atom which has gained translational energy. Symbolically we write

$$H + H + H \rightarrow H_2 + H$$

or (2-2)

$$3H \rightarrow H_2 + H$$

Here $N = 2$, and

$$M_1 = H, \qquad M_2 = H_2$$

$$\nu_1' = 3, \qquad \nu_2' = 0$$

$$\nu_1'' = 1, \qquad \nu_2'' = 1$$

Observe that in the present notation, no distinction is made between hydrogen atoms having different energies. Consider a second reaction, the overall chemical reaction between hydrogen and oxygen to form water:

$$2H_2 + O_2 \rightarrow 2H_2O \tag{2-3}$$

Here $N = 3$, and

$$M_1 = H_2, \qquad M_2 = O_2, \qquad M_3 = H_2O$$

$$\nu_1' = 2, \qquad \nu_2' = 1, \qquad \nu_3' = 0$$

$$\nu_1'' = 0, \qquad \nu_2'' = 0, \qquad \nu_3'' = 2$$

More complicated examples in which the ν_i' and ν_i'' are not necessarily integers can be given without difficulty. It is clear, however, that Eq. (2-1) is adequate to describe all possible chemical reactions.

The *law of mass action*,† which is confirmed by numerous experimental observations, states that the rate of disappearance of a chemical species is proportional to the products of the concentrations of the reacting chemical species, each concentration being raised to a power equal to the corresponding stoichiometric coefficient. Thus the reaction rate is given as

$$\mathrm{RR} = k \prod_{i=1}^{N} (C_{M_i})^{\nu_i'} \tag{2-4}$$

where k is the proportionality constant called the *specific reaction-rate constant*. For a given chemical reaction, k is independent of the concentrations C_{M_i} and depends only on the temperature. In general, k is expressed as

$$k = BT^{\alpha} \exp\left(-\frac{E_a}{R_u T} \right) \tag{2-5}$$

where BT^{α} represents the collision frequency and the exponential term is the Boltzmann factor, specifying the fraction of collisions that have an energy greater than the activation energy E_a. The values of B, α, and E_a are based on the nature of the elementary reaction.[3] For given chemical changes, these parameters are neither functions of the concentrations nor of temperature. In the following, we shall discuss the specific reaction rate constant k in more detail, in terms of theoretical chemical kinetics.

1.1 Total Collision Frequency

In the study of molecular collisions, one must consider the effect of the size of the molecules. Two identical spherical molecules collide when their centers come within a distance σ, the molecular diameter. If the distance between the two centers is greater than σ, the molecules do not collide. This is illustrated in Fig. 2.1*a* and *b*.

The larger the value of σ, the greater the probability that two molecules will collide. The collision of two molecules, each with a diameter of σ, can be considered equivalent to the collision of a molecule with diameter 2σ and another molecule represented as a point. In each representation, collision is judged to have occurred if the distance between the two centers of the molecules is less than σ (see Fig. 2.1*c*). Calculation of the collision frequency is simpler when based upon the second representation. As illustrated in Fig. 2.2, the average collision frequency for pairs of molecules in a gas is approximately equal to the total number of point molecules in the volume swept out in unit

† The origin of the law of mass action and its relation to first principles are discussed by Penner in Section 1.6 of Ref. 3.

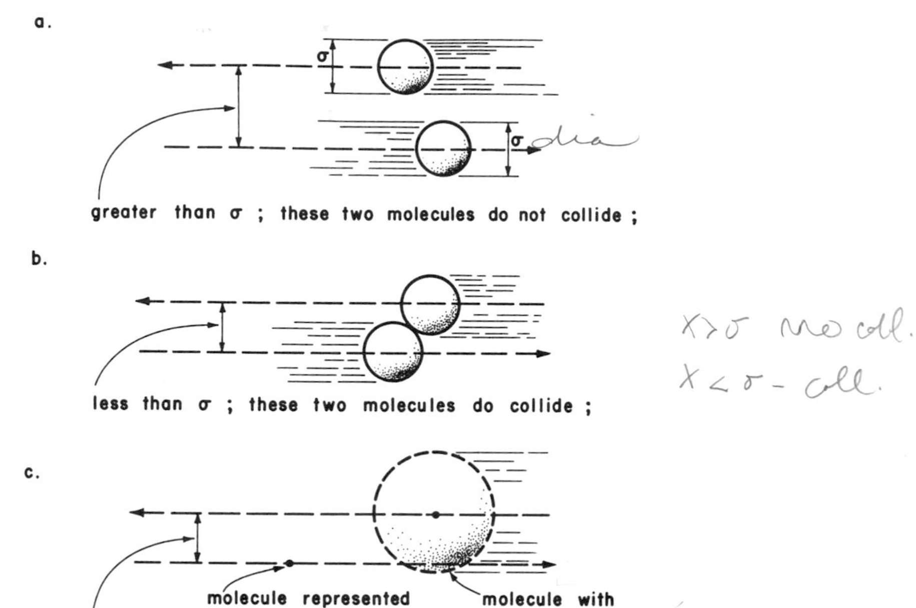

Figure 2.1 Illustration of the distance required between molecules for a collision to occur.

time by a molecule of radius σ moving with average velocity. A molecule sweeps through a cylinder of length $\bar{u}$ in 1 sec. For a molecule with $\sigma = 3.5 \times 10^{-8}$ cm and molecular weight of 130, the volume of this cylinder at 773 K (500°C) is

$$V = \pi r^2 l = \pi \sigma^2 \bar{u} \cdot 1$$

$$= \pi (3.5 \times 10^{-8})^2 \left[1.455 \times 10^4 \left(\tfrac{773}{130}\right)^{1/2}\right]$$

$$= 1.365 \times 10^{-10} \text{ cm}^3$$

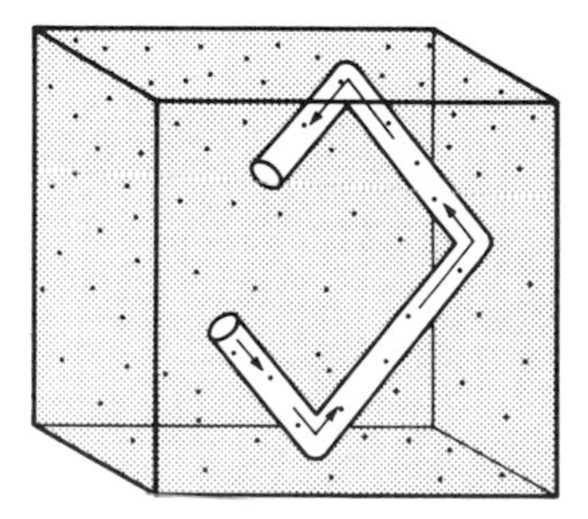

Figure 2.2 Collisions of a moving molecule of radius σ with stationary point molecules. In the volume swept out by the moving molecule, there are twelve point molecules. This is the number of collisions in the time interval in question according to this simple model.

Note: The average molecular velocity $\bar{u}$ can be calculated from

$$\bar{u} = \left(\frac{8R_uT}{\pi W}\right)^{1/2}$$

$$= 1.455 \times 10^4\left(\frac{T}{W}\right)^{1/2} \text{ cm/sec}$$

[Also see Eq. (3-15).]

If the concentration of gas molecules is 10^{-6} mole/cm^3, there are (10^{-6} mole/cm^3)(1.365×10^{-10} cm^3)(6.02×10^{23} molecules/mole) $= 8.2 \times 10^7$ molecules present in this volume. This is the number of collisions which the molecule in question makes in 1 sec., based upon the assumption that all molecules are stationary. If motion of all molecules is taken into account, the calculation leads to a $\sqrt{2}$-fold higher collision frequency. The total frequency of collisions of pairs of molecules in 1 cm^3 of gas is obtained by multiplying the collision frequency for one molecule by the number of molecules in 1 cm^3 and then dividing by 2, since each collision of two molecules has been counted twice. For molecules with $\sigma = 3.5 \times 10^{-8}$ cm and molecular weight 130 at a concentration of 10^{-6} mole/cm^3, the collision frequency per cm^3 at 773 K is therefore

$$\text{collision frequency} = \left(\tfrac{1}{2}\sqrt{2} \times 8.2 \times 10^7 \frac{\text{collisions}}{\text{molecules sec}}\right)\left(10^{-6}\frac{\text{mole}}{\text{cm}^3}\right)$$

$$\times\left(6.02 \times 10^{23}\frac{\text{molecules}}{\text{mole}}\right)$$

$$= 3.49 \times 10^{25} \text{ collision/cm}^3 \text{ sec}$$

If the molecules react upon every collision, the collision frequency will be equal to the reaction rate. In more familiar terms, the rate is

$$\frac{3.49 \times 10^{25}}{6.02 \times 10^{23}} = 58 \frac{\text{mole}}{\text{cm}^3 \text{ sec}}$$

—certainly a very high value. This is the instantaneous value of the rate at this concentration if reaction occurs at each collision.

In the calculation of the total number of collisions of two molecules in 1 cm^3 in 1 sec, the concentration of the gas appears twice as a factor: the first time in the evaluation of the number of stationary point molecules, the second time in the calculation of the overall collision frequency. The frequency of collisions of two gaseous molecules of the same kind is therefore proportional to the square of the concentration of the gas, that is

$$\text{collision frequency} \propto C_A^2$$

A reexamination of the derivation indicates that the collision frequency of two gas molecules of different kinds is proportional to the product of the concentrations of the two gases

$$\text{collision frequency} \propto C_B C_C$$

We therefore conclude that a second-order rate law will be associated with a gas-phase reaction whose mechanism is bimolecular collisions. Notice the use of the terms "second-order" and "bimolecular." The two are not synonymous. The reaction order pertains to the number of concentration factors in the experimentally established rate law; the molecularity pertains to the number of species taking part in a postulated reaction step. Although it may seem that reaction order and molecularity must be uniquely associated, later discussion will show that this assumption is unwarranted.

The collision frequency is equal to

$$Z = \sigma_{BC}^2 \left(\frac{8\pi KT}{\mu} \right)^{1/2} C_B C_C \tag{2-6}$$

where K is the Boltzmann's constant, μ is the reduced mass of molecules B and C ($\mu = m_B m_C/(m_B + m_C)$], σ_{BC} is the collision radius of species B and C, and m_B and m_C are the molecular masses.

1.2 Equation of Arrhenius

Svante Arrhenius (1859–1927) stated that only those molecules which possess energy greater than a certain amount E_a will react,[2] and these high-energy, active molecules lead to products. Since Arrhenius was also the first (1889) to introduce the Boltzmann factor $\exp(-E_a/R_u T)$ to calculate chemical reaction rates, the equation

$$k = A \exp\left(-\frac{E_a}{R_u T} \right) \tag{2-7}$$

is called the Arrhenius law; here A is assumed to include the effect of the collision terms, the steric factor associated with the orientation of the colliding molecules, and the mild temperature dependence of the preexponential factor. The parameter A corresponds to BT^α in Eq. (2-5), where the exponent α lies between 0 and 1.

For a bimolecular reaction, the reaction occurs in a collision only if the relative translational energy along the line of the centers of the two molecules at the moment of impact is in excess of E_a. For the second-order reaction

$$B + C \xrightarrow{k} \text{products}$$

the rate law is given by

$$\frac{dC_B}{dt} = -kC_B C_C = -AC_B C_C \exp\left(-\frac{E_a}{R_u T}\right) \tag{2-8}$$

According to Arrhenius,

$$\frac{dC_B}{dt} = -Z_{BC}\underline{P}\exp\left(-\frac{E_a}{R_u T}\right) \tag{2-9}$$

where T is the absolute temperature, Z_{BC} is the total collision frequency, and $\underline{P}$ is called the steric factor. The steric factor depends on the orientation of the colliding molecules. Its value is less than unity if some special orientation of colliding molecules is required (in addition to the necessary activation energy) for the reaction to occur. The meaning of the steric factor can be illustrated by considering the collision of two hydrogen iodide molecules: if the orientation of the colliding molecules is that shown in Fig. 2.3*a*, hydrogen and iodine molecules will form; if, however, the orientation is that shown in Fig. 2.3*b*, the colliding molecules will simply bounce off each other.

The numerical value of the steric factor for hydrogen iodide decomposition is found to be approximately 0.2. For more complex molecules the steric factor may be in the order of 0.01.

Comparing Eqs. (2-8) and (2-9) gives

$$AC_B C_C = Z_{BC}\underline{P}$$

Solving for A and substituting the value for Z_{BC} from Eq. (2-6) leads to

$$A = \sigma_{BC}^2 \left(\frac{8\pi KT}{\mu}\right)^{1/2} \underline{P} \tag{2-10}$$

Theoretical treatment indicates that the activation energy E_a obtained from the use of Eq. (2-7) applied to reactions of other orders has significance analogous

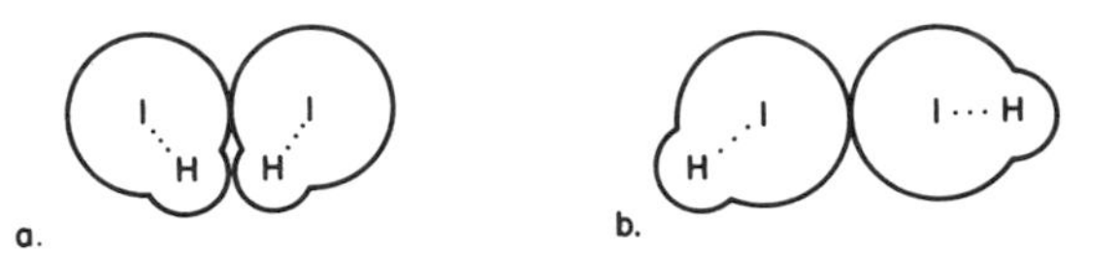

Figure 2.3 Two possible orientations for the collision of two HI molecules.

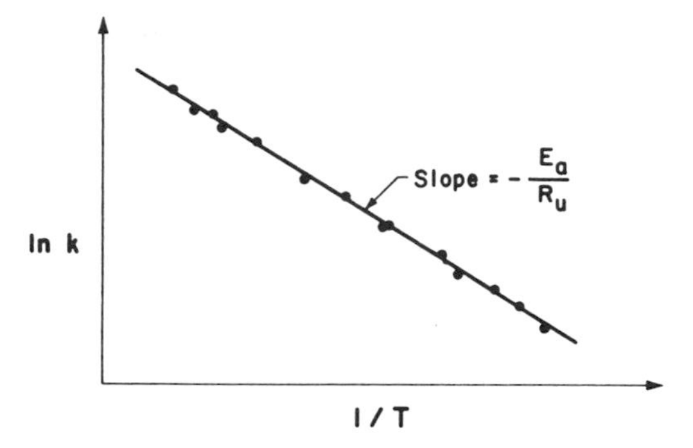

Figure 2.4 Graph showing temperature dependence of the specific reaction rate constant k.

to that already indicated for bimolecular reactions; hence Eq. (2-7) can be applied to reactions of all orders.

It should be mentioned that the specific rates of many reactions follow the Arrhenius law. For these reactions, the kinetic data plotted on a graph of ln k versus T^{-1} follow a straight line.

Figure 2.4 shows that in a given chemical reaction, the specific reaction rate constant k is independent of the concentrations C_{M_i} and depends only on the temperature. The equation for ln k, as shown in Fig. 2.4, can be derived from the natural logarithm of the Arrhenius equation (2-7), which gives

$$\ln k = \ln A - \frac{E_a}{R_u T} \tag{2-11}$$

It is important to note that the specific reaction rate constant depends on both temperature and temperature range. The Arrhenius equation generally cannot describe the combustion process over a wide temperature range. For example, a set of reactions which matches with test data at low temperatures may provide erroneous results at high temperature. There, however, another set of reactions may match the experimental results well. This is illustrated in Fig. 2.5. Caution must therefore be exercised in extrapolating the specific reaction rate to broader temperature ranges.

Although many reactions follow the Arrhenius law, there are two classes of reactions for which Eq. (2-7) does not hold. These involve:

A. Low activation-energy free radical reactions. In these reactions, temperature dependence in the preexponential term assumes greater importance and the so-called absolute theory of reaction appears to provide better correlation of kinetic data with temperature.[4] For further information, see Benson.[5]

B. Radical recombination. When simple radicals recombine to form a single product, energy must be removed from the product upon its formation in order to stabilize it. A third body is necessary to remove this energy. The

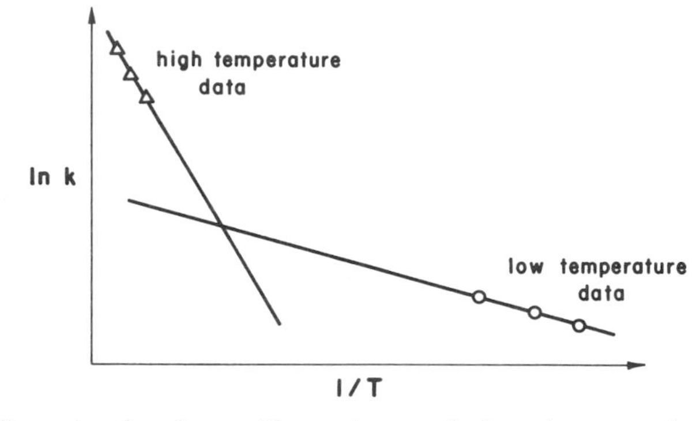

Figure 2.5 Plot illustrating that the specific reaction rate is dependent upon the temperature range as well as temperature.

pressure dependence of third-body recombination reactions can be quite pronounced. Hence, the specific reaction rate does not follow Eq. (2-7).

We shall always use k to denote a reaction-rate constant and shall employ frequently the subscript f to identify a forward reaction, that is, a reaction in which the reactants appear on the left side of the equation and the reaction products on the right side.

1.3 Apparent Activation Energy

The theory of absolute reaction rates begins with molecular collisions. Under favorable conditions, the collisions lead to the formation of a transitory chemical species, the *activated complex*. By using the methods of quantum mechanics, it is possible, in principle, to calculate the forces between atoms and molecules. A detailed description of the activated complex will provide insight into the nature of the changes in electronic and nuclear arrangement which characterize the chemical reaction. Consider the reaction between reactants A and B to form the reaction products C and D. The formation of reaction products is preceded by the production of an aggregate, the activated complex, which is designated by the symbol $X^{\ddagger}$. Thus,

$$A + B \rightleftharpoons X^{\ddagger} \rightleftharpoons C + D \tag{2-12}$$

The energy between two atoms of a diatomic molecule A_2 is a function of the distance between the two atoms. (This function was plotted and discussed in Chapter 1.) When a free atom B is approaching the molecule A_2, the distance between A and A changes. This is shown in Fig. 2.6. Immediately after the collision of B with A_2, the activated complex $BA_2^{\ddagger}$ may be formed. The activated complex has a much higher reactivity than normal atoms, and may separate in such a way as to give either the components BA and A or the initial components B and A_2.

A *reaction coordinate* may be associated with the path from reactants to reaction products. Depending on the choice of the reaction coordinate, differ-

Initial condition :

As B approaches A_2 the configuration becomes

Figure 2.6 The distance between two atoms in a molecule may change as a free atom approaches.

ent amounts of energy will be required for the reaction described by Eq. (2-2). The activated complex is located at the point of highest energy on the most favorable reaction path, and the activation energy E^* per molecule is required to form $X^{\ddagger}$. For the most favorable reaction path, the highest point ($X^{\ddagger}$) has lower total energy above the reactants (namely, $E_a = NE^*$ per mole) than any other reaction path. Figure 2.7 illustrates the activation complex and the required activation energy for exothermic and endothermic reactions in both forward and reverse directions.

An examination of Fig. 2.7 reveals that $E(\text{reactants}) > E(\text{product})$ for an exothermic reaction, while $E(\text{product}) > E(\text{reactants})$ for an endothermic reaction. Also note that the activation energy for forward and reverse reactions is not equal; that is, the forward and reverse reactions have different specific reaction rate constants.

In summary, the activation energy is the energy required for the reaction to occur; that is, it is the energy required to move the reactants over the energy barrier in order for reaction to begin. The activation energy is usually recovered by the heat released by the reaction process. Referring to Fig. 2.7, E_a is the activation energy and ΔH_r is the energy release observed thermodynamically.

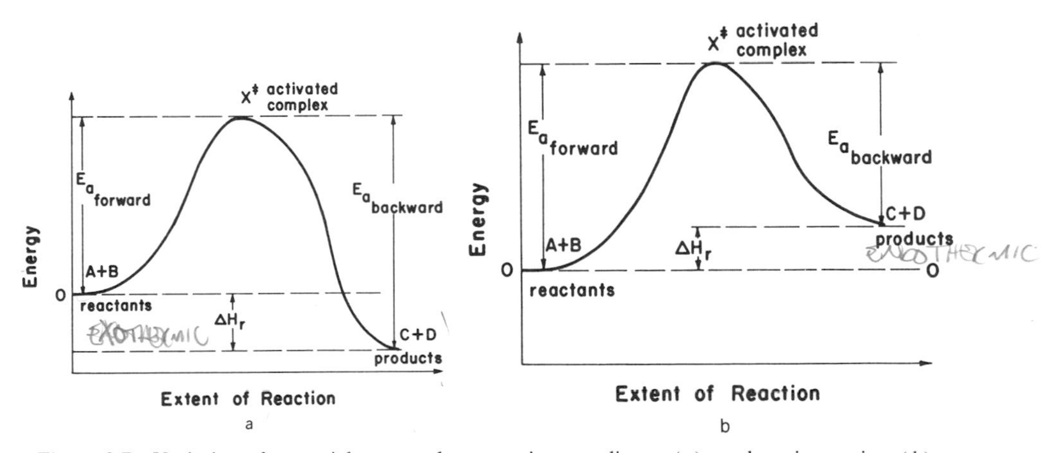

Figure 2.7 Variation of potential energy along reaction coordinate: (*a*) exothermic reaction, (*b*) endothermic reaction.

1.4 Rates of Reaction

The only observable results of a chemical reaction are net rates of change for the chemical components. It is clear from Eqs. (2-1) and (2-4) that the net rate of production of M_i is

$$\frac{dC_{M_i}}{dt} = (\nu_i'' - \nu_i')\text{RR} = (\nu_i'' - \nu_i')k_f \prod_{i=1}^{N} C_{M_i}^{\nu_i'} \tag{2-13}$$

Since species M_i may appear on both sides of Eq. (2-1) with different values for ν_i'' and ν_i', the difference $\nu_i'' - \nu_i'$ is multiplied by the reaction rate in the above equation.

The result of applying Eq. (2-13) to the reaction (2-2) is

$$\frac{dC_{\text{H}}}{dt} = (1 - 3)k_f C_{\text{H}}^3 = -2k_f C_{\text{H}}^3$$

and

$$\frac{dC_{\text{H}_2}}{dt} = (1 - 0)k_f C_{\text{H}}^3 = k_f C_{\text{H}}^3$$

Therefore,

$$\frac{dC_{\text{H}}}{dt} = -2\frac{dC_{\text{H}_2}}{dt} = -2k_f C_{\text{H}}^3$$

which states that the rate of depletion of H is twice as fast as the rate of formation of H_2.

The process represented by Eq. (2-1) is said to be *of order* ν_i' with respect to M_i. The overall order of the reaction is

$$m = \sum_{i=1}^{N} \nu_i' \tag{2-14}$$

That is, the overall order of the reaction is equal to the sum of the exponents in the reactant concentration terms. Thus the reaction (2-2) is of third order with respect to H and of zeroth order with respect to H_2, and it has an overall order of three.

The hypothetical chemical reaction described by Eq. (2-3) represents a third-order process which is of second order with respect to H_2 and of first order with respect to O_2. Equation (2-3) is an example of a valid overall chemical reaction which does *not* describe the reaction mechanism correctly. Thus the overall result for reaction between two moles of H_2 and one mole of O_2 is the production of two moles of H_2O; however, the conversion of H_2 and

O_2 occurs by means of a series of successive, interdependent, elementary chemical-reaction steps. For this reason, the rate expression obtained by applying Eq. (2-13) to Eq. (2-3)—

$$\frac{dC_{H_2}}{dt} = -\frac{dC_{H_2O}}{dt} = 2\frac{dC_{O_2}}{dt} = -2k_f C_{H_2}^2 C_{O_2}$$

—has no physical significance. The law of mass action, as expressed by Eq. (2-13), may be applied in a meaningful way only to elementary reaction steps which describe the correct reaction mechanism. For a number of simple chemical processes, a plausible reaction mechanism has been deduced by chemical kineticists; for some technically important reactions, chemical kineticists provide intelligent conjectures regarding the probable reaction mechanism. Detailed studies[6] show that the production of H_2O from H_2 and O_2 involves, among others, the elementary reaction steps

$$OH + H_2 \rightarrow H_2O + H$$

$$H + O_2 \rightarrow OH + O$$

The law of mass action applies to each step.

The stoichiometric coefficients for elementary reactions give information about the numbers of reacting moles, but not about the weights or volumes which are changing. If the cubic centimeter is chosen as the unit of volume, it is apparent that the units of the rate constant k_f are

$$\frac{\text{mole}}{\text{cm}^3\ \text{sec}} \frac{1}{(\text{mole}/\text{cm}^3)^m} = \text{mole}^{1-m}\ \text{cm}^{3m-3}\ \text{sec}^{-1}$$

Thus, for a first-order reaction, k_f is a frequency.

Example 2.1

The decomposition of nitrogen dioxide in the absence of inert gas is

$$2NO_2 \xrightarrow{k_f} 2NO + O_2$$

What is the order of this reaction? What is the rate law? And what are the units of the specific reaction rate constant?

Solution:

The decomposition of nitrogen dioxide is a second-order reaction, as can be seen by examining the exponent in the reaction concentration term in the rate law, given as

$$\frac{dC_{NO_2}}{dt} = -2k_f C_{NO_2}^2$$

Therefore, the units of the specific reaction rate constant can be determined from the above as

$$\text{moles}^{1-2}\,\text{cm}^{6-3}\,\text{sec}^{-1} = \text{mole}^{-1}\,\text{cm}^{3}\,\text{sec}^{-1} = \text{cm}^{3}/\text{mole sec}$$

Example 2.2

For the decomposition of nitrogen dioxide:

a. At 592 K, the value of k_f is 498 cm^3/mole sec. What is the rate of decomposition of nitrogen dioxide at this temperature if the concentration of nitrogen dioxide is 0.0030 mole/liter?
b. Values of this rate constant at other temperatures are

T (K)	k_f (cm^3/mole sec)
603.5	775
627	1810
651.5	4110
656	4740

What is the activation energy for this reaction?

Solution:

a. Converting the units of concentration to mole/cm^3, since 1 liter = 1000 ml = 1000 cm^3, we have

$$0.0030\ \text{mole/liter} = 3 \times 10^{-6}\ \text{mole/cm}^3$$

From Example 2.1, the rate of decomposition of nitrogen dioxide is given as

$$\frac{dC_{NO_2}}{dt} = -2k_f C_{NO_2}^2$$

which in turn gives

$$\frac{dC_{NO_2}}{dt} = -498\frac{\text{cm}^3}{\text{mole sec}} \times 2\left(3 \times 10^{-6}\frac{\text{mole}}{\text{cm}^3}\right)^2$$

$$= -8.964 \times 10^{-9}\ \text{mole/cm}^3\ \text{sec}$$

b. The activation energy for a reaction can be calculated directly from values of rate constants measured at two different temperatures. At T_1,

$$\ln k_1 = \ln A - \frac{E_a}{R_u T_1}$$

which was obtained by taking the natural logarithm of both sides of the Arrhenius law as given in Eq. (2-7). At T_2,

$$\ln k_2 = \ln A - \frac{E_a}{R_u T_2}$$

Subtracting these two equations yields

$$\ln \frac{k_1}{k_2} = \frac{E_a}{R_u}\left(\frac{1}{T_2} - \frac{1}{T_1}\right)$$

Upon rearranging,

$$E_a = \frac{R_u T_2 T_1}{T_1 - T_2} \ln \frac{k_1}{k_2}$$

Substituting in any two temperature values and the appropriate specific reaction rate constant, we have

$$E_a = \frac{(1.9872 \text{ cal/mole K})(627 \text{ K})(592 \text{ K})}{627 \text{ K} - 592 \text{ K}} \ln \frac{1810}{498} = 27.2 \text{ kcal/mole}$$

Experimentally, E_a is found by plotting measured values of $\ln k$ against $1/T$ and computing the slope of the best-fit straight line through the data points (slope $= -E_a/R_u$). After E_a is obtained, the factor A in the Arrhenius equation (2-7) may be calculated, using this equation and the measured values of k.

In the kinetic study of a reaction, there is no simple way to measure the rate directly; normally the concentration of a reactant or product is determined at various times. As shown in Fig. 2.8, a smooth curve should be obtained if concentration is plotted against time. At any given time, the rate of reaction is given by the slope $-d(C_a - C_x)/dt$ of the instantaneous-reactant-concentration $(C_a - C_x)$ curve, or by the rate of generation of product species concentration (dC_x/dt). In particular, the initial slope, at $t = 0$, provides the rate of the

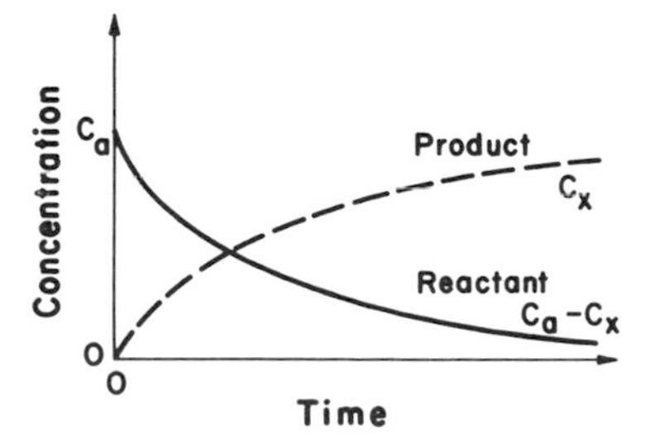

Figure 2.8 Plot of time variation of concentrations of reactant and product in terms of C_x, representing the portion of reactant consumed in chemical reaction.

reaction corresponding to the concentrations with which the experiment was begun. Another method commonly employed involves the use of expressions relating concentration to time for reactions of various orders. Some of the more important expressions will be derived in the next section.

2 ONE-STEP CHEMICAL REACTIONS OF VARIOUS ORDERS

2.1 First-Order Reactions

The rate law for the first-order reaction

$$A_2 \xrightarrow{k_f} 2A$$

is

$$\frac{dC_A}{dt} = 2k_f C_{A_2} = -2\frac{dC_{A_2}}{dt} \tag{2-15}$$

After separating the variables and performing integration from time 0 to t, we have

$$-\ln C_{A_2}\Big|_{C_{A_2,0}}^{C_{A_2,t}} = k_f(t-0)$$

or

$$\ln\left(\frac{C_{A_2,0}}{C_{A_2,t}}\right) = k_f t \tag{2-16}$$

which gives the concentration of A_2 as a function of time. The rate expression given in Eq. (2-15) also applies formally to the process

$$A + C \xrightarrow{k_f} D, \qquad \text{where} \quad C_C \gg C_A$$

Since $C_C \gg C_A$, the rate expression is given as

$$\frac{dC_A}{dt} = -\frac{dC_D}{dt} = -k_f C_A C_C = k' C_A$$

where k' is a new specific rate constant, which can be formulated because the concentration of C is approximately constant (i.e., C_C = constant). The decomposition of A_2 is a unimolecular reaction which obeys first-order kinetics, and the reaction $A + C \rightarrow D$ is a bimolecular reaction which obeys first-order kinetics. Thus, all unimolecular reactions are of order one, but not all first-order reactions are unimolecular.

Another example of a first-order reaction is the dissociation of molecule AB

$$AB \xrightarrow{k_f} A + B$$

where the rate law is given as

$$\frac{dC_{AB}}{dt} = -k_f C_{AB}$$

Example 2.3

Denoting the concentration of the products by C_x and the concentration of the reactants by $C_a - C_x$ (where C_a is the initial reactant concentration), develop an expression for the specific reaction rate constant. Also develop an expression for C_x. See Fig. 2.8.

Solution:

From Eq. (2-16)

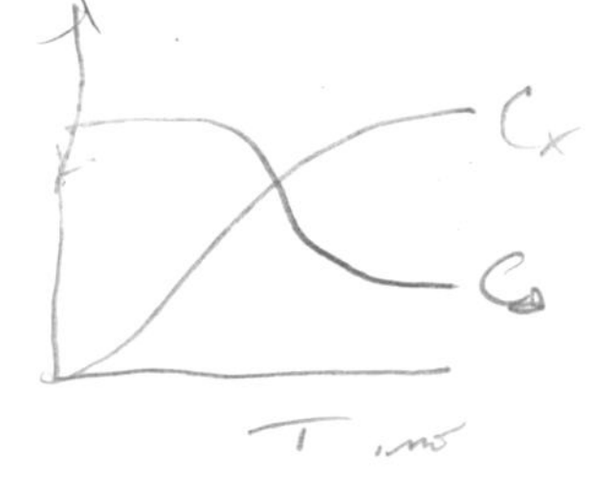

$$\ln\left(\frac{C_a}{C_a - C_x}\right) = k_f t$$

Therefore,

$$k_f = \frac{1}{t}\ln\left(\frac{C_a}{C_a - C_x}\right)$$

The concentration of x is

$$C_x = C_a(1 - e^{-k_f t})$$

2.2 Second-Order Reactions

Most chemical reactions are bimolecular and proceed as the result of reactions following binary collisions. It is therefore not surprising to note that chemical reactions frequently follow second-order kinetics. In complex processes, second-order kinetics may be considered as an indication that one of the bimolecular processes constitutes the slow or rate-determining step.

For the second-order bimolecular reaction

$$A + B \xrightarrow{k_f} AB$$

the rate law becomes

$$\frac{dC_A}{dt} = \frac{dC_B}{dt} = -\frac{dC_{AB}}{dt} = -k_f C_A C_B$$

In this reaction, the concentration of A is equal to the concentration of B, that is, $C_A = C_B$. The differential equation for this second-order reaction can be solved readily.

For the second-order bimolecular reaction

$$2A \overset{k_f}{\rightarrow} C + D$$

the rate law is

$$\frac{dC_A}{dt} = -2\frac{dC_C}{dt} = -2\frac{dC_D}{dt} = -2k_f C_A^2$$

Some representative second-order processes which have been postulated[6] as reactions occurring in flames are given below:

$$Cl + H_2 \rightarrow HCl + H$$

$$OH + H_2 \rightarrow H_2O + H$$

$$H + O_2 \rightarrow OH + O$$

$$O + H_2 \rightarrow OH + H$$

$$O_3 + CO \rightarrow CO_2 + 2O$$

$$OH + CH_4 \rightarrow H_2O + CH_3$$

The rate law can also be expressed in terms of the concentration of the reactant which was consumed in the reaction. For example, consider the following second-order reaction (sometimes called the atom-transfer reaction)

$$A + B \overset{k_f}{\rightarrow} C + D$$

The concentrations of species A and B are given as

$$C_A = C_{A0} - C_x$$

$$C_B = C_{B0} - C_x$$

where C_{A0} and C_{B0} are the initial concentrations and C_x is the portion of A and B which is consumed in the reaction. The rate law for this reaction is

$$\frac{dC_x}{dt} = k_f(C_{A0} - C_x)(C_{B0} - C_x) \tag{2-17}$$

If $C_{B0} \neq C_{A0}$, we can multiply both sides of the equation by

$$\frac{(C_{B0} - C_{A0})\, dt}{(C_{A0} - C_x)(C_{B0} - C_x)}$$

to obtain

$$\frac{(C_{B0} - C_{A0})\, dC_x}{(C_{A0} - C_x)(C_{B0} - C_x)} = k_f (C_{B0} - C_{A0})\, dt$$

Splitting the denominator of the fraction of the left-hand side of the above equation and integrating yields

$$\int \frac{dC_x}{C_{A0} - C_x} - \int \frac{dC_x}{C_{B0} - C_x} = \int k_f (C_{B0} - C_{A0})\, dt$$

$$\ln\left(\frac{C_x - C_{B0}}{C_x - C_{A0}}\right) = k_f (C_{B0} - C_{A0}) t + \text{constant}$$

To solve for k_f, use the fact that at $t = 0$, $C_x = 0$ from the definition of C_x:

$$\ln\left(\frac{0 - C_{B0}}{0 - C_{A0}}\right) = k_f (C_{B0} - C_{A0})(0) + \text{constant}$$

$$\ln\left(\frac{C_{B0}}{C_{A0}}\right) = \text{constant}$$

which gives

$$\ln\left(\frac{C_x - C_{B0}}{C_x - C_{A0}}\right) = k_f (C_{B0} - C_{A0}) t + \ln\left(\frac{C_{B0}}{C_{A0}}\right)$$

Solving for k_f, the result is

$$k_f = \frac{1}{t(C_{A0} - C_{B0})} \ln\left(\frac{C_{B0}(C_{A0} - C_x)}{C_{A0}(C_{B0} - C_x)}\right) \tag{2-18}$$

If $C_{B0} = C_{A0}$, one can show that

$$k_f = \frac{C_x}{t\, C_{A0}\,(C_{A0} - C_x)}$$

2.3 Third-Order Reactions

An example of a third-order and termolecular reaction is

$$2NO + O_2 \rightarrow 2NO_2$$

Another example is

$$M + 2A \xrightarrow{k_f} A_2 + M^*$$

M is called a third body and causes the reaction $2A \rightarrow A_2$. M^* has slightly different characteristics from those of M, since the nature of M is changed by the heat of the reaction process. Some particles may even emit radiation (light). The rate law for the above reaction is given as

$$\frac{dC_{A_2}}{dt} = k_f C_M C_A^2 = -\frac{1}{2}\frac{dC_A}{dt}$$

$$\frac{dC_A}{dt} = -2k_f C_M C_A^2$$

In case the concentration of M is truly a constant, it can then be combined with k to give

$$\frac{dC_A}{dt} = -k' C_A^2$$

where k' is a new rate constant. The order of the reaction is then reduced from three to two. If C_M is not truly constant, it is then a function of time, and the order of the reaction is still three.

For every reaction process, a rate equation can be used to replace the equilibrium-constant equation. Since the rate of a reaction is strongly dependent upon the path, the thermodynamic state functions cannot be used. Indeed, the detailed reaction mechanism for many chemical reactions is unknown. It is usually very difficult to measure the concentrations of every significant species.

Most reactions we shall deal with are second- and third-order. An example of a complicated reaction involving both second- and third-order reactions is

$$OH + H_2 \rightarrow H_2O + H$$

$$H + O_2 \rightarrow OH + O$$

$$H + H + H \rightarrow H_2 + H$$

3 CONSECUTIVE REACTIONS

Another type of complication that can occur in a reaction process is that in which the products of one reaction undergo further reaction to give other products. A simple example of this type is

$$A + B \xrightarrow{k_1} AB \xrightarrow{k_2} C + D \tag{2-19}$$

As can be seen from this equation, a consecutive reaction is a series reaction in

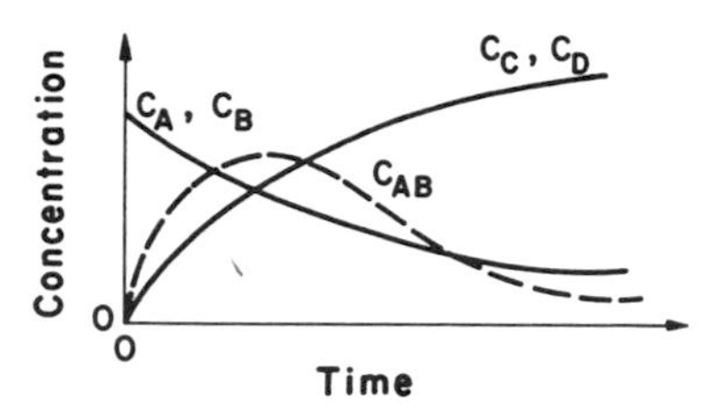

Figure 2.9 Illustration of the possibility of a concentration peak of the intermediate product AB in a consecutive reaction.

which k_1 and k_2 are the two specific reaction rate constants (the reverse reactions are neglected). The rate laws for the first and second reactions can be determined as

First reaction:

$$\frac{dC_{AB}}{dt} = k_1 C_A C_B = -\frac{dC_A}{dt} = -\frac{dC_B}{dt}$$

Second reaction:

$$\frac{dC_{AB}}{dt} = -k_2 C_{AB} = -\frac{dC_C}{dt} = -\frac{dC_D}{dt}$$

The net rate of change of C_{AB} is obtained by adding the first and second rate laws:

$$\left(\frac{dC_{AB}}{dt}\right)_{\text{net}} = k_1 C_A C_B - k_2 C_{AB} \tag{2-19a}$$

As the reaction proceeds, the concentrations of A and B decrease, and the concentrations of C and D increase. The concentration of AB, therefore, may have a peak at a particular time, as shown in Fig. 2.9.

Example 2.4

Determine an equation for the concentration of the intermediate product B in a simple consecutive reaction in which both reactions are of first order as given by the equation

$$A \xrightarrow{k_1} B \xrightarrow{k_2} C + D \tag{2-20}$$

where k_1 and k_2 are the two specific rate constants.

Solution:

The rate of disappearance of A is given by

$$\frac{dC_A}{dt} = -k_1 C_A$$

$\ln C_A = -k_1 t + C$

which integrates to

$$C_A = C_{A0}e^{-k_1 t} \tag{2-20a}$$

where C_{A0} is the initial concentration of A. The rate of formation of C is given by

$$\frac{dC_C}{dt} = k_2 C_B$$

while the net rate of production of B, the rate of its formation from A minus that of its destruction to give C and D, is

$$\frac{dC_B}{dt} = k_1 C_A - k_2 C_B \tag{2-20b}$$

Introduction of Eq. (2-20a) into (2-20b) gives

$$\frac{dC_B}{dt} = k_1 C_{A0} e^{-k_1 t} - k_2 C_B$$

which contains only the variables C_B and t. It integrates to

$$C_B = C_{A0}\frac{k_1}{k_2 - k_1}(e^{-k_1 t} - e^{-k_2 t}) \tag{2-20c}$$

The rates of change of C_A and C_B are thus given by (2-20a) and (2-20b); that of C_C is easily deduced by means of the relationship (note that $C_C = C_D$)

$$C_A + C_B + 2C_C = C_{A0} \tag{2-20d}$$

4 COMPETITIVE REACTIONS

A competitive reaction occurs when two or more sets of combustion products are produced from the same set of reactants. This situation is illustrated by

$$A + B \overset{k_1}{\rightarrow} AB$$

$$A + B \overset{k_2}{\rightarrow} E + F$$

The rate laws for the first and second reactions can be determined as follows:

First reaction:

$$\frac{dC_A}{dt} = -k_1 C_A C_B$$

Second reaction:

$$\frac{dC_A}{dt} = -k_2 C_A C_B$$

The net rate of disappearance of species A is found by summing the two equations given above:

$$\frac{dC_A}{dt} = -(k_1 + k_2) C_A C_B$$

The extrapolation of the rate law to a higher temperature range can lead to incorrect results because the specific reaction rate constants are very dependent on temperature. One reaction may be dominant at a given temperature, whereas at higher temperatures other competitive reactions may also need to be considered.

5 OPPOSING REACTIONS

In general, chemical reactions can proceed in both the forward direction (reactants forming products, rate constant k_f) and the reverse direction (reaction products reforming the reactants, rate constant k_b). At thermodynamic equilibrium there is no net change in composition. The rate constants k_f and k_b must therefore be related through the equilibrium constant, K_C, expressed in terms of the ratio of concentrations raised to appropriate powers:

$$K_C = \prod_{i=1}^{N} C_{M_i}^{(\nu_i'' - \nu_i')} \tag{2-21}$$

The general set of opposing chemical reactions is

$$\sum_{i=1}^{N} \nu_i' M_i \underset{k_b}{\overset{k_f}{\rightleftharpoons}} \sum_{i=1}^{N} \nu_i'' M_i \tag{2-22}$$

For simultaneous chemical reactions, the basic rate law of Eq. (2-13) must be applied to each reaction step; dC_{M_i}/dt represents the sum of the changes produced by the individual simultaneous reaction steps. Thus for the reaction symbolized by Eq. (2-22),

$$\frac{dC_{M_i}}{dt} = (\nu_i'' - \nu_i') k_f \prod_{j=1}^{N} C_{M_j}^{\nu_j'} + (\nu_i' - \nu_i'') k_b \prod_{j=1}^{N} C_{M_j}^{\nu_j''} \tag{2-23}$$

At thermodynamic equilibrium

$$\frac{dC_{M_i}}{dt} = 0 \quad \text{and} \quad C_{M_j} = C_{M_{j,e}} \tag{2-24}$$

where $C_{M_{j,e}}$ denotes the thermodynamic equilibrium value for species M_j. From Eqs. (2-23) and (2-24) it follows that

$$\frac{k_f}{k_b} = \prod_{j=1}^{N} C_{M_{j,e}}^{(\nu_j'' - \nu_j')} \equiv K_C \tag{2-25}$$

Here K_C represents the usual equilibrium constant defined in terms of concentration ratios. It is evident that Eq. (2-25) relates the ratio of the kinetic parameters k_f and k_b to the thermodynamic equilibrium constant K_C, which can be calculated quite accurately, for example, by quantum-statistical methods[7] from molecular properties. Equation (2-23) can be rewritten in terms of K_C as follows:

$$\frac{dC_{M_i}}{dt} = (\nu_i'' - \nu_i')k_f \prod_{j=1}^{N} C_{M_j}^{\nu_j'}\left(1 - \frac{1}{K_C}\prod_{j=1}^{N} C_{M_j}^{\nu_j'' - \nu_j'}\right) \tag{2-26}$$

Knowing K_C and the measured value of dC_{M_i}/dt, it is possible to calculate the forward rate constant from Eq. (2-26).

For a third-body reaction, the third-body concentration always cancels out in the equilibrium constant expression. For example, consider

$$\mathrm{H} + \mathrm{H} + M \rightleftharpoons \mathrm{H}_2 + M$$

The equilibrium constant for this expression is

$$K_C = \frac{C_{\mathrm{H}_2}C_M}{C_{\mathrm{H}}^2 C_M} = \frac{C_{\mathrm{H}_2}}{C_{\mathrm{H}}^2}.$$

5.1 First-Order Reaction Opposed by a First-Order Reaction

Consider the first-order reaction

$$A \underset{k_b}{\overset{k_f}{\rightleftharpoons}} B$$

for which Eq. (2-23) becomes

$$\frac{dC_x}{dt} = k_f(C_{A0} - C_x) - k_b C_x$$

where C_x is the portion of A converted to B. Therefore

$$C_A = C_{A0} - C_x$$

$$C_B = C_x$$

$$C_{B0} = 0$$

where the subscript 0 denotes initial conditions. Also,

$$\frac{k_f}{k_b} = K_C = \frac{C_{xe}}{C_{A0} - C_{xe}} \tag{2-27}$$

where the subscript e identifies the value of C_x at thermodynamic equilibrium. If $C_x = 0$ at $t = 0$, then substituting k_b of Eq. (2-27) into the ordinary differential equation for C_x and performing integration gives

$$k_f = \frac{C_{xe}}{C_{A0}t} \ln\left(\frac{C_{xe}}{C_{xe} - C_x}\right) \tag{2-28}$$

Hence if the equilibrium concentration $C_{xe} = C_{Be}$ is known, we can determine both k_f and k_b from experimental measurements of C_x as a function of time. Equation (2-27) can also be written in the form

$$\frac{C_{xe}}{C_{A0}} = \frac{k_f}{k_f + k_b}$$

from which Eq. (2-28) becomes

$$k_f + k_b = \frac{1}{t} \ln\left(\frac{C_{xe}}{C_{xe} - C_x}\right) \tag{2-29}$$

Equation (2-29) is formally identical with the first-order rate law for the forward reaction alone (see Example 2.3),

$$k_f = \frac{1}{t} \ln\left(\frac{C_{A0}}{C_{A0} - C_x}\right) \tag{2-30}$$

However, it is easy to distinguish between the chemical rate processes corresponding to Eq. (2-29) and Eq. (2-30) by studying the reaction rate as a function of the initial concentration C_{A0}. Equations (2-29) and (2-30) are written in a form suitable for the interpretation of experimental data.

5.2 First-Order Reaction Opposed by a Second-Order Reaction

A first-order reaction opposed by a second-order reaction is represented by the relation

$$A \underset{k_b}{\overset{k_f}{\rightleftharpoons}} B + C$$

Equation (2-26) becomes

$$\frac{dC_x}{dt} = k_f(C_{A0} - C_x) - k_b C_x^2$$

if $C_{B0} = C_{C0} = 0$. The integration equation is

$$k_f = \frac{C_{xe}}{t(2C_{A0} - C_{xe})} \ln\left(\frac{C_{A0}C_{xe} + C_x(C_{A0} - C_{xe})}{C_{A0}(C_{xe} - C_x)}\right) \tag{2-31}$$

5.3 Second-Order Reaction Opposed by a Second-Order Reaction

For the process

$$A + B \underset{k_b}{\overset{k_f}{\rightleftharpoons}} C + D$$

it is found that

$$k_f = \frac{C_{xe}}{2at(a - C_{xe})} \ln\left(\frac{C_x(a - 2C_{xe}) + aC_{xe}}{a(C_{xe} - C_x)}\right) \tag{2-32}$$

if $C_{A0} = C_{B0} = a$ and $C_{C0} = C_{D0} = 0$.

Example 2.5

Develop an equation for the specific reaction rate constant for the backward reaction of a first-order reaction opposed by a first-order reaction, and a first-order reaction opposed by a second-order reaction.

Solution:

For the first-order reaction opposed by a first-order reaction,

$$A \underset{k_b}{\overset{k_f}{\rightleftharpoons}} B$$

the rate law can be written as

$$\frac{dC_x}{dt} = k_f(C_{A0} - C_x) - k_b C_x$$

If C_{xe} is the equilibrium concentration of B at equilibrium when the net rate of reaction is zero, we have

$$k_f(C_{A0} - C_{xe}) = k_b C_{xe}$$

which rearranges to

$$k_b = \frac{k_f(C_{A0} - C_{xe})}{C_{xe}}$$

Using Eq. (2-30), we have

$$k_b = \frac{C_{A0} - C_{xe}}{tC_{xe}} \ln\left(\frac{C_{A0}}{C_{A0} - C_x}\right). \tag{2-33}$$

For the first-order reaction opposed by a second-order reaction,

$$A \underset{k_b}{\overset{k_f}{\rightleftharpoons}} B + C$$

the rate law is

$$\frac{dC_x}{dt} = k_f(C_{A0} - C_x) - k_b C_x^2$$

For the case in which the initial concentrations of B and C are a, while that of A is zero, the rate equation integrates to

$$k_b = \frac{C_{xe}}{t(a^2 - C_{xe}^2)} \ln\left(\frac{C_x(a^2 - C_x C_{xe})}{a^2(C_{xe} - C_x)}\right)$$

where C_{xe} is now the equilibrium concentration of A, and k_b is the specific rate formation of A.

6 CHAIN REACTIONS

Chain reactions are the most common type of chemical reactions. They consist of a series of consecutive, competitive, and opposing reaction steps with different reaction rate constants. These complex chemical reactions occur in all combustion processes. Processes which are understood in considerable detail will be discussed. For many combustion processes the specific rate constants for separate reaction steps either are not known or have been approximated only roughly.[6]

6.1 Free Radicals

In a reaction process, the most active species are called free radicals. In chemical terminology, free radicals are characterized by unpaired electrons or "bonds-paired" electrons. The hydrogen atom is a free radical, as is illustrated below, where the dots symbolize electrons:

$$\mathrm{H:H} \rightarrow \mathrm{H}\cdot + \mathrm{H}\cdot$$

If one hydrogen atom $\mathrm{H}\cdot$ is taken away from CH_4, two free radicals are formed:

$$\begin{array}{c}\mathrm{H}\\ \ddot{}\\ \mathrm{H:C:H}\\ \ddot{}\\ \mathrm{H}\end{array} \rightarrow \begin{array}{c}\mathrm{H}\\ \ddot{}\\ \mathrm{H:C}\cdot\\ \ddot{}\\ \mathrm{H}\end{array} + \cdot\mathrm{H}$$

Electromagnetic theory can be used to study the nature of free radicals in the reaction process.

Elementary reactions are called chain-initiating or chain-terminating reactions according as they produce or destroy free radicals. Also, with regard to the ratio of the number of free radicals in the product to that in the reactant, elementary reactions are called chain-propagating (or chain-carrying) reactions if the ratio is equal to 1, and chain-branching reactions if the ratio is greater than 1. Some elementary reactions and their denominations are given below:

$$A_2 \rightarrow 2A\} \quad \text{chain-initiating } (A_2 \text{ has low dissociation energy})$$

$$\left.\begin{array}{r} A + B_2 \rightarrow AB + B \\ B + A_2 \rightarrow AB + A \\ A + AB \rightarrow A_2 + B \\ B + AB \rightarrow B_2 + A \end{array}\right\} \quad \text{chain-propagating (usually very fast)}$$

$$\left.\begin{array}{r} M + 2A \rightarrow A_2 + M \\ M + 2B \rightarrow B_2 + M \end{array}\right\} \quad \text{chain-terminating}$$

A and B are called chain carriers or free radicals and seldom build up in high concentrations.

The elementary reaction

$$\mathrm{H} + \mathrm{O_2} \rightarrow \mathrm{OH} + \mathrm{O}$$

is a chain-branching reaction, since the number of chain carriers formed is more than the number of chain carriers used up in the reaction. More detailed discussion of chain-branching reactions is given in later sections.

6.2 Lindemann's Theory for First-Order Reaction

According to Lindemann,[8] the first-order chain-initiating reaction in the above set of elementary reactions occurs as the result of two-step reaction described below:

$$A + A \underset{k_b}{\overset{k_f}{\rightleftharpoons}} A^* + A \qquad \begin{matrix}(\text{fast})\\(\text{fast})\end{matrix} \tag{2-34}$$

$$A^* \xrightarrow{k_f'} \text{reaction products} \qquad (\text{slow}) \tag{2-35}$$

He proposed that reactant molecules receive energy by collisions with one another, and at any given time a small fraction of the molecules have sufficient energy to pass into the reaction products without having to receive any additional energy; such molecules will be referred to as *energized* molecules, A^*. The concentration of A^* depends upon the net rate of generation (rate of energization of A minus the rate of deenergization of A^*) by collision, as well as on the rate of destruction of A^* by decomposition into products. If the energized molecules are converted into products at a rate that is small compared to the rate at which they are energized by collision or that at which they are deenergized, a stationary concentration of them may be achieved. Also, since these energized molecules are in equilibrium with the normal molecules, their concentration is proportional to that of the normal molecules. The rate of reaction is proportional to the concentration of energized molecules, and is therefore proportional to the concentration of normal molecules; the reaction is therefore of first order. That is, the overall process will obey a first-order rate law as long as the formation of A^*, according to Eq. (2-34), is sufficiently rapid to maintain an equilibrium concentration of A^*. Since the frequency of binary collisions decreases as the pressure is reduced, it is reasonable to expect that the process symbolized by Eq. (2-34) will cease to be fast at reduced pressures; therefore, a first-order reaction should become of second order at sufficiently low pressures. This change of overall order with pressure is, in fact, observed in many first-order reactions.

The differential equations corresponding to the reaction process of Eqs. (2-34) and (2-35) are

$$\frac{dC_{A^*}}{dt} = k_f C_A^2 - k_b C_{A^*} C_A - k_f' C_{A^*} \tag{2-36}$$

and

$$\frac{dC_A}{dt} = -k_f C_A^2 + k_b C_{A^*} C_A \tag{2-37}$$

From Eq. (2-37) it is apparent that

$$C_{A*} = \frac{k_f C_A^2 + dC_A/dt}{k_b C_A} \tag{2-38}$$

By differentiating Eq. (2-38), we obtain

$$\frac{dC_{A*}}{dt} = \frac{k_f}{k_b}\frac{dC_A}{dt} + \frac{1}{k_b C_A}\frac{d^2 C_A}{dt^2} - \frac{1}{k_b C_A^2}\left(\frac{dC_A}{dt}\right)^2 \tag{2-39}$$

By combining Eqs. (2-36), (2-38), and (2-39) we obtain a single second-order differential equation in C_A and t which can be solved exactly only by numerical methods. One could also solve the two simultaneous equations (2-36) and (2-37) on a computer by a standard Runge–Kutta integration routine.

We have not yet utilized the result of our postulate that the reaction according to Eq. (2-35) is slow compared to the forward and reverse reactions given in Eq. (2-34). This physical notion leads to the conclusion that dC_{A*}/dt must be small compared to dC_A/dt, and suggests a simpler mathematical procedure for the solution of Eqs. (2-36) and (2-37). The first approximation is the classical steady-state approximation, which is expressed by the relation

$$\frac{dC_{A*}}{dt} = 0 \tag{2-40}$$

The steady-state assumption has often been used in work on combustion, especially in the early days before the development of computers and numerical techniques. The solution of a set of coupled first-order ordinary differential equations by computer should be familiar to all modern engineers and scientists.

If the steady-state assumption is valid for A^*, then Eq. (2-36) can be expressed by

$$C_{A*} = \frac{k_f C_A^2}{k_b C_A + k_f'} \tag{2-41}$$

which can be combined with Eq. (2-37) to yield the result

$$-\frac{dC_A}{dt} = \frac{k_f k_f' C_A^2}{k_b C_A + k_f'} \tag{2-41a}$$

This last relation can be integrated directly. Thus, for the chemical process described by Eqs. (2-36) and (2-37), introduction of the steady-state approximation of Eq. (2-40) leads to a straightforward solution of the problem.

A steady-state postulate for reaction intermediates is sometimes justified as a first approximation for chemical reaction in flow systems. However, it is necessary to examine the conditions in any given problem in order to at least verify the fact that the steady state is possible. The limitations of the steady-state assumption can be assessed most simply by comparing the complete solution with results derived from the steady-state treatment. Before applying the steady-state approximation to a given combustion problem, considerable ingenuity and physical insight may be required in order to obtain a rational estimate for the limits of reliability of the treatment.

6.3 Hydrogen – Bromine Reaction

A classical example of a complex reaction mechanism is provided by the formation of HBr from H_2 and Br_2. The global reaction for the generation of hydrogen bromide is

$$H_2 + Br_2 \rightarrow 2HBr$$

The rate of production of HBr does not follow the law of mass action given by Eq. (2-13). The experimentally determined rate law for the reaction is

$$\frac{dC_{HBr}}{dt} = \frac{a_1 C_{H_2} C_{Br_2}^{1/2}}{1 + C_{HBr}/\left(a_2 C_{Br_2}\right)}$$

where a_1 and a_2 are constants at a given temperature.

In the following, we shall first consider the detailed reaction mechanism, which consists of an interplay of various elementary reactions, then apply the steady-state treatment to free H and Br radicals, and finally derive a rate expression in the same form as that obtained experimentally. The H_2–Br_2 reaction also serves as an example of how a complex reaction mechanism can be proposed and verified.

To initiate this chemical reaction, heat is added. Br_2 begins to decompose first, since H_2 is more stable than Br_2 (note: $\Delta\mathscr{H}^\circ_{f,\mathrm{Br}} = 6.71$ kcal/mole, $\Delta\mathscr{H}^\circ_{f,\mathrm{H}} = 52$ kcal/mole). Bromine atoms are free radicals which can react with H_2. Therefore, a series of reactions are followed:

$$M + Br_2 \xrightarrow{k_1} 2Br + M \qquad \text{chain-initiating} \qquad (2\text{-}42)$$

$$Br + H_2 \xrightarrow{k_2} HBr + H \qquad (2\text{-}43)$$

$$H + Br_2 \xrightarrow{k_3} HBr + Br \qquad \text{chain-carrying} \qquad (2\text{-}44)$$

$$H + HBr \xrightarrow{k_4} H_2 + Br \qquad (2\text{-}45)$$

$$M + Br + Br \xrightarrow{k_5} Br_2 + M \qquad \text{chain-terminating} \qquad (2\text{-}46)$$

In Eqs. (2-42) and (2-46), the symbol M represents a third body, that is, any of the chemical species H, Br, H_2, Br_2, or HBr which may be present.

Equation (2-42) is a chain-initiating step. Equations (2-43) and (2-44) represent the chain-carrying reactions in which an atom (either Br or H) is produced for each atom which reacts. Equation (2-45) is the inverse of (2-43); the inverse of (2-44) is relatively slow and is therefore unimportant. Equation (2-46) represents the chain-breaking step. The chain-breaking step according to the process

$$2H + M \rightarrow H_2 + M$$

is not important in the present case, since the concentration of H atoms is generally small compared with that of Br atoms. However, at higher temperatures both of the equations

$$Br + HBr \rightarrow Br_2 + H$$

$$2H + M \rightarrow H_2 + M$$

may become quite significant. After examining the above opposing and consecutive reactions, it is easy to understand why the rate law derived from the equation

$$H_2 + Br_2 \rightarrow 2HBr$$

has very little significance.

The set of equations for the rate of change of concentration are

$$\frac{dC_{Br}}{dt} = 2k_1 C_M C_{Br_2} - k_2 C_{Br} C_{H_2} + k_3 C_H C_{Br_2} + k_4 C_H C_{HBr} - 2k_5 C_M C_{Br}^2 \tag{2-47}$$

$$\frac{dC_H}{dt} = k_2 C_{Br} C_{H_2} - k_3 C_H C_{Br_2} - k_4 C_H C_{HBr} \tag{2-48}$$

$$\frac{dC_{Br_2}}{dt} = -k_1 C_{Br_2} C_M - k_3 C_H C_{Br_2} + k_5 C_{Br}^2 C_M \tag{2-49}$$

$$\frac{dC_{H_2}}{dt} = -k_2 C_{Br} C_{H_2} + k_4 C_H C_{HBr} \tag{2-50}$$

$$\frac{dC_{HBr}}{dt} = k_2 C_{Br} C_{H_2} + k_3 C_H C_{Br_2} - k_4 C_H C_{HBr} \tag{2-51}$$

Applying the steady-state assumption that the mean concentrations of the free

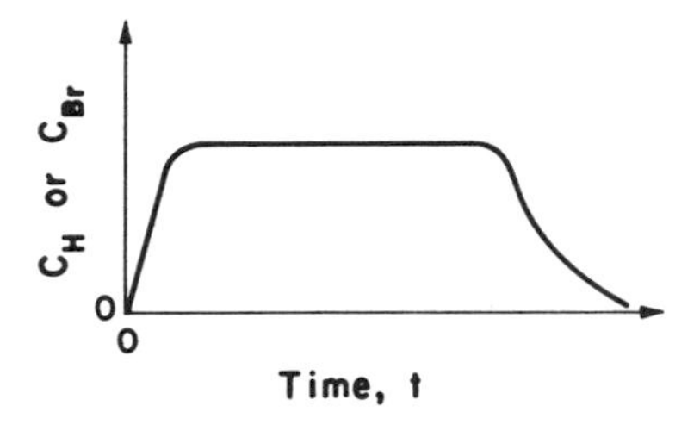

Figure 2.10 Throughout the bulk of the reaction the concentration of the free radicals is generally maintained as a constant.

radicals H and Br remain nearly constant leads to

$$\frac{dC_{\mathrm{H}}}{dt} = \frac{dC_{\mathrm{Br}}}{dt} = 0 \tag{2-52}$$

In actuality, the concentrations of H and Br will not remain constant throughout the reaction process, but they will remain constant throughout the bulk of the reaction. This situation is illustrated in Fig. 2.10, where, except for short initial and final periods, the concentration of the free radicals is nearly constant.

Using Eq. (2-52) to equate Eqs. (2-47) and (2-48), and then simplifying, we have

$$2k_5 C_{\mathrm{Br}}^2 C_M = 2k_1 C_{\mathrm{Br}_2} C_M$$

Therefore,

$$C_{\mathrm{Br}} = \sqrt{\frac{k_1}{k_5}}\sqrt{C_{\mathrm{Br}_2}} \tag{2-53}$$

Solving Eq. (2-48) for C_{H} gives

$$C_{\mathrm{H}} = \frac{k_2 C_{\mathrm{Br}} C_{\mathrm{H}_2}}{k_3 C_{\mathrm{Br}_2} + k_4 C_{\mathrm{HBr}}} \tag{2-54}$$

Equations (2-53) and (2-54) were obtained under the steady-state assumption. If the equilibrium assumption were used instead of the steady-state assumption, Eq. (2-54) would be different, since equilibrium-constant equations would be used to replace the rate expressions. It is quite obvious that these two assumptions are not interchangeable. Under either the steady-state assumption or the equilibrium assumption, the total number of unknowns is equal to six, namely

$$T_f,\ C_{\mathrm{H}},\ C_{\mathrm{Br}},\ C_{\mathrm{H}_2},\ C_{\mathrm{Br}_2},\ C_{\mathrm{HBr}}$$

In addition to Eqs. (2-47) through (2-51), we have one enthalpy-balance equation to make the system completely defined. By solving these six simulta-

$K = AT^b \exp\left(-\frac{E}{RT}\right)$

neous equations as a function of time, the reaction history of this combustion problem is obtained.

Now, if we follow the steady-state treatment and substitute Eqs. (2-53) and (2-54) into Eq. (2-51), we have

$$\frac{dC_{\mathrm{HBr}}}{dt} = k_2\sqrt{\frac{k_1 C_{\mathrm{Br}_2}}{k_5}}\,C_{\mathrm{H}_2} + \frac{k_3 C_{\mathrm{Br}_2} - k_4 C_{\mathrm{HBr}}}{k_3 C_{\mathrm{Br}_2} + k_4 C_{\mathrm{HBr}}} k_2 C_{\mathrm{Br}} C_{\mathrm{H}_2}$$

or

$$\frac{dC_{\mathrm{HBr}}}{dt} = k_2\sqrt{\frac{k_1 C_{\mathrm{Br}_2}}{k_5}}\,C_{\mathrm{H}_2}\left(\frac{2k_3 C_{\mathrm{Br}_2}}{k_3 C_{\mathrm{Br}_2} + k_4 C_{\mathrm{HBr}}}\right)$$

which simplifies to

$$\frac{dC_{\mathrm{HBr}}}{dt} = \frac{2k_2\sqrt{k_1/k_5}\,\sqrt{C_{\mathrm{Br}_2}}\,C_{\mathrm{H}_2}}{1 + (k_4/k_3)C_{\mathrm{HBr}}/C_{\mathrm{Br}_2}} \tag{2-55}$$

Equation (2-55) matches the empirical relation obtained from experimental measurements,

$$\frac{dC_{\mathrm{HBr}}}{dt} = \frac{2k_t C_{\mathrm{H}_2}\sqrt{C_{\mathrm{Br}_2}}}{1 + C_{\mathrm{HBr}}/(10C_{\mathrm{Br}_2})} \tag{2-56}$$

At the beginning of the reaction process, the concentration of HBr is very small, that is,

$$1 \gg \frac{C_{\mathrm{HBr}}}{10C_{\mathrm{Br}_2}}$$

In this case, Eq. (2-56) reduces to the Arrhenius form in which

$$\frac{dC_{\mathrm{HBr}}}{dt} = kC_{\mathrm{H}_2}C_{\mathrm{Br}_2}^{1/2} \tag{2-57}$$

The overall order of the reaction is $1\frac{1}{2}$. For the case in which

$$\frac{C_{\mathrm{HBr}}}{10C_{\mathrm{Br}_2}} \gg 1$$

the Arrhenius form is again obtained. In complex reactions, the order of reaction changes as a function of time.

7 CHAIN-BRANCHING EXPLOSIONS

In a mixture of hydrogen and oxygen it is presumed plausible that the presence of a free valence in the form of an OH radical or H atom should result in the following reaction cycle:

$$\left.\begin{array}{l} H + O_2 \rightarrow OH + O \\ O + H_2 \rightarrow OH + H \end{array}\right\} \quad \text{chain-branching}$$

$$OH + H_2 \rightarrow H_2O + H$$

The first reaction is endothermic by 17 kcal/mole. Thus, at room temperature and even at somewhat higher temperatures, a mixture of hydrogen and oxygen is very stable even if hydrogen atoms are introduced from another source. The free valence ultimately terminates at the wall through recombination processes. Above some temperatures, however, the chain-branching reaction becomes sufficiently frequent, when compared to the rate of removal of H atoms, to cause multiplication of free valences and explosion.

Explosions are conveniently classified into two distinct categories: *branched-chain explosions*, in which the reaction rate increases without limit because of chain branching; and *thermal explosions*, in which there is an exponential increase in reaction rate resulting from exothermic chemical reaction, heating of reactants, and an increase in the magnitude of the specific reaction rate constants.

Consider a 1-cm^3 container which contains initially one chain particle, that is, 1 free radical per cm^3. Assume that the number density in the container is 10^{19} molecules/cm^3 and the average collision rate is 10^8 collisions/sec. If the reaction in the volume is a chain-carrying reaction [i.e., one free radical can generate another free radical in the reaction ($\alpha' = 1.0$)], then the time required for all of the molecules to react (i.e., 10^{19} collisions) will be

$$t = \frac{10^{19}\ \text{collisions}}{10^8\ \text{collisions/sec}} = 10^{11}\ \text{sec} \approx 30{,}000\ \text{years}$$

Such a slow process cannot be called combustion. If the reaction in the volume is a chain-branching reaction [i.e., one free radical or chain particle can generate two chain particles in the reaction ($\alpha' = 2.0$)], then the time required for all of the molecules to react can be estimated as follows:

$$\frac{2^{N+1} - 1}{2 - 1} = 10^{19}\ \text{molecules}$$

where

$$N = 64\ \text{generations}$$

or

$$t = 64 \times 10^{-8} \text{ sec} \approx 10^{-6} \text{ sec} = 1\ \mu\text{sec}$$

This is certainly a very rapid combustion process. In an actual combustion process, not all reactions are chain-branching. However, the reaction rate is still very fast even for a very small portion of chain-branching reactions. For a combustion process in which 1% ($\alpha' = 1.01$) of the reactions are chain-branching, the time required for all the molecules in the volume to react would be only

$$\frac{\alpha'^{N+1} - 1}{\alpha' - 1} = \frac{1.01^{N+1} - 1}{0.01} = 10^{19} \text{ molecules/cm}^3$$

$$N = 3934$$

$$t = 3934 \times 10^{-8} \text{ sec} \approx 40\ \mu\text{sec}$$

This is still a very fast reaction.

In general, branched-chain reactions and explosions can be studied by considering the following chemical kinetics:

$$M \xrightarrow{k_1} R \qquad \text{chain-initiating}$$

$$R + M \xrightarrow{k_2} \alpha' R + M^* \qquad \text{chain-branching}$$

$$\left.\begin{array}{l} R + M \xrightarrow{k_3} P \\ R \underset{\text{(on wall)}}{\xrightarrow{k_4}} M \\ R \xrightarrow{k_5} \text{nonreactive species} \end{array}\right\} \text{chain-terminating} \tag{2-58}$$

The rate equation (applying the steady-state assumption) is

$$\frac{dC_R}{dt} = 0 = k_1 C_M + (\alpha' - 1) k_2 C_R C_M - k_3 C_R C_M - k_4 C_R - k_5 C_R \tag{2-59}$$

Solving for C_R gives

$$C_R = \frac{k_1 C_M}{k_3 C_M + k_4 + k_5 - (\alpha' - 1) k_2 C_M} \tag{2-60}$$

The rate of change of the product concentration is given as

$$\frac{dC_p}{dt} = k_3 C_R C_M = \frac{k_1 k_3 C_M^2}{k_3 C_M + k_4 + k_5 - k_2(\alpha' - 1) C_M} \tag{2-61}$$

The quantity $k_2(\alpha' - 1)C_M$ is positive; as its value increases it tends to decrease the denominator in Eq. (2-61). The critical value of α' is given as

$$\alpha'_{\text{critical}} = 1 + \frac{k_3 C_M + k_4 + k_5}{k_2 C_M} \tag{2-62}$$

and we have

$\alpha' \geq \alpha'_{\text{critical}} \quad \Rightarrow \quad$ chain-branching explosion

$\alpha' < \alpha'_{\text{critical}} \quad \Rightarrow \quad$ no explosion

However, it is important to note that for some actual explosion processes, because the concentration of R does not remain small, the steady-state approximation may not be valid. Other reaction steps may also become important. The postulated reaction kinetics in Eq. (2-58) may not always be applicable during an explosion.

8 EXPLOSION LIMITS

8.1 H_2-O_2 System

It is experimentally observed that a pressure vessel containing hydrogen and oxygen under the conditions shown in Fig. 2.11 will explode as the pressure is raised. Intuitively, one would assume that as the pressure is raised, the concentration of free radicals would be increased, which would lead to an explosion. However, an explosion is also experimentally observed as the pressure is lowered.

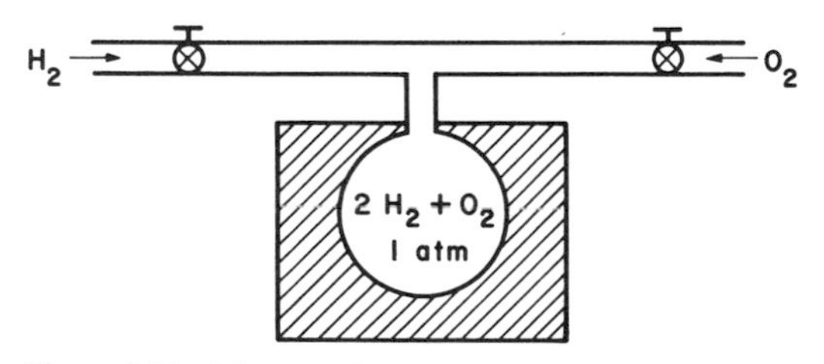

Figure 2.11 Mixture of H_2 and O_2 in a pressure vessel.

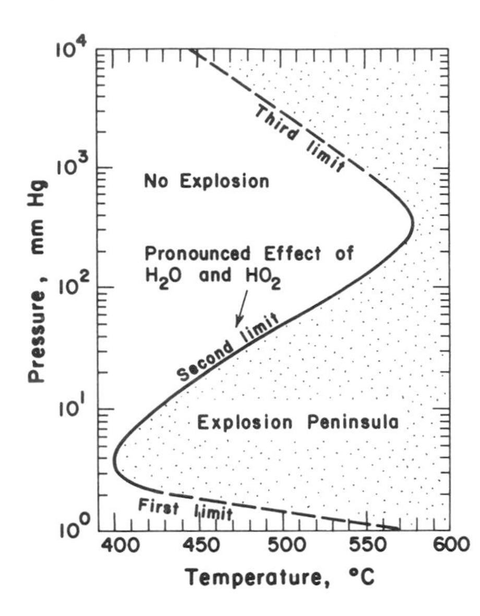

Figure 2.12 Explosion limits of a stoichiometric hydrogen–oxygen mixture in a spherical vessel. First and third limits are partly extrapolated. First limit is subject to erratic changes. (From Ref. 6.)

The existence of explosion limits in a closed vessel can be understood very simply from qualitative considerations of competition between chain-breaking and chain-branching reactions on surfaces and in the gas phase. Typical experimental results for hydrogen–oxygen mixtures are plotted schematically in Fig. 2.12. The first, or lower, explosion limit occurs at roughly the same pressure over a relatively large temperature range. The lower explosion limit is determined by a balance between the removal of chain carriers on the surface (wall effect) and production of chain carriers by gas-phase reactions. In this low pressure range, the number of collisions and the rate of production of chain carriers are both low. From Eq. (2-62) we know that

$$\alpha'_{\text{critical}} = (1 + C_1) + \frac{C_2}{C_M} \quad \text{and} \quad C_M \propto p$$

The lower the pressure, the larger the value of $\alpha'_{\text{critical}}$, and hence the smaller the chance for explosion. However, none of these analyses can predict the exact location of the explosion limits; they can only explain the mechanism of reaction in each region.

As the pressure is raised, the rate of production of chain carriers by gas-phase reactions increases to the point at which surface destruction is no longer sufficient to prevent a branching explosion. The lower explosion limit defines the condition at which chain branching in the gas phase is balanced by chain breaking at the surface.

As the pressure is raised in the explosion limit, chain branching in the gas phase becomes important. There are two postulations about the reaction kinetics:

Recommended by a number of investigators in the 1920s:

$$H_2 \rightarrow 2H + 106\ \text{kcal/mole} \qquad \text{(dissociation)} \tag{A}$$

Suggested by Lewis and von Elbe:[6]

$$H_2 + O_2 + M \rightarrow H_2O_2 + M^*$$
$$\hookrightarrow 2OH + 51\ \text{kcal/mole} \tag{B}$$

and after the OH radical is generated,

$$OH + H_2 \rightarrow H_2O + H - 15\ \text{kcal/mole}$$

$$H + O_2 \rightarrow OH + O + 16\ \text{kcal/mole}$$

$$O + H_2 \rightarrow OH + H + 2\ \text{kcal/mole}$$

The reaction proposed in (A) is more endothermic than the reaction in (B). However, reaction (B) requires a third-body reaction which is not as likely as the dissociation reaction. At low temperatures, therefore, reaction (B) is more likely, and at high temperatures reaction (A) is more likely.

At high pressures the second explosion limit is approached. The existence of the second explosion limit is readily explained if the three-body reaction

$$H + O_2 + M \rightarrow HO_2 + M \tag{2-63}$$

is added to the scheme. In this reaction the symbol M denotes any third molecule that stabilizes the combination of H and O_2. Because the metastable intermediate hydroperoxide radical (HO_2) is thought to be relatively unreactive, it is able to diffuse to the wall. HO_2 becomes a vehicle for the destruction of free valences, and therefore the above reaction is considered a chain-breaking reaction. With increasing pressure, the frequency of ternary collisions $H + O_2 + M$ increases relative to the frequency of binary collisions $H + O_2$. There is therefore a pressure above which the rate of removal of free valences exceeds the rate of formation of free valences by chain-branching reactions, and the second explosion limit is established. The destruction of the HO_2 molecule on the wall can be expressed by the reaction

$$HO_2 \xrightarrow{\text{wall}} \tfrac{1}{2}H_2 + O_2$$

$$HO_2 \xrightarrow{\text{wall}} \tfrac{1}{2}H_2O + \tfrac{3}{4}O_2$$

Up to this point, HO_2 is assumed to have no part in chain propagation or chain branching and to be destroyed at the wall.

At some pressure above the second explosion limit, however, HO_2 is assumed to participate in the chain propagation process according to the reaction

$$HO_2 + H_2 \rightarrow H_2O_2 + H$$
$$\hookrightarrow 2\,OH$$

Therefore, above a critical pressure, there is a rapid increase in the number of radicals. This critical pressure defines the third explosion limit. Now H_2O has a bond frequency very close to that of HO_2, the structures being

$$\mathrm{H{-}O{-}H}$$

$$\mathrm{O{-}O{-}H}$$

so H_2O is an excellent third body for the reaction given in Eq. (2-63). The region for pronounced effect of H_2O and HO_2 is indicated in the explosion-limit diagram. It is useful to note that for $T > 600°C$, HO_2 cannot be stabilized and therefore explosion is observed at all pressures.

The treatment of explosion limits in flow systems can be worked out through an extension of the methods developed for closed reaction vessels.

8.2 $CO-O_2$ System

As shown in Fig. 2.13, mixtures of carbon monoxide and oxygen also exhibit the phenomenon of explosion limits. The chain-initiating reaction is

$$CO + O_2 \rightarrow CO_2 + O, \qquad \Delta H_r = -9 \text{ kcal/mole} \quad \text{(exothermic)}$$

This initiating reaction is hard to achieve without H_2. Lewis and von Elbe[6]

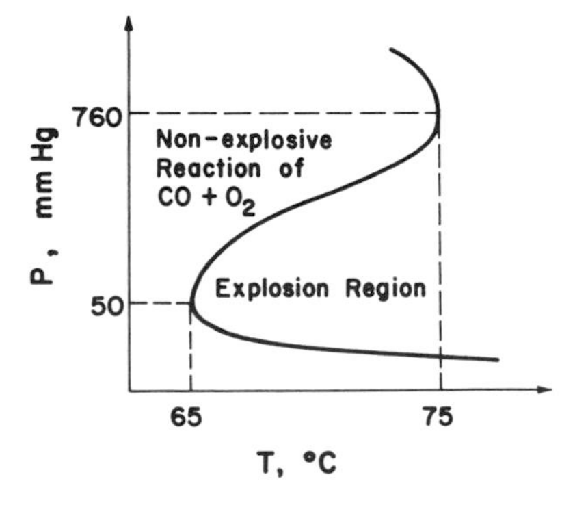

Figure 2.13 Explosion limits of a stoichiometric carbon monoxide–oxygen mixture.

have suggested that the explosion limit is essentially controlled by the reactions

$$M + CO + O \rightarrow CO_2 + M$$

$$M + O + O_2 \rightarrow O_3 + M^* \qquad \text{(exothermic)}$$

$$O_3 + CO \rightarrow CO_2 + 2O \qquad \text{(very rapid)}$$

$$O_3 + CO + M \rightarrow CO_2 + O_2 + M^*$$

It should be noted that the behavior of the CO–O_2 system is changed radically by the admixture of small amounts of H_2 or H_2O, and that the rate-controlling reaction mechanism now involves H, OH, H_2, HO_2, and H_2O as well as O, O_2, CO, CO_2, and O_3.

The water-gas reaction is most probably surface-catalyzed:

$$CO + H_2O \rightleftharpoons CO_2 + H_2 \qquad \text{(surface)}$$

and should be followed by the surface reaction in the hydrogen–oxygen reaction

$$H_2 + O_2 \rightleftharpoons H_2O_2 \qquad \text{(surface)}$$

Chain carriers are then provided by the dissociation at H_2O_2 in the gas phase:

$$H_2O_2 \rightarrow 2OH$$

$$OH + CO \rightarrow CO_2 + H$$

$$H + O_2 \rightarrow OH + O$$

9 RATE LAWS FOR ISOTHERMAL REACTIONS UTILIZING DIMENSIONLESS PARAMETERS[3]

For the sake of brevity, the following discussion will be restricted to a pair of opposing chemical reactions. Generalization to chain reactions can be made without difficulty.

For the most general opposing chemical reactions,

$$\sum_{i=1}^{N} \nu_i' M_i \underset{k_b}{\overset{k_f}{\rightleftharpoons}} \sum_{i=1}^{N} \nu_i'' M_i$$

As seen in Section 2.5, the net rate of production of species M_i is given by the phenomenological relation

$$\frac{dC_{M_i}}{dt} = (\nu_i'' - \nu_i')k_f \prod_{j=1}^{N} (C_{M_j})^{\nu_j'} \left\{ 1 - \frac{1}{K_c} \prod_{j=1}^{N} (C_{M_j})^{\nu_j'' - \nu_j'} \right\} \tag{2-26}$$

where K_C was defined by

$$K_C = \frac{k_f}{k_b} = \prod_{j=1}^{N} (C_{M_j,e})^{(\nu_j'' - \nu_j')} \tag{2-25}$$

A set of relations equivalent to that given above can now be formulated by replacing the concentrations with partial pressures, mole fractions, weight fractions, and so on. The transformations can be accomplished by assuming the validity of the ideal gas law, which constitutes an adequate approximation for a large class of combustion problems.

9.1 Equilibrium Constants

For ideal gases

$$C_{M_j,e} = \frac{p_{j,e}}{R_u T} \tag{2-64}$$

where $p_{j,e}$ denotes the equilibrium partial pressure of the chemical species identified by the symbol M_j. If the equilibrium constant K_p is defined by the relation

$$K_p = \prod_{j=1}^{N} (p_{j,e})^{(\nu_j'' - \nu_j')} \tag{2-65}$$

it is then apparent from Eqs. (2-25), (2-64), and (2-65) that

$$\frac{k_f}{k_b} = K_C = K_p (R_u T)^{-\Delta n} \tag{2-66}$$

where

$$\Delta n = \sum_{j=1}^{N} (\nu_j'' - \nu_j')$$

For ideal gases, the equilibrium mole fraction $X_{j,e}$ of species j is expressed as

$$X_{j,e} = \frac{p_{j,e}}{p} = \frac{C_{M_j,e}}{C_M} \tag{2-67}$$

where p represents the system pressure and C_M is the total number of moles per unit volume of gas mixture. From Eqs. (2-25), (2-64), (2-65), and (2-66) it can be seen that

$$\frac{k_f}{k_b} = K_C = K_p(R_uT)^{-\Delta n} = K_X\left(\frac{p}{R_uT}\right)^{\Delta n} = K_X(C_M)^{\Delta n} \qquad (2\text{-}68)$$

where

$$K_X = \prod_{j=1}^{N} (X_{j,e})^{(\nu_j'' - \nu_j')} \qquad (2\text{-}69)$$

For an ideal gas, the equilibrium mass of species j per unit volume, $\rho_{j,e}$, is

$$\rho_{j,e} = \left(\frac{p_{j,e}}{R_uT}\right)W_j = C_{M_j,e}W_j \qquad (2\text{-}70)$$

where W_j represents the molecular weight of species j. The equilibrium weight fraction at species j is then

$$Y_{j,e} = \frac{\rho_{j,e}}{\rho_e} \qquad (2\text{-}71)$$

where ρ_e denotes the equilibrium density of the fluid mixture. Using the preceding relations, it can be readily seen that

$$\frac{k_f}{k_b} = K_C = K_p(R_uT)^{-\Delta n} = K_X\left(\frac{p}{R_uT}\right)^{\Delta n} = K_Y F_w^{-1}\rho_e^{\Delta n} = K_\rho F_w^{-1} \qquad (2\text{-}72)$$

where

$$K_\rho \equiv \prod_{j=1}^{N} (\rho_{j,e})^{(\nu_j'' - \nu_j')} \qquad (2\text{-}73)$$

$$F_w \equiv \prod_{j=1}^{N} (W_j)^{(\nu_j'' - \nu_j')} \qquad (2\text{-}74)$$

$$K_Y \equiv \prod_{j=1}^{N} (Y_{j,e})^{(\nu_j'' - \nu_j')} \qquad (2\text{-}75)$$

Referring to the definitions of the various quantities involved, it is clear that only for $\Delta n = 0$, K_p, k_f/k_b, K_C, and F_w are dimensionless quantities. However, the equilibrium constants defined in terms of mole fractions K_X or in terms of weight fractions K_Y are always dimensionless. For use in Eq. (2-26), all of the expressions given in Eqs. (2-72) to (2-75) are equivalent, since the group

$$\left(\frac{k_b}{k_f}\right)\prod_{j=1}^{N}(C_{M_j})^{(\nu_j''-\nu_j')}$$

is always dimensionless. If we consider quantities K_C^*, K_p^*, K_X^*, K_ρ^*, and K_Y^* as analogous to the equilibrium constants except that concentrations, partial pressures, and so on refer to the prevailing local concentrations, partial pressure, and so on, rather than to equilibrium for the local conditions at T and p, then it is clear that

$$\left(\frac{k_b}{k_f}\right)\prod_{j=1}^{N}(C_{M_j})^{(\nu_j''-\nu_j')} = \frac{K_C^*}{K_C} = \frac{K_p^*}{K_p} = \frac{K_X^*}{K_X}$$

$$= \frac{K_\rho^*}{K_\rho} = \frac{K_Y^*}{K_Y} \equiv \frac{K^*}{K} \tag{2-76}$$

9.2 Net Rate of Production of Chemical Species

Using the definitions introduced in the preceding paragraph, it is easy to obtain the following equivalent expressions for dC_{M_i}/dt:

$$\frac{dC_{M_i}}{dt} = (\nu_i'' - \nu_i')k_f\left[1 - \left(\frac{K^*}{K}\right)\right]\prod_{j=1}^{N}(C_{M_j})^{\nu_j'}$$

$$= (\nu_i'' - \nu_i')\, k_f\,(R_uT)^{-m}\left[1 - \left(\frac{K^*}{K}\right)\right]\prod_{j=1}^{N}(p_j)^{\nu_j'}$$

$$= (\nu_i'' - \nu_i')k_f(C_M)^m\left[1 - \left(\frac{K^*}{K}\right)\right]\prod_{j=1}^{N}(X_j)^{\nu_j'}$$

$$= (\nu_i'' - \nu_i')k_f\left[1 - \left(\frac{K^*}{K}\right)\right]\prod_{j=1}^{N}\left(\frac{\rho_j}{W_j}\right)^{\nu_j'}$$

$$= (\nu_i'' - \nu_i')k_f\left[1 - \left(\frac{K^*}{K}\right)\right]\rho^m\prod_{j=1}^{N}\left(\frac{Y_j}{W_j}\right)^{\nu_j'} \tag{2-77}$$

where $m = \sum_{j=1}^{N}\nu_j'$, and the quantities p_j, X_j, ρ_j, and Y_j denote, respectively, the partial pressure, mole fraction, mass per unit volume, and weight fraction of species; and these quantities are the instantaneous local properties of the reacting mixture.

10 CONTEMPORARY METHODS FOR SOLVING COMPLEX CHEMICAL KINETIC SYSTEMS

In the above, we have discussed some simple kinetic systems which can be solved by using analytical methods. However, many real chemical systems, such as those which occur in nature or are involved in industrial processes, are often exceedingly complex. Depending on the question(s) asked, these systems may be too complex to solve analytically. In the past, chemists and engineers used their experience and intuition to fit the observed chemical processes with simple relations. When simple relations did not fit, more complex mechanisms were proposed, such as simultaneous or consecutive reactions. Quite often, in order to render the mathematical problem tractable, the steady-state approximation was used for intermediate species and free radicals. This provided a simplification, since by so doing, the concentrations of these intermediate species are obtained from algebraic rather than differential equations. In some cases, the steady-state approximations lead to remarkably satisfactory solutions. The success in achieving a mathematical solution based on the existence of steady states has led to a proliferation of such schemes. However, for a kinetically complex system, especially those involved in highly transient processes such as ignition, extinction, and the deflagration-to-detonation transition, the steady-state assumption often is not justified. Nowadays, because of recent developments in the numerical solution of systems of ODEs and PDEs, one may be able to solve the entire network of reactions. The use of numerical methods in obtaining solutions needs no apology; on the contrary it can be more economical and efficient to use computers as tools for solving these complicated problems.[10]

10.1 Numerical Programs for Solving ODEs

With modern numerical techniques, the complete time-dependent solutions of systems of ordinary differential equations (ODEs) can be obtained. This means that it is possible to test a proposed network of reactions in a quantitative fashion without the need to simplify either the reaction mechanism or the mathematics. In recent years, several ODE packages have been developed.[11] A well-known package subroutine called DIFSUB, based on fixed-step-size formulas, was developed by Gear[12,13] in 1971. It has since been modified for several special cases by Hindmarsh[14] and coworkers at Lawrence Livermore Laboratory. These modified codes are called the GEAR subroutines, after the first developer W. C. Gear. A program entitled EPISODE, based on variable-step-size formulas, was also developed by Hindmarsh.[15] For handling extremely fast reaction kinetics, the *K*-integrator[16,17] subroutine was developed at the U.S. Army Ballistic Research Laboratories. The special features of these programs will be discussed in Section 10.1.5. All these packages have been developed to handle a set of "stiff" simultaneous chemical kinetic equations.

10.1.1 Definition of the Time Constant

Let us consider the first-order decomposition reaction of species A:

$$A + M \rightarrow \text{products}$$

where M represents any collision partner and its concentration is constant. The rate of depletion of A can be written as

$$\frac{dC_A}{dt} = -kC_M C_A$$

whose solution is

$$C_A(t) = C_A(0)\exp(-kC_M t)$$

The time constant t_c is defined as the time it takes for C_A to decay to e^{-1} of the original concentration. In this case, inspection shows that

$$t_c = (kC_M)^{-1}$$

10.1.2 Common Error Types

Use of computing machinery to find solutions for mathematical problems entails an understanding of two very common sources of error.

The first is roundoff error, and it occurs because the computing device can represent a number only up to some maximum number of digits. The result of an operation (e.g., multiplication) can yield no more than this maximum. Generally the least significant digits are dropped and a roundoff error occurs.

The second is truncation error, and it occurs because a limiting process is broken off (truncated) before the limiting value is reached. Such errors occur when we approximate (a) an infinite series by a finite number of terms, (b) a derivative by a finite difference or (c) a nonlinear function by a linear one.

10.1.3 The Origin of Stiffness

A kinetic system is composed of several species whose concentrations can decay (or grow) at different rates; thus a kinetic system usually has a broad range of time constants. Mathematically this reaction network is expressed by a system of ODEs. It can happen that the numerical solution is dominated by the species that have the shortest time constants. This can be the case even after these species have decayed so that their relative concentrations are negligible. Such a system is termed *stiff*.

Mathematically the problem of stiffness lies in assuring the numerical stability of the solution algorithm. Stability here refers to the requirement that any errors (roundoff or truncation errors) introduced during the computation should be damped by the algorithm.

10.1.4 Implicit and Explicit Numerical Schemes

Numerical schemes to solve initial-value problems commonly involve multistep formulas of the form

$$y_n = \sum_{i=1}^{q} \alpha_i y_{n-i} + h_n \sum_{i=0}^{q} \beta_i y'_{n-i}$$

where $y_n = y(t_n)$, $y' = dy/dt$, q is the order of the multistep formula, and α_i and β_i are weighting coefficients. For $\beta_0 = 0$ the right-hand side is a combination of functions evaluated in previous time steps. Thus, y_n is given explicitly, and such formulas are termed *explicit*. On the other hand, for $\beta_0 \neq 0$ the right-hand side contains the term $h_0 \beta_0 y'_n$. Thus, y_n is a function of y'_n. Such formulas are termed *implicit*.

For explicit schemes, it can be shown that the time step h must satisfy the relation

$$h \leq 2t_c$$

in order that roundoff and truncation error shall decay. Thus, the smaller the time constant, the smaller the step size. This is the reason that the solution for a system of ODEs can be dominated by the smallest time constant.

In principle, as long as the step size meets the above inequality, an explicit scheme will remain stable. In practice, however, an explicit scheme applied to a stiff system of ODEs can require such intolerably small values for the step size h that roundoff and/or computation time can become critical factors. On the other hand, in implicit methods stability imposes no restrictions on the step size. For this reason all package programs for solving systems of stiff ODEs are based upon implicit numerical schemes.

10.1.5 Some Special Features of the Commonly Used ODE Solvers

Table 2.1 will familiarize the reader with some of the special features of the commonly used ODE solvers.

10.2 Package Programs as PDE Solvers

Besides the ODE solvers mentioned above, there are some package programs developed in recent years as PDE solvers for transient one-dimensional chemical kinetic calculations. A well-known PDE solver is the PDECOL package developed by Madsen and Sincovec.[20] This general package employs the method of lines. The spatial discretization is accomplished by finite-element collocation methods based on B-splines.[21] The overall solution strategy is to reduce PDEs to a system of ODEs in order to use standard ODE packages. PDECOL uses a finite-element method in approximating the solutions piecewise by polynomials for the consecutive regions whose boundaries are defined by

TABLE 2.1 Comparison of Special Features of the Commonly Used ODE Solvers

Program Name	Δt	Order of Numerical Scheme	Handling of Jacobian Matrix	Remarks
DIFSUB	Fixed	Variable: 1 to 6	↑	Suitable for fairly smooth problems
GEAR	Fixed	Variable: 1 to 5	Attempt to use previous Jacobian elements up to 10 Δt	Suitable for fairly smooth problems
EPISODE	Variable	Variable: 1 to 5	↓	Greater stability than GEAR—more suitable for rapid changes
K-integrator	Variable	Fixed 3	Update approx. Jacobian subset each Δt	Suitable for extremely rapid changes

the break points. The detailed solution procedure is beyond the scope of this chapter; interested readers are referred to Ref. 20 for details. A good example, using PDECOL to treat laminar steady-state flames, is given by Heimerl and Coffee.[22]

REFERENCES

1. K. J. Laidler, *Chemical Kinetics*, 2nd ed., McGraw-Hill Book Co., New York, 1965.
2. I. Glassman, *Combustion*, Academic Press, New York, 1977.
3. S. S. Penner, *Introduction to the Study of Chemical Reactions in Flow Systems*, Agardograph No. 7, Butterworth Scientific Publications, London, 1955.
4. F. A. Williams, *Combustion Theory*, Addison-Wesley Publishing Co., Reading, Mass., 1965.
5. S. W. Benson, *The Foundations of Chemical Kinetics*, McGraw-Hill Book Co., New York, 1960.
6. B. Lewis and G. von Elbe, *Combustion, Flames, and Explosions of Gases*, Academic Press, New York, 1951.
7. J. E. Mayer and M. G. Mayer, *Statistical Mechanics*, John Wiley & Sons, New York, 1940.
8. F. A. Lindemann, *Trans. Faraday Soc.*, Vol. 17, p. 598, 1922.

9. E. L. King, *How Chemical Reactions Occur*, W. A. Benjamin, New York, 1964.
10. D. Edelson, "The New Look In Chemical Kinetics," *J. Chem. Ed.*, Vol. 52, pp. 642–644, 1975.
11. D. Edelson, "Computer Simulation in Chemical Kinetics," *Science*, Vol. 214, pp. 981–986, 1981.
12. C. W. Gear, "The Automatic Integration of Ordinary Differential Equations," *Comm. ACM*, Vol. 14, pp. 176–179, 1971.
13. C. W. Gear, "DIFSUB for Solution of Ordinary Differential Equations," *Comm. ACM*, Vol. 14, pp. 185–189, 1971 (Algorithm 407).
14. A. C. Hindmarsh, "The LLL Family of Ordinary Differential Equation Solvers," Preprint UCRL-78129, April 1976.
15. A. C. Hindmarsh and E. D. Byrne, "Applications of EPISODE: An Experimental Package for the Integration of Systems of Ordinary Differential Equations," Preprint UCRL-75868, June 1975.
16. M. D. Kregel and E. L. Lortie, *Descriptions and Comparisons of the K-Method for Performing Numerical Integration of Stiff Ordinary Differential Equations*, BRL Report No. 1733, Aberdeen Proving Ground, Md, July 1974.
17. T. P. Coffee, J. M. Heimerl, and M. D. Kregel, *A Numerical Method to Integrate Stiff Systems of Ordinary Differential Equations*, ARBRL-TR-02206, Aberdeen Proving Ground, Md, January 1980.
18. G. Dahlquist and A. Björch, *Numerical Methods*, translated by N. Anderson, Prentice-Hall, Englewood Cliffs, N.J., 1974.
19. C. W. Gear, *Numerical Initial Value Problems in Ordinary Differential Equations*, Prentice-Hall, Englewood Cliffs, N.J., 1971.
20. B. K. Madsen and R. F. Sincovec, "PDECOL, General Collocation Software for Partial Differential Equations," *ACM Trans. Math. Software*, Vol. 5, pp. 326–351, 1979.
21. C. de Boor, "Package for Calculating with *B*-Splines," *SIAM J. Numer. Anal.*, Vol. 14, pp. 441–472, 1977.
22. J. M. Heimerl and T. P. Coffee, "The Detailed Modeling of Premixed, Laminar, Steady-State Flames. I. Ozone," *Combustion and Flame*, Vol. 39, pp. 301–315, 1980.

HOMEWORK

1. For the rate expression

$$-\frac{dC_A}{dt} = kC_A^2C_M$$

where M represents all of the constituents of the mixture:

(a) What is the order of the reaction?

(b) How would the rate of reaction depend on pressure? Show this explicitly.

2. The rate of a given reaction doubles for a 10°C rise in temperature from T_0 to $T_0 + 10$. Develop an expression for the activation energy of the reaction. Do not solve for a numerical answer.

3. Develop an expression for the time (half-life) at which an initial concentration C(0) decreases to $\frac{1}{2}$ its original value for a first-order reaction.

4. For the reaction sequence

$$N_2O_5 \xrightarrow{k_1} NO_2 + NO_3$$

$$NO_2 + NO_3 \xrightarrow{k_2} N_2O_5$$

$$NO_2 + O_3 \xrightarrow{k_3} NO_3 + O_2$$

$$2NO_3 \xrightarrow{k_4} 2NO_2 + O_2$$

(a) Write rate equations for the rates of formation of $N_2O_5, NO_2, NO_3, O_2, O_3$.

(b) Assume that the species NO_2 and NO_3 are in steady state. Solve for their concentrations.

(c) What do the results of the steady-state assumption imply with respect to the rate of N_2O_5 formation?

5. At low pressure, assuming no wall reaction, the H_2–O_2 reaction may be accounted for by the mechanism

$$H_2 + O_2 \xrightarrow{k_1} 2OH, \qquad k_1 \approx 10^{12}\exp(-39.0/R_uT)\ \text{cm}^3/\text{mole sec}$$

$$OH + H_2 \xrightarrow{k_2} H_2O + H, \quad k_2 \approx 10^{13.8}\exp(-5.9/R_uT)\ \text{cm}^3/\text{mole sec}$$

$$H + O_2 \xrightarrow{k_3} OH + O, \quad k_3 \approx 2.2 \times 10^{14}\exp(-16.5/R_uT)\ \text{cm}^3/\text{mole sec}$$

$$O + H_2 \xrightarrow{k_4} OH + H, \quad k_4 \approx 1.1 \times 10^{13}\exp(-9.4/R_uT)\ \text{cm}^3/\text{mole sec}$$

$$H + OH + M \xrightarrow{k_5} H_2O + M, \quad k_5 \approx 10^{17}\ \text{cm}^6/\text{mole}^2\ \text{sec} \quad (E_a = 0\ \text{kcal/mole})$$

Using the steady-state hypothesis, derive the differential equations expressing dC_{H_2O}/dt in terms of C_{H_2} and C_{O_2}.

ANSWER: $dC_{H_2O}/dt = 2k_2C_{O_2}C_{H_2}[k_1/k_2 + k_3/(k_5C_M)]$.

6. The combustion chamber in a certain rocket contains H atoms and OH radicals at equal concentrations of 4×10^{-6} mole/cm^3 and 3000 K. The total gas concentration is 4×10^{-4} mole/cm^3. Supposing the gases to be exhausted from the chamber at 1000 K (where H_2O is essentially undissociated) and at a density equal to $\frac{1}{40}$ of the density in the chamber, calculate how far in the exhaust stream, moving at 3048 m/sec, one will have to probe before finding that the H and OH are 99% recombined. Ignore recombination reactions other than $H + OH + M \rightarrow H_2O + M$, for which k is given as 10^{16} cm^6/mole^2 sec, independent of temperature (this is a typical figure for a three-body recombination rate constant).

ANSWER: $x = 30.8$ m.

7. One of the mechanisms of atom and radical combinations, called the *energy-transfer mechanism*, is described by the following chemical steps:

$$2R \underset{k_2}{\overset{k_1}{\rightleftharpoons}} R_2^*$$

$$R_2^* + X \xrightarrow{k_3} R_2 + X$$

where R represents a radical or an atom and X is a third body. Use the steady-state treatment for R_2^* to determine the rate of consumption of R. What is the order of the recombination reaction when the concentration of X is sufficiently large such that $k_3C_X \gg k_2$? What is the order of the recombination reaction at very low pressures?

8. Another mechanism of atom and radical combination is called the *atom–molecule* or *radical–molecule-complex mechanism*. This mechanism is particularly likely for atom combinations when the third body X is a molecule that can readily form a complex with one of the atoms. The reaction steps may be represented as

$$R + X \underset{k_2}{\overset{k_1}{\rightleftharpoons}} RX^*$$

The resulting complex retains the energy released in its formation, and may be deenergized by collision with another molecule X:

$$RX^* + X \underset{k_4}{\overset{k_3}{\rightleftharpoons}} RX + X + \underbrace{\Delta H_r}_{\text{heat release}}$$

Finally RX may react with R, giving rise to

$$RX + R \xrightarrow{k_5} R_2 + X$$

Apply the steady-state treatment to determine the rate of consumption of R. What is the order of the combination reaction when the concentration of R is sufficiently small? Does the reaction become second-order at high pressures (high concentrations of X)?

9. A mixture of high-temperature gases flows over a horizontal graphite slab. Some of the species $H_2O_{(g)}$, $CO_{2(g)}$, $H_{2(g)}$ in the mixture can react with graphite surface by heterogeneous reaction and cause thermochemical erosion of the solid. Consider the following heterogeneous reactions:

$$H_2O_{(g)} + C_{(s)} \rightarrow CO_{(g)} + H_{2(g)} \tag{1}$$

$$CO_{2(g)} + C_{(s)} \rightarrow 2CO_{(g)} \tag{2}$$

$$H_{2(g)} + 2C_{(s)} \rightarrow C_2H_{2(g)} \tag{3}$$

Determine the heats of reaction of the above three reactions at 2000 K. Are these reactions exothermic or endothermic?

10. Suppose the order of reaction of a chemical reacting system is n, that is,

$$\frac{dC}{dt} = -kC^n$$

where C is the concentration of the reacting species. Show that the rate of change of mole fraction of reacting species, dX/dt, is proportional to $-kX^n\rho^{n-1}$. What is the pressure dependence of dX/dt? List your assumptions. Can we say that

$$\frac{d\varepsilon}{dt} \propto -k\varepsilon^n\rho^{n-1}$$

where ε is the reaction progress variable defined by Eq. (1-12) in Chapter 1?

11. Express the concentration of the ith species, C_i, in terms of following parameters:

(a) ρ_i, W_i

(b) ρ, Y_i, W_i

(c) C, X_i

(d) p, Y_i, T, W_i

(e) p, T, X_i.

CHAPTER

3 The Conservation Equations for Multicomponent Reacting Systems

ADDITIONAL SYMBOLS

Symbol	Description	Dimension
B_i	Body force per unit volume in i-direction	F/L^3
C	Molar concentration	N/L^3
d	Molecular diameter	L
$\mathscr{D}_{AB}$	Binary diffusivity for A–B system	L^2/t
α_i	Thermal diffusion coefficient for species i	L^2/t
e_{ij}	Strain rate tensor	t^{-1}
Ea_k	Activation energy for the kth reaction	Q/N
$\mathbf{f}_i$	External force per unit mass on species i	F/M
$\mathbf{F}$	Force	F
$\mathbf{F}_S$	Surface force	F
h	Enthalpy per unit mass	Q/M
h_t	Total enthalpy per unit mass [Eq. (3-80)]	Q/M
$\mathbf{I}$	Identity matrix or vector form of Kronecker delta δ_{ij}	—
$\mathbf{J}_i$	Mass flux of species i relative to mass-average velocity	M/L^2t
$\mathbf{J}_i^*$	Molar flux of species i relative to molar-average velocity	N/L^2t
K	Boltzmann constant	(Q/T)/molecule

l	Mean free path	L
$\dot{\mathbf{m}}$	Mass flux	M/L^2t
$\dot{\mathbf{n}}$	Molar flux	N/L^2t
N_i	Number of moles of species i	—
$\tilde{N}$	Avogadro's number	—
$\mathbf{q}$	Heat-flux vector	Q/L^2t
$T°$	Fixed standard reference temperature	T
$\bar{u}$	Arithmetic-mean molecular speed	L/t
u_i	Velocity component in i-direction	L/t
$\mathbf{v}$	Mass-average velocity	L/t
v	Control volume	L^3
$\mathbf{v}_i$	Velocity of ith species with respect to stationary coordinate axes	L/t
$\mathbf{v}^*$	Molar-average velocity	L/t
$\mathbf{V}_i$	Mass diffusion velocity of ith species	L/t
$\mathbf{V}_i^*$	Molar diffusion velocity of ith species	L/t
$\mathbf{v}_{ik}$	Mass diffusion velocity of kth component in i-direction	L/t
y	Space coordinate in y-direction	L
z	Space coordination in z-direction	L
Z_{ij}	Number of collisions between ith- and jth-species molecules per unit volume	t^{-1}
$\boldsymbol{\Gamma}_i$	Net rate of change of momentum of molecules of type i per unit volume	Ft/L^2
α	Thermal diffusivity	L^2/t
λ	Thermal conductivity or second viscosity	Q/tLT or Ft/L^2
μ	Dynamic viscosity or first viscosity	Ft/L^2
μ'	Bulk viscosity	Ft/L^2
μ_{ij}	Reduced mass of molecules of species i and j	m
σ_{ij}, $\tilde{\sigma}$	Total stress tensor	F/L^2
τ_{ij}	Viscous shear tensor	F/L^2
Ω_i	Molar rate of production of species i	N/tL^3
ω_i	Mass rate of production of species i	M/tL^3

Subscripts

A, B	Species in binary system
i, j, k	Species in multicomponent systems
x, y	Fluxes in x and y directions
$\mathbf{i}, \mathbf{j}, \mathbf{k}$	Unit vectors in x, y, and z directions

1 DEFINITIONS OF CONCENTRATIONS, VELOCITIES, AND MASS FLUXES

The concentration of the various species in a multicomponent system may be expressed in numerous ways. This study will consider four such ways: *mass concentration* ρ_i, which is the mass of species i per unit volume of solution; *molar concentration* $C_i = \rho_i / W_i$, which is the number of moles of species i per unit volume; *mass fraction* $Y_i = \rho_i / \rho$, which is the mass concentration of species i divided by the total mass density of the solution; and *mole fraction* $X_i = C_i / C$, which is the molar concentration of species i divided by the total molar density of the solution.

The various chemical species in a diffusing mixture move at different velocities. $\mathbf{v}_i$ denotes the velocity of the ith species with respect to stationary coordinate axes. Thus, for a mixture of n species, the local *mass-average velocity* $\mathbf{v}$ is defined as

$$\mathbf{v} = \frac{\sum_{i=1}^{N} \rho_i \mathbf{v}_i}{\sum_{i=1}^{N} \rho_i} \tag{3-1}$$

Similarly, a local *molar-average velocity* $\mathbf{v}^*$ may be defined as

$$\mathbf{v}^* = \frac{\sum_{i=1}^{N} C_i \mathbf{v}_i}{\sum_{i=1}^{N} C_i} \tag{3-2}$$

In flow systems, one is frequently interested in the velocity of a given species with respect to $\mathbf{v}$ or $\mathbf{v}^*$ rather than with respect to stationary coordinates. This leads to the definition of the *diffusion velocities* (see diagram):

mass diffusion velocity

$$\mathbf{V}_i = \mathbf{v}_i - \mathbf{v} \tag{3-3}$$

molar diffusion velocity

$$\mathbf{V}_i^* = \mathbf{v}_i - \mathbf{v}^* \tag{3-4}$$

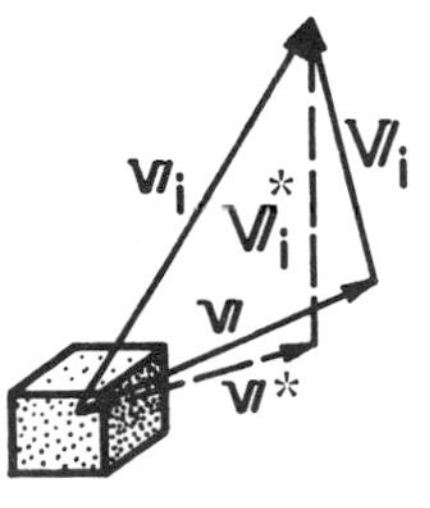

Diagram of velocity vectors.

These diffusion velocities indicate the motion of component i relative to the local motion of the fluid stream.

To illustrate the meaning of the various kinds of velocities, we consider a simple case of a binary system in which $X_A = \frac{1}{3}$ and velocity vectors are colinear; $|\mathbf{v}^*| = 10$, $|\mathbf{v}_A - \mathbf{v}^*| = 2$, and $W_A = 3W_B$:

$$|\mathbf{v}^*| = \frac{\sum_{i=1}^{N} C_i |\mathbf{v}_i|}{\sum_{i=1}^{N} C_i} = \sum_{i=1}^{N} X_i |\mathbf{v}_i| = \tfrac{1}{3} \times 12 + \tfrac{2}{3} |\mathbf{v}_B|$$

$$\sum_{i=1}^{N} X_i = \frac{\sum_{i=1}^{N} C_i}{C} = \frac{C}{C} = 1$$

Using the relations given above, one obtains $|\mathbf{v}_B| = 9$. Further, from the definition of molar concentration and mass-average velocity, one obtains $|\mathbf{v}| = 10.8$. Thus, in general, $\mathbf{v}$ and $\mathbf{v}^*$ are not equal.

Now that concentrations and velocities have been discussed, we are in a position to define mass and molar fluxes. The mass (or molar) flux of species i is a vector quantity denoting the mass (or number of moles) of species i that passes through a unit area per unit time. Thus, the mass and molar fluxes relative to stationary coordinates are

$$\dot{\mathbf{m}}_i = \rho_i \mathbf{v}_i \qquad \text{(mass flux)} \tag{3-5}$$

$$\dot{\mathbf{n}}_i = C_i \mathbf{v}_i \qquad \text{(molar flux)} \tag{3-6}$$

The relative mass and molar fluxes are defined as

$$\mathbf{J}_i = \rho_i(\mathbf{v}_i - \mathbf{v}) = \rho_i \mathbf{V}_i \tag{3-7}$$

$$\mathbf{J}_i^* = C_i(\mathbf{v}_i - \mathbf{v}^*) = C_i \mathbf{V}_i^* \tag{3-8}$$

Example 3.1

Relations among the molar fluxes:

a. How are the fluxes $\mathbf{J}_i^*$ and $\dot{\mathbf{n}}_i$ related in an N-component system?
b. Show that the sum of the fluxes $\mathbf{J}_i^*$ is zero.

Solution:

Using the definitions of $\mathbf{v}^*$ and $\mathbf{J}_i^*$, one obtains

$$\mathbf{J}_i^* = C_i(\mathbf{v}_i - \mathbf{v}^*)$$

$$= C_i \mathbf{v}_i - \frac{C_i}{C} \sum_{j=1}^{N} C_j \mathbf{v}_j$$

The definitions of $\dot{\mathbf{n}}_i$ and X_i yield

$$\mathbf{J}_i^* = \dot{\mathbf{n}}_i - X_i \sum_{j=1}^{N} \dot{\mathbf{n}}_j \tag{3-9}$$

Summation of Eq. (3-9) from $i = 1$ to $i = N$ gives

$$\sum_{i=1}^{N} \mathbf{J}_i^* = 0 \tag{3-10}$$

2 FICK'S LAW OF DIFFUSION

The mass diffusivity $\mathscr{D}_{AB} = \mathscr{D}_{BA}$ in a binary system is defined as

$$\mathbf{J}_A^* = -C\mathscr{D}_{AB}\nabla X_A \tag{3-11}$$

This is Fick's first law of diffusion, written in terms of the molar diffusion flux $\mathbf{J}_A^*$. This equation states that species A diffuses (moves relative to the mixture) in the direction of decreasing mole fraction of A, just as heat flows by conduction in the direction of decreasing temperature. The units of $\mathscr{D}_{AB}$ are ft^2/hr or cm^2/sec.

The molar flux relative to stationary coordinates can now be defined as

$$\dot{\mathbf{n}}_A = C_A\mathbf{v}^* - C\mathscr{D}_{AB}\nabla X_A \tag{3-12}$$

This equation shows that the diffusion flux $\dot{\mathbf{n}}_A$ relative to stationary coordinates is the result of two vector quantities: the molar flux of A resulting from the bulk motion of the fluid, and the vector $\mathbf{J}_A^*$, which is the molar flux of A resulting from the diffusion superimposed on the bulk flow. In terms of mass flux, Fick's first law is

$$\mathbf{J}_A = -\rho\mathscr{D}_{AB}\nabla Y_A \tag{3-13}$$

The mass flux of A relative to the stationary coordinates is

$$\dot{\mathbf{m}}_A = \rho_A\mathbf{v} - \rho\mathscr{D}_{AB}\nabla Y_A \tag{3-14}$$

The similarity between mass transport, momentum transport, and energy transport in the y-direction of a binary system can be seen from

$$J_{Ay} = -\mathscr{D}_{AB}\frac{\partial}{\partial y}(\rho_A) \qquad \text{(Fick's law for constant } \rho)$$

$$\tau_{yx} = -\nu\frac{\partial}{\partial y}(\rho v_x) \qquad \text{(Newton's law for constant } \rho)$$

$$q_y = -\alpha\frac{\partial}{\partial y}(\rho C_p T) \qquad \text{(Fourier's law for constant } \rho C_p)$$

3 THEORY OF ORDINARY DIFFUSION IN GASES AT LOW DENSITY

The mass diffusivity $\mathscr{D}_{AB}$ for binary mixtures of nonpolar gases is predictable within about 5% by kinetic theory.

Consider a pure gas containing two molecular species A and A^*, both species having the same m_A and the same size and shape. Assume temperature T and molar concentration C to be constant.

From kinetic theory, the molecular velocity relative to fluid velocity $\mathbf{v}$ has an average magnitude $\bar{u}$:

$$\bar{u} = \sqrt{\frac{8KT}{\pi m_A}} \tag{3-15}$$

K is the Boltzmann constant = (universal gas constant)/(Avogadro's number). The frequency of molecular collisions on a stationary surface exposed to the gas, per unit area, is

$$Z = \tfrac{1}{4}\tilde{n}\bar{u} \tag{3-16}$$

where $\tilde{n}$ represents the molecules per unit volume. $\tilde{n}$ is a constant, since the molar concentration C is a constant. The mean free path l, from kinetic theory, is

$$l = \frac{1}{\sqrt{2}\,\pi d^2 \tilde{n}} \tag{3-17}$$

where d is the diameter of the molecules.

The molecules reaching any plane in the gas have, on the average, had their last collision at a distance a from the plane (see Fig. 3.1):

$$a = \tfrac{2}{3} l \tag{3-18}$$

The relative molar flux $\mathbf{J}_A^*$ of species A across any plane of constant y is found by counting the molecules of A that cross a unit area of the plane in the positive y-direction and subtracting the number that cross in the negative y-direction. Thus

$$\mathbf{J}_A^* = \frac{ZX_A|_{y-a} - ZX_A|_{y+a}}{\tilde{N}}\mathbf{j} \tag{3-19}$$

if the concentration profile $X_A(y)$ is assumed to be essentially linear. ($\tilde{N}$ is

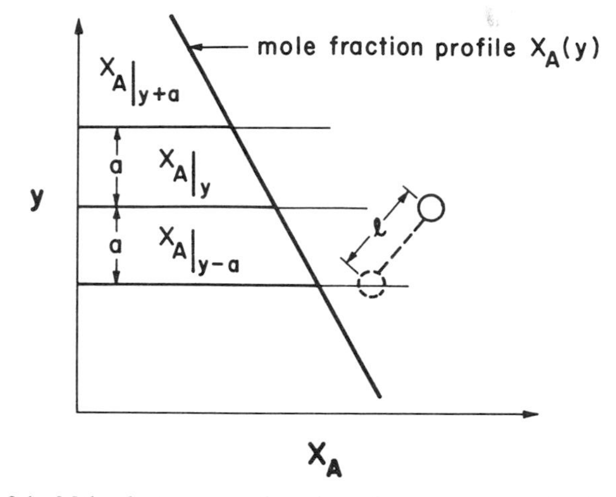

Figure 3.1 Molecular transport of species A from plane at y-a to plane at y.

Avogadro's number.) As a further consequence of this assumption,

$$\frac{dX_A}{dy} = \frac{X_A|_y - X_A|_{y-a}}{a} = \frac{X_A|_{y+a} - X_A|_y}{a}$$

$$X_A|_{y-a} = X_A|_y - \tfrac{2}{3}l\frac{dX_A}{dy} \tag{3-20}$$

$$X_A|_{y+a} = X_A|_y + \tfrac{2}{3}l\frac{dX_A}{dy} \tag{3-21}$$

Substituting Eqs. (3-20) and (3-21) into Eq. (3-19), we have

$$\mathbf{J}_A^* = \frac{Z}{\tilde{N}}\left(-\tfrac{4}{3}l\frac{dX_A}{dy}\right)\mathbf{j} = -\left(\tfrac{1}{3}\frac{\tilde{n}}{\tilde{N}}\bar{u}l\right)\frac{dX_A}{dy}\mathbf{j}$$

$$\mathbf{J}_A^* = -C\left(\tfrac{1}{3}\bar{u}l\right)\frac{dX_A}{dy}\mathbf{j} \tag{3-22}$$

This equation corresponds to the y-component of Fick's law, with the following approximate value for $\mathscr{D}_{AA}^*$:

$$\mathscr{D}_{AA}^* = \tfrac{1}{3}\bar{u}l \tag{3-23}$$

We substitute for $\bar{u}$ and l to obtain

$$\mathscr{D}_{AA}^{*} = \frac{1}{3}\sqrt{\frac{8KT}{\pi m_A}}\,\frac{1}{\sqrt{2}\,\pi d_A^2 \tilde{n}} = \frac{2}{3}\sqrt{\frac{K^3}{\pi^3 m_A}}\,\frac{T^{3/2}}{d_A^2}\,\frac{1}{\tilde{n}KT}$$

$$= \frac{2}{3}\sqrt{\frac{K^3}{\pi^3 m_A}}\,\frac{T^{3/2}}{d_A^2}\,\frac{1}{\tilde{n}KT}$$

Using the perfect-gas law $p = CR_uT = \tilde{n}KT$,

$$\mathscr{D}_{AA}^{*} = \frac{2}{3}\sqrt{\frac{K^3}{\pi^3 m_A}}\,\frac{T^{3/2}}{d_A^2}\,\frac{1}{CR_uT}$$

$$= \frac{2}{3}\sqrt{\frac{K^3}{\pi^3 m_A}}\,\frac{T^{3/2}}{pd_A^2}$$

$$\mathscr{D}_{AA}^{*} = \frac{2}{3}\sqrt{\frac{K^3}{\pi^3 m_A}}\,\frac{T^{3/2}}{pd_A^2} \tag{3-24}$$

$\mathscr{D}_{AA}^{*}$ represents the mass diffusivity of a mixture of two species of rigid spheres of identical mass and diameter. The calculation of $\mathscr{D}_{AB}$ for rigid spheres of unequal mass and diameter is considerably more difficult; the corresponding result is

$$\mathscr{D}_{AB} = \frac{2}{3}\left(\frac{K^3}{\pi^3}\right)^{1/2}\left(\frac{1}{2m_A} + \frac{1}{2m_B}\right)^{1/2}\frac{T^{3/2}}{p\left(\dfrac{d_A + d_B}{2}\right)^2} \tag{3-25}$$

4 EQUATIONS OF CONTINUITY

To derive the equation of continuity for each species in a multicomponent mixture, we begin by making a mass balance over an arbitrary differential fluid element in a binary mixture. We apply the law of conservation of mass of species A to a volume element $\Delta x\,\Delta y\,\Delta z$ fixed in space through which a binary mixture of A and B is flowing. (See Fig. 3.2.) Within this element, A may be produced by chemical reaction at a rate ω_A (kg m^{-3} sec^{-1}). The various contributions to the mass balance are:

Time rate of change of mass of A in volume element,

$$\frac{\partial \rho_A}{\partial t}\,\Delta x\,\Delta y\,\Delta z$$

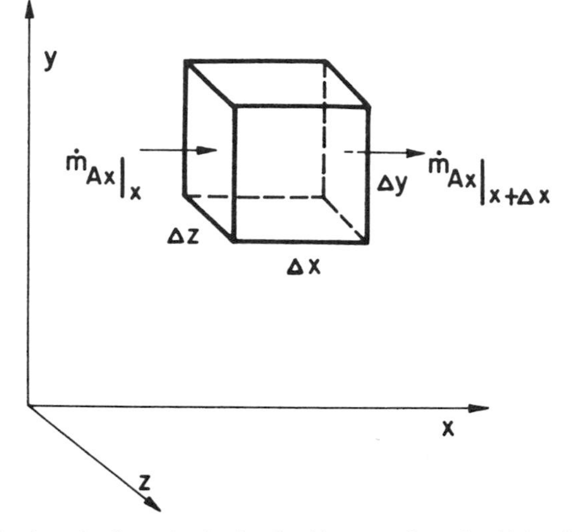

Figure 3.2 Region of volume $\Delta x\, \Delta y\, \Delta z$ fixed in space through which a fluid is flowing.

Input of A across face at x,

$$\dot{m}_{Ax}|_x \Delta y \Delta z$$

Output of A across face at $x + \Delta x$,

$$\dot{m}_{Ax}|_{x+\Delta x}\, \Delta y \Delta z = \dot{m}_{Ax}|_x\, \Delta y \Delta z + \frac{\partial \dot{m}_{Ax}}{\partial x}\, \Delta x\, \Delta y \Delta z$$

Rate of production of A by chemical reaction,

$$\omega_A\, \Delta x\, \Delta y \Delta z$$

There are also input and output terms in the y and z directions. When the entire mass balance is divided through by $\Delta x\, \Delta y\, \Delta z$, one obtains

$$\frac{\partial \rho_A}{\partial t} + \left(\frac{\partial \dot{m}_{Ax}}{\partial x} + \frac{\partial \dot{m}_{Ay}}{\partial y} + \frac{\partial \dot{m}_{Az}}{\partial z} \right) = \omega_A \tag{3-26}$$

This is the *equation of continuity for component A* in a binary mixture. The quantities $\dot{m}_{Ax}$, $\dot{m}_{Ay}$, $\dot{m}_{Az}$ are the rectangular components of the mass flux vector $\dot{\mathbf{m}}_A$. In vector notation, the equation may be rewritten as

$$\frac{\partial \rho_A}{\partial t} + (\nabla \cdot \dot{\mathbf{m}}_A) = \omega_A \tag{3-27}$$

Similarly, the equation of continuity for component B is

$$\frac{\partial \rho_B}{\partial t} + (\nabla \cdot \dot{\mathbf{m}}_B) = \omega_B \tag{3-28}$$

Addition of Eqs. (3-27) and (3-28) gives

$$\frac{\partial \rho}{\partial t} + (\nabla \cdot \rho \mathbf{v}) = 0 \tag{3-29}$$

which is the *equation of continuity for the mixture*. In arriving at Eq. (3-29), we have made use of both the relation $\dot{\mathbf{m}}_A + \dot{\mathbf{m}}_B = \rho \mathbf{v}$ and the law of conservation of mass in the form $\omega_A + \omega_B = 0$. For a fluid of constant mass density ρ, Eq. (3-29) becomes

$$\nabla \cdot \mathbf{v} = 0 \tag{3-30}$$

The development given above could have been made equally well in terms of molar units. If Ω_A is the molar rate of production of A per unit volume, then the molar analog of Eq. (3-27) is

$$\frac{\partial C_A}{\partial t} + \nabla \cdot \dot{\mathbf{n}}_A = \Omega_A \tag{3-31}$$

Substituting Eq. (3-14) into Eq. (3-27), we obtain

$$\frac{\partial \rho_A}{\partial t} + \nabla \cdot \rho_A \mathbf{v} = \nabla \cdot \rho \mathscr{D}_{AB} \nabla Y_A + \omega_A \tag{3-32}$$

Substituting Eq. (3-12) into Eq. (3-31), we obtain

$$\frac{\partial C_A}{\partial t} + \nabla \cdot C_A \mathbf{v}^* = \nabla \cdot C \mathscr{D}_{AB} \nabla X_A + \Omega_A \tag{3-33}$$

If no chemical reactions occur, ω_A, ω_B, Ω_A, and Ω_B are all zero. If in addition $\mathbf{v}$ is zero in Eq. (3-32) or $\mathbf{v}^*$ is zero in Eq. (3-33), we get

$$\frac{\partial C_A}{\partial t} = \mathscr{D}_{AB} \nabla^2 C_A \tag{3-34}$$

which is called *Fick's second law of diffusion*. This equation is generally used for diffusion in solids or stationary liquids and for equimolar counterdiffusion in gases. This equation is similar to the heat-conduction equation

$$\frac{\partial T}{\partial t} = \alpha \nabla^2 T$$

Equations of Continuity

For a multicomponent system, Eq. (3-27) (using the rela $\rho_i = Y_i\rho$ and $\mathbf{v}_i = \mathbf{v} + \mathbf{V}_i$) becomes

$$\frac{\partial Y_i\rho}{\partial t} + \nabla \cdot \rho Y_i(\mathbf{v} + \mathbf{V}_i) = \omega_i \tag{3-35}$$

or

$$\rho\frac{\partial Y_i}{\partial t} + Y_i\frac{\partial \rho}{\partial t} + Y_i\nabla \cdot (\rho\mathbf{v}) + \rho\mathbf{v} \cdot \nabla Y_i + \nabla \cdot \rho Y_i\mathbf{V}_i = \omega_i$$

From continuity equation,

$$Y_i\frac{\partial \rho}{\partial t} + Y_i\nabla \cdot \rho\mathbf{v} = 0$$

Thus Eq. (3-35) simplifies to

$$\rho\frac{\partial Y_i}{\partial t} + \rho\mathbf{v} \cdot \nabla Y_i + \nabla \cdot \rho Y_i\mathbf{V}_i = \omega_i$$

or

$$\frac{\partial Y_i}{\partial t} + \mathbf{v} \cdot \nabla Y_i + \frac{1}{\rho}\nabla \cdot \rho Y_i\mathbf{V}_i = \frac{\omega_i}{\rho} \qquad i = 1, 2, \ldots, N \tag{3-36}$$

In a general multicomponent system, there are N equations of this kind. The addition of these equations gives the equation of continuity for the mixture. Any one of these N equations may be replaced by the equation of continuity

TABLE 3.1 Equation of Continuity in Several Coordinate Systems

Rectangular coordinates (x, y, z):

$$\frac{\partial \rho}{\partial t} + \frac{\partial}{\partial x}(\rho u_x) + \frac{\partial}{\partial y}(\rho u_y) + \frac{\partial}{\partial z}(\rho u_z) = 0 \tag{A}$$

Cylindrical coordinates (r, θ, z):[a]

$$\frac{\partial \rho}{\partial t} + \frac{1}{r}\frac{\partial}{\partial r}(\rho r u_r) + \frac{1}{r}\frac{\partial}{\partial \theta}(\rho u_\theta) + \frac{\partial}{\partial z}(\rho u_z) = 0 \tag{B}$$

Spherical coordinates (r, θ, ϕ):[b]

$$\frac{\partial \rho}{\partial t} + \frac{1}{r^2}\frac{\partial}{\partial r}(\rho r^2 u_r) + \frac{1}{r\sin\theta}\frac{\partial}{\partial \theta}(\rho u_\theta \sin\theta) + \frac{1}{r\sin\theta}\frac{\partial}{\partial \phi}(\rho u_\phi) = 0 \tag{C}$$

[a] $r \geq 0,\ 2\pi \geq \theta \geq 0$.

[b] $r \geq 0,\ 2\pi > \phi \geq 0,\ \pi \geq \theta \geq 0$.

for the mixture in any given problem. The fact that these are N-1 independent equations for Y_i corresponds to the fact that only N-1 of the Y_i are independent.

It may be noted that the above derivations have for simplicity been made in rectangular coordinates. However, rectangular coordinates are not always the most convenient for solving combustion problems. To facilitate the use of the

TABLE 3.2 Equation of Continuity for *i*th Species in Several Coordinate Systems

Rectangular coordinates (x, y, z):

$$\rho\left(\frac{\partial Y_i}{\partial t} + u_x\frac{\partial Y_i}{\partial x} + u_y\frac{\partial Y_i}{\partial y} + u_z\frac{\partial Y_i}{\partial z}\right) + \frac{\partial}{\partial x}(\rho Y_i V_{ix}) + \frac{\partial}{\partial y}(\rho Y_i V_{iy}) + \frac{\partial}{\partial z}(\rho Y_i V_{iz}) = \omega_i \tag{A}$$

[where

$$V_{ix} = -\frac{\mathscr{D}}{Y_i}\frac{\partial Y_i}{\partial x}, \qquad V_{iy} = -\frac{\mathscr{D}}{Y_i}\frac{\partial Y_i}{\partial y}, \quad \text{and} \quad V_{iz} = -\frac{\mathscr{D}}{Y_i}\frac{\partial Y_i}{\partial z}$$

according to Fick's law of mass diffusion]

Cylindrical coordinates (r, θ, z):

$$\rho\left(\frac{\partial Y_i}{\partial t} + u_r\frac{\partial Y_i}{\partial r} + \frac{u_\theta}{r}\frac{\partial Y_i}{\partial \theta} + u_z\frac{\partial Y_i}{\partial z}\right) + \frac{1}{r}\frac{\partial}{\partial r}(r\rho Y_i V_{ir}) + \frac{1}{r}\frac{\partial}{\partial \theta}(\rho Y_i V_{i\theta}) + \frac{\partial}{\partial z}(\rho Y_i V_{iz}) = \omega_i \tag{B}$$

[where

$$V_{ir} = -\frac{\mathscr{D}}{Y_i}\frac{\partial Y_i}{\partial r}, \qquad V_{i\theta} = -\frac{\mathscr{D}}{Y_i r}\frac{\partial Y_i}{\partial \theta}, \quad \text{and} \quad V_{iz} = -\frac{\mathscr{D}}{Y_i}\frac{\partial Y_i}{\partial z}$$

according to Fick's law of mass diffusion]

Spherical coordinates (r, θ, ϕ):

$$\rho\left(\frac{\partial Y_i}{\partial t} + u_r\frac{\partial Y_i}{\partial r} + \frac{u_\theta}{r}\frac{\partial Y_i}{\partial \theta} + \frac{u_\phi}{r\sin\theta}\frac{\partial Y_i}{\partial \phi}\right) + \frac{1}{r^2}\frac{\partial}{\partial r}(r^2\rho Y_i V_{ir}) + \frac{1}{r\sin\theta}\frac{\partial}{\partial \theta}(\sin\theta\,\rho Y_i V_{i\theta}) + \frac{1}{r\sin\theta}\frac{\partial}{\partial \phi}(\rho Y_i V_{i\phi}) = \omega_i \tag{C}$$

[where

$$V_{ir} = -\frac{\mathscr{D}}{Y_i}\frac{\partial Y_i}{\partial r}, \qquad V_{i\theta} = -\frac{\mathscr{D}}{Y_i r}\frac{\partial Y_i}{\partial \theta}, \quad \text{and} \quad V_{i\phi} = -\frac{\mathscr{D}}{Y_i r\sin\theta}\frac{\partial Y_i}{\partial \phi}$$

according to Fick's law of mass diffusion]

governing conservation equations, several forms of the equation of continuity for the gas mixture are listed in Table 3.1, and those for individual species in Table 3.2.

5 CONSERVATION OF MOMENTUM

In this section we shall derive and discuss the partial differential momentum equation. The basic assumption is that we are dealing with continuous, isotropic, and homogeneous media. We shall consider the special case of a Newtonian fluid, that is, a fluid exhibiting a linear relationship between shear stress and rate of deformation, resulting in the Navier–Stokes equation.

We derive first the momentum equation in terms of stress. Then, we shall consider the relationship between stress and fluid deformation.

5.1 Momentum Equation in Terms of Stress

We shall derive the momentum equation in terms of stress using three different approaches; each of these exists in the literature, although it is not always clear which approach is being followed. The basis for *any* derivation of the momentum conservation equations is Newton's second law of motion,

$$\sum \mathbf{F} = \frac{d(m\mathbf{v})}{dt} \tag{3-37}$$

Basically, the three approaches are:

1. *Infinitesimal Particle.* Consider a fluid particle as it moves through space relative to some fixed coordinate system. Equation (3-37) describes the motion of the particle. The acceleration of that particle is then related to various particles at "fixed points" in the flow field. Thus, we relate the motion of a particle to the observation of conditions of various particles at a fixed point in space.

2. *Infinitesimal Control Volume.* Consider a cubical infinitesimal volume element fixed in space. Then, equate the net momentum flux out of the control volume plus the time rate of change of momentum in the control volume to the net force on the mass within the control volume.

3. *Finite Control Volume.* Consider a gas-permeable control volume of finite size. This control volume can have any arbitrary shape and is fixed in space. Gauss's theorem is used to relate the surface and volume integrals, resulting in an equation involving only volume integrals. It is argued that the conditions must be satisfied for the integrand, since the integration is arbitrary. This results in the desired differential momentum equation.

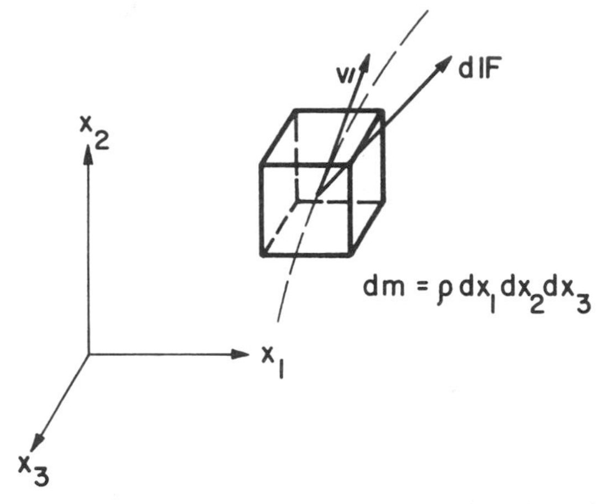

Figure 3.3 An infinitesimal fluid particle moving in a three-dimensional space.

Each of these three approaches may be found in a number of texts on chemically inert fluid mechanics. For example:

1. The first approach is presented in Hinze,[5] Shames,[6] Li and Lam,[7] and Aris.[8]
2. The second approach is presented in Bird, Stewart, and Lightfoot.[1]
3. The third approach is presented in Landau and Liftshitz.[9]

Let us consider each of these in order.

5.1.1 Infinitesimal Particle

Consider a particle of mass dm moving through space with velocity $\mathbf{v}$, as shown in Fig. 3.3. If a net force $d\mathbf{F}$ is acting on the particle, the momentum is

$$d\mathbf{F} = \frac{d}{dt}[(dm)\mathbf{v}]$$

For a particle of fixed mass

$$d\mathbf{F} = dm\,\frac{d}{dt}(\mathbf{v}) \tag{3-37a}$$

This equation describes the motion of the particle as it moves through space relative to the fixed coordinate system. Thus, $d\mathbf{v}/dt$ represents the acceleration of a particular fluid particle as it moves from point to point through space.

Next, we shall relate the derivative of the velocity to conditions at a fixed point in space. In considering the description of fluid motion, we want to describe the position, velocity, and acceleration throughout the flow for all fluid particles of interest. Fluid motion can be described by (1) considering the motion of distinct particles as they move through space, or by (2) looking at each point in the flow and establishing conditions as a function of time. As previously mentioned, the first is called the "particle" or "Lagrangian" approach, and the second is called the "field" or "Eulerian" approach.

Considering the Lagrangian viewpoint in Eq. (3-37a), we follow a particle and observe changes as it moves in space. Then the expression for velocity $\mathbf{v}(x_1, x_2, x_3, t)$ does not refer to fixed coordinates, but allows for changes in coordinates to locate a distinct particle at various times. The positions are given by

$$x_1 = x_1(t), \qquad x_2 = x_2(t), \qquad x_3 = x_3(t)$$

because the time t is the only independent variable in the Lagrangian approach. Based upon the above relationships, we have

$$\begin{aligned} \mathbf{v} &= \mathbf{v}[x_1(t), x_2(t), x_3(t), t] \\ &= \mathbf{v}(t) \end{aligned}$$

where $\mathbf{v}$ refers to the velocity of a distinct particle for different times (as a function of time). Similarly, for acceleration of the particle we have

$$\begin{aligned} \mathbf{a} &= \mathbf{a}[x_1(t), x_2(t), x_3(t), t] \\ &= \mathbf{a}(t) \end{aligned}$$

or

$$\mathbf{a} = \frac{d\mathbf{v}}{dt} = \frac{\partial \mathbf{v}}{\partial x_1}\frac{dx_1}{dt} + \frac{\partial \mathbf{v}}{\partial x_2}\frac{dx_2}{dt} + \frac{\partial \mathbf{v}}{\partial x_3}\frac{dx_3}{dt} + \frac{\partial \mathbf{v}}{\partial t}$$

However,

$$\frac{dx_1}{dt} = u_1$$

$$\frac{dx_2}{dt} = u_2$$

$$\frac{dx_3}{dt} = u_3$$

where

$$\mathbf{v} = \mathbf{i}u_1 + \mathbf{j}u_2 + \mathbf{k}u_3 = [u_1, u_2, u_3]$$

The acceleration then becomes

$$\frac{du_i}{dt} = u_1 \frac{\partial u_i}{\partial x_1} + u_2 \frac{\partial u_i}{\partial x_2} + u_3 \frac{\partial u_i}{\partial x_3} + \frac{\partial u_i}{\partial t}$$

The above quantity is the acceleration of a distinct particle moving through space. However, the quantities u_1, u_2, u_3, $\partial u_i/\partial x_1$, etc., represent the conditions of a particle at a fixed point in space at a fixed time. That is, they represent conditions from a field point of view as

$$u_i = u_i(x_1, x_2, x_3, t) = u_i(\mathbf{x}, t)$$

The operator

$$\frac{d}{dt} \equiv \frac{D}{Dt} \equiv u_1 \frac{\partial}{\partial x_1} + u_2 \frac{\partial}{\partial x_2} + u_3 \frac{\partial}{\partial x_3} + \frac{\partial}{\partial t}$$

is called the *substantial derivative*.

Equation (3-37) can now be written as

$$dF_i = dm \left[u_1 \frac{\partial u_i}{\partial x_1} + u_2 \frac{\partial u_i}{\partial x_2} + u_3 \frac{\partial u_i}{\partial x_3} + \frac{\partial u_i}{\partial t} \right]$$

or

$$dF_1 = dm \frac{Du_1}{Dt}$$

$$dF_2 = dm \frac{Du_2}{Dt}$$

$$dF_3 = dm \frac{Du_3}{Dt}$$

and in Cartesian tensor notation we have

$$dF_i = dm \frac{Du_i}{Dt}$$

$$dF_i = dm \left[u_j \frac{\partial u_i}{\partial x_j} + \frac{\partial u_i}{\partial t} \right] \tag{3-38}$$

If we consider a particle of mass dm, which is cubical in shape and has sides of length dx_1, dx_2, and dx_3, then

$$dm = \rho \, dx_1 \, dx_2 \, dx_3$$

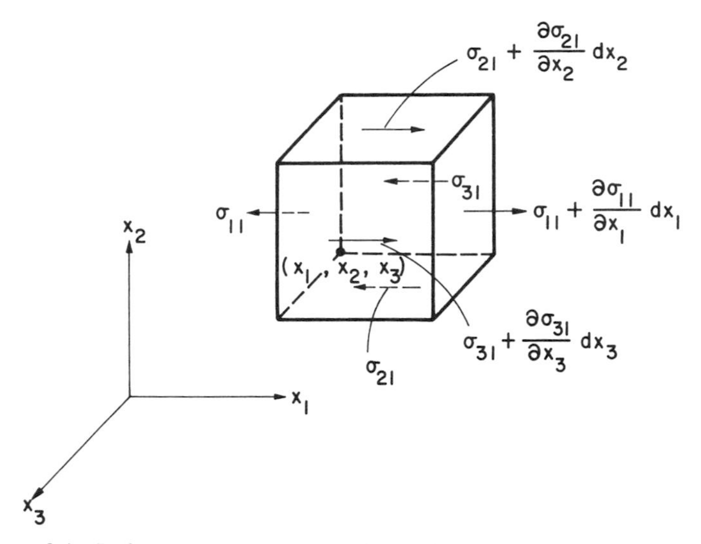

Figure 3.4 Surface-stress components acting on the fluid particle in the x_1-direction.

The force acting on the particle is split into a surface force df_i and a body force (per unit volume) B_i. In a mixture of N species, the body forces acting on the species may differ. Thus for a multicomponent system

$$B_i = \rho \sum_{k=1}^{N} (Y_k f_k)_i \tag{3-39}$$

where f_k is the force per unit mass on the kth species, and

$$dF_i = df_i + B_i \, dx_1 \, dx_2 \, dx_3 \tag{3-40}$$

Now we shall express the net surface force in terms of the stress acting on the different faces of the particle. Consider the cube of Fig. 3.4. In this case, the surface forces for the three directions are

$$df_1 = \left(\frac{\partial \sigma_{11}}{\partial x_1} + \frac{\partial \sigma_{21}}{\partial x_2} + \frac{\partial \sigma_{31}}{\partial x_3} \right) dx_1 \, dx_2 \, dx_3$$

$$df_2 = \left(\frac{\partial \sigma_{12}}{\partial x_1} + \frac{\partial \sigma_{22}}{\partial x_2} + \frac{\partial \sigma_{32}}{\partial x_3} \right) dx_1 \, dx_2 \, dx_3$$

$$df_3 = \left(\frac{\partial \sigma_{13}}{\partial x_1} + \frac{\partial \sigma_{23}}{\partial x_2} + \frac{\partial \sigma_{33}}{\partial x_3} \right) dx_1 \, dx_2 \, dx_3$$

After dividing by $dx_1\, dx_2\, dx_3$, the momentum equation becomes

$$\rho\left[u_j\frac{\partial u_i}{\partial x_j}+\frac{\partial u_i}{\partial t}\right]=\frac{\partial \sigma_{ji}}{\partial x_j}+B_i$$

$$=\frac{\partial \sigma_{ji}}{\partial x_j}+\rho\sum_{k=1}^{N}(Y_k f_k)_i \tag{3-41}$$

which is the equation of motion in terms of the stress tensor.

5.1.2 Infinitesimal Control Volume

Consider Fig. 3-4. However, instead of considering a cubical particle which moves through space, assume that this figure represents a cubical volume, fixed in space, through which the fluid moves. For this volume we have

$$\left\{\begin{matrix}\text{rate of}\\\text{momentum}\\\text{increase}\\\text{in control}\\\text{volume}\end{matrix}\right\}=\left\{\begin{matrix}\text{net rate}\\\text{of momentum}\\\text{in control}\\\text{volume}\end{matrix}\right\}+\left\{\begin{matrix}\text{sum of forces}\\\text{acting on}\\\text{system}\end{matrix}\right\}$$

Considering only the x_1-direction, the net momentum flux is

$$\left\{\rho u_1 u_1-\left[\rho u_1 u_1+\frac{\partial(\rho u_1 u_1)}{\partial x_1}\,dx_1\right]\right\}dx_2\,dx_3$$

$$+\left\{\rho u_2 u_1-\left[\rho u_2 u_1+\frac{\partial(\rho u_2 u_1)}{\partial x_2}\,dx_2\right]\right\}dx_1\,dx_3$$

$$+\left\{\rho u_3 u_1-\left[\rho u_3 u_1+\frac{\partial(\rho u_3 u_1)}{\partial x_3}\,dx_3\right]\right\}dx_1\,dx_2$$

which becomes

$$-\left[\frac{\partial(\rho u_1 u_1)}{\partial x_1}+\frac{\partial(\rho u_2 u_1)}{\partial x_2}+\frac{\partial(\rho u_3 u_1)}{\partial x_3}\right]dx_1\,dx_2\,dx_3$$

The increase of the rate of momentum in the volume for the x_1-direction is

$$\frac{\partial(\rho u_1)}{\partial t}\,dx_1\,dx_2\,dx_3$$

Then combining with the stress derived previously, we have for the x_1-direction

$$\left[\frac{\partial(\rho u_1)}{\partial t} + \frac{\partial(\rho u_1 u_1)}{\partial x_1} + \frac{\partial(\rho u_2 u_1)}{\partial x_2} + \frac{\partial(\rho u_3 u_1)}{\partial x_3}\right] dx_1\, dx_2\, dx_3$$

$$= \left(\frac{\partial \sigma_{11}}{\partial x_1} + \frac{\partial \sigma_{21}}{\partial x_2} + \frac{\partial \sigma_{31}}{\partial x_3}\right) dx_1\, dx_2\, dx_3 + B_1\, dx_1\, dx_2\, dx_3$$

The equations for the x_2 and x_3 directions may be obtained in a similar manner. The final result after dividing by $dx_1\, dx_2\, dx_3$ is

$$\frac{\partial(\rho u_i)}{\partial t} + \frac{\partial(\rho u_j u_i)}{\partial x_j} = \frac{\partial \sigma_{ji}}{\partial x_j} + B_i \tag{3-42}$$

After differentiating the left-hand-side terms, Eq. (3-42) becomes

$$u_i\left[\frac{\partial \rho}{\partial t} + \frac{\partial(\rho u_j)}{\partial x_j}\right] + \rho\left[\frac{\partial u_i}{\partial t} + u_j \frac{\partial u_i}{\partial x_j}\right] = \frac{\partial \sigma_{ji}}{\partial x_j} + B_i$$

But the first term on the left side is zero by the continuity equation. The final equation becomes

$$\rho\left[u_j \frac{\partial u_i}{\partial x_j} + \frac{\partial u_i}{\partial t}\right] = \frac{\partial \sigma_{ji}}{\partial x_j} + B_i$$

$$= \frac{\partial \sigma_{ji}}{\partial x_j} + \rho \sum_{k=1}^{N} (Y_k f_k)_i \tag{3-43}$$

which is the same result as that of the previous section.

5.1.3 Finite Control Volume

If one starts with Newton's second law for a particle,

$$\mathbf{F} = \frac{d(m\mathbf{V})}{dt}$$

the momentum equation for a finite control volume in space may be derived (see undergraduate texts such as Shames,[6] giving

$$\frac{\partial}{\partial t}\int_v \mathbf{V}\rho\, dv + \int_A \mathbf{V}(\rho\mathbf{V} \cdot d\mathbf{A}) = \mathbf{F}_s + \int_v \mathbf{B}\, dv \tag{3-44}$$

which states that

$$\begin{Bmatrix}\text{rate of}\\ \text{momentum}\\ \text{increase}\\ \text{in volume}\end{Bmatrix} + \begin{Bmatrix}\text{net rate}\\ \text{of momentum}\\ \text{flux out of}\\ \text{volume}\end{Bmatrix} = \begin{Bmatrix}\text{sum of surface}\\ \text{and body forces}\\ \text{acting on fluid}\\ \text{within volume}\end{Bmatrix}$$

In tensor notation, Eq. (3-44) is

$$\frac{\partial}{\partial t}\int_v u_i\rho\, dv + \int_A \rho u_i u_j\, dA_j = \int_A \sigma_{ij}\, dA_j + \int_v B_i\, dv \tag{3-45}$$

Applying Gauss's theorem to the second term to change it to a volume integral, we have

$$\int_A \rho u_i u_j\, dA_j = \int_v \frac{\partial(\rho u_i u_j)}{\partial x_j}\, dv$$

Similarly, applying Gauss's theorem to the third term, we have

$$\int_A \sigma_{ij}\, dA_j = \int_v \frac{\partial \sigma_{ij}}{\partial x_j}\, dv$$

Combining these with Eq. (3-45), we now obtain a result which involves only volume integrals:

$$\int_v \left[\frac{\partial(\rho u_i)}{\partial t} + \frac{\partial(\rho u_i u_j)}{\partial x_j}\right] dv = \int_v \left[\frac{\partial \sigma_{ij}}{\partial x_j} + B_i\right] dv \tag{3-46}$$

Since the integrands are continuous functions and there is arbitrary volume integration, we have (note $\sigma_{ij} = \sigma_{ji}$)

$$\frac{\partial(\rho u_i)}{\partial t} + \frac{\partial(\rho u_i u_j)}{\partial x_j} = \frac{\partial \sigma_{ji}}{\partial x_j} + B_i \tag{3-47}$$

which is the same result as that given in the previous section [Eq. (3-42)]. Then, following the method of the previous section, Eq. (3-47) or (3-42) becomes

$$\rho\left[u_j \frac{\partial u_i}{\partial x_j} + \frac{\partial u_i}{\partial t}\right] = \frac{\partial \sigma_{ji}}{\partial x_j} + B_i$$

$$= \frac{\partial \sigma_{ji}}{\partial x_j} + \rho \sum_{k=1}^{N} (Y_k f_k)_i \tag{3-48}$$

Thus, with three similar but basically different approaches, we have arrived at the momentum equation in terms of stress. Perhaps three derivations are no more convincing than one, but the fact that all three appear in the literature is often confusing. A better understanding (or a lesser misunderstanding) may be achieved by looking at the three different approaches.

Our next task is to establish a relationship between stress and mean velocity.

5.2 Stress – Strain-Rate Relationship (Constitutive Relationship)

Before we develop the constitutive relationship, it is important to point out a fundamental difference between solids and fluids. For solid elastic bodies, shear stress is proportional to the *magnitude* of angular deformation through Hooke's law; whereas for a fluid, shear stress is proportional to the *rate of angular deformation* through Stokes's law.

A fluid tends to deform when subjected to a shear stress. The rate of deformation varies for different fluids, and it is dependent upon the thermodynamic state for a given fluid. Resistance to deformation is a property of the fluid.

There are several ways to establish stress–strain relationships. The approach used in classical texts such as Schlichting[3] is to write the stress–strain relationships for solid elastic bodies and then replace the displacement vector by a time derivative of the displacement. This approach is lengthy but has a strong physical basis.

We want to relate the term $\partial\sigma_{ji}/\partial x_j$ in some way to the mean velocity. In other words, we want to relate the stress tensor σ_{ij} to the rate of deformation of a fluid element.

In the stress tensor

$$\sigma_{ij} = \begin{Bmatrix} \sigma_{11} & \sigma_{12} & \sigma_{13} \\ \sigma_{21} & \sigma_{22} & \sigma_{23} \\ \sigma_{31} & \sigma_{32} & \sigma_{33} \end{Bmatrix} = -p\delta_{ij} + \tau_{ij} \tag{3-49}$$

the components with $i = j$ (i.e., $\sigma_{11}, \sigma_{22}, \sigma_{33}$) represent normal stress and are thus related to pressure and linear deformation. The components with $i \neq j$ represent tangential stress. (Note that σ_{21}, for example, is a stress on the constant-x_2 face [first index] in the x_1-direction [second index].)

The stress tensor is symmetric. That is,

$$\sigma_{ij} = \sigma_{ji}$$

or specifically,

$$\sigma_{12} = \sigma_{21}$$

$$\sigma_{13} = \sigma_{31}$$

$$\sigma_{23} = \sigma_{32}$$

This can be shown easily by summing moments and is left as an exercise for the reader.

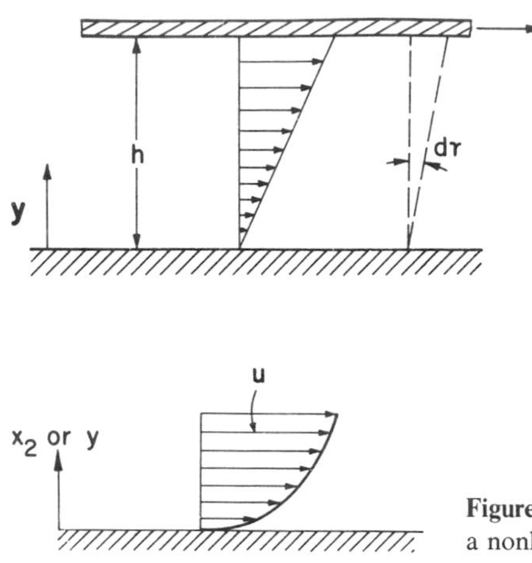

Figure 3.5 Fluid between two parallel plates with the lower one stationary and the upper one moving at constant velocity (the linear velocity profile corresponding to zero pressure gradient in the flow direction).

Figure 3.6 Velocity distribution of a parallel flow having a nonlinear velocity profile.

5.2.1 Strain Rate

Consider the special case of a fluid undergoing strain as shown in Fig. 3.5. The fluid is contained between parallel walls with the lower plate stationary and the upper plate moving at a constant velocity U. The fluid experiences angular deformation at the rate of

$$\frac{d\gamma}{dt} = \frac{U}{h}$$

In the case of a nonlinear velocity distribution as shown in Fig. 3.6, an infinitesimal fluid element experiences a strain rate as shown in Fig. 3.7. The

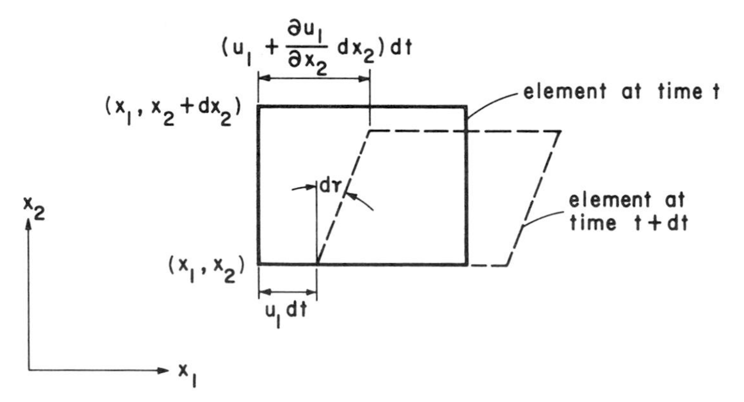

Figure 3.7 Deformation of an infinitesimal fluid element in a two-dimensional parallel flow.

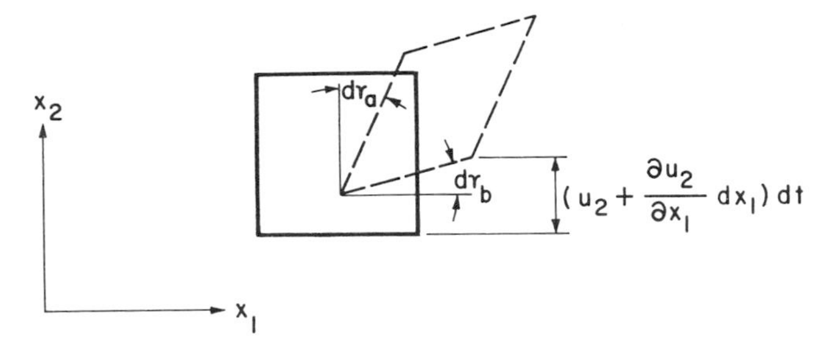

Figure 3.8 Deformation of an infinitesimal fluid element in a two-dimensional flow (angular deformations are allowed for both vertical and horizontal segments).

total angular strain in time dt is

$$d\gamma = \frac{\left(u_1 + \dfrac{\partial u_1}{\partial x_2} dx_2\right) dt - u_1\, dt}{dx_2}$$

so the strain rate is

$$\frac{d\gamma}{dt} = \frac{\partial u_1}{\partial x_2}$$

The above result is for the case of one-dimensional deformation of an element. In the case of two-dimensional deformation (i.e., one in which there is a nonzero velocity in the x_2-direction, u_2), we have angular deformation as shown in Fig. 3.8. The total angular deformation is

$$d\gamma = d\gamma_a + d\gamma_b = \frac{\left(u_1 + \dfrac{\partial u_1}{\partial x_2} dx_2\right) dt - u_1\, dt}{dx_2} + \frac{\left(u_2 + \dfrac{\partial u_2}{\partial x_1} dx_1\right) dt - u_2\, dt}{dx_1}$$

so

$$\frac{d\gamma}{dt} = \frac{\partial u_1}{\partial x_2} + \frac{\partial u_2}{\partial x_1} \equiv 2e_{12} = 2e_{21}$$

where the general strain-rate tensor is

$$e_{ij} \equiv \frac{1}{2}\left(\frac{\partial u_i}{\partial x_j} + \frac{\partial u_j}{\partial x_i}\right).$$

5.2.2 Stress

Now we make the very important assumption that the shear stress is linearly proportional to the rate of angular deformation, that is, the fluid is Newtonian. For a parallel-flow system,

$$\tau \propto \frac{d\gamma}{dt} = \frac{\partial u_1}{\partial x_2}$$

or

$$\sigma_{21} = \sigma_{12} = \mu \frac{\partial u_1}{\partial x_2} \tag{3-50}$$

For the case of both velocities u_1 and u_2 nonzero, we have

$$\tau_{12} = \mu \left(\frac{\partial u_1}{\partial x_2} + \frac{\partial u_2}{\partial x_1} \right). \tag{3-51}$$

Further, if we consider the x_1–x_3 and x_2–x_3 planes in the same way, we obtain the general shear stress tensor for a Newtonian fluid

$$\sigma_{ij} = \mu \left(\frac{\partial u_i}{\partial x_j} + \frac{\partial u_j}{\partial x_i} \right), \qquad i \neq j \tag{3-52}$$

The evaluation of the off-diagonal elements of the general stress tensor σ_{ij} is now complete. This leaves the task of determining the diagonal elements of the stress tensor in terms of pressure and velocity.

We have considered angular deformation. Let us consider next the linear deformation of a fluid particle as shown in Fig. 3.9. The linear deformation in time dt is

$$\frac{\left(u_1 + \dfrac{\partial u_1}{\partial x_1} dx_1 \right) dt - u_1 \, dt}{dx_1} = \frac{\partial u_1}{\partial x_1} dt$$

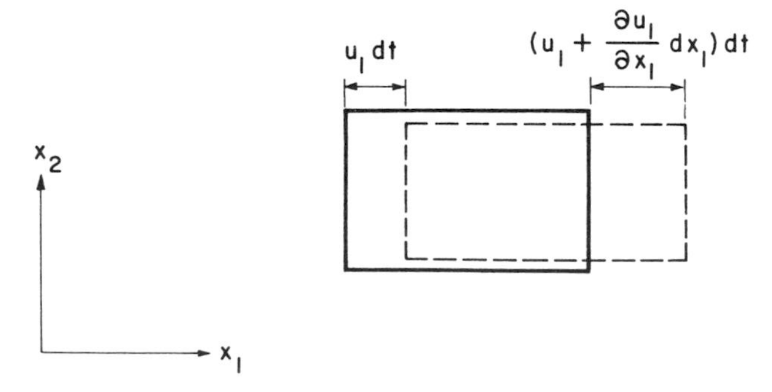

Figure 3.9 Linear deformation of a fluid particle.

The rate of linear deformation is

$$e_{11} = \frac{\partial u_1}{\partial x_1}$$

and for the x_2 and x_3 directions, we have

$$e_{22} = \frac{\partial u_2}{\partial x_2}$$

$$e_{33} = \frac{\partial u_3}{\partial x_3}$$

The total rate-of-deformation tensor is

$$\frac{\partial u_i}{\partial x_j}$$

where we have linear deformation for $i = j$ and angular deformation for $i \neq j$. We write the deformation tensor in two parts:

$$\frac{\partial u_i}{\partial x_j} = \frac{1}{2}\left(\frac{\partial u_i}{\partial x_j} + \frac{\partial u_j}{\partial x_i}\right) + \frac{1}{2}\left(\frac{\partial u_i}{\partial x_j} - \frac{\partial u_j}{\partial x_i}\right)$$

The first term is a symmetric tensor representing $\frac{1}{2}$ of the angular-deformation tensor. The second term is an antisymmetric tensor representing rotation of a fluid without deformation.

Now, we consider the relationship for stress and strain for the normal stresses to be

$$\begin{aligned}
\sigma_{11} &= -p + ce_{11} + \lambda e_{22} + \lambda e_{33} \\
\sigma_{22} &= -p + \lambda e_{11} + ce_{22} + \lambda e_{33} \\
\sigma_{33} &= -p + \lambda e_{11} + \lambda e_{22} + ce_{33}
\end{aligned} \tag{3-53}$$

This states that there is linear relationship between stress, say σ_{11}, and strain in the same direction, e_{11} (see Fig. 3.10). Also, the stress σ_{11} causes a strain in the other two directions, e_{22} and e_{33}. Since there is no preferred direction, they are related by the same coefficient λ; that is, the fluid is assumed to be isotropic.

In Eq. (3-53), p represents the hydrostatic pressure, which must be included, since when there is no linear deformation ($e_{11} = e_{22} = e_{33} = 0$), we still have a normal stress in all directions expressed as

$$\sigma_{11} = -p, \qquad \sigma_{22} = -p, \quad \text{and} \quad \sigma_{33} = -p$$

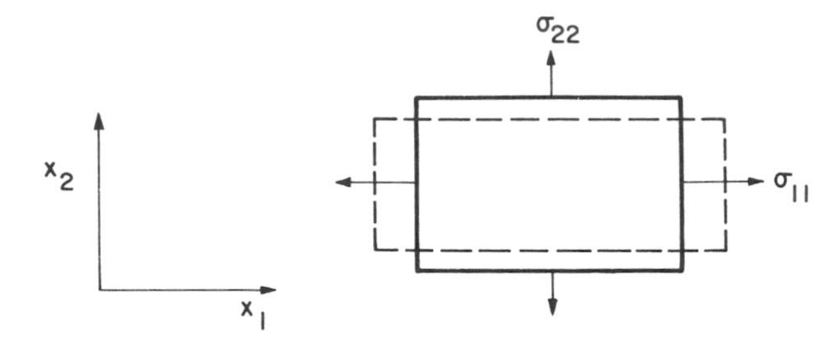

Figure 3.10 Normal stresses acting on a two-dimensional fluid particle.

The normal stress–strain relationships may be written as

$$\sigma_{11} = -p + \lambda(e_{11} + e_{22} + e_{33}) + (c - \lambda)e_{11}$$

$$\sigma_{22} = -p + \lambda(e_{11} + e_{22} + e_{33}) + (c - \lambda)e_{22}$$

$$\sigma_{33} = -p + \lambda(e_{11} + e_{22} + e_{33}) + (c - \lambda)e_{33} \tag{3-53a}$$

The above two coefficients λ and c can be replaced by two other coefficients μ and μ' according to the following relations:

$$c - \lambda = 2\mu$$

$$\lambda = \mu' - \tfrac{2}{3}\mu \tag{3-54}$$

of which the first can be shown to be valid for an isotropic Newtonian fluid. Here μ is usually called the dynamic viscosity or the first viscosity, μ' is sometimes called the bulk viscosity,[2] and λ is called the second viscosity. (It should be noted that some authors call λ the bulk viscosity, since it is associated with volume expansion.) For monatomic gas mixtures, kinetic theory shows that $\mu' = 0$. Up to now, there are no direct and conclusive data on μ'. The usual practice is to employ the hypothesis made by Stokes in 1845:

$$\lambda + \tfrac{2}{3}\mu = 0, \qquad \text{or} \quad \mu' = 0 \tag{3-55}$$

Physically, the nonzero values of μ' for polyatomic gases, as explained by Williams,[2] are caused by the relaxation effects between the translational motion and the various internal degrees of freedom. These effects lead to a positive value of μ'. So far, no theoretical computations or reliable experimental measurements of μ' exist, and in fact μ' is usually negligible in combustion processes.

Substituting Eq. (3-54) into Eq. (3-53), we have

$$\sigma_{11} = -p + \left(\mu' - \tfrac{2}{3}\mu\right)e_{kk} + 2\mu e_{11}$$

$$\sigma_{22} = -p + \left(\mu' - \tfrac{2}{3}\mu\right)e_{kk} + 2\mu e_{22}$$

$$\sigma_{33} = -p + \left(\mu' - \tfrac{2}{3}\mu\right)e_{kk} + 2\mu e_{33}$$

and the general stress–strain-rate relationship becomes

$$\sigma_{ij} = -p\delta_{ij} + \left(\mu' - \tfrac{2}{3}\mu\right)e_{kk}\delta_{ij} + 2\mu e_{ij} \tag{3-56}$$

The stress tensor may also be written as

$$\boxed{\sigma_{ij} = -p\delta_{ij} + \left(\mu' - \tfrac{2}{3}\mu\right)\frac{\partial u_k}{\partial x_k}\delta_{ij} + \mu\left(\frac{\partial u_i}{\partial x_j} + \frac{\partial u_j}{\partial x_i}\right)} \tag{3-57}$$

5.3 Navier – Stokes Equation

The stress–strain relationship (3-57) may be substituted into the momentum equation (3-48) to give

$$\begin{aligned}\rho\left[u_j\frac{\partial u_i}{\partial x_j} + \frac{\partial u_i}{\partial t}\right] &= \frac{\partial}{\partial x_j}\left[-p\delta_{ij} + \left(\mu' - \tfrac{2}{3}\mu\right)\frac{\partial u_k}{\partial x_k}\delta_{ij} + \mu\left(\frac{\partial u_i}{\partial x_j} + \frac{\partial u_j}{\partial x_i}\right)\right] + B_i \\ &= \frac{\partial}{\partial x_j}\left[-p\delta_{ij} + \left(\mu' - \tfrac{2}{3}\mu\right)\frac{\partial u_k}{\partial x_k}\delta_{ij} + \mu\left(\frac{\partial u_i}{\partial x_j} + \frac{\partial u_j}{\partial x_i}\right)\right] \\ &\quad + \rho\left(\sum_{k=1}^{N} Y_k f_k\right)_i\end{aligned} \tag{3-58}$$

If we assume that $\sigma_{ii} = -3p$, the bulk viscosity is zero. The same result is obtained from kinetic theory for a perfect monatomic gas. The momentum equation becomes

$$\rho\left[u_j\frac{\partial u_i}{\partial x_j} + \frac{\partial u_i}{\partial t}\right] = \frac{\partial}{\partial x_j}\left\{-p\delta_{ij} + \mu\left[\frac{\partial u_i}{\partial x_j} + \frac{\partial u_j}{\partial x_i}\right] - \tfrac{2}{3}\mu\left[\delta_{ij}\frac{\partial u_k}{\partial x_k}\right]\right\} + B_i \tag{3-59}$$

TABLE 3.3 The Equation of Motion in Rectangular Coordinates (x, y, z)[a]

In terms of τ:

x-component,
$$\rho\left(\frac{\partial u_x}{\partial t} + u_x\frac{\partial u_x}{\partial x} + u_y\frac{\partial u_x}{\partial y} + u_z\frac{\partial u_x}{\partial z}\right) = -\frac{\partial p}{\partial x} + \left(\frac{\partial \tau_{xx}}{\partial x} + \frac{\partial \tau_{yx}}{\partial y} + \frac{\partial \tau_{zx}}{\partial z}\right) + B_x \quad \text{(A)}$$

y-component,
$$\rho\left(\frac{\partial u_y}{\partial t} + u_x\frac{\partial u_y}{\partial x} + u_y\frac{\partial u_y}{\partial y} + u_z\frac{\partial u_y}{\partial z}\right) = -\frac{\partial p}{\partial y} + \left(\frac{\partial \tau_{xy}}{\partial x} + \frac{\partial \tau_{yy}}{\partial y} + \frac{\partial \tau_{zy}}{\partial z}\right) + B_y \quad \text{(B)}$$

z-component,
$$\rho\left(\frac{\partial u_z}{\partial t} + u_x\frac{\partial u_z}{\partial x} + u_y\frac{\partial u_z}{\partial y} + u_z\frac{\partial u_z}{\partial z}\right) = -\frac{\partial p}{\partial z} + \left(\frac{\partial \tau_{xz}}{\partial x} + \frac{\partial \tau_{yz}}{\partial y} + \frac{\partial \tau_{zz}}{\partial z}\right) + B_z \quad \text{(C)}$$

In terms of velocity gradients for a Newtonian fluid with constant ρ and μ:

x-component,
$$\rho\left(\frac{\partial u_x}{\partial t} + u_x\frac{\partial u_x}{\partial x} + u_y\frac{\partial u_x}{\partial y} + u_z\frac{\partial u_x}{\partial z}\right) = -\frac{\partial p}{\partial x} + \mu\left(\frac{\partial^2 u_x}{\partial x^2} + \frac{\partial^2 u_x}{\partial y^2} + \frac{\partial^2 u_x}{\partial z^2}\right) + B_x \quad \text{(D)}$$

y-component,
$$\rho\left(\frac{\partial u_y}{\partial t} + u_x\frac{\partial u_y}{\partial x} + u_y\frac{\partial u_y}{\partial y} + u_z\frac{\partial u_y}{\partial z}\right) = -\frac{\partial p}{\partial y} + \mu\left(\frac{\partial^2 u_y}{\partial x^2} + \frac{\partial^2 u_y}{\partial y^2} + \frac{\partial^2 u_y}{\partial z^2}\right) + B_y \quad \text{(E)}$$

z-component,
$$\rho\left(\frac{\partial u_z}{\partial t} + u_x\frac{\partial u_z}{\partial x} + u_y\frac{\partial u_z}{\partial y} + u_z\frac{\partial u_z}{\partial z}\right) = -\frac{\partial p}{\partial z} + \mu\left(\frac{\partial^2 u_z}{\partial x^2} + \frac{\partial^2 u_z}{\partial y^2} + \frac{\partial^2 u_z}{\partial z^2}\right) + B_z \quad \text{(F)}$$

[a]Adapted from Ref. 1.

TABLE 3.4 The Equation of Motion in Cylindrical Coordinate

In terms of τ:

r-component,[b]

$$\rho\left(\frac{\partial u_r}{\partial t} + u_r\frac{\partial u_r}{\partial r} + \frac{u_\theta}{r}\frac{\partial u_r}{\partial \theta} - \frac{u_\theta^2}{r} + u_z\frac{\partial u_r}{\partial z}\right)$$

$$= -\frac{\partial p}{\partial r} + \left(\frac{1}{r}\frac{\partial}{\partial r}(r\tau_{rr}) + \frac{1}{r}\frac{\partial \tau_{r\theta}}{\partial \theta} - \frac{\tau_{\theta\theta}}{r} + \frac{\partial \tau_{rz}}{\partial z}\right)$$

θ-component,[c]

$$\rho\left(\frac{\partial u_\theta}{\partial t} + u_r\frac{\partial u_\theta}{\partial r} + \frac{u_\theta}{r}\frac{\partial u_\theta}{\partial \theta} + \frac{u_r u_\theta}{r} + u_z\frac{\partial u_\theta}{\partial z}\right)$$

$$= -\frac{1}{r}\frac{\partial p}{\partial \theta} + \left(\frac{1}{r^2}\frac{\partial}{\partial r}(r^2\tau_{r\theta}) + \frac{1}{r}\frac{\partial \tau_{\theta\theta}}{\partial \theta} + \frac{\partial \tau_{\theta z}}{\partial z}\right) + \rho B_\theta \qquad \text{(B)}$$

z-component,

$$\rho\left(\frac{\partial u_z}{\partial t} + u_r\frac{\partial u_z}{\partial r} + \frac{u_\theta}{r}\frac{\partial u_z}{\partial \theta} + u_z\frac{\partial u_z}{\partial z}\right)$$

$$= -\frac{\partial p}{\partial z} + \left(\frac{1}{r}\frac{\partial}{\partial r}(r\tau_{rz}) + \frac{1}{r}\frac{\partial \tau_{\theta z}}{\partial \theta} + \frac{\partial \tau_{zz}}{\partial z}\right) + \rho B_z \qquad \text{(C)}$$

In terms of velocity gradients for a Newtonian fluid with constant ρ and μ:

r-component,[b]

$$\rho\left(\frac{\partial u_r}{\partial t} + u_r\frac{\partial u_r}{\partial r} + \frac{u_\theta}{r}\frac{\partial u_r}{\partial \theta} - \frac{u_\theta^2}{r} + u_z\frac{\partial u_r}{\partial z}\right)$$

$$= -\frac{\partial p}{\partial r} + \mu\left[\frac{\partial}{\partial r}\left(\frac{1}{r}\frac{\partial}{\partial r}(ru_r)\right) + \frac{1}{r^2}\frac{\partial^2 u_r}{\partial \theta^2} - \frac{2}{r^2}\frac{\partial u_\theta}{\partial \theta} + \frac{\partial^2 u_r}{\partial z^2}\right] + \rho B_r \qquad \text{(D)}$$

θ-component,[c]

$$\rho\left(\frac{\partial u_\theta}{\partial t} + u_r\frac{\partial u_\theta}{\partial r} + \frac{u_\theta}{r}\frac{\partial u_\theta}{\partial \theta} + \frac{u_r u_\theta}{r} + u_z\frac{\partial u_\theta}{\partial z}\right)$$

$$= -\frac{1}{r}\frac{\partial p}{\partial \theta} + \mu\left[\frac{\partial}{\partial r}\left(\frac{1}{r}\frac{\partial}{\partial r}(ru_\theta)\right) + \frac{1}{r^2}\frac{\partial^2 u_\theta}{\partial \theta^2} + \frac{2}{r^2}\frac{\partial u_r}{\partial \theta} + \frac{\partial^2 u_\theta}{\partial z^2}\right] + \rho B_\theta \qquad \text{(E)}$$

z-component,

$$\rho\left(\frac{\partial u_z}{\partial t} + u_r\frac{\partial u_z}{\partial r} + \frac{u_\theta}{r}\frac{\partial u_z}{\partial \theta} + u_z\frac{\partial u_z}{\partial z}\right)$$

$$= -\frac{\partial p}{\partial z} + \mu\left[\frac{1}{r}\frac{\partial}{\partial r}\left(r\frac{\partial u_z}{\partial r}\right) + \frac{1}{r^2}\frac{\partial^2 u_z}{\partial \theta^2} + \frac{\partial^2 u_z}{\partial z^2}\right] + \rho B_z \qquad \text{(F)}$$

---

[a]Adapted from Ref. 1.

[b]The term $\rho u_\theta^2/r$ is the *centrifugal force*. It gives the effective force in the r-direction resulting from fluid motion in the θ-direction. This term arises automatically on transformation from rectangular to cylindrical coordinates; it does not have to be added on physical grounds.

[c]The term $\rho u_r v_\theta/r$ is the *Coriolis force*. It is an effective force in the θ-direction when there is flow in both the r and θ directions. This term also arises automatically in the coordinate transformation. The Coriolis force arises in the problem of flow near a rotating disk (see, for example, Ref. 3).

.erms of τ:

r-component,
$$\rho\left(\frac{\partial u_r}{\partial t} + u_r\frac{\partial u_r}{\partial r} + \frac{u_\theta}{r}\frac{\partial u_r}{\partial \theta} + \frac{u_\phi}{r\sin\theta}\frac{\partial u_r}{\partial \phi} - \frac{u_\theta^2 + u_\phi^2}{r}\right)$$
$$= -\frac{\partial p}{\partial r} + \left(\frac{1}{r^2}\frac{\partial}{\partial r}(r^2\tau_{rr}) + \frac{1}{r\sin\theta}\frac{\partial}{\partial\theta}(\tau_{r\theta}\sin\theta) + \frac{1}{r\sin\theta}\frac{\partial\tau_{r\phi}}{\partial\phi} - \frac{\tau_{\theta\theta}+\tau_{\phi\phi}}{r}\right) + \rho B_r \quad \text{(A)}$$

θ-component,
$$\rho\left(\frac{\partial u_\theta}{\partial t} + u_r\frac{\partial u_\theta}{\partial r} + \frac{u_\theta}{r}\frac{\partial u_\theta}{\partial \theta} + \frac{u_\phi}{r\sin\theta}\frac{\partial u_\theta}{\partial \phi} + \frac{u_r u_\theta}{r} - \frac{u_\phi^2\cot\theta}{r}\right)$$
$$= -\frac{1}{r}\frac{\partial p}{\partial \theta} + \left(\frac{1}{r^2}\frac{\partial}{\partial r}(r^2\tau_{r\theta}) + \frac{1}{r\sin\theta}\frac{\partial}{\partial\theta}(\tau_{\theta\theta}\sin\theta) + \frac{1}{r\sin\theta}\frac{\partial\tau_{\theta\phi}}{\partial\phi} + \frac{\tau_{r\theta}}{r} - \frac{\cot\theta}{r}\tau_{\phi\phi}\right) + \rho B_\theta \quad \text{(B)}$$

ϕ-component,
$$\rho\left(\frac{\partial u_\phi}{\partial t} + u_r\frac{\partial u_\phi}{\partial r} + \frac{u_\theta}{r}\frac{\partial u_\phi}{\partial \theta} + \frac{u_\phi}{r\sin\theta}\frac{\partial u_\phi}{\partial \phi} + \frac{u_\phi u_r}{r} + \frac{u_\theta u_\phi}{r}\cot\theta\right)$$
$$= -\frac{1}{r\sin\theta}\frac{\partial p}{\partial \phi} + \left(\frac{1}{r^2}\frac{\partial}{\partial r}(r^2\tau_{r\phi}) + \frac{1}{r}\frac{\partial\tau_{\theta\phi}}{\partial\theta} + \frac{1}{r\sin\theta}\frac{\partial\tau_{\phi\phi}}{\partial\phi} + \frac{\tau_{r\phi}}{r} + \frac{2\cot\theta}{r}\tau_{\theta\phi}\right) + \rho B_\phi \quad \text{(C)}$$

In terms of velocity gradients for a Newtonian fluid with constant ρ and μ:[b]

r-component,
$$\rho\left(\frac{\partial u_r}{\partial t} + u_r\frac{\partial u_r}{\partial r} + \frac{u_\theta}{r}\frac{\partial u_r}{\partial \theta} + \frac{u_\phi}{r\sin\theta}\frac{\partial u_r}{\partial \phi} - \frac{u_\theta^2 + u_\phi^2}{r}\right)$$
$$= -\frac{\partial p}{\partial r} + \mu\left(\nabla^2 u_r - \frac{2}{r^2}u_r - \frac{2}{r^2}\frac{\partial u_\theta}{\partial\theta} - \frac{2}{r^2}u_\theta\cot\theta - \frac{2}{r^2\sin\theta}\frac{\partial u_\phi}{\partial\phi}\right) + \rho B_r \quad \text{(D)}$$

θ-component,
$$\rho\left(\frac{\partial u_\theta}{\partial t} + u_r\frac{\partial u_\theta}{\partial r} + \frac{u_\theta}{r}\frac{\partial u_\theta}{\partial \theta} + \frac{u_\phi}{r\sin\theta}\frac{\partial u_\theta}{\partial \phi} + \frac{u_r u_\theta}{r} - \frac{u_\phi^2\cot\theta}{r}\right)$$
$$= -\frac{1}{r}\frac{\partial p}{\partial \theta} + \mu\left(\nabla^2 u_\theta + \frac{2}{r^2}\frac{\partial u_r}{\partial\theta} - \frac{u_\theta}{r^2\sin^2\theta} - \frac{2\cos\theta}{r^2\sin^2\theta}\frac{\partial u_\phi}{\partial\phi}\right) + \rho B_\theta \quad \text{(E)}$$

ϕ-component,
$$\rho\left(\frac{\partial u_\phi}{\partial t} + u_r\frac{\partial u_\phi}{\partial r} + \frac{u_\theta}{r}\frac{\partial u_\phi}{\partial \theta} + \frac{u_\phi}{r\sin\theta}\frac{\partial u_\phi}{\partial \phi} + \frac{u_\phi u_r}{r} + \frac{u_\theta u_\phi}{r}\cot\theta\right)$$
$$= -\frac{1}{r\sin\theta}\frac{\partial p}{\partial \phi} + \mu\left(\nabla^2 u_\phi - \frac{u_\phi}{r^2\sin^2\theta} + \frac{2}{r^2\sin\theta}\frac{\partial u_r}{\partial\phi} + \frac{2\cos\theta}{r^2\sin^2\theta}\frac{\partial u_\theta}{\partial\phi}\right) + \rho B_\phi \quad \text{(F)}$$

[a]Adapted from Ref. 1.

[b]In these equations $\nabla^2 = \frac{1}{r^2}\frac{\partial}{\partial r}(r^2\frac{\partial}{\partial r}) + \frac{1}{r^2 \sin\theta}\frac{\partial}{\partial\theta}(\sin\theta\frac{\partial}{\partial\theta}) + \frac{1}{r^2 \sin^2\theta}(\frac{\partial^2}{\partial\phi^2})$

TABLE 3.6 Components of the Stress Tensor for Newtonian Fluids in Rectangular Coordinates (x, y, z)[a]

$$\tau_{xx} = \mu\left[2\frac{\partial u_x}{\partial x} - \tfrac{2}{3}(\nabla \cdot \mathbf{v})\right] \quad \text{(A)}$$

$$\tau_{yy} = \mu\left[2\frac{\partial u_y}{\partial y} - \tfrac{2}{3}(\nabla \cdot \mathbf{v})\right] \quad \text{(B)}$$

$$\tau_{zz} = \mu\left[2\frac{\partial u_z}{\partial z} - \tfrac{2}{3}(\nabla \cdot \mathbf{v})\right] \quad \text{(C)}$$

$$\tau_{xy} = \tau_{yx} = \mu\left[\frac{\partial u_x}{\partial y} + \frac{\partial u_y}{\partial x}\right] \quad \text{(D)}$$

$$\tau_{yz} = \tau_{zy} = \mu\left[\frac{\partial u_y}{\partial z} + \frac{\partial u_z}{\partial y}\right] \quad \text{(E)}$$

$$\tau_{zx} = \tau_{xz} = \mu\left[\frac{\partial u_z}{\partial x} + \frac{\partial u_x}{\partial z}\right] \quad \text{(F)}$$

$$\nabla \cdot \mathbf{v} = \frac{\partial u_x}{\partial x} + \frac{\partial u_y}{\partial y} + \frac{\partial u_z}{\partial z} \quad \text{(G)}$$

[a]Adapted from Ref. 1.

TABLE 3.7 Components of the Stress Tensor for Newtonian Fluids in Cylindrical Coordinates (r, θ, z)[a]

$$\tau_{rr} = \mu\left[2\frac{\partial u_r}{\partial r} - \tfrac{2}{3}(\nabla \cdot \mathbf{v})\right] \quad \text{(A)}$$

$$\tau_{\theta\theta} = \mu\left[2\left(\frac{1}{r}\frac{\partial u_\theta}{\partial \theta} + \frac{u_r}{r}\right) - \tfrac{2}{3}(\nabla \cdot \mathbf{v})\right] \quad \text{(B)}$$

$$\tau_{zz} = \mu\left[2\frac{\partial u_z}{\partial z} - \tfrac{2}{3}(\nabla \cdot \mathbf{v})\right] \quad \text{(C)}$$

$$\tau_{r\theta} = \tau_{\theta r} = \mu\left[r\frac{\partial}{\partial r}\left(\frac{u_\theta}{r}\right) + \frac{1}{r}\frac{\partial u_r}{\partial \theta}\right] \quad \text{(D)}$$

$$\tau_{\theta z} = \tau_{z\theta} = \mu\left[\frac{\partial u_\theta}{\partial z} + \frac{1}{r}\frac{\partial u_z}{\partial \theta}\right] \quad \text{(E)}$$

$$\tau_{zr} = \tau_{rz} = \mu\left[\frac{\partial u_z}{\partial r} + \frac{\partial u_r}{\partial z}\right] \quad \text{(F)}$$

$$\nabla \cdot \mathbf{v} = \frac{1}{r}\frac{\partial}{\partial r}(r u_r) + \frac{1}{r}\frac{\partial u_\theta}{\partial \theta} + \frac{\partial u_z}{\partial z} \quad \text{(G)}$$

[a]Adapted from Ref. 1.

TABLE 3.8 Components of the Stress Tensor for Newtonian Fluids in Spherical Coordinates (r, θ, ϕ)[a]

$$\tau_{rr} = \mu\left[2\frac{\partial u_r}{\partial r} - \tfrac{2}{3}(\nabla \cdot \mathbf{v})\right] \tag{A}$$

$$\tau_{\theta\theta} = \mu\left[2\left(\frac{1}{r}\frac{\partial u_\theta}{\partial \theta} + \frac{u_r}{r}\right) - \tfrac{2}{3}(\nabla \cdot \mathbf{v})\right] \tag{B}$$

$$\tau_{\phi\phi} = \mu\left[2\left(\frac{1}{r\sin\theta}\frac{\partial u_\phi}{\partial \phi} + \frac{u_r}{r} + \frac{u_\theta \cot\theta}{r}\right) - \tfrac{2}{3}(\nabla \cdot \mathbf{v})\right] \tag{C}$$

$$\tau_{r\theta} = \tau_{\theta r} = \mu\left[r\frac{\partial}{\partial r}\left(\frac{u_\theta}{r}\right) + \frac{1}{r}\frac{\partial u_r}{\partial \theta}\right] \tag{D}$$

$$\tau_{\theta\phi} = \tau_{\phi\theta} = \mu\left[\frac{\sin\theta}{r}\frac{\partial}{\partial \theta}\left(\frac{u_\phi}{\sin\theta}\right) + \frac{1}{r\sin\theta}\frac{\partial u_\theta}{\partial \phi}\right] \tag{E}$$

$$\tau_{\phi r} = \tau_{r\phi} = \mu\left[\frac{1}{r\sin\theta}\frac{\partial u_r}{\partial \phi} + r\frac{\partial}{\partial r}\left(\frac{u_\phi}{r}\right)\right] \tag{F}$$

$$\nabla \cdot \mathbf{v} = \frac{1}{r^2}\frac{\partial}{\partial r}(r^2 u_r) + \frac{1}{r\sin\theta}\frac{\partial}{\partial \theta}(u_\theta \sin\theta) + \frac{1}{r\sin\theta}\frac{\partial u_\phi}{\partial \phi} \tag{G}$$

[a]Adapted from Ref. 1.

This equation is known as the Navier–Stokes equation. If the flow is assumed incompressible, the equation then becomes

$$\rho\left[\frac{\partial u_i}{\partial t} + u_j\frac{\partial u_i}{\partial x_j}\right] = -\frac{\partial p}{\partial x_i} + \frac{\partial}{\partial x_j}\left[\mu\left(\frac{\partial u_i}{\partial x_j} + \frac{\partial u_j}{\partial x_i}\right)\right] + \rho\sum_{k=1}^{N}(Y_k f_k)_i \tag{3-60}$$

Various forms of the equation of motion written in rectangular, cylindrical, and spherical coordinates are given in Tables 3.3–3.5. The components of the stress tensor for Newtonian fluids in these coordinates are given in Tables 3.6–3.8.

To facilitate readers' application of the conservation equations in vector form and to enhance their understanding of vector operations in various coordinate systems, the differential operations involving the ∇-operator in rectangular, cylindrical, and spherical coordinates are listed in Tables 3.9–3.11, respectively. All operations involving the stress tensor $\boldsymbol{\tau}$ are valid for symmetrical $\boldsymbol{\tau}$ only. The expressions for $(\boldsymbol{\tau} : \nabla \mathbf{v})$ are useful for the energy equation to be discussed in the following section.

TABLE 3.9 Summary of Differential Operations Involving the ∇-Operator in Rectangular Coordinates (x, y, z)[a]

$$\nabla \cdot \mathbf{v} = \frac{\partial u_x}{\partial x} + \frac{\partial u_y}{\partial y} + \frac{\partial u_z}{\partial z} \tag{A}$$

$$\nabla^2 s = \frac{\partial^2 s}{\partial x^2} + \frac{\partial^2 s}{\partial y^2} + \frac{\partial^2 s}{\partial z^2} \tag{B}$$

$$\boldsymbol{\tau} : \nabla \mathbf{v} = \tau_{xx}\left(\frac{\partial u_x}{\partial x}\right) + \tau_{yy}\left(\frac{\partial u_y}{\partial y}\right) + \tau_{zz}\left(\frac{\partial u_z}{\partial z}\right) + \tau_{xy}\left(\frac{\partial u_x}{\partial y} + \frac{\partial u_y}{\partial x}\right) + \tau_{yz}\left(\frac{\partial u_y}{\partial z} + \frac{\partial u_z}{\partial y}\right) + \tau_{zx}\left(\frac{\partial u_z}{\partial x} + \frac{\partial u_x}{\partial z}\right) \tag{C}$$

$$[\nabla s]_x = \frac{\partial s}{\partial x} \tag{D}$$

$$[\nabla s]_y = \frac{\partial s}{\partial y} \tag{E}$$

$$[\nabla s]_z = \frac{\partial s}{\partial z} \tag{F}$$

$$[\nabla \times \mathbf{v}]_x = \frac{\partial u_z}{\partial y} - \frac{\partial u_y}{\partial z} \tag{G}$$

$$[\nabla \times \mathbf{v}]_y = \frac{\partial u_x}{\partial z} - \frac{\partial u_z}{\partial x} \tag{H}$$

$$[\nabla \times \mathbf{v}]_z = \frac{\partial u_y}{\partial x} - \frac{\partial u_x}{\partial y} \tag{I}$$

$$[\nabla \cdot \boldsymbol{\tau}]_x = \frac{\partial \tau_{xx}}{\partial x} + \frac{\partial \tau_{xy}}{\partial y} + \frac{\partial \tau_{xz}}{\partial z} \tag{J}$$

$$[\nabla \cdot \boldsymbol{\tau}]_y = \frac{\partial \tau_{yx}}{\partial x} + \frac{\partial \tau_{yy}}{\partial y} + \frac{\partial \tau_{yz}}{\partial z} \tag{K}$$

$$[\nabla \cdot \boldsymbol{\tau}]_z = \frac{\partial \tau_{zy}}{\partial x} + \frac{\partial \tau_{zy}}{\partial y} + \frac{\partial \tau_{zz}}{\partial z} \tag{L}$$

$$[\nabla^2 \mathbf{v}]_x = \frac{\partial^2 u_x}{\partial x^2} + \frac{\partial^2 u_x}{\partial y^2} + \frac{\partial^2 u_x}{\partial z^2} \tag{M}$$

$$[\nabla^2 \mathbf{v}]_y = \frac{\partial^2 u_y}{\partial x^2} + \frac{\partial^2 u_y}{\partial y^2} + \frac{\partial^2 u_y}{\partial z^2} \tag{N}$$

$$[\nabla^2 \mathbf{v}]_z = \frac{\partial^2 u_z}{\partial x^2} + \frac{\partial^2 u_z}{\partial y^2} + \frac{\partial^2 u_z}{\partial z^2} \tag{O}$$

TABLE 3.9 (Continued).

$$[\mathbf{v}\cdot\nabla\mathbf{v}]_x = u_x\frac{\partial u_x}{\partial x} + u_y\frac{\partial u_x}{\partial y} + u_z\frac{\partial u_x}{\partial z} \tag{P}$$

$$[\mathbf{v}\cdot\nabla\mathbf{v}]_y = u_x\frac{\partial u_y}{\partial x} + u_y\frac{\partial u_y}{\partial y} + u_z\frac{\partial u_y}{\partial z} \tag{Q}$$

$$[\mathbf{v}\cdot\nabla\mathbf{v}]_z = u_x\frac{\partial u_z}{\partial x} + u_y\frac{\partial u_z}{\partial y} + u_z\frac{\partial u_z}{\partial z} \tag{R}$$

For Newtonian fluids

$$\frac{\boldsymbol{\tau}:\nabla\mathbf{v}}{\mu} = 2\left[\left(\frac{\partial u_x}{\partial x}\right)^2 + \left(\frac{\partial u_y}{\partial y}\right)^2 + \left(\frac{\partial u_z}{\partial z}\right)^2\right] + \left[\frac{\partial u_y}{\partial x} + \frac{\partial u_x}{\partial y}\right]^2 + \left[\frac{\partial u_z}{\partial y} + \frac{\partial u_y}{\partial z}\right]^2 + \left[\frac{\partial u_x}{\partial z} + \frac{\partial u_z}{\partial x}\right]^2 - \frac{2}{3}\left[\frac{\partial u_x}{\partial x} + \frac{\partial u_y}{\partial y} + \frac{\partial u_z}{\partial z}\right]^2 \tag{S}$$

[a]Adapted from Ref. 1. Operations involving the tensor $\boldsymbol{\tau}$ are given for symmetrical $\boldsymbol{\tau}$ only.

TABLE 3.10 Summary of Differential Operations Involving the ∇-Operator in Cylindrical Coordinates (r, θ, z)[a]

$$\nabla\cdot\mathbf{v} = \frac{1}{r}\frac{\partial}{\partial r}(ru_r) + \frac{1}{r}\frac{\partial u_\theta}{\partial \theta} + \frac{\partial u_z}{\partial z} \tag{A}$$

$$\nabla^2 s = \frac{1}{r}\frac{\partial}{\partial r}\left(r\frac{\partial s}{\partial r}\right) + \frac{1}{r^2}\frac{\partial^2 s}{\partial \theta^2} + \frac{\partial^2 s}{\partial z^2} \tag{B}$$

$$\boldsymbol{\tau}:\nabla\mathbf{v} = \tau_{rr}\left(\frac{\partial u_r}{\partial r}\right) + \tau_{\theta\theta}\left(\frac{1}{r}\frac{\partial u_\theta}{\partial \theta} + \frac{u_r}{r}\right) + \tau_{zz}\left(\frac{\partial u_z}{\partial z}\right) + \tau_{r\theta}\left[r\frac{\partial}{\partial r}\left(\frac{u_\theta}{r}\right) + \frac{1}{r}\frac{\partial u_r}{\partial \theta}\right] + \tau_{\theta z}\left(\frac{1}{r}\frac{\partial u_z}{\partial \theta} + \frac{\partial u_\theta}{\partial z}\right) + \tau_{rz}\left(\frac{\partial u_z}{\partial r} + \frac{\partial u_r}{\partial z}\right) \tag{C}$$

$$[\nabla s]_r = \frac{\partial s}{\partial r} \tag{D}$$

$$[\nabla s]_\theta = \frac{1}{r}\frac{\partial s}{\partial \theta} \tag{E}$$

$$[\nabla s]_z = \frac{\partial s}{\partial z} \tag{F}$$

$$[\nabla\times\mathbf{v}]_r = \frac{1}{r}\frac{\partial u_z}{\partial \theta} - \frac{\partial u_\theta}{\partial z} \tag{G}$$

$$[\nabla\times\mathbf{v}]_\theta = \frac{\partial u_r}{\partial z} - \frac{\partial u_z}{\partial r} \tag{H}$$

$$[\nabla\times\mathbf{v}]_z = \frac{1}{r}\frac{\partial}{\partial r}(ru_\theta) - \frac{1}{r}\frac{\partial u_r}{\partial \theta} \tag{I}$$

TABLE 3.10 (Continued).

$$[\nabla \cdot \boldsymbol{\tau}]_r = \frac{1}{r}\frac{\partial}{\partial r}(r\tau_{rr}) + \frac{1}{r}\frac{\partial}{\partial \theta}\tau_{r\theta} - \frac{1}{r}\tau_{\theta\theta} + \frac{\partial \tau_{rz}}{\partial z} \tag{J}$$

$$[\nabla \cdot \boldsymbol{\tau}]_\theta = \frac{1}{r}\frac{\partial \tau_{\theta\theta}}{\partial \theta} + \frac{\partial \tau_{r\theta}}{\partial r} + \frac{2}{r}\tau_{r\theta} + \frac{\partial \tau_{\theta z}}{\partial z} \tag{K}$$

$$[\nabla \cdot \boldsymbol{\tau}]_z = \frac{1}{r}\frac{\partial}{\partial r}(r\tau_{rz}) + \frac{1}{r}\frac{\partial \tau_{\theta z}}{\partial \theta} + \frac{\partial \tau_{zz}}{\partial z} \tag{L}$$

$$[\nabla^2 \mathbf{v}]_r = \frac{\partial}{\partial r}\left(\frac{1}{r}\frac{\partial}{\partial r}(ru_r)\right) + \frac{1}{r^2}\frac{\partial^2 u_r}{\partial \theta^2} - \frac{2}{r^2}\frac{\partial u_\theta}{\partial \theta} + \frac{\partial^2 u_r}{\partial z^2} \tag{M}$$

$$[\nabla^2 \mathbf{v}]_\theta = \frac{\partial}{\partial r}\left(\frac{1}{r}\frac{\partial}{\partial r}(ru_\theta)\right) + \frac{1}{r^2}\frac{\partial^2 u_\theta}{\partial \theta^2} + \frac{2}{r^2}\frac{\partial u_r}{\partial \theta} + \frac{\partial^2 u_\theta}{\partial z^2} \tag{N}$$

$$[\nabla^2 \mathbf{v}]_z = \frac{1}{r}\frac{\partial}{\partial r}\left(r\frac{\partial u_z}{\partial r}\right) + \frac{1}{r^2}\frac{\partial u_z}{\partial \theta^2} + \frac{\partial^2 u_z}{\partial z^2} \tag{O}$$

$$[\mathbf{v} \cdot \nabla \mathbf{v}]_r = u_r\frac{\partial u_r}{\partial r} + \frac{u_\theta}{r}\frac{\partial u_r}{\partial \theta} - \frac{u_\theta^2}{r} + u_z\frac{\partial u_r}{\partial z} \tag{P}$$

$$[\mathbf{v} \cdot \nabla \mathbf{v}]_\theta = u_r\frac{\partial u_\theta}{\partial r} + \frac{u_\theta}{r}\frac{\partial u_\theta}{\partial \theta} + \frac{u_r u_\theta}{r} + u_z\frac{\partial u_\theta}{\partial z} \tag{Q}$$

$$[\mathbf{v} \cdot \nabla \mathbf{v}]_z = u_r\frac{\partial u_z}{\partial r} + \frac{u_\theta}{r}\frac{\partial u_z}{\partial \theta} + u_z\frac{\partial u_z}{\partial z} \tag{R}$$

For Newtonian fluids

$$\begin{aligned}
\frac{\boldsymbol{\tau} : \nabla \mathbf{v}}{\mu} &= 2\left[\left(\frac{\partial u_r}{\partial r}\right)^2 + \left(\frac{1}{r}\frac{\partial u_\theta}{\partial \theta} + \frac{u_r}{r}\right)^2 + \left(\frac{\partial u_z}{\partial z}\right)^2\right] \\
&+ \left[r\frac{\partial}{\partial r}\left(\frac{u_\theta}{r}\right) + \frac{1}{r}\frac{\partial u_r}{\partial \theta}\right]^2 + \left[\frac{1}{r}\frac{\partial u_z}{\partial \theta} + \frac{\partial u_\theta}{\partial z}\right]^2 \\
&+ \left[\frac{\partial u_r}{\partial z} + \frac{\partial u_z}{\partial r}\right]^2 - \frac{2}{3}\left[\frac{1}{r}\frac{\partial}{\partial r}(ru_r) + \frac{1}{r}\frac{\partial u_\theta}{\partial \theta} + \frac{\partial u_z}{\partial z}\right]^2
\end{aligned} \tag{S}$$

[a]Adapted from Ref. 1. Operations involving the tensor $\boldsymbol{\tau}$ are given for symmetrical $\boldsymbol{\tau}$ only.

TABLE 3.11 Summary of Differential Operations Involving the ∇-Operator in Spherical Coordinates (r, θ, ϕ)[a]

$$(\nabla \cdot \mathbf{v}) = \frac{1}{r^2}\frac{\partial}{\partial r}(r^2 u_r) + \frac{1}{r\sin\theta}\frac{\partial}{\partial \theta}(u_\theta \sin\theta) + \frac{1}{r\sin\theta}\frac{\partial u_\phi}{\partial \phi} \tag{A}$$

$$\nabla^2 s = \frac{1}{r^2}\frac{\partial}{\partial r}\left(r^2 \frac{\partial s}{\partial r}\right) + \frac{1}{r^2\sin\theta}\frac{\partial}{\partial \theta}\left(\sin\theta \frac{\partial s}{\partial \theta}\right) + \frac{1}{r^2\sin^2\theta}\frac{\partial^2 s}{\partial \phi^2} \tag{B}$$

$$\begin{aligned}
\boldsymbol{\tau} : \nabla \mathbf{v} = {} & \tau_{rr}\left(\frac{\partial u_r}{\partial r}\right) + \tau_{\theta\theta}\left(\frac{1}{r}\frac{\partial u_\theta}{\partial \theta} + \frac{u_r}{r}\right) + \tau_{\phi\phi}\left(\frac{1}{r\sin\theta}\frac{\partial u_\phi}{\partial \phi} + \frac{u_r}{r} + \frac{u_\theta \cot\theta}{r}\right) \\
& + \tau_{r\theta}\left(\frac{\partial u_\theta}{\partial r} + \frac{1}{r}\frac{\partial u_r}{\partial \theta} - \frac{u_\theta}{r}\right) + \tau_{r\phi}\left(\frac{\partial u_\phi}{\partial r} + \frac{1}{r\sin\theta}\frac{\partial u_r}{\partial \phi} - \frac{u_\phi}{r}\right) \\
& + \tau_{\theta\phi}\left(\frac{1}{r}\frac{\partial u_\phi}{\partial \theta} + \frac{1}{r\sin\theta}\frac{\partial u_\theta}{\partial \phi} - \frac{\cot\theta}{r} u_\phi\right)
\end{aligned} \tag{C}$$

$$[\nabla s]_r = \frac{\partial s}{\partial r} \tag{D}$$

$$[\nabla s]_\theta = \frac{1}{r}\frac{\partial s}{\partial \theta} \tag{E}$$

$$[\nabla s]_\phi = \frac{1}{r\sin\theta}\frac{\partial s}{\partial \phi} \tag{F}$$

$$[\nabla \times \mathbf{v}]_r = \frac{1}{r\sin\theta}\frac{\partial}{\partial \theta}(u_\phi \sin\theta) - \frac{1}{r\sin\theta}\frac{\partial u_\theta}{\partial \phi} \tag{G}$$

$$[\nabla \times \mathbf{v}]_\theta = \frac{1}{r\sin\theta}\frac{\partial u_r}{\partial \phi} - \frac{1}{r}\frac{\partial}{\partial r}(r u_\phi) \tag{H}$$

$$[\nabla \times \mathbf{v}]_\phi = \frac{1}{r}\frac{\partial}{\partial r}(r u_\theta) - \frac{1}{r}\frac{\partial u_r}{\partial \theta} \tag{I}$$

$$[\nabla \cdot \boldsymbol{\tau}]_r = \frac{1}{r^2}\frac{\partial}{\partial r}(r^2 \tau_{rr}) + \frac{1}{r\sin\theta}\frac{\partial}{\partial \theta}(\tau_{r\theta}\sin\theta) + \frac{1}{r\sin\theta}\frac{\partial \tau_{r\phi}}{\partial \phi} - \frac{\tau_{\theta\theta} + \tau_{\phi\phi}}{r} \tag{J}$$

$$\begin{aligned}
[\nabla \cdot \boldsymbol{\tau}]_\theta = {} & \frac{1}{r^2}\frac{\partial}{\partial r}(r^2 \tau_{r\theta}) + \frac{1}{r\sin\theta}\frac{\partial}{\partial \theta}(\tau_{\theta\theta}\sin\theta) + \frac{1}{r\sin\theta}\frac{\partial \tau_{\theta\phi}}{\partial \phi} \\
& + \frac{\tau_{r\theta}}{r} - \frac{\cot\theta}{r}\tau_{\phi\phi}
\end{aligned} \tag{K}$$

$$[\nabla \cdot \boldsymbol{\tau}]_\phi = \frac{1}{r^2}\frac{\partial}{\partial r}(r^2 \tau_{r\phi}) + \frac{1}{r}\frac{\partial \tau_{\theta\phi}}{\partial \theta} + \frac{1}{r\sin\theta}\frac{\partial \tau_{\phi\phi}}{\partial \phi} + \frac{\tau_{r\phi}}{r} + \frac{2\cot\theta}{r}\tau_{\theta\phi} \tag{L}$$

$$[\nabla^2 \mathbf{v}]_r = \nabla^2 u_r - \frac{2u_r}{r^2} - \frac{2}{r^2}\frac{\partial u_\theta}{\partial \theta} - \frac{2u_\theta \cot\theta}{r^2} - \frac{2}{r^2\sin\theta}\frac{\partial u_\phi}{\partial \phi} \tag{M}$$

$$[\nabla^2 \mathbf{v}]_\theta = \nabla^2 u_\theta + \frac{2}{r^2}\frac{\partial u_r}{\partial \theta} - \frac{u_\theta}{r^2\sin^2\theta} - \frac{2\cos\theta}{r^2\sin^2\theta}\frac{\partial u_\phi}{\partial \phi} \tag{N}$$

$$[\nabla^2 \mathbf{v}]_\phi = \nabla^2 u_\phi - \frac{u_\phi}{r^2\sin^2\theta} + \frac{2}{r^2\sin\theta}\frac{\partial u_r}{\partial \phi} + \frac{2\cos\theta}{r^2\sin^2\theta}\frac{\partial u_\theta}{\partial \phi} \tag{O}$$

TABLE 3.11 (Continued).

$$[\mathbf{v}\cdot\nabla\mathbf{v}]_r = u_r\frac{\partial u_r}{\partial r} + \frac{u_\theta}{r}\frac{\partial u_r}{\partial\theta} + \frac{u_\phi}{r\sin\theta}\frac{\partial u_r}{\partial\phi} - \frac{u_\theta^2 + u_\phi^2}{r} \qquad \text{(P)}$$

$$[\mathbf{v}\cdot\nabla\mathbf{v}]_\theta = u_r\frac{\partial u_\theta}{\partial r} + \frac{u_\theta}{r}\frac{\partial u_\theta}{\partial\theta} + \frac{u_\phi}{r\sin\theta}\frac{\partial u_\theta}{\partial\phi} + \frac{u_r u_\theta}{r} - \frac{u_\phi^2\cot\theta}{r} \qquad \text{(Q)}$$

$$[\mathbf{v}\cdot\nabla\mathbf{v}]_\phi = u_r\frac{\partial u_\phi}{\partial r} + \frac{u_\theta}{r}\frac{\partial u_\phi}{\partial\theta} + \frac{u_\phi}{r\sin\theta}\frac{\partial u_\phi}{\partial\phi} + \frac{u_\phi u_r}{r} + \frac{u_\theta u_\phi\cot\theta}{r} \qquad \text{(R)}$$

For Newtonian fluids

$$\begin{aligned}\frac{\boldsymbol{\tau}:\nabla\boldsymbol{v}}{\mu} = {} & 2\left[\left(\frac{\partial u_r}{\partial r}\right)^2 + \left(\frac{1}{r}\frac{\partial u_\theta}{\partial\theta} + \frac{u_r}{r}\right)^2 + \left(\frac{1}{r\sin\theta}\frac{\partial u_\phi}{\partial\phi} + \frac{u_r}{r} + \frac{u_\theta\cot\theta}{r}\right)^2\right] \\ & + \left[r\frac{\partial}{\partial r}\left(\frac{u_\theta}{r}\right) + \frac{1}{r}\frac{\partial u_r}{\partial\theta}\right]^2 + \left[\frac{\sin\theta}{r}\frac{\partial}{\partial\theta}\left(\frac{u_\phi}{\sin\theta}\right) + \frac{1}{r\sin\theta}\frac{\partial u_\theta}{\partial\phi}\right]^2 \\ & + \left[\frac{1}{r\sin\theta}\frac{\partial u_r}{\partial\phi} + r\frac{\partial}{\partial r}\left(\frac{u_\phi}{r}\right)\right]^2 \\ & - \frac{2}{3}\left[\frac{1}{r^2}\frac{\partial}{\partial r}(r^2u_r) + \frac{1}{r\sin\theta}\frac{\partial}{\partial\theta}(u_\theta\sin\theta) + \frac{1}{r\sin\theta}\frac{\partial u_\phi}{\partial\phi}\right]^2 \qquad \text{(S)}\end{aligned}$$

[a]Adapted from Ref. 1. Operations involving the tensor $\boldsymbol{\tau}$ are given for symmetrical $\boldsymbol{\tau}$ only.

6 CONSERVATION OF ENERGY

Before deriving the conservation-of-energy equation for a multicomponent system, we must be aware of the various effects which contribute to the heat flux. There are two additional effects besides conduction which contribute to the heat flux in a multicomponent system. When the average velocity $\mathbf{v}_i$ of any component i differs from the mass-average velocity of the mixture, there is then a mass flux $\rho_i\mathbf{V}_i$ of component i across a surface moving with the mass-average velocity of the gas mixture.

If the average enthalpy (per unit mass) associated with the ith component is h_i, its molecules will carry across the surface an extra enthalpy per unit area per unit time equal to $h_i\rho_i\mathbf{V}_i$ or $\rho h_i Y_i\mathbf{V}_i$. The total enthalpy (of all the components) per unit area per unit time flowing relative to the mass average motion of the mixture is $\rho\Sigma_{i=1}^N h_i Y_i\mathbf{V}_i$. This energy-flux term, caused by interdiffusion processes, constitutes an additional component to q in binary and multicomponent systems.

Onsager's reciprocal relations for the thermodynamics of irreversible processes imply that if temperature gives rise to diffusion velocities (the "thermal-diffusion effect" or "Soret effect"), concentration gradients must produce a heat flux. This reciprocal effect, known as the *Dufour effect*, provides an additional contribution to the heat flux $\mathbf{q}$.

It is conventional to express the concentration gradients in terms of differences in diffusion velocities. In terms of diffusion velocities, the Dufour heat flux can be expressed as

$$R_u T \sum_{i=1}^{N} \sum_{j=1}^{N} \left(\frac{X_j \alpha_i}{W_i \mathscr{D}_{ij}} \right) (\mathbf{V}_i - \mathbf{V}_j).$$

In most cases, the Dufour effect is so small that it is negligible even when thermal diffusion is not negligible. Although it is omitted in most applications, the term is retained in the general equation for completeness.

Conduction and the above two effects yield the heat-flux vector as

$$\mathbf{q} = -\lambda \nabla T + \rho \sum_{i=1}^{N} h_i Y_i \mathbf{V}_i + R_u T \sum_{i=1}^{N} \sum_{j=1}^{N} \left(\frac{X_j \alpha_i}{W_i \mathscr{D}_{ij}} \right) (\mathbf{V}_i - \mathbf{V}_j) \tag{3-61}$$

We can now proceed to derive the conservation of energy equation. The stored energy per unit mass e_t is defined as

$$e_t = e + \frac{u_i u_i}{2} \tag{3-62}$$

where e, the specific internal energy, is defined as

$$e = h - \frac{p}{\rho} = \sum_{i=1}^{N} h_i Y_i - \frac{p}{\rho} \tag{3-63}$$

where

$$h_i = \Delta h_{f,i}^{\circ} + \int_{T^{\circ}}^{T} C_{p,i}\, dT \tag{3-64}$$

Beginning with a stationary differential volume element through which a pure fluid is flowing, we then write the law of conservation of energy for the

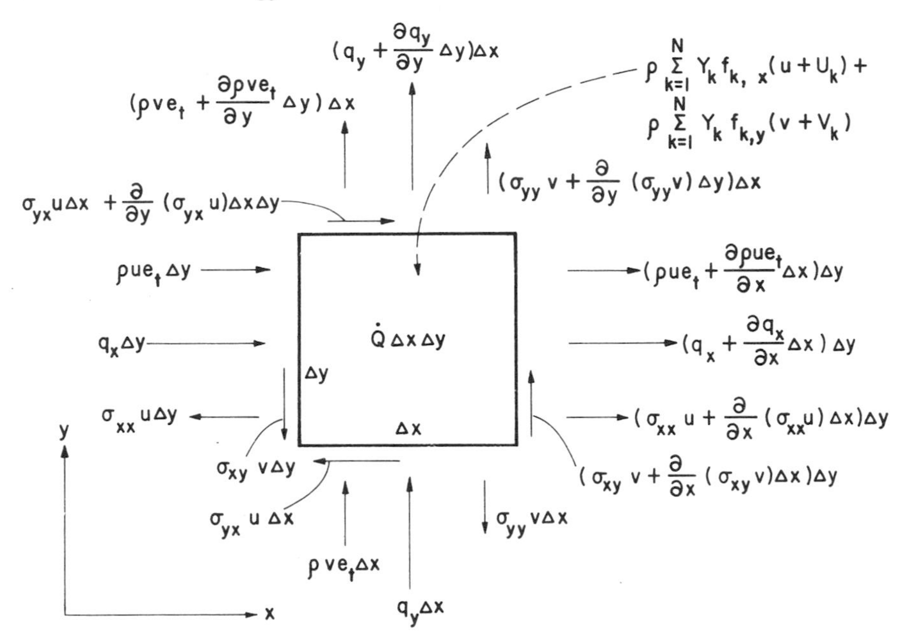

Figure 3.11 Terms in the energy-flux balance for a two-dimensional flow.

fluid contained within the volume element of any given time:

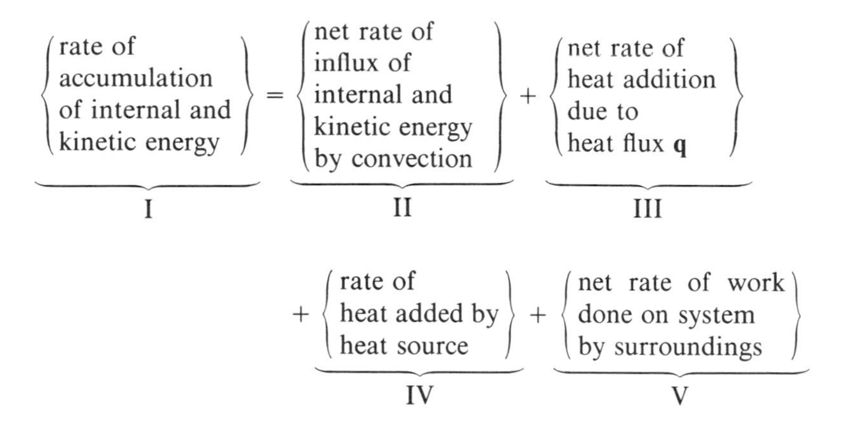

$$\underbrace{\begin{Bmatrix}\text{rate of}\\ \text{accumulation}\\ \text{of internal and}\\ \text{kinetic energy}\end{Bmatrix}}_{\text{I}} = \underbrace{\begin{Bmatrix}\text{net rate of}\\ \text{influx of}\\ \text{internal and}\\ \text{kinetic energy}\\ \text{by convection}\end{Bmatrix}}_{\text{II}} + \underbrace{\begin{Bmatrix}\text{net rate of}\\ \text{heat addition}\\ \text{due to}\\ \text{heat flux } \mathbf{q}\end{Bmatrix}}_{\text{III}}$$

$$+ \underbrace{\begin{Bmatrix}\text{rate of}\\ \text{heat added by}\\ \text{heat source}\end{Bmatrix}}_{\text{IV}} + \underbrace{\begin{Bmatrix}\text{net rate of work}\\ \text{done on system}\\ \text{by surroundings}\end{Bmatrix}}_{\text{V}}$$

The rate-of-work term V consists of two parts. One is due to the total surface stress tensor acting on the boundary of the control volume (such as $\sigma_{xx}u\,\Delta y \cdot 1$, $\sigma_{xy}v\,\Delta y \cdot 1$, $\sigma_{yx}u\,\Delta x \cdot 1$ shown in Fig. 3.11); the other is due to the body force. The body force acting on the kth species in the x-direction is $\rho Y_k f_{k,x}$. The rate of work done on the fluid due to this force component is the product of this force component with the x-component of the average velocity of the kth species, that is, $\rho Y_k f_{k,x} u_k$, or $\rho Y_k f_{k,x}(u + U_k)$. Summing over all species in the mixture, we have $\rho\Sigma_{k=1}^{N} Y_k f_{k,x}(u + U_k)$ as the total contribution of the body force in the x-direction.

The reason for considering a two-dimensional reacting flow in Fig. 3.11 is to be able to fit all the terms around the infinitesimal control volume $\Delta x\, \Delta y \cdot 1$ in the same diagram. After obtaining the energy-balance equation, we shall then generalize it to a three-dimensional space. Based upon the energy-balance statement, the energy equation made of the corresponding terms I–V for a two-dimensional unsteady reacting flow is shown below:

$$\underbrace{\frac{\partial}{\partial t}(\rho e_t)}_{\text{I}} = \underbrace{-\frac{\partial}{\partial x}(\rho u e_t) - \frac{\partial}{\partial y}(\rho v e_t)}_{\text{II}} \underbrace{- \frac{\partial q_x}{\partial x} - \frac{\partial q_y}{\partial y}}_{\text{III}} + \underbrace{\dot{Q}}_{\text{IV}}$$

$$\underbrace{+\rho \sum_{k=1}^{N} Y_k f_{k,x}(u + U_k) + \rho \sum_{k=1}^{N} Y_k f_{k,y}(v + V_k) + \frac{\partial \sigma_{xx} u}{\partial x} + \frac{\partial \sigma_{yx} u}{\partial y} + \frac{\partial \sigma_{yy} v}{\partial y} + \frac{\partial \sigma_{xy} v}{\partial x}}_{\text{V}} \tag{3-65}$$

where U_k, V_k are relative mass-diffusion velocity components.

To generalize Eq. (3-65) to a three-dimensional space, we have

$$\begin{aligned}
\frac{\partial}{\partial t}(\rho e_t) &+ \frac{\partial}{\partial x}(\rho u e_t) + \frac{\partial}{\partial y}(\rho v e_t) + \frac{\partial}{\partial z}(\rho w e_t) \\
&= -\frac{\partial q_x}{\partial x} - \frac{\partial q_y}{\partial y} - \frac{\partial q_z}{\partial z} + \dot{Q} + \frac{\partial \sigma_{xx} u}{\partial x} + \frac{\partial \sigma_{yx} u}{\partial y} + \frac{\partial \sigma_{zx} u}{\partial z} \\
&\quad + \frac{\partial \sigma_{xy} v}{\partial x} + \frac{\partial \sigma_{yy} v}{\partial y} + \frac{\partial \sigma_{zy} v}{\partial z} + \frac{\partial \sigma_{xz} w}{\partial x} + \frac{\partial \sigma_{yz} w}{\partial y} + \frac{\partial \sigma_{zz} w}{\partial z} \\
&\quad + \rho \sum_{k=1}^{N} Y_k f_{k,x}(u + U_k) + \rho \sum_{k=1}^{N} Y_k f_{k,y}(v + V_k) \\
&\quad + \rho \sum_{k=1}^{N} Y_k f_{k,z}(w + W_k)
\end{aligned} \tag{3-66}$$

Writing the equation in vector–tensor notation, it becomes

$$\frac{\partial}{\partial t}(\rho e_t) + \frac{\partial}{\partial x_i}(\rho e_t u_i) = -\frac{\partial q_i}{\partial x_i} + \dot{Q} + \frac{\partial \sigma_{ji} u_i}{\partial x_j} + \rho \sum_{k=1}^{N} Y_k f_{k,i}(u_i + V_{k,i}) \tag{3-67}$$

This is called the divergence form of the energy equation. The left-hand side of Eq. (3-67) may be written as

$$\rho\frac{\partial e_t}{\partial t} + e_t\frac{\partial \rho}{\partial t} + e_t\frac{\partial \rho u_i}{\partial x_i} + \rho u_i\frac{\partial e_t}{\partial x_i}$$

From continuity,

$$e_t\left(\frac{\partial \rho}{\partial t} + \frac{\partial \rho u_i}{\partial x_i}\right) = 0$$

Thus, Eq. (3-67) simplifies to

$$\boxed{\begin{aligned}\rho\frac{\partial e_t}{\partial t} + \rho u_i\frac{\partial e_t}{\partial x_i} = -\frac{\partial q_i}{\partial x_i} + \dot{Q} + \frac{\partial \sigma_{ji}u_i}{\partial x_j}\\ +\rho\sum_{k=1}^{N} Y_k f_{k,i}(u_i + V_{k,i})\end{aligned}} \tag{3-68}$$

This is called the Euler form of the energy equation.

The momentum equation (3-48) is

$$\rho\frac{\partial u_i}{\partial t} + \rho u_j\frac{\partial u_i}{\partial x_j} = \frac{\partial \sigma_{ji}}{\partial x_j} + \rho\sum_{k=1}^{N} Y_k f_{k,i}$$

Multiplying it by u_i, one obtains

$$\boxed{\rho\frac{\partial\left(\frac{1}{2}u_i u_i\right)}{\partial t} + \rho u_j\frac{\partial\left(\frac{1}{2}u_i u_i\right)}{\partial x_j} = u_i\frac{\partial \sigma_{ji}}{\partial x_j} + \rho\sum_{k=1}^{N} Y_k f_{k,i}u_i} \tag{3-69}$$

The above equation is the mechanical-energy equation.

Subtracting Eq. (3-69) from Eq. (3-68), we get

$$\boxed{\rho\frac{De}{Dt} = -\frac{\partial q_i}{\partial x_i} + \sigma_{ji}\frac{\partial u_i}{\partial x_j} + \dot{Q} + \rho\sum_{k=1}^{N} Y_k f_{k,i}V_{k,i}} \tag{3-70}$$

The above equation can be rewritten in vector form as

$$\rho\frac{\partial e}{\partial t} + \rho\mathbf{v}\cdot\nabla e = -\nabla\cdot\mathbf{q} + \tilde{\boldsymbol{\sigma}}:\nabla\mathbf{v} + \dot{Q} + \rho\sum_{k=1}^{N} Y_k\mathbf{f}_k\cdot\mathbf{V}_k \tag{3-71}$$

Since

$$\frac{\partial u_i}{\partial x_j} = \frac{1}{2}\left(\frac{\partial u_i}{\partial x_j} + \frac{\partial u_j}{\partial x_i}\right) + \frac{1}{2}\left(\frac{\partial u_i}{\partial x_j} - \frac{\partial u_j}{\partial x_i}\right)$$

$$= e_{ij} + \hat{\omega}_{ij} \tag{3-72}$$

it follows that

e_{ij}, the strain-rate tensor, which is symmetric

$\hat{\omega}_{ij}$, the angular-velocity tensor, which is antisymmetric

$\sigma_{ji}\hat{\omega}_{ij} =$ (symmetric tensor) $\times$ (antisymmetric tensor) $= 0$

$$\sigma_{ji}\frac{\partial u_i}{\partial x_j} = \sigma_{ji}(e_{ij} + \hat{\omega}_{ij}) = \sigma_{ij}e_{ij}$$

$$= \frac{1}{2}(-p - \tfrac{2}{3}\mu\frac{\partial u_k}{\partial x_k})\delta_{ij}(\frac{\partial u_i}{\partial x_j} + \frac{\partial u_j}{\partial x_i}) + 2\mu e_{ij}e_{ij} \tag{3-73}$$

[In the above operations, we have neglected the bulk-viscosity (μ') term.] Thus Eq. (3-71) reduces to

$$\underbrace{\rho\frac{De}{Dt}}_{\substack{\text{increase of}\\ \text{internal}\\ \text{energy}}} = -\underbrace{\frac{\partial q_i}{\partial x_i}}_{\substack{\text{heat influx}\\ \text{due to}\\ \text{conduction,}\\ \text{diffusion,}\\ \text{and Dufour effect}}} - \underbrace{p\frac{\partial u_j}{\partial x_j}}_{\substack{\text{flow}\\ \text{work}}} \underbrace{- \tfrac{2}{3}\mu\left(\frac{\partial u_j}{\partial x_j}\right)^2 + 2\mu e_{ij}e_{ij}}_{\substack{\Phi\text{, dissipation by}\\ \text{viscous stress}}}$$

$$+ \underbrace{\dot{Q}}_{\substack{\text{heat}\\ \text{input}}} + \underbrace{\rho\sum_{k=1}^{N} Y_k f_{k,i}V_{k,i}}_{\substack{\text{body-force}\\ \text{work}}} \tag{3-74}$$

$$\boxed{\rho\frac{De}{Dt} + p\frac{\partial u_j}{\partial x_j} = -\nabla\cdot\mathbf{q} + \Phi + \dot{Q} + \rho\sum_{k=1}^{N} Y_k\mathbf{f}_k\cdot\mathbf{V}_k} \tag{3-75}$$

$$h = e + \frac{p}{\rho}$$

$$\frac{De}{Dt} = \frac{Dh}{Dt} - \frac{1}{\rho}\frac{Dp}{Dt} + \frac{p}{\rho^2}\frac{D\rho}{Dt} \tag{3-76}$$

From continuity

$$\frac{1}{\rho}\frac{D\rho}{Dt} = -\frac{\partial u_j}{\partial x_j} \tag{3-77}$$

Using Eqs. (3-76) and (3-77), Eq. (3-75) becomes

$$\boxed{\rho\frac{Dh}{Dt} - \frac{Dp}{Dt} = -\nabla\cdot\mathbf{q} + \Phi + \dot{Q} + \rho\sum_{k=1}^{N} Y_k \mathbf{f}_k \cdot \mathbf{V}_k} \tag{3-78}$$

Adding the mechanical-energy equation (3-69) to the above equation, we obtain

$$\rho\frac{D}{Dt}\left(h + \frac{u_i u_i}{2}\right) - \frac{Dp}{Dt} = u_i\frac{\partial \sigma_{ji}}{\partial x_j} + \dot{Q} - \nabla\cdot\mathbf{q} + \rho\sum_{k=1}^{N} Y_k \mathbf{f}_k \cdot (\mathbf{u} + \mathbf{V}_k) + \Phi \tag{3-79}$$

Using Eq. (3-57) and

$$h_t = h + \frac{u_i u_i}{2} \tag{3-80}$$

we have

$$\rho\frac{Dh_t}{Dt} - \frac{\partial p}{\partial t} - u_i\frac{\partial p}{\partial x_i} = -u_i\frac{\partial p}{\partial x_j}\delta_{ij} + u_i\frac{\partial \tau_{ji}}{\partial x_j} + \Phi$$

$$+\dot{Q} - \nabla\cdot\mathbf{q} + \rho\sum_{k=1}^{N} Y_k \mathbf{f}_k \cdot (\mathbf{u} + \mathbf{V}_k) \tag{3-81}$$

where

$$\tau_{ij} = -\tfrac{2}{3}\mu\frac{\partial u_k}{\partial x_k}\delta_{ij} + \mu\left(\frac{\partial u_i}{\partial x_j} + \frac{\partial u_j}{\partial x_i}\right) = \sigma_{ij} + p\delta_{ij}$$

Now,

$$u_i\frac{\partial p}{\partial x_j}\delta_{ij} = u_i\frac{\partial p}{\partial x_i}$$

BLE 3.12 Energy Equation[a] in Several Coordinate Systems

Rectangular coordinates (x, y, z):

$$\rho C_p\left(\frac{\partial T}{\partial t} + u_x\frac{\partial T}{\partial x} + u_y\frac{\partial T}{\partial y} + u_z\frac{\partial T}{\partial z}\right) - \left(\frac{\partial p}{\partial t} + u_x\frac{\partial p}{\partial x} + u_y\frac{\partial p}{\partial y} + u_z\frac{\partial p}{\partial z}\right)$$

$$= \lambda\left(\frac{\partial^2 T}{\partial x^2} + \frac{\partial^2 T}{\partial y^2} + \frac{\partial^2 T}{\partial z^2}\right) - \sum_{i=1}^{N} \omega_i \Delta h_{f,i}^{\circ}$$

$$- \left[\frac{\partial}{\partial x}\left(\rho T \sum_{i=1}^{N} C_{pi} Y_i V_{ix}\right) + \frac{\partial}{\partial y}\left(\rho T \sum_{i=1}^{N} C_{pi} Y_i V_{iy}\right) + \frac{\partial}{\partial z}\left(\rho T \sum_{i=1}^{N} C_{pi} Y_i V_{iz}\right)\right]$$

$$+ \mu\left\{2\left[\left(\frac{\partial u_x}{\partial x}\right)^2 + \left(\frac{\partial u_y}{\partial y}\right)^2 + \left(\frac{\partial u_z}{\partial z}\right)^2\right] + \left[\frac{\partial u_y}{\partial x} + \frac{\partial u_x}{\partial y}\right]^2\right.$$

$$\left. + \left[\frac{\partial u_z}{\partial y} + \frac{\partial u_y}{\partial z}\right]^2 + \left[\frac{\partial u_x}{\partial z} + \frac{\partial u_z}{\partial x}\right]^2 - \frac{2}{3}\left[\frac{\partial u_x}{\partial x} + \frac{\partial u_y}{\partial y} + \frac{\partial u_z}{\partial z}\right]^2\right\}$$

$$+ \rho \sum_{i=1}^{N} Y_i (f_{ix} V_{ix} + f_{iy} V_{iy} + f_{iz} V_{iz}) \tag{A}$$

Cylindrical coordinates (r, θ, z):

$$\rho C_p\left(\frac{\partial T}{\partial t} + u_r\frac{\partial T}{\partial r} + \frac{u_\theta}{r}\frac{\partial T}{\partial \theta} + u_z\frac{\partial T}{\partial z}\right) - \left(\frac{\partial p}{\partial t} + u_r\frac{\partial p}{\partial r} + \frac{u_\theta}{r}\frac{\partial p}{\partial \theta} + u_z\frac{\partial p}{\partial z}\right)$$

$$= \lambda\left[\frac{1}{r}\frac{\partial}{\partial r}\left(r\frac{\partial T}{\partial r}\right) + \frac{1}{r^2}\frac{\partial^2 T}{\partial \theta^2} + \frac{\partial^2 T}{\partial z^2}\right] - \sum_{i=1}^{N} \omega_i \Delta h_{f,i}^{\circ}$$

$$- \left[\frac{1}{r}\frac{\partial}{\partial r}\left(r\rho T \sum_{i=1}^{N} C_{pi} Y_i V_{ir}\right) + \frac{1}{r}\frac{\partial}{\partial \theta}\left(\rho T \sum_{i=1}^{N} C_{pi} Y_i V_{i\theta}\right) + \frac{\partial}{\partial z}\left(\rho T \sum_{i=1}^{N} C_{pi} Y_i V_{iz}\right)\right]$$

$$+ \mu\left\{2\left[\left(\frac{\partial u_r}{\partial r}\right)^2 + \left(\frac{1}{r}\frac{\partial u_\theta}{\partial \theta} + \frac{u_r}{r}\right)^2 + \left(\frac{\partial u_z}{\partial z}\right)^2\right] + \left[r\frac{\partial}{\partial r}\left(\frac{u_\theta}{r}\right) + \frac{1}{r}\frac{\partial u_r}{\partial \theta}\right]^2\right.$$

$$+ \left[\frac{1}{r}\frac{\partial u_z}{\partial \theta} + \frac{\partial u_\theta}{\partial z}\right]^2 + \left[\frac{\partial u_r}{\partial z} + \frac{\partial u_z}{\partial r}\right]^2$$

$$\left. - \frac{2}{3}\left[\frac{1}{r}\frac{\partial}{\partial r}(r u_r) + \frac{1}{r}\frac{\partial u_\theta}{\partial \theta} + \frac{\partial u_z}{\partial z}\right]^2\right\}$$

$$+ \rho \sum_{i=1}^{N} Y_i (f_{ir} V_{ir} + f_{i\theta} V_{i\theta} + f_{iz} V_{iz}) \tag{B}$$

TABLE 3.12 (Continued).

Spherical coordinates (r, θ, ϕ):

$$\rho C_p\left(\frac{\partial T}{\partial t} + u_r\frac{\partial T}{\partial r} + \frac{u_\theta}{r}\frac{\partial T}{\partial \theta} + \frac{u_\phi}{r\sin\theta}\frac{\partial T}{\partial \phi}\right) - \left(\frac{\partial p}{\partial t} + u_r\frac{\partial p}{\partial r} + \frac{u_\theta \partial p}{r\partial \theta} + \frac{u_\phi}{r\sin\theta}\frac{\partial p}{\partial \phi}\right)$$

$$= \lambda\left[\frac{1}{r^2}\frac{\partial}{\partial r}\left(r^2\frac{\partial T}{\partial r}\right) + \frac{1}{r^2\sin\theta}\frac{\partial}{\partial \theta}\left(\sin\theta\frac{\partial T}{\partial \theta}\right) + \frac{1}{r^2\sin^2\theta}\frac{\partial^2 T}{\partial \phi^2}\right] - \sum_{i=1}^{N}\omega_i \Delta h^\circ_{f,i}$$

$$-\left[\frac{1}{r^2}\frac{\partial}{\partial r}\left(r^2\rho T\sum_{i=1}^{N} C_{pi}Y_iV_{ir}\right) + \frac{1}{r\sin\theta}\frac{\partial}{\partial\theta}\left(\rho T\sin\theta\sum_{i=1}^{N} C_{pi}Y_iV_{i\theta}\right)\right.$$

$$\left.+\frac{1}{r\sin\theta}\frac{\partial}{\partial\phi}\left(\rho T\sum_{i=1}^{N} C_{pi}Y_iV_{i\phi}\right)\right]$$

$$+\mu\left\{2\left[\left(\frac{\partial u_r}{\partial r}\right)^2 + \left(\frac{1}{r}\frac{\partial u_\theta}{\partial\theta} + \frac{u_r}{r}\right)^2 + \left(\frac{1}{r\sin\theta}\frac{\partial u_\phi}{\partial\phi} + \frac{u_r}{r} + \frac{u_\theta\cot\theta}{r}\right)^2\right]\right.$$

$$+\left[r\frac{\partial}{\partial r}\left(\frac{u_\theta}{r}\right) + \frac{1}{r}\frac{\partial u_r}{\partial\theta}\right]^2 + \left[\frac{\sin\theta}{r}\frac{\partial}{\partial\theta}\left(\frac{u_\phi}{\sin\theta}\right) + \frac{1}{r\sin\theta}\frac{\partial u_\theta}{\partial\phi}\right]^2$$

$$+\left[\frac{1}{r\sin\theta}\frac{\partial u_r}{\partial\phi} + r\frac{\partial}{\partial r}\left(\frac{u_\phi}{r}\right)\right]^2$$

$$\left.-\frac{2}{3}\left[\frac{1}{r^2}\frac{\partial}{\partial r}(r^2u_r) + \frac{1}{r\sin\theta}\frac{\partial}{\partial\theta}(u_\theta\sin\theta) + \frac{1}{r\sin\theta}\frac{\partial u_\phi}{\partial\phi}\right]^2\right\}$$

$$+\rho\sum_{i=1}^{N} Y_i(f_{ir}V_{ir} + f_{i\theta}V_{i\theta} + f_{i\phi}V_{i\phi}) \qquad \text{(C)}$$

[a] Based upon the following assumptions: (1) all species in the mixture follow the perfect-gas law, (2) constant C_p and λ, (3) Dufour effect negligible, (4) no radiation effect.

and from Eq. (3-73) and from the definition of τ_{ij} it follows that

$$\Phi = \tau_{ij}\frac{\partial u_i}{\partial x_j} \tag{3-82}$$

Substituting in Eq. (3-81), it follows that

$$\boxed{\rho\frac{Dh_t}{Dt} - \frac{\partial p}{\partial t} = \frac{\partial(u_i\tau_{ji})}{\partial x_j} + \dot{Q} - \nabla\cdot\mathbf{q} + \rho\sum_{k=1}^{N} Y_k\mathbf{f}_k\cdot(\mathbf{u} + \mathbf{V}_k)} \tag{3-83}$$

From the above equation it follows that if (1) flow is inviscid and steady, (2) there is no heat flux due to conduction, diffusion, or Dufour effect, (3) there is no body force, and (4) there is no heat input, then h_t will be constant.

The energy equation written in terms of temperature in several coordinate systems is given in Table 3.12.

7 PHYSICAL DERIVATION OF THE MULTICOMPONENT DIFFUSION EQUATION

In order to render the conservation equations soluble for a multicomponent system, the diffusion velocities $\mathbf{V}_i$ appearing in these equations must be determined. In general, there are two ways to evaluate the diffusion velocities. The approximate way is to employ Fick's law of mass diffusion and to replace the diffusion velocities by the concentration gradient times some appropriate coefficient [see Eqs. (3-5) to (3-8) and Eqs. (3-11) to (3-14)]. The more exact method is to find $\mathbf{V}_i$ from the multicomponent diffusion equation, which is derived in the following through physical arguments.

The binary-collision problem has been studied extensively in classical mechanics.[4] The three-dimensional dynamical problem of binary collision between two particles with masses m_i and m_j is found to be mathematically equivalent to a one-body problem in a plane, with the reduced mass

$$\mu_{ij} \equiv \frac{m_i m_j}{m_i + m_j}.$$

The average momentum transferred from a molecule of type i to a molecule of type j must be equal to the negative of that transferred from j to i molecules. It is expected that on the average, the momentum transferred to an i-molecule in a collision between i and j molecules is approximately

$$\mu_{ij}(\mathbf{V}_j - \mathbf{V}_i)$$

where μ_{ij} is the reduced mass and $\mathbf{V}_j - \mathbf{V}_i$ is the average relative velocity between molecules of species i and j. The body force acting on molecules of type i in a unit volume element is $\rho Y_i \mathbf{f}_i$, or $\rho Y_i \mathbf{f}_i \Sigma_{j=1}^{N} Y_j$, since $\Sigma_{i=1}^{N} Y_j = 1$. Hence, the net rate of change of momentum of molecules of type i per unit volume is

$$\boldsymbol{\Gamma}_i = \sum_{j=1}^{N} \mu_{ij} Z_{ij}(\mathbf{V}_j - \mathbf{V}_i) + \sum_{j=1}^{N} \rho Y_i Y_j \mathbf{f}_i \qquad \text{for} \quad i = 1, 2, \ldots, N \quad (3\text{-}84)$$

where Z_{ij} is the total number of collisions per unit volume per second between the molecules of types i and j.

This change of momentum of species i is evident in changes in both the random velocity and ordered velocity of the molecules. The rate of change of the ordered momentum is $\rho Y_i \, D\mathbf{v}/Dt$. The partial pressure p_i of species i represents physically the momentum of molecules of type i transported per second across a surface of unit area, traveling with the mass-average velocity of the fluid. Thus, ∇p_i is the rate of change of momentum of random motion of molecules of type i per unit volume. The quantity $\boldsymbol{\Gamma}_i$ is therefore also given by the relation

$$\Gamma_i = \rho Y_i \frac{D\mathbf{v}}{Dt} + \nabla p_i, \qquad i = 1, 2, \ldots, N \tag{3-85}$$

According to Dalton's law of partial pressures,

$$p_i = X_i p, \qquad i = 1, 2, \ldots, N$$

Hence

$$\nabla p_i = p \nabla X_i + X_i \nabla p, \qquad i = 1, 2, \ldots, N \tag{3-86}$$

Neglecting the viscous forces, the momentum equation can be used to show that

$$\rho \frac{D\mathbf{v}}{Dt} = -\nabla p + \rho \sum_{j=1}^{N} Y_j \mathbf{f}_j \tag{3-87}$$

Substituting Eqs. (3-85), (3-86), and (3-87) into Eq. (3-84) yields

$$\nabla X_i = \sum_{j=1}^{N} \frac{\mu_{ij} Z_{ij}}{p} (\mathbf{V}_j - \mathbf{V}_i) + (Y_i - X_i) \frac{\nabla p}{p} + \frac{\rho}{p} \sum_{j=1}^{N} Y_i Y_j (\mathbf{f}_i - \mathbf{f}_j) \tag{3-88}$$

The product $\mu_{ij} Z_{ij}$ may be related to the binary diffusion coefficient by considering the limiting case of a constant-pressure process in a two-component system with no body forces, for which Eq. (3-88) reduces to

$$\nabla X_A = \frac{\mu_{AB} Z_{AB}}{p} (\mathbf{V}_B - \mathbf{V}_A) \tag{3-89}$$

Using Eq. (3-9) one can show that the relative molar flux J_A^* can be expressed in terms of the mole fractions of species A and B, and the velocity-vector difference between B and A as

$$\mathbf{J}_A^* = -X_A X_B C (\mathbf{V}_B - \mathbf{V}_A) \tag{3-90}$$

Substituting $\mathbf{V}_B - \mathbf{V}_A$ of Eq. (3-90) into Eq. (3-89) and comparing it with Eq. (3-11), we have $\mathscr{D}_{AB} = (X_A X_B p)/(\mu_{AB} Z_{AB})$, which implies that, in general,

$$\mathscr{D}_{ij} = \frac{X_i X_j p}{\mu_{ij} Z_{ij}} \tag{3-91}$$

Utilizing Eq. (3-91), Eq. (3-88) becomes

$$\nabla X_i = \sum_{j=1}^{N} \frac{X_i X_j}{\mathscr{D}_{ij}}(\mathbf{V}_j - \mathbf{V}_i) + (Y_i - X_i)\frac{\nabla p}{p} + \frac{\rho}{p}\sum_{j=1}^{N} Y_i Y_j(\mathbf{f}_i - \mathbf{f}_j), \qquad i=1,2,\ldots,N \tag{3-88a}$$

The multicomponent diffusion equation obtained through more rigorous derivation from the kinetic theory is

$$\boxed{\begin{aligned}\nabla X_i = {} & \sum_{j=1}^{N} \frac{X_i X_j}{\mathscr{D}_{ij}}(\mathbf{V}_j - \mathbf{V}_i) + (Y_i - X_i)\frac{\nabla p}{p} \\ & + \frac{\rho}{p}\sum_{j=1}^{N} Y_i Y_j(\mathbf{f}_i - \mathbf{f}_j) + \sum_{j=1}^{N} \frac{X_i X_j}{\rho\mathscr{D}_{ij}}\left(\frac{\alpha_j}{Y_j} - \frac{\alpha_i}{Y_i}\right)\frac{\nabla T}{T}, \\ & i = 1,2,\ldots,N\end{aligned}} \tag{3-88b}$$

where α_j is the thermal diffusion coefficient of species j. It may be noted that the last summation of the above equation was absent in Eq. (3-88a), since we did not consider the thermal-diffusion (Soret) effect[4] in the derivation. Opposite to the Dufour effect, the Soret effect describes the mass diffusion resulting from temperature gradients. Physically, Eq. (3-88b) states that concentration gradients may be supported by diffusion velocities, pressure gradients, the differences in the body force per unit mass on molecules of different species, and thermal-diffusion effects.

The diffusion equation reduces to Fick's law of diffusion for binary mixtures when (a) thermal diffusion is negligible, (b) the body force per unit mass is the same for each species ($\mathbf{f}_A = \mathbf{f}_B$), and (c) either the pressure is a constant or the molecular weights of both species are the same.

For a multicomponent system which assumes that thermal diffusion, body forces, and pressure-induced diffusions are negligible, the diffusion equation simplifies to

$$\nabla X_i = \sum_{j=1}^{N} \frac{X_i X_j}{\mathscr{D}_{ij}}(\mathbf{V}_j - \mathbf{V}_i) \tag{3-92}$$

By using Eqs. (3-3) and (3-6), it can be shown easily that the above equation can be written as

$$\nabla X_i = \sum_{j=1}^{N} \frac{1}{C\mathscr{D}_{ij}}(X_i \dot{\mathbf{n}}_j - X_j \dot{\mathbf{n}}_i) \tag{3-93}$$

In performing calculations for multicomponent systems, it is often convenient to define an effective binary diffusion $\mathscr{D}_{im}$ for the diffusion of species i in a mixture. Equation (3-12) then becomes

$$\dot{\mathbf{n}}_i = C_i \mathbf{v}^* - C\mathscr{D}_{im} \nabla X_i \tag{3-94}$$

This equation, simpler than Eq. (3-93), also relates $\dot{\mathbf{n}}_i$ to ∇X_i.

8 OTHER NECESSARY EQUATIONS IN MULTICOMPONENT SYSTEMS

The ω_i in each species continuity equation is determined by the phenomenological chemical kinetic expression

$$\omega_i = W_i \sum_{k=1}^{M} (\nu''_{i,k} - \nu'_{i,k}) B_k T^{\alpha_k} \, exp - \frac{E_{ak}}{R_u T} \prod_{j=1}^{N} \left(\frac{X_j p}{R_u T} \right)^{\nu'_{j,k}} \tag{3-95}$$

where M is the total number of chemical reactions occurring and N is the total number of chemical species present.

The ideal-gas equation states that

$$p = \rho R_u T \sum_{i=1}^{N} \frac{Y_i}{W_i} \tag{3-96}$$

The relationship between X_i and Y_i is

$$X_i = \frac{Y_i / W_i}{\sum_{j=1}^{N} (Y_j / W_j)}, \qquad i = 1, 2, \ldots, N \tag{3-97}$$

9 SOLUTION OF A MULTICOMPONENT-SPECIES SYSTEM

If $\mathbf{V}_i$ can be substituted by Fick's law in the species and energy equations, then in a system with N species there are $N + 6$ unknowns. They are

$$Y_1, Y_2, Y_3, \ldots, Y_N, \rho, T, p, u, v, w$$

The $N + 6$ equations to be solved are

1 overall mass continuity	Eq. (3-29)
3 momentum equations	Eq. (3-60)
1 energy equation	Eq. (3-75)
$N - 1$ species equations	Eq. (3-36)
1 equation of state	Eq. (3-96)
1 equation relating all Y_i	$Y_1 + Y_2 + \cdots + Y_N = 1$

If $\mathbf{V}_i$ must be solved from the diffusion equation for a multicomponent system, there will be $5N + 6$ unknowns. They are

$$Y_1, Y_2, Y_3 \ldots Y_N, \rho, T, p, u, v, w$$

$$X_1, X_2, X_3 \ldots X_N$$

and

$$V_{1,x}, V_{1,y}, V_{1,z}$$

$$V_{2,x}, V_{2,y}, V_{2,z}$$

$$\vdots$$

$$V_{N,x}, V_{N,y}, V_{N,z}$$

There are $5N + 6$ equations. Of those, $N + 6$ equations are listed above. Further there are

$3N$ diffusion equations	Eq. (3-88b)
N equations relating X_i to Y_i	Eq. (3-97)

10 SHVAB – ZEL'DOVICH FORMULATION

A number of effects in the governing equations are unimportant in a large class of combustion problems and can be omitted in many applications. These include

a. Body forces ($\mathbf{f}_i$ terms)
b. Soret and Dufour effects (terms involving α_i)
c. Pressure gradient diffusion [$(Y_i - X_i)\nabla p/p$ term in Eq. (3-88b)]
d. Bulk viscosity (μ')

Additional restrictive assumptions will be made to obtain the Shvab–Zel'dovich formulation of the conservation equations.

For *steady flow*, the overall continuity equation (Eq. (3-29)) reduces to

$$\nabla \cdot \rho \mathbf{v} = 0 \tag{3-98}$$

For *low-speed* steady-flow problems, viscous effects are often negligible, and the momentum equation (3-60) reduces to

$$\frac{\partial p}{\partial x_i} \approx 0 \tag{3-99}$$

In other words, the pressure

$$p = \text{constant} \tag{3-100}$$

The energy equation (3-78) reduces to

$$\rho \mathbf{v} \cdot \nabla h = \nabla \cdot (\lambda \nabla T) - \nabla \cdot \left(\rho \sum_{i=1}^{N} h_i Y_i \mathbf{V}_i \right) + \dot{Q} \tag{3-101}$$

because $\Phi = 0$ and $p =$ constant.

Adding $\sum_{i=1}^{N} h_i Y_i \nabla \cdot \rho \mathbf{v} = 0$ on the left-hand side of Eq. (3-101), noting that $h = \sum_{i=1}^{N} h_i Y_i$, and setting the external heat addition $\dot{Q} = 0$, one obtains

$$\nabla \cdot \left[\rho \sum_{i=1}^{N} h_i Y_i (\mathbf{v} + \mathbf{V}_i) - \lambda \nabla T \right] = 0 \tag{3-102}$$

The conservation-of-species equation (3-35) for steady flow reduces to

$$\nabla \cdot [\rho Y_i (\mathbf{v} + \mathbf{V}_i)] = \omega_i, \qquad i = 1, 2, \ldots, N \tag{3-103}$$

When the *binary-diffusion coefficients of all pairs of species are equal*, the diffusion equation (3-92) reduces to

$$\mathscr{D} \nabla X_i = X_i \sum_{j=1}^{N} X_j \mathbf{V}_j - X_i \mathbf{V}_i, \qquad i = 1, 2, \ldots, N \tag{3-104}$$

Multiplying the above equation by Y_i / X_i, and summing over i, we obtain

$$\sum_{i=1}^{N} \frac{Y_i}{X_i} \mathscr{D} \nabla X_i = \sum_{i=1}^{N} Y_i \sum_{j=1}^{N} X_j \mathbf{V}_j - \sum_{i=1}^{N} Y_i \mathbf{V}_i \tag{3-105}$$

Using the fact that

$$\sum_{i=1}^{N} Y_i = 1 \quad \text{and} \quad \sum_{i=1}^{N} Y_i \mathbf{V}_i = 0$$

we obtain

$$\sum_{j=1}^{N} X_j \mathbf{V}_j = \sum_{i=1}^{N} Y_i \mathscr{D} \nabla \ln X_i = \sum_{j=1}^{N} Y_j \mathscr{D} \nabla \ln X_j \tag{3-106}$$

Substituting Eq. (3-106) into Eq. (3-104) and dividing by X_i, we obtain

$$\mathscr{D} \left[\nabla \ln X_i - \sum_{j=1}^{N} Y_j \nabla \ln X_j \right] = -\mathbf{V}_i, \qquad i = 1, 2, \ldots, N \tag{3-107}$$

Substituting Eq. (3-97) for X_i in the above equation, we obtain Fick's law

$$\mathbf{V}_i = -\mathscr{D}\nabla \ln Y_i, \qquad i = 1, 2, \ldots, N \tag{3-108}$$

Substituting Eqs. (3-103), (3-108) and (3-64) into Eq. (3-102), we obtain

$$\nabla \cdot \left[\rho\mathbf{v}\int_{T^\circ}^{T}\left(\sum_{i=1}^{N} Y_i C_{p,i}\right) dT - \rho\mathscr{D}\sum_{i=1}^{N}\nabla Y_i \int_{T^\circ}^{T} C_{p,i}\, dT - \rho\mathscr{D}\frac{\lambda}{\rho C_p \mathscr{D}} C_p \nabla T\right]$$

$$= -\sum_{i=1}^{N}\Delta h_{f,i}^{\circ}\omega_i \tag{3-109}$$

Assuming that the Lewis number is unity, that is,

$$\mathrm{Le} = \frac{\lambda}{\rho C_p \mathscr{D}} = 1 \tag{3-110}$$

and using the fact that

$$\sum_{i=1}^{N} Y_i C_{p,i} = C_p \tag{3-111}$$

Eq. (3-109) then becomes

$$\nabla \cdot \left\{\rho\mathbf{v}\int_{T^\circ}^{T} C_p\, dT - \rho\mathscr{D}\overbrace{\left[\sum_{i=1}^{N}\nabla Y_i \int_{T^\circ}^{T} C_{p,i}\, dT + C_p\nabla T\right]}^{\nabla \int_{T^\circ}^{T} C_p\, dT}\right\} = -\sum_{i=1}^{N}\Delta h_{f,i}^{\circ}\omega_i \tag{3-112}$$

Note that

$$\nabla\int_{T^\circ}^{T} C_p\, dT = \nabla\sum_{i=1}^{N} Y_i\int_{T^\circ}^{T} C_{p,i}\, dT$$

$$= \sum_{i=1}^{N}(\nabla Y_i)\int_{T^\circ}^{T} C_{p,i}(T^*)\, dT^* + \sum_{i=1}^{N} Y_i\nabla\int_{T^\circ}^{T} C_{p,i}(T^*)\, dT^*$$

$$= \sum_{i=1}^{N}(\nabla Y_i)\int_{T^\circ}^{T} C_{p,i}\, dT + \sum_{i=1}^{N} Y_i C_{p,i}(T)\nabla T$$

$$= \sum_{i=1}^{N}(\nabla Y_i)\int_{T^\circ}^{T} C_{p,i}\, dT + C_p\nabla T \tag{3-113}$$

Substituting Eq. (3-113) into Eq. (3-112), we obtain the Shvab–Zel'dovich energy equation

$$\nabla \cdot \left\{ \rho \mathbf{v} \int_{T^\circ}^{T} C_p \, dT - \rho \mathscr{D} \nabla \int_{T^\circ}^{T} C_p \, dT \right\} = - \sum_{i=1}^{N} \Delta h_{f,i}^\circ \omega_i \qquad (3\text{-}114)$$

Substituting Eq. (3-108) in Eq. (3-103), we obtain the Shvab–Zel'dovich species equation.

$$\nabla \cdot \{ \rho \mathbf{v} Y_i - \rho \mathscr{D} \nabla Y_i \} = \omega_i \qquad (3\text{-}115)$$

Note the similarity in Eqs. (3-114) and (3-115). In deriving these equations, we have *not* assumed that the specific heat (or any transport coefficient) of the mixture is a constant or that the specific heats of all the species are equal.

The Shvab–Zel'dovich form for the conservation equations is particularly useful when one may assume that chemical changes occur by a single reaction step

$$\sum_{i=1}^{N} \nu_i' M_i \rightarrow \sum_{i=1}^{N} \nu_i'' M_i \qquad (3\text{-}116)$$

In this case

$$\frac{\omega_i}{W_i(\nu_i'' - \nu_i')} \equiv \omega, \qquad i = 1, 2, \ldots, N \qquad (3\text{-}117)$$

Defining two new parameters η_T and η_i such that

$$\eta_T \equiv \frac{\int_{T^\circ}^{T} C_p \, dT}{\sum_{i=1}^{N} \Delta h_{f,i}^\circ W_i (\nu_i' - \nu_i'')} \qquad (3\text{-}118)$$

$$\eta_i \equiv \frac{Y_i}{W_i(\nu_i'' - \nu_i')}, \qquad i = 1, 2, \ldots, N \qquad (3\text{-}119)$$

Then the energy equation (3-114) becomes

$$\nabla \cdot [\rho \mathbf{v} \eta_T - \rho \mathscr{D} \nabla \eta_T] = \omega \qquad (3\text{-}120)$$

and the species equation (3-115) becomes

$$\nabla \cdot [\rho \mathbf{v} \eta_i - \rho \mathscr{D} \nabla \eta_i] = \omega \qquad (3\text{-}121)$$

The nonlinear rate term ω may be eliminated from N of the $N+1$ equations corresponding to

$$L(\eta) = \omega \tag{3-122}$$

where L is the linear operator (if $\rho\mathscr{D}$ is independent of η)

$$L(\eta) = \nabla \cdot [\rho \mathbf{v} \eta - \rho\mathscr{D}\nabla\eta] \tag{3-123}$$

and where η can be $\eta_T, \eta_1, \eta_2, \ldots, \eta_N$. If it is assumed that $\eta = \eta_1$, then η_1 must be found from

$$L(\eta_1) = \omega \tag{3-124}$$

The other flow variables are determined by the linear function

$$L(\beta) = 0 \tag{3-125}$$

where β can be $\beta_T, \beta_2, \beta_3, \ldots, \beta_N$:

$$\beta_T = \eta_T - \eta_1$$

$$\beta_2 = \eta_2 - \eta_1$$

$$\beta_N = \eta_N - \eta_1 \tag{3-126}$$

This result is most important for systems in which the reactants are initially unmixed; in these systems one often finds that by solving the linear equations for relations between flow variables, burning rates may eventually be determined without solving the nonlinear equation.

11 DIMENSIONLESS RATIOS OF TRANSPORT COEFFICIENTS

In flow problems, the conservation equations can often be written in forms involving dimensionless ratios of various transport coefficients. The *Lewis number* (or Lewis–Seminov number) is defined as

$$\boxed{\mathrm{Le} \equiv \frac{\lambda}{\rho C_p \mathscr{D}} = \frac{\alpha}{\mathscr{D}} = \frac{\text{rate of energy transport}}{\text{rate of mass transport}}} \tag{3-127}$$

The *Prandtl number* is defined as

$$\boxed{\mathrm{Pr} \equiv \frac{C_p \mu}{\lambda} = \frac{C_p \rho(\mu/\rho)}{\lambda} = \frac{\nu}{\alpha} = \frac{\text{rate of momentum transport}}{\text{rate of energy transport}}} \tag{3-128}$$

The *Schmidt number* is defined as

$$\boxed{\mathrm{Sc} \equiv \frac{\nu}{\mathscr{D}} = \frac{\text{rate of momentum transport}}{\text{rate of mass transport}}} \tag{3-129}$$

Le and Sc may be defined for each pair of species in multicomponent mixtures. From the above we see that

$$\boxed{\mathrm{Le} \equiv \frac{\mathrm{Sc}}{\mathrm{Pr}}} \tag{3-130}$$

In many systems, Le is very nearly unity. It is often slightly less then unity in combustible gas mixtures. The approximation Le = 1 is frequently very helpful in theoretical combustion analyses.

Note: Le is sometimes defined as the reciprocal of Eq. (3-127); that is, $\mathrm{Le} = \mathscr{D}/\alpha$.

12 BOUNDARY CONDITIONS AT AN INTERFACE

In the formulation of a combustion problem, besides the consideration of the governing differential equations, one also has to formulate the boundary and initial conditions adequately. The specification of boundary conditions is not always trivial. The number of boundary conditions required for a set of governing differential equations depends quite often upon the flow conditions of the reactive mixture, in other words, upon the nature of the governing equations. Either under- or overspecifying the boundary conditions will make the overall formulation ill posed. The boundary-condition specification for various types of combustion problems is discussed in many places in later chapters. The purpose for describing boundary conditions here is to familiarize readers with the formal way of specifying them through the use of conservation conditions at material interfaces, where the boundary conditions are usually more complex than those for far fields.

Relations expressing the conservation of mass, momentum, and energy at an interface may be derived by converting conservation equations in differential form to integral form and then passing to the limit in which the volume of integration approaches a surface.

The time rate of change of the integral of an arbitrary function $f(x, t)$ over a given volume $\mathcal{v}$ can be written as the sum of two parts:

$$\frac{d}{dt}\left(\iiint_{\mathcal{v}} f\,d\mathcal{v}\right) = \iiint_{\mathcal{v}} \frac{\partial f}{\partial t}\,d\mathcal{v} + \iint_{\mathcal{S}} \mathbf{v}_B \cdot \mathbf{n} f\,d\mathcal{S} \tag{3-131}$$

The first term represents the contribution due to the time rate of change of the integrand. The second term represents the contribution due to the motion of the control surface $\mathcal{S}$ bounding the volume $\mathcal{v}$. Here $\mathbf{n}$ is the local outward normal unit vector to the surface $\mathcal{S}$, and $\mathbf{v}_B$ is the velocity of the boundary of the control volume. In general, both $\mathbf{n}$ and $\mathbf{v}_B$ can vary from point to point on $\mathcal{S}$. Equation (3-131) can be proved[2] and is used here as a mathematical identity.

Now, let us derive the mass conservation condition at an interface by first considering the continuity equation (3-29):

$$\frac{\partial \rho}{\partial t} + \nabla \cdot (\rho \mathbf{v}) = 0.$$

Upon integration of this equation over $\mathcal{v}$, we have

$$\iiint_{\mathcal{v}} \left[\frac{\partial \rho}{\partial t} + \nabla \cdot (\rho \mathbf{v}) \right] d\mathcal{v} = 0.$$

Using Eq. (3-131) and the divergence theorem,

$$\iiint_{\mathcal{v}} \nabla \cdot (\rho \mathbf{v})\, d\mathcal{v} = \iint_{\mathcal{S}} \rho \mathbf{v} \cdot \mathbf{n}\, d\mathcal{S},$$

we obtain

$$\frac{d}{dt} \iiint_{\mathcal{v}} \rho\, d\mathcal{v} - \iint_{\mathcal{S}} \mathbf{v}_B \cdot \mathbf{n} \rho\, d\mathcal{S} + \iint_{\mathcal{S}} \rho \mathbf{v} \cdot \mathbf{n}\, d\mathcal{S} = 0$$

or

$$\frac{d}{dt} \left(\iiint_{\mathcal{v}} \rho\, d\mathcal{v} \right) + \iint_{\mathcal{S}} \rho (\mathbf{v} - \mathbf{v}_B) \cdot \mathbf{n}\, d\mathcal{S} = 0 \tag{3-132}$$

Equation (3-132) represents the integral form of the continuity equation.

Example 3.1

If the right surface of a solid material is burning at a rate of v_b, show that [from Eq. (3-131)] the mass burning rate is

$$\frac{dM}{dt} = -v_b \rho_s A_b$$

where A_b is the burning surface area and ρ_s is the density of the solid. If one lets the control volume on both sides move at the same rate as v_b, show that the rate of change of mass in the control volume is zero.

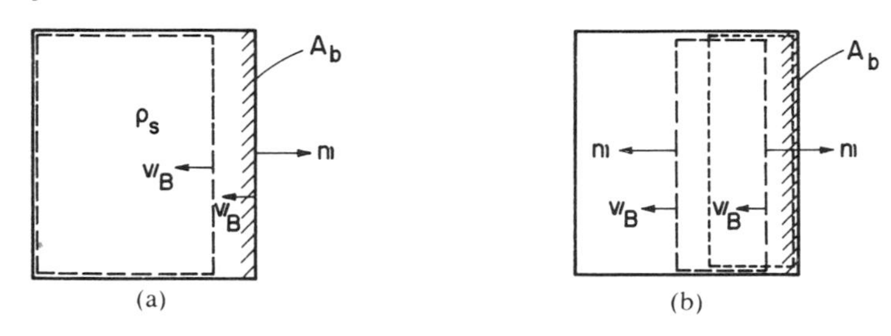

Solution:

We have

$$\underbrace{\frac{d}{dt}\iiint_{v} \rho_s \, dv}_{M} = \iiint_{v} \frac{\partial \rho_s}{\partial t}^{\,0} dv + \underbrace{\iint_{S} \mathbf{v}_B \cdot \mathbf{n} \rho_s \, dS}_{-v_b \rho_s A_b}$$

$$\therefore \quad \frac{dM}{dt} = -v_b \rho_s A_b.$$

If the solid material is burning away, but we avoid changing the size of the control volume by moving the boundaries of the control volume at the same rate as the burning rate, then we have

$$\frac{d}{dt}\iiint_{v} \rho_s \, dv = \iiint_{v} \frac{\partial \rho_s}{\partial t}^{\,0} dv + \iint_{S} \mathbf{v}_B \cdot \mathbf{n} \rho_s \, dS$$

$$= +v_{B\,\text{left}} \rho_s A_b - v_{B\,\text{right}} \rho_s A_b$$

$$= v_b \rho_s A_b - v_b \rho_s A_b = 0$$

The momentum equation (3-48) is

$$\rho \frac{\partial \mathbf{v}}{\partial t} + \rho \mathbf{v} \cdot \nabla \mathbf{v} = \nabla \cdot \tilde{\boldsymbol{\sigma}} + \rho \sum_{i=1}^{N} Y_i \mathbf{f}_i \tag{3-133}$$

Multiplying the continuity equation (3-36) by **v**, we obtain

$$\mathbf{v} \frac{\partial \rho}{\partial t} + \mathbf{v} \nabla \cdot (\rho \mathbf{v}) = 0 \tag{3-134}$$

Adding Eqs. (3-133) and (3-134) gives

$$\frac{\partial}{\partial t}(\rho \mathbf{v}) + \nabla \cdot (\rho \mathbf{v} \mathbf{v}) = \nabla \cdot \tilde{\boldsymbol{\sigma}} + \rho \sum Y_i \mathbf{f}_i \tag{3-135}$$

We may integrate this over $\mathcal{v}$ to show that

$$\frac{d}{dt}\iiint_{\mathcal{v}} \rho\mathbf{v}\, d\mathcal{v} + \iint_{\mathcal{S}} \rho\mathbf{v}[(\mathbf{v}-\mathbf{v}_B)\cdot\mathbf{n}]\, d\mathcal{S} = \iint_{\mathcal{S}} \tilde{\boldsymbol{\sigma}}\cdot\mathbf{n}\, d\mathcal{S} + \iiint_{\mathcal{v}} \rho\sum_{i=1}^{N} Y_i\mathbf{f}_i\, d\mathcal{v} \tag{3-136}$$

The above equation represents the integral form of the momentum equation.

The energy equation (3-67) can be written in vector notation as

$$\frac{\partial}{\partial t}(\rho e_t) + \nabla\cdot(\rho e_t\mathbf{v}) = -\nabla\cdot\mathbf{q} + \dot{Q} + \nabla\cdot(\tilde{\boldsymbol{\sigma}}\cdot\mathbf{v}) + \rho\sum_{i=1}^{N} Y_i\mathbf{f}_i\cdot(\mathbf{v}+\mathbf{V}_i) \tag{3-137}$$

After integration, we obtain

$$\frac{d}{dt}\iiint_{\mathcal{v}} \rho e_t\, d\mathcal{v} + \iint_{\mathcal{S}} \rho e_t[(\mathbf{v}-\mathbf{v}_B)\cdot\mathbf{n}]\, d\mathcal{S}$$

$$= \iiint_{\mathcal{v}} \dot{Q}\, d\mathcal{v} - \iint_{\mathcal{S}} \mathbf{q}\cdot\mathbf{n}\, d\mathcal{S} + \iint_{\mathcal{S}} \mathbf{v}\cdot\tilde{\boldsymbol{\sigma}}\cdot\mathbf{n}\, d\mathcal{S} + \iiint_{\mathcal{v}} \rho\sum_{i=1}^{N} Y_i\mathbf{f}_i\cdot(\mathbf{v}+\mathbf{V}_i)\, d\mathcal{v} \tag{3-138}$$

Equation (3-138) represents the integral form of the energy equation.

The continuity-of-species equation (3-35) is

$$\frac{\partial}{\partial t}(\rho Y_i) + \nabla\cdot[\rho Y_i(\mathbf{v}+\mathbf{V}_i)] = \omega_i$$

On integrating, we obtain

$$\frac{d}{dt}\left(\iiint_{\mathcal{v}} \rho Y_i\, d\mathcal{v}\right) + \iint_{\mathcal{S}} \rho Y_i(\mathbf{v}+\mathbf{V}_i-\mathbf{v}_B)\cdot\mathbf{n}\, d\mathcal{S} = \iiint_{\mathcal{v}} \omega_i\, d\mathcal{v} \tag{3-139}$$

We now consider the control volume $\mathcal{v}$ to be a thin slab, the thickness of which is made to approach zero (see Fig. 3.12). In the limit, the integrals over $\mathcal{S}$ are composed of two parts, namely integrals over each face of the slab.

Let us arbitrarily choose one face as the "positive" side of the slab, identifying quantities on this side by the subscript + and those on the other side by the subscript −. It is clear that n_- tends to $-n_+$ at corresponding points on the two faces; in the limit, the integral over the surface of $\mathcal{v}$ may be replaced by an integral over the interface area $\mathcal{S}_I$, and hence Eq. (3-132)

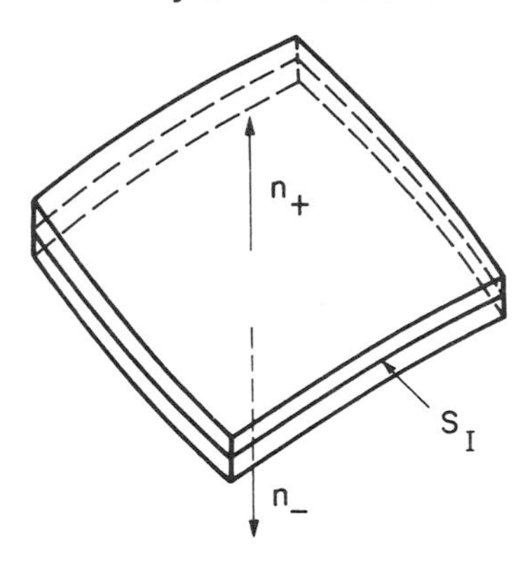

Figure 3.12 Control volume for the derivation of interface conditions.

becomes

$$\iint_{\mathcal{S}_I} [\rho_+(\mathbf{v}_+ - \mathbf{v}_{B+}) - \rho_-(\mathbf{v}_- - \mathbf{v}_{B-})] \cdot \mathbf{n}_+ \, d\mathcal{S} = -\lim_{v \to 0} \left[\frac{d}{dt} \iiint_v \rho \, dv \right] \tag{3-140}$$

Equation (3-136) becomes

$$\iint_{\mathcal{S}_I} \{ \rho_+ \mathbf{v}_+ [(\mathbf{v}_+ - \mathbf{v}_{B+}) \cdot \mathbf{n}_+] - \rho_- \mathbf{v}_- [(\mathbf{v}_- - \mathbf{v}_{B-}) \cdot \mathbf{n}_+] - (\tilde{\boldsymbol{\sigma}}_+ - \tilde{\boldsymbol{\sigma}}_-) \cdot \mathbf{n}_+ \} \, d\mathcal{S}$$

$$= \lim_{v \to 0} \left[\iiint_v \rho \sum_{i=1}^{N} Y_i \mathbf{f}_i \, dv - \frac{d}{dt} \iiint_v \rho \mathbf{v} \, dv \right] \tag{3-141}$$

The energy equation (3-138) becomes

$$\iint_{\mathcal{S}_I} [\rho_+ e_{t+} (\mathbf{v}_+ - \mathbf{v}_{B+}) - \rho_- e_{t-} (\mathbf{v}_- - \mathbf{v}_{B-}) + \mathbf{q}_+ - \mathbf{q}_- - \mathbf{v}_+ \cdot \tilde{\boldsymbol{\sigma}}_+ + \mathbf{v}_- \cdot \tilde{\boldsymbol{\sigma}}_-] \cdot \mathbf{n}_+ \, d\mathcal{S}$$

$$= \lim_{v \to 0} \left\{ \iiint_v \dot{Q} \, dv + \iiint_v \rho \sum_{i=1}^{N} Y_i \mathbf{f}_i \cdot (\mathbf{v} + \mathbf{V}_i) \, dv - \frac{d}{dt} \iiint_v \rho e_t \, dv \right\} \tag{3-142}$$

The species equation (3-139) becomes

$$\iint_{\mathcal{S}_I} [\rho_+ Y_{i+}(\mathbf{v}_+ + \mathbf{V}_{i+} - \mathbf{v}_{B+}) - \rho_- Y_{i-}(\mathbf{v}_- + \mathbf{V}_{i-} - \mathbf{v}_{B-})] \cdot \mathbf{n}_+ \, d\mathcal{S}$$

$$= \lim_{v \to 0} \left(\iiint_v \omega_i \, dv - \frac{d}{dt} \iiint_v \rho Y_i \, dv \right), \qquad i = 1, 2, \ldots, N \quad (3\text{-}143)$$

The last term in the equation represents surface reactions as well as the nonzero time rate of accumulation of the ith chemical species at the interface.

For example, if the interface is taken to be a solid surface at which chemical reactions are proceeding at a finite rate, then

$$\omega_i \to \omega_i' \delta(y - y_I)$$

where ω_i' is the surface reaction rate, y is the coordinate normal to the surface at $y = y_I$, and δ is the Dirac delta function. Then,

$$\lim_{v \to 0} \iiint_v \omega_i \, dv = \iint_{\mathcal{S}} \omega_i' \, d\mathcal{S}$$

The right-hand side represents the mass rate of production of species i on the surface.

If ρ_i' is the mass of species i per unit area, then

$$\rho Y_i \to \rho_i' \delta(y - y_I)$$

$$\lim_{v \to 0} \frac{d}{dt} \iiint_v \rho Y_i \, dv = + \frac{d}{dt} \lim_{v \to 0} \iiint_v \rho_i' \delta(y - y_I) \, dv$$

$$= + \frac{d}{dt} \iint_{\mathcal{S}_I} \rho_i' \, d\mathcal{S} \qquad (3\text{-}144)$$

Thus Eq. (3-143) expresses the following physically obvious condition: the difference between the mass flux of species i leaving the surface and that entering the surface equals the mass rate of production of species i at the surface minus the mass rate of accumulation of species i at the surface [the term on the right-hand side of Eq. (3-144)]. Similar interpretations can be given for Eqs. (3-140), (3-141), and (3-142).

In many problems, all terms in Eqs. (3-140) through (3-143) involving $\lim_{v \to 0}$ are zero. If it is assumed that the interface is stationary, i.e.,

$$\mathbf{v}_B = 0$$

and that viscosity is negligible, then Eqs. (3-140) through (3-143) simplify considerably:

$$\rho_+ \mathbf{v}_+ \cdot \mathbf{n}_+ = \rho_- \mathbf{v}_- \cdot \mathbf{n}_+ \tag{3-145}$$

$$\rho_+ \mathbf{v}_+ (\mathbf{v}_+ \cdot \mathbf{n}_+) + p_+ \mathbf{n}_+ = \rho_- \mathbf{v}_- (\mathbf{v}_- \cdot \mathbf{n}_+) + p_- \mathbf{n}_+ \tag{3-146}$$

$$\left\{\rho_+ \left[h_{t+}\mathbf{v}_+ + \sum_{i=1}^{n} h_{i+} Y_{i+} \mathbf{V}_{i+}\right] - \lambda_+ (\nabla T)_+ + \mathbf{q}_{R+}\right\} \cdot \mathbf{n}_+$$

$$= \left\{\rho_- \left[h_{t-}\mathbf{v}_- + \sum_{i=1}^{n} h_{i-} Y_{i-} \mathbf{V}_{i-}\right] - \lambda_- (\nabla T)_- + \mathbf{q}_{R-}\right\} \cdot \mathbf{n}_+ \tag{3-147}$$

where $\mathbf{q}_{R+}$ and $\mathbf{q}_{R-}$ represent the radiative heat-flux contribution. This term has been included because radiative losses from surfaces are often substantial. If there is no surface reaction to generate species i, then the mass-flux balance at the surface gives

$$\rho_+ Y_{i+} (\mathbf{v}_+ + \mathbf{V}_{i+}) \cdot \mathbf{n}_+ = \rho_- Y_{i-} (\mathbf{v}_- + \mathbf{V}_{i-}) \cdot \mathbf{n}_+ \tag{3-148}$$

Example 3.2

Write the momentum flux balance at a gas–solid interface.

Solution:

If the viscous effect is negligible, the total stress tensor $\tilde{\boldsymbol{\sigma}}$ in the gas phase becomes

$$\tilde{\boldsymbol{\sigma}} = -p\mathbf{I}$$

$$\because \quad \sigma_{ij} = -p\delta_{ij} + \left(\mu' - \tfrac{2}{3}\mu\right)\frac{\partial u_k}{\partial x_k}\delta_{ij} + \mu\left(\frac{\partial u_i}{\partial x_j} + \frac{\partial u_j}{\partial x_i}\right)$$

or

$$\tilde{\boldsymbol{\sigma}} = -p\mathbf{I} + \left(\cancel{\mu'}^{\,0} - \tfrac{2}{3}\cancel{\mu}^{\,0}\right)(\nabla \cdot \mathbf{v})\mathbf{I} + \cancel{\mu}^{\,0}\left[(\nabla \mathbf{v}) + (\nabla \mathbf{v})^T\right]$$

where the superscript T means the transpose. The stress tensor in a linear viscoelastic solid material can be written as

$$\sigma_{ij}(t) = \int_{-\infty}^{t} G_1(t - t')\frac{d}{dt'}\varepsilon_{ij}(t')\,dt' - \tfrac{1}{3}\delta_{ij}\int_{-\infty}^{t} G_1(t - t')\frac{d\varepsilon_{kk}}{dt'}\,dt' + \delta_{ij}K\varepsilon_{kk}$$

where $G_1(t - t')$ is called the relaxation modulus in shear, ε_{ij} are the strain-

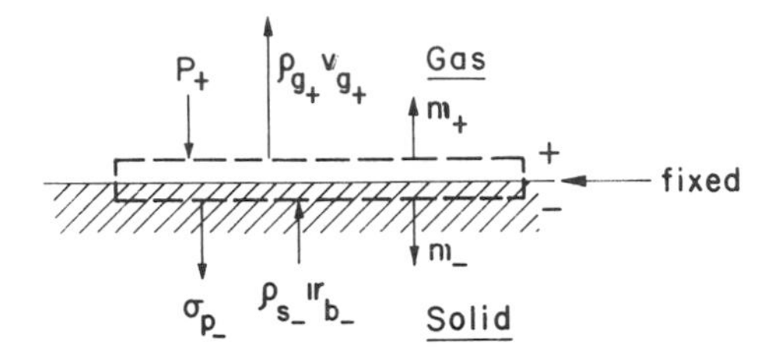

tensor components, and K is the bulk modulus. (Note: solid propellants usually can be assumed to be viscoelastic in shear and elastic in bulk.)

For most viscoelastic materials

$$G_1 = g_0 + \sum_{m=1}^{M} g_m e^{-\beta_m t}$$

For simplicity, let us denote the normal stress in the solid phase by $\sigma_p \mathbf{I}$. σ_p is positive for tension. Then the normal-stress tensor in the solid phase becomes

$$\tilde{\boldsymbol{\sigma}} = \sigma_p \mathbf{I}$$

Now, we shall assume the interface is stationary and the body-force and viscous effects are negligible. Following Eq. (3-141), we have

$$A\rho_{g+}\mathbf{v}_{g+}(\mathbf{v}_{g+}\cdot \mathbf{n}_+) - A\rho_{s-}\mathbf{r}_{b-}(\mathbf{r}_{b-}\cdot \mathbf{n}_+) - (-p_+\mathbf{I} - \sigma_{p-}\mathbf{I})\cdot \mathbf{n}_+ A$$

$$= \lim_{v\to 0}\left[\iiint_v \rho \sum_{i=1}^{N} Y_i \mathbf{f}_i \, dv - \frac{d}{dt}\iiint_v \rho \mathbf{v}\, dv\right] \to 0$$

or

$$\underbrace{\underbrace{A\rho_{g+}\mathbf{v}_{g+}\,\mathbf{v}_{g+}}_{\left(\frac{M}{t}\right)\left(\frac{L}{t}\right)}}_{\substack{\text{Momentum}\\\text{flux leaving}\\\text{control}\\\text{surface}}} - \underbrace{\underbrace{A\rho_{s-}\mathbf{r}_{b-}\,\mathbf{r}_{b-}}_{\left(\frac{M}{t}\right)\left(\frac{L}{t}\right)}}_{\substack{\text{Momentum}\\\text{flux entering}\\\text{control}\\\text{surface}}} + \underbrace{\underbrace{Ap_+\mathbf{n}_+}_{(F)}}_{\substack{\text{Force}\\\text{exerted}\\\text{on}\\\text{top surface}\\\text{of control}\\\text{volume}}} + \underbrace{\underbrace{A\sigma_{p-}\mathbf{n}_+}_{(F)}}_{\substack{\text{Force exerted}\\\text{on bottom}\\\text{surface of}\\\text{control volume}}} = 0$$

$\mathbf{n}_+$ can be eliminated from the above equation because

$$\mathbf{v}_{g+} = v_g \mathbf{n}_+, \qquad \mathbf{r}_{b-} = r_b \mathbf{n}_+$$

Thus we have

$$\rho_{g+}v_{g+}^2 + p_+ = \rho_{s-}r_{b-}^2 - \sigma_{p-}$$

Now $\rho_{g+}v_{g+}^2 \gg \rho_{s-}r_{b-}^2$ (because $\rho_{g+}v_{g+} = \rho_{s-}r_{b-}$ and $v_{g+} \gg r_{b-}$); then

$$-\sigma_{p-} = p_+ + \rho_{g+}v_{g+}^2$$

Therefore the normal compressive stress $(-\sigma_{p-})$ in the solid phase immediately below the interface is approximately equal to the sum of hydrostatic pressure and the momentum of the gas leaving the burning surface.

Example 3.3

Write the species mass flux balance at a gas–solid interface. Assumptions:

1. The interface location is fixed in space.
2. Fick's law of mass diffusion is valid.
3. No subsurface diffusion.

Solution:

See the accompanying diagram.

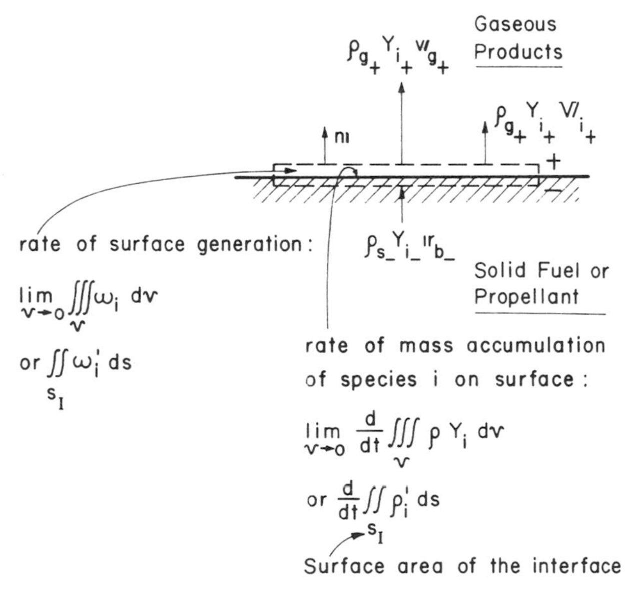

Mass-flux balance of ith species at the gas–solid interface.

Following Eq. (3-143), we have

$$A\rho_{g+} Y_{i+}(\mathbf{v}_{g+} + \mathbf{V}_{i+}) \cdot \mathbf{n}_{+} - A\rho_{s-} Y_{i-}\mathbf{r}_{b-} \cdot \mathbf{n}_{+} = \iint_{s_I} \omega_i' \, ds - \frac{d}{dt} \iint_{s_I} \rho_i' \, ds$$

where

$$\mathbf{V}_{i+} = -\frac{\mathscr{D}}{Y_{i+}} \frac{\partial Y_i}{\partial y}\bigg|_{+} \mathbf{n}_{+}$$

so that

$$\rho_{g+} Y_{i+} v_{g+} - \rho_{g+}\mathscr{D}\frac{\partial Y_i}{\partial y}\bigg|_{+} - \rho_{s-} Y_{i-} r_{b-} = \omega_i' - \frac{1}{A}\frac{d}{dt} \iint_{s_I} \rho_i' \, ds$$

If we further assume that the mass-accumulation term is negligible, then we have

$$\rho_{g+} Y_{i+} v_{g+} - \rho_{g+}\mathscr{D}\frac{\partial Y_i}{\partial y}\bigg|_{+} = \rho_{s-} Y_{i-} r_{b-} + \omega_i'$$

In general, a gaseous species i in the product does not exist in the same form in the solid; hence $Y_{i-} = 0$. Therefore

$$\underbrace{\rho_{g+} Y_{i+} v_{g+}}_{\substack{\text{mass flux of} \\ \text{species } i \\ \text{convected out}}} - \underbrace{\rho_{g+}\mathscr{D}\frac{\partial Y_i}{\partial y}\bigg|_{+}}_{\substack{\text{mass flux of} \\ \text{species } i \\ \text{diffused out}}} = \underbrace{\omega_i'}_{\substack{\text{mass flux of species } i \\ \text{generated on the surface}}}$$

Example 3.4

Write the energy-flux balance at a gas–liquid interface.

Solution:

Following Eq. (3-142), if the body-force work is zero and if there is no accumulation of total energy with respect to time, we have

$$\begin{aligned} &\iint_{s_I} \rho_{g+}\left(e_{g+} + \frac{v_{g+}^2}{2}\right) v_{g+} \, ds - \iint_{s_I} \rho_{l-}\left(e_{l-} + \frac{r_{b-}^2}{2}\right) r_{b-} \, ds \\ &\quad + \iint_{s_I} \mathbf{q}_{+} \cdot \mathbf{n}_{+} \, ds - \iint_{s_I} \mathbf{q}_{-} \cdot \mathbf{n}_{+} \, ds \\ &\quad - \iint_{s_I} v_{g+}(-p_{g+} + \tau_{ii+}) \, ds - \iint_{s_I} r_{b-} p_{l-} \, ds = \lim_{v \to 0} \iiint_{v} \dot{Q} \, dv \qquad \text{(A)} \end{aligned}$$

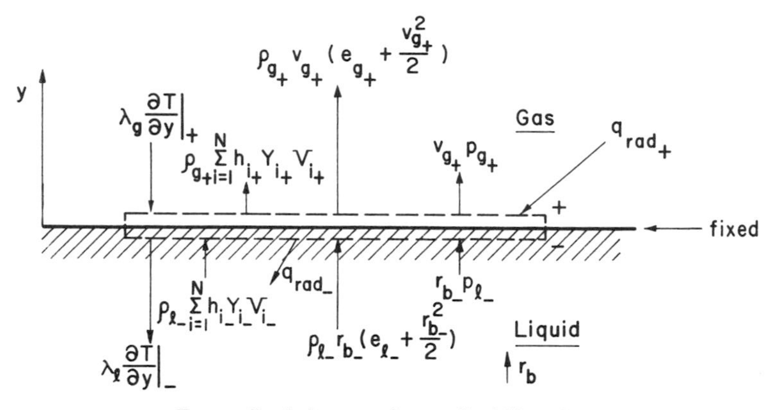

Energy-flux balance at the gas–liquid interface.

(see accompanying diagram). Here

$$\mathbf{q}_{+} = -\lambda \nabla T + \rho \sum_{i=1}^{N} h_i Y_i \mathbf{V}_i + \mathbf{q}_{\text{Dufour}}$$

where $\mathbf{q}_{\text{Dufour}}$ can be neglected; also, the nonhydrostatic normal stresses, τ_{ii+}, can be neglected in comparison with p_{g+}.

Now $\lim_{v \to 0} \iiint_v \dot{Q}\, dv$ is equal to $\iint_{s_I} (\mathbf{q}_{\text{rad}-} - \mathbf{q}_{\text{rad}+}) \cdot \mathbf{n}_+ \, ds$, which represents the net rate of energy absorption at the interface. Thus the energy-flux balance becomes

$$\rho_{g+} v_{g+} \left(e_{g+} + \frac{p_{g+}}{\rho_{g+}} + \frac{v_{g+}^2}{2} \right) + \rho_{g+} \sum_{i=1}^{N} h_{i+} Y_{i+} V_{i+} - \lambda_g \frac{\partial T}{\partial y}\bigg|_{+} - q_{\text{rad}+}$$

$$= \rho_{l-} r_{b-} \left(e_{l-} + \frac{p_{l-}}{\rho_{l-}} + \underbrace{\frac{r_b^2}{2}}_{\text{small}} \right) + \rho_{l-} \sum_{i=1}^{N} h_{i-} Y_{i-} V_{i-} - \lambda_l \frac{\partial T}{\partial y}\bigg|_{-} - q_{\text{rad}-} \tag{B}$$

Note that

$$e \equiv h - \frac{p}{\rho} = \sum_{i=1}^{N} h_i Y_i - \frac{p}{\rho} \tag{C}$$

Substituting Eq. (C) into (B), we have

$$\rho_{g+}\Bigg[\underbrace{v_{g+}\left(h_{g+}+\frac{v_{g+}^{2}}{2}\right)}_{\substack{\text{total enthalpy flux of}\\ \text{gaseous mixture}\\ \text{leaving surface by}\\ \text{bulk motion of}\\ \text{gases at surface}}} + \underbrace{\sum_{i=1}^{N} h_{i+}Y_{i+}V_{i+}}_{\substack{\text{enthalpy flux of}\\ \text{gaseous mixture}\\ \text{leaving surface}\\ \text{by mass diffusion}}}\Bigg] - \underbrace{\lambda_{g}\frac{\partial T}{\partial y}\bigg|_{+}}_{\substack{\text{energy flux}\\ \text{to surface}\\ \text{due to heat}\\ \text{conduction}}} - \underbrace{q_{\text{rad}+}}_{\substack{\text{energy flux}\\ \text{to surface}\\ \text{due to}\\ \text{radiation}}}$$

$$= \rho_{l-}\Bigg[\underbrace{r_{b-}\left(h_{l-}+\frac{r_{b-}^{2}}{2}\right)}_{\substack{\text{total enthalpy flux}\\ \text{of liquid material}\\ \text{arriving at interface}\\ \text{by bulk motion}\\ \text{of liquid at surface}}} + \underbrace{\sum_{i=1}^{N} h_{i-}Y_{i-}V_{i-}}_{\substack{\text{enthalpy flux}\\ \text{of liquid}\\ \text{arriving at}\\ \text{interface}\\ \text{by mass diffusion}}}\Bigg] - \underbrace{\lambda_{l}\frac{\partial T}{\partial y}\bigg|_{-}}_{\substack{\text{energy flux}\\ \text{to}\\ \text{subsurface}\\ \text{due to heat}\\ \text{conduction}}} - \underbrace{q_{\text{rad}-}}_{\substack{\text{energy flux}\\ \text{leaving}\\ \text{interface by}\\ \text{radiation}}} \tag{D}$$

It should be noted that the enthalpy contains not only the sensible enthalpy but also the chemical enthalpy, that is,

$$h_{i} = \Delta h_{f,i}^{\circ} + \int_{T^{\circ}}^{T} C_{pi}\, dT \tag{E}$$

The species mass-flux balance can be written as

$$\rho_{g+}Y_{i+}v_{g+} - \rho_{g+}\mathscr{D}\frac{\partial Y_{i}}{\partial y}\bigg|_{+} = \rho_{l-}Y_{i-}r_{b-} - \rho_{l-}\mathscr{D}\frac{\partial Y_{i}}{\partial y}\bigg|_{-} + \omega_{i}' \tag{F}$$

or

$$\rho_{g+}\left[Y_{i+}v_{g+} + Y_{i+}V_{i+}\right] = \rho_{l-}\left[Y_{i-}r_{b-} + Y_{i-}V_{i-}\right] + \omega_{i}' \tag{F$'$}$$

Since usually $h_{g+} \gg v_{g+}^{2}/2$ and $h_{l-} \gg r_{b-}^{2}/2$, we can drop the kinetic-energy terms from Eq. (D). After using Eqs. (C) and (E), Eq. (D) becomes

$$\rho_{g+}\sum_{i=1}^{N} h_{i+}Y_{i+}\left(v_{g+} + V_{i+}\right) - \lambda_{g}\frac{\partial T}{\partial y}\bigg|_{+} - q_{\text{rad}+}$$

$$= \rho_{l-}\sum_{i=1}^{N} h_{i-}Y_{i-}\left(r_{b-} + V_{i-}\right) - \lambda_{l}\frac{\partial T}{\partial y}\bigg|_{-} - q_{\text{rad}-} \tag{G}$$

or

$$\sum_{i=1}^{N}\left[\Delta h_{f,i}^{\circ} + \int_{T^{\circ}}^{T} C_{pi}\, dT\right]_{+}\rho_{g+}Y_{i+}\left(v_{g+} + V_{i+}\right) - \lambda_{g}\frac{\partial T}{\partial y}\bigg|_{+} - q_{\text{rad}+}$$

$$= \sum_{i=1}^{N}\left[\Delta h_{f,i}^{\circ} + \int_{T^{\circ}}^{T} C_{pi}\, dT\right]_{-}\rho_{l-}Y_{i-}\left(r_{b-} + V_{i-}\right) - \lambda_{l}\frac{\partial T}{\partial y}\bigg|_{-} - q_{\text{rad}-} \tag{H}$$

or

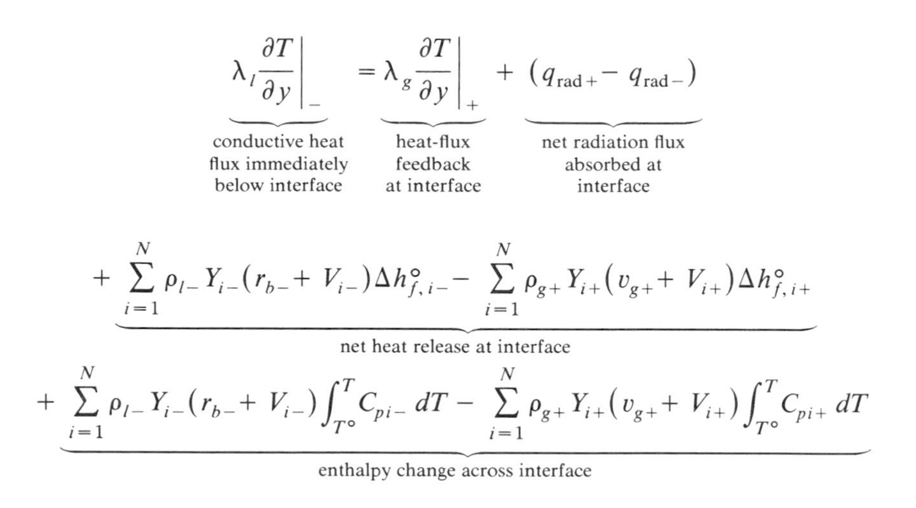

the net heat release at the interface is $\approx r_{b-}\rho_{l-}Q_s$ if V_i are small for all i. The enthalpy change across the interface is equal to

$$\sum_{i=1}^{N} \rho_{l-} Y_{i-}(r_{b-} + V_{i-}) T_s (C_{pi-} - C_{pi+})$$

if $C_{pi}T_s = \int_{T^\circ}^{T} C_{pi}\, dT$ for all i; this term also can be further simplified to $\rho_{l-} r_{b-} T_s (C_{pl} - C_{pg})$ if $V_i = 0$ for all i.

REFERENCES

1. R. B. Bird, W. E. Stewart, and E. N. Lightfoot, *Transport Phenomena*, John Wiley & Sons, New York, 1960.
2. F. A. Williams, *Combustion Theory*, Addison-Wesley Publishing Co., Reading, Mass., 1965.
3. H. Schlichting, *Boundary-Layer Theory*, 6th ed., Chapter 5, McGraw-Hill Book Company, New York, 1968.
4. J. O. Hirschfelder, C. F. Curtiss, and R. B. Bird, *Molecular Theory of Gases and Liquids*, Chapter 1, John Wiley & Sons, New York, 1954.
5. J. O. Hinze, *Turbulence*, McGraw-Hill Book Company, New York, 1975.
6. I. H. Shames, *Mechanics of Fluids*, McGraw-Hill Book Company, New York, 1962.
7. W. H. Li and S. H. Lam, *Principles of Fluid Mechanics*, Addison-Wesley Publishing Co., Reading, Mass., 1964.
8. R. Aris, *Vectors, Tensors, and the Basic Equations of Fluid Mechanics*, Prentice-Hall, Englewood Cliffs, N.J., 1962.
9. L. C. Landau and E. M. Liftshitz, *Fluid Mechanics*, Pergamon Press, New York, 1959.

HOMEWORK

1. Use Eq. (3-9) to show that the molar flux of species A can be expressed as

$$\mathbf{J}_A^* = -CX_A X_B(\mathbf{V}_B - \mathbf{V}_A)$$

in a binary system with species A and B.

2. Evaluate the following sums for a multicomponent system with N species:

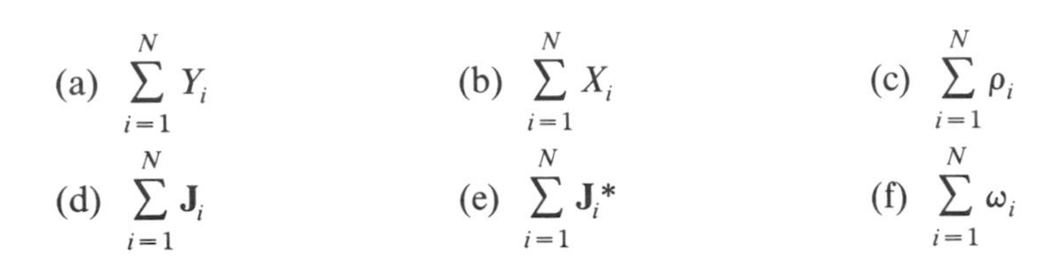

(a) $\sum_{i=1}^{N} Y_i$ (b) $\sum_{i=1}^{N} X_i$ (c) $\sum_{i=1}^{N} \rho_i$

(d) $\sum_{i=1}^{N} \mathbf{J}_i$ (e) $\sum_{i=1}^{N} \mathbf{J}_i^*$ (f) $\sum_{i=1}^{N} \omega_i$

3. For a multicomponent flow of reactive mixture with N species, show that

$$\sum_{i=1}^{N} Y_i \mathbf{V}_i = 0$$

where Y_i represents the mass fraction of species i, and $\mathbf{V}_i$ represents the relative mass diffusion velocity of species i. If the x-component of $\mathbf{V}_i$, V_{X_i}, is substituted for $\mathbf{V}_i$ in the above equation, will it still be valid?

4. Show that the total stress tensor τ_{ij} for an isotropic Newtonian fluid is symmetric.

5. Show that for any isotropic Newtonian fluid, the relationship $c - \lambda = 2\mu$ given in Eq. (3-54) is valid, where μ is the dynamic viscosity of the fluid. Hint: Consider a coordinate transformation between the principal stress axes and some arbitrary axes x_1, x_2, x_3 for which the shear stresses are not zero.

6. Express the quantity $\dot{Q}$ in the energy equation in terms of the heat of formation and the rate of local production of all the species considered in reacting flow systems, if the internal energy e in the following equation contains *no* chemical energy:

$$\rho \frac{\partial e}{\partial t} + \rho \mathbf{v} \cdot \nabla e = -\nabla \cdot \mathbf{q} + \tilde{\sigma} : \nabla \mathbf{v} + \dot{Q} + \rho \sum_{k=1}^{N} Y_k \mathbf{f}_k \cdot \mathbf{V}_k$$

ANSWER: $\dot{Q} = -\Sigma_{i=1}^{N} \omega_i \, \Delta h_{f,i}^{\circ}$

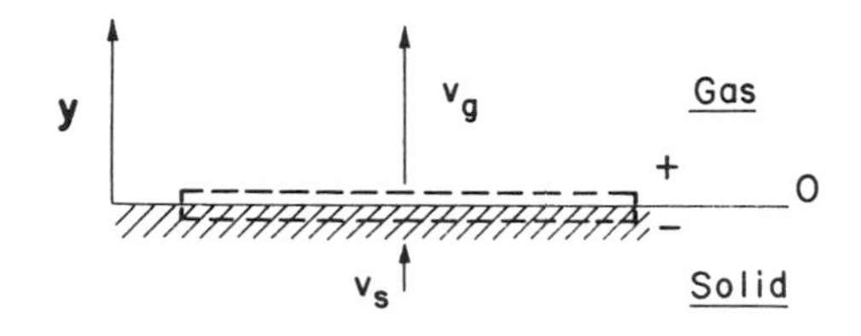

7. If a solid fuel is gasifying at a fixed surface, write down the boundary conditions for the balance of mass flux at the interface.

8. If the interface control volume moves at the regression rate of the solid in the above problem, will the mass-flux balance equation be the same?

9. A mixture of high-temperature gases flows over a horizontal graphite slab. Some of species [$H_2O_{(g)}$, $CO_{2(g)}$ and $H_{2(g)}$] in the mixture can react with graphite surface by heterogeneous reaction and cause thermochemical erosion of the solid. Consider the following heterogeneous reactions:

$$H_2O_{(g)} + C_{(s)} \rightarrow CO_{(g)} + H_{2(g)} \tag{1}$$

$$CO_{2(g)} + C_{(s)} \rightarrow 2CO_{(g)} \tag{2}$$

$$H_{2(g)} + 2C_{(s)} \rightarrow C_2H_{2(g)} \tag{3}$$

   (a) Write the species mass-flux balance at the gas–solid interface for species $H_2O_{(g)}$, $CO_{2(g)}$, $CO_{(g)}$, and $H_{2(g)}$. Give the physical meanings of each term in the mass-flux balance for $H_2O_{(g)}$. The rates of consumption of graphite due to H_2O, CO_2, and H_2 can be denoted as $-\omega_{C,H_2O}$, $-\omega_{C,CO_2}$, and $-\omega_{C,H_2}$ respectively. These rates are assumed to be known functions of T_{wall} and the partial pressure of each reacting species.

   (b) What is the energy-flux balance at the gas–solid interface? Give the physical meaning of each term.

   (c) Sketch the vertical concentration profiles of the gaseous species near the gas–solid interface.

PROJECT

In the process of modeling certain complex combustion problems, it is often helpful to consider some simpler problems that are similar or closely related to the combustion problems under investigation. For example, in order to formulate the pyrolysis process of a solid fuel under radiative heating, one could start with a simpler problem by considering the sublimation of ice under radiative heating. In this project, let us consider a very thick layer of ice on a large pond.

1. Assuming the air is still (no wind), formulate the ice melting and evaporation problem under solar heating. The initial temperature of ice, T_i, could be much below its melting point (273 K). Write the governing equations for the temperature profiles in ice and air. Specify the initial and boundary conditions so that the problem can be solved numerically. Please list all assumptions and define all important variables. (The radiation heat flux could be solved from a two-flux model.)
2. Reformulate the above problem, allowing air to flow over the ice surface.

(6)

e HAS ONLY THERMAL ENERGY

e^C CHEMICAL ENERGY

$$e^C = \sum_{i=1}^{N} h_i^c Y_i - P/\rho = \sum_{i=1}^{N}\left(\Delta h^\circ_{f,i} + \int_{T_0}^{T} C_{p,i}\,dT\right) Y_i - \frac{P}{\rho} = e + \sum_{i=1}^{N} \Delta h^\circ_{f,i} Y_i$$

(3-64) $\quad h - \frac{P}{\rho} = e$

$$h \equiv \sum_{i=1}^{N} Y_i \int_{T_0}^{T} C_{p,i}\,dT$$

TAKE PARTIAL

$$\frac{\partial e^C}{\partial t} = \frac{\partial e}{\partial t} + \sum_{i=1}^{N} \Delta h^\circ_{f,i} \frac{\partial Y_i}{\partial t}$$

$$\nabla e^C = \nabla e + \sum_{i=1}^{N} \Delta h^\circ_{f,i} \nabla Y_i$$

DEFINE

$$q^C \equiv -\lambda \nabla T + \rho \sum h_i^c Y_i V_i + R_u T \sum_{i=1}^{N}\sum_{j=1}^{N} \frac{X_j D_{T,i}}{W_i D_{ij}} (V_i - V_j) \quad (h_i^c = h_i + \Delta h^\circ_{f,i})$$

$$= q + \rho \sum_{i=1}^{N} Y_i V_i \Delta h^\circ_{f,i}$$

$$\nabla \cdot q^C = \nabla \cdot q + \nabla \cdot \left(\rho \sum_{i=1}^{N} Y_i V_i \Delta h^\circ_{f,i}\right)$$

NO EXT. ENERGY INPUT TO THE CONTROL VOL. (3-71), $\dot{Q} = 0$

SUB. 3-71 w/ $\dot{Q} \neq 0$ FROM STATED EQNS

$$\dot{Q} = \rho\left(\frac{\partial e}{\partial t} - \frac{\partial e^C}{\partial t}\right) + \rho \bar{V} \cdot \nabla e - \rho \bar{V} \cdot \bar{\nabla} e^C + \bar{\nabla} \cdot q - \bar{\nabla} \cdot q^C$$

$$= -\rho \sum_{i=1}^{N} \Delta h^\circ_{f,i} \frac{\partial Y_i}{\partial t} - \rho \bar{V} \cdot \sum_{i=1}^{N} \Delta h^\circ_{f,i} \bar{\nabla} Y_i - \bar{\nabla} \cdot \rho \sum_{i=1}^{N} Y_i \bar{V}_i \Delta h^\circ_{f,i} \quad (3)$$

SPECIES EQN

$$\rho \frac{\partial Y_i}{\partial t} + \rho \bar{V} \cdot \bar{\nabla} Y_i = \dot{w}_i - \nabla \cdot (\rho Y_i \bar{V}_i) \quad i = 1, \ldots, N$$

MULT. BY $\Delta h^\circ_{f,i}$ + SUM ALL SPECIES

$$\rho \sum_{i=1}^{N} \Delta h^\circ_{f,i} \frac{\partial Y_i}{\partial t} + \rho \bar{V} \cdot \sum_{i=1}^{N} \Delta h^\circ_{f,i} \nabla Y_i + \bar{\nabla} \cdot \rho \sum_{i=1}^{N} Y_i \bar{V}_i \Delta h^\circ_{f,i} = \sum_{i=1}^{N} \dot{w}_i \Delta h^\circ_{f,i} \quad (4)$$

SUB. (4) INTO (3): $\dot{Q} = -\sum_{i=1}^{N} \dot{w}_i \Delta h^\circ_{f,i}$

$\dot{w}_i$ = RATE PROD OF ith SPECIES

$\Delta h^\circ_{f,i}$ IS NEG. FOR EXOTHERMIC REACTION IN FORMATION OF SPECIES i

CHAPTER

4 Detonation and Deflagration Waves of Premixed Gases

ADDITIONAL SYMBOLS

Symbol	Description	Dimension
c	Local sonic velocity	L/t
$\overline{C}_p$	Average specific heat	Q/MT
v_w	Wave velocity (Fig. 4.3)	L/t
f	Effective mole fraction [Eq. (4-159)]	—
h	Enthalpy per unit mass (thermal plus chemical)	Q/M
M	Mach number	—
q	Heat per unit mass [(Eq. (4-16)]	Q/M
q_{cond}	Heat per unit mass by conduction	Q/M
R	Specific gas constant	Q/T
s	Entropy per unit mass	Q/MT
u	Velocity (Fig. 4.1)	L/t
v	Velocity (Fig. 4.3)	L/t
Δe_f°	Specific energy of formation	Q/M
Δh_c	Enthalpy increase per unit mass due to combustion	Q/M
Δh_s	Enthalpy increase per unit mass due to shock heating	Q/M

$\Delta h^{\circ}_{f,i}$	Specific enthalpy of formation	Q/M
Δu	Velocity difference [Eq. (4-48)]	L/t
α	Constant [Eq. (4-164)]	—
β	Constant [Eq. (4-164)]	—
γ	Specific-heat ratio	—
λ	Thermal Conductivity	Q/LTt

Superscript

$\circ$	Standard state condition

Subscripts

C–J	Taken at Chapman–Jouguet points
HC	Taken along Hugoniot curve
j	jth species
L	Taken at lower Chapman–Jouguet point J
k	Index of summation
min	Minimum value of quantity
s	Taken at constant entropy
t	Total
U	Taken at upper Chapman–Jouguet point Y
0	Initial state
$1/\rho$	Taken at constant $1/\rho$
1	Condition of unburned gases in region I (Fig. 4.1)
2	Condition of burned gases in region II (Fig. 4.1)

The chemical-reaction zone is often called the "flame zone," "flame front," "reaction wave," or the like. Within the flame zone, rapid reactions take place and light is usually (but not always) emitted from the flame. In general, there are two types of flames:

a. *Premixed Flame*. Reactants are perfectly mixed before chemical reaction.
b. *Diffusion Flame*. Reactants diffuse into each other during the chemical reaction.

In this chapter, we shall concentrate on the premixed flame.

Depending upon the existence of a combustion wave and the speed of wave propagation through a reacting mixture, reactions of premixed gases are generally divided into three categories:

1. *Explosion.* Rate of heat generation is extremely fast, but it does not require the passage of a combustion wave through the exploding medium.
2. *Deflagration.* A combustion wave propagating at subsonic speed.
3. *Detonation.* A combustion wave propagating at supersonic speed.

In the following, we shall discuss the characteristics of the deflagration and detonation waves, and derive the Rankine–Hugoniot equations to relate the properties on either side of the wave. The method of calculating the speed and structure of the detonation wave will also be discussed.

1 QUALITATIVE DIFFERENCES BETWEEN DETONATION AND DEFLAGRATION

The waves we consider in the first part of this chapter will be limited to one-dimensional (1D) planar waves. A schematic diagram of a one-dimensional combustion wave is shown in Fig. 4.1 to help visualize the physical situation. In this figure, we are following the motion of a one-dimensional planar combustion wave in a very long duct with constant area. The combustion wave is moving to the left at a constant velocity u_1. In a reference frame following the wave motion, the stationary unburned gases ahead of the wave can be considered to move at velocity u_1 toward the wave front. The selection of this frame of reference is convenient for working with a stationary wave. In Fig. 4.1, the subscript 1 designates conditions of the unburned gases ahead of the wave, and subscript 2 indicates conditions of the burned gases behind the wave. Velocities u_1 and u_2 are defined with respect to the coordinate system fixed relative to the stationary wave.

Typical data on the upstream and downstream conditions can give us a good picture of the physical situation during a deflagration and detonation

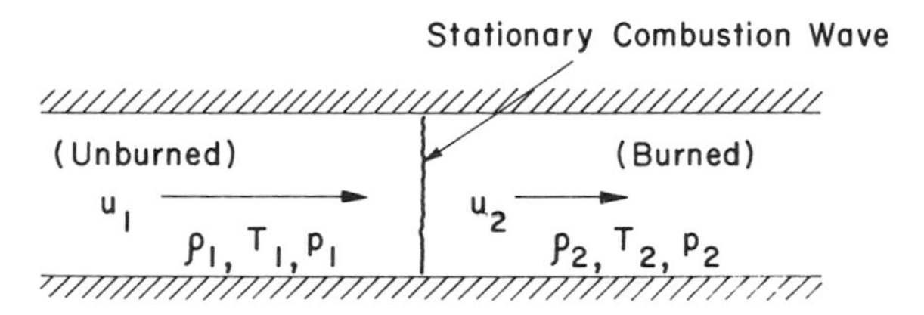

Figure 4.1 Schematic diagram of a stationary one-dimensional combustion wave (deflagration or detonation wave).

TABLE 4.1 Qualitative Differences between Detonation and Deflagration in Gases[a]

	Detonation	Deflagration
u_1/c_1	5–10	0.0001–0.03
u_2/u_1	0.4–0.7 (deceleration)	4–6 (acceleration)
p_2/p_1	13–55 (compression)	≈ 0.98 (slight expansion)
T_2/T_1	8–21 (heat addition)	4–16 (heat addition)
ρ_2/ρ_1	1.7–2.6	0.06–0.25

[a] Data taken by Friedman.[1]

wave. A comparison of data between deflagration and detonation, taken from Friedman,[1] is shown in Table 4.1.

A tube containing an explosive gas mixture having one or both ends open will permit a combustion wave to propagate when ignited at the open end. The wave can attain a steady velocity and will not accelerate to a detonation wave. If the mixture is ignited at the closed end, the reacted hot gases act as a piston which pushes the reaction front to the unburned gas. This wave can accelerate and lead to detonation.

2 THE HUGONIOT CURVE

The conservation equations for steady one-dimensional flow, with no body forces, no external heat addition or heat loss, negligible Dufour and species interdiffusion effects, are as follows:

Continuity.

$$\frac{d(\rho u)}{dx} = 0 \tag{4-1}$$

Momentum.

$$\rho u \frac{du}{dx} = -\frac{dp}{dx} + \frac{d}{dx}\left[\left(\tfrac{4}{3}\mu + \mu'\right)\frac{du}{dx}\right] \tag{4-2}$$

Energy.

$$\rho u\left[\frac{d}{dx}\left(h + \frac{u^2}{2}\right)\right] = -\frac{d}{dx}q_{\text{cond}} + \frac{d}{dx}\left[u\left(\tfrac{4}{3}\mu + \mu'\right)\frac{du}{dx}\right] \tag{4-3}$$

where

$$q_{\text{cond}} = -\lambda\frac{dT}{dx} \tag{4-4}$$

$$h = C_p T + h^\circ \tag{4-5}$$

The viscosity $(\frac{4}{3}\mu + \mu')$ in the momentum equation is due to

$$\tau_{ij} = \mu\left(\frac{\partial u_i}{\partial x_j} + \frac{\partial u_j}{\partial x_i}\right) + \left(\mu' - \tfrac{2}{3}\mu\right)\frac{\partial u_k}{\partial x_k}\delta_{ij} \tag{4-6}$$

$$\tau_{11} = \left(2\mu + \mu' - \tfrac{2}{3}\mu\right)\frac{du}{dx} = \left(\tfrac{4}{3}\mu + \mu'\right)\frac{du}{dx} \tag{4-7}$$

The bulk viscosity μ' is usually very small and can be neglected.

By integrating Eq. (4-1) to Eq. (4-3) we have

$$\rho u = \text{constant} \equiv \dot{m} \tag{4-8}$$

$$\rho u \frac{du}{dx} + u\frac{d}{dx}(\rho u) = -\frac{dp}{dx} + \frac{d}{dx}\left(\tfrac{4}{3}\mu\frac{du}{dx}\right) \tag{4-2}$$

From continuity

$$\frac{d}{dx}(\rho u) = 0 \tag{4-1}$$

Equation (4-2) becomes

$$\frac{d}{dx}\left[\rho u^2 + p - \tfrac{4}{3}\mu\frac{du}{dx}\right] = 0 \tag{4-9}$$

Integrating,

$$\rho u^2 + p - \tfrac{4}{3}\mu\frac{du}{dx} = \text{constant} \tag{4-10}$$

The energy equation becomes

$$\rho u\left(C_p T + h^\circ + \tfrac{1}{2}u^2\right) - \lambda\frac{dT}{dx} - u\left(\tfrac{4}{3}\mu\frac{du}{dx}\right) = \text{constant}' \tag{4-11}$$

Because du/dx and dT/dx are both equal to zero in the burned and unburned regions, the following conservation equations provide the relationships between the flow properties in these two regions:

$$\rho_1 u_1 = \rho_2 u_2 = \dot{m} \tag{4-12}$$

$$p_1 + \rho_1 u_1^2 = p_2 + \rho_2 u_2^2 \tag{4-13}$$

$$C_p T_1 + \tfrac{1}{2}u_1^2 + q = C_p T_2 + \tfrac{1}{2}u_2^2 \tag{4-14}$$

or

$$h_1 + \tfrac{1}{2}u_1^2 = h_2 + \tfrac{1}{2}u_2^2 \tag{4-14a}$$

and

$$p_2 = \rho_2 R_2 T_2 \tag{4-15}$$

where

$$q \equiv h_1^\circ - h_2^\circ \tag{4-16}$$

and

$$h^\circ = \sum_{i=1}^{N} Y_i \Delta h_{f,i}^\circ \tag{4-17}$$

Equation (4-15) is simply the equation of state. There are four equations, but five unknowns: $u_1, u_2, \rho_2, T_2, p_2$.

Note that Eqs. (4-12) to (4-15) can be obtained by considering the physics of the problem directly rather than from the general fluid-dynamic equations.

The four equations can be reduced to one equation in two unknowns, p_2 and ρ_2. Combining Eq. (4-12) and Eq. (4-13),

$$p_2 - p_1 = \rho_1 u_1^2 - \rho_2 u_2^2 = \frac{(\rho_1 u_1)^2}{\rho_1} - \frac{(\rho_2 u_2)^2}{\rho_2} = \left(\frac{1}{\rho_1} - \frac{1}{\rho_2}\right)\dot{m}^2 \tag{4-18}$$

$$\therefore \quad \rho_1^2 u_1^2 = \frac{p_2 - p_1}{\dfrac{1}{\rho_1} - \dfrac{1}{\rho_2}} = \dot{m}^2 \tag{4-19}$$

Equation (4-19) is generally referred to as the *Rayleigh-line relation*. This can also be written in terms of the Mach number $M_1 \equiv u_1/c_1$. We have

$$c_1 \equiv \sqrt{\gamma R_1 T_1} = \sqrt{\gamma\left(\frac{p_1}{\rho_1}\right)} \tag{4-20}$$

$$\frac{\gamma}{\gamma}\frac{\rho_1^2 u_1^2}{\rho_1 p_1} = \frac{\dfrac{p_2}{p_1} - 1}{1 - \dfrac{\rho_1}{\rho_2}} \tag{4-21}$$

Combining, the Rayleigh-line relation becomes

$$\gamma M_1^2 = \frac{\dfrac{p_2}{p_1} - 1}{1 - \dfrac{\rho_1}{\rho_2}} \tag{4-22}$$

Using the definition for C_p and the equation of state, Eq. (4-14) can be rewritten in the form of Eq. (4-25) by using

$$C_p - C_v = R \tag{4-23}$$

or

$$C_p = \frac{R}{1 - 1/\gamma} = \frac{\gamma}{\gamma - 1} R \tag{4-24}$$

We obtain

$$\frac{\gamma}{\gamma - 1}\left(\frac{p_2}{\rho_2} - \frac{p_1}{\rho_1}\right) - \tfrac{1}{2}\left(u_1^2 - u_2^2\right) = q \tag{4-25}$$

Rearranging Eq. (4-13) to obtain expressions for u_1^2 and u_2^2 and substituting these expressions into Eq. (4-25), we have

$$\frac{\gamma}{\gamma - 1}\left(\frac{p_2}{\rho_2} - \frac{p_1}{\rho_1}\right) - \frac{1}{2}\left(\underbrace{\frac{p_2 - p_1}{\rho_1} + \frac{\rho_2}{\rho_1}u_2^2}_{u_1^2} + \underbrace{\frac{p_2 - p_1}{\rho_2} - \frac{\rho_1}{\rho_2}u_1^2}_{-u_2^2}\right) = q \tag{4-26}$$

Rearranging Eq. (4-12) and combining with Eq. (4-26), we have

$$\boxed{\frac{\gamma}{\gamma - 1}\left(\frac{p_2}{\rho_2} - \frac{p_1}{\rho_1}\right) - \tfrac{1}{2}(p_2 - p_1)\left(\frac{1}{\rho_1} + \frac{1}{\rho_2}\right) = q} \tag{4-27}$$

Equation (4-27) is known as the (*Rankine–*) *Hugoniot relation*. The plot of p_2 versus $1/\rho_2$ for a fixed heat release per unit mass, q, is called the *Hugoniot curve*, and is shown in Fig. 4.2. Alternatively, the Hugoniot relation can be expressed in terms of total (thermal plus chemical) enthalpy h.

Combining Eq. (4-5) with Eq. (4-24) and the perfect-gas law, we have

$$h = \left(\frac{\gamma}{\gamma - 1}\right)\frac{p}{\rho} + h^\circ \tag{4-28}$$

Combining Eq. (4-16) and Eq. (4-28) gives

$$h_2 - h_1 = \frac{\gamma}{\gamma - 1}\frac{p_2}{\rho_2} + h_2^\circ - \frac{\gamma}{\gamma - 1}\frac{p_1}{\rho_1} - h_1^\circ = \frac{\gamma}{\gamma - 1}\left(\frac{p_2}{\rho_2} - \frac{p_1}{\rho_1}\right) - q \tag{4-29}$$

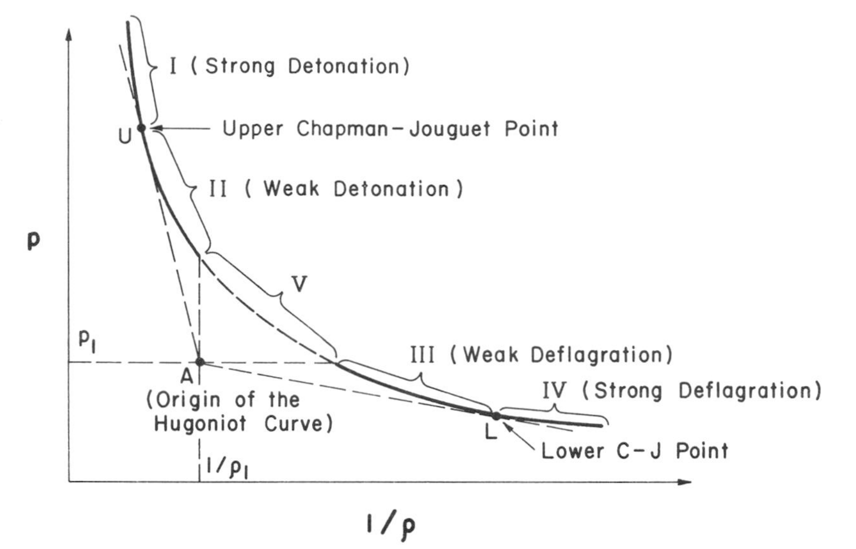

Figure 4.2 Hugoniot curve on p-versus-$1/\rho$ plane, showing various sections of the curve corresponding to the solution domains of various combustion conditions.

Substituting into Eq. (4-27) gives the equivalent Hugoniot relation:

$$h_2 - h_1 = \tfrac{1}{2}(p_2 - p_1)\left(\frac{1}{\rho_1} + \frac{1}{\rho_2}\right) \tag{4-30}$$

3 PROPERTIES OF THE HUGONIOT CURVE

The Hugoniot is essentially a plot of all the possible values of $(1/\rho_2, p_2)$ for a given value of $(1/\rho_1, p_1)$ and q. The point $(1/\rho_1, p_1)$, which is usually called the *origin* of the Hugoniot plot, is designated by the symbol A.

Regions of possible solutions are constructed by drawing tangents to the curve through the point A, and vertical and horizontal lines from A. The curve is thus divided into five regions, specified as Roman numerals in Fig. 4.2. The two tangent points to the curve are called Chapman–Jouguet points,[2,3] generally referred to as C–J points, and denoted by U for the upper C–J point and L for the lower C–J point.

It must be noted that although the curve represents all the *possible* solutions of the Hugoniot equation, not all the solutions are valid, for physical reasons. We shall establish which regions of the curve represent valid solutions.

In region V, $p_2 > p_1$ and $1/\rho_2 > 1/\rho_1$. The Rayleigh-line expression (4-19) implies that u_1 is imaginary. Thus, region V is shown to be a physically impossible region.

To study the characteristic nature at the C–J points, let us consider q to be fixed and differentiate the Hugoniot relation (4-27) with respect to $1/\rho_2$:

$$\frac{1}{\rho_2}\left(\frac{\gamma}{\gamma-1}\right)\frac{dp_2}{d(1/\rho_2)}+\left(\frac{\gamma}{\gamma-1}\right)p_2$$

$$-\tfrac{1}{2}(p_2-p_1)-\frac{1}{2}\frac{dp_2}{d(1/\rho_2)}\left(\frac{1}{\rho_1}+\frac{1}{\rho_2}\right)=0 \tag{4-31}$$

After rearranging, we have

$$\frac{dp_2}{d(1/\rho_2)}=\frac{(p_2-p_1)-\left(\dfrac{2\gamma}{\gamma-1}\right)p_2}{\left(\dfrac{2\gamma}{\gamma-1}\right)\dfrac{1}{\rho_2}-\left(\dfrac{1}{\rho_1}+\dfrac{1}{\rho_2}\right)} \tag{4-32}$$

The slopes at the tangent points U and L can also be represented as

$$\left[\frac{dp_2}{d(1/\rho_2)}\right]_{\text{C-J}}=\frac{p_2-p_1}{\dfrac{1}{\rho_2}-\dfrac{1}{\rho_1}} \tag{4-33}$$

Equating the right-hand side of Eq. (4-32) to that of Eq. (4-33) and rearranging, we have

$$\frac{p_2-p_1}{\left(\dfrac{1}{\rho_2}-\dfrac{1}{\rho_1}\right)}=-\gamma\rho_2 p_2 \qquad \text{at C–J points} \tag{4-34}$$

Combining Eq. (4-34) with Eq. (4-19), we have

$$u_2^2=\frac{\gamma p_2}{\rho_2}=c_2^2, \qquad \text{or} \quad |u_2|=c_2 \tag{4-35}$$

Therefore, at the C–J points $M_2=1$.

In the detonation branch of the Hugoniot curve (regions I and II), $1/\rho_2 < 1/\rho_1$; therefore

$$u_2-u_1=\dot{m}\left(\frac{1}{\rho_2}-\frac{1}{\rho_1}\right)<0$$

or

$$u_1>u_2 \tag{4-36}$$

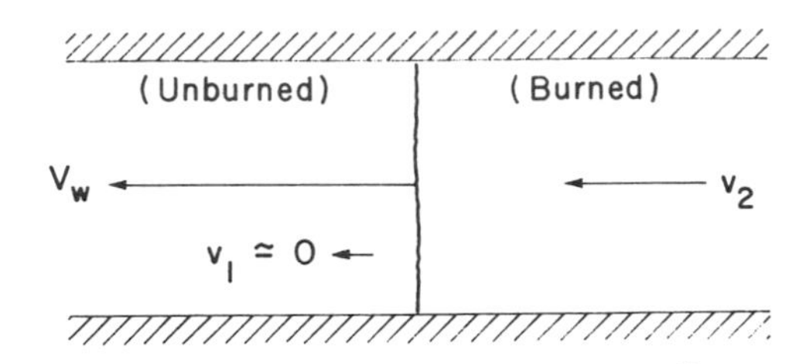

Figure 4.3 Detonation wave in the laboratory coordinate system.

where u_1 and u_2 are the velocities of the unburned and burned gases relative to the detonation wave (Fig. 4.1). If we now consider a laboratory coordinate system in which the detonation wave is moving at velocity V_w relative to the stationary duct, the direction of the absolute velocities of the gases before and behind the detonation wave is shown in Fig. 4.3. The algebraic relationships between the absolute and relative velocities are

$$v_1 = V_w - u_1$$

$$v_2 = V_w - u_2 \tag{4-37}$$

While u_1 and u_2 are positive in the direction shown in Fig. 4.1, v_1 and v_2 are positive in the direction shown in Fig. 4.3.

Since the absolute velocity of the unburned gases is assumed to be zero, $V_w = u_1$, and from the inequality (4-36) we have

$$v_2 = V_w - u_2 = u_1 - u_2 > 0 \tag{4-38}$$

Physically, this means that the burned gases behind a detonation wave try to follow the wave motion. We shall examine the magnitude of v_2 with respect to V_w to see whether the burned gases can catch up with the wave motion or not. First of all, the detonation-wave speed V_w can be expressed as the sum of u_2 and v_2 by rearranging Eq. (4-37):

$$V_w = u_2 + v_2$$

For the upper C–J point, $u_2 = c_2$ and we have

$$V_w = c_2 + v_2 > c_2 \tag{4-39}$$

This implies that the C–J detonation wave is traveling at supersonic speed. Also, because $V_w > v_2$, the burned gases, although traveling in the same direction as the detonation wave, can never catch up with it.

Region I is called the strong-detonation region; within it the pressure of the burned gases is greater than that of the C–J detonation wave, that is, $p_2 > p_U$. In passing through a strong detonation wave, the gas velocity relative to the wave front is slowed down substantially from supersonic speed to subsonic

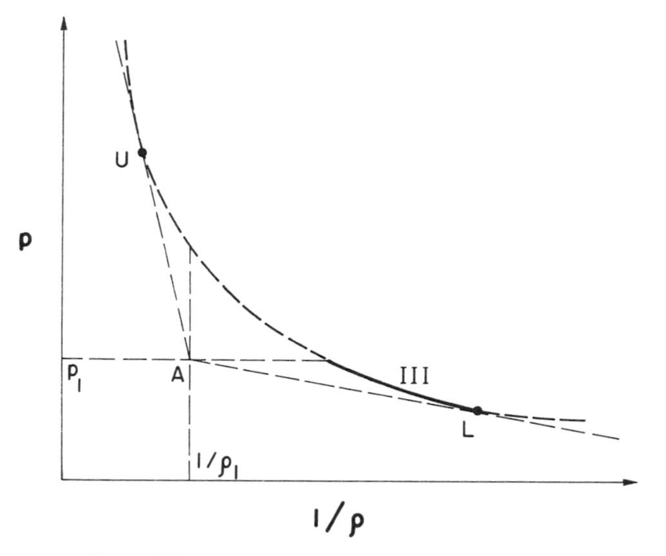

Figure 4.4 Solution regions on the Hugoniot curves.

(shown later). At the same time, the pressure and density increase significantly. It should be noted that the strong detonation with $p_2 \to \infty$ is physically unattainable. As a matter of fact, a strong detonation wave is seldom observed, since it requires a special experimental setup for generating overdriven shock waves in a very strong confinement.

Region II is called the weak-detonation region; within it the pressure of the burned gases is smaller than that of the C–J detonation wave, that is, $p_2 < p_U$. In passing through a weak detonation wave, the gas velocity relative to the wave front is slowed down, but the burned mixture still has a velocity greater than that of sound. It is interesting to note that the isochoric ($1/\rho_2 \approx 1/\rho_1$) weak detonation corresponding to infinite wave velocity [see Eq. (4-19)] is physically unattainable. In general, a weak detonation wave is seldom observed; it requires a gas mixture chosen specially to have extremely fast chemical kinetics.

Under most experimental conditions, detonations are Chapman–Jouguet waves. A part of the reason for ruling out regions I and II stems from the structure of the detonation wave, which will be discussed in a later section. The usual solution on the detonation branch of the Hugoniot curve is therefore represented by the upper C–J point, U in Fig. 4.4. The wave speed at U corresponds to the minimum detonation wave speed, since a straight line through the origin ($1/\rho_1$, p_1) will not intersect the detonation branch anywhere if the magnitude of its slope is less than that of the tangent line shown in Fig. 4.4.

Opposite to the upper C–J point, the C–J deflagration (corresponding to L) has the maximum wave speed of all deflagrations [see Eq. (4-19)]. This is because a straight line through the origin A fails to intersect the deflagration branch if the magnitude of its slope exceeds that of the tangent line, and it intersects the deflagration branch at two points (once in Region III and once in Region IV) if its slope is less than that of the tangent line (see Fig. 4.2).

Region III is called the weak-deflagration region; within it $p_1 \geq p_2 > p_L$ and $1/\rho_2 < 1/\rho_L$. In passing through a weak deflagration wave, the gas velocity relative to the wave front is accelerated from a subsonic velocity to a higher subsonic velocity. The isobaric weak deflagration ($p_s = p_1$) corresponds to zero propagation velocity. The solution in region III is often observed; in most experimental conditions the pressure in the burned-gas zone is slightly lower than that of the unburned gases. The detailed wave structure and methods for determining the wave speed (laminar flame speed of premixed gases) are discussed in detail in Chapter 5.

Region IV is called the strong-deflagration region. In passing through a strong deflagration wave, the gas velocity relative to the wave front must be accelerated substantially from subsonic to supersonic. Considerations of wave structure forbid a change from subsonic to supersonic flow in a constant-area duct.[4] Consequently, strong deflagration is never observed experimentally; the possible solution in the deflagration branch is shown in Fig. 4.4 as the solid line on the Hugoniot curve. The C–J deflagration, corresponding to point L, is also not observed experimentally; hence it is excluded from the solid curve.

In the realizable deflagration-solution region, the burned gases move away from the combustion wave front. This can be seen by using the continuity equation (4-12) and the conditions $1/\rho_2 > 1/\rho_1$ and $\dot{m} > 0$: we have

$$u_2 - u_1 = \dot{m}\left(\frac{1}{\rho_2} - \frac{1}{\rho_1}\right) > 0.$$

Using Eq. (4-37) and substituting V_w for u_1, the above inequality becomes

$$v_2 < 0.$$

This shows that the burned gases, unlike those shown in Fig. 4.3 for a detonation wave, flow away from the deflagration wave. This is one of the characteristic differences between deflagration and detonation.

3.1 Entropy Distribution along the Hugoniot Curve

Now we shall consider the entropy variation along the Hugoniot curve and show that the entropy has a minimum at U and a maximum at L. Let us recall that

$$h \equiv e + \frac{p}{\rho} \tag{4-40}$$

where e represents internal energy including chemical energy:

$$h_2 - h_1 = (e_2 - e_1) + \left(\frac{p_2}{\rho_2} - \frac{p_1}{\rho_1}\right) \tag{4-41}$$

From Eq. (4-30) we know that

$$e_2 - e_1 = \tfrac{1}{2}(p_2 - p_1)\left(\frac{1}{\rho_1} + \frac{1}{\rho_2}\right) - \frac{p_2}{\rho_2} + \frac{p_1}{\rho_1} \tag{4-42}$$

After rearranging, we have

$$e_2 - e_1 = \tfrac{1}{2}(p_2 + p_1)\left(\frac{1}{\rho_1} - \frac{1}{\rho_2}\right) \tag{4-43}$$

From the first and second laws of thermodynamics we have

$$T_2\, ds_2 = de_2 + p_2\, d\left(\frac{1}{\rho_2}\right) \tag{4-44}$$

Differentiating Eq. (4-43), we have

$$de_2 = \tfrac{1}{2}(dp_2)\left(\frac{1}{\rho_1} - \frac{1}{\rho_2}\right) - \frac{1}{2}\left(d\frac{1}{\rho_2}\right)(p_2 + p_1) \tag{4-45}$$

Substitute Eq. (4-45) into Eq. (4-44):

$$T_2\, ds_2 = \frac{1}{2}\left(\frac{1}{\rho_1} - \frac{1}{\rho_2}\right) dp_2 + \tfrac{1}{2}(p_2 - p_1)\, d\left(\frac{1}{\rho_2}\right)$$

or

$$T_2 \frac{ds_2}{d(1/\rho_2)} = \frac{1}{2}\left(\frac{1}{\rho_1} - \frac{1}{\rho_2}\right)\left(\frac{dp_2}{d(1/\rho_2)} - \frac{p_2 - p_1}{1/\rho_2 - 1/\rho_1}\right) \tag{4-46}$$

Using Eq. (4-33) for the slopes at the C–J points, Eq. (4-46) gives

$$\left[\frac{ds_2}{d(1/\rho_2)}\right]_{\text{C-J}} = 0 \tag{4-47}$$

We conclude that the entropy has a maximum or minimum at the C–J points. In order to determine which it is, we have to obtain the second derivative of s_2 with respect to $1/\rho_2$. Differentiating Eq. (4-46), we have

$$T_2\left[\frac{d^2 s_2}{d(1/\rho_2)^2}\right]_{\text{C-J}} = \frac{1}{2}\left(\frac{1}{\rho_1} - \frac{1}{\rho_2}\right)$$
$$\times\left[\frac{d^2 p_2}{d(1/\rho_2)^2} - \frac{1}{1/\rho_2 - 1/\rho_1}\frac{dp_2}{d(1/\rho_2)} + \frac{p_2 - p_1}{(1/\rho_2 - 1/\rho_1)^2}\right]_{\text{C-J}} \tag{4-48}$$

or

$$\left[\frac{d^2 s_2}{d(1/\rho_2)^2}\right]_{C-J} = \frac{1}{2T_2}\left(\frac{1}{\rho_1} - \frac{1}{\rho_2}\right)\left[\frac{d^2 p_2}{d(1/\rho_2)^2}\right]_{C-J} + \frac{1}{2}\left\{\frac{dp_2}{d(1/\rho_2)} - \frac{p_2 - p_1}{1/\rho_2 - 1/\rho_1}\right\}_{C-J}\frac{1}{T_2}$$

Using Eq. (4-33), the last term in the above equation vanishes, so we have

$$\left[\frac{d^2 s_2}{d(1/\rho_2)^2}\right]_{C-J} = \frac{1}{2T_2}\left(\frac{1}{\rho_1} - \frac{1}{\rho_2}\right)\left[\frac{d^2 p_2}{d(1/\rho_2)^2}\right]_{C-J} \tag{4-49}$$

From the shape of the Hugoniot curve (Fig. 4.2), we observe that

$$\frac{d^2 p_2}{d(1/\rho_2)^2} > 0 \qquad \text{for every point} \tag{4-50}$$

This inequality will be shown later after we have introduced the asymptotes of Hugoniot curve.

For the upper C–J point U, we have $1/\rho_1 > 1/\rho_2$, and Eq. (4-49) gives

$$\frac{d^2 s_2}{d(1/\rho_2)^2} > 0 \tag{4-51}$$

This shows that entropy has a minimum at point U. For the lower C–J point L, we have $1/\rho_1 < 1/\rho_2$, and Eq. (4-49) gives

$$\left[\frac{d^2 s_2}{d(1/\rho_2)^2}\right]_L < 0 \tag{4-52}$$

This shows that entropy has a maximum at point L. The entropy distribution thus appears as shown in Fig. 4.5.

3.2 Comparison of the Burned-Gas Velocity behind a Detonation Wave with the Local Speed of Sound

Since the entropy s is a function of p and $1/\rho$,

$$ds = \left[\frac{\partial s}{\partial(1/\rho)}\right]_p d\left(\frac{1}{\rho}\right) + \left[\frac{\partial s}{\partial p}\right]_{1/\rho} dp \tag{4-53}$$

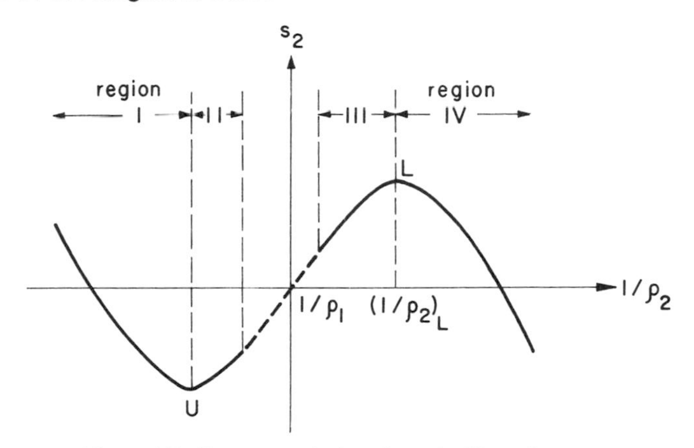

Figure 4.5 Entropy variation along the Hugoniot curve.

The speed of sound of the burned gases can be obtained from the consideration of an adiabatic reversible (isentropic) process. By setting $ds = 0$, Eq. (4-53) reduces to

$$\left[\frac{\partial p}{\partial(1/\rho)}\right]_s = -\frac{\left[\frac{\partial s}{\partial(1/\rho)}\right]_p}{\left[\frac{\partial s}{\partial p}\right]_{1/\rho}} \tag{4-54}$$

We use the subscript HC to represent that the derivative is being taken along the Hugoniot curve. From Eq. (4-53) we know that

$$\left[\frac{\partial s}{\partial(1/\rho)}\right]_{HC} = \left[\frac{\partial s}{\partial(1/\rho)}\right]_p + \left[\frac{\partial s}{\partial p}\right]_{1/\rho}\left[\frac{\partial p}{\partial(1/\rho)}\right]_{HC} \tag{4-55}$$

After rearranging, we have

$$\left[\frac{\partial p}{\partial(1/\rho)}\right]_{HC} = \frac{1}{[\partial s/\partial p]_{1/\rho}}\left\{\left[\frac{\partial s}{\partial(1/\rho)}\right]_{HC} - \left[\frac{\partial s}{\partial(1/\rho)}\right]_p\right\} \tag{4-55a}$$

Subtracting Eq. (4-54) from Eq. (4-55a) yields

$$\left[\frac{\partial p}{\partial(1/\rho)}\right]_{HC} - \left[\frac{\partial p}{\partial(1/\rho)}\right]_s = \frac{\left[\frac{\partial s}{\partial(1/\rho)}\right]_{HC}}{\left[\frac{\partial s}{\partial p}\right]_{1/\rho}}. \tag{4-56}$$

At C–J points, we have

$$\left[\frac{\partial s}{\partial(1/\rho)}\right]_{HC} = 0 \tag{4-57}$$

Using Eqs. (4-56) and (4-57), we have

$$\left[\frac{\partial p}{\partial(1/\rho)}\right]_{HC} = \left[\frac{\partial p}{\partial(1/\rho)}\right]_s \quad \text{(at C–J points)} \tag{4-58}$$

This implies that the tangent lines to the Hugoniot curve on a p-versus-$1/\rho$ plot at C–J points are also the tangent to the constant-entropy line.

After rearranging Eq. (4-46) and combining with Eq. (4-56), we have

$$\frac{2T_2}{1/\rho_1 - 1/\rho_2}\left[\frac{\partial s}{\partial(1/\rho)}\right]_{HC} + \frac{p_2 - p_1}{1/\rho_2 - 1/\rho_1} - \left[\frac{\partial p}{\partial(1/\rho)}\right]_s = \frac{\left[\frac{\partial s}{\partial(1/\rho)}\right]_{HC}}{\left[\frac{\partial s}{\partial p}\right]_{1/\rho}} \tag{4-59}$$

Using Eq. (4-19) and rearranging the above equation, we get

$$-\rho_2^2 u_2^2 + \rho_2^2\left[\frac{\partial p}{\partial \rho}\right]_s = \left[\frac{\partial s}{\partial(1/\rho)}\right]_{HC}\left[\frac{-2T_2}{1/\rho_1 - 1/\rho_2} + \frac{1}{[\partial s/\partial p]_{1/\rho}}\right] \tag{4-60}$$

Based upon the definition of the speed of sound, we know that

$$\rho_2^2\left[\frac{\partial p_2}{\partial \rho_2}\right]_s = \rho_2^2 c_2^2 \tag{4-61}$$

Substituting Eq. (4-61) into Eq. (4-40), we have

$$-\rho_2^2 u_2^2 + \rho_2^2 c_2^2 = \left[\frac{\partial s}{\partial(1/\rho)}\right]_{HC}\left[\frac{-2T_2}{1/\rho_1 - 1/\rho_2} + \frac{1}{[\partial s/\partial p]_{1/\rho}}\right] \tag{4-62}$$

We note that if the right-hand side of Eq. (4-62) is positive in region I, then

$$c_2^2 > u_2^2 \tag{4-63}$$

which implies that the burned-gas velocity is subsonic.

From Fig. 4.5 we known that in region I

$$\left[\frac{\partial s}{\partial(1/\rho)}\right]_{HC} < 0 \tag{4-64}$$

and the right-hand side of Eq. (4-62) will be positive if

$$\frac{2T_2}{1/\rho_1 - 1/\rho_2} > \frac{1}{[\partial s/\partial p]_{1/\rho}} \tag{4-65}$$

In order to prove the above inequality, the partial derivative term on the right-hand side of Eq. (4-65) must be first replaced by an algebraic term involving the properties of the gases at the burned end, since $[\partial s/\partial p]_{1/\rho}$ represents the change of entropy of the burned gases with respect to p_2 when $1/\rho_2$ is held constant. Differentiating Eq. (4-40), we have

$$dh = de + p\,d\left(\frac{1}{\rho}\right) + \frac{1}{\rho}\,dp \tag{4-66}$$

The first law of thermodynamics gives

$$dh = T\,ds + \frac{1}{\rho}\,dp \tag{4-67}$$

For mixtures with constant C_p, we have

$$C_p\,dT = T\,ds + \frac{1}{\rho}\,dp \tag{4-68}$$

From the perfect-gas law,

$$T = \frac{p}{\rho R} \tag{4-69}$$

$$dT = \frac{1}{\rho R}\,dp + \frac{p}{R}\,d\left(\frac{1}{\rho}\right) \tag{4-70}$$

$$C_p = \frac{\gamma}{\gamma - 1}R \tag{4-24}$$

Substituting Eq. (4-70) and Eq. (4-24) into Eq. (4-68) yields

$$\frac{\gamma}{\gamma - 1}R\left[\frac{1}{\rho R}\,dp + \frac{p}{R}\,d\left(\frac{1}{\rho}\right)\right] = T\,ds + \left(\frac{1}{\rho}\right)dp \tag{4-71}$$

Substituting Eq. (4-69) into Eq. (4-71) to eliminate T, and solving for ds, we have

$$ds = \frac{\gamma R}{\gamma - 1}\rho\, d\left(\frac{1}{\rho}\right) + \frac{1}{(\gamma - 1)\rho T}\, dp \tag{4-72}$$

$$\left[\frac{\partial s}{\partial p}\right]_{1/\rho} = \frac{1}{(\gamma - 1)\rho T} \tag{4-73}$$

Using Eq. (4-73), the second factor on the right-hand side of Eq. (4-62) becomes

$$\frac{-2T_2}{1/\rho_1 - 1/\rho_2} + \frac{1}{[\partial s/\partial p]_{1/\rho}} = \frac{-2T_2}{1/\rho_1 - 1/\rho_2} + (\gamma - 1)\rho_2 T_2 \tag{4-74}$$

It can also be written in the following form after some algebraic manipulations:

$$\frac{-2T_2}{1/\rho_1 - 1/\rho_2} + \frac{1}{[\partial s/\partial p]_{1/\rho}} = \frac{p_2}{R}\left[\gamma - \frac{1 + (\rho_1/\rho_2)}{1 - (\rho_1/\rho_2)}\right] \tag{4-75}$$

In order to prove that $c_2 > u_2$ in region I, we need to show that

$$\gamma - \frac{1 + (\rho_1/\rho_2)}{1 - (\rho_1/\rho_2)} < 0 \tag{4-76}$$

The Rankine–Hugoniot relation given by Eq. (4-27) can be arranged to give

$$\frac{\gamma + 1}{2(\gamma - 1)}\frac{p_2}{\rho_2} - \frac{1}{2}\frac{p_2}{\rho_1} + \frac{p_1}{2\rho_2} - \frac{\gamma + 1}{2(\gamma - 1)}\frac{p_1}{\rho_1} - q = 0 \tag{4-77}$$

This is the equation of a hyperbola in the p_2–$1/\rho_2$ plane. To represent p_2 in terms of $1/\rho_2$, we have

$$p_2 = \frac{q + \dfrac{\gamma + 1}{2(\gamma - 1)}\dfrac{p_1}{\rho_1} + \left[-\dfrac{p_1}{2}\right]\dfrac{1}{\rho_2}}{-\dfrac{1}{2}\dfrac{1}{\rho_1} + \left[\dfrac{\gamma + 1}{2(\gamma - 1)}\right]\dfrac{1}{\rho_2}} = \frac{a + \dfrac{b}{\rho_2}}{c + \dfrac{d}{\rho_2}} \tag{4-78}$$

where

$$a = q + \frac{\gamma + 1}{2(\gamma - 1)} \frac{p_1}{\rho_1} \tag{4-79}$$

$$b = -\frac{p_1}{2} \tag{4-80}$$

$$c = -\frac{1}{2\rho_1} \tag{4-81}$$

$$d = \frac{\gamma + 1}{2(\gamma - 1)} \tag{4-82}$$

Solving Eq. (4-78) for $1/\rho_2$, we have

$$\frac{1}{\rho_2} = \frac{a - cp_2}{-b + dp_2} \tag{4-83}$$

The vertical asymptote is determined by

$$c + \frac{d}{\rho_2} = 0$$

or

$$\left(\frac{1}{\rho_2}\right)_{\min} = -\frac{c}{d} = \left(\frac{\gamma - 1}{\gamma + 1}\right)\frac{1}{\rho_1} \tag{4-84}$$

The horizontal asymptote is determined by $p_2 = b/d$, or

$$(p_2)_{\min} = -\left(\frac{\gamma - 1}{\gamma + 1}\right)p_1 \tag{4-85}$$

From the vertical asymptote, we have

$$\frac{1}{\rho_2} > \left(\frac{\gamma - 1}{\gamma + 1}\right)\frac{1}{\rho_1} \tag{4-86}$$

or

$$(\gamma + 1)\frac{\rho_1}{\rho_2} > \gamma - 1 \tag{4-86a}$$

or

$$\frac{\rho_1}{\rho_2} + 1 > \gamma\left(1 - \frac{\rho_1}{\rho_2}\right) \tag{4-86b}$$

For region I,

$$1 - \frac{\rho_1}{\rho_2} > 0 \tag{4-87}$$

Dividing both sides of Eq. (4-86b) by this positive quantity we have

$$\frac{1 + (\rho_1/\rho_2)}{1 - (\rho_1/\rho_2)} > \gamma \qquad \text{(for region I)} \tag{4-88}$$

The above inequality together with Eqs. (4-75) and (4-62) shows that

$$c_2 > u_2 \qquad \text{(in region I)} \tag{4-89}$$

This means that the velocity of the burned gas at any point above the upper C–J point U is subsonic. If we consider the laboratory coordinates, then V_w is the speed of the detonation wave relative to the tube, and $V_w = v_2 + u_2$ [see Eq. (4-37)]. Using this relationship, Eq. (4-89) becomes

$$c_2 > V_w - v_2$$

or

$$c_2 + v_2 > V_w \qquad \text{(in region I)} \tag{4-90}$$

The physical meaning of the terms in the above inequality are given below:

v_2 = the gas particle velocity following the wave.
c_2 = the propagation velocity of any disturbance.
$c_2 + v_2$ = the resultant propagation velocity of any disturbance in the wave motion direction.

Equation (4-90) implies that the rarefaction disturbances propagate at a higher speed than the combustion wave. They catch up with the combustion wave and reduce its strength. Therefore any solution in region I falls back to U.

Note that previously we assumed from the shape of the Hugoniot curve that

$$\frac{d^2 p_2}{d(1/\rho_2)^2} > 0 \tag{4-50}$$

This inequality can be proved by using the asymptotic limits. By differentiating Eq. (4-78) with respect to $1/\rho_2$, we have

$$\frac{dp_2}{d(1/\rho_2)} = \frac{bc - ad}{(c + d/\rho_2)^2} \tag{4-91}$$

and

$$\frac{d^2 p_2}{d(1/\rho_2)^2} = \frac{-2d(bc - ad)}{(c + d/\rho_2)^3} \tag{4-92}$$

From the vertical asymptote (4-84), it can be shown that

$$\left(c + \frac{d}{\rho_2}\right)^3 > 0 \tag{4-93}$$

Also,

$$\begin{aligned} bc - ad &= \frac{p_1}{4\rho_1} - \frac{\gamma + 1}{2(\gamma - 1)}q - \frac{(\gamma + 1)^2}{4(\gamma - 1)^2}\frac{p_1}{\rho_1} \\ &= \frac{-\gamma}{(\gamma - 1)^2}\frac{p_1}{\rho_1} - \frac{\gamma + 1}{2(\gamma - 1)}q \\ &= -\left[\frac{\gamma}{(\gamma - 1)^2}\frac{p_1}{\rho_1} + \frac{\gamma + 1}{2(\gamma - 1)}q\right] \end{aligned} \tag{4-94}$$

Since the quantity in the square brackets is always positive for exothermic reactions, $bc - ad$ is less than zero. Based upon Eqs. (4-93), (4-94), and (4-82), we have

$$\frac{d^2 p_2}{d(1/\rho_2)^2} > 0 \tag{4-95}$$

The solution in region II is ruled out in view of the ZND detonation wave structure,[5] which will be discussed in Section 5. In summary, the commonly observed detonation waves usually correspond to the upper C–J point, and the commonly observed deflagration waves lie in region III of the Hugoniot curve as shown in Fig. 4.4. Chapter 5 is devoted to the determination of the eigenvalue (the actual laminar flame speed) on the weak-deflagration branch.

4 DETERMINATION OF CHAPMAN – JOUGUET DETONATION-WAVE VELOCITY

The detonation velocity u_1 is obviously of great interest to us. Many articles describe methods for the calculation of C–J detonation-wave velocities.[5-15] The calculation methods can be divided into two categories. First are trial-and-error methods,[5-9] and second are Newton–Raphson iteration methods.[10-15] The trial-and-error calculations are generally inferior to Newton–Raphson calculations for two reasons. First, the former give less precision for an equivalent amount of calculation. Second, their success often depends on the intuition of the person performing the calculations in obtaining successive approximations to the solution. To understand their differences, we shall discuss both methods in the following.

4.1 Trial-and-Error Method

The speed of sound behind a detonation wave is given by

$$c_2 = \sqrt{\left[\frac{\partial p_2}{\partial \rho_2}\right]_s} \tag{4-96}$$

Since $u_2 = c_2$ at the upper C–J point, using Eqs. (4-12) and (4-96), we have

$$u_1 = \frac{1}{\rho_1} c_2 \rho_2 = \frac{1}{\rho_1}\sqrt{\rho_2^2\left[\frac{\partial p_2}{\partial \rho_2}\right]_s} = \frac{1}{\rho_1}\sqrt{-\left[\frac{\partial p_2}{\partial(1/\rho_2)}\right]_s} \tag{4-97}$$

The isentropic relation for the burned gas can be written as

$$p_2\left(\frac{1}{\rho_2}\right)^{\gamma_2} = \text{constant} \tag{4-98}$$

Differentiating the above equation and rearranging, we obtain

$$-\left[\frac{\partial p_2}{\partial(1/\rho_2)}\right]_s = \frac{\gamma_2 p_2}{1/\rho_2} \tag{4-99}$$

Substituting this expression into Eq. (4-97), we have

$$u_1 = \frac{1/\rho_1}{1/\rho_2}\sqrt{\gamma_2 p_2 \frac{1}{\rho_2}} = \frac{\rho_2}{\rho_1}\sqrt{\gamma_2 R_2 T_2} \tag{4-100}$$

This equation is useful, but insufficient to determine u_1, since the properties at the burned end are not known. A different form of Eq. (4-100) can be written as

$$\rho_1^2 u_1^2 = \gamma_2 p_2 \rho_2 \tag{4-101}$$

Substituting into the Rayleigh-line relation (4-19), we get

$$\frac{1}{\rho_1} - \frac{1}{\rho_2} = \frac{p_2 - p_1}{\gamma_2 p_2 \rho_2} \tag{4-102}$$

Substituting Eq. (4-102) into Eq. (4-43), we have

$$e_2 - e_1 = \tfrac{1}{2}\left(p_2^2 - p_1^2\right)\frac{1}{\gamma_2 p_2 \rho_2} \tag{4-103}$$

Making an acceptable approximation for the detonation wave that

$$p_2^2 \gg p_1^2$$

then Eq. (4-103) becomes

$$e_2 - e_1 \approx \tfrac{1}{2} p_2 \frac{1}{\gamma_2 \rho_2} = \frac{R_2 T_2}{2\gamma_2} \tag{4-104}$$

The molar form of Eq. (4-104) can be obtained by multiplying the above equation by the molecular weight W_2:

$$e_2 W_2 - \frac{W_2}{W_1} e_1 W_1 = \frac{1}{2}\frac{R_u T_2}{\gamma_2} \tag{4-105}$$

where $e_2 W_2$ and $e_1 W_1$ represent the molar internal energy of the products and reactants, respectively.

Multiplying Eq. (4-102) by a factor of $p_2 + p_1$, we have

$$(p_1 + p_2)\left(\frac{1}{\rho_1} - \frac{1}{\rho_2}\right) = \left(p_2^2 - p_1^2\right)\frac{1}{\gamma_2 p_2 \rho_2} \tag{4-106}$$

Again assuming that $p_2^2 \gg p_1^2$, and expanding the left-hand side, Eq. (4-106) becomes

$$\frac{p_1}{\rho_1} - \frac{p_1}{\rho_2} + \frac{p_2}{\rho_1} - \frac{p_2}{\rho_2} = \frac{p_2}{\gamma_2 \rho_2} \tag{4-107}$$

Using the equation of state to eliminate p_1 and p_2, the above equation becomes

$$R_1 T_1 - \frac{\rho_1}{\rho_2} R_1 T_1 + \frac{\rho_2}{\rho_1} R_2 T_2 - R_2 T_2 = \frac{R_2 T_2}{\gamma_2} \tag{4-108}$$

Multiplying by $\rho_2/(\rho_1 R_2 T_2)$ and rearranging, the approximate Rayleigh-line relation gives

$$\left(\frac{\rho_2}{\rho_1}\right)^2 - \left(\frac{1}{\gamma_2} + 1 - \frac{R_1 T_1}{R_2 T_2}\right)\left(\frac{\rho_2}{\rho_1}\right) - \frac{R_1 T_1}{R_2 T_2} = 0 \tag{4-109}$$

Equation (4-109) allows us to solve for ρ_2/ρ_1 in term of T_2. Also, using the equation of state, p_2 is related to ρ_2/ρ_1 and T_2 by

$$p_2 = \left(\frac{\rho_2}{\rho_1}\right)\left(\frac{R_2 T_2}{R_1 T_1}\right) p_1 \tag{4-110}$$

Using the above relations, the C–J detonation velocity u_1 can be determined according to the following iteration procedure:

1. Assume p_2.
2. Assume T_2.
3. Calculate the equilibrium composition based upon T_2 and p_2.
4. From the equilibrium composition, find γ_2, R_2, e_2.
5. Check Eq. (4-104) to see whether the calculated e_2, γ_2, and R_2 and the assumed T_2 satisfy the equation. If not, go to step 2 and reassume another T_2. If equation is satisfied, proceed to step 6.
6. Solve for ρ_2/ρ_1 from Eq. (4-109).
7. Find p_2 from Eq. (4-110). If the calculated p_2 equals the assumed p_2, then the iteration sequence has converged. Otherwise, go to step 1 and assume a new p_2.
8. Calculate the detonation velocity u_1 from Eq. (4-100).

Note: A quick way to have a good guess of the value of ρ_2/ρ_1 is to assume that $p_2 \gg p_1$. By so doing, Eq. (4-102) reduces to

$$\frac{1}{\rho_1} - \frac{1}{\rho_2} = \frac{1}{\gamma_2 \rho_2} \tag{4-111}$$

or

$$\frac{\rho_2}{\rho_1} \approx \frac{\gamma_2 + 1}{\gamma_2} \tag{4-112}$$

This equation is useful, since an excellent guess of the value of γ_2 can be easily made. One can thus obtain a good guess for p_2.

4.2 The Newton – Raphson Iteration Method

The general iterative method for obtaining the C–J detonation-wave velocity to be described below is based on those of Refs. 14 and 15. The method of Eisen[10,11] also uses the Newton–Raphson iteration procedure[16] for obtaining successive approximations to roots of algebraic equations. However, Eisen's method requires at least four, and often more, separate solutions of the detonation equations to obtain the desired C–J solution. Furthermore, his interpolation scheme can easily lead to estimates for the detonation velocity for which no solution exists. The methods used by McGill[12] and Luker,[13] although they follow the same general Newton–Raphson iteration method, encounter some severe oscillations due to the omission of the changes in composition and heats of reaction with respect to temperature and pressure. The method developed by Zeleznik and Gordon,[14] and programmed by Gordon and McBride,[15] is considered to be the best among all the Newton–Raphson iteration methods and is described below.

There are three steps in the Gordon–McBride calculation procedure. The first step is an initial estimation of the detonation pressure and temperature. The second is an improved estimation of these parameters by means of a recursion formula. The third is obtaining the correct values by means of a Newton–Raphson iteration procedure. The derivation of the required equations is given in Ref. 14, and they are summarized here in slightly modified form.

The conservation equations (4-12), (4-13), and (4-14a) for continuity, momentum, and energy, with the additional constraint $u_2 = c_2$, were reduced to the following two equations:

$$\frac{p_1}{p_2} = \underbrace{1 - \gamma_{s,2}\left(\frac{\rho_2}{\rho_1} - 1\right)}_{p''} \tag{4-113}$$

$$h_2 = h_1 + \frac{R_u \gamma_{s,2} T_2}{2W_2}\left[\left(\frac{\rho_2}{\rho_1}\right)^2 - 1\right] = h'' \tag{4-114}$$

where γ_s represents the exponent used in Eq. (4-98) for an isentropic process; it can be expressed as

$$\gamma_s = \left[\frac{\partial \ln p}{\partial \ln \rho}\right]_s \tag{4-115}$$

It should be noted that γ_s in general is not equal to the ratio of specific heats. Equations (4-113) and (4-114) can be rearranged to give the well-known Rankine–Hugoniot equation (4-30). However, in order to facilitate numerical calculations of detonation velocity, Zeleznik, Gordon, and McBride[14,15] chose to use Eqs. (4-113) and (4-114) rather than use the Hugoniot equation directly.

For convenience in writing the iteration equations the symbols p'' and h'' are used to represent the right sides of Eqs. (4-113) and (4-114), respectively. These equations then become

$$p'' - \frac{p_1}{p_2} = 0 \tag{4-116}$$

$$h'' - h_2 = 0 \tag{4-117}$$

Applying the Newton–Raphson method to Eqs. (4-116) and (4-117) divided by R_u, and using for independent variables the logarithms of the temperature ratio and the pressure ratio across the detonation wave, we have

$$\frac{\partial(p'' - p_1/p_2)}{\partial \ln(p_2/p_1)} \Delta \ln \frac{p_2}{p_1} + \frac{\partial(p'' - p_1/p_2)}{\partial \ln(T_2/T_1)} \Delta \ln \frac{T_2}{T_1} = \frac{p_1}{p_2} - p'' \tag{4-118}$$

$$\frac{\partial\left(\dfrac{h'' - h_2}{R_u}\right)}{\partial \ln \dfrac{p_2}{p_1}} \Delta \ln \frac{p_2}{p_1} + \frac{\partial\left(\dfrac{h'' - h_2}{R_u}\right)}{\partial \ln \dfrac{T_2}{T_1}} \Delta \ln \frac{T_2}{T_1} = \frac{h_2 - h''}{R_u} \tag{4-119}$$

where $\Delta \ln(p_2/p_1)$ represents the difference $\ln(p_2/p_1)_{k+1} - \ln(p_2/p_1)_k$, and the subscript k stands for the kth iteration. Similar expression is used for $\Delta \ln(T_2/T_1)$.

The partial derivatives appearing in Eqs. (4-118) and (4-119) can be evaluated if γ_s is taken to be independent of temperatures and pressures. Within the accuracy of this assumption, the partial derivatives are

$$\frac{\partial(p'' - p_1/p_2)}{\partial \ln(p_2/p_1)} = \frac{p_1}{p_2} - \gamma_{s,2}\left(\frac{\rho_2}{\rho_1}\right)\left(\frac{\partial \ln \rho}{\partial \ln p}\right)_{T,2} \tag{4-120}$$

$$\frac{\partial(p'' - p_1/p_2)}{\partial \ln(T_2/T_1)} = -\gamma_{s,2}\left(\frac{\rho_2}{\rho_1}\right)\left(\frac{\partial \ln \rho}{\partial \ln T}\right)_{p,2} \tag{4-121}$$

$$\frac{\partial\left(\dfrac{h'' - h_2}{R_u}\right)}{\partial \ln \dfrac{p_2}{p_1}} = \frac{\gamma_{s,2} T_2}{2W_2}\left\{\left(\frac{\rho_2}{\rho_1}\right)^2 - 1 + \left(\frac{\partial \ln \rho}{\partial \ln p}\right)_{T,2}\left[1 + \left(\frac{\rho_2}{\rho_1}\right)^2\right]\right\} - \frac{T_2}{W_2}\left[\left(\frac{\partial \ln \rho}{\partial \ln T}\right)_{p,2} + 1\right] \tag{4-122}$$

$$\frac{\partial\left(\dfrac{h'' - h_2}{R_u}\right)}{\partial \ln \dfrac{T_2}{T_1}} = \frac{\gamma_{s,2} T_2}{2W_2}\left[\left(\frac{\rho_2}{\rho_1}\right)^2 + 1\right]\left(\frac{\partial \ln \rho}{\partial \ln T}\right)_{p,2} - \frac{T_2(C_p)_2}{R_u} \tag{4-123}$$

To perform calculations, the equation of state for the products of reaction must be known. In Refs. 14 and 15, consideration is given only to the case where the products of reaction obey the perfect-gas law $pW = \rho R_u T$. From the equation of state, it follows that

$$\left[\frac{\partial \ln \rho}{\partial \ln p}\right]_T = \left[\frac{\partial \ln W}{\partial \ln p}\right]_T + 1 \tag{4-124}$$

$$\left[\frac{\partial \ln \rho}{\partial \ln T}\right]_p = \left[\frac{\partial \ln W}{\partial \ln T}\right]_p - 1 \tag{4-125}$$

The evaluation of the derivatives of the molecular weight appearing in the above equations is discussed in detail in Ref. 17. It should be noted that derivatives of the molecular weight generally have some effect on the convergence and can not be neglected unless they are small relative to unity. Zeleznik and Gordon[14] showed that the omission of these derivatives can result in oscillations.

Gordon and McBride[15] reported that a good initial estimate for the pressure ratio is not as important as a good estimate for the temperature ratio. For a number of chemical systems that were investigated, an initial estimate of $(p_2/p_1)_0 = 15$ has been found to be satisfactory. An initial estimate for the temperature ratio is found by calculating the flame temperature T_2 corresponding to the following enthalpy (see Ref. 14 for details):

$$h_2 = h_1 + \frac{3}{4}\frac{R_u T_1}{W_1}\left(\frac{p_2}{p_1}\right)_0 \tag{4-126}$$

Improved initial assumed values of $(p_2/p_1)_0$ and estimated values of $(T_2/T_1)_0$ corresponding to h_2 in Eq. (4-126) can be obtained by the use of the following recursion formulas:

$$\left(\frac{p_2}{p_1}\right)_{k+1} = \frac{1+\gamma_{s,2}}{2\gamma_{s,2}\alpha_k}\left\{1 + \left[1 - \frac{4\gamma_{s,2}\alpha_k}{(1+\gamma_{s,2})^2}\right]^{1/2}\right\} \tag{4-127}$$

$$\left(\frac{T_2}{T_1}\right)_{k+1} = \left(\frac{T_2}{T_1}\right)_0 - \frac{3}{4}\frac{R_u}{W_1(C_p)_2}\left(\frac{p_2}{p_1}\right)_0 + \frac{R_u\gamma_{s,2}}{2W_1(C_p)_2}\left(\frac{r_{k+1}^2 - 1}{r_{k+1}}\right)\left(\frac{p_2}{p_1}\right)_{k+1} \tag{4-128}$$

where

$$\alpha_k \equiv \left(\frac{T_1}{T_2}\right)_k \left(\frac{W_2}{W_1}\right) \tag{4-129}$$

$$r_{k+1} \equiv \alpha_k \left(\frac{p_2}{p_1}\right)_{k+1} \tag{4-130}$$

The quantities W_2, $\gamma_{s,2}$, and $(C_p)_2$ in Eqs. (4-127) to (4-129) are the equilibrium values for $(p_2/p_1)_0$ and $(T_2/T_1)_0$. Repeating the use of Eqs. (4-127) to (4-130) three times in the program CEC76 generally provides excellent initial estimates for the Newton–Raphson iteration.

An arbitrary control factor f_c is applied to the corrections obtained from solution of Eqs. (4-118) and (4-119) before getting improved estimates of $\ln(p_2/p_1)$ and $\ln(T_2/T_1)$ for the new iteration step. The value of f_c used by Gordon and McBride permits a maximum correction of 1.5 of the previous estimates of p_2/p_1 and T_2/T_1. This is the same as permitting a maximum absolute correction of 0.4054652 on $\ln(p_2/p_1)$ and $\ln(T_2/T_1)$. The control factor is determined by the following equation:

$$f_c = \min\left[\frac{0.4054652}{|\Delta \ln(p_2/p_1)|}, \frac{0.4054652}{|\Delta \ln(T_2/T_1)|}, 1\right] \tag{4-131}$$

Improved estimates are then obtained by using f_c and the following equations:

$$\ln\left(\frac{p_2}{p_1}\right)_{k+1} = \ln\left(\frac{p_2}{p_1}\right)_k + f_c\left[\Delta \ln\left(\frac{p_2}{p_1}\right)\right]_k$$

$$\ln\left(\frac{T_2}{T_1}\right)_{k+1} = \ln\left(\frac{T_2}{T_1}\right)_k + f_c\left[\Delta \ln\left(\frac{T_2}{T_1}\right)\right]_k \tag{4-132}$$

The iteration process is continued until corrections obtained from Eqs. (4-118) and (4-119) meet the following criteria:

$$\left|\Delta \ln\left(\frac{p_2}{p_1}\right)\right| < 0.00005$$

$$\left|\Delta \ln\left(\frac{T_2}{T_1}\right)\right| < 0.00005 \tag{4-133}$$

Convergence of Gordon and McBride's program can generally be reached in three to five iterations.

4.3 Comparison of Calculated Detonation-Wave Velocities with Experimental Data

The first critical experimental test of the C–J theory for its prediction of the detonation velocity was carried out by Lewis and Friauf[18] on mixtures of hydrogen and oxygen diluted with various gases. The dissociation equilibria

TABLE 4.2 Comparison of Lewis and Friauf's Detonation-Velocity Data with the Calculated Results Based upon Chapman–Jouguet Theory

Explosive Mixture	p_2 (atm)	T_2 (K)	u_1 (calc.) (m/sec)	u_1 (expt.) (m/sec)	Dissociation (mole %)
$(2H_2 + O_2)$	18.0	3583	2806	2819	32
$(2H_2 + O_2) + O_2$	17.4	3390	2302	2314	30
$(2H_2 + O_2) + N_2$	17.4	3367	2378	2407	18
$(2H_2 + O_2) + 5N_2$	14.4	2685	1850	1822	2
$(2H_2 + O_2) + 6H_2$	14.2	2650	3749	3532	1

considered were

$$H_2O \rightleftharpoons H_2 + \tfrac{1}{2}O_2$$

$$H_2O \rightleftharpoons \tfrac{1}{2}H_2 + OH$$

$$H_2 \rightleftharpoons 2H$$

Table 4.2 contains a summary of the results for $p_1 = 1$ atm, $T_1 = 298$ K. Except for large excess of hydrogen, the agreement between the calculated and experimental velocities is very good.

The influence of dissociation on the u_1 cannot be neglected. If dissociation is neglected (no H or OH species present), the calculated velocities are considerably larger, by several hundred meters per second, than the experimental data. The equilibrium concentrations of radicals (amount of dissociation shown in the last column) suggest that the chain-carrying concentration in the reaction zone must be very large indeed. This, together with the high temperature, provides an explanation for the extreme rapidity of the reaction (which supports the supersonic wave).

It is seen from Table 4.2 that dilution of the stoichiometric mixture with hydrogen results in an increase in the detonation velocity, despite the reduced temperature T_2. This reflects the influence of the much reduced density of the mixture, a result that suggests the performance of experiments with helium and argon as diluents. The addition of helium to a stoichiometric mixture should increase the velocity because of the decreased density. On the other hand, the addition of argon should decrease the velocity. These two inert gases are identical in their thermal effect, but affect the detonation velocity in opposite directions. How well this prediction is fulfilled can be seen from Table 4.3, obtained by Lewis and Friauf.[18] The experimental fact that dilution with the inert gas helium increases the detonation velocity provides a particularly clear confirmation of the predictability of the detonation wave velocity by the simple theory.

Berets, Greene, and Kistiakowsky[19] repeated the calculations made by Lewis and Friauf,[18] using more accurate spectroscopic data, and obtained

TABLE 4.3 Effect of Diluent Concentration and Initial Density on Detonation-Wave Velocity[a]

Explosive Mixture	p_2 (atm)	T_2 (K)	u_1 (calc.) (m/sec)	u_1 (expt.) (m/sec)
$(2H_2 + O_2) + 5Ar$	16.3	3097	1762	1700
$(2H_2 + O_2) + 3Ar$	17.1	3265	1907	1800
$(2H_2 + O_2) + 1.5Ar$	17.6	3412	2117	1950
$(2H_2 + O_2)$	18.0	3583	2806	2819
$(2H_2 + O_2) + 1.5He$	17.6	3412	3200	3010
$(2H_2 + O_2) + 3He$	17.1	3265	3432	3130
$(2H_2 + O_2) + 5He$	16.3	3097	3613	3160

[a] From Lewis and Friauf.[18] $P_1 = 1$ atm; $T_1 = 291$ K.

good agreement with the theoretical values of detonation velocities reported in Ref. 18 (about 1% difference). They further showed that the measured and calculated u_1 agree closely except for lean mixtures in the neighborhood of the limits of detonability. Under those conditions, the measured data are significantly below the theoretical values. These discrepancies were attributed[20] in early days to the lateral heat losses, which are not considered in the one-dimensional theory. More comparisons between the experimentally measured detonation velocities and theoretical results based upon the classical C–J theory can be found in the books by Lewis and von Elbe[20] and Strehlow.[46] In general, the classical C–J theory predicts properties of the bulk flow which ordinarily are in good agreement with measured values.[21] However, poor agreement has been found for initial pressures, mixture compositions, and tube dimensions which are near the limit values for the occurrence of a constant-velocity propagating wave, as indicated by Strehlow.[21] This disagreement has been attributed to multidimensional transverse structure, which is much more pronounced under these marginal conditions. Such detonation is termed *marginal*, and average frontal velocities as low as 85–90% of the C–J velocity are observed.

5 DETONATION-WAVE STRUCTURE

The understanding of detonation-wave structure has gone through a revolution due to great advances made in the 1960s by various researchers. The evolution from the classical one-dimensional point of view to the modern concepts of multidimensional wave structure can be found in several excellent comprehensive review papers, including those of Oppenheim, Manson, and Wagner,[22] Strehlow,[23] Lee, Soloukhin, and Oppenheim,[24] Schott,[25] Shchelkin,[26] and van Tiggelen and de Soete.[27] To introduce some commonly used terminology in the field of detonation and also to appreciate the efforts involved in reaching this current understanding of detonation, we shall discuss both the classical and modern views.

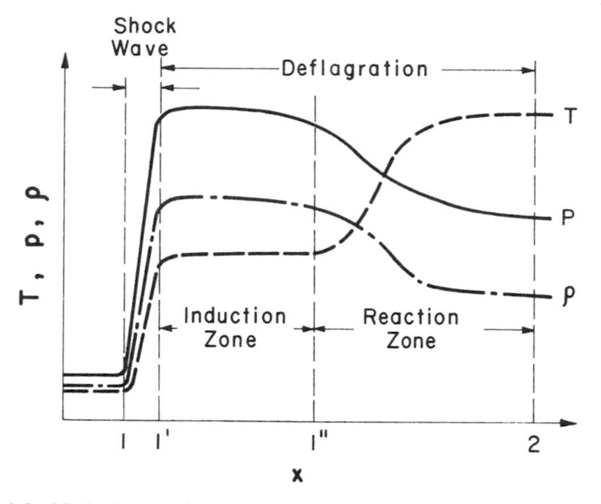

Figure 4.6 Variation of physical properties through a ZND detonation wave.

5.1 ZND One-Dimensional Wave Structure

Extending the classical C–J theory[2,3] (Chapman, 1899; Jouguet, 1905, 1906), Zel'dovich[28] (1940), von Neumann[29] (1942), and Döring[30] (1943) independently assumed that the flow is strictly one-dimensional and steady relative to the detonation front. They postulated that the detonation wave consisted of a shock wave moving at the detonation velocity, with chemical reactions occurring behind the shock in a region much thicker than that of a typical shock wave. They considered that the shock wave initially heats the reactants to a temperature at which they can react at a rate high enough for the ensuing deflagration to propagate as fast as the shock wave. Their assumption of very limited reactions in the shock-wave region itself was based on the fact that the thickness of a shock wave is usually very small, in the order of a few mean free paths of the gas molecules. In view of very small number of collisions between molecules within the shock wave, the major amount of heat release in a detonation wave was believed to occur in a thick region behind the shock wave.

Figure 4.6 shows the variation of physical properties through a one-dimensional ZND detonation wave. The magnitude of the pressure, temperature, and density immediately behind the shock (at station 1′) depends upon the fraction of gaseous mixture reacted. If the reaction rate follows the Arrhenius law, then the rate of reaction increases slowly in the region immediately behind the shock, where the temperature is not extremely high. As a result, the pressure, temperature, and density profiles are relatively flat in the region immediately behind the shock front. This region is therefore called the induction zone. After the induction period, the gas properties change sharply as the reaction rate increases drastically to high values. When the reaction is completed, the thermodynamic properties reach their equilibrium values. The physical dis-

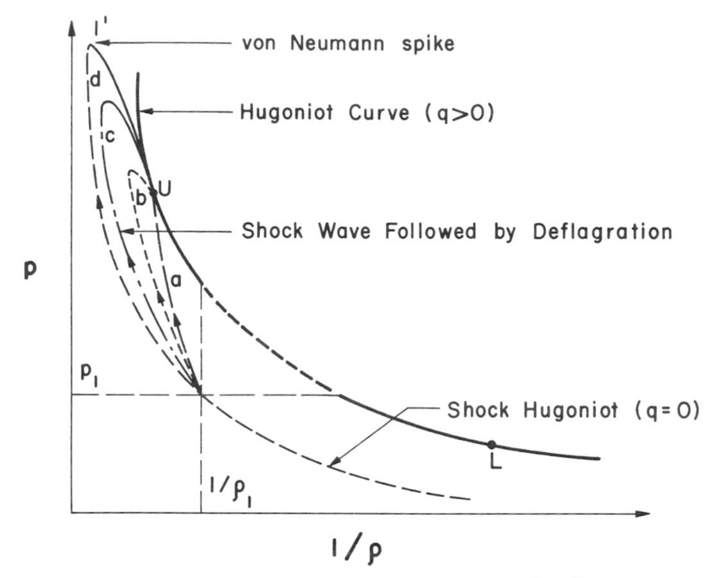

Figure 4.7 ZND detonation structure on $(p, 1/\rho)$ diagram.

tance between the shock front and the fully reacted station is in the order of 1 cm.

The locus of the reacting mixture is shown on the Hugoniot plot in Fig. 4.7. As shown in the figure, there are many paths by which a reacting mixture may pass through the detonation wave from the unreacted state to the fully reacted state. All paths (a, b, c, etc.) can satisfy the conservation equations.

A gaseous mixture may enter the detonation wave in the state corresponding to the initial point $(1/\rho_1, p_1)$ and move directly to the upper C–J point by path a. However, this path would demand that reaction occur everywhere along the path. Since there is no significant compression in the beginning part of this path, there cannot be sufficient temperature to initiate any reaction. Thus, path a is very unlikely to have enough energy release to sustain the wave. Path b represents a possible locus for mixtures with fast chemical kinetics. For slow chemical kinetics, the path may be c. In the limit of zero chemical-energy release in the shock wave, the initial portion of path c will coincide with the shock Hugoniot curve. The peak pressure behind the shock wave in the ZND model is called the *von Neumann spike*.

According to the ZND model, there exist sharp pressure and density gradients at the front of the detonation wave. For detonations propagating into subatmospheric initial pressures, the sharp rises in pressure and density have been confirmed experimentally by several investigators, notably Kistiakowsky and his students,[31,32] Wagner and his coworkers,[33-36] Edwards and his coworkers,[37-39] White,[40-42] and Fay.[43,44]

However, the crucial question, whether the so-called von Neumann spike has been attained or not, could not be answered by these experiments, because their effective space resolution was inadequate. An extrapolation of the density profile, measured by the x-ray absorption technique,[31,32] to the front of the reaction zone demonstrated that a peak density of about 70% of the expected von Neumann spike was probably attained. Just and Wagner[45] measured, by means of a schlieren technique, a peak density that was 70 to 90% of the von Neumann spike value. The preliminary records of the pressure profile, obtained by Edwards et al.,[39] can be extrapolated to a pressure peak that seems to be in fair agreement with the von Neumann spike.

It is worth pointing out that the state at the von Neumann spike is usually evaluated assuming complete equilibrium for different degrees of freedom. However, it is known in particular that the vibrational relaxation times of several species appearing in the vicinity of the von Neumann spike are on the order of a microsecond. In these cases, lower peak densities and higher pressures than those of the equilibrium von Neumann state can be expected to occur.

5.2 Multidimensional Detonation-Wave Structure

According to Strehlow,[46] the first evidence of multidimensional wave structure was obtained in 1926 by Campbell and Woodhead.[47] By observing spin in limit mixtures in circular tubes, they showed that detonation waves might prefer to travel in a manner which is locally three-dimensional and nonsteady. For more than 25 years after their discovery, spinning detonation was thought to be a phenomena peculiar to limit mixtures. The detailed wave structure of self-sustaining detonation waves was not studied until Denisov and Troshin[48] adapted an experimental technique used initially by Antolik[49] in 1875. Antolik used soot-coated plates near a spark discharge and recorded many thin lines on the smoked foil. These lines were interpreted much later as being caused by the intersection of "sound waves." Denisov and Troshin revived the technique in 1959 and first applied it to the observation of transverse waves in detonations. In effect, the smoked-foil method exploits the fact (realized by Denisov and Troshin) that the detonation, unlike any other singular wave process, could actually "write on the walls"; that is, it could leave a record of its passage in the form of an imprint on a film-coated wall, the most convenient material for this purpose being a thin layer of soot. Since 1959, the smoked-foil method has been extensively applied to the study of detonation structure. According to Strehlow,[46] wood smoke deposited in an almost opaque layer on the surface produces the best smoked-foil records. The foil may be "frozen" after firing by spraying it with a clear lacquer.

A few examples of smoked-foil records obtained with self-sustaining propagating detonation waves[46,50] are shown in Figs. 4.8, 4.9, and 4.10. The reason that a propagating detonation wave is able to write on the walls is the presence of triple points at the intersecting lines of triple shock waves. The triple point (line) can be described using Fig. 4.11. In this figure, a steady supersonic

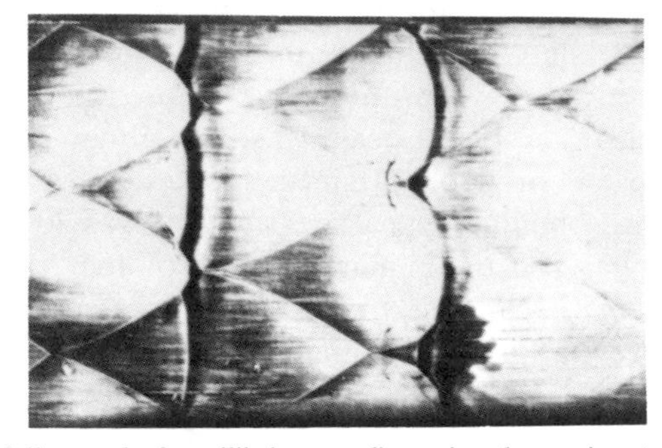

Figure 4.8 Smoked-foil record of equilibrium-configuration detonation (rectangular mode). The active width of this smoked foil is 8.26 cm. The detonation is propagating from left to right. H_2–O_2 system, 40% argon, $p_1 = 60$ Torr, $\phi = 1$ (after R. A. Strehlow[50]).

two-dimensional flow, passing through a convergent ramp section from right to left, produces a shock-wave pattern as shown in the schematic diagram, when the ramp angle is very steep. The triple point (line) is the intersection point (line) of Mach-stem, incident, and reflected shocks. A slipstream is also generated in this situation. A detailed description of such flow configurations can be found in the book by Shapiro.[51]

The detonation-wave structure is characterized by a nonplanar leading shock wave which at every instant consists of many curved shock sections which are convex toward the incoming flow. The lines of intersection of these curved shock segments are propagating in various directions at high velocities (see Fig. 4.10). The third shock (R in Fig. 4.11) of these intersections extends back into the reactive flow regime and is required for the flow to be balanced at the intersection of the two convex leading shock waves. In general, the flow in the neighborhood of the shock front is quite complex. The schematic diagram of symmetric planar interaction is shown in Fig. 4.10.

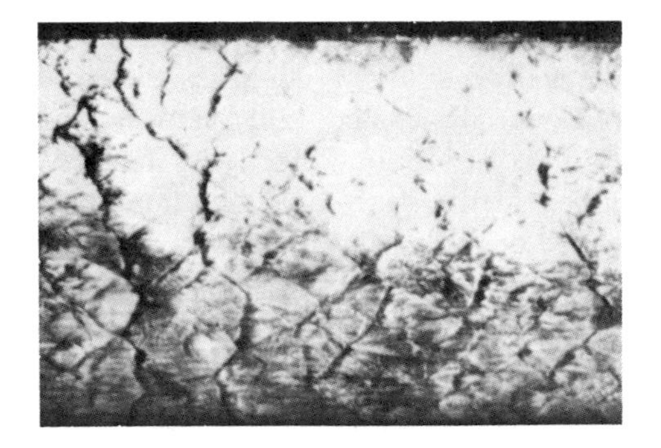

Figure 4.9 Smoked-foil record of a time-average steady detonation, illustrating the lack of an equilibrium configuration. The active width of this smoked foil is 8.26 cm. The detonation is propagating from left to right. 20% C_3H_8–80% O_2 system, $p_1 = 50$ Torr (after R. A. Strehlow[50]).

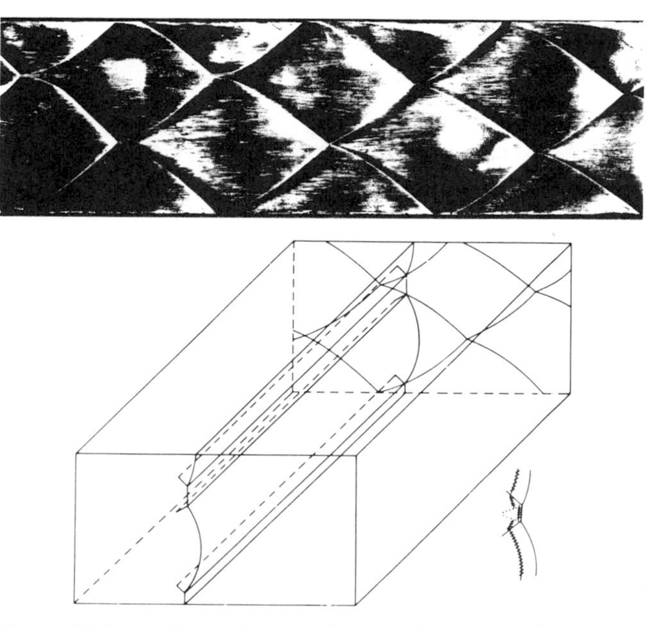

Figure 4.10 Smoked-foil record and schematic diagram of symmetric planar interaction (after R. A. Strehlow[46]).

In agreement with the above wave-structure description, the interferogram of a self-sustained detonation (Fig. 4.12) obtained by White[41] (1961) shows an irregular pattern of the fringes in and behind the reaction zone, whereas the shock front itself is quite smooth and planar. Flash schlieren pictures taken by Fay[43,52] and Opel[52] demonstrated also that the shape of the reaction zone is irregular. On the basis of optical reflectivity measurements, Sastri et al.[53] concluded that detonations in $H_2 + 3O_2$ mixture show the wave front is neither planar nor smoothly curved.

6 THE MECHANISM OF DEFLAGRATION-TO-DETONATION TRANSITION (DDT) IN GASEOUS MIXTURES

The current understanding of the DDT mechanism is presented in an excellent review by Oppenheim, Manson, and Wagner.[22] This understanding has been achieved due largely to the recent development of laser schlieren photography, which has permitted the attainment of extremely short (less than 0.01 μsec) light pulses at ultrahigh (up to 10^6/sec) repetition rate. This technique generates a stroboscopic set of still photographs which reveal many details of the DDT processes which could not be observed by any other means. Some detailed experimental observations of DDT in explosive gases were reported by Urtiew and Oppenheim.[55]

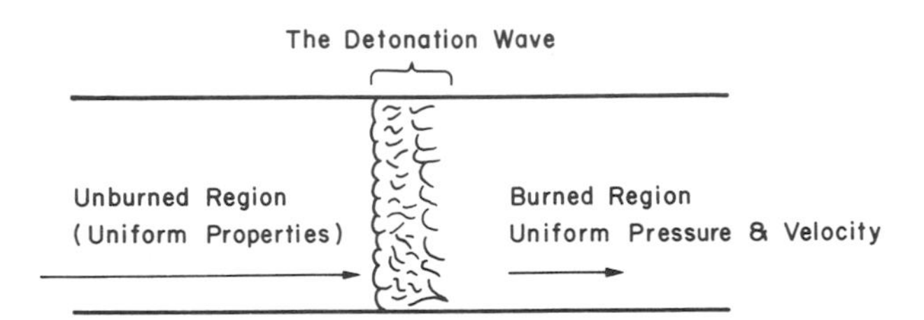

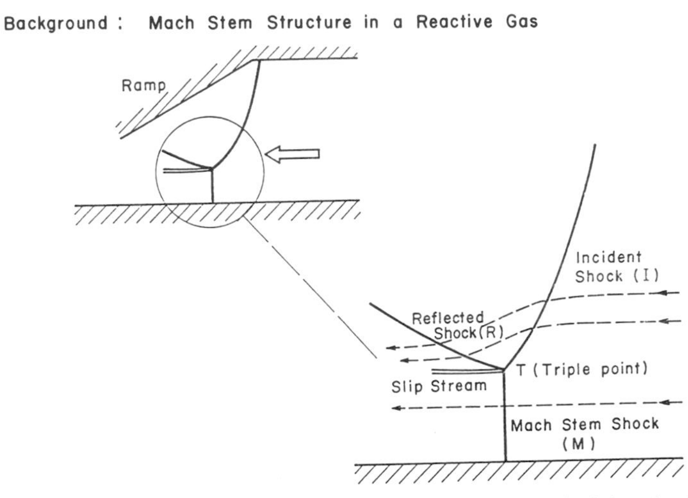

Figure 4.11 Schematic diagram showing the shock-wave pattern and triple point in a two-dimensional supersonic flow passing through a convergent ramp section.

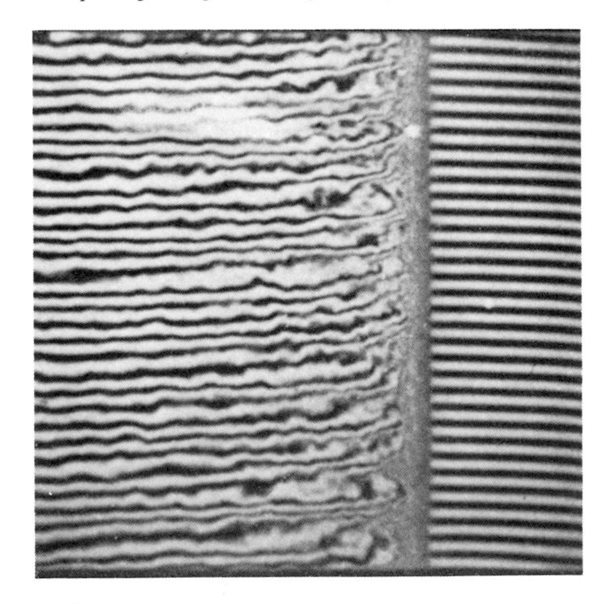

Figure 4.12 Interferogram, obtained by White,[41] of a self-sustained detonation in a $2H_2 + O_2 + 2CO$ mixture initially at 0.3-atm pressure and room temperature. Increase in density shifts fringes upwards.

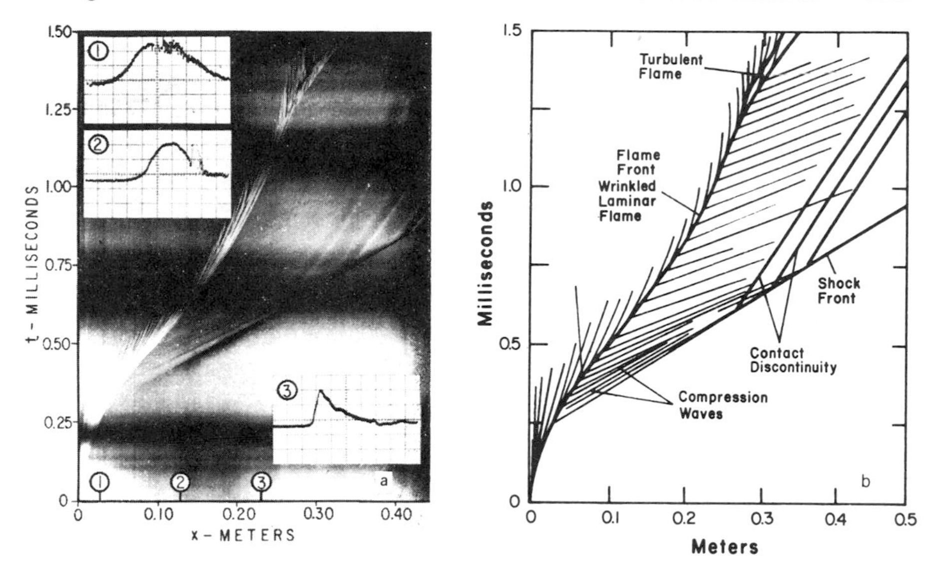

Figure 4.13 Streak schlieren photograph of the development of detonation in stoichiometric hydrogen–oxygen mixture initially at normal temperature and pressure, showing the generation of pressure waves ahead of the accelerating flame. Spark ignition by discharging 1.0 mJ across a $\frac{1}{32}$-in. gap. Electrodes located at closed end of a $1 \times 1\frac{1}{2}$-in.-cross-section tube. Pressure records at positions 1, 2, and 3 shown as inserts. Vertical grid: 5.2 (lb/in.^{2})/div for insert 1, 10.4 (lb/in.^{2})/div for inserts 2 and 3. Horizontal grid: 0.10 msec/div for inserts 1 and 2, 0.20 msec/div for insert 3, from left to right.[22]

The sequence of events leading to detonation in a tube containing explosive gases can be seen in Figs. 4.13–4.16, which are streak and flash schlieren photographs with interpretations. The development of detonation can be summarized as follows:

1. Generation of compression waves ahead of an accelerating laminar flame (see Fig. 4.13). The laminar flame front is wrinkled at this stage.
2. Formation of a shock front due to coalescence of compression waves (see Fig. 4.13).
3. Movement of gases induced by shock, causing the flame to break into a turbulent brush (see Fig. 4.13).
4. Onset of "an explosion in an explosion" at a point within the turbulent reaction zone, producing two strong shock waves in opposite directions and transverse oscillations in between. These oscillations are called transverse waves (see Fig. 4.14). The forward shock is referred to as superdetonation and moves into unburned gases. In the opposite direction, a shock moves into the burned gases and is known as *retonation*.

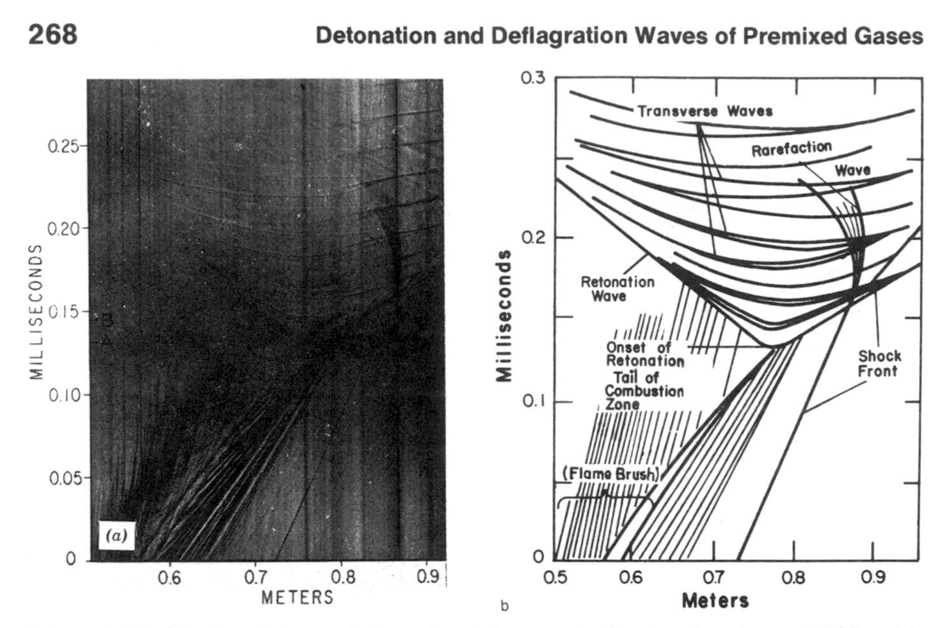

Figure 4.14 Streak schlieren photograph of the onset of retonation in a stoichiometric hydrogen–oxygen mixture initially at normal temperature and pressure. Hot-wire ignitor, made up of a $\frac{1}{2}$-in.-long by $\frac{1}{8}$-in.-diam electrically heated coil, located at closed end of a $1 \times 1\frac{1}{2}$-in.-cross-section tube. The abscissa scale denotes the distance from the end of the tube. Symbols *A* and *B* mark instants at which the accompanying flash schlieren photographs were obtained.[22]

5. Development of spherical shock wave at the onset of the "explosion in an explosion," with a center located in the vicinity of the boundary layer (see Fig. 4.15).
6. Interaction of transverse waves with shock front, retonation wave, and reaction zone (see Fig. 4.16).
7. Establishment of a final "steady wave" as a result of a long sequence of wave interaction processes that lead finally to the shock-deflagration ensemble: the self-sustained C–J detonation.

A typical example of DDT in a $2H_2 + O_2$ mixture ignited by a glow coil in a 1-m-long tube is shown in Fig. 4.17, obtained by Urtiew and Oppenheim.[55]

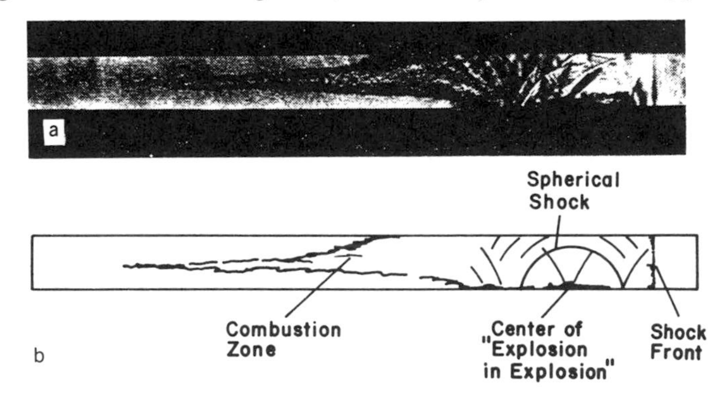

Figure 4.15 Flash schlieren photograph of the onset of retonation in a stoichiometric hydrogen–oxygen mixture initially at normal temperature and pressure at an instant marked by *A* on the streak schlieren photograph (Fig. 4.14).

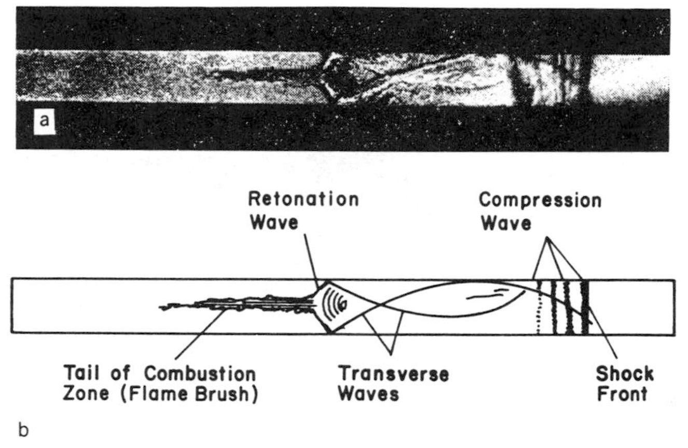

Figure 4.16 Flash schlieren photograph of transverse waves set up at the onset of retonation in a stoichiometric hydrogen–oxygen mixture initially at normal temperature and pressure, shown at an instant marked by B on the streak schlieren photograph (Fig. 4.14).

Occupying the major part of the figure is a streak recorded with a slit located along the center line of the tube, and the knife edge oriented in the direction perpendicular to the main wave motion. Above it is a single-flash photograph which, as demonstrated by the two chain-dotted lines, matches the events of the streaks as if it were taken concurrently. A pressure record, obtained simultaneously with the streak photography by means of a transducer located at a position corresponding to about 0.24 m of the distance scale of Fig. 4.17, is included as an insert.

The nature of the transverse waves can be seen clearly in Fig. 4.17. The minima of the hyperbolic traces indicate the instant the transverse waves crossed the tube center line. The minima of the hyperbolic traces remain nearly at the same spatial location in Fig. 4.17. This implies that the center of the "explosion in an explosion" must have been brought practically to rest. The forward-moving wave was studied using wall imprints of the detonation process, an example of which is shown in Fig. 4.18. The characteristic fish-scale pattern, which corresponds to inception of the forward shock, is a distinguishing feature of a self-sustained detonation front.

In giving these general descriptions of the detonation development, it must be pointed out that there are several different modes of transition processes. The detailed mechanism of each mode depends strongly on the action of the pressure waves that are generated in the course of the transition. Four different modes are generally observed and classified, based upon the location of the onset of an "explosion in an explosion," as follows:

Mode 1. Between flame and shock front (see Figs. 4.19 and 4.20*a*).
Mode 2. At flame front (see Fig. 4.20*b*).
Mode 3. At shock front (see Fig. 4.20*c*).
Mode 4. At contact discontinuity (see Fig. 4.20*d*).

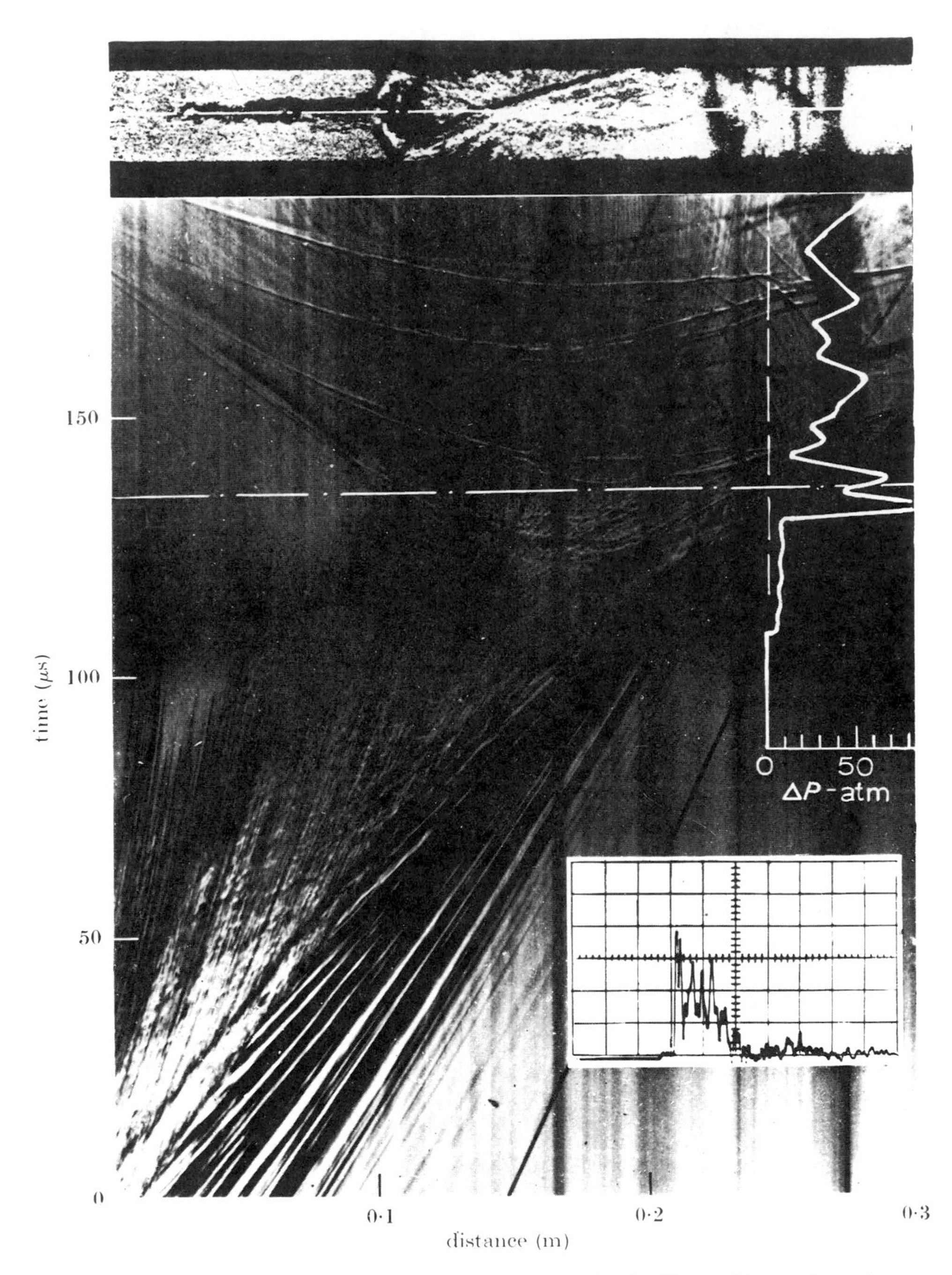

Figure 4.17 Streak schlieren record of wave processes associated with transition to detonation, and a matching single-flash photograph of the wave pattern. $2H_2 + O_2$ initially at atmospheric pressure. Pressure record shown in insert. Vertical scale: 1 div = 260 lb/in.^{2}. Horizontal scale: 1 div = 50 μsec. Oscilloscope sweep leads the streak photograph by 18 μsec.[55]

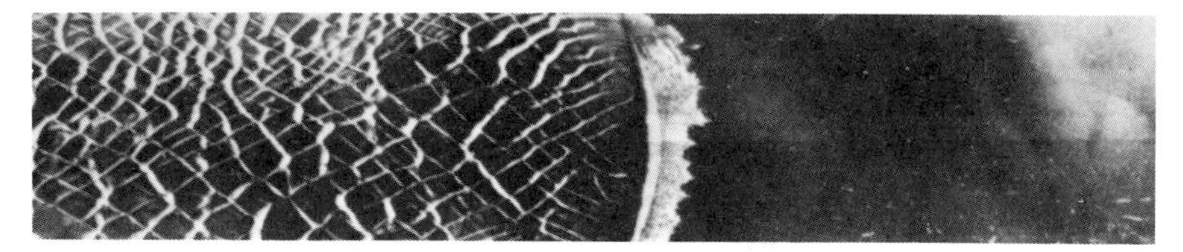

Figure 4.18 Wall imprints of the transition process.

In view of the existence of various modes, the onset of detonation depends on the particular pattern of shock fronts created by the accelerating flame. Since the generation of any particular pattern depends on some minute inhomogeneities in its development, the process of transition to detonation is nonreproducible in its detailed sequence of events.

7 DETONABILITY AND CHEMICAL KINETICS: LIMITS OF DETONABILITY

Belles[54] combined kinetic data with the observed wave behavior during the transition from flame to detonation, and developed a useful technique for predicting the limits of detonability of hydrogen–oxygen mixtures. His analysis was based upon the premise that the transition to detonation occurs when the preflame shock becomes strong enough to cause the mixture to explode. The prediction of limits of detonability were based upon the following considerations and procedures:

A. The chemical kinetics condition for branched-chain explosion of hydrogen was defined in terms of temperature, pressure, and mixture composition.

B. Standard shock-wave equations were used to express the temperature and pressure of the shock-heated gas in terms of the shock strength, represented by the Mach number.

C. Steps A and B were combined so as to express the explosion condition in terms of shock strength and mixture composition. Critical shock strengths for explosion were then determined.

D. Conservation of energy was used to show that some mixtures, on burning, can produce enough energy to support a shock of the critical strength, whereas others cannot. Thus, the limits of detonability can be found.

The following gas-phase kinetics for the oxidation of hydrogen were considered

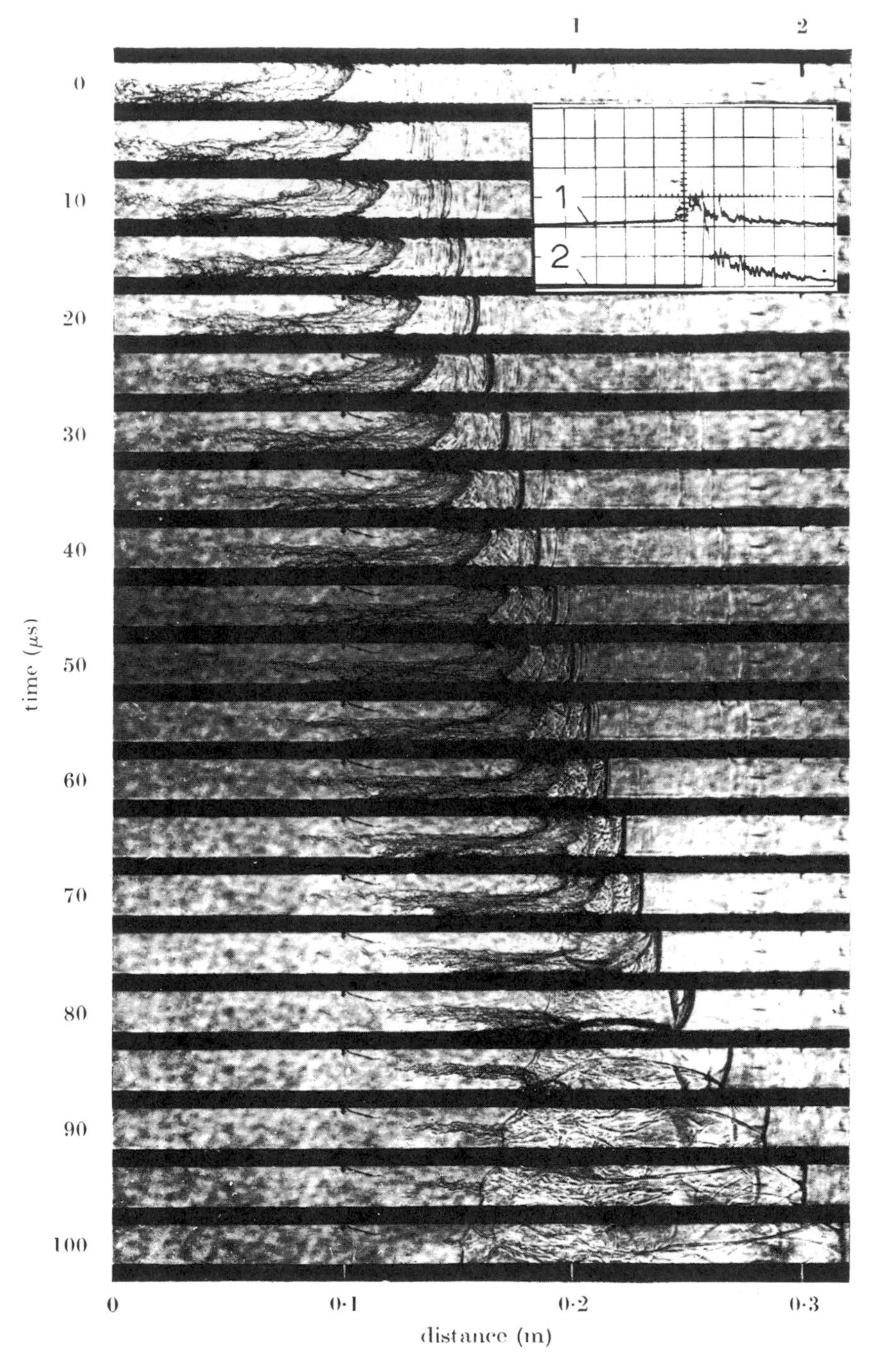

Figure 4.19 Stroboscopic schlieren record of the transition to detonation with onset between flame and shock. $2H_2 + O_2$ initially at a pressure of 554 Torr. Pressure record shown in insert. Vertical scale: 1 div = 200 lb/in.^{2}. Horizontal scale: 1 div = 50 μsec. Oscilloscope sweep leads the photographic record by 180 μsec. (After Urtiew and Oppenheim.[55])

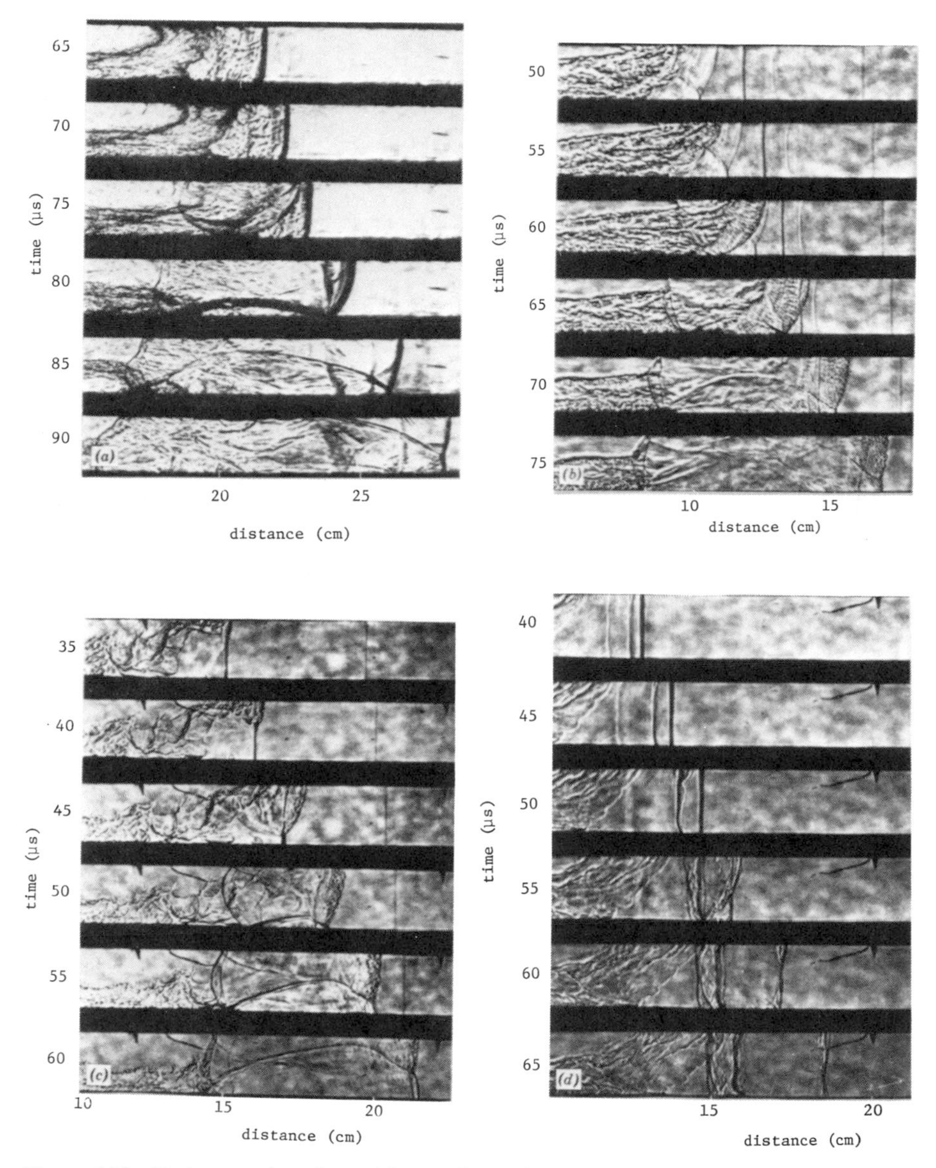

Figure 4.20 Various modes of transition to detonation observed in $2H_2 + O_2$ mixtures (after Urtiew and Oppenheim[55]): (*a*) mode 1 with onset occurring between flame and shock, (*b*) mode 2 with onset occurring at flame front, (*c*) mode 3 with onset occurring at shock front, (*d*) mode 4 with onset occurring at contact discontinuity.

by Belles:

$$OH + H_2 \xrightarrow{k_1} H_2O + H \tag{4-134}$$

$$H + O_2 \xrightarrow{k_2} OH + O \tag{4-135}$$

$$O + H_2 \xrightarrow{k_3} OH + H \tag{4-136}$$

$$H + O_2 + M \xrightarrow{k_4} HO_2 + M \tag{4-137}$$

where k_1, k_2, k_3, k_4 are the rate constants for the reactions represented by Eqs. (4-134) to (4-137) respectively, and M represents the third body in the formation of HO_2. Since the diffusion time of a free radical to the cold wall is much longer than the characteristic time of detonation, the chain-killing reactions on the wall were neglected.

The steady-state assumption was used for the free radicals. Thus

$$\frac{dC_{H_2O}}{dt} = k_1 \frac{C_{H_2}^2 C_X}{k_4 C_M - 2k_2} \tag{4-138}$$

The branched-chain explosion limit condition is therefore

$$\frac{2k_2}{k_4 C_M} = 1 \tag{4-139}$$

Using the rate constants of Lewis and von Elbe,[20] an expression for the concentration C_M in terms of temperature and pressure, and the gas law, Eq. (4-139) becomes

$$\frac{(3.11) T \exp(-8550/T)}{X_M p} = 1 \tag{4-140}$$

where X_M is the effective mole fraction of the third bodies in the formation of HO_2 in Eq. (4-137). Recall that the mole fraction X_M is related to the concentration by the following equation:

$$C_M = \frac{X_M p}{R_u T} \tag{4-141}$$

where X_M is an effective mole fraction of third bodies for the formation of

HO_2 in Eq. (4-137), and is given by the following empirical equation[20]:

$$X_M = X_{H_2} + 0.35X_{O_2} + 0.43X_{N_2} + 0.2X_{Ar} = 1.47X_{CO_2} \tag{4-142}$$

The numerical factors in Eq. (4-142) express ratios of the rates of reaction for Eq. (4-137) with various third bodies, to the rate with H_2 as the third body.

Rewriting Eq. (4-140) in the logarithmic form, we have

$$\frac{3.71}{T} - \log_{10}\left(\frac{T}{p}\right) = \log_{10}\left(\frac{3.11}{X_M}\right) \tag{4-143}$$

As soon as a given H_2–O_2 mixture (having a characteristic value of X_M dependent upon its composition) is raised to a temperature and pressure that satisfy Eq. (4-143), the mixture will explode. The pressure and temperature behind an incident shock may be calculated from the following equations:

$$\frac{p}{p_1} = \frac{1}{\alpha}\left(\frac{M^2}{\beta} - 1\right) \tag{4-144}$$

$$\frac{T}{T_1} = \frac{1}{\alpha^2\beta M^2}\left(\frac{M^2}{\beta} - 1\right)\left(\beta M^2 + \frac{1}{\gamma}\right) \tag{4-145}$$

where

$$\alpha = \frac{\gamma + 1}{\gamma - 1}, \qquad \beta = \frac{\gamma - 1}{2\gamma} \tag{4-146}$$

and T_1 and p_1 are the initial temperature and pressure. Equations (4-144) and (4-145) apply to the ideal case in which C_p = constant. However, the shock strengths required to cause explosion in hydrogen mixtures are sufficiently low that real-gas effects are usually not very important.

The critical Mach number may be found by combining Eqs. (4-143) through (4-145). One obtains

$$\frac{3.710\alpha^2\beta M^2}{T_1\left(\frac{M^2}{\beta} - 1\right)\left(\beta M^2 + \frac{1}{\gamma}\right)} - \log_{10}\left[\frac{T_1\left(\beta M^2 + \frac{1}{\gamma}\right)}{p_1\alpha\beta M^2}\right] = F(T_1, p_1, \gamma, M)$$

$$= \log_{10}\left(\frac{3.11}{X_M}\right) \tag{4-147}$$

Equation (4-147) may be solved graphically. Figure 4.21 shows typical plots of

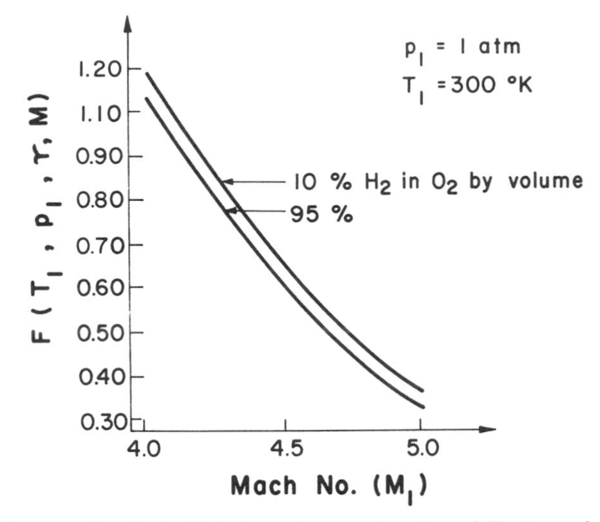

Figure 4.21 Critical Mach number as a function of T_1, P_1, and γ.

the left-hand side of Eq. (4-147) against Mach number, for different hydrogen–oxygen mixtures.

Conservation of energy provides the link between the calculated critical shock Mach numbers for explosion and the prediction of limits of detonability. The preflame shock provides a mechanism whereby some of the energy released by combustion is distributed to unburned gas ahead of the flame. Ultimately, this energy is recovered by the burned gas, when the flame consumes the shock-heated mixture. But if the enthalpy increase in the shock-heated state corresponding to the critical shock Mach number is greater than the heat of combustion, an impossible situation arises. No transition to detonation could ever occur, even if the duct were infinitely long. The criterion for detonability is therefore

$$\Delta h_s \leq \Delta h_c \tag{4-148}$$

where Δh_s is the enthalpy rise across a shock of critical Mach number, and Δh_c the heat of combustion. Both Δh_s and Δh_c are expressed per unit mass of mixture. The energy increase of the shocked state is obtained from

$$\Delta h_s = \overline{C}_p(T_t - T_0) \tag{4-149}$$

where the total temperature T_t is given by

$$T_t = T_0\left(1 + \frac{\gamma - 1}{2} M_c^2\right) \tag{4-150}$$

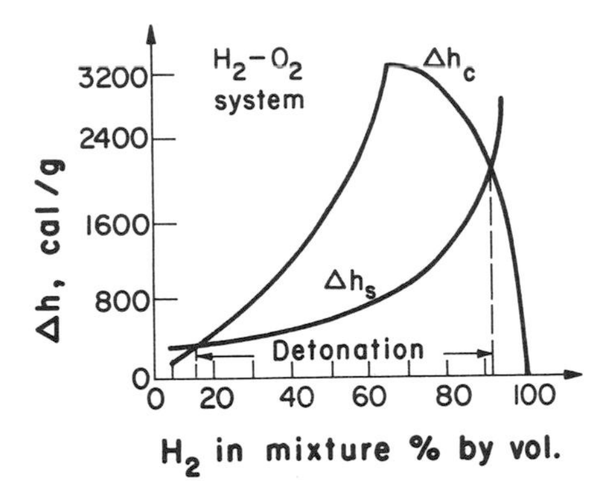

Figure 4.22 Δh_s and Δh_c versus percentage of H_2 in mixture, and the limits of detonability.

TABLE 4.4 Comparison of Belles's Theory[54] and Experimental Results for Hydrogen Mixtures

	Lean Limit, % H_2[a]		Rich Limit, % H_2[a]	
	Exptl.	Theory (Belles)	Exptl.	Theory (Belles)
Detonation in air	18.3	15.8	59	59.7
Flammability limit	14	–	70	–
Detonation in O_2	15	16.3	90	92.3

[a] By volume.

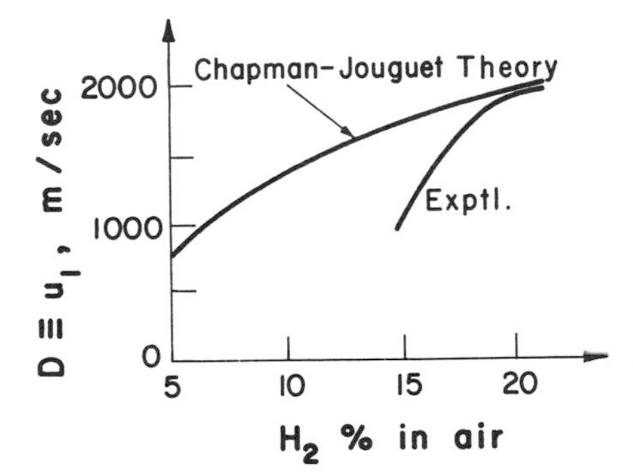

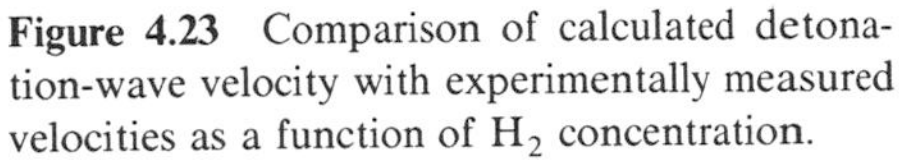
Figure 4.23 Comparison of calculated detonation-wave velocity with experimentally measured velocities as a function of H_2 concentration.

TABLE 4.5 Experimental Deflagration and Detonation Limits for Different Fuels

Mixture	Deflagration Lean Limit (% Fuel)[a]	Detonation Lean Limit (% Fuel)[a]	Detonation Rich Limit (% Fuel)[a]	Deflagration Rich Limit (% Fuel)[a]
H_2–O_2	4.6	15	90	93.9
H_2–air	4	18.3	59	74
CO–O_2 (moist)	15.5	38	90	93.9
(CO + H_2)–O_2	12.5	17.2	91	92
(CO + H_2)–air	6.05	19	59	71.8
NH_3–O_2	13.5	25.4	75	79
C_3H_8–O_2	2.4	3.2	37	55
C_2H_2–O_2	2.8	3.5	92	93
$C_4H_{10}O$–air	1.85	2.8	4.5	36.5

[a] By volume.

and $\overline{C}_p$, the average specific heat of the mixture, is defined as

$$\overline{C}_p = \sum_{i=1}^{N} X_i \overline{C}_{p,i} \tag{4-151}$$

where $\overline{C}_{p,i}$ is the average molar heat capacity at constant pressure of species i.

Figure 4.22 is a plot of Δh_s and Δh_c versus the percentage of H_2 in the mixture. The two curves intersect at two concentrations. The criterion of detonability [Eq. (4-148)] is satisfied only between these two concentrations. Hence, they are the predicted limits of detonability. The calculated results of Belles's study[54] showed good agreements with measured detonability limits (see Table 4.4). This implies that the limits of detonability of H_2 are well accounted for by the kinetic requirements for explosion.

Since the calculated detonation velocity is usually very accurate, many researchers have used the detonation-wave velocity to check the thermodynamic species and also to determine the bond energy. It should be noted that the departure between theory and experiment becomes significant when the concentration of the fuel is decreased below the lean limit as shown in Fig. 4.23. As shown in Table 4.5, there exist limits of detonability, analogous to limits of flammability, beyond which stable detonation is not observed. In comparison with the deflagration limits, one can see that the deflagration limits are wider than the detonation limits. It is also interesting to note that hydrocarbon mixtures such as C_3H_8 (see Fig. 4.24) and C_2H_2 mixtures (see Fig. 4.25) show maximum detonation velocities that are widely displaced from those for the stoichiometric composition for combustion to CO_2 and H_2O; instead they are fairly close to compositions corresponding to combustion to CO and H_2O. This is because the reaction time in the detonation process is very short; the time to form stoichiometric CO and H_2O mixtures is much

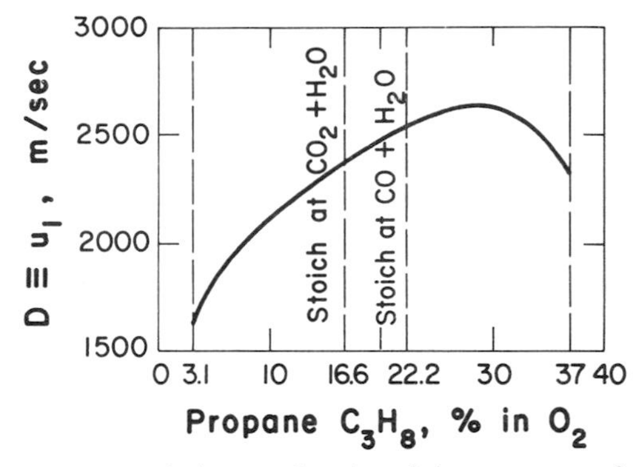

Figure 4.24 Detonation-wave velocity as a function of the percentage of propane in O_2.

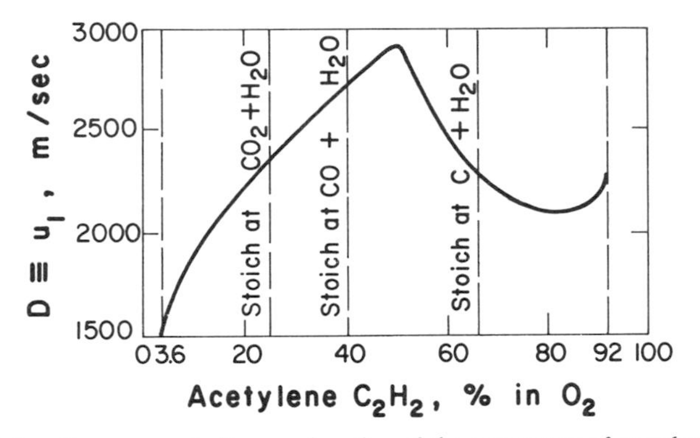

Figure 4.25 Detonation-wave velocity as a function of the percentage of acetylene in O_2.

shorter than that for CO_2 and H_2O. The oxidation process from CO to CO_2 is relatively slow. The change of detonation with mixture composition is further illustrated in Figs. 4.24 and 4.25.

REFERENCES

1. R. Friedman, *Am. Rocket Soc. J.*, Vol. 24, p. 349, November 1953.
2. D. L. Chapman, "On the Rate of Explosion of Gases," *Phil. Mag.*, Vol. 47, pp. 90–103, 1899.
3. E. Jouguet, *J. Mathematique*, p. 347, 1905; p. 6, 1906; *Mécanique des Explosifs*, O. Doin, Paris, 1917.
4. F. A. Williams, *Combustion Theory*, Addison-Wesley Publishing Co., Reading, Mass., 1965.
5. I. Glassman, *Combustion*, Academic Press, New York, 1977.
6. B. T. Wolfson and R. G. Dunn, "The Calculation of Detonation Parameters for Gaseous Mixtures," *Proc. Propellant Thermodynamics and Handling Conf.*, Special Rept. 12, pp. 397–440, Eng. Exp. Station, Ohio State University, June 1960.
7. M. P. Moyle, *Effect of Temperature on the Detonation Characteristics of Hydrogen–Oxygen Mixtures*, Ph.D. Thesis, University of Michigan, 1956.

8. R. L. Gealer, *Influence of High Pressure on the Properties of Hydrogen–Oxygen Detonation Waves*, Ph.D. Thesis, University of Michigan, 1958.

9. L. E. Bollinger and R. Edse, "Thermodynamic Calculations of Hydrogen–Oxygen Detonation Parameters for Various Initial Pressures," *ARS J.*, Vol. 31, pp. 251–256, February 1961.

10. C. L. Eisen, R. A. Gross, and T. J. Rivlin, *Theoretical Calculations in Gaseous Detonations*, TN 58-326, Office of Scientific Research, March 15, 1958.

11. C. L. Eisen, "Theoretical Calculations in Gaseous Detonations," *Proc. Propellant Thermodynamics and Handling Conf.*, Special Rept. 12, pp. 345–356, Eng. Exp. Station, Ohio State University, June 1960.

12. P. L. McGill and J. A. Luker, *Detonation Pressures of Stoichiometric Hydrogen-Oxygen Mixtures Saturated with Water at High Initial Temperatures and Pressures*, Rept. ChE273-5611F, Research Institute, Syracuse University, December 1956.

13. J. A. Luker, L. B. Adler, and E. C. Hobaica, *The Formation of Detonation in Saturated Mixtures of Knallgas–Steam and in Stoichiometric Mixtures of Deuterium–Oxygen (Heavy Knallgas) Saturated with Deuterium Oxide (Heavy Water)*, Rept. ChE273-591F, Research Institute, Syracuse University, January 23, 1959.

14. F. J. Zeleznik and S. Gordon, "Calculation of Detonation Properties and Effect of Independent Parameters on Gaseous Detonations," *ARS J.*, Vol. 32, No. 4, pp. 606–615, April 1962.

15. S. Gordon and B. J. McBride, *Computer Program for Calculation of Complex Chemical Equilibrium Compositions, Rocket Performance, Incident and Reflected Shocks, and Chapman–Jouguet Detonations*, NASA SP-273, 1971.

16. F. B. Hildebrand, *Advanced Calculus for Applications*, Chapter 7, Prentice-Hall, Englewood Cliffs, N.J., 1962.

17. S. Gordon, F. J. Zeleznik, and V. N. Huff, *A General Method for Automatic Computation of Equilibrium Compositions and Theoretical Rocket Performance of Propellants*, NASA TN D-132, 1959.

18. B. Lewis and J. B. Friauf, *J. Am. Chem. Soc.*, Vol. 52, p. 3905, 1930.

19. D. G. Berets, E. F. Greene, and G. B. Kistiakowsky, *J. Am. Chem. Soc.*, Vol. 72, p. 1080, 1950.

20. B. Lewis and G. von Elbe, *Combustion, Flames and Explosion of Gases*, 2nd ed., Chapter VIII, Academic Press, New York, 1961.

21. R. A. Strehlow, "Detonation Structure and Gross Properties," *Combustion Sci. and Technol.*, Vol. 4, pp. 65–71, 1971.

22. A. K. Oppenheim, N. Manson, and H. Gg. Wagner, "Recent Progress in Detonation Research," *AIAA J.*, Vol. 1, pp. 2243–2252, 1963.

23. R. A. Strehlow, "Gas Phase Detonations: Recent Developments," *Combustion and Flame*, Vol. 12, No. 2, pp. 81–101, 1968.

24. J. H. Lee, R. I. Soloukhin, and A. K. Oppenheim, "Current Views on Gaseous Detonation," *Astronautica Acta*, Vol. 14, pp. 565–584, 1969.

25. G. L. Schott, "Structure, Chemistry and Instability of Detonation in Homogeneous Low-Density Fluids—Gases," paper presented at the Fourth International Symposium on High Explosive Detonation, White Oak, Maryland, 1964.

26. K. I. Shchelkin, *Usp. Akad. Nauk*, Vol. 80, p. 525, 1963; Translation, *Soviet Physics Uspekhi*, Vol. 6, p. 523, 1964.

27. A. van Tiggelen and G. de Soete, *Rev. Inst. Franc. Petrole, Ann. Combust. Liquides*, Vol. 21, pp. 239–284, 455–486, 604–678, 1966.

28. Y. B. Zel'dovich, "The Theory of the Propagation of Detonation in Gaseous Systems," *Expt. Theor. Phys. S.S.S.R.*, Vol. 10, p. 542, 1940; Translation, NACA TM 1261, 1950.

29. J. von Neumann, *Theory of Detonation Waves*, Proj. Report No. 238, April 1942; OSRD Report No. 549, 1942.

30. W. Döring, "Über den Detonationvorgang in Gasen," *Ann. Phys. Lpz.*, Vol. 43, pp. 421–436, 1943.

31. J. P. Chesick and G. B. Kistiakowsky, "Gaseous Detonations X—Study of Reaction Zones," *J. Chem. Phys.*, Vol. 28, pp. 956–961, 1958.

32. G. B. Kistiakowsky and J. P. Kydd, "Gaseous Detonations. IX. A Study of the Reaction Zone by Gas Density Measurements," *J. Chem. Phys.*, Vol. 25, p. 824, 1956.

33. W. Jost, Th. Just, and H. Gg. Wagner, "Investigation of the Reaction Zone of Gaseous Detonations," *Eighth Symposium (International) on Combustion*, pp. 582–588, 1962.

34. Th. Just, F. J. Luig, and H. Gg. Wagner, "Untersuchungen der Reaktionszone von Detonationen in Knallgas verschiedener Zusammensetzung," *Z. Elektrochem.*, Vol. 65, p. 403, 1961.

35. W. Pusch and H. Gg. Wagner, "Investigation of the Dependence of the Limits of Detonability on Tube Diameter," *Combustion and Flame*, Vol. 6, pp. 157–162, 1962.

36. H. Gg. Wagner, "Reaction Zone and Stability of Gaseous Detonations," *Ninth Symposium (International) on Combustion*, pp. 454–460, 1963.

37. D. H. Edwards, G. T. Williams, and J. C. Breeze, "Pressure and Velocity Measurements on Detonation Waves in Hydrogen–Oxygen Mixtures," *J. Fluid Mech.*, Vol. 6, pp. 497–517, 1959.

38. D. H. Edwards and T. G. Jones, "Vibration Phenomena in Detonation Waves in Hydrogen–Oxygen Mixtures," *Brit. J. Appl. Phys.*, Vol. 11, pp. 190–194, 1960.

39. D. H. Edwards, G. T. Williams, and B. Price, "Pressure Measurements on Detonation Waves in Hydrogen–Oxygen Mixtures," *Les Ondes de Detonation*, pp. 249–256, CNRS, Paris, 1962.

40. F. J. Martin and D. White, "The Formation and Structure of Gaseous Detonation Waves," *Seventh Symposium (International) on Combustion*, pp. 856–865, 1959.

41. D. R. White, "Turbulent Structure of Gaseous Detonation," *Phys. Fluids*, Vol. 4, pp. 465–480, 1961.

42. D. R. White and K. H. Cary, "Structure of Gaseous Detonation, II. Generation of Laminar Detonation," *Phys. Fluids*, Vol. 6, pp. 749–750, 1963.

43. J. A. Fay, "Two-Dimensional Gaseous Detonations: Velocity Deficit," *Phys. Fluids*, Vol. 2, pp. 283–289, 1959.

44. J. A. Fay, "The Structure of Gaseous Detonation Waves," *Eighth Symposium (International) on Combustion*, pp. 30–40, 1962.

45. Th. Just and H. Gg. Wagner, "Die Reaktionszone in Gasdetonationen, I," *Z. Physik. Chem.*, Vol. 13, pp. 241–243, 1957. Th. Just and H. Gg. Wagner, "Reaktionszone von Knallgasdetonationen," *Z. Physik. Chem.*, Vol. 19, p. 250, 1959. Th. Just and H. Gg. Wagner, "Untersuchung der Reaktionszone von Detonationen in Knallgas," *Z. Elektrochem.*, Vol. 64, p. 501, 1960.

46. R. A. Strehlow, *Fundamentals of Combustion*, Chapter 9, International Textbook Company, 1968.

47. C. Campbell and D. W. Woodhead, *J. Chem. Soc.*, p. 3010, 1926; p. 1572, 1927.

48. Y. H. Denisov and Y. K. Troshin, *Dokl. Akad. Sci. SSSR*, Vol. 125, p. 110, 1959; translation, *Phys. Chem. Sect.*, Vol. 125, p. 217, 1960.

49. K. Antolik, *Pogg. Ann.* 230, Series 2, Vol. 154, p. 14, 1875.

50. R. A. Strehlow, "Multi-Dimensional Detonation Wave Structure," *Astronautica Acta*, Vol. 15, pp. 345–357, 1970.

51. A. H. Shapiro, *Compressible Fluid Flow*, Vol. I, Chapter 16, The Ronald Press Company, New York, 1953.

52. J. A. Fay and G. L. Opel, "Two-Dimensional Effects in Gaseous Detonation Waves," *J. Chem. Phys.*, Vol. 29, pp. 955–958, 1958.
53. M. L. N. Sastri, L. M. Schwartz, B. F. Myers, and D. F. Horning, "Optical Studies of the Structure of Gaseous Detonation Waves," *Ninth Symposium (International) on Combustion*, pp. 470–473, 1963.
54. Belles, F. E., *Seventh Symposium on Combustion (International)*, Butterworths, London, p. 745, 1959.
55. P. A. Urtiew and A. K. Oppenheim, "Experimental Observations of the Transition to Detonation in an Explosive Gas," *Proc. Roy. Soc.* A, Vol. 295, pp. 13–28, 1966.

HOMEWORK

1. Calculate the steady-state detonation-wave velocity in gaseous ozone at initial pressure of 1 atm and initial temperature of 298 K:

$$O_3 \rightarrow \nu''_{O_2} O_2 + \nu''_O O$$

Compare your solution with numerical results obtained from the NASA/Lewis code (TRAN76).

ANSWER: $U_1 = 1911$ m/sec.

2. Suppose the perfect-gas law is replaced by the Noble–Abel dense-gas law

$$p\left(\frac{1}{\rho} - b\right) = RT,$$

where b represents the covolume (the volume occupied by gaseous molecules themselves). Develop a Rankine–Hugoniot relation using the above equation of state for the gases behind the detonation wave front.

3. A mixture of hydrogen, oxygen, and nitrogen, having partial pressures in the ratio 2 : 1 : 5 in the order listed, is observed to detonate with a speed of 1890 m/sec when the initial temperature is 293 K and the initial pressure is 1 atm. Assuming fully relaxed conditions, calculate the peak pressure in the detonation wave, and calculate the pressure and temperature of the gas just after the passage of the wave. Prove that the observed speed corresponds to the C–J condition. Reasonable approximations are allowed, such as neglect of dissociation, the assumption that the pressure after the wave passes is much greater than the initial pressure, the use of existing gas-dynamics tables designed for pure air to analyze processes inside the wave, and specific heats independent of pressure.

4. Estimate the steady-state detonation-wave velocity for the premixed gaseous mixture $2H_2 + O_2 + 3N_2$ (assuming no dissociation of the product gases). The initial temperature and pressure of the reactants are $T_1 = 298.15$ K and $p_1 = 1$ atm, respectively.

5. Derive the equations expressing the pressure ratio p_2/p_1, the temperature ratio T_2/T_1, and the velocity ratio u_2/u_1 in terms of the Mach number M_1, M_2 and the specific-heat ratio γ for the combustion wave shown in Fig. 4.1. Assume (1) ideal-gas behavior, (2) steady state, (3) no body force, (4) no external heat addition or heat loss, (5) Dufour effect negligible, (6) interdiffusion negligible.

6. Consider a stationary combustion wave in a constant-area tube. Assume that there is no external heat transfer across the walls and that the problem can be treated as one-dimensional and steady state. There are no body forces or Dufour effect.
   (a) Derive an expression to relate the change of entropy to the pressure ratio in this problem, for example, $\Delta S = \mathscr{F}(p_2/p_1)$.
   (b) Use the above relationship to comment about the difference between detonation and deflagration waves.

7. Verify the following C–J condition: *The C–J condition implies that when the Hugoniot curve is tangent to the Rayleigh line passing through the origin, it is also tangent to the isentrope. Hence the local Mach number at the upper C–J point and the lower C–J point is unity irrespective of the thermodynamic properties of the medium.*

8. Assume the conditions at the unburned side of a stationary one-dimensional combustion wave are:

$$p_1 = 1 \text{ atm} \quad \text{and} \quad \rho_1 = 1.185 \text{ kg/m}^3$$

   (a) Derive an equivalent Hugoniot relation in terms of compression ratio, $\varepsilon(\equiv v_1/v_2)$, and pressure ratio, $\pi_r(\equiv p_2/p_1)$.
   (b) Calculate the heat release per unit mass using the data given in Table 4.1 for both detonation and deflagration waves, and compare the calculated results.

PROJECT

The interferograms of both Chapman–Jouguet and slightly overdriven detonations of initial pressures up to 0.3 atm show that without exception the shock front reaction zone structure is not planar and that the following flow has variations in refractive index that we can interpret as due to turbulence.† Show, by resolving the burned gas properties U_2 (velocity), ρ_2 (density), p_2 (pressure), and E_2 (energy) into mean and fluctuating quantities, the form of the Rayleigh and Rankine–Hugoniot functions when turbulence is considered. You may

† D. R. White, "Turbulent Structure of Gaseous Detonation," *The Physics of Fluids*, Vol. 4, No. 4, pp. 465–480, April 1961.

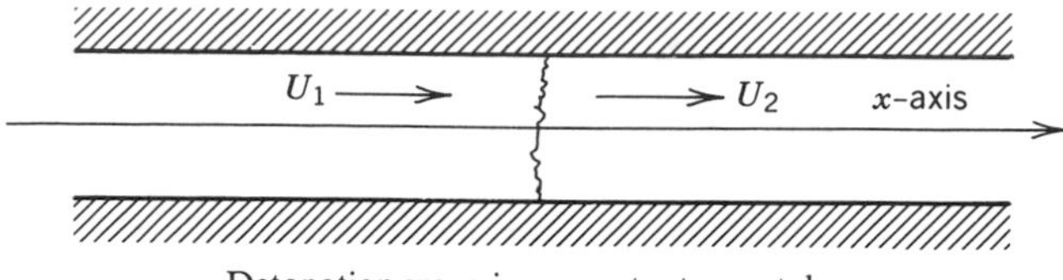

Detonation wave in a constant-area tube.

make the following assumptions:

a. The detonation wave occurs as a statistically steady process in a constant area tube (see figure above).
b. The flow ahead of the detonation wave is nonturbulent.
c. The mean gas velocity U is parallel to the axis of the tube (x-axis) and the transverse derivatives of any time-averaged quantity vanishes.
d. Neglect velocity fluctuation terms of third order or higher.
e. Assume isotropic turbulence $\overline{u_i'u_j'} = \frac{1}{3}u^2\delta_{ij}$.

CHAPTER

5 Premixed Laminar Flames

ADDITIONAL SYMBOLS

Symbol	Description	Dimension
a	Number of molecules of reactant per unit volume	L^{-3}
A_v	Avogadro's number	N^{-1}
C_t	Total number of moles per unit volume	NL^{-3}
d_p	Penetration distance	L
d_q	Quenching distance	L
g	Boundary velocity gradient	t^{-1}
G	Total mass velocity per unit area	$M/t \cdot L^2$
H_{R_A}	Standard heat of reaction of A	Q/M
n_r	Number of moles of reactant	N
n_p	Number of moles of product	N
r_{BA}	Stoichiometric ratio	—
R_A	Consumption rate of A	$M/L^3 \cdot t$
S_L	Laminar flame speed	L/t
T_f	Flame temperature	T
T_{inf}	Temperature at point of inflection	T
δ_{ph}	Length of preheat zone	L
δ_r	Length of reaction zone	L

1 INTRODUCTION

From the Hugoniot plot (Fig. 4.2) we have learned that there are two types of combustion waves: detonation waves (regions I and II of the Hugoniot curve)

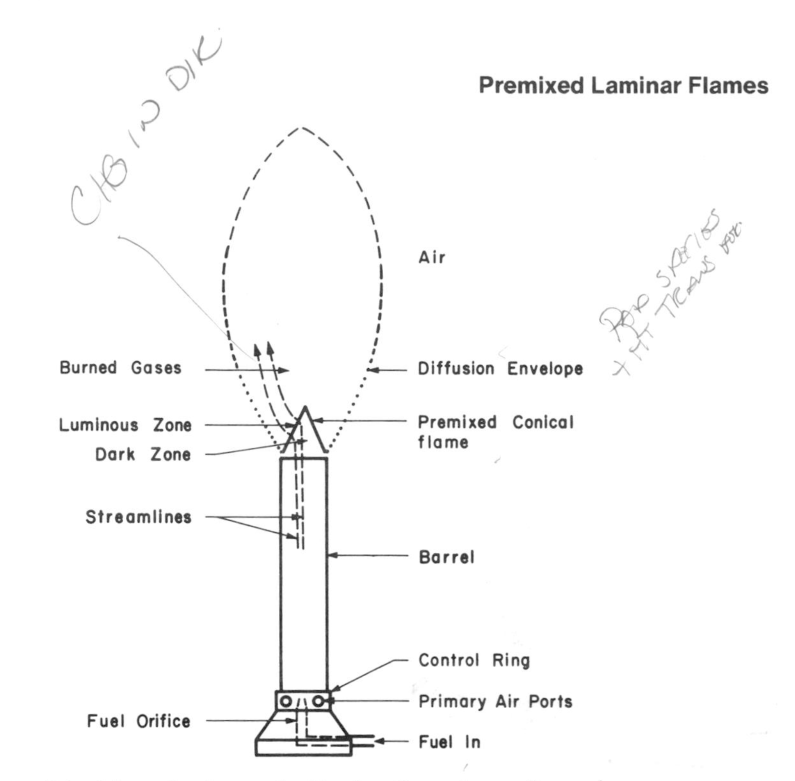

Figure 5.1 Schematic of a premixed laminar flame above a Bunsen burner.

and deflagration waves (regions III and IV). Detonation waves are supersonic waves supported by combustion, while deflagration waves are subsonic waves propagated by heat release from chemical reactions. As mentioned earlier, solutions in region IV do not exist, because combustion products cannot depart from the wave at supersonic speeds. The unique solution in region III is the main subject of this chapter. In other words, the laminar-flame theories are essentially studies to obtain a unique solution for a deflagration wave under a given set of conditions.

The first laboratory premixed-flame burner was invented by Bunsen around 1855. Figure 5.1 shows a schematic diagram of the Bunsen burner. The fuel gas enters the burner through a feedline near the bottom. The gaseous fuel jet induces the entrainment of primary air through a number of ports on the control ring. As fuel and air flow up the barrel tube they mix, and before reaching the top of the burner, the gases are well mixed and can be considered to be homogeneous. At the port, the flame is anchored near the top of the burner. As long as the fuel feeding rate remains constant, the premixed flame in a quiescent ambient atmosphere will remain stationary and steady.

A picture of a premixed ethylene–air flame is shown in Fig. 5.2. A shadow graph of a conical premixed laminar flame is shown in Fig. 5.3. A simultaneous photograph of visible (outer cone) and schlieren (inner cone) cones of a premixed flame is shown in Fig. 5.4. These photographs indicate clearly that there is a luminous conical region within which reaction and heat release are taking place. Underneath the luminous cone, there is a dark zone (marked in Fig. 5.1) within which the unburned gas molecules change their flow direction

Figure 5.2 Photograph of a premixed ethylene–air flame (from Gaydon and Wolfhard[30]).

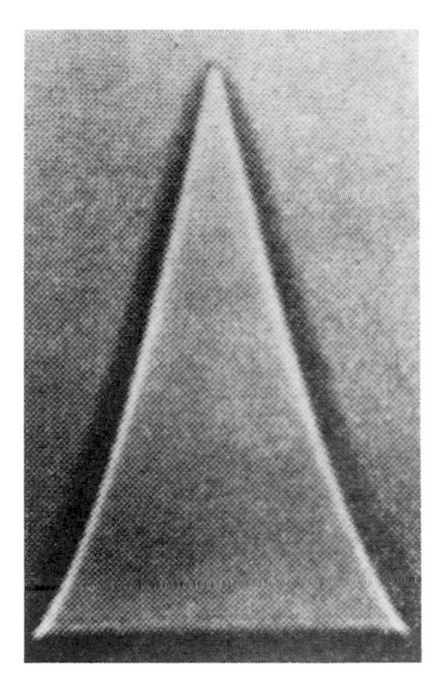

Figure 5.3 Shadowgraph of conical premixed laminar flame (from Weinberg;[31] photographed by J. W. Anderson).

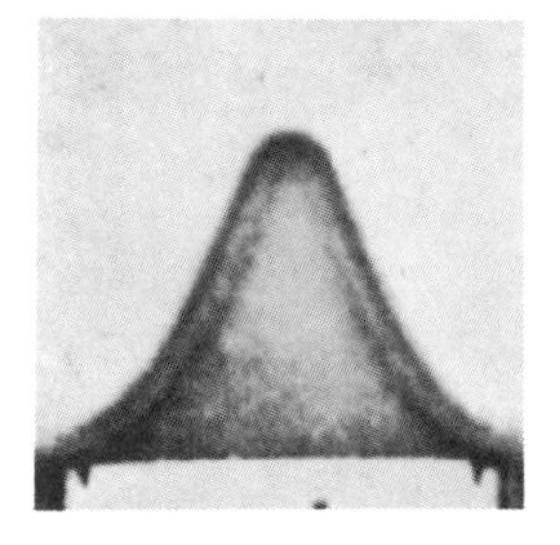

Figure 5.4 Simultaneous photographs of visible (outer) and schlieren (inner) cones of a premixed flame (from Weinberg;[31] photographed by J. W. Linnett).

from the initially vertical direction to outward directions. Immediately beneath the luminous cone, the unburned gases are heated to a critical temperature at which rapid chemical reaction starts. The burned gases expand as they leave the flame zone and are diluted and cooled by the surrounding air. A few streamlines are sketched in Fig. 5.1 to show the general flow characteristics near a Bunsen burner flame.

The luminous zone is usually no more than 1 mm thick. The color of the luminous zone, according to Glassman,[1] changes with the fuel–air ratio. When the mixture is fuel-lean, the color of the cone surface is deep violet due to large concentrations of excited CH radicals. When the mixture is fuel-rich, the color of the cone surface is green due to large concentrations of C_2 molecules. The high-temperature burned gases usually show a reddish glow which arises from radiation from CO_2 and water vapor. When the mixture is highly fuel-rich, carbon particles form and an intense yellow radiation appears. It is yellow because the blackbody radiation curve peaks in the yellow for the temperatures that normally exist in these premixed flames.

It should be noted that a laminar flame does not have to be stationary like the Bunsen-burner flame shown above; it can be a traveling flame in a tube or other devices to be discussed later. Furthermore, a deflagration wave does not have to have a planar flame front. As a matter of fact, most laminar flames are nonplanar. A realistic shape of the laminar flame front propagating in a horizontal tube is shown in Fig. 5.19 which will be discussed in a later section.

In laminar-flame theory, we are mostly interested in determining the flame speed (also called burning velocity or flame velocity or normal combustion velocity). The laminar-flame speed is defined as the velocity of the *unburned* gases *normal* to the combustion wave surface as these gases move into the combustion front.

In recent years, many physical and chemical processes involved in steady-state flame propagation have been studied. A variety of methods of attack have been developed. An extensive review of the classical laminar-flame theories was conducted by Evans.[2] On the basis of the major assumptions, theories are divided into three groups:

1. Thermal theory:
   Mallard and Le Chatelier's Development[3]
   The Damköhler theory[4]

The theory of Bartholomé[5]
The theory of Emmons, Harr, and Strong[6]
The theory of Bechert[7]

2. Comprehensive theory:
The theory of Lewis and von Elbe[8]
The theory of Zel'dovich, Frank-Kamenetsky, and Semenov[9]
The theory of Boys and Corner[10]
The theory of Hirschfelder and Curtiss[11]
The theory of von Karman and Penner[12]

3. Diffusion theory:
The theory of Tanford and Pease[13]
The theory of Van Tiggelen[14]
The theory of Manson[15]
The theory of Gaydon and Wolfhard[16]

For detailed information of these theories, the reader should refer to the original paper of each theory or the discussions in Evans's review paper. In the following, one theory is selected from each group to show the special considerations in the development of each group.

2 THERMAL THEORY: MALLARD AND LE CHATELIER'S DEVELOPMENT[2] (1883)

The objective of laminar-flame theory is to find the laminar-flame speed S_L. The temperature variation across the flame can be plotted as shown in Fig. 5.5. Mallard and Le Chatelier divided the flame into two zones. Zone I is the preheat zone, in which the gases are heated by conduction and reach ignition at the ignition boundary. Zone II is the chemical reaction zone, in which chemical enthalpy is converted into sensible enthalpy.

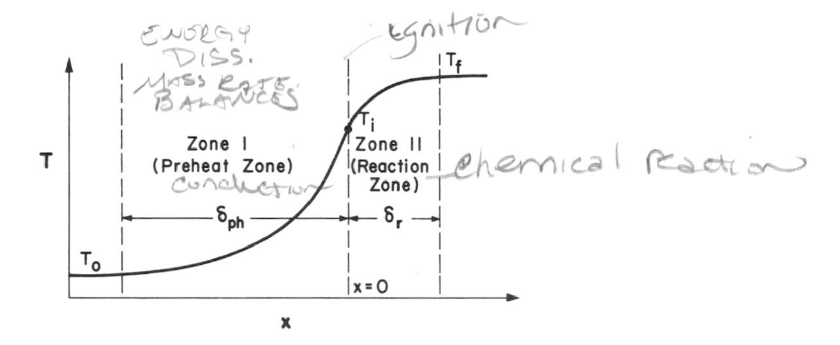

Figure 5.5 Schematic diagram of the temperature variation across a typical laminar flame.

The energy balance of zone I gives

$$\dot{m}C_p(T_i - T_0) = \lambda\frac{T_f - T_i}{\delta_r} \tag{5-1}$$

where the left-hand side of the equation represents the amount of energy absorbed as the unburned mixture flows into the entrance of the preheat zone at T_0 and is heated to T_i at the exit of the preheat zone, and the right-hand side is the heat flux to the interface. The mass flow rate per unit area, $\dot{m}$, is defined as

$$\dot{m} = \rho S_L \tag{5-2}$$

where ρ is the density of unburned gas and S_L is the laminar flame speed. From (5-1) and (5-2) we have

$$S_L = \frac{\lambda}{\rho C_p}\frac{T_f - T_i}{T_i - T_0}\frac{1}{\delta_r} \tag{5-3}$$

If we define τ_r to be the reaction time, then

$$\delta_r = S_L\tau_r = S_L\frac{1}{d\varepsilon/dt} \tag{5-4}$$

where $d\varepsilon/dt$ is the rate of reaction (RR) and ε is the reaction-progress variable discussed in Chapter 1. Then

$$\delta_r \propto S_L\frac{1}{RR}$$

Substituting Eq. (5-4) into Eq. (5-3), we get

$$S_L = \sqrt{\left(\frac{\lambda}{\rho C_p}\right)\frac{T_f - T_i}{T_i - T_0}\left(\frac{d\varepsilon}{dt}\right)} \propto \sqrt{\alpha \cdot \text{RR}} \tag{5-5}$$

where T_i represents the ignition temperature. The reaction rate RR was not specified by Mallard and Le Chatelier at any particular temperature. However, their analysis suggests that the flame speed is proportional to the square root of the product of thermal diffusivity and reaction rate. This result is one of the most important relationships in laminar-flame theories.

As seen in Problem 2.10, for an nth-order chemical reaction we have

$$\frac{d\varepsilon}{dt} = k\varepsilon^n p^{n-1} = Ae^{-E_a/R_uT}\varepsilon^n p^{n-1} \tag{5-6}$$

For the pressure dependence, we have

$$S_L \propto \sqrt{\frac{1}{\rho} p^{n-1}} \propto \sqrt{p^{n-2}} \tag{5-7}$$

This equation implies that for second-order chemical reactions, the laminar-flame speed should be independent of pressure. Based upon Eq. (5-5) and the Arrhenius law for RR, the temperature dependence can be expressed:

$$S_L \propto \sqrt{e^{-E_a/R_u T}} = e^{-E_a/2R_u T} \tag{5-8}$$

As we shall see later, in order to obtain realistic values of S_L, one needs to replace the temperature T in Eq. (5-8) by T_f, since most chemical reaction takes place around T_f. It is useful to point out here that in the study of laminar flames, the conduction (thermal diffusion) term is of the same order as the convection term. However, this is not true for all combustion problems.

3 COMPREHENSIVE THEORY: THE THEORY OF ZEL'DOVICH, FRANK-KAMENETSKY, AND SEMENOV

These investigators have adopted Mallard and Le Chatelier's idea[2,3] of dividing the flame into two zones (preheat and reaction zones). However, instead of considering the energy equation alone, they have used the species-conservation equation together with the energy equation. They propose that the ignition temperature is very close to the adiabatic flame temperature and consequently replace T_i with T_f in their estimation of reaction rates. Their basic assumptions are:

a. The pressure is constant.
b. The number of moles does not vary during the reaction. This restriction is later removed to allow the number to change in the constant ratio

$$\frac{n_r}{n_p} = \frac{\text{moles of reactant}}{\text{moles of product}}$$

(according to the stoichiometric reaction)

c. C_p and λ are constant. DIFF. COEFF.
d. $\lambda/C_p = \mathscr{D}\rho$. This implies that Le = 1, and has the effect of reducing the number of differential equations from two to one by replacing one of the differential equations with an algebraic equation. Later this restriction is modified to allow Le = constant. $Le = \frac{\lambda}{C_p \rho \mathscr{D}}$
e. The flame is one-dimensional and steady-state.

The equations of species continuity and energy conservation are first written in their most general forms and then simplified according to the above assumptions:

The general species conservation equation derived in Chapter 3 can be written as

$$\frac{\partial \rho_r}{\partial t} + \nabla \cdot \rho_r \mathbf{v} = \nabla \cdot (\rho \mathscr{D} \nabla Y_r) - \omega_r \tag{5-9}$$

ω_r is positive for rate of destruction of reactant, and ω_r has dimensions of $Mt^{-1}L^{-3}$. Under assumption e, Eq. (5-9) becomes

$$\frac{d}{dx}\left(\frac{\rho_r}{\rho} \cdot \rho u\right) = \frac{d}{dx}\left[\rho \mathscr{D} \frac{d}{dx}\left(\frac{\rho_r}{\rho}\right)\right] - \omega_r$$

or

$$\frac{\rho_r}{\rho}\frac{d(\rho u)}{dx} + \rho u \frac{d(\rho_r/\rho)}{dx} = \rho \mathscr{D} \frac{d^2}{dx^2}\left(\frac{\rho_r}{\rho}\right) - \omega_r \tag{5-10}$$

From continuity equation $d(\rho u)/dx = 0$, then Eq. (5-10) becomes

$$\rho u \frac{d(\rho_r/\rho)}{dx} = \rho \mathscr{D} \frac{d^2}{dx^2}\left(\frac{\rho_r}{\rho}\right) - \omega_r \tag{5-11}$$

ρ_r can be written in terms of the number density a of reactant molecules (cm^{-3}) as

$$\rho_r = \frac{a}{A_v} \cdot W_r \tag{5-12}$$

where A_v is Avogadro's number, and W_r is the reactant molecular weight (g/mole). Equation (5-11) then becomes

$$\rho u \frac{d(aW_r/\rho A_v)}{dx} = \rho \mathscr{D} \frac{d^2}{dx^2}\left(\frac{aW_r/A_v}{\rho}\right) - \omega_r \tag{5-13}$$

Multiplying through by A_v/W_r, we have

$$\rho u \frac{d(a/\rho)}{dx} = \rho \mathscr{D} \frac{d^2}{dx^2}\left(\frac{a}{\rho}\right) - \frac{A_v}{W_r}\omega_r \tag{5-14}$$

The inhomogeneous term in Eq. (5-14), denoted by

$$\frac{A_v}{W_r} \cdot \omega_r \equiv \omega \tag{5-15}$$

equals the number of molecules of reactant per cm^3 per sec. Hence,

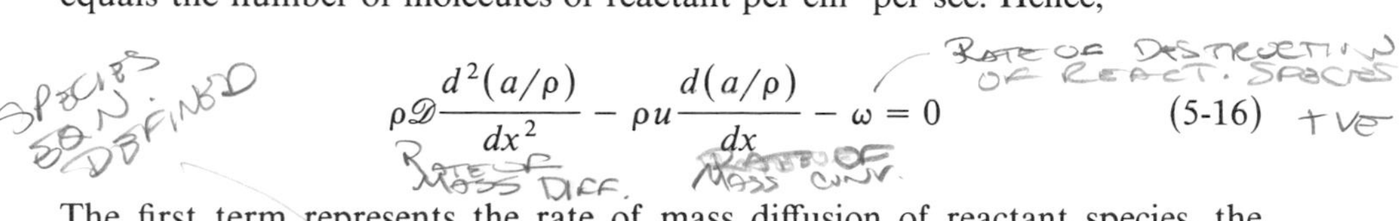

$$\rho\mathscr{D}\frac{d^2(a/\rho)}{dx^2} - \rho u\frac{d(a/\rho)}{dx} - \omega = 0 \tag{5-16}$$

The first term represents the rate of mass diffusion of reactant species, the second term the rate of mass convection, and the third term the rate of destruction of reactant species.

From Chapter 3 we have learned that the general form of the energy equation in terms of enthalpy is

$$\rho\frac{Dh}{Dt} - \frac{Dp}{Dt} = -\nabla\cdot\mathbf{q} + \Phi + \dot{Q} + \rho\sum_{K=1}^{N} Y_K\mathbf{f}_K\cdot\mathbf{v}_K \tag{3-78}$$

Neglecting dissipation, body-force work, and unsteady terms, and also using

$$q = -\lambda\frac{dT}{dx}, \qquad h = C_pT$$

we have the following energy equation in one-dimensional form:

$$\rho C_p u\frac{dT}{dx} = \lambda\frac{d^2T}{dx^2} + \dot{Q}$$

or

$$\frac{\lambda}{\rho C_p}\frac{d^2T}{dx^2} = u\frac{dT}{dx} - \frac{\dot{Q}}{\rho C_p} \tag{5-17}$$

where $\dot{Q} = -\sum_{i=1}^{N}\omega_i\Delta h_{f,i}^{\circ} = \omega Q$ (Q represents the heat of reaction per molecule of reactant). Equation (5-17) can be written as

$$\frac{\lambda}{C_p}\frac{d^2T}{dx^2} - \rho u\frac{dT}{dx} + \frac{\omega Q}{C_p} = 0 \tag{5-18}$$

where ρu is the eigenvalue of the problem.

Following the Shvab–Zel'dovich formulation (see Chapter 3), two new variables θ and α are defined as

$$\theta \equiv C_p\frac{T - T_0}{Q} \tag{5-19}$$

$$\alpha \equiv \frac{a_0}{\rho_0} - \frac{a}{\rho} \tag{5-20}$$

where the subscript 0 designates initial values at an undisturbed station. Substituting Eqs. (5-19) and (5-20) into Eqs. (5-18) and (5-16), we have

$$\frac{\lambda}{C_p}\frac{d^2\theta}{dx^2} - \rho u \frac{d\theta}{dx} + \omega = 0 \tag{5-21}$$

$$\rho \mathscr{D} \frac{d^2\alpha}{dx^2} - \rho u \frac{d\alpha}{dx} + \omega = 0 \tag{5-22}$$

These two equations are similar if Le = 1.

The boundary conditions for (5-21) and (5-22) are prescribed as follows: For $x = -\infty$ (cold boundary),

$$\alpha = 0, \qquad \theta = 0$$

For $x = +\infty$ (hot boundary),

$$\alpha = \frac{a_0}{\rho_0}, \qquad \theta = C_p \frac{T_f - T_0}{Q}$$

It is interesting to note that the solutions of Eqs. (5-21) and (5-22) are identical (i.e., $\alpha = \theta$) over the entire combustion wave zone if

$$\frac{a_0}{\rho_0} = C_p \frac{T_f - T_0}{Q}$$

The above relationship can be satisfied if the flame is adiabatic. Under this assumption, the following algebraic equation can be introduced to replace one of the two differential equations [say, Eq. (5-22)]:

$$C_p T + \frac{aQ}{\rho} = C_p T_0 + \frac{a_0 Q}{\rho_0} = C_p T_f \tag{5-23}$$

The meaning of this equation is that the sum of the thermal (sensible) and chemical energies per unit mass of mixture is constant throughout the combustion zone. This implies that only one ordinary differential equation, (5-21) or (5-17), need be solved. In region I ($\omega = 0$), the energy equation can be written as

$$\frac{d^2T}{dx^2} - \frac{(\rho u) C_p}{\lambda}\frac{dT}{dx} = 0 \tag{5-24}$$

with the boundary conditions:

$$x = \begin{cases} -\infty, & T = T_0 \\ 0^-, & T = T_i \end{cases}$$

In region II, the convective term was considered by these researchers to be negligible, since T_i is usually very close to T_f. Then the energy equation can be written as

$$\frac{d^2T}{dx^2} + \frac{\omega Q}{\lambda} = 0 \tag{5-25}$$

with boundary conditions

$$x = \infty, \qquad T = T_f, \quad \frac{dT}{dx} = 0$$

$$x = 0^+, \qquad T = T_i$$

In order to have heat-flux balance at the interface of the two zones, it is required that

$$\left(\frac{dT}{dx}\right)_{x=0^-} = \left(\frac{dT}{dx}\right)_{x=0^+} \tag{5-26}$$

By integrating Eq. (5-24) with respect to x, we have:

$$\frac{dT}{dx} = \frac{(\rho u)C_p}{\lambda}T + \text{constant}$$

Since at $x = -\infty$ we have $T = T_0$, $dT/dx = 0$, it follows that

$$\frac{dT}{dx} = \frac{(\rho u)C_p}{\lambda}(T - T_0)$$

This gives

$$\left(\frac{dT}{dx}\right)_{x=0^-} = \frac{(\rho u)C_p(T_i - T_0)}{\lambda} \approx \frac{(\rho u)C_p(T_f - T_0)}{\lambda} \tag{5-27}$$

if T_i is very close to T_f.

In order to integrate Eq. (5-25), we multiply it by 2 dT/dx and rearrange it to give

$$\frac{d}{dx}\left(\frac{dT}{dx}\right)^2 = -2\left(\frac{dT}{dx}\right)\frac{\omega Q}{\lambda}$$

Integrating from $x = 0^+$ to $x = \infty$, and applying the boundary conditions, we have

$$0 - \left(\frac{dT}{dx}\right)^2_{x=0^+} = -\int_{T_i}^{T_f}\frac{2Q\omega}{\lambda}\,dT$$

or

$$\left(\frac{dT}{dx}\right)^2_{x=0^+} = \int_{T_i}^{T_f} \frac{2Q\omega}{\lambda}\, dT \tag{5-28}$$

Substituting both (5-27) and (5-28) into (5-26), we have

$$\frac{(\rho u)^2 C_p^2 (T_f - T_0)^2}{\lambda^2} = \frac{2Q}{\lambda} \int_{T_i}^{T_f} \omega\, dT$$

or

$$(\rho u)^2 = \frac{2\lambda Q}{C_p^2 (T_f - T_0)^2} \int_{T_i}^{T_f} \omega\, dT$$

Also we know that $a_0 Q/\rho_0 = C_p(T_f - T_0)$; then

$$(\rho u)^2 = \rho_0^2 S_L^2 = \rho_0^2 u_0^2 = \frac{2\lambda}{C_p(T_f - T_0)} \frac{\rho_0}{a_0} \int_{T_i}^{T_f} \omega\, dT$$

Therefore,

$$S_L = u_0 = \sqrt{\left(\frac{\lambda}{\rho_0 C_p}\right) \frac{2}{T_f - T_0} \left(\frac{1}{a_0} \int_{T_i}^{T_f} \omega\, dT\right)} = \sqrt{\left(\frac{\lambda}{\rho_0 C_p}\right) \frac{2}{T_f - T_0} I} \tag{5-29}$$

where

$$I = \frac{1}{a_0} \int_{T_i}^{T_f} \omega\, dT \approx \frac{1}{a_0} \int_{T_0}^{T_f} \omega\, dT \tag{5-30}$$

These two quantities are nearly equal because there is no reaction between T_0 and T_i.

Now let ω be a function of T, according to the expression for a zero-order chemical reaction for simplicity:

$$\omega = A e^{-E_a/R_u T} \tag{5-31}$$

where A, the Arrhenius factor, is considered a constant which does not depend upon the concentration. For most hydrocarbon fuels $E_a/R_u T_f \gg 1$. Since $E_a \approx$ 30,000–40,000 cal/mole, $T_f \approx$ 1500–2000 K, and $R_u =$ 1.987 cal/mole K, then

$$\frac{E_a}{R_u T_f} \sim 10 \gg 1$$

Set $\sigma \equiv T_f - T$; its values range from $\sigma_1 = T_f - T_i$ down to 0 for zone II. Then

$$T = T_f - \sigma = T_f\left(1 - \frac{\sigma}{T_f}\right)$$

Since σ/T_f is generally small, we have

$$e^{-E_a/R_uT} = \exp\left(-\frac{E_a}{R_uT_f(1-\sigma/T_f)}\right) \approx \exp\left[-\frac{E_a}{R_uT_f}\left(1+\frac{\sigma}{T_f}\right)\right]$$

$$= e^{-E_a/R_uT_f}e^{-E_a\sigma/R_uT_f^2} \tag{5-32}$$

The expression for ω is now introduced into (5-30) to give

$$I = \frac{Ae^{-E_a/R_uT_f}}{a_0}\int_{T_i}^{T_f} e^{-E_a\sigma/R_uT_f^2}\,dT = -\frac{Ae^{-E_a/R_uT_f}}{a_0}\int_{\sigma_1}^{0} e^{-E_a\sigma/R_uT_f^2}\,d\sigma \tag{5-33}$$

Let

$$\beta \equiv \frac{E_a\sigma}{R_uT_f^2} \quad \text{and} \quad \beta_1 \equiv \frac{E_a\sigma_1}{R_uT_f^2} \tag{5-34}$$

Substitute (5-34) into (5-33):

$$I = \frac{Ae^{-E_a/R_uT_f}}{a_0}\,\frac{R_uT_f^2}{E_a}\int_0^{\beta_1} e^{-\beta}\,d\beta \approx \frac{Ae^{-E_a/R_uT_f}}{a_0}\,\frac{R_uT_f^2}{E_a}\,j$$

where $j = \int_0^{\beta_1} e^{-\beta}\,d\beta = 1 - e^{-\beta_1} \approx 1$. Therefore,

$$I \approx \frac{Ze^{-E_a/R_uT_f}}{a_0}\left(\frac{R_uT_f^2}{E_a}\right) \tag{5-35}$$

According to Eq. (5-29), u_0 for a zero-order reaction can be written as

$$S_L = u_0 = \sqrt{\frac{2\lambda}{C_p\rho_0}\,\frac{1}{a_0}\left(\frac{Ae^{-E_a/R_uT_f}}{T_f - T_0}\right)\frac{R_uT_f^2}{E_a}} \tag{5-36}$$

Since $a_0 \propto p$ and $\rho_0 \propto p$, then $S_L \propto \sqrt{p^{-2}}$ for a zero-order reaction. This agrees with the result obtained before, that $S_L \propto \sqrt{p^{n-2}}$.

Note: This theory is valid only when $E_a/R_uT_f \gg 1$, in other words for high-activation-energy reactions. Reactions in which ω is a function of reactant concentration as well as temperature have been considered. For first-order (unimolecular) reactions

$$\omega = CAe^{-E_a/R_uT} \qquad (C = \text{concentration}) \tag{5-37}$$

and for second-order (bimolecular) reactions

$$\omega = C^2Ae^{-E_a/R_uT} \tag{5-38}$$

Some of the restricting assumptions may now be removed. If we allow the number of molecules to change in the ratio n_r/n_p, assign λ and C_p the mean values $\bar{\lambda}$ and $\bar{C}_p$ in region I and the value for the burned gases λ_f and C_{pf} in region II, and set $(\lambda/C_p)/\rho\mathscr{D} = \text{Le}$, then the equation for a first-order reaction becomes

$$S_L = u_0 = \sqrt{\frac{2\lambda_f C_{pf} A}{\rho_0 \bar{C}_p^2}\left(\frac{T_0}{T_f}\right)\left(\frac{n_r}{n_p}\right)(\text{Le})\left(\frac{R_u T_f^2}{E_a}\right)^2 \frac{e^{-E_a/R_uT_f}}{(T_f - T_0)^2}} \tag{5-39}$$

and for a second-order reaction

$$S_L = u_0 = \sqrt{\frac{2\lambda_f (C_{pf})^2 A a_0}{\rho_0 \bar{C}_p^3}\left(\frac{T_0}{T_f}\right)\left(\frac{n_r}{n_p}\right)^2(\text{Le})^2\left(\frac{R_u T_f^2}{E_a}\right)^3 \frac{e^{-E_a/R_uT_f}}{(T_f - T_0)^3}} \tag{5-40}$$

The theory does not give very accurate results, but it does predict the trend of the flame propagation speed. Note that we still have the basic relationship

$$S_L \propto \sqrt{\alpha(\text{RR})_{T_f}} \tag{5-41}$$

The difference between this theory and that of Mallard and Le Chatelier is that the reaction rate is based upon the flame temperature.

4 DIFFUSION THEORY: THE THEORY OF TANFORD AND PEASE

Tanford and Pease[13] assumed that for certain reactions in laminar flames the rate of diffusion of active radicals into the unburned gas determines the magnitude of the flame velocity. Their work was reported in three parts. The first part[13(a)] gave calculations of the equilibrium atom and free-radical concentrations in a moist carbon monoxide flame. The results indicated that the equilibrium concentration of the hydrogen atoms is an important factor in determining the flame velocity. In the second part, Tanford[13(b)] presented calculations to establish the relative importance of mass diffusion and heat conduction in creating the hydrogen atoms in the flame zone. The conclusion was that diffusion is the controlling process. The third paper[13(c)] developed an equation for the flame velocity based on this conclusion.

They plotted their calculated equilibrium concentration of H in the moist $CO-O_2-N_2$ flame versus the flame velocities reported by Jahn[17] and showed a linkage between S_L and C_H, suggesting that S_L is a function of hydrogen-atom concentration. Two mechanisms were considered by which radicals appear throughout the flame zone, and an evaluation of the relative importance of the two mechanisms was made. The mechanisms considered were:

1. Local production by thermal dissociation—the concentrations being functions of temperature and hence dependent upon the heat-conduction process.
2. Supply by diffusion from a station at which the reaction has reached equilibrium.

The concentration of hydrogen atoms, C_{H}, as a function of x was evaluated for the first mechanism by setting up an energy equation which upon solution gives T as a function of x. From the $T(x)$ curve the maximum possible values of C_{H} as a function of x were calculated. The distribution of C_{H} throughout the combustion wave was evaluated according to the second mechanism by solving the species-continuity equation and applying boundary conditions at both burned and unburned stations.

Comparison was made between the two C_{H} distributions obtained by using the two different mechanisms. It was found that the concentration of hydrogen atoms from thermal dissociation is less than 10% of that due to mass diffusion at a small distance away from the fully reacted station for both the carbon monoxide flame and hydrogen flame. On the basis of these results a formula for the flame velocity was developed on the assumption that the determining factor is the diffusion of active species, chiefly hydrogen atoms, from the station at which the combustion has reached equilibrium. Several further assumptions were used in the development of their diffusion theory: (1) zero activation energy for radical species, (2) exponential concentration distribution for all active free radicals, (3) a constant mean temperature = $0.7T_f$ for the whole combustion zone, (4) a constant mass diffusion coefficient for gases in the combustion zone. (5) a first-order chemical-reaction expression for the source term in the species equation, and (6) no chain-branching reactions.

Using these assumptions, an expression for the laminar-flame speed was derived as

$$S_L \equiv u_0 = \sqrt{\frac{C_r}{X_P} \sum_i \frac{k_i p_i \mathscr{D}_{i,0}}{B_i'}} \tag{5-42}$$

where p_i is the equilibrium partial pressure of free-radical species i; $\mathscr{D}_{i,0}$ is the mass diffusivity of ith species at the initial temperature of the unburned mixture; k_i is the specific rate constant for the consumption of ith species; C_r is the concentration of the reactant mixture; X_P is the mole fraction of products; and B_i' is a function of the mass diffusivity of the gases in the reaction zone, the kinetic parameter, and the laminar flame speed. One obvious defect of the above equation is that the flame velocity of the dry carbon monoxide mixture goes to zero with free-radical concentration or partial pressure. Tanford and Pease therefore assumed that there is a constant contribution to the flame velocity independent of the radical concentration, and this constant (17 cm/sec) is added to the Eq. (5-42). By considering only H and OH radicals in moist carbon monoxide flames, they obtained

$$S_L = 17\ \text{cm/sec} + \sqrt{\frac{C_r}{X_P}\left(\frac{k_{\text{H}} p_{\text{H}} \mathscr{D}_{\text{H},0}}{B'_{\text{H}}} + \frac{k_{\text{OH}} p_{\text{OH}} \mathscr{D}_{\text{OH},0}}{B'_{\text{OH}}}\right)} \tag{5-43}$$

Since both B'_{H} and B'_{OH} are functions of S_L, Eq. (5-43) is an implicit equation for calculating S_L.

Flame velocities for moist CO–O_2–N_2 mixtures typically vary from 25 to 106 cm/sec. The calculated values of S_L from Eq. (5-43) generally have errors less than 25%. Equation (5-42) was applied to the combustion of H_2 in a similar way, with results of the same degree of accuracy.

Dugger[18] (1951) measured the velocity as a function of initial temperature for propane–air and ethylene–air flames, and compared the curves for the experimental data with curves calculated by the Zel'dovich–Frank-Kamenetsky theory as well as with curves calculated according to the Tanford–Pease theory. In this comparison, Dugger assumed that the controlling step was a bimolecular reaction. To examine the temperature dependence, he rearranged the second-order Zel'dovich–Frank-Kamenetsky–Semenov equation and approximated the temperature-dependent terms with those for air. In this way, he reduced Eq. (5-40) to the form:

$$S_L \propto \sqrt{\frac{T_0^2 T_f^{4.9} e^{-E_a/R_u T_f}}{(T_f - T_0)^3}} \tag{5-44}$$

In using the Tanford–Pease equation, Dugger assumed that only p_i, $\mathscr{D}_{i,m}$, and the number density of the reactant molecules are temperature-dependent. Dugger's expression for predicting the relative effect of temperature on flame speed according to the Tanford–Pease theory is

$$S_L \propto \sqrt{\left(\sum k_i p_i \mathscr{D}_{i,m}\right) T_0^2 T_f^{-1.33}} \tag{5-45}$$

where $\mathscr{D}_{i,m}$ is the mass diffusivity at $0.7T_f$. Figure 5.6 shows the results

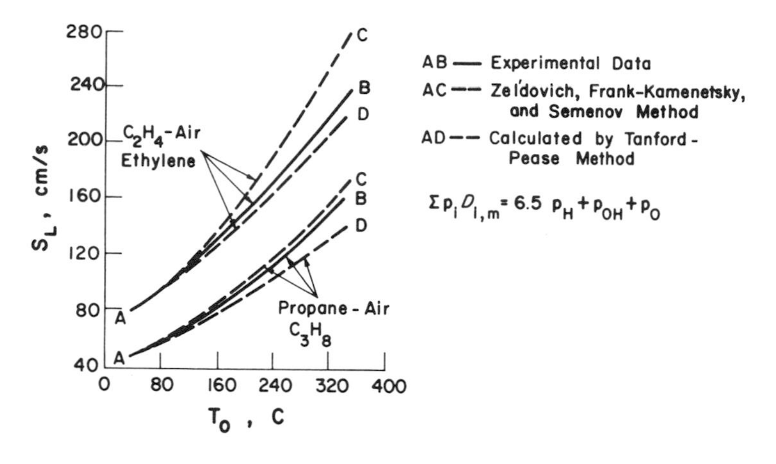

Figure 5.6 Comparison of calculated S_L versus T_0 with the measured data (adapted from Evans[2]).

for the Tanford–Pease theory as well as for the Zel'dovich–Frank-Kamenetsky–Semenov theory. The effect of the initial temperature on the flame velocity appears to be explained equally well or equally badly by the two theories.

Simon[19] found the calculated flame velocities of 35 hydrocarbons from the diffusion flame theory to be consistent with those measured by Gerstein, Levine, and Wong.[20] Equilibrium concentrations of H, O, and OH radicals were calculated at compositions giving the maximum flame velocity. The sum of the active radical concentrations each multiplied by the respective diffusion coefficient was plotted against S_L as shown in Fig. 5.7. It should be noted that Eq. (5-42) was modified for application to hydrocarbons by including a factor r for the total number of moles of water vapor and carbon dioxide which form per mole of hydrocarbon, and by assuming that H, OH, and O are equally

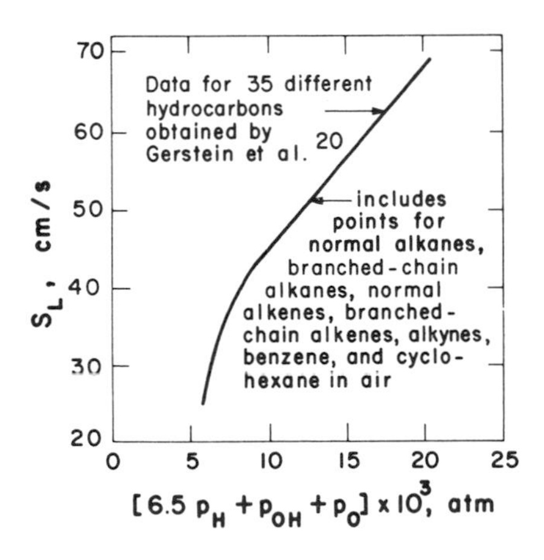

Figure 5.7 S_L versus radical concentrations as calculated by Simon.[19]

effective as chain carriers. Then Eq. (5-42) becomes

$$S_L = \sqrt{\frac{rC_r}{X_P} k \left(\frac{\mathscr{D}_{\mathrm{H}} p_{\mathrm{H}}}{B'_{\mathrm{H}}} + \frac{\mathscr{D}_{\mathrm{OH}} p_{\mathrm{OH}}}{B'_{\mathrm{OH}}} + \frac{\mathscr{D}_{\mathrm{O}} p_{\mathrm{O}}}{B'_{\mathrm{O}}} \right)} \tag{5-46}$$

From Eq. (5-46) k was found for all hydrocarbons except ethylene to have a value of $(1.4 \pm 0.1) \times 10^{11}$ cm^3/mole sec. This constancy of k suggests that the rate constants for the oxidation of the hydrocarbons either are the same or are unimportant in the mechanism of flame propagation.

Simon[19] points out that a correlation between flame velocity and flame temperature exists for these hydrocarbons, and that a flame mechanism which depended strongly on flame temperature might have given an equally good correlation with flame speed. Therefore a thermal mechanism cannot be ruled out.

The consistency of the calculated results with experimental data does not prove the validity of these simple theories developed in the 1940s. Indeed, the researchers then were handicapped by not having advanced computers and sophisticated numerical schemes to solve nonlinear coupled partial or ordinary differential equations. The current approach to this type of problem is quite different from those in the pioneering age. A formulation using crude assumptions like those of Tanford and Pease would be inexcusable in today's research work. We shall follow one example to show a modern approach to laminar-flame problems.

5 CONTEMPORARY METHOD FOR SOLVING LAMINAR-FLAME PROBLEMS

With the advent of high-speed digital computers, the solution of laminar-flame problems has changed significantly from the traditional methods discussed previously. We shall use one of the more recent papers by Heimerl and Coffee[21] as an example of a contemporary method. A one-dimensional, premixed, laminar, steady-state ozone flame was considered in their theoretical model. One reason for choosing ozone was its simplicity; it involves only three species: O, O_2, and O_3. The chemical reactions between these species are

$$O_3 + M \underset{k_{b_1}}{\overset{k_{f_1}}{\rightleftharpoons}} O + O_2 + M \tag{5-47}$$

$$O + O_3 \underset{k_{b_2}}{\overset{k_{f_2}}{\rightleftharpoons}} 2O_2 \tag{5-48}$$

$$2O + M \underset{k_{b_3}}{\overset{k_{f_3}}{\rightleftharpoons}} O_2 + M \tag{5-49}$$

TABLE 5.1 Kinetic Coefficients

Coefficient	Expression	Reference	Remarks
k_{f1}	$4.31 \times 10^{14} \exp\left(-\frac{11{,}161}{T}\right) \frac{\text{cm}^3}{\text{mole s}}$	Heimerl and Coffee[22] (1979)	$M = O_3$ $300 < T < 3000$ K
k_{b1}	$1.2 \times 10^{13} \exp\left(\frac{+976}{T}\right) \frac{\text{cm}^6}{\text{mole}^2\ \text{s}}$	—	Derived from equilibrium constant
k_{f2}	$1.14 \times 10^{13} \exp\left(-\frac{2300}{T}\right) \frac{\text{cm}^3}{\text{mole s}}$	Hampson[23] (1973)	$200 < T < 1000$ K
k_{b2}	$1.19 \times 10^{13} \exp\left(-\frac{50{,}600}{T}\right) \frac{\text{cm}^3}{\text{mole s}}$	—	Derived from equilibrium constant
k_{f3}	$1.38 \times 10^{18} T^{-1} \exp\left(-\frac{171}{T}\right) \frac{\text{cm}^6}{\text{mole s}}$	Johnston[24] (1968)	$M = O_2$ $1000 < T < 8000$ K
k_{b3}	$2.75 \times 10^{19} T^{-1} \exp\left(-\frac{59{,}732}{T}\right) \frac{\text{cm}^3}{\text{mole s}}$	Johnston[24] (1968)	$M = O_2$ $1000 < T < 8000$ K

where M represents the third body, which could be either O, O_2, or O_3. The kinetic coefficients for these reactions are given as a function of temperature in Table 5.1.

The rate coefficient may vary with the identity of the chaperon third body M. The chaperon relationships used are

$$k_{f_1}(M = \text{O}) = k_{f_1}(M = \text{O}_2) = 0.44 k_{f_1}(M = \text{O}_3) \tag{5-50}$$

$$k_{f_3}(M = \text{O}) = 3.6 k_{f_3}(M = \text{O}_2) = 3.6 k_{f_3}(M = \text{O}_3) \tag{5-51}$$

Equation (5-50) is based upon the work of Johnston,[24] and Eq. (5-51) is based on the work of Baulch et al.[25]

The rate of production of species i for Eqs. (5-47) to (5-49) can be given as

$$\omega_i = W_i \sum_{k=1}^{6} (\nu''_{i,k} - \nu'_{i,k}) \underbrace{B_k T^{\alpha_k} e^{-E_a/R_u T}}_{k_k} \prod_{j=1}^{3} \left(\frac{X_j}{R_u} \frac{p}{T} \right)^{\nu'_{j,k}} \tag{5-52}$$

which considers all six chemical reactions [Eqs. (5-47) to (5-49)] and all three chemical species. This equation has the same form as Eq. (3-95).

In addition to the kinetic coefficients, there are two sets of data required as inputs to the model: the thermodynamic and transport coefficients. By means of the thermodynamic coefficients, the sensible enthalpy and constant-pressure specific heat are related to the temperature by the following equations:

$$h = R_u\left(a_0 + \sum_{n=1}^{5} a_n \frac{T^n}{n}\right) \tag{5-53}$$

$$C_p = R_u \sum_{n=1}^{5} a_n T^{n-1} \tag{5-54}$$

For each species, two sets of coefficients are given in Ref. 21 for two temperature ranges. The transport coefficients λ and μ are given as functions of temperature. The mass diffusivity $\mathscr{D}_{ij}$ is assumed to be a function of both temperature and pressure.

The governing equations used in the model are given in one-dimensional and time-dependent forms as follows:

Overall continuity equation:

$$\frac{\partial \rho}{\partial t} + \frac{\partial(\rho u)}{\partial x} = 0 \tag{5-55}$$

Species continuity equation:

$$\rho\frac{\partial Y_k}{\partial t} + \rho u \frac{\partial Y_k}{\partial x} = -\frac{\partial}{\partial x}(\rho Y_k V_k) + \omega_k \tag{5-56}$$

where $k = 1$, 2, or 3 for O, O_2, or O_3 respectively.

Energy equation:

$$\rho C_p \frac{\partial T}{\partial t} + \rho u C_p \frac{\partial T}{\partial x} = \frac{\partial}{\partial x}\left(\lambda \frac{\partial T}{\partial x}\right) - \sum_{k=1}^{N} \omega_k \Delta h^\circ_{f_k} - \rho \sum_{k=1}^{N} C_{p,k} Y_k V_k \frac{\partial T}{\partial x} \tag{5-57}$$

Diffusion equation:

$$\frac{\partial X_k}{\partial x} = \sum_{j=1}^{N} \frac{X_k X_j}{\mathscr{D}_{kj}}(V_j - V_k) \tag{5-58}$$

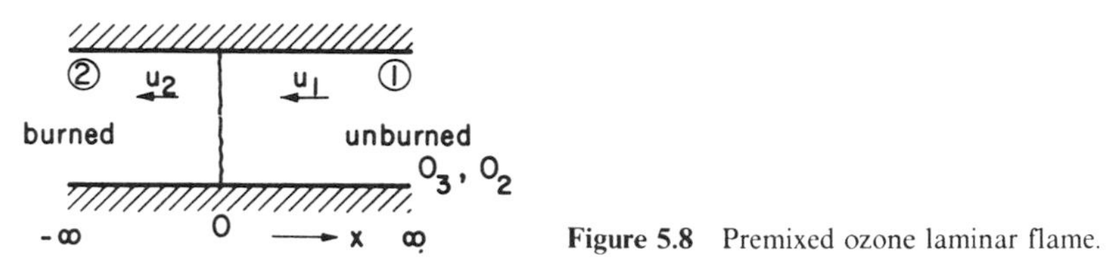

Figure 5.8 Premixed ozone laminar flame.

The energy equation was arrived at under the assumptions that the pressure is constant in the reaction zone, the viscous dissipation is negligible, and there is no body-force work.

The boundary conditions for the totally unburned end are

$$x \to \infty: \qquad T = T_1$$

$$Y_k = Y_{k1} \qquad \text{for} \quad k = 1,2,3 \tag{5-59}$$

The boundary conditions for the totally burned end are

$$x \to -\infty: \qquad \frac{\partial T}{\partial x} = \frac{\partial Y_k}{\partial x} = 0 \tag{5-60}$$

(See Fig. 5.8.)

In order to avoid solving Eq. (5-55) along with the other equations, Heimerl and Coffee[21] used a Lagrangian coordinate ψ and a new time coordinate τ defined by

$$\psi \equiv \int_0^x \rho(x', t)\, dx'$$

$$\tau \equiv t \tag{5-61}$$

By the chain rule, we have

$$\frac{\partial}{\partial t}(\ ) = \frac{\partial \psi}{\partial t}\frac{\partial(\ )}{\partial \psi} + \frac{\partial \tau}{\partial t}\frac{\partial(\ )}{\partial \tau} = [(\rho u)_0 - (\rho u)]\frac{\partial(\ )}{\partial \psi} + \frac{\partial(\ )}{\partial \tau} \tag{5-62}$$

$$\frac{\partial}{\partial x}(\ ) = \frac{\partial \psi}{\partial x}\frac{\partial(\ )}{\partial \psi} + \frac{\partial \tau}{\partial x}\frac{\partial(\ )}{\partial \tau} = \rho\frac{\partial(\ )}{\partial \psi} \tag{5-63}$$

In arriving at the final expressions in Eqs. (5-62) and (5-63), we have used the following relationships which come directly from the definition given in Eq. (5-61):

$$\frac{\partial \psi}{\partial x} = \rho, \qquad \frac{\partial \tau}{\partial x} = 0$$

$$\frac{\partial \psi}{\partial t} = (\rho u)_{x=0} - (\rho u)_x, \qquad \frac{\partial \tau}{\partial t} = 1 \tag{5-64}$$

One can easily show that by integrating the overall continuity equation (5-55) from 0 to x, one obtains

$$\frac{\partial}{\partial t}\int_0^x \rho\, dx + [(\rho u)_x - (\rho u)_0] = 0 \tag{5-65}$$

which automatically satisfies one of the relationships given in Eq. (5-64). Therefore, the overall continuity equation need not be solved in the (ψ, τ) coordinate system.

It is important to point out that the product ρu at $x = 0$ is the eigenvalue of the problem. It is defined as

$$m_0 \equiv (\rho u)_0 \tag{5-66}$$

After the transformation and replacement of τ by t, we have

$$\frac{\partial Y_k}{\partial t} + m_0 \frac{\partial Y_k}{\partial \psi} = -\frac{\partial}{\partial \psi}(\rho Y_k V_k) + \frac{\omega_k}{\rho} \qquad \text{for} \quad k = 1, 3 \tag{5-67}$$

For O_2, the mass fraction Y_2 can be determined from

$$Y_2 = 1 - Y_1 - Y_3 \tag{5-68}$$

The energy equation in the new coordinate system becomes

$$\frac{\partial T}{\partial t} = m_0 \frac{\partial T}{\partial \psi} = \frac{1}{C_p}\frac{\partial}{\partial \psi}\left(\rho\lambda \frac{\partial T}{\partial \psi}\right) - \frac{1}{\rho C_p}\sum_{k=1}^{N} \omega_k \Delta h_{fk}^{\circ} - \sum_{k=1}^{N} \frac{C_{p,k}}{C_p}\rho Y_k V_k \frac{\partial T}{\partial \psi} \tag{5-69}$$

The eigenvalue of m_0 or S_L was determined from the steady-state solution of the problem, that is,

$$\frac{\partial Y_k}{\partial t} = 0 \quad \text{and} \quad \frac{\partial T}{\partial t} = 0. \tag{5-70}$$

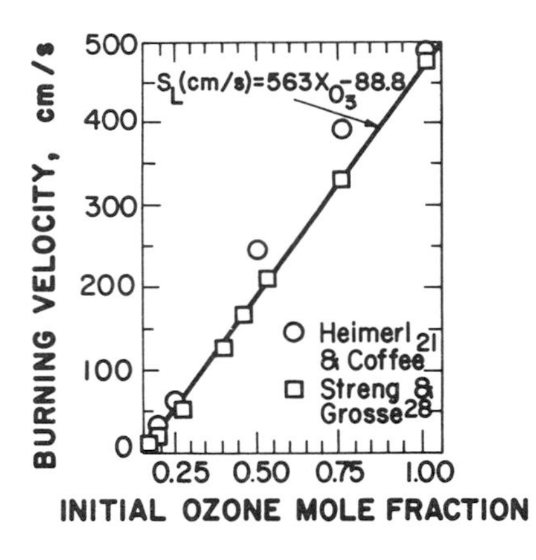

Figure 5.9 Comparison of theoretical results of Heimerl and Coffee[21] with experimental data of Streng and Grosse.[28]

Under the steady-state condition,

$$\rho u = \text{constant} = \rho(\infty)u(\infty) = -\rho_1 S_L \tag{5-71}$$

Integrating Eq. (5-56) over any interval (a, b), one obtains

$$\rho u[Y_k(b) - Y_k(a)] = \int_a^b \omega_k \, dx - \rho Y_k V_k \Big|_a^b \tag{5-72}$$

or

$$S_L = -\frac{\int_a^b \omega_k \, dx - \rho Y_k V_k \Big|_a^b}{\rho_1[Y_k(b) - Y_k(a)]} = -\frac{\int_{x=a}^{x=b} \frac{\omega_k}{\rho} \, d\psi - \rho Y_k V_k \Big|_a^b}{\rho_1[Y_k(b) - Y_k(a)]} \tag{5-73}$$

In the numerical solution, Heimerl and Coffee[21] used a general PDE solver (PDECOL), developed by Madsen and Sincovec[26] and discussed in Chapter 2. In

TABLE 5.2 Comparison of Computed Burning Velocities by Heimerl and Coffee[21] with the Measurements of Streng and Grosse[28]

Initial Ozone Mole Fraction	Burning Velocity (cm/sec)		Ratio
	Heimerl	Streng and Grosse[a]	
1.00	497	474	1.05
0.75	396	333	1.19
0.50	248	193	1.28
0.25	64	52	1.23
0.20	33	24	1.38

[a] Determined from Streng and Grosse's equation in Fig. 5.9.

TABLE 5.3 Comparison of Burning Velocities Computed by Warnatz[29] and by Heimerl and Coffee[21]

Initial Ozone Mole Fraction	Burning Velocity (cm/sec) Warnatz	Burning Velocity (cm/sec) Heimerl and Coffee	Ratio
1.00	445	497	0.90
0.75	350	396	0.88
0.50	225	248	0.91
0.25	65	64	1.02
0.20	37	33	1.12

the solution procedure, the spatial discretization is accomplished by finite-element collocation methods based on *B*-splines, developed by de Boor.[27] The solution was written in the form of a finite series

$$Y_k \approx \sum_{i=1}^{\mathrm{NC}} C_k^{(i)}(t) B_i(\psi), \qquad k = 1, 3$$

$$T \approx \sum_{i=1}^{\mathrm{NC}} C_T^{(i)}(t) B_i(\psi) \tag{5-74}$$

where $C_k^{(i)}$ and $C_T^{(i)}$ are time-dependent coefficients, $B_i(\psi)$ are basis functions of the *B*-splines, and NC is the number of collocation points. The time-dependent coefficients are determined uniquely by requiring that the preceding expansion satisfy the given boundary conditions and that they satisfy Eqs. (5-67) and (5-69) exactly at interior collocation points.

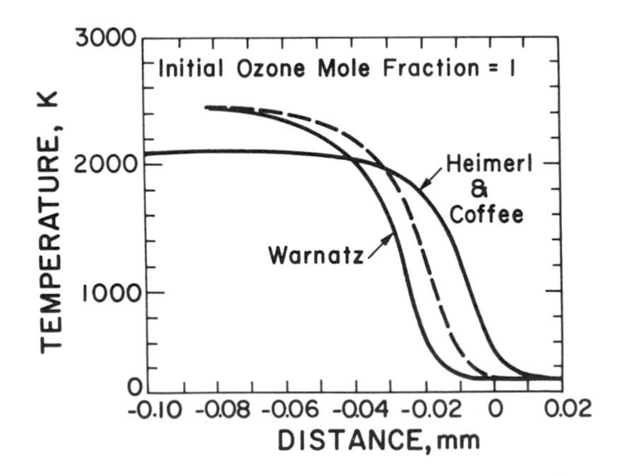

Figure 5.10 Comparison of calculated temperature profiles by Warnatz[29] and Heimerl et al.[21] (Dashed profile represents the result obtained by substituting Warnatz's expressions for k_{f1} and k_{f2} into the Heimerl–Coffee model.)

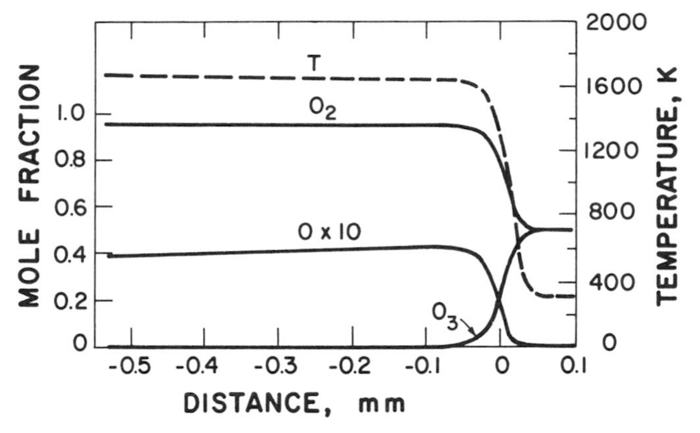

Figure 5.11 Calculated temperature and concentration profiles.

The predicted flame velocities are in reasonable agreement with the measurements of Streng and Grosse,[28] as shown in Fig. 5.9 and Table 5.2. The calculated results of Heimerl and Coffee[21] are also in good agreement (within $\pm 12\%$) with the computed results of Warnatz,[29] even though he used quite different expressions for some of the kinetic coefficients (see Table 5.3). The calculated temperature and atomic oxygen profiles are quite different from those of Warnatz. The comparison of temperature profiles is shown in Fig. 5.10.

A typical set of calculated profiles of temperature and concentration are shown in Fig. 5.11. As depicted in this figure, the major variations of temperature and concentrations are within a distance of 0.1 mm.

Heimerl and Coffee[21] stressed that profile measurements for temperature and concentrations are vital to test the model's input coefficients and so validate the model. They further indicated that agreement with burning velocities, even over a wide range of initial ozone mole fractions, is a necessary but insufficient condition to insure that the input coefficients are realistic.

6 EFFECT OF CHEMICAL AND PHYSICAL VARIABLES ON FLAME SPEED

6.1 Chemical Variables

6.1.1 Effect of Mixture Ratio

The variation of the laminar-flame speed with fuel–oxidant ratio is governed predominantly by the variation of the temperature with the mixture ratio. For hydrocarbon fuels, the peak of the flame speed occurs at stoichiometric or slightly fuel-rich mixtures. Some typical results on flame speeds (measured by Hartmann[32]) as a function of fuel volume percentage in air are shown in Fig. 5.12.

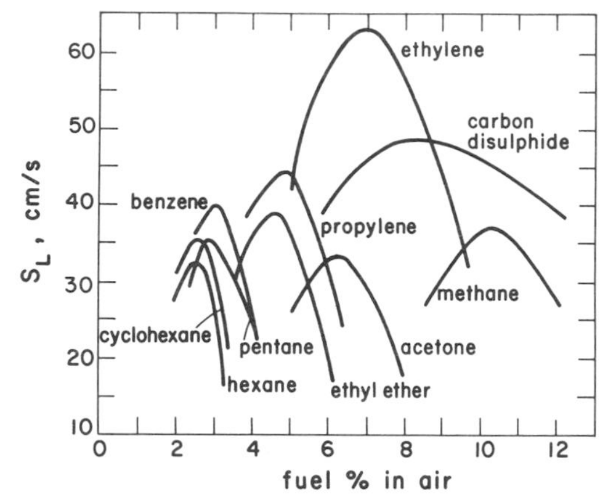

Figure 5.12 The variation of flame speed with mixture ratio (after Kanury[33]).

It is generally acceptable to assume that a mixture with maximum flame temperature is also a mixture with maximum flame speed.[33] In very lean or very rich mixtures, it is impossible to propagate a flame because there is too little fuel or oxidant to maintain a steady deflagration wave. Thus, there exist upper and lower flammability limits. The deflagration limits for some fuel–oxidant mixtures are listed in Table 4.5.

6.1.2 Effect of Fuel Molecular Structure

Attempts have been made to correlate the flame speed with the fuel structure.[20,34] The variation of flame speed with respect to the number of substituted methyl groups is shown in Fig. 5.13.

Gerstein et al.[20] and Reynolds[34] found that as the fuel molecular weight increases, the range of flammability becomes narrower (also shown in Fig. 5.12). Figure 5.14 shows the maximum flame speed as a function of the number of carbon atoms in the fuel molecule for three families. For saturated hydro-

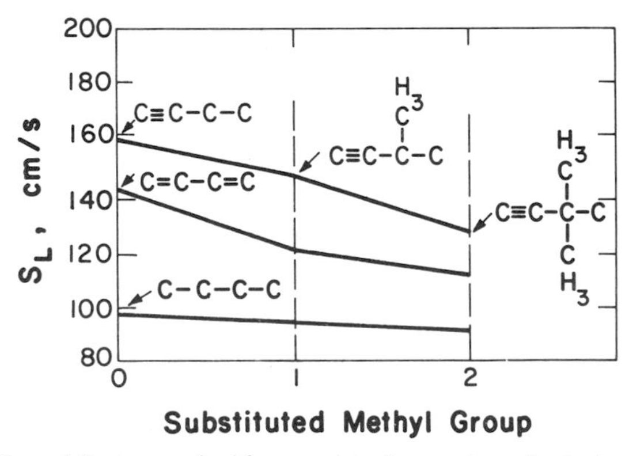

Figure 5.13 Variation of flame speed with respect to the number of substituted methyl groups.

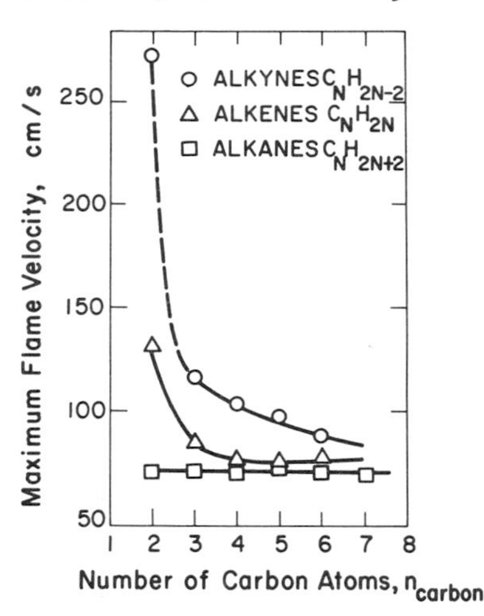

Figure 5.14 Effect of the number of carbon atoms in the chain on maximum flame velocity (after Reynolds and Gerstein[34]).

carbons [alkanes (also known as paraffins) such as C_2H_6 (ethane), C_3H_8 (propane), C_4H_{10} (butane), C_5H_{12} (pentane), and C_6H_{14} (hexane)], the maximum flame speed (70 cm/sec) is nearly independent of the number of carbon atoms in the molecule. For unsaturated hydrocarbons [either alkenes (also known as olefins) such as C_2H_4 (ethylene) and C_3H_6 (propylene), or alkynes (also known as the acetylene series) such as C_2H_2 (acetylene), C_3H_4 (propyne), C_4H_6 (butyne), and C_5H_8 (pentyne)], the laminar flame speed is higher for fuels with a smaller number of carbon atoms. The values of S_L fall steeply as n_{carbon} increases to 4, and then falls slowly with further increases in n_{carbon}, approaching the value of S_L of saturated fuels when $n_{\text{carbon}} \geq 8$.

From the above results, it seems that the number of substituted methyl groups and the structure of the fuel have a strong influence on S_L (Fig. 5.13 and Fig. 5.14). It is concluded by Gerstein et al.[20] that the variation of flame speed with respect to different substitutions of methyl groups is believed to be due to the flame temperature rather than the molecular structure. The influence of the number of carbon atoms in various fuels on laminar flame speed, however, is not due to the flame temperature, since the adiabatic flame temperature of most fuels is around 2200 K; the activation energies for most fuels reacting with oxygen are within a narrow range around 40,000 cal/mole. The difference in S_L for fuels containing different numbers of carbon atoms is mainly due to the changes in thermal diffusivity, which is a function of the fuel molecular weight.

So far, we have kept saying that $S_L \propto \sqrt{\alpha\, \text{RR}}$. However, this is true only for similar chemical compounds. The reactivity of the fuel may affect the flame velocity drastically. For example, there is a significant difference between the reactivity of a C–Si bond and a Si–H bond. This can be seen in Table 5.4. Therefore, for different chemical compounds, the reactivity of the combustion system should be considered.

TABLE 5.4 Effect of Fuel Reactivity on Flame Speed

Fuel	S_L (cm/sec)	T_F (K)	Reactivity
$(CH_3)_4C$	33	2254	
$(CH_3)_4Si$	60	2247	Greater than that of $(CH_3)_4C$
$(C_2H_5)_2SiH_2$	111	2278	Higher reactivity
$(C_4H_9)SiH_3$	148	2290	↓
SiH_4	Spontaneously flammable		Extremely reactive

6.1.3 *Effects of Additives*

The main purpose of using an additive is to raise the ignition temperature and reduce the tendency to preignition, as well as knocking. It only slightly affects the flame speed. However, studies with moist carbon monoxide show that its flame speed is raised appreciably by addition of small amounts of hydrogen or hydrogen-containing fuels. The pronounced effect of small amounts of H_2O on the reaction of CO gas was noticed by Tanford and Pease[13] and has been discussed in the diffusion theories (see Section 5.4). The antiknocking compound is used to slow down the low-temperature oxidation and is also used as a diluent. It does not have a strong effect on the flame speed, which is mainly controlled by high-temperature reactions. Mixing of different fuels usually does not have a significant effect on S_L. However, if the inert substance added does change the thermal diffusivity α of the fuel, then it will affect the flame speed significantly.

6.2 Physical Variables

6.2.1 *Effect of Pressure*

Lewis[35] studied the effect of pressure on the laminar flame speed of various hydrocarbon–O_2 mixtures containing N_2, Ar, or He. A power law, $S_L \propto p^n$, was developed, where the exponent n is referred to as the Lewis pressure index n. He observed that when $S_L < 50$ cm/sec, the exponent n is usually negative, implying that S_L increases with decreasing pressure; for $50 < S_L < 100$ cm/sec, S_L is independent of pressure; and when $S_L > 100$ cm/sec, S_L increases with increasing pressure.

In view of the thermal and the comprehensive theories discussed in Sections 5.2 and 5.3, we know that $S_L \propto p^{(n-2)/2}$. The pressure dependence observed by Lewis implies that the overall order of the reaction is less than 2 for flames with $S_L < 50$ cm/sec, equal to 2 for those with $50 < S_L < 100$ cm/sec, and greater than 2 for those with $S_L > 100$ cm/sec. The relationship $S_L \propto p^{(n-2)/2}$

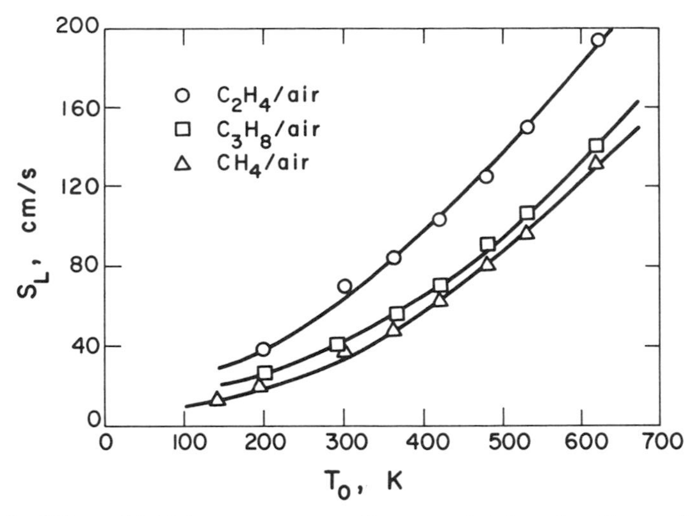

Figure 5.15 Effect of initial temperature on flame speed (adapted, with permission of the American Chemical Society, from Dugger, Heimel, and Weast[36]).

is supported by much experimental evidence. For example, the first-order decomposition reaction for ethylene oxide monopropellant (C_2H_4O) is shown to follow $S_L \propto \sqrt{p^{-1}}$. For many hydrocarbon fuels undergoing second-order reactions, the laminar-flame speeds were found to be independent of pressure.

6.2.2 *Effect of Initial Temperature*

Dugger et al.[36] conducted a series of experiments showing the dependence of S_L on T_0 for three mixtures. Figure 5.15 shows that S_L (or u_0) increases with T_0 for all three hydrocarbon–air mixtures. The results can be represented by the following relationship:

$$S_L \propto T_0^m \tag{5-75}$$

where the exponent m ranges between 1.5 and 2. The increase of S_L with T_0 is mainly due to the preheating effect. In general, the change of T_f caused by preheating the initial mixture is not significant. This is due to the fact that the heat release term a_0Q/ρ_0 is more or less fixed, and this term is also much larger that C_pT_0. From Eq. (5-23), we can see that T_f cannot change much by the preheating effect. Nevertheless, even a slight change in T_f may alter S_L significantly, as shown below.

6.2.3 *Effect of Flame Temperature*

The maximum flame speed for several mixtures as a function of flame temperature is shown in Fig. 5.16. The effect of T_f on S_L is obviously very strong. From these data, we can conclude that the laminar-flame speed S_L is essentially determined by the value of T_f; this is consistent with the idea of Zel'dovich[9a–9c] that the reaction rate is based upon T_f. At high temperatures,

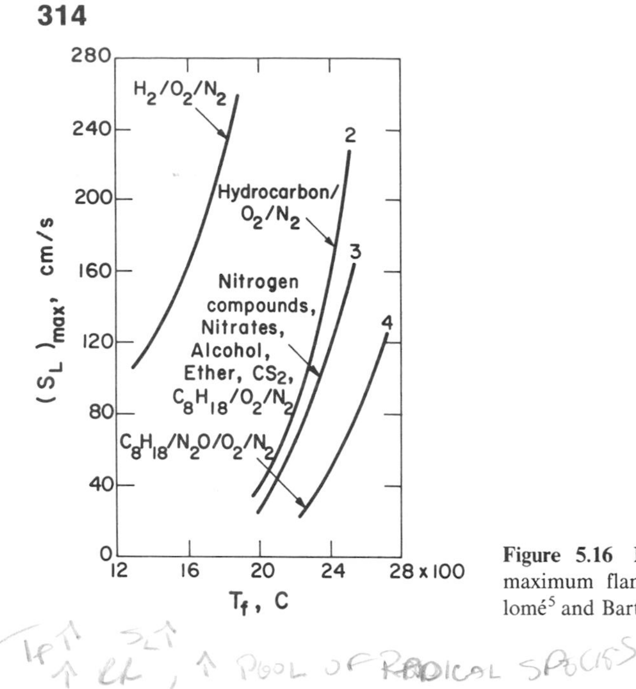

Figure 5.16 Effect of flame temperature on maximum flame speed (adapted from Bartholomé[5] and Bartholomé and Sachsse[37]).

dissociation reactions are favored. These reactions can introduce numerous free radicals into the flame, and these free radicals act as chain carriers to promote the reaction and hence the propagation of the flame. The lighter these free radicals are, the easier they are able to diffuse to the preflame zone. This is why H atoms can significantly enhance flame speeds. This description of radical-species diffusion is exactly that of Tanford and Pease.[13]

6.2.4 Effect of Thermal Diffusivity and Specific Heat

Clingman et al.[38] performed a set of experiments to elucidate the dominant effects of the thermal diffusivity and reaction rates on S_L. They measured the flame propagation speed of methane in various oxygen–inert-gas mixtures. The volumetric ratio of oxygen to inert gas was always set at 0.21 : 0.79. The inert gases chosen were nitrogen (N_2), helium (He), and argon (Ar). Their results are shown in Fig. 5.17.

The results can be explained as follows: By comparing the two top curves, we can see that S_L in He mixtures is higher than that in Ar mixtures. This is due to the fact that the thermal diffusivity of helium is much larger than that of argon because the molecular weight of helium is much smaller. Recall the relationship given by Eq. (5-41): if the flame temperatures are the same in both cases, then the higher the thermal diffusivity, the higher the S_L. Helium and argon are both monatomic gases, and thus their flame temperatures are equal. Consequently, S_L for He mixtures is greater than that for Ar mixtures.

In comparing the middle curve with the bottom curve of Fig. 5.17, we can see that S_L in Ar mixtures is higher than that in N_2 mixtures (regular air). This

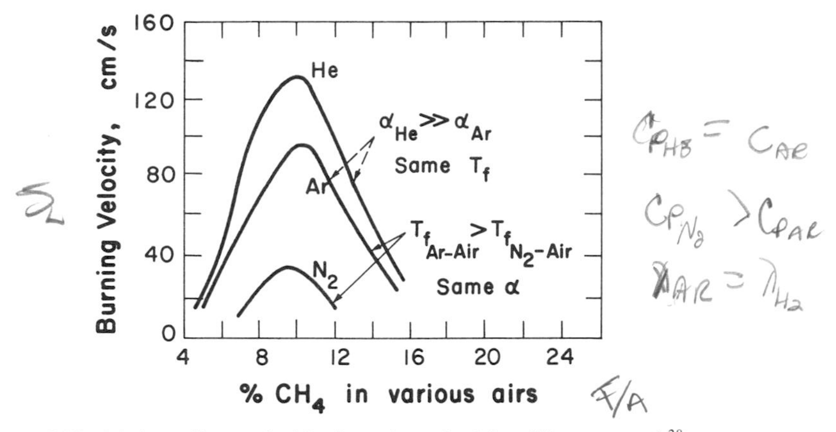

Figure 5.17 Methane flame velocities in various airs (after Clingman et al.[38]).

is due to the fact that Ar and N_2 have nearly the same values of thermal diffusivity. However, Ar is a monatomic gas which has a lower specific heat $[(C_p)_{\mathrm{Ar}} = \frac{5}{2}R]$ than the diatomic gas N_2 $[(C_p)_{\mathrm{N}_2} = \frac{9}{2}R]$. Since the heat release on all systems is the same, the flame temperature will be higher in the Ar mixture than that in the N_2 mixture. Thus, based upon Eq. (5-41), S_L is higher in Ar mixtures than in N_2 mixtures.

7 METHODS OF MEASURING FLAME VELOCITY

The flame velocity has been defined as the velocity of the unburned gases through the combustion wave in the direction normal to the wave surface. Various measurement techniques for this quantity have been developed. Some common techniques are briefly discussed below. For more detailed information on techniques for measuring the speed and structure of laminar premixed flames, the reader is referred to the books by Glassman,[1] Gaydon and Wolfhard,[30] Beer and Chigier,[39] Weinberg,[31] and Kanury.[33]

There are six basic types of experiments to measure laminar-flame speeds. These are (1) the Bunsen-burner method, (2) the transparent-tube method, (3) the constant-volume bomb method, (4) the soap-bubble method, (5) the particle-track method, and (6) the flat-flame burner method. In some of these methods, the flame remains stationary; in others, the flame front moves with respect to the laboratory frame of reference. These methods are discussed in the following.

7.1 Bunsen-Burner Method

The earliest premixed-flame burner was invented by Bunsen around 1855. Previously, simple flames of the diffusion type had been used; these diffusion flames were luminous and sometimes very smoky, and tended to form carbon deposits on surfaces which came in contact with the flame. The effective temperatures of these diffusion flames were low. The premixed flames from Bunsen burners are relatively clean and give much more intense combustion with higher effective temperatures. The simple design principle of the Bunsen burner has been incorporated in many gas appliances such as cooking stoves and gas burners.

In the simple type of Bunsen burner, as shown in Fig. 5.1, the fuel gas issues from a small nozzle or orifice and entrains some air (primary air). The mixture of gas and primary air passes up the burner tube at a speed which is sufficient to prevent the flame stroking back down the tube. The mixture thus burns at the top of the burner, the combustion being assisted by the surrounding (secondary) air. For good air entrainment, the air holes must be of sufficient size. The amount of air entrained is usually well below that needed for complete combustion of the gas.

There are limits to the size of Bunsen burners. The lower limit is set by the quenching distance of the gas–air mixture. The upper limit is set by the increasing tendency toward flashback (a term to be discussed later) with large burners. To maintain the flame at the port, there is a maximum gas velocity for a given supply pressure. To obtain sufficient velocity at all points in the tube and to prevent flashback, the average velocity must be increased for larger burners.

To obtain good mixing and allow time for the turbulence created by the gas jet to die down, we obviously require a long burner, but if it is too long, it will cause unnecessary resistance to flow and reduce the air entrainment. For natural gases, the length should be about 6 times the diameter of the mouth. For manufactured gases with higher burning velocities, a longer burner is required, usually 10 times the diameter. The diameter of the tube is usually on the order of 1 cm.

A sketch of the flow configuration is shown in Fig. 5.18*a*. Let us call the conical surface area of the inner cone A_f, the tube cross-sectional area A_t, and the average flow velocity in the tube V_t. Then the laminar flame velocity S_L can be determined from the mass continuity:

$$S_L = V_t \frac{A_t}{A_f} \tag{5-76}$$

Since the conical surface area A_f is greater than A_t, S_L must be less than V_t.

Although Eq. (5-76) is a very simple equation, there are some difficulties in determining S_L. This is due to the uncertainty of the surface area A_f. Depending upon the optical method used, the conical area varies in the manner shown in Fig. 5.18*b*. The shadow cone is used by many experimentalists because it is simpler than the Schlieren techniques. The shadow cone, being cooler, gives a more correct result than the visible cone.

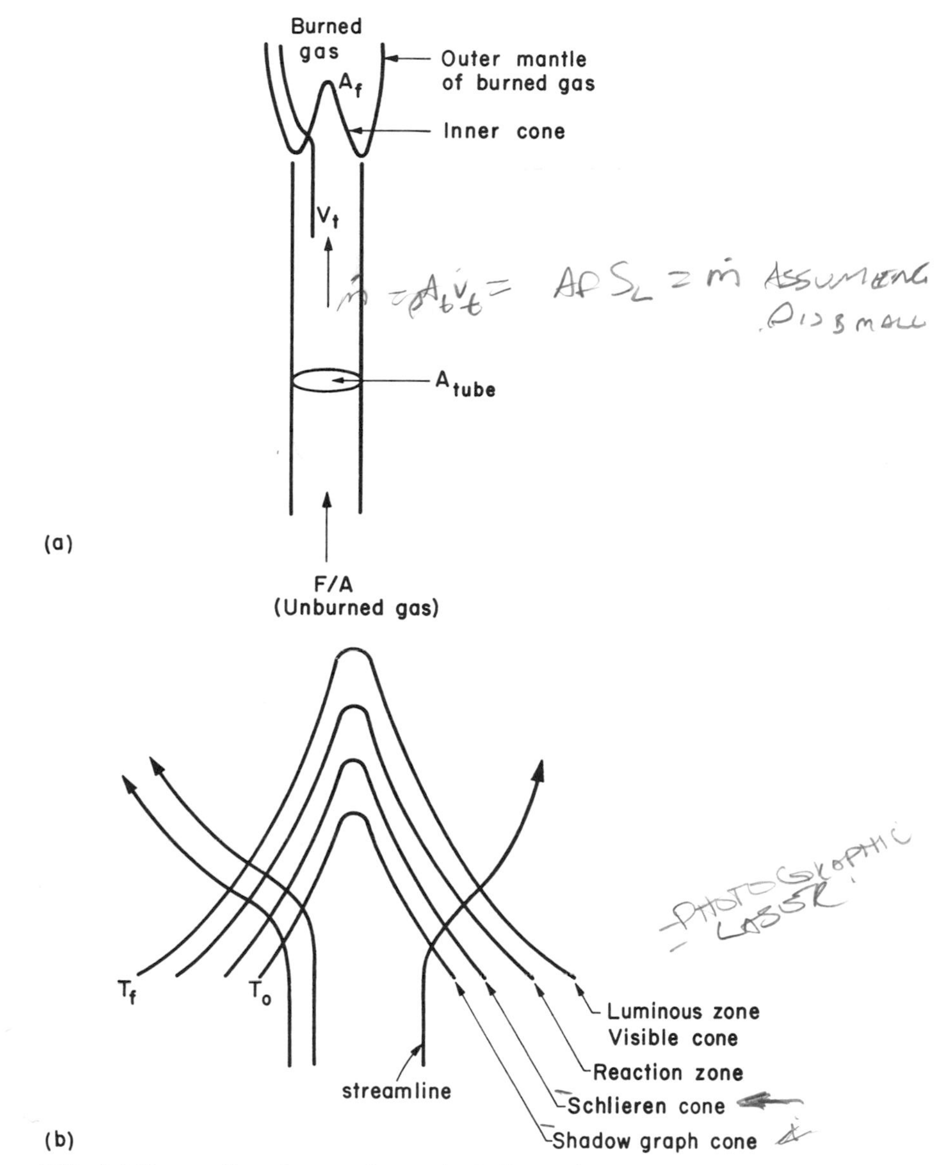

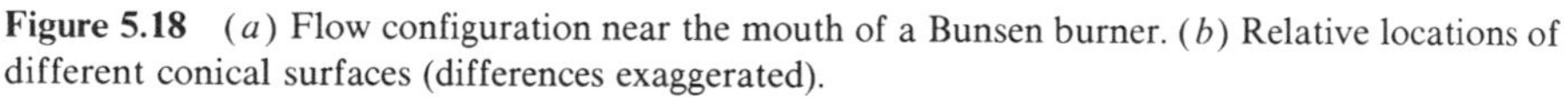
Figure 5.18 (*a*) Flow configuration near the mouth of a Bunsen burner. (*b*) Relative locations of different conical surfaces (differences exaggerated).

Experiments with fine magnesium oxide particles entrained in the gaseous stream have shown that the flow streamlines remain fairly straight until the Schlieren cone is reached, and then the flow diverges from the burner axis before reaching the visible cone. These experiments have led many investigators to use the Schlieren cone for flame-speed evaluation. In general, the wider the flame zone, the greater the resulting error in the determination of S_L.

The chief advantages of burner methods are that the equipment is simple, flexible, and easily adapted for measurements at varying temperatures and pressures. However, there are many disadvantages:

1. The diffusional interchange with the surrounding atmosphere alters the fuel–oxident ratio, so that the flame velocity observed may not represent the measured fuel–air ratio.
2. One can never completely eliminate wall-quenching effects.
3. The flame cone can act as a lens in a shadow measurement; this causes uncertainties in the proper cone size.
4. There may be inadequate entrainment of air and resultant tendency toward flashback with large-diameter burners.

The accuracy of the Bunsen-burner method is on the order of $\pm 20\%$.

7.2 Transparent-Tube Method

If a horizontal transparent tube (see Fig. 5.19), having an inside diameter greater than quenching diameter (a critical diameter below which the flame is quenched), is filled with a homogeneous combustible gas mixture and then ignited at one end of the tube, a flame will travel through the tube. By placing a suitable orifice at the closed end of the tube to reduce reflected pressure waves, uniform linear flame movement over a good portion of the tube and a flame of constant shape for that portion may be obtained. In order to evaluate the mass flow rate through the orifice, a coating of soap solution may be applied to the outer surface of the tube. The growth of the soap bubble at the end of the tube can be used to deduce the value of V_g from

$$S_L = \left(V_0 - V_g\right) \frac{A_{\text{tube}}}{A_f} \tag{5-77}$$

where V_0 is the linear velocity of the uniform flame movement, and V_g is the velocity of unburned gas ahead of flame.

The flame surface is roughly parabolic. This distortion from planarity is due to two factors:

1. The gas outflow at the tube axis is considerably more rapid than that at the tube walls, so that the pressure drop between the fresh and burned gas is

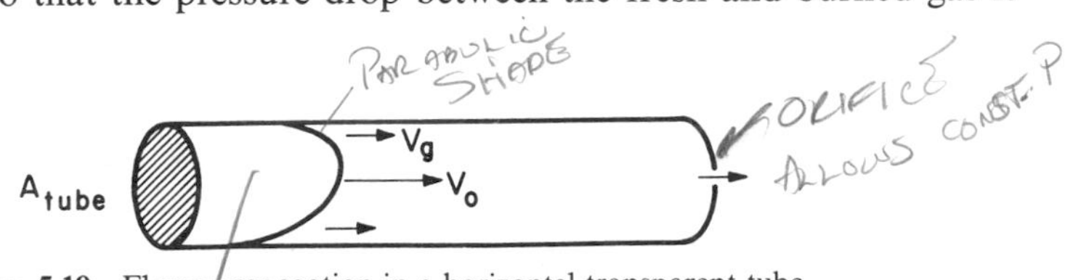

Figure 5.19 Flame propagation in a horizontal transparent tube.

greater along the tube axis. This tends to make the flame front parabolic in shape.

2. The difference in the densities of the burned and fresh gases produces natural convection, which stretches the flame front further.

In general, the results obtained from the transparent-tube method are close to those obtained with a Bunsen burner.

7.3 Constant-Volume Bomb (Closed Spherical Bomb) Method

In the constant-volume bomb method, the combustible mixture is ignited at the center of a rigid spherical vessel, which is usually on the order of 30 cm in diameter. As the flame progresses, the expansion of the burned gas causes both the pressure and temperature of the unburned gas to increase because of adiabatic compression. The temperature increase causes the flame velocity to increase continuously from the center towards the wall. In this method, simultaneous records of the size of the spherical zone of the burned gas and of the pressure in the vessel must be made to determine S_L.

It can be shown that

$$S_L = \left(1 - \frac{R^3 - r^3}{3p\gamma_u r^2}\frac{dp}{dr}\right)\frac{dr}{dt}$$

$$= \frac{dr}{dt} - \frac{R^3 - r^3}{3p\gamma_u r^2}\frac{dp}{dt} \tag{5-78}$$

where

$\gamma_u \equiv \dfrac{C_p}{C_v}$ (unburned gas)

p = pressure at time t

R = sphere radius

r = instantaneous radius of spherical flame

In Eq. (5-78), S_L is obtained as a difference of two quantities which are of comparable magnitude; therefore, errors in these derivatives can be magnified in the S_L calculation.

An alternative method of evaluating the burning velocity is to determine the rate of change of the mass fraction Y of burned gas. That is,

$$S_L = \frac{1}{3}\frac{R^3}{r^2}\left(\frac{p_i}{p}\right)^{1/\gamma_u}\frac{dY}{dt} \tag{5-79}$$

where p_i is the initial pressure. For small values of Y,

$$Y \equiv \frac{p - p_i}{p_e - p_i}$$

where p_e is the pressure corresponding to combustion at constant volume and may be computed thermochemically.

The flame speed deduced according to the above method assumes complete equilibrium behind the flame front and no heat losses. Any delay in attaining equilibrium in a rather large volume immediately behind the flame front will cause error; the value of S_L calculated using the above expression is usually less than the actual value.

7.4 Soap-Bubble (Constant-Pressure Bomb) Method

In the soap-bubble method, a homogeneous combustible mixture is used to blow a soap bubble around a pair of spark electrodes. The flame velocity is calculated from

$$S_L = \frac{V_{\text{flame}} r_i^3}{r_f^3} \tag{5-80}$$

where

V_{flame} is the average spatial velocity of the spherical flame front.
r_i is the initial radius of the soap bubble.
r_f is the final radius of the sphere of burned gas.

If the effective mean reaction-zone temperature is constant and the reaction mechanism does not change with composition, the true effects of fuel and oxidant concentrations on flame velocity and hence on the overall reaction rate can be expressed by

$$S_L^2 \propto X_F^a X_0^c \tag{5-81}$$

where

X_F is the mole fraction of fuel in unburned gas.
X_0 is the mole fraction of oxidant in unburned gas.
a and c are empirical exponents.

At time zero, the gas mixture contained in the spherical soap bubble is ignited by the spark (Fig 5.20). In general, we can assume that:

1. The spherical flame spreads radially through the gas.

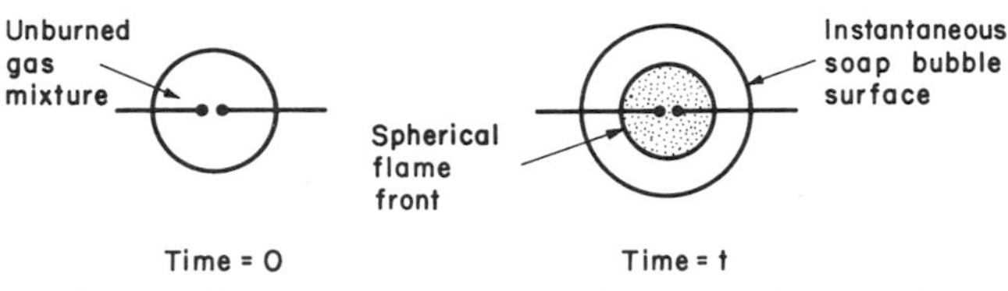

Figure 5.20 Experimental setup of the soap-bubble method.

2. The pressure remains constant.
3. The growth of the flame front can be followed by some photographic means.

By equating the mass flux in front of the flame front to that behind the flame front, we have

$$S_L A \rho_u = u_r A \rho_b \tag{5-82}$$

Hence,

$$S_L = u_r \frac{\rho_b}{\rho_u} = u_r \frac{T_u}{T_b} \tag{5-83}$$

where u_r represents the recorded flame velocity.

One obvious disadvantage of this method is the large uncertainty in the temperature ratio T_u/T_b needed for obtaining ρ_b/ρ_u. Although one can assume that the gases reach the theoretical flame temperature, comparison of measured and calculated expansion ratios often reveals serious discrepancies. Also, for use in Eq. (5-80), the initial and final sizes of the bubble must be known very accurately because they are raised to the third power in the calculation. In fact, however, the final size is difficult to measure. There are several other problems in using the soap-bubble method, namely:

a. The method is not suitable for studying the flame spread of dry mixtures, since they can be moistened by evaporation from the soap solution.
b. There is inevitably some heat loss to the electrodes.
c. For slow flames, the flame front may not remain spherical; also, the thickness of the reaction zone becomes large.
d. For very fast flames, the flame front is no longer smooth, due to the formation of cellular flame structure.

7.5 Particle-Track Method

With conical flames on round burners, we always have curved flame surfaces and difficulty in photography. To overcome these difficulties, Lewis and

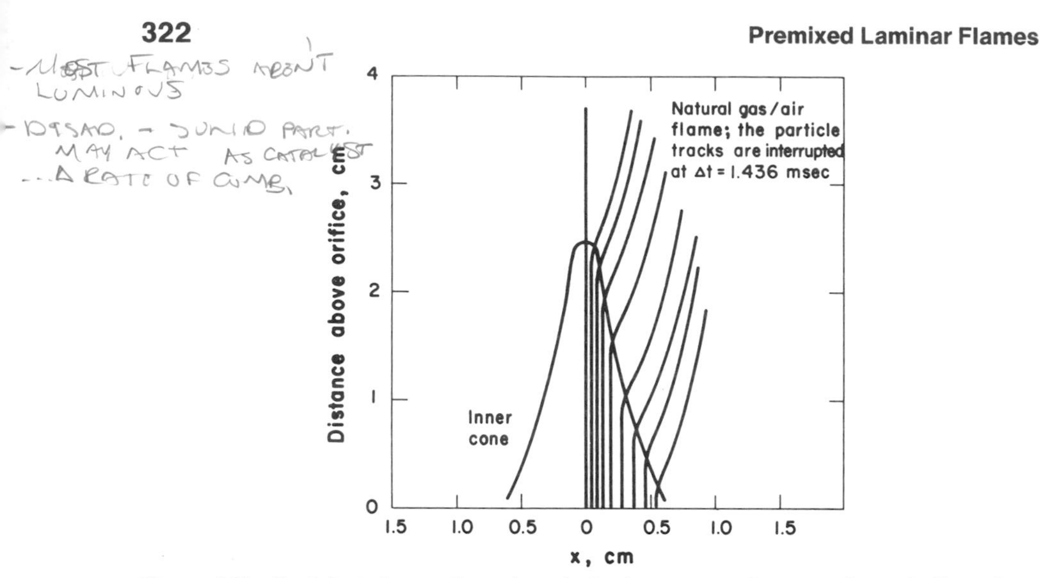

Figure 5.21 Particle trajectory lines through the inner cone of a natural-gas–air flame from particle-track measurements.

von Elbe[40] made an important investigation using rectangular burners. A particle-track method was devised in which small magnesium oxide particles in a gas stream were illuminated intermittently from the side. A photograph of the track of a particle then shows its direction. Typical results are shown in Fig. 5.21. The velocity of the particle can also be deduced from consecutive photographs.

The width of their burner was 0.755 cm, which was rather small; and the propagation of the flame front with uniform burning velocity would be faster if larger burners were employed. They showed that the burning velocity is a genuine physical constant (observe the flat region in Fig. 5.22).

A distinct weakness of this method is that the introduction of solid particles may, by catalytic effects at the surface, modify the combustion processes and hence alter S_L. Errors may also occur if the particles are too large to follow the gas flow accurately. The particle-track method is generally too laborious for regular measurement of burning velocities.

7.65 Flat-Flame Burner Method

The flat-flame burner method is usually attributed to Powling.[41] It is probably the most accurate method, because it offers the simplest flame front and one in which the areas of the shadow, Schlieren, and visible fronts are all the same. It is achieved (Fig. 5.23) by placing a porous metal disk or a series of small tubes of 1-mm diameter or less at the exit of the larger flow tube. A modern flat-flame burner usually consists of a sintered, porous bronze (or 316 stainless steel) water-cooled burner surrounded by a sintered, porous shroud ring for the introduction of a shielding gas (usually nitrogen), both enclosed in a

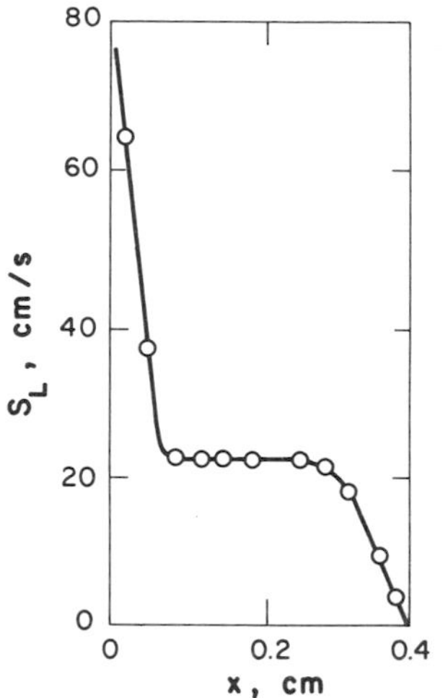

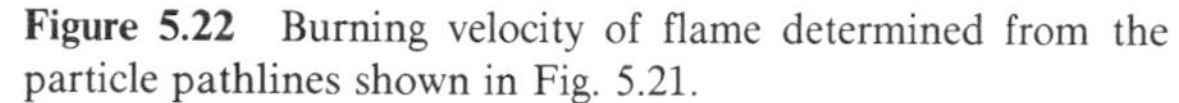
Figure 5.22 Burning velocity of flame determined from the particle pathlines shown in Fig. 5.21.

precision-machined housing assembly with connections for water cooling, combustion gas, and shielding gas.

The gaseous mixture is usually ignited at high flow rate and adjusted until the flame is flat. By controlling the rate of efflux of burned gases with a grid, a more stable flame is obtained. This method is applicable only to mixtures having low burning velocities, of the order of 15 cm/sec or less. At higher S_L, the flame front positions itself far from the burner and forms conical shapes. Spalding and Botha[42] extended the method to higher flame speeds by cooling the plug. The cooling brings the flame front closer to the plug and stabilizes it.

A plot of flame speed versus cooling rate is made and is extrapolated to zero cooling to obtain the adiabatic flame speed S_L (Fig. 5.24). This procedure can be used for all mixture ratios within the flammability limits.

8 PRINCIPLE OF STABILIZATION OF COMBUSTION WAVES IN LAMINAR STREAMS

A common example of the fixation of a combustion wave in a gas stream is afforded by a Bunsen-burner flame. We shall now examine the mechanism by

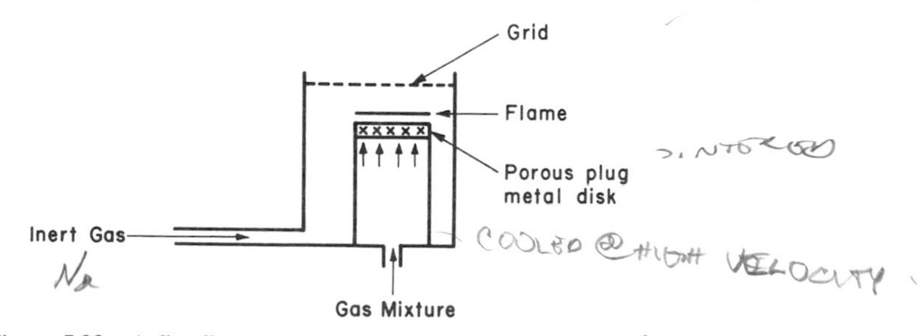

Figure 5.23 A flat-flame apparatus (adapted from Glassman[1]).

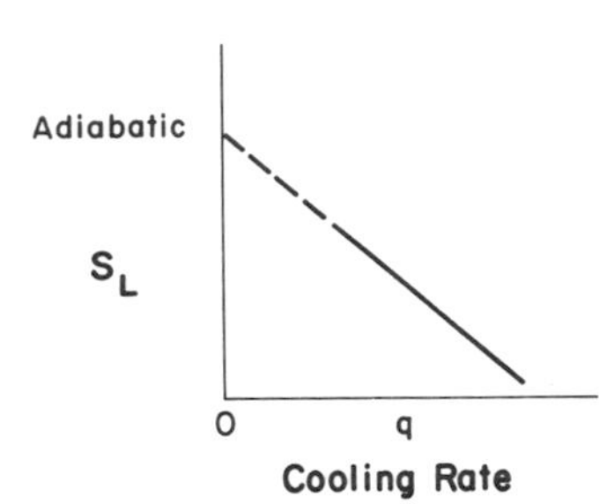

Figure 5.24 The effect of cooling on laminar flame speed (adapted from Glassman[1]).

which the inner cone maintains a fixed position with respect to the burner rim. Near the fringe of the combustion wave in the vicinity of the burner rim, the mixture gas velocity can be represented by an approximately linear profile. Figure 5.25 shows the mechanism which determines the shape of a Bunsen flame. There are four different situations. The comparison of profiles of approach velocity and flame propagation velocity near the rim shows how the flame is anchored to the burner rim. When the premixed-gas flow is weak, S_L is greater than u_{supply} over almost the entire cross-section of the barrel; hence, the flame propagates into the barrel to produce *flashback*. On the other hand, when the flow is very strong, u_{supply} is greater than S_L over the entire cross-section, producing *blowoff*. The critical condition for blowoff arises when the S_L and u_{supply} profiles are tangent at a certain radial location. Stable combustion is shown in Fig. 5.25*d*.

Consider a velocity profile slightly inside the barrel (Fig. 5.26); the critical flashback condition is reached if u_{supply} is represented by u_2. The critical value of boundary velocity gradient g at which this condition is first realized is

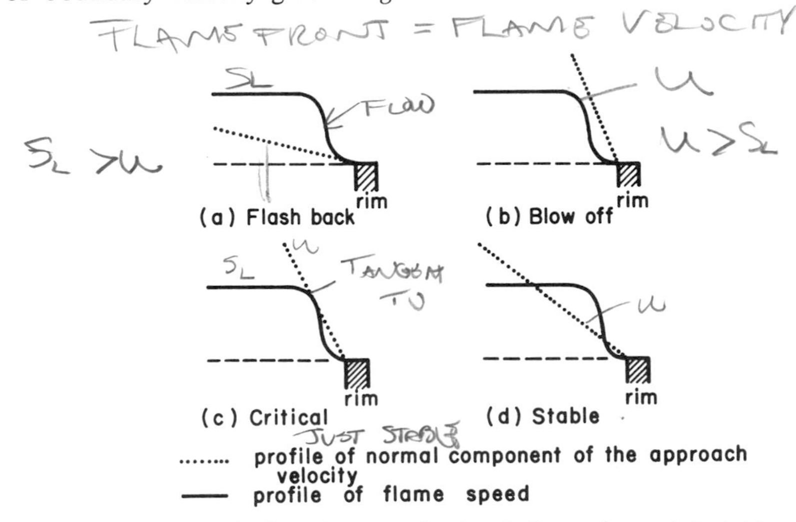

Figure 5.25 Stability of a flame front near the rim of a Bunsen burner (adapted from Kanury[33]).

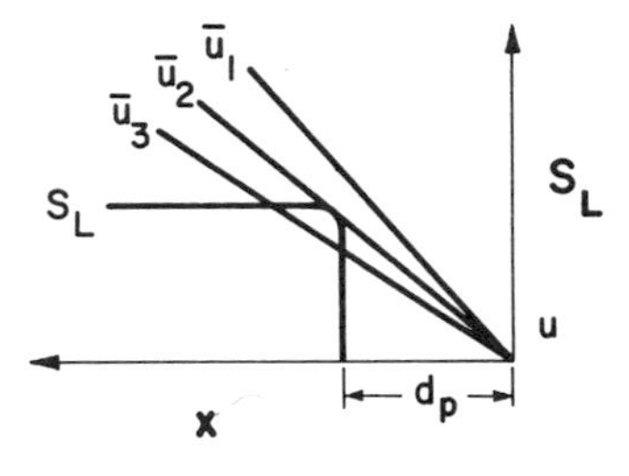

Figure 5.26 Burning velocity and gas velocity inside a Bunsen tube (after Lewis and von Elbe[40]).

denoted by g_F. That is,

$$g_F = \frac{S_L}{d_p} \tag{5-84}$$

where d_p, the distance between the flame edge and the burner wall, is called the *penetration distance* and is equal to half the quenching distance (or diameter; see Section 9) d_q.

When the gas flow is increased, the equilibrium position shifts upward from the burner rim. Then the premixed, unburned gas becomes progressively diluted by interdiffusion of surrounding gases (see Fig. 5.27). The critical value of g corresponding to blowoff is denoted by g_B and given by

$$g_B = \lim_{r \to R} \left(-\frac{du}{dr} \right)_{\bar{u}_{\text{average}}} \tag{5-85}$$

The critical boundary velocity gradients g_F and g_B are given by the shapes of curves 1 and 3, respectively, in Fig. 5.27. The values of g_F and g_B which define the flame stability on laminar jets are determined experimentally. It is clear from this discussion that the velocity gradients have definite effects on flame stability. A qualitative sketch of critical boundary velocity gradients versus fuel–oxidant ratio is shown in Fig. 5.28.

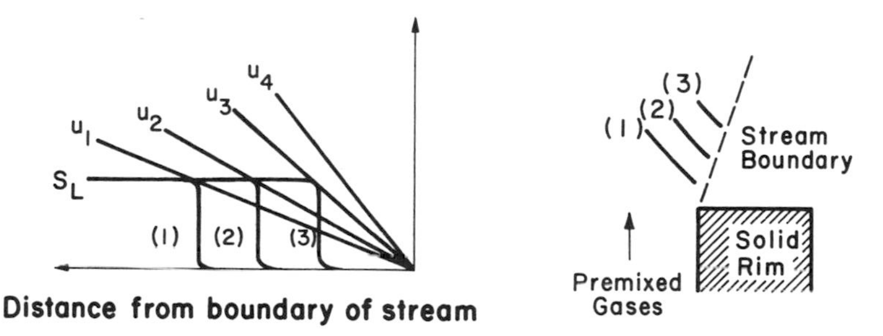

Figure 5.27 Burning velocity and gas velocity above a Bunsen-tube rim (after Lewis and von Elbe[40]).

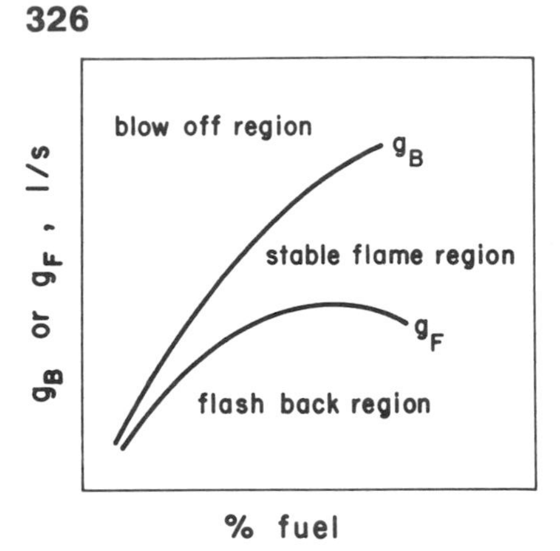

Figure 5.28 Effect of critical boundary velocity gradient and fuel percentage on flame stability (adapted from Kanury[33]).

9 FLAME QUENCHING

Flame quenching can significantly affect the process of flame propagation. Let us first define a quenching distance by considering the following phenomena. After a flame has been established on the Bunsen-burner port, if the mixture is suddenly cut off, the flame flashes back and propagates down the burner tube. We then replace the barrel with a narrower one. The experiment is repeated until the tube just barely permits the flame to propagate. The diameter of the last tube is called the quenching distance (or quenching diameter) for the given fuel–oxidant mixture under the specified conditions of temperature and pressure. Below this diameter, the flame is quenched by the tube wall and the flame cannot flash back.

For the two-dimensional problem of a pair of parallel plates, the quenching distance d can be defined as the distance between the plates such that rate of heat generation, $\dot{q}_R$, is exactly equal to the rate of heat removal, $\dot{q}_L$:

$$\dot{q}_R = \phi_0(\text{RR})(Ad)Q_R \tag{5-86}$$

where

d is the distance between the two parallel plates (L).
A is a small element of area (L^2).
ϕ_0 is the stoichiometric mole fraction of the combustible mixture.
Q_R is the heat of reaction per mole of stoichiometric mixture (Q/mole).
RR is the reaction rate (mole$/tL^3$).

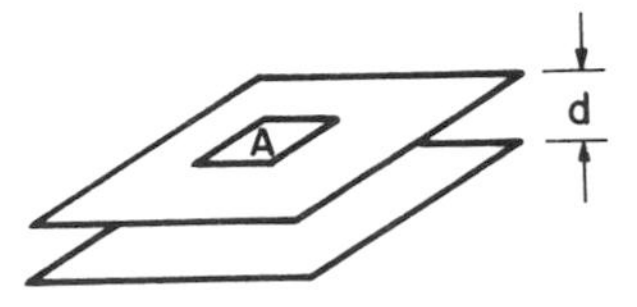

Also

$$\dot{q}_L = \lambda A \frac{dT}{dx} \tag{5-87}$$

Let T_q signify the lowest temperature at which the flame can propagate. Let T_0 signify the temperature at the cold boundary. Approximating the temperature distribution in the gas phase by two linear straight-line segments with a maximum temperature T_q at the center, and equating $\dot{q}_R$, we get

$$2A\lambda \frac{T_q - T_0}{d/2} = \phi_0 Q_R \, \text{RR} \, Ad$$

or

$$d^2 = \frac{4\lambda(T_q - T_0)}{\phi_0 \, \text{RR} \, Q_R} \tag{5-88}$$

From the energy balance, we have

$$Q_R = C_p(T_f - T_0) \tag{5-89}$$

Substituting Eq. (5-89) into Eq. (5-88), we have

$$d^2 = \frac{4\lambda}{\phi_0 C_p \, \text{RR}} \left(\frac{T_q - T_0}{T_f - T_0} \right) \tag{5-90}$$

Since α can be expressed as $\lambda / C_p \rho_0$, then

$$\boxed{d \propto \sqrt{\frac{\alpha}{\text{RR}}}} \tag{5-91}$$

Combining Eq. (5-91) with Eq. (5-41), we have

$$\frac{S_L}{d} \propto \text{RR} \tag{5-92}$$

and

$$S_L d \propto \alpha \tag{5-93}$$

It is important to note that Eq. (5-90) is derived under the assumption of negligible heat loss from the flame to the cold gas. Because of this assumption, Eq. (5-90) cannot satisfy the limiting condition at RR $\rightarrow$ 0. If one sets the

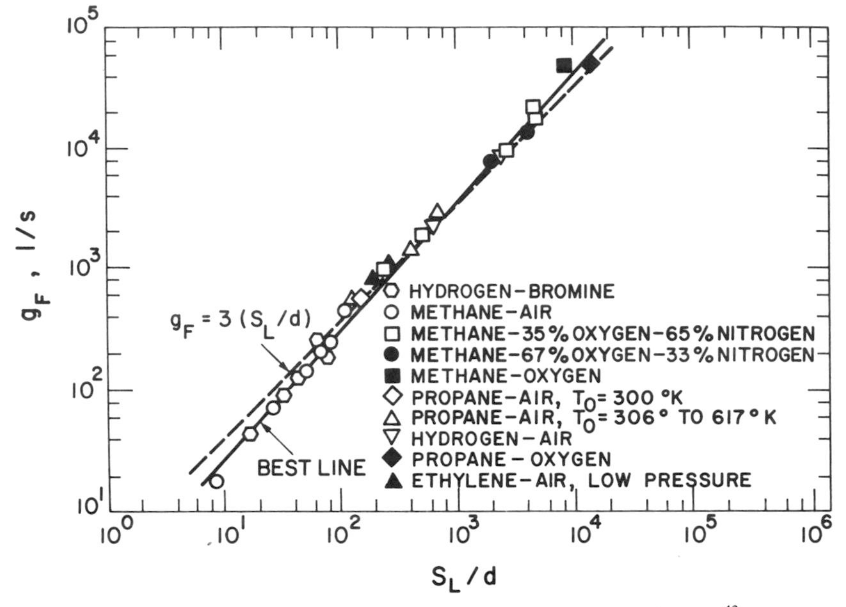

Figure 5.29 Correlation of g_F with S_L/d (after Berlad and Potter[43]).

reaction rate equal to zero in Eq. (5-90), the quenching diameter approaches infinity; however, from experiments, we know that the quenching diameter is finite.

Equation (5-92), in general, is useful in determining trends in reaction rate. The ratio S_L/d can also be related to the boundary velocity gradient at flashback [see Eq. (5-84)]. Lewis and von Elbe[40] suggest that twice the depth of penetration of quenching ($2d_p$) should be equal to the quenching distance between parallel plates. For cylindrical tubes (the usual experimental configuration), d_p should be about $\frac{1}{3}$ the quenching distance.[43,44] One can thus write

$$g_F = 3\frac{S_L}{d} \tag{5-94}$$

for cylindrical tubes. The results of Berlad and Potter[43] in an examination of a number of fuel–oxidant mixtures is shown in Fig. 5.29. The best line through the data has the equation

$$g_F = 14.125\left(\frac{S_L}{d}\right)^{1.168} \tag{5-95}$$

Equation (5-94) is also sketched in Fig. 5.29. Considering the variety of mixtures included in the correlation, the agreement with Eq. (5-94) is good. It would be difficult to decide whether the deviations are due to the lack of precision of the experimental data or of the analytical expression. Both are probably involved. In general, the quenching distance is smaller for more reactive fuels. The typical quenching distance for hydrocarbon fuel between parallel plates is around 0.18 cm. It has been found that a very small centerbody has a large quenching action in the annular case. For a given mixture, the quenching distance depends on the inert diluent used in the order

$$d_{He} > d_{Ar} > d_{N_2} > d_{CO_2}$$

This implies that the flame is, in effect, more insulated from the walls by CO_2 than by helium. The reason is that

$$\alpha_{He} > \alpha_{Ar} > \alpha_{N_2} > \alpha_{CO_2}$$

As the initial temperature of the unburned mixture is increased, the flame is able to pass through smaller openings, and hence the quenching distance is decreased. The quenching distance also depends upon pressure; it increases as pressure decreases. We can see this dependence from the following equations:

$$\alpha = \frac{\lambda}{\rho C_p} \propto \frac{1}{p} \tag{5-96}$$

$$\mathrm{RR} \propto p^n \tag{5-97}$$

$$d_q = \sqrt{\frac{\alpha}{\mathrm{RR}}} = \sqrt{\frac{1}{p^{n+1}}} \propto \frac{1}{p^b} \tag{5-98}$$

In other words, the quenching effect of walls on flame becomes greater when the pressure is reduced. A rough approximation for hydrocarbon burning in air is

$$d_q \propto \frac{1}{p} \qquad \text{(second-order reaction)} \tag{5-99}$$

10 FLAMMABILITY LIMITS OF PREMIXED LAMINAR FLAMES

In this section, we shall discuss the stability limits of premixed laminar flames. On this subject, there are two groups of studies. One concentrates on the laminar flame propagation by the mixture ratio; the other studies the relationship of flammability limits with flow conditions. The latter addresses flame stabilization phenomena such as flashback, blowoff, onset of turbulence, and so on. Some of this is discussed in Section 8.

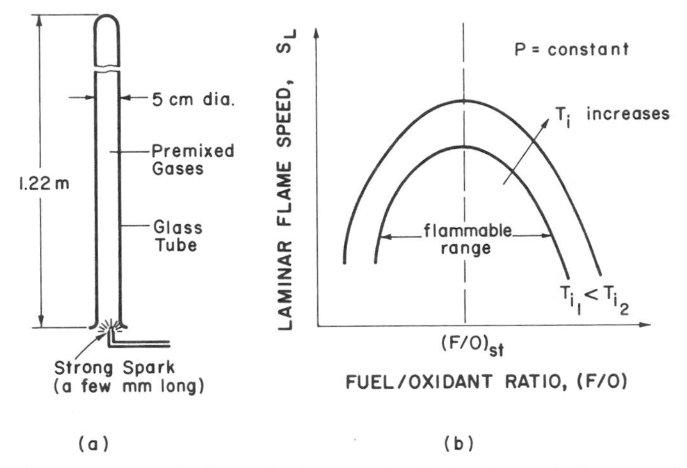

Figure 5.30 Standard glass tube for testing flammability limits of premixed gases and the laminar flame speed as functions of fuel/oxidant ratio and initial temperature.

10.1 Flammability Limits Determined from a Standard Glass Tube

To measure the lean and rich concentration limits of flammability, we need to exclude all external effects so that it is a property of the fuel–oxidant mixture and test conditions which usually are specified in terms of temperature and pressure. In the early 1950s, a standard vertical glass tube of 1.22-m (4-ft) length with an inner diameter of 5 cm was selected as a standard tube for observing whether a mixture is truly capable of propagating a flame indefinitely away from the ignition source. The ignition source is a spark plug located at the bottom of the tube and can generate a strong spark of a few millimeters in length (see Fig. 5.30*a*).

Typical variations of laminar flame speed as functions of F/O and initial temperature of the gaseous mixture are shown in Fig. 5.30*b*. When the initial temperature of the mixture is increased, the flammable F/O range becomes wider, or the flammability limits are broadened. The upper and lower flammability limits of mixtures of certain fuels with air are listed in Table 5.5. These are assembled from data of Smith and Stinson,[45] Kanury[46] and NACA Report 1300.[47]

10.2 Effect of Pressure and Temperature on Flammability Limit

Pressure has a definite effect on flammability limits. As discussed in Chapter 2 (Section 2.6), the rich limits become much wider with increasing pressure (see Fig. 5.37). The lean limits are not appreciably affected by the pressure. This is especially valid for hydrocarbon–air mixtures. The change of slope at the upper (rich) limit in Fig. 5.37 is caused by the change of second-order reaction at low pressures to first-order reaction at high pressures.

TABLE 5.5 Flammability Limits of Certain Fuel–Air Mixtures

	Lower (Lean) Limits, Vol. (%) in Air	Upper (Rich) Limits, Vol. (%) in Air	Stoichiometric, Vol. (%) in Air
Methane, CH_4	5	15	9.47
Heptane, C_7H_{16}	1	6	1.87
Hydrogen, H_2	4	74.2	29.2
Carbon monoxide, CO	12.5	74.2	29.5
Acetylene	2.5	80	7.7
Ethane, C_2H_4	2.82	15.34	5.64
Ethylene oxide, C_2H_4O	3.0	100	7.72
Propane, C_3H_8	2.05	11.38	4.02
Methanol (g), CH_3OH	5.88	49.94	12.24

For simple hydrocarbons such as ethane, propane, butane and pentane, it appears that the rich limits extend linearly with increasing pressure. The lean limits, on the other hand, are at first extended slightly and approach asymptotic values thereafter as pressure is increased. A sketch of the above dependence is shown in Fig. 5.31. The measured flammability data[40] of natural gas–air mixtures are shown in Fig. 5.32. It is interesting to note that the limits shown in this figure are as wide at subatmospheric pressures as those at 1 atm. These low-pressure limits are sometimes called "flame propagation limits in a tube with specified diameter."

It is useful to point out that the size of the test vessel has a nonnegligible effect on the relationship between pressure and flammability limits. As the size of the test vessel decreases, the effect of pressure on flammability limits becomes more important; this is due to the quenching process which is governed by the density or pressure level of the gaseous mixture. Similarly, as pressure is reduced, the mixture becomes more difficult to ignite unless the test vessel is increased to a larger size (or diameter). The minimum diameter of the test vessel is therefore found to increase with a decrease in pressure. This relationship is given by Eq. (5-99).

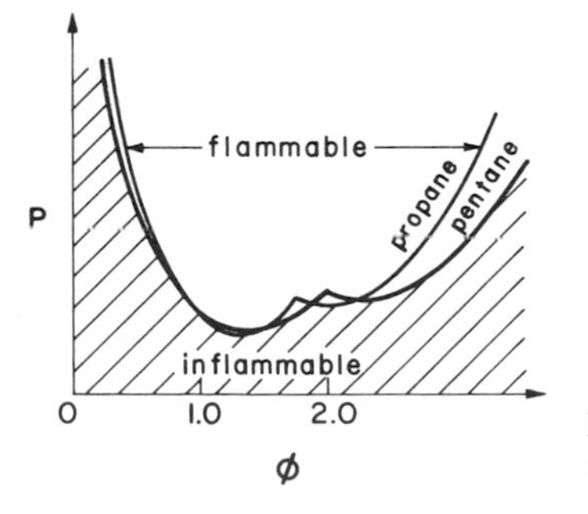

Figure 5.31 Dependence of flammability limits on pressure and stoichiometric ratio of hydrocarbon–air mixture.

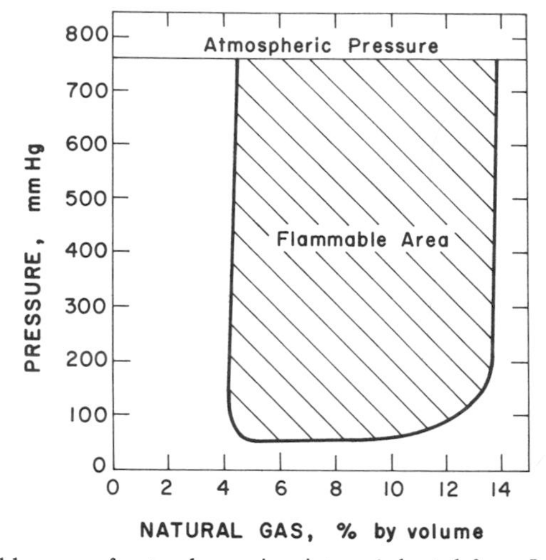

Figure 5.32 Flammable zone of natural gas–air mixture (adapted from Lewis and von Elbe,[40] based upon the original work of Jones and Kennedy of Bureau of Mines, Report of Investigation No. 3798, 1945).

If the temperature is increased, the flammability limits will be broadened. That is, the lean limit will lie at a lower F/O equivalence ratio whereas the rich limit will lie at a higher one (Fig. 5.30*b*). It is observed experimentally that the limits vary linearly with temperature. For example, the lean limit of *n*-pentane drops from 1.53% in volume at room temperature to 1.22% at 572 K at a slope of 11.4×10^{-4} vol.%/K. The rich limit increases at a steeper slope of 30.6×10^{-4} vol.%/K. Hence, by raising the temperature of hydrocarbon–air mixture, the flammability limits become broader; however, the temperature effect on flammability limits is usually less significant than that of pressure.

10.3 Spalding's Theory of Flammability Limits and Flame Quenching

A theoretical approach of predicting the flammability limit trends was developed by Spalding.[48] This theory was supposed to be able to predict the following:

a. The fact that below certain concentrations of combustible gas or oxygen, flame will be unable to propagate.
b. The lower limit concentration.
c. The burning velocity at the limit.

In his model, heat exchange between the gas and its solid surroundings is considered. It was demonstrated that combustible mixtures have, in general, two possible burning velocities. Only the upper one is stable in ordinary circumstances. At the inflammability limits, two burning velocities coincide; beyond the limits, they are just imaginary. Spalding claimed that the conclusions of the theory are independent of whether the reaction proceeds by a single step or by a chain mechanism. They are also independent of particular assumptions as to the dependence of reaction rate on concentration and temperature, other than that the latter dependence should be steeper than that of the heat-transfer process.

In Spalding's mathematical model, a single-step forward reaction is considered:

$$A + B \rightarrow C \tag{5-100}$$

The mass flow rate of the unburned mixture stream $A + B$ is assumed to maintain at a constant value of G in the positive x-direction. In case the species B is in excess, a part of unreacted B will exist. Even though B may exist in the product, it is still general enough to define C as our product. The conservation equations under the one-dimensional steady-state assumption have the following forms:

Energy equation:

$$\frac{d}{dx}\left(\lambda \frac{dT}{dx}\right) - C_p G \frac{dT}{dx} + H_{r_A}^\circ \dot{R}_A = \dot{Q}_{\text{loss}} \tag{5-101}$$

Species equation:

$$\frac{d}{dx}\left(\frac{\lambda}{C_p}\frac{dY_A}{dx}\right) - G\frac{dY_A}{dx} - \dot{R}_A = 0 \tag{5-102}$$

Stoichiometric equation:

$$Y_{B_u} - Y_B = \left(Y_{A_u} - Y_A\right) r_{BA} \tag{5-103}$$

where G = total mass velocity per unit area (g/sec m^2) in x-direction
Y = mass fraction of component in mixture
$\dot{R}_A$ = mass consumption rate of species A (g/m^3 sec)
$H_{R_A}^\circ$ = standard heat of reaction of A (J/g of A)
$\dot{Q}_{\text{loss}}$ = rate of heat loss per unit volume (J/sec m^3)
r_{BA} = ratio of reacted amount of B to that of A (g/g)
λ = thermal conductivity (J/m sec K)

The subscripts u and b denote completely unburned and completely burned situations respectively.

For convenience, the following dimensionless forms are employed:

$$\tau \equiv \frac{T - T_u}{T_b - T_u} \tag{5-104}$$

$$\xi \equiv \exp\left(C_p \frac{Gx}{\lambda} \right) \tag{5-105}$$

$$\alpha \equiv \frac{Y_A}{Y_{A_u}} \tag{5-106}$$

$$\lambda^+ = \frac{H_{r_A}^\circ \dot{R}_A^* \lambda_b}{(T_b - T_u)(C_p G)^2} \tag{5-107}$$

$$\phi = \frac{\lambda \dot{R}_A}{\lambda_b \dot{R}_A^*} \tag{5-108}$$

$$\psi = \frac{\lambda \dot{Q}_{\text{loss}}}{\lambda_b \dot{Q}_{\text{loss}}^*} \tag{5-109}$$

$$K = \frac{\dot{Q}_{\text{loss}}^*}{\dot{R}_A^* H_{r_A}^\circ} \tag{5-110}$$

where $\dot{R}_A^*$ is the value of $\dot{R}_A$ when $T = T_b$ and $Y_A = Y_{A_u}$
$\dot{Q}_{\text{loss}}^*$ is the value of $\dot{Q}_{\text{loss}}$ at $T = T_b$

Upon substitutions, Eqs. (5-101) and (5-102) become two second-order ordinary differential equations:

$$\frac{d^2\tau}{d\xi^2} = -\frac{\lambda^+}{\xi^2}(\phi - K\psi) \tag{5-111}$$

$$\frac{d^2\alpha}{d\xi^2} = \frac{\lambda^+}{\xi^2}\phi \tag{5-112}$$

Here, ξ is a distorted space variable chosen so as to eliminate the first derivatives in Eqs. (5-101) and (5-102). From the definition of ξ, we have $\xi = 0$ corresponding to the cold unburned condition at $x \to -\infty$. Boundary conditions in terms of ξ are:

$$\xi = 0: \quad \tau = 0, \quad \alpha = 1$$
$$\xi = \infty: \quad \frac{d\tau}{d\xi} = 0, \quad \alpha = 0 \quad \text{(assuming } B \text{ is in excess)} \tag{5-113}$$

The question now is to find the eigenvalue λ^+ that satisfies these conditions. With λ^+ known, the burning velocity G can be evaluated from Eq. (5.107). We shall consider two cases: $K = 0$, corresponding to the adiabatic case, and $K > 0$, corresponding to the nonadiabatic case.

Case I: $K = 0$ (i.e., Zero Heat Loss)

Equations (5-111) and (5-112) become almost identical except having opposite signs for the right-hand side terms. Using the boundary conditions described by Eq. (5-113) and the definitions of α and τ given by Eqs. (5-104) and (5-106), α and τ can be related by the following algebraic equation:

$$\alpha + \tau = 1 \tag{5-114}$$

Therefore, only one of the two ordinary differential equations needs to be solved. The boundary condition that $\xi = \infty$ at $\tau = 1$ is somewhat inconvenient mathematically, since ξ approaches ∞. Thus, a new boundary condition was taken by Spalding as

$$\xi = 1: \qquad \alpha = 0, \quad \frac{d\tau}{d\xi} = -\frac{d\alpha}{d\xi} = \left(\frac{d\tau}{d\xi}\right)_1 \tag{5-115}$$

This corresponds physically to supposing that an adiabatic porous catalyst plug is situated at $\xi = 1$ and reduces the concentration of species A to zero, with the result that it has the temperature $\tau = 1$ and there is a finite gradient at the catalyst surface. A set of solutions for τ versus ξ and α versus ξ is sketched in Fig. 5.33.

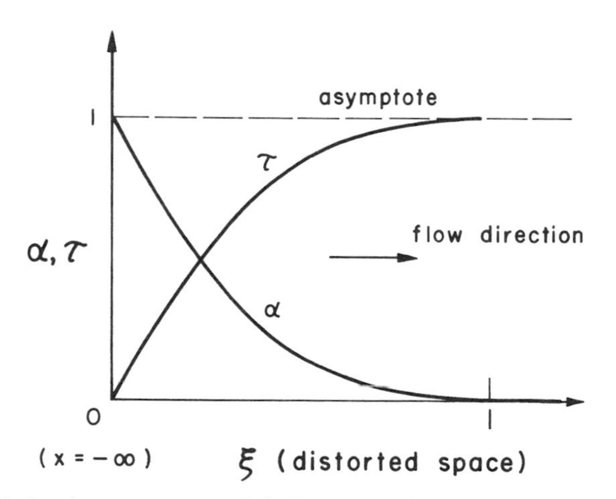

Figure 5.33 Calculated temperature and fuel concentration profiles in flame (adapted from Spalding[48]).

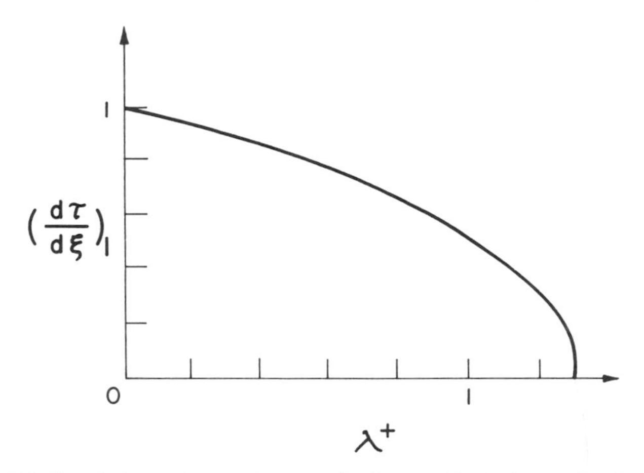

Figure 5.34 Relation between temperature gradient at catalyst plug and value of λ^+ for a particular flame (adapted from Spalding[48]).

This catalyst plug postulation restricts the infinite domain to a finite domain with $0 < \xi \leqq 1$. Solution of Eq. (5-111) yields a different profile of τ for each value of $(d\tau/d\xi)_1$. The required value of λ^+ for $(d\tau/d\xi)_1 = 0$ can then be obtained by extrapolating the curve in Fig. 5.34 to $(d\tau/d\xi)_1 = 0$.

Case II: $K > 0$ (Heat Losses Considered)

The solution differs from Case I in that τ, after reaching a peak at $\xi = 1$, falls gradually to zero as ξ increases beyond the reaction zone (see Fig. 5.35). In this case τ and α profiles are not similar; both Eqs. (5-111) and (5-112) must be

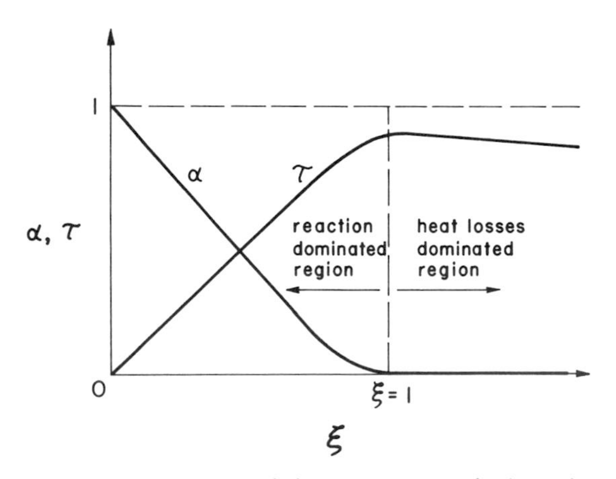

Figure 5.35 τ and α profiles in flame with heat losses from the burned gas (adapted from Spalding[48]).

solved simultaneously. The line $\xi = 1$ serves to separate the region where reaction rate is dominant from that where heat losses are dominant.

In order to permit an analytical solution, attention will be restricted to reacting gas mixtures characterized by

$$\phi = \alpha \tau^n \tag{5-116}$$

which signifies that the reaction rate is proportional to the concentration of A (fuel). According to Spalding, this is usually close to the truth for limit mixtures in which the other reactant, B, is in large excess. Usually, the value of temperature exponent n is between 6 and 15. The heat-loss parameter ψ was taken as another power law of temperature τ for the heat loss zone, for example,

$$\begin{aligned} \psi &= \tau^m \qquad \text{for } \xi > 1 \\ \psi &= 0 \qquad \text{for } \xi \leqq 1 \end{aligned} \tag{5-117}$$

The exponent m varies between approximately 1 and 5 according to whether the heat transfer is predominantly by conduction or radiation. The heat-loss term ψ is neglected upstream of the line $\xi = 1$ in comparison with chemical reaction. This assumption will be justified later. In the following, we shall see that solutions are developed in two parts and patched at $\xi = 1$.

For the region $0 < \xi \leqq 1$, we have

$$\frac{d^2\tau}{d\xi^2} = -\lambda^+ \frac{\alpha\tau^n}{\xi^2} \tag{5-118}$$

$$\frac{d^2\alpha}{d\xi^2} = \lambda^+ \frac{\alpha\tau^n}{\xi^2} \tag{5-119}$$

with boundary conditions:

$$\begin{aligned} \xi = 0: &\qquad \alpha = 1, \quad \tau = 0 \\ \xi = 1: &\qquad \alpha = 0, \quad \frac{d\alpha}{d\xi} = 0 \\ &\qquad \tau = \tau_1, \quad \frac{d\tau}{d\xi} = \left(\frac{d\tau}{d\xi}\right)_1 \end{aligned} \tag{5-120}$$

From the similarity of Eqs. (5-118) and (5-119), we can see that α and τ must be linearly related by

$$\alpha = -\tau + a\xi + b \tag{5-121}$$

One can easily verify this by differentiating Eq. (5-121) twice with respect to ξ and then comparing the results with Eqs. (5-118) and (5-119). It is interesting to note that the boundary conditions for α and τ are not similar. The coefficients a and b must be determined from the boundary conditions:

$$\text{at } \xi = 0: \quad 1 = 0 + 0 + b \quad \therefore\ b = 1$$

$$\text{at } \xi = 1: \quad 0 = -\tau_1 + a + 1 \quad \therefore\ a = \tau_1 - 1$$

Substituting a and b into Eq. (5-121), we have

$$\alpha = 1 - \tau - (1 - \tau_1)\xi \tag{5-122}$$

After rearranging and differentiating with respect to ξ and setting $\xi = 1$, we have

$$\left(\frac{d\tau}{d\xi}\right)_1 = (\tau_1 - 1) \tag{5-123}$$

Multiplying Eq. (5-119) by $(d\tau/d\xi)\,d\xi$, we have

$$\frac{d\tau}{d\xi}\frac{d^2\alpha}{d\xi^2}\,d\xi = \lambda^+ \frac{\alpha\tau^n}{\xi^2}\frac{d\tau}{d\xi}\,d\xi$$

Substituting Eqs. (5-122) and (5-123) into the above equation and integrating from $\xi = 0$ to 1, we have

$$\int_0^1 \frac{d^2\alpha}{d\xi^2}(\tau_1 - 1)\,d\xi - \int_0^1 \frac{d\alpha}{d\xi}\frac{d^2\alpha}{d\xi^2}\,d\xi = \lambda^+ \int_0^{\tau_1} \frac{[1 - \tau - (1 - \tau_1)\xi]}{\xi^2}\tau^n\,d\tau$$

$$(\tau_1 - 1)\int_{\xi=0}^{\xi=1} d\left(\frac{d\alpha}{d\xi}\right) - \frac{1}{2}\int_{\xi=0}^{\xi=1} d\left(\frac{d\alpha}{d\xi}\right)^2 = \lambda^+ \int_0^{\tau_1} \frac{[1 - \tau - (1 - \tau_1)\xi]}{\xi^2}\tau^n\,d\tau$$

$$\therefore\ \lambda^+ = -\frac{\dfrac{1}{2}\left[\left(\dfrac{d\alpha}{d\xi}\right)_1^2 - \left(\dfrac{d\alpha}{d\xi}\right)_0^2\right] + (1 - \tau_1)\left[\left(\dfrac{d\alpha}{d\xi}\right)_1 - \left(\dfrac{d\alpha}{d\xi}\right)_0\right]}{\displaystyle\int_0^{\tau_1} \frac{1 - \tau - (1 - \tau_1)\xi}{\xi^2}\tau^n\,d\tau} \tag{5-124}$$

The value of n is so large that the reaction is confined to the upper temperature levels. This means that where the integrand in Eq. (5-124) is finite, ξ may be taken as unity without appreciable error. In the region beyond the small neighborhood of $\xi = 1$, the heat release term is nearly zero. From Eq. (5-119), we have

$$\left(\frac{d^2\alpha}{d\xi^2}\right) = 0 \quad \text{or} \quad \frac{d\alpha}{d\xi} = \text{constant}$$

This means that α is linear with respect to ξ. Connecting $\alpha = 1$ at $\xi = 0$ to $\alpha = 0$ at $\xi = 1$, a rough approximation for the slope at $\xi = 0$ can be given as

$$\left(\frac{d\alpha}{d\xi}\right)_0 \cong -1 \tag{5-125}$$

Also, since

$$\left(\frac{d\alpha}{d\xi}\right)_1 = 0 \tag{5-126}$$

and

$$\tau_1 - 1 = \left(\frac{d\tau}{d\xi}\right)_1 \cong 0 \tag{5-127}$$

Eq. (5.124), after the substitution of the above relationships, becomes

$$\lambda^+ \cong \frac{1}{2\int_0^{\tau_1}(1-\tau)\tau^n\,d\tau} \cong \frac{1}{2\int_0^{\tau_1}(\tau_1-\tau)\tau^n\,d\tau} \tag{5-128}$$

Before evaluating the integral, it is worth noting that for the particular case when $\tau_1 = 1$, this is the well-known formula of Zel'dovich and Frank-Kamenetsky.[9(a),48] Evaluating Eq. (5-128) gives

$$\lambda^+ \cong \frac{(n+1)(n+2)}{2\tau_1^{(n+2)}} \tag{5-129}$$

When τ_1 is known, as for example, when heat losses are zero so that $\tau_1 = 1$, λ^+ and therefore the flame speed can be evaluated immediately. When heat losses are present, τ_1 must first be determined by considering the region beyond $\xi = 1$. Thus we have the following equation:

$$\xi \geqq 1: \qquad \frac{d^2\tau}{d\xi^2} = \lambda^+ \frac{K\tau^m}{\xi^2} \tag{5-130}$$

The boundary conditions for the heat-loss region are:

$$\xi = 1: \qquad \tau = \tau_1, \quad \frac{d\tau}{d\xi} = \left(\frac{d\tau}{d\xi}\right)_1 \tag{5-131}$$

$$\xi = +\infty: \qquad \frac{d\tau}{d\xi} = 0 \tag{5-132}$$

Equation (5-130) has a simple solution if $m = 1$. However, if $\lambda^+ K$ is small, τ^m can be treated as a constant close to $\xi = 1$ (i.e., $\tau^m \simeq \tau_1^m$) and upon integration, we have

$$\frac{d\tau}{d\xi} = -\lambda^+ K \frac{\tau_1^m}{\xi} \tag{5-133}$$

Putting $\xi = 1$ in Eq. (5-133) and postulating that the gradients on both sides of the $\xi = 1$ line are equal, we have from Eq. (5-123)

$$\lambda^+ K = \tau_1^{-m}(1 - \tau_1) \tag{5-134}$$

and so from Eq. (5-129) we have the required relation between K and τ_1 as:

$$K = \frac{2\tau_1^{n+2-m}}{(n+1)(n+2)}(1 - \tau_1) \tag{5-135}$$

When K exceeds a critical value K_c, however, the τ_1 values are imaginary. A relation between K and λ^+ can be obtained by eliminating τ_1 from Eq. (5-135) and (5-129). A relationship for $m = 4$, $n = 11$ is plotted in Fig. 5.36.

It can be seen that for any value of the heat-loss parameter less than K_c, two values exist for λ^+ and thus two possible flame speeds. If K exceeds K_c, however, there are no real values of λ^+, which means no steadily propagating flame is possible. K_c therefore represents the magnitude of the ratio of heat-loss rate to chemical-reaction rate which corresponds to the critical condition at the flammability limit. The magnitude of the flame speed at flammability limit $S_{L,c}$ may be obtained using Eqs. (5-107) and (5-110) leading to:

$$S_{L,c} = \frac{1}{C_p \rho_u}\sqrt{\frac{H_{r_A}^\circ \dot{R}_{A_c}^* \lambda_b}{(T_b - T_u)\lambda_c^+}} = \frac{1}{C_p \rho_u}\sqrt{\frac{\lambda_b \dot{Q}_{\text{loss}}^*}{(T_b - T_u) K_c \lambda_c^+}} \tag{5-136}$$

For the particular flame under consideration, K_c and λ_c^+ are fixed. Their product, $K_c\lambda_c^+$, equals 0.169. For a weak hydrocarbon–air flame with an adiabatic temperature of 1500 K, a value of $\dot{Q}_{\text{loss}}^* = 0.1$ cal/(cm^3 sec) was reported by Fishenden and Saunders.[49,48] Inserting the gas properties and the

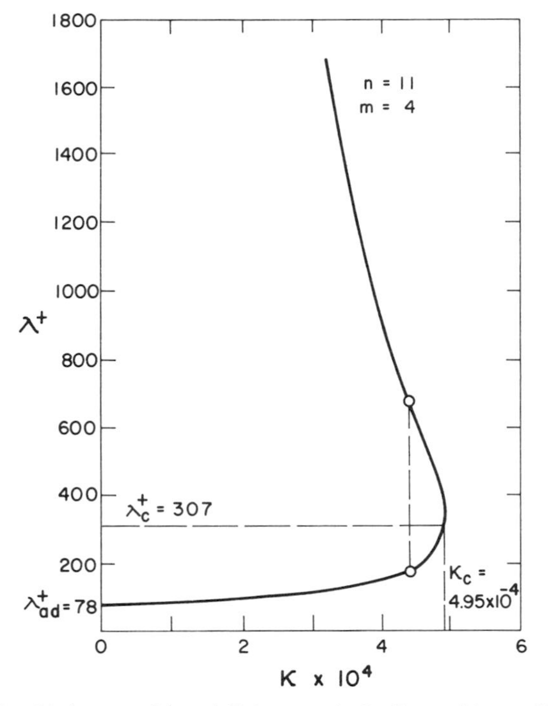

Figure 5.36 Relationship between λ^+ and K for a particular flame with $n = 11$ and $m = 4$ (after Spalding[48]).

above value of $K_c\lambda_c^+$ into Eq. (5-136), the magnitude of $S_{L,c}$ is found to be 1.2 cm/sec.

The effect of pressure on $S_{L,c}$ can also be seen from Eq. (5-136). Since both ρ_u and $\dot{Q}^*_{\text{loss}}$ are proportional to P, we obtain the relation

$$S_{L,c} \propto \frac{1}{p^{1/2}} \propto p^{-1/2} \tag{5-137}$$

It is evident that $S_{L,c}$ does not depend upon the order of reaction whereas S_L does depend upon the order of reaction [see Eq. (5.7)]. This explains the generally observed trend in Fig. 5.37 of the boundary variation of the flammable region with respect to pressure. It is clear that in order to have flame propagation, we must have $S_L > S_{L,c}$ or $S_L/S_{L,c} > 1$:

$$\begin{aligned} &\text{For } n = 1, \quad \frac{S_L}{S_{L,c}} \text{ is independent of } p \\ &\text{For } n = 2, \quad \frac{S_L}{S_{L,c}} = \text{constant}/p^{-1/2} \sim \sqrt{p} \end{aligned} \tag{5-138}$$

Hence, the flammable region is widened with increasing pressure in the case of second-order reactions and maintains constant width for first-order reactions.

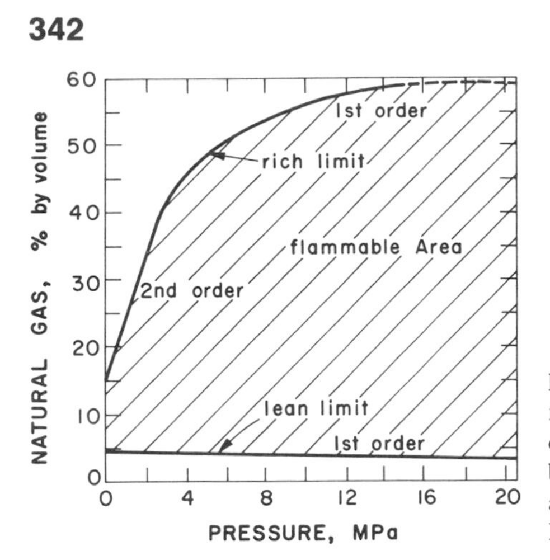

Figure 5.37 Dependence of the width of flammable region on pressure and the order of chemical reactions (after Lewis and von Elbe[40], based upon the original work of Jones, Kennedy, and Spolan of Bureau of Mines, Report of Investigation No. 4557, 1949).

The above theory was also extended by Spalding to consider the wall-quenching effect as a flame travels through a stagnant gaseous mixture in a duct. Readers are referred to the original paper[48] for detailed discussions on this aspect.

REFERENCES

1. I. Glassman, *Combustion*, Academic Press, New York, 1977.
2. M. W. Evans, "Current Theoretical Concepts of Steady-State Flame Propagation," *Chem. Rev.*, Vol. 51, pp. 363–429, 1952.
3. E. Mallard and H. L. LeChatelier, *Ann. Mines*, Vol. 4, p. 379, 1883.
4. G. Damköhler, *Z. Elektrochem.*, Vol. 46, p. 601, 1940; translation, N.A.C.A. Tech. Memo. 1112.
5. E. Bartholomé, *Naturwissenschaften*, Vol. 36, p. 171, 1949; Vol. 36, p. 206, 1949.
6. H. W. Emmons, J. A. Harr, and P. Strong, *Thermal Flame Propagation*, Computation Laboratory of Harvard University, December 1949.
7. K. Bechert, *Z. Naturforsch.*, Vol. 3A, p. 584, 1948; *Portugaliae Phys.*, Vol. 3, p. 29, 1949; *Ann. Physik*, Vol. 6, No. 4, p. 191, 1949; *Naturwissenschaften*, Vol. 37, p. 112, 1950.
8. B. Lewis and G. von Elbe, *J. Chem. Phys.*, Vol. 2, p. 537, 1934.
9. (a) Ya. B. Zel'dovich and D. A. Frank-Kamenetsky, *Compt. Rend. Acad. Sci. URSS*, Vol. 19, p. 693, 1938. (b) Ya. B. Zel'dovich and N. Semenov, *J. Expt. Theoret. Phys. USSR*, Vol. 10, p. 1116, 1940; translation, N.A.C.A. Tech. Memo. 1084. (c) Ya. B. Zel'dovich, *J. Phys. Chem. USSR*, Vol. 22, No. 1, 1948.
10. S. F. Boys and J. Corner, *Proc. Roy. Soc. London*, Vol. A197, p. 90, 1949.
11. (a) J. O. Hirschfelder and C. F. Curtiss, *Third Symposium on Combustion, Flame, and Explosion Phenomena*, p. 121, The Williams and Wilkins Company, Baltimore, 1949. (b) J. O. Hirschfelder and C. F. Curtiss, *J. Chem. Phys.*, Vol. 17, p. 1076, 1949.
12. Th. von Karman and S. S. Penner, *Selected Combustion Problems, Fundamentals, and Aeronautical Applications*, pp. 5–41, AGARD, Butterworths Scientific Publications, London, 1954.

13. (a) C. Tanford and R. N. Pease, *J. Chem. Phys.*, Vol. 15, p. 431, 1947. (b) C. Tanford, *J. Chem. Phys.*, Vol. 15, p. 433, 1947. (c) C. Tanford and R. N. Pease, *J. Chem. Phys.*, Vol. 15, p. 861, 1947.

14. (a) A. Van Tiggelen, *Bull. Soc. Chim. Belg.*, Vol. 55, p. 202, 1946. (b) A. Van Tiggelen, *Bull. Soc. Chim. Belg.*, Vol. 58, p. 259, 1949.

15. N. Manson, *J. Chem. Phys.*, Vol. 17, p. 837, 1949.

16. (a) A. G. Gaydon and H. G. Wolfhard, *Proc. Roy. Soc. London*, Vol. A194, p. 169, 1948. (b) A. G. Gaydon and H. G. Wolfhard, *Fuel*, Vol. 29, p. 15, 1950.

17. G. Jahn, *Der. Zündvorgang in Gasgemischen*, Oldenbourg, Berlin, 1942.

18. G. L. Dugger, *J. Am. Chem. Soc.*, Vol. 73, p. 2308, 1951.

19. D. M. Simon, *J. Am. Chem. Soc.*, Vol. 73, p. 422, 1951.

20. M. Gerstein, O. Levine, and E. L. Wong, *J. Am. Chem. Soc.*, Vol. 73, p. 418, 1951.

21. J. M. Heimerl and T. P. Coffee, "The Detailed Modeling of Premixed, Laminar, Steady-State Flames. I. Ozone," *Combustion and Flame*, Vol. 39, pp. 301–315, 1980.

22. J. M. Heimerl and T. P. Coffee, "The Unimolecular Ozone Decomposition Reaction," *Combustion and Flame*, Vol. 35, pp. 117–123, 1979.

23. R. F. Hampson (Ed.), "Survey of Photochemical Rate Data for Twenty-Eight Reactions of Interest in Atmospheric Chemistry," *J. Phys. Chem.* Ref. Data 2, pp. 267–312, 1973.

24. H. S. Johnston, *Gas Phase Reaction Kinetics of Neutral Oxygen Species*, NSRDS-NBS-20, September 1968.

25. D. L. Baulch, D. D. Drysdale, J. Duxbury, and S. J. Grant, *Evaluated Kinetic Data for High Temperature Reactions, Vol. 3, Homogeneous Gas Phase Reactions of the* O_2/O_3 *System, the* $CO/O_2/H_2$ *System and of Sulphur Containing Species*, Buttersworths, Boston, 1976.

26. B. K. Madsen and R. F. Sincovec, "PDECOL: General Collocation Software for Partial Differential Equations," Preprint UCRL-78263 (Rev 1), Lawrence Livermore Laboratory, 1977.

27. C. de Boor, "Package for Calculating with B-Splines," *SIAM J. Numer. Anal.*, Vol. 14, pp. 441–472, 1977.

28. A. G. Streng and A. V. Grosse, "The Ozone to Oxygen Flame," *Sixth Symposium (International) on Combustion*, pp. 264–273, Reinhold, New York, 1957.

29. J. Warnatz, *Berechnung der Flammemgeschwindigkeit und der Struktur von laminaren flachen Flammen*, Habilitationsschrift, Darmstadt, 1977.

30. A. G. Gaydon and H. G. Wolfhard, *Flames: Their Structure, Radiation and Temperature*, 2nd ed., Chapman and Hall, London, 1960.

31. F. J. Weinberg, *Optics of Flames*, Butterworths, London, 1963.

32. E. Hartmann, Thesis, Karlsruhe, 1931.

33. A. M. Kanury, *Introduction to Combustion Phenomena*, Chapter 8, Gordon and Breach, New York, 1975.

34. T. W. Reynolds and M. Gerstein, "Influence of Molecular Structure of Hydrocarbons on Rate of Flame Propagation," *Third Symposium (International) on Combustion*, pp. 190–194, 1949.

35. B. Lewis, "Discussion," *Selected Combustion Problems, AGARD*, p. 177, Butterworths, London, 1954.

36. G. L. Dugger, S. Heimel, and R. C. Weast, *Ind. and Engrg. Chem.*, Vol. 47, p. 114, 1955.

37. E. Bartholomé and H. Sachsse, *Z. Electrochem.*, Vol. 53, p. 183, 1949.

38. W. H. Clingman, Jr., R. S. Brokaw, and R. N. Pease., *Fourth Symposium on Combustion (International)*, p. 310, Combustion Inst., Pittsburgh, 1953.

39. J. M. Beer and N. A. Chigier, *Combustion Aerodynamics*, Applied Science Publishers, London, 1974.
40. B. Lewis and G. von Elbe, *Combustion, Flames and Explosions of Gases*, 2nd ed., Chapter V, Academic Press, New York, 1961.
41. J. Powling, *Fuel*, Vol. 28, p. 25, 1949.
42. D. B. Spalding and J. P. Botha, *Proc. Roy. Soc. London*, Vol. A225, p. 71, 1954.
43. A. L. Berlad and A. E. Potter, "Relation of Boundary Velocity Gradient for Flash-back to Burning Velocity and Quenching Distance," *Combustion and Flame*, Vol. 1, No. 1, pp. 127–128, 1957.
44. M. Gerstein, "Review of Some Recent Combustion Experiments," *Third AGARD Symposium*, pp. 307–332, 1958.
45. M. L. Smith and K. W. Stinson, *Fuels and Combustion*, McGraw-Hill, New York, 1952.
46. A. M. Kanury, *Introduction to Combustion Phenomenon*, Chapter 4, Gordon and Breach Science Publishers, New York, 1975.
47. NACA Report 1300, "Basic Considerations in the Combustion of Hydrocarbon Fuels in Air," National Advisory Committee for Aeronautics, 1957.
48. D. B. Spalding, "A Theory of Inflammability Limits and Flame Quenching," *Proc. Roy. Soc. London*, Vol. A240, pp. 83–100, 1957.
49. M. Fishenden and O. A. Saunders, *Introduction to Heat Transfer*, p. 21, Oxford University Press, Oxford, 1950.

HOMEWORK

1. A long cylindrical tube of length L and diameter D is initially filled with a premixed combustible gaseous mixture of methane and oxygen. Before ignition, the mixture is stagnant with a pressure of 1 atm and a temperature of 298 K. Ignition at the left end of the tube is achieved by a spark. The flame front then moves toward the right. If the heat loss to the tube wall is *not* negligible, the flame front will not be a planar surface. The shape of the flame front is approximately parabolic, as shown in the following figure:
   (a) If we are interested in studying the transient flame propagation process and the detailed combustion-wave structure, list the assumptions you would make in order to formulate this problem.
   (b) Give the governing equations.
   (c) What chemical reaction equations should you consider in order to provide sufficient information for the source terms in your governing equations?

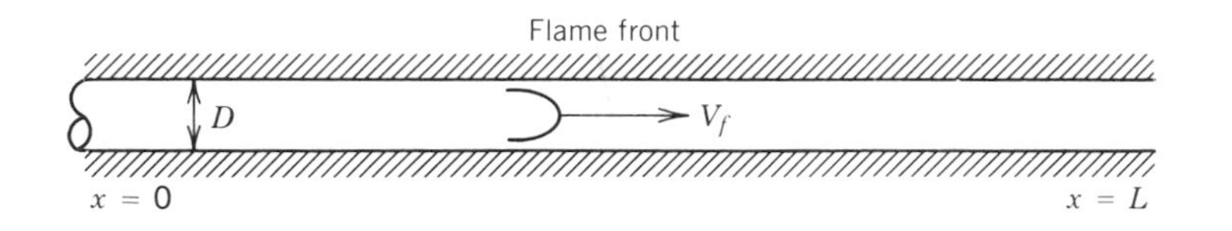

(d) Give initial and boundary conditions necessary to complete the model.

(e) What kind of empirical inputs and/or correlations are needed to complete the formulation?

(f) Sketch your anticipated results for:

(i) Axial temperature distributions at various times along the centerline of the tube.

(ii) Axial concentration distributions at various times along the centerline of the tube.

(iii) Radial temperature and concentration profiles near the flame front.

(g) If the tube diameter becomes very small, what will happen to the flame propagation process?

2. A numerical analysis is to be done for a laminar premixed flame in a Bunsen burner to predict the flashback characteristics of a stoichiometric mixture of air and methane. Give the governing equations and initial and boundary conditions required to solve the problem. List all assumptions.

3. Consider the one-step hydrazine decomposition reaction

$$N_2H_4 \rightarrow 2\,NH_2$$

in a laminar flame. Formulate the governing equations for species and temperature, and solve for the laminar burning velocity as the eigenvalue of the problem.

4. A gas mixture is contained in a soap bubble and ignited by a spark so that a spherical flame spreads radially. It is assumed that the soap bubble can expand. The growth of a flame front along a radius is followed by some photographic means. Relate the velocity of the flame front as determined from the photographs to the laminar flame speed. If this method were used to measure flame speeds, what would you consider its advantages and disadvantages?

5. In each of the cases below, properly order the flame speeds:

(a) Stoichiometric H_2 with

$$21\%\ O_2\text{–}79\%\ He$$

$$21\%\ O_2\text{–}79\%\ N_2$$

$$21\%\ O_2\text{–}70\%\ Ar$$

(b) Wet CO–O_2 and dry CO–O_2, both at the same mixture ratio.

(c) A monopropellant decomposition flame and a hydrocarbon–oxygen flame.

(d) A hydrocarbon–air flame at high pressure and the same flame at low pressure.

6. A constant-volume spherical combustion chamber of radius R is filled initially with premixed combustible gaseous mixture. At time zero, a spark is produced at the center of the combustion chamber and a spherical flame propagates radially outward. If the instantaneous radius of the flame is r, derive the laminar flame speed given by Eq. (5-78), of which the instantaneous pressure is measured from a pressure transducer mounted on the wall of a spherical combustion chamber.
(*Hint*: The compression of unburned gas by the burned gas can be considered to be adiabatic.)

7. Show that the energy and species conservation equations [(5-101) and (5-102)] in Spalding's flammability model can be transformed into Eqs. (5-111) and (5-112) after using the parameters defined in Eqs. (5-104) through (5-110).

CHAPTER

6 Gaseous Diffusion Flames and Combustion of a Single Liquid Fuel Droplet

In the broadest sense, a diffusion flame may be defined as any flame in which the fuel and oxidizer are initially separated (non-premixed). For example, a pan of oil burning in air, a lighted candle, a fuel droplet burning in oxygen, and pure solid fuel burning in a ramjet engine all produce diffusion flames. In a restricted sense, a diffusion flame may be defined as a non-premixed, steady or unsteady, nearly isobaric flame in which most of the reaction occurs in a narrow zone that can be approximated as a surface. The mixing rate is lower than the chemical-reaction rate; the latter is often of negligible importance for diffusion flame. Diffusion flames can either be laminar or turbulent; in this chapter, we shall concentrate on laminar ones.

1 BURKE AND SCHUMANN'S THEORY OF LAMINAR DIFFUSION FLAMES

A classical example of a laminar diffusion flame, which was first described quantitatively by Burke and Schumann,[1] is provided by a system in which fuel and air flow with the same linear flow velocity in coaxial cylindrical tubes.

The observed shapes of diffusion flames may be divided into two classes. If the ratio of the duct radii r_s to r_j is such that more air is available than what is required for complete combustion, then an *overventilated* flame is formed and the flame boundary converges to the cylinder axis. On the other hand, if the air supply is insufficient for complete burning, then an *underventilated* flame is produced in which the flame surface expands to the outer tube wall. (See Fig. 6.1.)

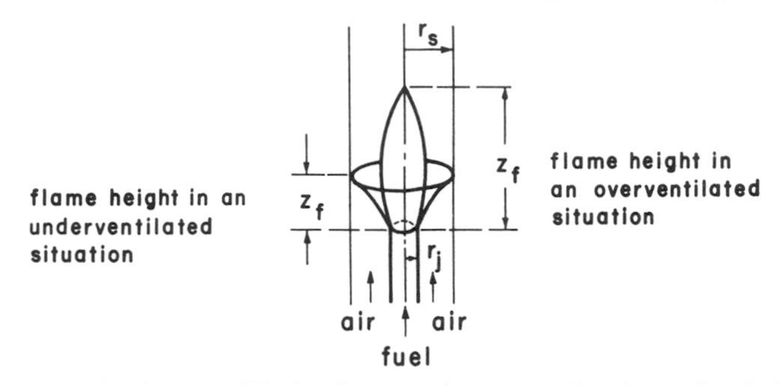

Figure 6.1 The shape of diffusion flames under over- and underventilated situations.

The gross features of diffusion flames can often be described by postulating the existence of a surface (assumed to be coincident with the luminous "combustion surface") at which chemical reactions occur instantaneously. A typical composition profile of a hydrogen–air diffusion flame is shown in Fig. 6.2.

Of particular interest is the fact that very little fuel and oxygen penetrate through the luminous flame boundary, a result which shows that the chemical reactions are completed in a very narrow region and which suggests that the flame surface may be identified, in first approximation, as the surface to which the fuel and oxygen rates of delivery are in stoichiometric proportions.

The nonsteady conservation-of-species equations given in Eq. (3-32) can be written as

$$\frac{\partial \rho_A}{\partial t} + \nabla \cdot (\rho_A \mathbf{v}) = \nabla \cdot \left[\rho \mathscr{D} \nabla \left(\frac{\rho_A}{\rho} \right) \right] - \omega_A$$

where ω_A has a negative sign because it now represents the rate of consumption of mass of species A. For cylindrical geometry under axisymmetric

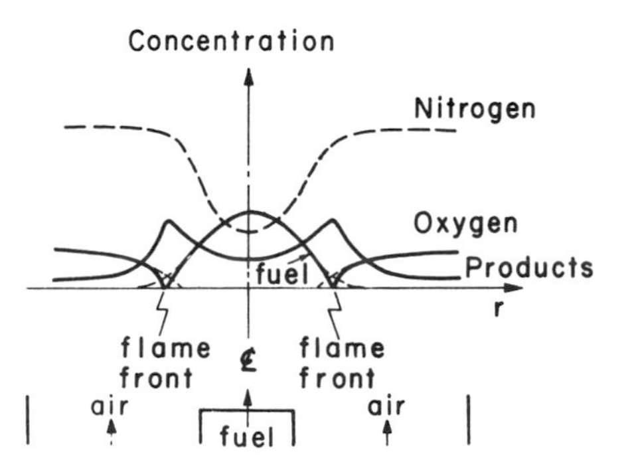

Figure 6.2 Species variation through a diffusion flame at a fixed height above the fuel jet tube.

condition, the above species continuity equation becomes

$$\underbrace{\frac{\partial \rho_A}{\partial t}}_{\substack{\text{rate of}\\ \text{increase}\\ \text{in } \rho_A}} = \underbrace{\frac{1}{r}\frac{\partial}{\partial r}\left[\mathscr{D}\rho r \frac{\partial Y_A}{\partial r}\right]}_{\substack{\text{mass rate}\\ \text{of } A \text{ diffused}\\ \text{in (from}\\ \text{Fick's law)}}} - \underbrace{\left[\frac{1}{r}\frac{\partial (r\rho_A v_r)}{\partial r} + \frac{\partial v_z \rho_A}{\partial z}\right]}_{\substack{\text{mass rate of } A\\ \text{convected out}}} - \underbrace{\omega_A}_{\substack{\text{mass}\\ \text{rate of}\\ \text{consumption}\\ \text{of } A}} \tag{6-1}$$

Here the axial diffusion is neglected, since it is considered to be small in comparison with radial diffusion term:

$$\frac{\partial^2 Y_A}{\partial z^2} \ll \frac{\partial^2 Y_A}{\partial r^2}$$

where

$$Y_A = \text{mass fraction of species } A$$

$$\rho = \text{total density}$$

$$\rho_A = \text{density of } A$$

The dependence of the product $\mathscr{D}\rho$ on the temperature can be deduced from the expressions for the random molecular velocity and mean free path:

$$\mathscr{D}\rho = \tfrac{1}{3}\bar{u}l\rho = \tfrac{1}{3}\bar{u}\frac{kTp}{\sqrt{2}\,\pi\sigma^2 pRT} = \frac{1}{3}\sqrt{\frac{8kT}{\pi m}}\,\frac{k}{\sqrt{2}\,\pi R\sigma^2}$$

Since the product $\mathscr{D}\rho$ is a weak function of temperature, as a first-order approximation it can be considered to be independent of temperature. The species continuity equation can thus be expressed in the same form as the energy equation. The convective terms of Eq. (6-1) can be rearranged to give

$$\frac{1}{r}\frac{\partial (r\rho_A v_r)}{\partial r} = \frac{1}{r}\frac{\partial [r\rho v_r(\rho_A/\rho)]}{\partial r} = \rho v_r \frac{\partial (Y_A)}{\partial r} + \frac{Y_A}{r}\frac{\partial (\rho v_r r)}{\partial r}$$

$$\frac{\partial v_z \rho_A}{\partial z} = \frac{\partial v_z \rho Y_A}{\partial z} = Y_A \frac{\partial v_z \rho}{\partial z} + v_z \rho \frac{\partial Y_A}{\partial z}$$

Considering the steady-state condition

$$\frac{\partial \rho_A}{\partial t} = 0,$$

and using the overall continuity equation

$$\frac{1}{r}\frac{\partial(\rho v_r r)}{\partial r}+\frac{\partial(\rho v_z)}{\partial z}=0$$

to simplify the convective terms, Eq. (6-1) becomes

$$\frac{\mathscr{D}\rho}{r}\frac{\partial}{\partial r}\left(r\frac{\partial Y_A}{\partial r}\right)-\cancelto{\text{small}}{\left(\frac{\rho v_r}{r}\right)}r\frac{\partial Y_A}{\partial r}-v_z\rho\frac{\partial Y_A}{\partial z}=\omega_A \tag{6-2}$$

If $A + F \rightarrow P$ represents the overall chemical reaction in the diffusion flame, then the species continuity equation for B can be written as

$$\frac{\mathscr{D}\rho}{r}\frac{\partial}{\partial r}\left(r\frac{\partial Y_F}{\partial r}\right)-(\rho v_z)\frac{\partial Y_F}{\partial z}=\omega_F=\frac{\omega_A}{\phi} \tag{6-3}$$

where ϕ is the stoichiometric mass ratio of A to F. The general form of the energy equation from Eq. (3-78) is

$$\rho\frac{Dh}{Dt}-\frac{Dp}{Dt}=-\nabla\cdot\mathbf{q}+\cancelto{0}{\Phi}+\dot{Q}+\rho\sum_{k=1}^{n}Y_k\cancelto{0}{\mathbf{F}_k}\cdot\mathbf{V}_k$$

Assuming steady state, constant pressure, small v_r, and negligible axial heat conduction, then

$$\rho\cancelto{0}{\frac{\partial h}{\partial t}}+\rho v_r\cancelto{\text{small}}{\frac{\partial h}{\partial r}}+\rho v_z\frac{\partial h}{\partial z}-\cancelto{0}{\frac{\partial p}{\partial t}}-v_r\cancelto{0}{\frac{\partial p}{\partial r}}-v_z\cancelto{0}{\frac{\partial p}{\partial z}}$$

$$=+\frac{\lambda}{r}\frac{\partial}{\partial r}\left(r\frac{\partial T}{\partial r}\right)+\lambda\underbrace{\frac{\partial}{\partial z}\left(\frac{\partial T}{\partial z}\right)}_{\text{small}}+\omega_F\Delta H_r$$

The energy equation after simplification becomes

$$\frac{\lambda}{C_p}\frac{1}{r}\frac{\partial}{\partial r}\left[r\frac{\partial C_pT}{\partial r}\right]-(\rho v_z)\frac{\partial(C_pT)}{\partial z}=-\omega_F\Delta H_r=-\frac{\omega_A}{\phi}\Delta H_r=\dot{H} \tag{6-4}$$

where ΔH_r is the heat of reaction per unit mass of fuel (F). Recall that in the

Shvab–Zel'dovich formulation we used

$$\mathrm{Le} = 1, \qquad \text{or} \quad \mathscr{D}\rho = \frac{\lambda}{C_p} \tag{6-5}$$

The energy equation, after setting the Lewis number equal to one, becomes

$$\frac{\mathscr{D}\rho}{r}\frac{\partial}{\partial r}\left[\frac{r\partial(C_pT)}{\partial r}\right] - \rho v_z \frac{\partial(C_pT)}{\partial z} = -\frac{\omega_A}{\phi}\Delta H_r \tag{6-6}$$

Multiplying Eq. (6-2) by $\Delta H_r/\phi$ and combining with Eq. (6-6), we get

$$\frac{1}{r}\frac{\partial}{\partial r}\left\{\mathscr{D}\rho\left[r\frac{\partial}{\partial r}\left(C_pT + \frac{Y_A\Delta H_r}{\phi}\right)\right]\right\} - \rho v_z\frac{\partial}{\partial z}\left(C_pT + \frac{Y_A\Delta H_r}{\phi}\right) = 0 \tag{6-7}$$

For more complex chemical reactions in diffusion flames, one can follow the Shvab–Zel'dovich procedure to obtain equations of the same form of Eq. (6-7). Some detailed procedures are given below.

In general coordinate systems, the energy and mass equations are

$$\nabla \cdot \left[(\rho\mathbf{v})(C_pT) - \frac{\lambda}{C_p}\nabla(C_pT)\right] = -\dot{H} = \sum_j \Delta h^\circ_{f,j}\omega_j \tag{6-8}$$

and

$$\nabla \cdot \left[(\rho\mathbf{v})Y_j - \rho\mathscr{D}\nabla Y_j\right] = -\omega_j \tag{6-9}$$

where ω_j is the rate of consumption of species j, and $\Delta h^\circ_{f,j}$ is the heat of formation per unit mass at the base temperature of species j.

Sometimes we can consider the global reaction as follows:

$$F + \phi_m O \rightarrow P \tag{6-10}$$

or

$$\nu'_F F + \nu'_O O \rightarrow \text{product}$$

with

$$\phi_m = \text{molar stoichiometric oxidizer-to-fuel ratio} = \frac{\nu'_O}{\nu'_F}$$

If we divide Eq. (6-9) by $W_j(\nu_j'' - \nu_j')$, we get the general species equation as

$$\nabla \cdot \left[(\rho \mathbf{v}) \frac{Y_j}{W_j(\nu_j'' - \nu_j')} - (\rho \mathscr{D}) \nabla \frac{Y_j}{W_j(\nu_j'' - \nu_j')} \right] = - \frac{\omega_j}{W_j(\nu_j'' - \nu_j')} \equiv -\dot{M} \tag{6-11}$$

Note that in the global reaction, if we let j represent the oxidizer, then

$$\nu_j'' - \nu_j' = -\nu_O' = -\phi_m$$

and if we let j represent fuel, we have

$$\nu_j'' - \nu_j' = -\nu_F' = -1$$

The mass diffusion equation for all the species present in a general system can be put into the same form by setting

$$\alpha_j \equiv \frac{Y_j}{W_j(\nu_j'' - \nu_j')} \tag{6-12}$$

Then Eq. (6-11) reduces to

$$\nabla \cdot \left[(\rho \mathbf{v}) \alpha_j - \rho \mathscr{D} \nabla \alpha_j \right] = -\dot{M} \tag{6-13}$$

Following the Shvab–Zel'dovich formulation given in Section 3.10, we can arrange the energy equation in the following form:

$$\nabla \cdot \left[\rho \mathbf{v} \frac{C_p T}{\Delta H_r W_j(\nu_j'' - \nu_j')} - \frac{\lambda}{C_p} \nabla \frac{C_p T}{\Delta H_r W_j(\nu_j'' - \nu_j')} \right] = \frac{-\omega_j}{W_j(\nu_j'' - \nu_j')} = -\dot{M}$$

Let

$$\alpha_T \equiv \frac{C_p T}{\Delta H_r W_j(\nu_j'' - \nu_j')}$$

then

$$\nabla \cdot \left[(\rho \mathbf{v}) \alpha_T - \rho \mathscr{D} \nabla \alpha_T \right] = -\dot{M} \tag{6-14}$$

Comparing Eq. (6-14) with Eq. (6-13), we notice that both α_j and α_T satisfy the same differential equation.

Equation (6-13) and Eq. (6-14) may both be expressed as

$$L(\alpha) = -\dot{M} \tag{6-15}$$

where the linear operator L is defined by

$$L(\alpha) \equiv \nabla \cdot [(\rho \mathbf{v})\alpha - \rho \mathscr{D} \nabla \alpha] \tag{6-16}$$

The inhomogeneous nonlinear rate term may be eliminated from all except one of the relations corresponding to Eq. (6-15). Selecting α_1 to be the dependent variable for the inhomogeneous equation, we have

$$L(\alpha_1) = -\dot{M} \tag{6-17}$$

Other flow variables are then determined by the linear homogeneous equation through the use of coupling functions β:

$$L(\beta) = 0 \tag{6-18}$$

with $\beta = \alpha_T - \alpha_1 \equiv \beta_T$ or $\beta = \alpha_j - \alpha_1 \equiv \beta_j$ $(j \neq 1)$. This procedure is quite important in solving diffusion-flame problems, since in these problems one often finds that by solving the linear equations for relations between flow variables, burning rates may eventually be determined without solving the nonlinear equation. Equation (6-18) is deceptively simple in appearance; it is hard to solve unless additional approximations are made. In general, $\rho \mathbf{v}$ and $\rho \mathscr{D}$ do depend on β_j or β_T; thus the operator L depends implicitly on β, and Eq. (6-18) is actually nonlinear.

1.1 Basic Assumptions and Solution Method

Burke and Schumann assumed that:

1. At port position, the velocities of air and fuel are constant, equal, and uniform across their respective tubes. This is accomplished by varying the radii of the tubes and also the molar fuel ratio which is given by $r_j^2/(r_s^2 - r_j^2)$.
2. The velocity of the fuel and air up the duct in the region of the flame is the same as the velocity at the port (no tube friction loss).
3. $\rho \mathscr{D}$ is constant.
4. Diffusion in the axial direction is negligible in comparison with that in the radial direction:

$$\frac{\partial^2 Y_i}{\partial r^2} \gg \frac{\partial^2 Y_j}{\partial z^2}$$

5. Mixing is caused by diffusion only, and the radial velocity component is equal to zero:

$$v_r = 0$$

6. Reaction takes place at $\phi = 1$ (at the flame surface).

The only differential equation that we need to consider for the mass-fraction distribution is Eq. (6-18):

$$L(\beta) = 0$$

with

$$\beta = \alpha_F - \alpha_O \tag{6-19}$$

$$\alpha_F = \frac{-Y_F}{W_F \nu'_F}, \qquad \alpha_0 = \frac{-Y_O}{W_O \nu'_O} \tag{6-20}$$

In cylindrical coordinates, this differential equation is

$$\left(\frac{v_z}{\mathscr{D}}\right)\frac{\partial \beta}{\partial z} - \frac{1}{r}\frac{\partial}{\partial r}\left(r\frac{\partial \beta}{\partial r}\right) = 0 \tag{6-21}$$

The boundary conditions for Eq. (6-21) are

$$\beta = -\frac{(Y_F)_{z=0}}{W_F \nu'_F} \qquad \text{at} \quad z = 0, \quad 0 \le r \le r_j \tag{6-22}$$

$$\beta = +\frac{(Y_O)_{z=0}}{W_O \nu'_O} \qquad \text{at} \quad z = 0, \quad r_j \le r \le r_s \tag{6-23}$$

and

$$\frac{\partial \beta}{\partial r} = 0 \qquad \text{at } r = 0, \ z > 0, \text{ and at } r = r_s, \ z > 0 \tag{6-24}$$

It is convenient to introduce the dimensionless coordinates

$$\xi \equiv \frac{r}{r_s}$$

$$\eta \equiv \frac{z\mathscr{D}}{v_z r_s^2} \tag{6-25}$$

and to define the reduced parameters

$$C \equiv \frac{r_j}{r_s}$$

$$\nu \equiv \frac{(Y_O)_{z=0} W_F \nu'_F}{(Y_F)_{z=0} W_O \nu'_O} \tag{6-26}$$

and the reduced dependent variable

$$\gamma \equiv \beta \frac{W_F \nu_F'}{(Y_F)_{z=0}} \tag{6-27}$$

In terms of these quantities, Eqs. (6-21) through (6-24) become

$$\frac{\partial \gamma}{\partial \eta} = \frac{1}{\xi} \frac{\partial}{\partial \xi}\left(\xi \frac{\partial \gamma}{\partial \xi}\right), \tag{6-28}$$

$$\gamma = \begin{cases} 1 & \text{at} \quad \eta = 0, \quad 0 \le \xi < C, \\ -\nu & \text{at} \quad \eta = 0, \quad C < \xi < 1, \end{cases} \tag{6-29}$$

$$\frac{\partial \gamma}{\partial \xi} = 0 \qquad \text{at } \xi = 1, \text{ and at } \xi = 0, \quad \eta > 0 \tag{6-30}$$

By the method of separation of variables, it can be shown that a part of the decomposed solution satisfies the Bessel's differential equation of order n:

$$\frac{d^2 y}{dx^2} + \frac{1}{x}\frac{dy}{dx} + \left(1 - \frac{n^2}{x^2}\right) y = 0.$$

The functions J_0 and J_1 shown in the accompanying graph are solutions of the above differential equation; they are called the Bessel functions of the first kind (of order 0 and 1 respectively). The mathematical forms of the Bessel functions are

$$J_0(x) = 1 - \frac{x^2}{2^2} + \frac{x^4}{2^4(2!)^2} - \cdots + (-1)^n \frac{x^{2n}}{2^{2n}(n!)^2}$$

$$= \sum_{n=0}^{\infty} (-1)^n \frac{(x/2)^{2n}}{n!n!}$$

and

$$J_m(x) = \sum_{n=0}^{\infty} \frac{(-1)^n x^{m+2n}}{2^{m+2n} n! \Gamma(m+n+1)}, \qquad m > 0$$

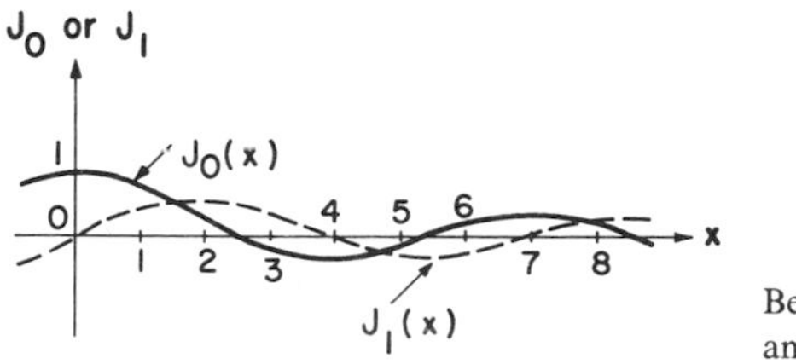

Bessel functions of the first kind of order 0 and 1.

The final solution for γ in power-series form is

$$\gamma = (1+\nu)C^2 - \nu + 2(1+\nu)C \sum_{n=1}^{\infty} \frac{1}{\phi_n} \frac{J_1(C\phi_n)}{[J_0(\phi_n)]^2} J_0(\phi_n \xi) e^{-\phi_n^2 \eta} \quad (6\text{-}31)$$

where ϕ_n represent successive roots of the equation $J_1(\phi) = 0$ (with ordering convention $\phi_n > \phi_{n-1}$, $\phi_0 = 0$).

1.2 Flame Shape and Flame Height

If the entire reaction is to occur at a flame surface, then

$$\beta = 0, \qquad \text{or} \quad \gamma = 0$$

at the flame surface. Hence, setting $\gamma = 0$ in Eq. (6-31) provides a relation between ξ and η that defines the locus of the flame surface [$\eta = f(\xi)$, or $z = g(r)$]. The shape of the surface obtained in this manner is shown in Fig. 6.1 for two different values of ν.

The flame height is obtained by solving Eq. (6-31) for η after setting $\xi = 0$ for overventilated flames or $\xi = 1$ for underventilated flames (and, of course, $\gamma = 0$ in either case). Since the flame heights are generally large enough to cause the factor $e^{-\phi_n^2 \eta}$ to decrease rapidly as n increases at these values of η, it usually suffices to retain only the first few terms of the sum in Eq. (6-31) for this calculation. Neglecting all the terms except $n = 1$, we obtain the rough approximation

$$\eta = \frac{1}{\phi_1^2} \ln \left\{ \frac{2(1+\nu)CJ_1(C\phi_1)}{[\nu - (1+\nu)C^2]\phi_1 J_0(\phi_1)} \right\} \quad (6\text{-}32)$$

for the dimensionless flame height of an underventilated flame. The first zero of $J_1(\phi)$ is $\phi_1 = 3.83$. The flame shapes and flame heights obtained from Eq. (6-32) by Burke and Schumann (Fig. 6.3) are in surprisingly good agreement with experiments, considering the drastic nature of some of the assumptions.

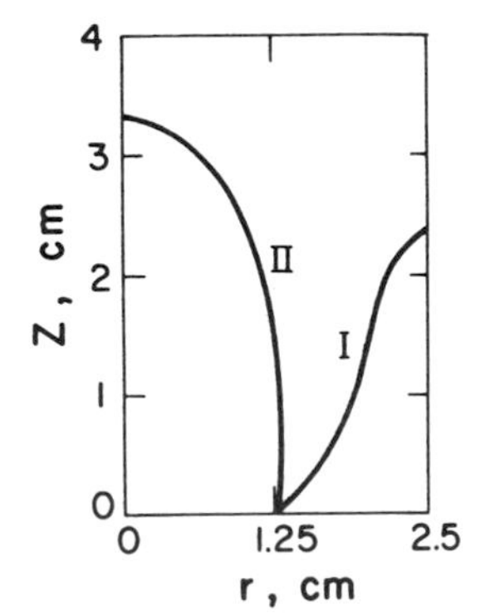

Figure 6.3 Calculated diffusion flame height: curve I, underventilated cylindrical flame; curve II, overventilated cylindrical flame.

In more recent work, Penner and Bahadori[24] and Chung and Law[25] reexamined the Burke-Schumann problem and retained the axial diffusion terms. In Penner and Bahadori's study,[24] they considered diffusion flames with arbitrary transport coefficients and chemical reactions. The closed-form solution obtained is slightly more complex than that of Burke and Schumann.[1] In Chung and Law's work,[25] the solution was obtained by a perturbation method under the assumption that the Lewis number is very close to unity. The zeroth order solution was obtained by separation of variables, and the 1st order problem was solved using Green's functions. Their results showed that the flame is made longer and narrower by including streamwise diffusion transport. They also showed that the Burke-Schumann solution is generally valid for large Padet numbers.

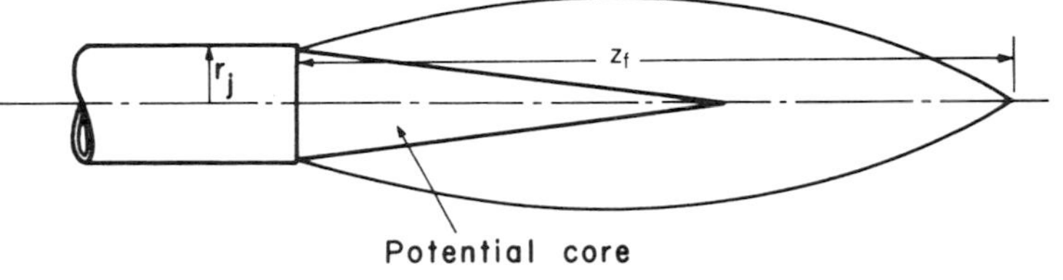

Horizontal laminar fuel jet.

2 PHENOMENOLOGICAL ANALYSIS OF FUEL JETS

In the following, we shall use a simple phenomenological approach to show that for a laminar flame, the flame height is proportional to the volumetric flow rate and inversely proportional to the mass diffusivity:

$$(z_f)_{\text{laminar}} \propto \frac{\text{volumetric flow rate}}{\mathscr{D}} \tag{6-33}$$

For a turbulent flame

$$(z_f)_{\text{turbulent}} \propto \text{port size} \tag{6-34}$$

Sometimes very useful information can be obtained from simple phenomenological reasoning. The fundamental assumption which has been made here is that the combustion process does not affect the mixing rate between the fuel jet and the surrounding oxidizer. In essence, as soon as the oxidizer mixes into the fuel, it reacts. We have considered a simple cylindrical problem; thus as soon as the oxidizer has diffused from the jet edge to the center, we have completely burned the fuel—or, in other words, we have the flame height.

Our problem is simply one of the laminar diffusional mixing of a jet—which in itself can be approached phenomenologically. Simple kinetic theory tells us that in molecular diffusion the average displacement of a molecule is given by

$$\frac{1}{2}\frac{d\overline{X}^2}{dt} = \mathscr{D} \tag{6-35}$$

where $\mathscr{D} = \frac{1}{2}\bar{u}l$ is the molecular diffusion coefficient and $\overline{X}^2$ is the mean square of the displacement of the particle in a specified direction in the time t. To arrive at an estimate of the flame length, it is proposed to use this equation from kinetic theory in the integrated form

$$\xi^2 = 2\mathscr{D}t \tag{6-36}$$

where ξ^2 denotes the mean squared displacement of a molecule from its initial location due to diffusion during time t. The length of the flame is assumed to correspond to the condition that at the point on the stream axis where combustion is complete, the average depth of penetration of air into fuel must be approximately equal (or proportional, if one likes) to the radius of the burner tube. As an approximation, ξ is identified with the average depth of penetration. The gas velocity $\overline{V}$ (inside the port) is taken as constant, so that the time t required for completion of the diffusion process—which is the time in which a gas element flows from the burner port to the flame tip—is given by

$$t = \frac{z_f}{\overline{V}} \qquad \text{(residence time)} \tag{6-37}$$

The radial diffusion time can be approximated by $\xi^2/2\mathscr{D}$ or $r_j^2/2\mathscr{D}$. Assuming that the residence time is of the same order of the diffusion time, we have

$$z_{f,L} \propto \frac{r_j^2\overline{V}}{2\mathscr{D}} \propto \frac{\left(\pi r_j^2\right)\overline{V}}{2\pi\mathscr{D}} \propto \frac{\text{volumetric flow rate}}{\mathscr{D}} \tag{6-38}$$

This result is significant; it clearly indicates that for laminar-diffusion flames the height is proportional to the volumetric flow rate, and inversely proportional to the mass diffusivity. The relationship is also helpful for studying turbulent fuel jets, as explained in the following simple analysis.

First, let us recall that the Schmidt number is defined as $\nu/\mathscr{D}$. For constant Schmidt number, we have $\mathscr{D} \propto \nu$. Hence, Eq. (6-38) can be rewritten as

$$z_{f,L} \propto \frac{r_j^2\overline{V}}{\nu}$$

For the turbulent case, we use the same reasoning to arrive at a relationship similar to the above, except that instead of the molecular viscosity ν, we must use the turbulent eddy viscosity ν_t in the denominator, since $\nu_t \gg \nu$. Thus,

$$z_{f,T} \propto \frac{r_j^2\overline{V}}{\nu_t} \tag{6-39}$$

Note that $\nu_t \propto l u'_{\text{rms}}$, where l is the scale of turbulence and is proportional to the tube diameter (or tube radius) and u'_{rms} is the intensity of turbulence, which is approximately proportional to the mean flow velocity at the axis; thus,

$$\nu_t \propto r_j\overline{V} \tag{6-40}$$

Combining the Eqs. (6-39) and (6-40), we have

$$z_{f,T} \propto \frac{r_j^2\overline{V}}{r_j\overline{V}} \propto r_j \tag{6-41}$$

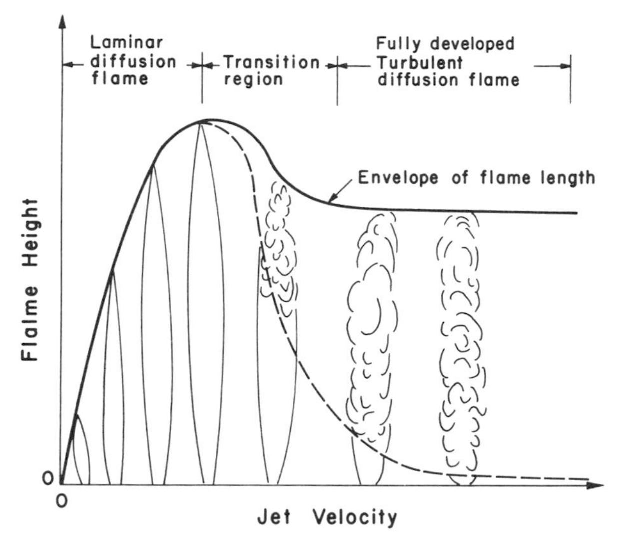

Figure 6.4 The variation of flame height and character as a function of jet velocity (after Hottel and Hawthorne[2]).

This says that the height of a turbulent-diffusion flame is proportional to the port radius only. It is a very important practical conclusion that has been verified in many ways. The variation of diffusion flame height as a function of jet velocity is shown in Fig. 6.4. The laminar-diffusion height follows the functional dependence shown in Eq. (6-38), while the turbulent-diffusion height follows Eq. (6-41). More discussion of turbulent diffusion-flame structure is given in Chapter 7.

3 LAMINAR DIFFUSION FLAME JETS

Injection of fuel jets into a combustor containing oxidizers is a common practice in many combustion systems such as diesel engines, gas-turbine engines, industrial furnaces, and ramjet engines. Under the condition when there is no cross-flow or buoyancy effect and also the fuel jet and the ambient air are parallel, the diffusion flame is essentially the same as the Burke–Schumann laminar-diffusion flame we discussed in Section 2. The only difference is that the effect of the outer tube becomes negligible when the combustor diameter becomes very large in comparison with the fuel-jet diameter. The schematic diagram of a single fuel jet under the above condition is shown in Fig. 6.5*a*. When the buoyancy or cross-flow effect is significant, the flame geometry and the boundary of the hot gases will not be axisymmetric (see Fig. 6.5*b*). Before we study the detailed structure of these diffusion flames, we shall provide some background on laminar jet mixing.

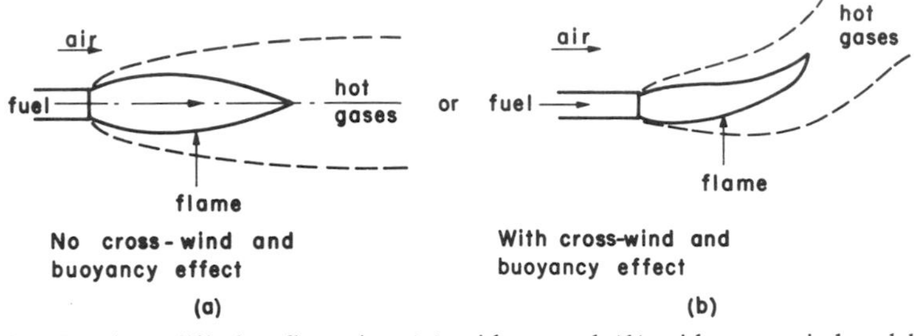

Figure 6.5 Laminar diffusion flame jets (*a*) without and (*b*) with cross-wind and buoyancy effect.

3.1 Laminar Jet Mixing

Consider a gaseous fuel jet with a uniform (top-hat) velocity profile issuing from a circular hole of radius r_0 into quiescent air. Mixing will occur between the gaseous fuel jet and ambient air as shown in Fig. 6.6. At some downstream location, the velocity profile will have a maximum at the centerline and a gradual decay to zero at the boundary of the mixing zone. The air is entrained through the boundary of the mixing zone. For simplicity of analysis, let us make the following assumptions:

1. The surrounding air far from the jet is at rest.
2. Chemical reaction is absent.
3. ρ, μ and other properties of the gases are uniform (uniform property flow).
4. The flow is steady.
5. The effect of buoyancy is absent.
6. The pressure in the fluid is uniform.

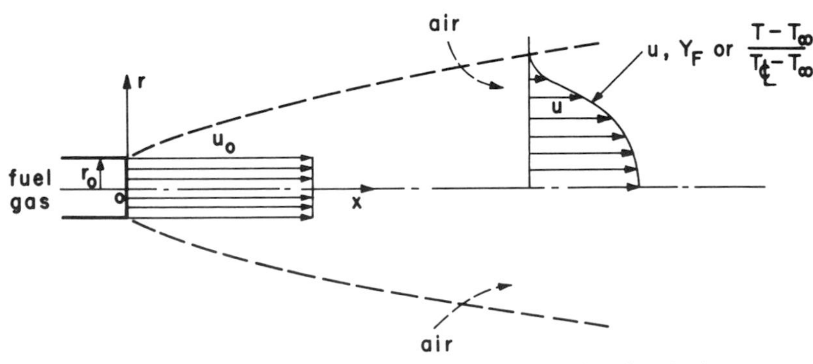

Figure 6.6 Velocity and concentration profile distributions in a laminar jet.

7. Mass diffusion, heat conduction, and viscous action in the axial direction are negligible.
8. Sc = Pr = 1.

The overall mass conservation equation can be written as

$$\frac{\partial}{\partial x}(\rho u r) + \frac{\partial}{\partial r}(\rho v r) = 0 \tag{6-42}$$

The x-momentum equation is

$$\frac{\partial}{\partial x}(\rho u r u) + \frac{\partial}{\partial r}(\rho v r u) = \frac{\partial}{\partial r}\left(\mu r \frac{\partial u}{\partial r}\right) \tag{6-43}$$

The fuel species conservation equation written in terms of mixture fraction, f, is

$$\frac{\partial}{\partial x}(\rho u r f) + \frac{\partial}{\partial r}(\rho v r f) = \frac{\partial}{\partial r}\left(\rho \mathscr{D} r \frac{\partial f}{\partial r}\right) \tag{6-44}$$

where the mixture fraction has been defined in Section 8 of Chapter 1. Taking ρ, μ and $\mathscr{D}$ to be uniform and applying the assumptions that Sc = Pr = Le = 1, the above equations can be reduced to

$$\frac{\partial u}{\partial x} + \frac{\partial v}{\partial r} + \frac{v}{r} = 0 \tag{6-45}$$

$$u\frac{\partial u}{\partial x} + v\frac{\partial u}{\partial r} = \frac{\nu}{r}\frac{\partial}{\partial r}\left(r\frac{\partial u}{\partial r}\right) \tag{6-46}$$

$$u\frac{\partial f}{\partial x} + v\frac{\partial f}{\partial r} = \frac{\nu}{r}\frac{\partial}{\partial r}\left(r\frac{\partial f}{\partial r}\right) \tag{6-47}$$

The boundary conditions at the entrance station are

$$x=0, \quad r \le r_0: \qquad u = u_0 \tag{6-48}$$

$$f = 1 \tag{6-49}$$

$$x=0, \quad r > r_0: \qquad u = 0 \tag{6-50}$$

$$f = 0 \tag{6-51}$$

At large radius, one has

$$r \to \infty: \qquad u = 0 \tag{6-52}$$

$$f = 0 \tag{6-53}$$

There are two jet invariants based upon the conservation of momentum and species. These jet invariants are defined as follows:

$$I_u \equiv \frac{1}{\nu}\int_0^\infty u^2 r\,dr = \frac{1}{\nu}\left(\tfrac{1}{2}u_0^2 r_0^2\right) \tag{6-54}$$

$$I_f = \frac{1}{\nu}\int_0^\infty ufr\,dr = \frac{1}{\nu}\left(\tfrac{1}{2}u_0 r_0^2\right) \tag{6-55}$$

According to Spalding[3] and Schlichting,[4] the following equations satisfy the governing partial differential equations and boundary conditions at the far field:

$$u = \frac{3}{4}\frac{I_u}{x}\left(1 + \frac{\xi^2}{4}\right)^{-2} \tag{6-56}$$

or

$$\frac{u}{u_0} = \frac{3}{32}\left[\frac{\mathrm{Re}_{d_0} d_0}{x}\right]\left(1 + \tfrac{1}{4}\xi^2\right)^{-2} \tag{6-56a}$$

$$v = \left(\tfrac{3}{8}I_u \nu\right)^{1/2}\frac{\xi}{x}\left(1 - \frac{\xi^2}{4}\right)\left(1 + \frac{\xi^2}{4}\right)^{-2} \tag{6-57}$$

or

$$\frac{v}{u_0} = \frac{\sqrt{3}}{8}\frac{d_0}{x}\left(\xi - \tfrac{1}{4}\xi^3\right)\left(1 + \tfrac{1}{4}\xi^2\right)^{-2} \tag{6-57a}$$

and

$$f = \frac{3}{4}\frac{I_f}{x}\left(1 + \frac{\xi^2}{4}\right)^{-2} \tag{6-58}$$

where the dimensionless variable ξ is defined as

$$\xi \equiv \left(\frac{3}{8}\frac{I_u}{\nu}\right)^{1/2}\frac{r}{x} = \frac{\sqrt{3}}{8}\left(\frac{r}{d_0}\right)\left[\frac{\mathrm{Re}_{d_0} d_0}{x}\right] \tag{6-59}$$

It is quite evident from the above solution that the velocity and concentration profiles are self-similar; that is, they depend upon r/x alone. Based upon Eqs. (6-56) and (6-58), the centerline values of u and f are

$$u_{\mathbb{C}} x = \tfrac{3}{4}I_u$$

$$f_{\mathbb{C}} x = \tfrac{3}{4}I_f \tag{6-60}$$

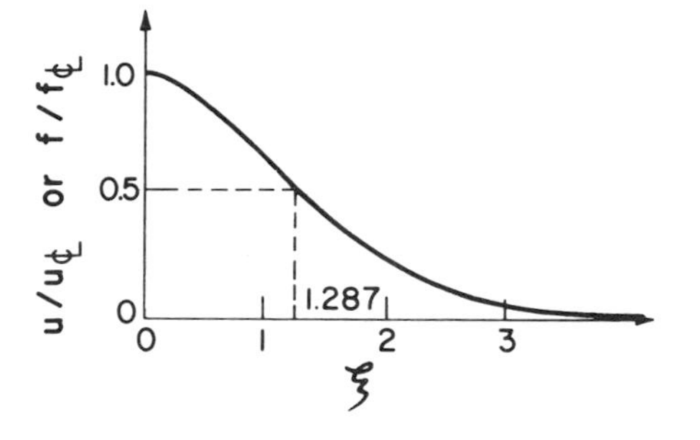

Figure 6.7 Similarity profile of velocity or concentration as a function of dimensionless variable.

Since the centerline velocity is less than the jet exit velocity ($u < u_0$) and also the mixture fraction at the centerline in the downstream location is less than one, we can conclude from Eq. (6-60) that the x-value satisfying the above far-field solution must obey the following inequality:

$$x > \frac{3}{4}\frac{I_u}{u_0}, \qquad \text{or} \qquad \frac{x}{r_0} > \frac{3}{8}\frac{u_0 r_0}{\nu} \tag{6-61}$$

The radial profiles of u and f, when normalized by their respective centerline values, are

$$\frac{u}{u_{℄}} = \frac{f}{f_{℄}} = \frac{1}{\left(1 + \frac{1}{4}\xi^2\right)^2} \tag{6-62}$$

The sketch shown in Fig. 6.7 represents the velocity or concentration profiles qualitatively. It should be noted that

$$\frac{u}{u_{℄}} = 0.5 \qquad \text{for} \quad \xi = 1.287.$$

The radius at which the velocity has one-half of its centerline velocity is denoted $r_{1/2}$. From Eq. (6-59), we have

$$\frac{r_{1/2}}{x} = 1.287\left(\frac{8\nu}{3I_u}\right)^{1/2} = 2.97\frac{\nu}{u_0 r_0} \tag{6-63}$$

This equation implies that the angle of the spreading of the jet is inversely proportional to the Reynolds number.

The total mass flow rate in the jet, $\dot{m}$, can be obtained from the following integration:

$$\dot{m} \equiv \int_0^{\infty} 2\pi r \rho u \, dr \tag{6-64}$$

Substituting Eq. (6-56) into Eq. (6-64) and carrying out the integration, we have

$$\dot{m} = 2\pi\rho x^2\left(\frac{8}{3}\frac{\nu}{I_u}\right)\left(\frac{3}{4}\frac{I_u}{x}\right)\int_0^\infty \frac{\xi\,d\xi}{\left(1+\frac{1}{4}\xi^2\right)^2} = 8\pi\mu x \tag{6-65}$$

The entrainment rate can be obtained by differentiating Eq. (6-65), i.e.,

$$\frac{d\dot{m}}{dx} = 8\pi\mu \tag{6-66}$$

Associated with the entrainment process, an entrainment velocity can be defined by

$$v_{\text{ent}} \equiv -\frac{1}{2\pi r_{jb}\rho}\frac{d\dot{m}}{dx} = \frac{-4\mu}{r_{jb}\rho} = -4\frac{\nu}{r_{jb}} \tag{6-67}$$

where r_{jb} represents the radius at the jet boundary. In dimensionless form, Eq. (6-57) can be written as

$$\frac{v}{v_{\text{ent}}} = -\frac{\sqrt{3}}{16}\frac{u_0 r_0}{\nu}\frac{\xi}{x}\left(1-\frac{\xi^2}{4}\right)\left(1+\frac{\xi^2}{4}\right)^{-2} r_{jb} \tag{6-68}$$

According to Schlichting,[4]

$$r_{jb} \approx \frac{16}{\sqrt{3}}\left(\frac{x}{Re_{d_0} d_0}\right) d_0 \tag{6-69}$$

Using the above expression, the radial velocity can be expressed as

$$\frac{v}{v_{\text{ent}}} = -\tfrac{1}{4}\xi^2\left(1-\frac{\xi^2}{4}\right)\left(1+\frac{\xi^2}{4}\right)^{-2} \tag{6-70}$$

It is interesting to note that at large ξ, $v/v_{\text{ent}} \to 1$ and at small ξ, $v/v_{\text{ent}} \sim -\xi^2/4$.

The equations describing lines of constant u and f can be found by rearranging Eq. (6-56) or (6-58).

$$r = \frac{16}{\sqrt{3}}\frac{x}{\text{Re}_{d_0}}\sqrt{\sqrt{\frac{3}{32}\left(\frac{\text{Re}_{d_0} d_0}{x}\right)\frac{u_0}{u}} - 1} \tag{6-71}$$

where

$$\text{Re}_{d_0} \equiv \frac{u_0 d_0}{\nu}$$

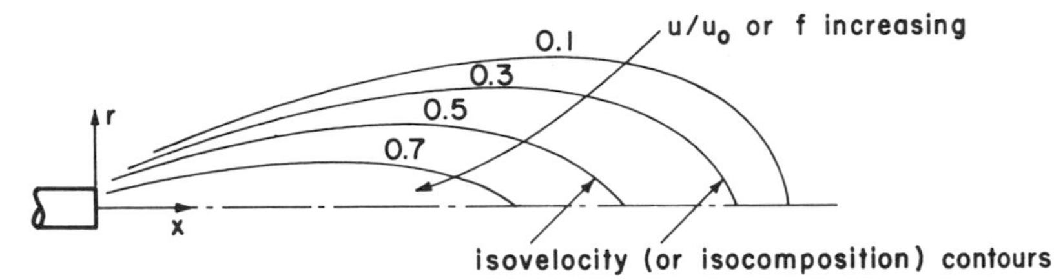

Figure 6.8 Isovelocity or isocomposition contours of a laminar jet.

Curves obeying Eq. (6-71), for fixed values of u/u_0 (or f), have the form shown in Fig. 6.8.

Since there is exact similarity between the processes of mass transfer and momentum transfer, we can write

$$f = \frac{u}{u_0} \tag{6-72}$$

If we further assume that Le = 1 and consider the energy equation, we find that the temperature field is related to the concentration field by

$$\frac{T - T_\infty}{T_0 - T_\infty} = f \tag{6-73}$$

Thus, the curves in Fig. 6.8 can also be interpreted as isotherms.

If the Schmidt number is not equal to unity but equal to some constant, then one can show that

$$\frac{f}{f_{\mathbb{C}}} = \left(\frac{u}{u_{\mathbb{C}}}\right)^{\mathrm{Sc}} \tag{6-74}$$

This equation implies that for Sc < 1, the concentration profile is broader than the velocity profile. It should also be noted that, under this condition, the centerline mixture fraction $f_{\mathbb{C}}$ is smaller than that with Sc = 1. Furthermore, the $f_{\mathbb{C}}$ for Sc ≠ 1 no longer obeys Eq. (6-60).

It is worthwhile to point out that when μ, ρ, and $\mathscr{D}$ vary with f and T, the analytical solution of the governing equations ceases to be possible, and we must adopt numerical methods. Under these circumstances, the quantitative values will be different from those with constant transport properties; however, no qualitative change in the behavior of the jet is anticipated.

The buoyancy effect may be important in some cases. For upward vertical jets with the density of the injected fluid lower than that of the surroundings, buoyancy effects can cause the momentum flux in the jet to increase with vertical distance. Also, when buoyancy forces have components at right angles to the jet axis, the flow will no longer be axisymmetric. The solutions of these problems are more complex and one usually has to rely upon numerical methods.

3.2 Laminar Jet with Chemical Reactions

In many industrial furnaces, flares, and combustors of propulsive devices, one needs to predict the shape and structure of the diffusion flame resulting from the injection of fuel gases into an oxidizing atmosphere. In order to achieve good predictive ability, the laminar diffusion flame jets must be modeled. Let us consider a steady, axisymmetric, vertical laminar jet with low Mach number but fast chemical reaction rates. We further assume that the pressure is uniform and the buoyancy effect is negligible. Under these specified conditions, the governing differential equations are given as follows:

Continuity (overall) equation:

$$\frac{\partial}{\partial x}(\rho u r) + \frac{\partial}{\partial r}(\rho v r) = 0 \tag{6-75}$$

Axial-momentum conservation equation:

$$\frac{\partial}{\partial x}(\rho u r u) + \frac{\partial}{\partial r}(\rho v r u) = \frac{\partial}{\partial r}\left(\mu r \frac{\partial u}{\partial r}\right) \tag{6-76}$$

Fuel species continuity equation:

$$\frac{\partial}{\partial x}(\rho u r Y_F) + \frac{\partial}{\partial r}(\rho v r Y_F) = \frac{\partial}{\partial r}\left(\rho \mathscr{D}_F r \frac{\partial Y_F}{\partial r}\right) + r\omega_F \tag{6-77}$$

Oxidant species continuity equation:

$$\frac{\partial}{\partial x}(\rho u r Y_O) + \frac{\partial}{\partial r}(\rho v r Y_O) = \frac{\partial}{\partial r}\left(\rho \mathscr{D}_O r \frac{\partial Y_O}{\partial r}\right) + r\omega_O \tag{6-78}$$

Energy-conservation equation:

$$\frac{\partial}{\partial x}(\rho u r h) + \frac{\partial}{\partial r}(\rho v r h) = \frac{\partial}{\partial r}\left(\lambda r \frac{\partial T}{\partial r}\right) + (\Delta H_r)\frac{\partial}{\partial r}\left(\rho \mathscr{D}_F r \frac{\partial Y_F}{\partial r}\right) \tag{6-79}$$

where

$$h \equiv C_p T + Y_F \Delta H_r \tag{6-80}$$

and ΔH_r is the heat of reaction per mass of fuel.

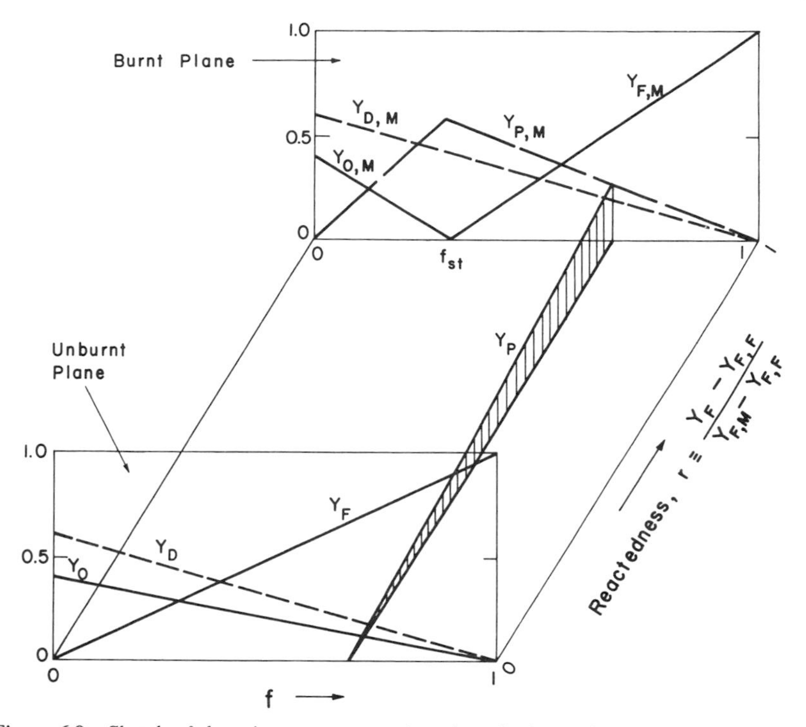

Figure 6.9 Sketch of the mixture state as a function of mixture fraction and reactedness.

To facilitate our discussion and theoretical solution, let us define a term introduced initially by Spalding[3] for many combustion problems: in a *simple chemically reacting system* (SCRS), the pure fuel and pure oxidant are imagined to always unite in fixed proportions, the specific heats of all components are equal, and the transport properties are equal at any point in the mixture but need not be uniform. For such systems, a sketch of mixture state as a function of mixture fraction and reactedness is shown in Fig. 6.9. The symbols $Y_{F,F}$ and $Y_{F,M}$ in Fig. 6.9 represent the mass fractions of fuel in the fuel stream and in the burned mixture stream respectively.

If the fuel and the oxidizer are presumed to form a SCRS, one can relate the rate of production of fuel to that of oxidizer by

$$\omega_F = \left(\frac{F}{O}\right)_{\text{st}} \omega_O \tag{6-81}$$

Since $\nu = \mathscr{D} = \alpha$ for a SCRS, so that Sc = Pr = Le = 1, then Eqs. (6-77) and (6-78) can be combined to form

$$\frac{\partial}{\partial x}(\rho u r \zeta) + \frac{\partial}{\partial r}(\rho v r \zeta) = \frac{\partial}{\partial r}\left(\mu r \frac{\partial \zeta}{\partial r}\right) \tag{6-82}$$

where

$$\zeta \equiv Y_F - \left(\frac{F}{O}\right)_{\text{st}} Y_O \tag{6-83}$$

Since ζ is a conserved property and related to the mixture fraction by Eq. (1-120), the mixture fraction f also satisfies Eq. (6-82). Namely,

$$\frac{\partial}{\partial x}(\rho u r f) + \frac{\partial}{\partial r}(\rho v r f) = \frac{\partial}{\partial r}\left(\mu r \frac{\partial f}{\partial r}\right) \tag{6-84}$$

One can also show that Eq. (6-79) can be reduced to

$$\frac{\partial}{\partial x}(\rho u r h) + \frac{\partial}{\partial r}(\rho v r h) = \frac{\partial}{\partial r}\left(\mu r \frac{\partial h}{\partial r}\right) \tag{6-85}$$

The similarity of the equations for u, f, and h permits a single form of solution to serve for all three equations. The solutions for the set of governing equations with uniform properties (μ = const, ρ = const′) are identical to those of inert laminar jet as discussed in Section 3.1.

The far-field solution, which satisfies the boundary conditions at large r

$$r \to \infty: \quad \begin{cases} u = 0 \\ f = 0 \\ h = h_\infty \end{cases} \tag{6-86}$$

can be written as

$$\frac{ux}{I_u} = \frac{fx}{I_f} = \frac{(h - h_\infty)x}{I_h} = \frac{3}{4}\frac{1}{\left(1 + \frac{1}{4}\xi^2\right)^2} \tag{6-87}$$

where ξ is defined as before by Eq. (6-59) and the third jet invariant I_h is defined by

$$I_h \equiv \frac{1}{\nu}\int_0^\infty u(h - h_\infty) r\, dr = \frac{u_0(h_0 - h_\infty) r_0^2}{2\nu} \tag{6-88}$$

In terms of the Reynolds number $\text{Re}_{d_0} (\equiv u_0 d_0/\nu)$ of the jet at the exit station, Eq. (6-87) can be written as

$$\frac{ux}{u_0 r_0}\frac{1}{\text{Re}_{d_0}} = \frac{fx}{r_0}\frac{1}{\text{Re}_{d_0}} = \frac{h - h_\infty}{h_0 - h_\infty}\frac{x}{r_0}\frac{1}{\text{Re}_{d_0}}$$

$$= \frac{\frac{3}{16}}{\left(1 + \frac{3}{256}\text{Re}_{d_0}^2 r^2/x^2\right)^2} \tag{6-89}$$

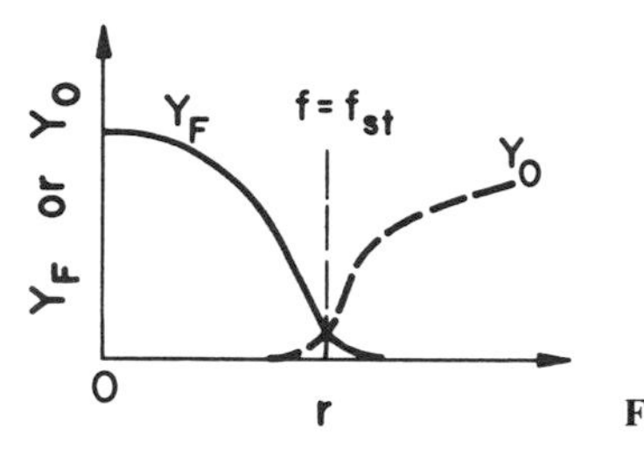

Figure 6.10 Radial distribution of mass fractions.

By examining Eqs. (6-87), (6-56), and (6-58), one can see that there is no primary effect of combustion on the distribution of velocity and of conserved properties such as f and h. The secondary effects are caused mainly by the influence of temperature on density and transport properties. It can be seen from Eq. (6-89) that the centerline values of velocity, mixture fraction, and enthalpy difference $h_{\mathbb{C}} - h_\infty$ fall off as the reciprocal of x. The radial profiles fall asymptotically to zero as r/x increases.

If the reaction rates are infinitely fast, fuel and oxidizer cannot be simultaneously present in finite concentration; then the reaction zone is a surface (no overlapping region for Y_F and Y_O). The surface is located at

$$Y_F - \left(\frac{F}{O}\right)_{\text{st}} Y_O = 0 \tag{6-90}$$

or

$$f = f_{\text{st}} = \frac{(F/O)_{\text{st}} Y_{O,A}}{1 + (F/O)_{\text{st}} Y_{O,A}} \tag{6-90a}$$

In practice, the overlapping region for many hydrocarbon fuels and air systems is very small, as shown in Fig. 6.10. The radius of the flame at any particular x can be found by inserting f_{st} into Eq. (6-89) and rearranging; the result is

$$\frac{r_{\text{flame}}}{x} = \frac{16}{\sqrt{3}\,\text{Re}_{d_0}} \sqrt{\sqrt{\frac{3}{16}\frac{\text{Re}_{d_0} r_0}{x f_{\text{st}}}} - 1} \tag{6-91}$$

and is sketched in Fig. 6.11. The flame length can be obtained by equating

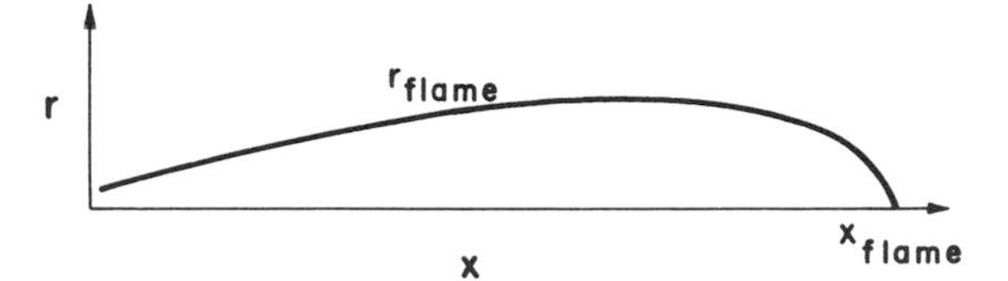

Figure 6.11 Contour of a laminar diffusion flame.

r_{flame} to zero in Eq. (6-91):

$$\frac{x_{\text{flame}}}{r_0} = \frac{3}{16} \frac{\text{Re}_{d_0}}{f_{\text{st}}} \tag{6-92}$$

$$x_{\text{flame}} = \frac{3}{8} \frac{u_0 r_0^2}{f_{\text{st}} \nu} = \frac{3}{8\pi} \frac{1}{f_{\text{st}}} \frac{\pi r_0^2 u_0}{\nu}$$

$$\propto \frac{\text{volumetric flow rate}}{\mathscr{D}} \tag{6-92a}$$

Again, we see that the flame height is proportional to the volumetric flow rate and inversely proportional to the mass diffusivity. Expressing f_{st} in terms of $Y_{O,A}$, Eq. (6-92) becomes

$$\frac{x_{\text{flame}}}{r_0} = \tfrac{3}{16} \text{Re}_{d_0} \left[1 + \frac{1}{(F/O)_{\text{st}} Y_{O,A}} \right] \tag{6-93}$$

This equation says that as $Y_{O,A}$ (or Y_{O_∞}) decreases, the flame length increases, which implies that fuel has farther to go to find its oxidizer. Short flame lengths are often preferred in rocket motors; to obtain them, Eq. (6-93) suggests the use of a small injector radius. This is why many liquid-fuel rocket-engine combustors have numerous tiny fuel-injector nozzles. In industrial furnaces, on the other hand, long flames are often desired for uniformity of heat transfer over the furnace length; and so the fuel is often supplied through a large single nozzle.

4 THE BURNING OF A SINGLE DROPLET IN A QUIESCENT ATMOSPHERE

In the burning of a single fuel droplet in an oxidizing atmosphere, fuel is evaporated from the liquid surface and diffuses to the flame front, while the oxygen moves from the surroundings to the flame front. The shape of the envelope flame can be either spherical or nonspherical (Fig. 6.12). Nonspherical flames are generally caused by the relative motion between the surrounding gases and droplet and by convection effects. When the droplet is very small, it can be entrained easily by the surrounding gases and the relative velocity between the droplet and nearby gases becomes small; then the diffusion flame surrounding the droplet becomes nearly spherical.

The rate at which the droplet evaporates and burns is generally considered to be determined by the rate of heat transfer from the flame front to the fuel surface. In the case of gaseous diffusion flames, chemical processes are assumed to occur so rapidly that the burning rates are determined solely by mass and heat transfer rates. Most theoretical models describing such a burning process

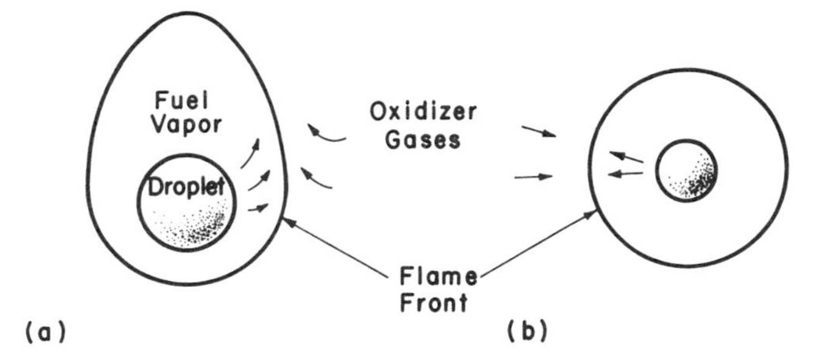

Figure 6.12 Shapes of diffusion flames surrounding a burning spherical fuel droplet: (*a*) nonspherical, (*b*) spherical.

consider a *double-film model* for the combustion of a single droplet. One film separates the droplet surface from the flame front, and the other separates the flame front from the surrounding oxidizer (see Fig. 6.13).

In most analytical developments, the liquid surface is assumed to be at the *normal boiling point* temperature of the fuel. Surveys of the temperature fields in burning liquids indicate that in fact the temperature is only a few degrees below the boiling temperature.[3,5] In the following analysis, the temperature distribution within the droplet is assumed to be uniform and is slightly below the normal boiling point. In *film I*, heat is conducted from the flame front to the liquid, where it vaporizes the fuel. Most analyses assume that the fuel is heated to the flame temperature before it chemically reacts and that the fuel does not react until it reaches the flame front. Most early investigators did not realize that it was *not* necessary to determine T_f in order to obtain the droplet burning rate. The value of T_f can, however, be determined easily by assuming an infinitely thin reaction zone at the stoichiometric position. In *film II*, oxygen diffuses to the flame front, and combustion products and heat are transported to the surrounding atmosphere. The position of the boundary,

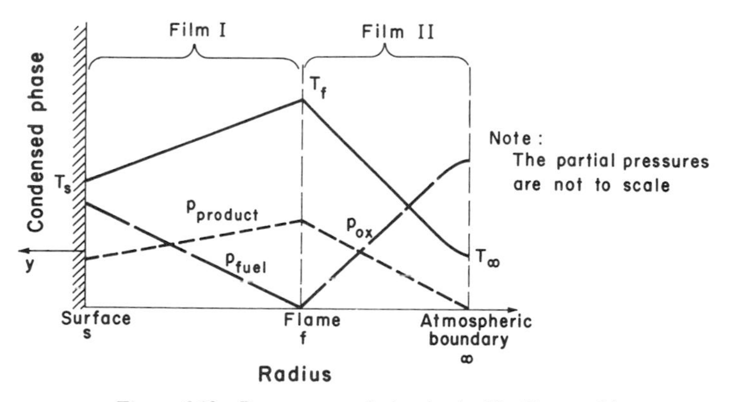

Figure 6.13 Parameter variation in double-film model.

designated by ∞, is determined by convection. A stagnant atmosphere places the boundary at an infinite distance from the fuel surface.

In droplet burning, there are several important questions to consider. How long is the lifetime of a droplet? How does one calculate the rate of consumption of a fuel droplet? How does the burning rate of a droplet depend upon the heat of reaction and flame temperature? What are the effects of relative motion between the droplet and surrounding gases? What is the effect of buoyancy? Is there an internal recirculation within the droplet? What is the effect of ambient pressure on the lifetime of droplet and other parameters? Three parameters are generally evaluated to answer a part of the above questions: the mass burning rate, the flame-front location, and the flame temperature. The most important parameter is the mass burning rate, since it depicts the heat release rate in a combustor. It also permits the evaluation of the so-called evaporation coefficient, which is most readily measured experimentally.

The use of the term "evaporation coefficient" comes about from mass- and heat-transfer experiments on evaporation without combustion, as generally used in spray drying and humidification. Basically the *evaporation coefficient* β_v is defined by the following d^2 evaporation law, which has been verified experimentally:

$$d^2 = d_0^2 - \beta_v t \tag{6-94}$$

where d_0 is the original drop diameter, and d the drop diameter after time t. It will be shown later that the same expression has been found to hold true for mass and heat transfer with chemical reaction.

Here we concern ourselves with the burning of droplets, but the concepts to be used are just as applicable to the evaporation of liquids, sublimination of solids, hybrid burning rates, ablation heat transfer, transpiration cooling, and so on. In all cases, we are interested in the mass consumption, or the rate of regression of condensed material, whereas in gaseous diffusion flames we had no specific property to measure. The condensed phase must be gasified, and consequently there must be an energy input into the condensed material. The heat flux at the surface determines the rate of regression. Thus, we may write

$$q_s = \dot{r}\rho_{\text{liq}} Q \tag{6-95}$$

where

q_s is the heat flux to the surface (Q/tL^2).

$\dot{r}$ is the regression rate (L/t).

Q is the sum of two terms representing the heat of vaporization, sublimation, or gasification and the enthalpy required to bring the material from its initial temperature to the temperature of vaporization, sublimation, or

gasification (Q/M), namely,

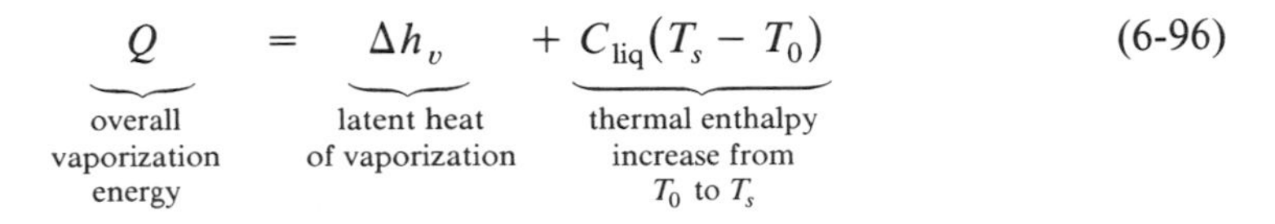

$$\underbrace{Q}_{\substack{\text{overall}\\ \text{vaporization}\\ \text{energy}}} = \underbrace{\Delta h_v}_{\substack{\text{latent heat}\\ \text{of vaporization}}} + \underbrace{C_{\text{liq}}(T_s - T_0)}_{\substack{\text{thermal enthalpy}\\ \text{increase from}\\ T_0 \text{ to } T_s}} \tag{6-96}$$

The heat flux at the surface (assuming no radiation) is

$$q_s = -\lambda_g \left(\frac{\partial T}{\partial y} \right)_s \tag{6-97}$$

where y is the distance below the droplet surface. This same equation holds true even if there is mass evolution from the surface and even if convection effects prevail. For convective cases, one can write the surface heat flux in terms of the heat-transfer coefficient $\bar{h}_c$, averaged over the whole surface of the sphere, as

$$q_s = \bar{h}_c (T_\infty - T_s) \tag{6-98}$$

By equating Eqs. (6-97) and (6-98), we have

$$q_s = -\lambda_g \left(\frac{\partial T}{\partial y} \right)_s = \bar{h}_c (T_\infty - T_s) \tag{6-99}$$

The heat-transfer coefficient is often expressed in terms of Nusselt number

$$\overline{\text{Nu}} \equiv \frac{\bar{h}_c d_0}{\lambda_g} \tag{6-100}$$

where d_0 is the initial diameter of the droplet. The Nusselt number can be considered as a dimensionless heat-transfer coefficient; it is usually correlated with the Reynolds number and Prandtl number as

$$\overline{\text{Nu}} = f(\text{Re}, \text{Pr}) \tag{6-101}$$

A specific form[6] suitable for spherical droplets is

$$\overline{\text{Nu}}_d = \frac{\bar{h}_c d}{\lambda_g} = 2 + 0.6 \left[\frac{|U_g - U_p| d}{\nu_g} \right]^{1/2} \left(\frac{\nu_g}{\alpha_g} \right)^{1/3} \tag{6-102}$$

where $U_g - U_p$ is the relative velocity between the nearby gases and the fuel particle.

When radiation becomes important, the surface heat flux can be written as the sum of two terms:

$$q_s = -\lambda_g \left(\frac{\partial T}{\partial y} \right)_s + q_{\text{rad}} \tag{6-103}$$

It is important to note that the mathematical procedure for obtaining the solution of the problem does not become much more difficult, since the radiation effects are only involved in the boundary condition and the differential equations describing the processes are not altered.

4.1 Evaporation of a Single Fuel Droplet

We shall follow Spalding's initial approach[7,8] and treat a single fuel droplet in a nonconvective atmosphere of a given temperature and pressure. A *quasi-steady assumption* is used in the following analysis; it implies that the droplet is of fixed size. One can imagine that the droplet evaporates so slowly that it can be replaced by a porous sphere into which fuel is being fed at a rate equal to the mass evaporation rate, so that the surface of the sphere is always wet. The steady-state species and energy equations for the single droplet can then be written as:

Fuel species continuity equation:

$$r^2 \rho v \frac{dY_F}{dr} = \frac{d}{dr}\left(r^2 \mathscr{D} \rho \frac{dY_F}{dr} \right) + r^2 \dot{\omega}_F \tag{6-104}$$

Energy equation:

$$r^2 \rho v \frac{dC_p T}{dr} = \frac{d}{dr}\left(\frac{\lambda}{C_p} r^2 \frac{dC_p T}{dr} \right) + r^2 \dot{Q} \tag{6-105}$$

Consider a droplet of a certain radius r_s. If the droplet is vaporizing, then the fluid will leave the surface by convection and diffusion. Since, just below the droplet surface, only fuel (F) exists, the total fuel mass flow rate at the surface is equal to the sum of the fuel mass flow rates due to convection and diffusion:

$$\underbrace{\rho_s v_s}_{\substack{\text{total gaseous}\\ \text{mass flux of}\\ \text{fuel species}\\ \text{leaving}\\ \text{surface}}} = \underbrace{\rho_s (Y_F)_s v_s}_{\substack{\text{mass flux of fuel}\\ \text{species due to}\\ \text{bulk velocity of}\\ \text{mixture at}\\ \text{surface}}} + \underbrace{-\rho_s \mathscr{D}_s \left(\frac{dY_F}{dr} \right)_s}_{\substack{\text{mass flux of}\\ \text{fuel species}\\ \text{due to}\\ \text{mass}\\ \text{diffusion}}} \tag{6-106}$$

After rearranging Eq. (6-106), we have

$$v_s = \frac{\mathscr{D}(dY_F/dr)_s}{(Y_F)_s - 1} \tag{6-107}$$

Spalding defined a new parameter

$$b \equiv \frac{Y_F}{Y_{Fs} - 1} \equiv -\frac{\text{mass fraction of } F \text{ at any location } r}{\text{mass fraction of species other than } F \text{ at surface } r = r_s} \tag{6-108}$$

Then v_s becomes

$$v_s = \mathscr{D}\left(\frac{db}{dr}\right)_s \tag{6-109}$$

For an evaporization process $\dot{\omega}_F = 0$ and the equation (6-104) becomes

$$r^2 \rho v \frac{db}{dr} = \frac{d}{dr}\left(r^2 \mathscr{D} \rho \frac{db}{dr}\right) \tag{6-110}$$

The boundary condition at $r = \infty$ is $Y_F = Y_{F\infty}$, or

$$r = \infty: \qquad b = b_\infty = \frac{Y_{F\infty}}{Y_{Fs} - 1} \tag{6-111}$$

From the integrated continuity equation, we have

$$r^2 \rho v = \text{constant} = r_s^2 \rho_s v_s \tag{6-112}$$

Using Eq. (6-112) and integrating Eq. (6-110) with respect to r, we get

$$r^2 \rho v b = r^2 \rho \mathscr{D} \frac{db}{dr} + \text{constant}$$

Applying the boundary condition (6-109) at the surface, we have

$$r_s^2 \rho_s v_s (b - b_s + 1) = r^2 \rho \mathscr{D} \frac{db}{dr}$$

By separating variables,

$$\frac{r_s^2 \rho_s v_s}{r^2 \rho \mathscr{D}}\, dr = \frac{db}{b - b_s + 1}$$

Taking the product $\rho\mathscr{D}$ to be constant ($= \rho_s\mathscr{D}_s$) and integrating again, we have

$$-\frac{r_s^2 v_s}{r\mathscr{D}_s} = \ln(b - b_s + 1) + \text{constant} \tag{6-113}$$

Using the boundary condition (6-111), we have

$$\frac{r_s^2 v_s}{r\mathscr{D}_s} = \ln\left(\frac{b_\infty - b_s + 1}{b - b_s + 1}\right) \tag{6-114}$$

At $r = r_s$,

$$\frac{r_s v_s}{\mathscr{D}_s} = \ln[(b_\infty - b_s) + 1] \tag{6-115}$$

The difference $b_\infty - b_s$ is called the *Spalding transfer number B*:

$$B \equiv b_\infty - b_s = \frac{Y_{Fs} - Y_{F\infty}}{1 - Y_{Fs}} \tag{6-116}$$

Then Eq. (6-115) becomes

$$r_s v_s = \mathscr{D}_s \ln(1 + B) \tag{6-117}$$

The mass flow rate per unit area is usually called G_F:

$$G_F \equiv \frac{\dot{m}_F}{4\pi r_s^2} = \rho_s v_s \tag{6-118}$$

The fuel evaporation-rate expression (6-117) can be rearranged to give

$$\boxed{G_F = \frac{\dot{m}_F}{4\pi r_s^2} = \rho_s \mathscr{D}_s \frac{\ln(1 + B)}{r_s}} \tag{6-119}$$

In order to calculate the mass evaporation rate from Eq. (6-119), the value of B must be evaluated; before that, Y_F (or p_F) must be determined. A reasonable assumption would be that the gas which surrounds the droplet is saturated by the vapor of the liquid constituting the droplet at the surface temperature. Thus the problem now becomes to determine T_s, since vapor-pressure data are available.

In order to determine the surface temperature T_s, we must consider the energy equation. For the case of evaporation (no combustion), $\dot{Q} = 0$ and

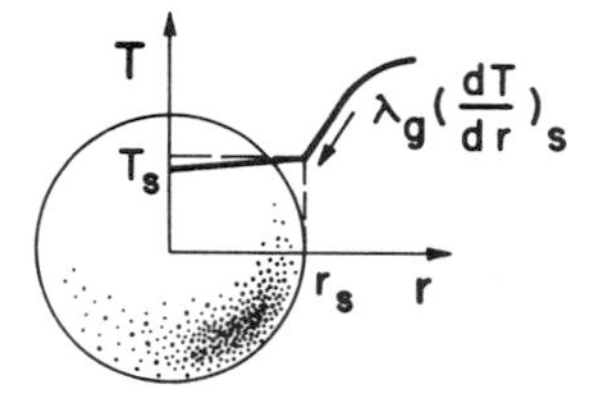

Figure 6.14 Temperature distribution of an evaporating liquid droplet.

Eq. (6-105) become

$$r_s^2 \rho_s v_s C_p \frac{dT}{dr} = \frac{d}{dr}\left(r^2 \lambda \frac{dT}{dr}\right) \tag{6-120}$$

Integrating the above equation with respect to r, we have

$$r_s^2 \rho_s v_s C_p T = r^2 \lambda \frac{dT}{dr} + \text{constant} \tag{6-121}$$

The boundary condition at the surface is

$$\lambda\left(\frac{dT}{dr}\right)_s = \rho_s v_s \Delta h_v, \qquad T(r = r_s) = T_s \tag{6-122}$$

(see Fig. 6.14), where Δh_v is the latent heat of vaporization at temperature T_s. Since the temperature distribution in the droplet is usually quite uniform, one can neglect the conduction heat flux below the surface. After applying the boundary condition at the droplet surface, we have

$$r_s^2 \rho_s v_s C_p \left(T - T_s + \frac{\Delta h_v}{C_p}\right) = r^2 \lambda \frac{dT}{dr} \tag{6-123}$$

After separating the variables and integrating the above equation, we have

$$-\frac{r_s^2 \rho_s v_s C_p}{r\lambda} = \ln\left(T - T_s + \frac{\Delta h_v}{C_p}\right) + \text{constant}' \tag{6-124}$$

Using the boundary condition that $T \to T_\infty$ as $r \to \infty$, we have

$$\frac{r_s^2 \rho_s v_s C_p}{r\lambda} = \ln\left(\frac{T_\infty - T_s + \Delta h_v/C_p}{T - T_s + \Delta h_v/C_p}\right) \tag{6-125}$$

Setting r equal to r_s at the surface, we have

$$\frac{r_s \rho_s v_s C_p}{\lambda} = \ln\left(\frac{C_p(T_\infty - T_s)}{\Delta h_v} + 1\right) \tag{6-126}$$

Since $\alpha \equiv \lambda/\rho C_p$,

$$r_s v_s = \alpha_s \ln\left[1 + \frac{C_p(T_\infty - T_s)}{\Delta h_v}\right] \equiv \alpha_s \ln[1 + B_T] \tag{6-127}$$

Comparing Eq. (6-127) with Eq. (6-117), we have

$$r_s v_s = \alpha_s \ln\left[1 + \frac{C_p(T_\infty - T_s)}{\Delta h_v}\right] = \mathscr{D}_s \ln\left[1 + \frac{Y_{F\infty} - Y_{Fs}}{Y_{Fs} - 1}\right]$$

or

$$r_s v_s = \alpha_s \ln[1 + B_T] = \mathscr{D}_s \ln[1 + B_M] \tag{6-128}$$

If $\alpha_s = \mathscr{D}_s$ (i.e., Le = 1), then

$$B_T = B_M$$

and

$$\frac{C_p(T_\infty - T_s)}{\Delta h_v} = \frac{Y_{F\infty} - Y_{Fs}}{Y_{Fs} - 1} \tag{6-129}$$

This equation relates the two unknowns T_s and Y_{Fs}; one additional equation is needed to solve for them. The following equation relates Y_F to the partial pressure p_F:

$$Y_F = \frac{\rho_F}{\rho} = \frac{n_F W_F}{nW} = \frac{p_F}{p}\frac{W_F}{W} \tag{6-130}$$

This equation is helpful, since we can further relate the partial pressure of fuel to the surface temperature T_s by the Clausius–Clapeyron vapor-pressure equation discussed in Section 17 of Chapter 1. In a form similar to Eq. (1-216), the Clausius–Clapeyron equation can be written as

$$\frac{d \ln p_F}{dT} = \frac{\Delta h_v}{RT_s^2} \tag{6-131}$$

or

$$\ln \frac{p_F}{p_{F\,\mathrm{ref}}} = \frac{\Delta h_v}{R}\left(\frac{1}{T_{\mathrm{ref}}} - \frac{1}{T_s}\right) \tag{6-132}$$

Now we have three equations [Eqs. (6-129), (6-130), and (6-132)] for three unknowns: Y_{Fs}, T_s, and p_F. Once the solution is obtained, the transfer number B and mass evaporation rate can be calculated from Eq. (6-116) and Eq. (6-119).

The lifetime or evaporation time, t_v, of a liquid droplet can be calculated using Eq. (6-119). The droplet evaporation time is an important parameter in combustion-chamber design, since the lifetime of the largest droplet in a spray determines the minimum time the droplet must be allowed to reside in the combustion chamber. Also, the residence time is related to other design parameters such as the air-stream velocity through the combustor, the velocity of the spray injection, the angle of injection, and the combustor geometry. Using the mass continuity at the droplet surface ($-\rho_l\, dr_s/dt = \rho_s v_s$) and rearranging Eq. (6-119), we have

$$\frac{dr_s}{dt} = -\frac{\rho_s \mathscr{D}_s}{\rho_l r_s}\ln(1+B) \tag{6-133}$$

After integration and expressing the result in terms of droplet diameter, we have

$$d^2 = d_0^2 - \left[\frac{8\rho_s \alpha_s}{\rho_l}\ln(1+B)\right]t \tag{6-134}$$

Comparing Eq. (6-134) with the d^2 evaporation law given by Eq. (6-94), we obtain

$$\beta_v = \frac{8\rho_s \alpha_s}{\rho_l}\ln(1+B) \tag{6-135}$$

where the evaporation coefficient β_v represents the magnitude of the negative slope of the straight line on the d^2–t plot shown in Figure 6.15. The lifetime of the droplet is therefore equal to

$$t_v = \frac{d_0^2}{\beta_v} = \frac{\rho_l d_0^2}{8\rho_s \alpha_s \ln(1+B)} \tag{6-136}$$

This implies that t_v is longer for larger droplets, and also that lighter fuels evaporate faster. The effect of gas density and thermal diffusivity on droplet evaporation time can also be seen clearly from Eq. (6-136). It should be noted that even though the above equation was derived for evaporating droplets, it is applicable to burning droplets. Godsave[11] and many other experimentalists have found in droplet-burning studies[5] that the time variation of the droplet diameter was the same as that for evaporation.

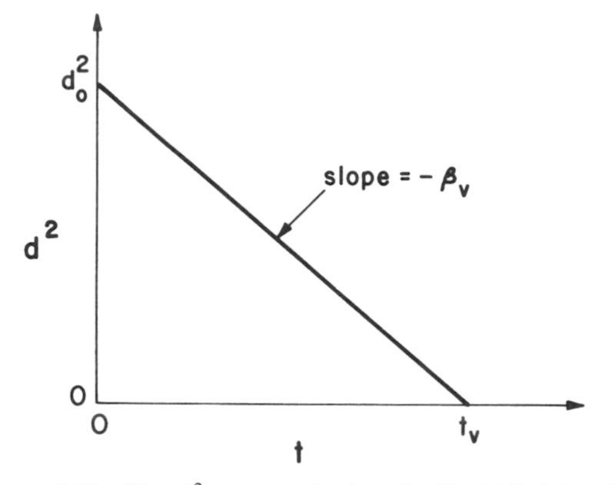

Figure 6.15 The d^2 evaporation law for liquid fuel droplets.

4.2 Mass Burning Rate of a Single Fuel Droplet

In droplet-burning analysis we assume that the fuel and oxidant depletion rates are related in stoichiometric proportions. Consequently, the following relationship holds:

$$\dot{\omega}_F = \dot{\omega}_O (F/O)_{\text{st}} \tag{6-137}$$

$$\dot{\omega}_F \Delta h_{r,F} = \dot{\omega}_O \Delta h_{r,F} (F/O)_{\text{st}} = -\dot{Q} \tag{6-138}$$

where $\Delta h_{r,F}$ is the heat of reaction per unit mass of fuel. If one defines the variables

$$b_{F,O} \equiv \frac{Y_F - Y_O (F/O)_{\text{st}}}{(Y_{Fs} - 1) + Y_{Os} (F/O)_{\text{st}}} \tag{6-139}$$

$$b_{F,T} \equiv \frac{Y_F \Delta h_{r,F} + C_p T}{\Delta h_v + \Delta h_{r,F} (Y_{Fs} - 1)} \tag{6-140}$$

$$b_{O,T} \equiv \frac{Y_O (F/O)_{\text{st}} \Delta h_{r,F} + C_p T}{\Delta h_v + (F/O)_{\text{st}} Y_{Os} \Delta h_{r,F}} \tag{6-141}$$

then the energy and species continuity equations can be combined (assuming Le = 1) to give

$$r^2 \rho v \frac{db}{dr} = \frac{d}{dr}\left(r^2 \rho \mathscr{D} \frac{db}{dr} \right) \tag{6-142}$$

where b can be either $b_{F,O}$ or $b_{F,T}$ or $b_{O,T}$ defined by Eqs. (6-139) through

(6-141). Since the product $\rho\mathscr{D}$ is considered independent of temperature and $r^2\rho v$ = constant, Eq. (6-142) may be integrated twice with the boundary conditions

$$r = r_s: \qquad v_s = \mathscr{D}\left(\frac{db}{dr}\right)_s \tag{6-143}$$

$$r \to \infty: \qquad b = b_\infty \tag{6-144}$$

to obtain the solution

$$\frac{r^2\rho v}{r\mathscr{D}\rho} = \ln\left(\frac{b_\infty - b_s + 1}{b - b_s + 1}\right) \tag{6-145}$$

At $r = r_s$

$$\frac{r_s^2\rho_s v_s}{(\mathscr{D}_s\rho_s)r_s} = \ln(b_\infty - b_s + 1) = \ln(1 + B)$$

or

$$G_F = \frac{\dot{m}_F}{4\pi r_s^2} = \mathscr{D}_s\rho_s\frac{\ln(1 + B)}{r_s} \tag{6-146}$$

The transfer number B can be obtained from one of the following:

$$B_{F,O} = \frac{(Y_{F\infty} - Y_{Fs}) + (Y_{Os} - Y_{O\infty})(F/O)_{\text{st}}}{(Y_{Fs} - 1) + (F/O)_{\text{st}}(Y_0)_s} = \frac{(F/O)_{\text{st}}Y_{O\infty} + Y_{Fs}}{1 - Y_{Fs}}$$

$$B_{F,T} = \underbrace{\frac{\Delta h_{r,F}(Y_{F\infty} - Y_{Fs}) + C_p(T_\infty - T_s)}{\Delta h_v + \Delta h_{r,F}(Y_{Fs} - 1)}}_{\text{with or without combustion}} = \underbrace{\frac{C_p(T_\infty - T_s) - Y_{Fs}\Delta h_{r,F}}{\Delta h_v + \Delta h_{r,F}(Y_{Fs} - 1)}}_{\text{with combustion}}$$

$$B_{O,T} = \underbrace{\frac{(F/O)_{\text{st}}(Y_{O\infty} - Y_{Os})\Delta h_{r,F} + C_p(T_\infty - T_s)}{\Delta h_v + (F/O)_{\text{st}}Y_{Os}\Delta h_{r,F}}}_{\text{with or without combustion}}$$

$$= \underbrace{\frac{C_p(T_\infty - T_s) + Y_{O\infty}(F/O)_{\text{st}}\Delta h_{r,F}}{\Delta h_v}}_{\text{with combustion}} \tag{6-147}$$

TABLE 6.1 Values of the Transfer Number for Various Condensed Combustible Substances[8–10]

Combustible in Air	B
iso-Octane	6.41
Benzene	5.97
n-Heptane	5.82
Toluene	5.69
Aviation gasoline	≈ 5.5
Automobile gasoline	≈ 5.3
Kerosene	≈ 3.4
Gas oil	≈ 2.5
Light fuel oil	≈ 2.0
Heavy fuel oil	1.7
Carbon	0.12

where for the combustion case $Y_{Os} = Y_{F\infty} = 0$. Because $B_{F,O} = B_{O,T}$, we have

$$\frac{Y_{O\infty}(F/O)_{\mathrm{st}} + Y_{Fs}}{1 - Y_{Fs}} = \frac{\Delta h_{r,F}(F/O)_{\mathrm{st}} Y_{O\infty} + C_p(T_\infty - T_s)}{\Delta h_v} \tag{6-148}$$

The unknowns in this equation are T_s and Y_{Fs}. These two unknowns together with the vapor pressure of the fuel can be calculated by the same method outlined above for a noncombusting liquid droplet. Once these parameters are known, $\dot{m}_F$ can be determined from Eq. (6-146).

Values of B taken from both Spalding[8] (1955) and Blackshear[9] (1960) for various condensed combustible substances are given in Table 6.1. The most convenient form of B is $B_{O,T}$, which can be expressed as

$$B_{O,T} = \frac{C_p(T_\infty - T_s) + (F/O)_{\mathrm{st}} Y_{O\infty}\,\Delta h_{r,F}}{\Delta h_v} \approx \frac{Y_{O\infty}\,\Delta h_{r,F}}{\Delta h_v}(F/O)_{\mathrm{st}} \tag{6-149}$$

As soon as $Y_{O\infty}$ is known, $B_{O,T}$ can be calculated from Eq. (6-149) and the burning rate of the droplet can be determined from Eq. (6-146). In general, the surface temperature has less effect on the burning rate than the mass fractions Y_{Fs} and $Y_{O\infty}$ have. It is interesting to note that for most hydrocarbon fuels, B does not change too much and is in the range of $1.7 < B < 5.5$. It can be seen clearly from Eq. (6-146) that the controlling term for the difference in mass burning rates between various fuels is due to the magnitude of the product $\mathscr{D}_s \rho_s$. Since $1 + B$ is inside a natural logarithm, a tenfold variation in B results only in an approximately twofold variation in burning rate. Therefore, the $\Delta h_{r,F}$ does not play an important role in determining the mass consumption rate.

If it is assumed that the flame exists at the position where $Y_O = Y_F = 0$, then the temperature of the flame and its position can be determined. However, it is important to show that the burning rate can be determined *without* knowing where the flame front is. Equation (6-145) may be written in the following form:

$$\frac{r^2\rho v}{\mathscr{D}\rho r} = \frac{r_s^2\rho_s v_s}{\mathscr{D}_s\rho_s r} = \ln\left[\frac{Y_{F\infty} - Y_{Fs} - (F/O)_{st}(Y_{O\infty} - Y_{Os}) + (Y_{Fs} - 1)}{Y_F - Y_{Fs} - (F/O)_{st}(Y_O - Y_{Os}) + (Y_{Fs} - 1)}\right] \tag{6-150}$$

At the flame surface $Y_F = Y_O = 0$ and $Y_{F\infty} = Y_{Os} = 0$; thus Eq. (6-150) becomes

$$\frac{r_s^2\rho_s v_s}{\mathscr{D}_s\rho_s r_{\text{stoich}}} = \ln[1 + (F/O)_{st}Y_{O\infty}] \tag{6-151}$$

Since $4\pi r_s^2\rho_s v_s = \dot{m}_F$ is known, r_{stoich} can be solved for. The flame temperature at r_{stoich} can be obtained by writing Eq. (6-145) with $b = b_{O,T}$ and making use of Eq. (6-151). It is interesting to note that the flame temperature T_f can be *higher* than the adiabatic flame temperature, since the nitrogen in film I is preheated.

5 FUEL DROPLET IN A CONVECTIVE STREAM

When the droplets are not at rest relative to the oxidizing atmosphere, the results for quiescent flow will no longer hold true and effects of forced convection must be considered; especially when $\text{Re}_d > 20$, there exists a boundary-layer flow region around the front of the sphere and a wake region behind the sphere. The boundary condition at the droplet surface can be expressed as

$$\bar{h}_c(\Delta T) = \rho_s v_s \Delta h_v = \frac{\dot{m}_f}{4\pi r_s^2}\Delta h_v = \frac{\mathscr{D}_s\rho_s}{r_s}\Delta h_v \ln(1 + B) \tag{6-152}$$

where

$$\Delta T \equiv T_\infty + \frac{\psi Y_{O\infty}\Delta h_r}{C_p} - T_s \tag{6-153}$$

and

$$\psi \equiv (F/O)_{st} \tag{6-154}$$

The thermal driving potential in Eq. (6-153) is expressed as the ambient temperature T_∞ plus the rise in temperature due to the chemical-energy release minus the surface temperature.

Substituting the ΔT expression of Eq. (6-153) into Eq. (6-152), we have

$$\bar{h}_c \frac{\psi Y_{O\infty}\,\Delta h_r + C_p(T_\infty - T_s)}{C_p} = \frac{\mathscr{D}_s \rho_s}{r_s}\,\Delta h_v \ln(1 + B)$$

$$= \frac{\lambda_s}{C_p r_s}\,\Delta h_v \ln(1 + B) \qquad (6\text{-}155)$$

After rearranging, we have

$$\frac{\bar{h}_c r_s}{\lambda_s} = \left[\frac{\psi Y_{O\infty}\,\Delta h_r + C_p(T_\infty - T_s)}{\Delta h_v}\right]^{-1} \ln(1 + B) \qquad (6\text{-}156)$$

Recalling the definition of B and the Nussett number, the above equation becomes

$$\mathrm{Nu} \equiv \frac{\bar{h}_c r_s}{\lambda_s} = \frac{\ln(1 + B)}{B} \qquad (6\text{-}157)$$

For problems with high Reynolds numbers, the quantity $[\ln(1 + B)]/B$ has been commonly used as an empirical correction factor, and a classical expression[10] for Nu with mass transfer is

$$\mathrm{Nu}_{r_s} = \frac{\ln(1 + B)}{B}\left[1 + 0.39\,\mathrm{Pr}^{1/3}\mathrm{Re}_{r_s}^{1/2}\right] \qquad (6\text{-}158)$$

It is interesting to note that as Re_{r_s} approaches 0, Eq. (6-158) reduces to Eq. (6-157). Also, if $\mathrm{Pr} = 1$ and $\mathrm{Re}_{r_s} \gg 1$, Eq. (6-158) becomes

$$\mathrm{Nu}_{r_s} = 0.39\left[\frac{\ln(1 + B)}{B}\right]\mathrm{Re}_{r_s}^{1/2} \qquad (6\text{-}159)$$

Using Eqs. (6-152), (6-153), and (6-147), we have

$$\frac{\bar{h}_c r_s}{\lambda_s} = \frac{\bar{h}_c \Delta T\, r_s}{\Delta T\, \lambda_s} = \frac{\rho_s v_s\,\Delta h_v\, r_s C_p \mu}{\Delta T\, \lambda_s C_p \mu}$$

$$= \frac{\rho_s v_s r_s}{\mu}\,\frac{\Delta h_v}{\psi Y_{O\infty}\,\Delta h_r + C_p(T_\infty - T_s)} = \frac{\rho_s v_s r_s}{\mu}\,\frac{1}{B}$$

Hence, Eq. (6-159) becomes

$$\frac{\rho_s v_s r_s}{\mu} = 0.39\,\mathrm{Re}_{r_s}^{1/2} \ln(1 + B) \tag{6-160}$$

Due to the fact that a wake region exists behind the droplet, this equation, deduced from Eq. (6-158), is not likely to give accurate quantitative predictions. However, it has been found that after multiplying the left-hand side of the above equation by $B^{0.15}$, the following two dimensionless parameters can often be used to correlate experimental data:

$$\left(\frac{\rho_s v_s r_s}{\mu}\right)\frac{B^{0.15}}{\ln(1 + B)} \quad \text{and} \quad \mathrm{Re}_{r_s}^{1/2}$$

The above relationship implies that droplet burning rate under a laminar convective flow condition follows a $d^{3/2}$ burning-rate law. As we shall discuss in a later section, droplet burning follows a power law in the diameter with exponent on the order of one in turbulent flows.

6 SUPERCRITICAL BURNING OF LIQUID DROPLETS IN A STAGNANT ENVIRONMENT

Quasisteady theories of particle combustion become inaccurate at high gas pressures, and are nearly useless above the critical pressure. The critical pressure and temperature are shown in a phase diagram in Fig. 6.16. There are two major reasons why quasisteady equations are inappropriate at the high pressures prevailing in some jet engines or combustors. First, as the pressure approaches the critical pressure of the injected fuel, the latent heat of vaporization (Δh_v) decreases to zero; the quasisteady equations (6-149) and (6-119) correspondingly predict a vaporization rate which rises to infinity. Second, the gaseous mass in the region influenced by the droplet increases with pressure; therefore the transient term in the energy equation is no longer negligible.

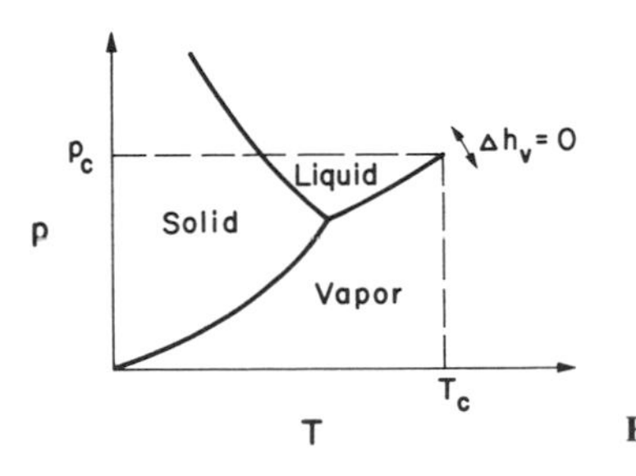

Figure 6.16 Phase diagram of a liquid fuel.

The transient burning of a composite solid propellant was studied by Summerfield.[12] Extending Summerfield's approach to transient-burning problems, Spalding[13] developed a theory to predict the rate of combustion of liquid droplets under high pressure. The basic equations used by Spalding were based on the conservation of elements. Rosner[14] and Dominicis[15] modified Spalding's analysis and used conservation of chemical compounds. The basic model consists of a pocket of fuel vapor surrounded by a spherical flame zone. Outside the flame zone is an environment containing an oxidant. Fuel diffuses outward to the flame zone, and oxidant diffuses inward to the flame, where they react instantly to form combustion products.

The basic assumptions of this model are:

1. The fluid properties are constant.
2. The mass diffusivities of all species are identical, and only concentration diffusion is present in the system.
3. The radial bulk velocity is zero.
4. The combustion is a single one-step reaction ($\nu_F M_F + \nu_O M_O \rightarrow \nu_P M_P$).
5. The reaction takes place in a thin flame zone.

The conservation of species i is given by

$$\rho \frac{\partial Y_i}{\partial t} = \omega_i - [\nabla \cdot (\rho Y_i \mathbf{V}_i)] \tag{6-161}$$

The constitutive diffusion relation is

$$\nabla X_i = \sum_{j=1}^{N} \frac{X_i X_j}{\mathscr{D}_{ij}} (\mathbf{V}_j - \mathbf{V}_i) \tag{6-162}$$

The phenomenological chemical-kinetic expression is

$$\omega_i = W_i(\nu_i'' - \nu_i')\omega \tag{6-163}$$

Using assumption 2, Eq. (6-162) can be rearranged to give

$$\mathbf{V}_i = -\mathscr{D} \frac{\nabla Y_i}{Y_i} \tag{6-164}$$

Substituting Eq. (6-164) into Eq. (6-161) gives

$$\rho \frac{\partial Y_i}{\partial t} = \omega_i + \nabla \cdot (\rho \mathscr{D} \nabla Y_i) \tag{6-165}$$

Substituting Eq. (6-163) into Eq. (6-165), we have

$$\rho \frac{\partial Y_i}{\partial t} - \nabla \cdot (\rho \mathscr{D} \nabla Y_i) = W_i(\nu_i'' - \nu_i')\omega \tag{6-166}$$

Defining a new dependent variable

$$\alpha_i \equiv \frac{Y_i}{W_i(\nu_i'' - \nu_i')} \tag{6-167}$$

and substituting α_i for Y_i, Eq. (6-166) becomes

$$\rho \frac{\partial \alpha_i}{\partial t} - \nabla \cdot (\rho \mathscr{D} \nabla \alpha_i) = \omega \tag{6-168}$$

for $i = F$, O, or P.

Let

$$\beta' \equiv \alpha_P - \alpha_F \tag{6-169}$$

which can also be written as

$$\beta' = \frac{Y_P}{\nu_P W_P} + \frac{Y_F}{\nu_F W_F} \tag{6-170}$$

Replacing α_P with β', Eq. (6-168) becomes

$$\rho \frac{\partial \beta'}{\partial t} - \nabla \cdot (\rho \mathscr{D} \nabla \beta') = 0 \tag{6-171}$$

Let us define β_0 as the value of β' for $r \leq r_s$ and $t = 0$; then

$$\beta_0' \equiv \frac{1}{\nu_F W_F} \tag{6-172}$$

Also define β_∞' as the value of β' as r approaches ∞; then

$$\beta_\infty' \equiv \frac{Y_{P\infty}}{\nu_P W_P} \tag{6-173}$$

β_∞' may be nonzero, because the gaseous mixture far from the droplet may have some constant concentration of products, $Y_{P\infty}$.

If we further define a dependent variable

$$\Gamma \equiv \frac{\beta' - \beta'_\infty}{\beta'_0 - \beta'_\infty} \tag{6-174}$$

and substitute Γ for β', Eq. (6-171) becomes

$$\rho \frac{\partial \Gamma}{\partial t} - \nabla \cdot (\rho \mathscr{D} \nabla \Gamma) = 0 \tag{6-175}$$

In spherical coordinates, the above equation can be written as

$$\rho \frac{\partial \Gamma}{\partial t} - \frac{\mathscr{D}\rho}{r^2} \frac{\partial}{\partial r}\left(r^2 \frac{\partial \Gamma}{\partial r} \right) = 0 \tag{6-176}$$

The boundary condition at the far field is

$$r \to \infty: \qquad \Gamma = 0 \tag{6-177}$$

The total amount of fuel in the system is constant; that is, the unreacted and reacted portions together form the overall fuel mass,

$$M_t = \int_0^\infty 4\pi r^2 \rho \left(Y_F + \frac{\nu_F W_F}{\nu_P W_P} Y_P \right) dr \tag{6-178}$$

If we denote M_f as the mass of fuel injected into the system at time $t = 0$, one can show that

$$M_f = \int_0^\infty 4\pi r^2 \rho \Gamma \, dr \tag{6-179}$$

The initial conditions are

$$t = 0 \text{ and } r \leq r_{s0}: \qquad \Gamma = 1 \tag{6-180}$$

$$t = 0 \text{ and } r > r_{s0}: \qquad \Gamma = 0 \tag{6-181}$$

The particular solution for Eq. (6-176) is

$$\Gamma = \frac{M_f}{\rho (4\pi \mathscr{D} t)^{3/2}} \exp\left(\frac{-r^2}{4\mathscr{D} t} \right) \tag{6-182}$$

It should be noted that for times of order greater than $r_{s0}^2/\mathscr{D}$ the fuel distribution is independent of whether the initial mass is concentrated at a point or is finitely distributed. It should also be noted that Eq. (6-182) is unrealistic at very short times.

To determine the location of the spherical flame front, we can set $Y_F = 0$, $Y_O = 0$, and $Y_P = 1$. Also note that at infinity we have

$$Y_{P\infty} + Y_{O\infty} = 1 \tag{6-183}$$

The value of Γ at the flame front can be calculated from Eq. (6-174). It can also be expressed as

$$\Gamma_{\text{fl}} = \frac{Y_{O\infty}}{(\nu_O W_O/\nu_F W_F) + Y_{O\infty}} \tag{6-184}$$

Solving Eq. (6-182) for r and setting $\Gamma = \Gamma_{\text{fl}}$, the flame position is given as a function of time:

$$\frac{r_{\text{fl}}^2}{4\mathscr{D}t} = \ln\left[\frac{M_f}{\rho\Gamma_{\text{fl}}(4\pi\mathscr{D}t)^{3/2}}\right] \tag{6-185}$$

In terms of dimensionless variables, one can define the following two parameters:

$$\phi \equiv \frac{r_{\text{fl}}}{(M_f/\rho\Gamma_{\text{fl}})^{1/3}} \tag{6-186}$$

$$\theta \equiv \frac{4\pi\mathscr{D}t}{(M_f/\rho\Gamma_{\text{fl}})^{2/3}} \tag{6-187}$$

The dimensionless flame radius ϕ is shown in Fig. 6.17 as a function of θ based upon the following equation deduced from Eq. (6-185):

$$\phi = \left[-\left(\frac{3}{2\pi}\right)\theta \ln \theta\right]^{1/2} \tag{6-188}$$

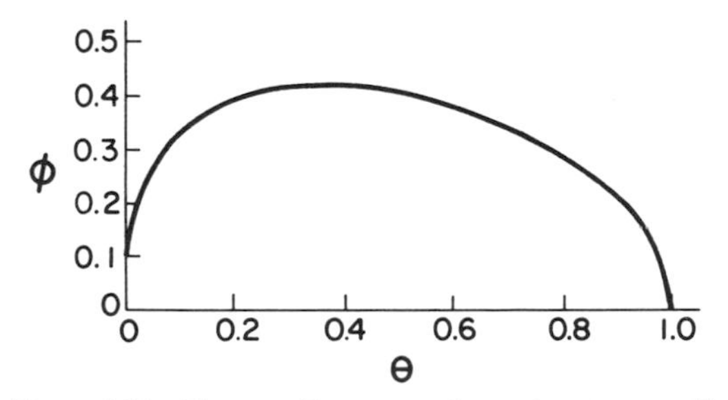

Figure 6.17 Flame radius versus time (after Spalding[13]).

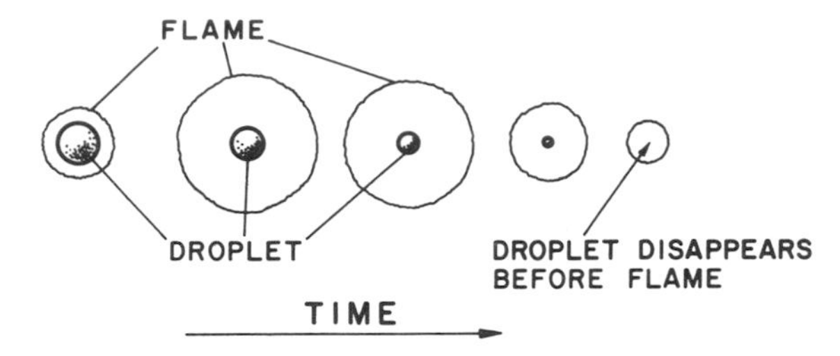

Figure 6.18 Appearance of burning droplets under transient conditions (after Spalding[13]).

The maximum flame radius can be shown to occur at $\theta = 0.368$ with $\phi_{max} = 0.419$. The maximum flame radius can also be expressed in terms of M_f, ρ, and Γ_{fl}, according to the following equation:

$$(r_{fl})_{max} = 0.419\left(\frac{M_f}{\rho\Gamma_{fl}}\right)^{1/3} \tag{6-189}$$

and the time variation of the size of diffusion flame is shown in Fig. 6.18. The total burning time, t_b, can be obtained by setting $r_{fl} = 0$ or $\phi = 0$. Corresponding to $\phi = 0$, θ must be equal to 1 in Eq. (6-188). Substituting $\theta = 1$ into Eq. (6-187), one obtains

$$t_b = \frac{(M_f/\rho\Gamma_{fl})^{2/3}}{4\pi\mathscr{D}} \tag{6-190}$$

To study the pressure dependence of t_b, one can rearrange Eq. (6-190) to give

$$t_b = \frac{\rho^{1/3}(M_f/\Gamma_{fl})^{2/3}}{4\pi(\rho\mathscr{D})} \propto p^{1/3} \tag{6-191}$$

where the product $\rho\mathscr{D}$ is assumed independent of pressure.

The result that the burning time was proportional to $p^{1/3}$, obtained from the modified analyses of Rosner,[14] Dominicis,[15] and Chervinsky,[16] is essentially unchanged from that of Spalding's early theory of high-pressure drop combustion.[13] Faeth et al.[17] conducted supercritical combustion tests at pressures on the order of 100 atm, under zero-gravity conditions in a free-fall apparatus to eliminate convection effects. Their experimental results generally confirm the transient model. The supercritical combustion lifetime of *n*-decane droplets in air as a function of pressure, measured by Faeth et al.,[17] is shown in Fig. 6.19. In this figure a comparison is also made between experimental data and two theoretical formulations due to Slattery and to Hirshfelder et al.[23] The pressure dependence of these two theories appears to be satisfactory within the accuracy of the data.

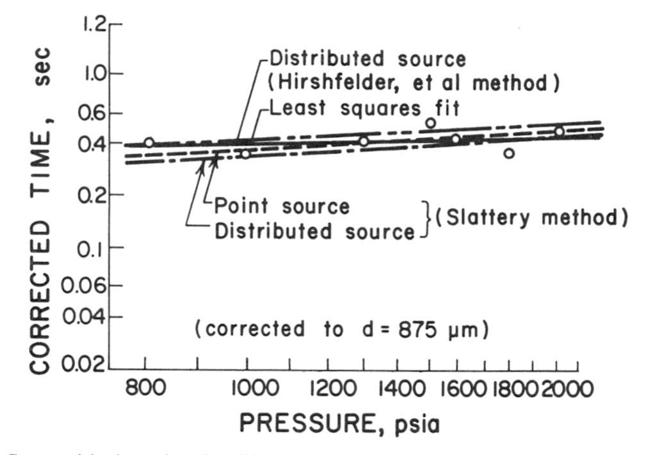

Figure 6.19 Supercritical combustion lifetimes for *n*-decane in air as a function of pressure (after Faeth et al.[17]).

7 EFFECT OF INTERNAL CIRCULATION ON DROPLET VAPORIZATION RATE

In many practical high-pressure combustors, the Reynolds number based on the relative velocity between gas and droplet is on the order of 100 or higher. This implies that the shear stress at the gas–liquid interface can be large enough to induce internal liquid-phase circulation. The liquid motion can then be important in determining the heat and mass transfer processes within the droplet and thereby modify the vaporization rate. Prakash[18-20] and Sirignano[19-20] developed a theoretical model for the prediction of droplet vaporization rate, considering the effect of internal circulation within a droplet.

They demonstrated that boundary layers exist in both the gas and liquid phases near the droplet surface over most of its lifetime, and that convection is an important transport mode within the droplet. They considered a thin boundary layer near the surface of the droplet in the liquid phase. The liquid motion in the core region was approximated by a Hill's spherical vortex as shown in Fig. 6.20*a*. The third liquid-phase region considered in their physical model was an invicid wake near the axis of symmetry shown in Fig. 6.20*b*. The gas-phase motion near the surface was analyzed for three different regions: a front stagnation region, a boundary-layer region, and a separation region. For the boundary layer, the Karman–Pohlhausen integral approach[4] was used. The matching with the front stagnation region was performed by requiring that the gradient of momentum and energy along the droplet surface be continuous. In the separation region, the boundary-layer equations cannot be used. The heat flux in this region was neglected in comparison with the fluxes in the boundary layer.

The gas-phase flow over a vaporizing droplet, rigorously speaking, is unsteady due to the temporal change in the size of the droplet. However, the

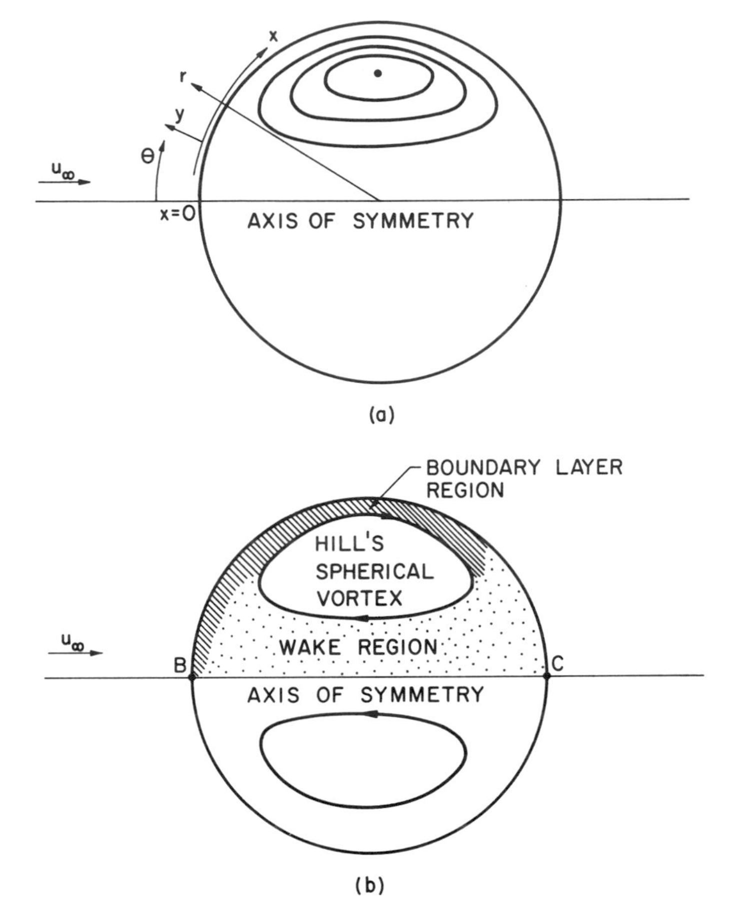

Figure 6.20 (*a*) Hill's spherical vortex and orthogonal boundary-layer coordinate system used by Prakash and Sirignano.[19-20] (*b*) Various regions considered in Prakash and Sirignano's model.[18-20]

characteristic time for changes in the gas phase is the residence time in the neighborhood of the droplet, and the time is usually much shorter than the droplet lifetime. Therefore, the quasisteady gas-phase assumption was employed in Prakash and Sirignano's model. It should be noted that although the conditions in many combustors are turbulent, a locally laminar boundary layer exists over the spherical droplet surface, since the typical size of droplet is much smaller than a typical large eddy size. This is why laminar gas-phase equations are considered in their model. (The reader is referred to the original publications[18-20] for complete formulation and solutions.)

In brief, for evaluating the heating of the droplet, the thermal-boundary-layer equations were solved in a way similar to the viscous-boundary-layer equations. The heating of the core has been shown to be essentially normal to the closed streamlines of the Hill's vortex, and the heating process is unsteady. In

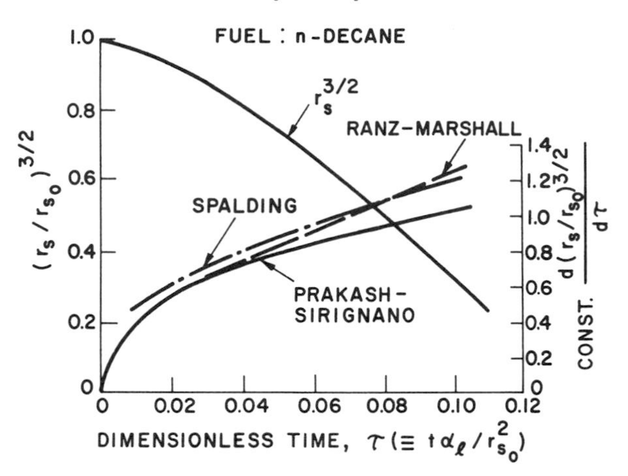

Figure 6.21 Droplet size and vaporization rate versus time (adapted from Prakash and Sirignano[20]).

the solution of this coupled gas–liquid problem, the gas-phase boundary layer is first treated using the initial guess or the relaxed previous iterated values for surface velocity and surface temperature. With the shear stress and the heat flux from the gas-phase solution, the liquid-phase viscous and thermal boundary layer and then the thermal core are resolved. The gas-phase and liquid-phase solutions are iterated until convergence is reached. The new droplet radius corresponding to the next time step is calculated, and the process repeated at the advanced time. Since the droplet becomes heated as time increases, the liquid viscosity and thus the strength of the Hill's vortex are updated in time. The overall vaporization rate at any instant is calculated by integrating the vaporizing flux over the droplet surface.

Three hydrocarbon fuels (n-hexane, n-decane, and n-hexadecane) evaporating in air were considered for $T = 1000$ K and $p = 10$ atm. The results for n-decane are shown in Fig. 6.21, which shows the variation of $(r_s/r_{s0})^{3/2}$ and the dimensionless average vaporizing mass flux. A comparison of the calculated results of Prakash with those based upon the Ranz–Marshall correlation[21] and the Spalding correlation[22] are shown in this figure. The Ranz–Marshall correlation is

$$\dot{m} = \dot{m}_{ss}\left(1 + 0.276\,\text{Re}_d^{1/2}\text{Pr}^{1/3}\right) \tag{6-192}$$

where $\dot{m}_{ss}$ is the mass vaporization rate with spherically symmetric vaporization. The Spalding correlation is given in terms of the Spalding transfer number B and the Reynolds number based upon droplet diameter, that is,

$$\frac{\dot{m}}{4\pi r_s \mu} = 0.265 B^{3/5}\text{Re}_d^{1/2} \qquad (\text{for} \quad 0.6 < B < 5) \tag{6-193}$$

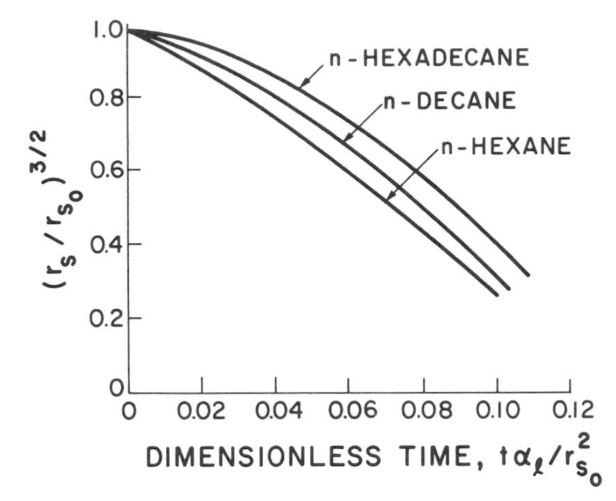

Figure 6.22 Comparison of droplet lifetimes of three hydrocarbon fuels (after Prakash and Sirignano[20]).

Spalding's correlation was based on experiments with Reynolds numbers between 800 and 4000, and the scatter of the experimental results was about 15–20%.

In using these correlations, the heat of vaporization was modified by Prakash and Sirignano to take into account the heat flux into the liquid phase. The modified heat of vaporization was calculated using their results for the average heat flux into the liquid phase. Figure 6.21 shows that the solution of Prakash and Sirignano is in good agreement with the Ranz–Marshall correlation during the initial part of the lifetime when the vaporization rate is small. The numerical results are lower than Spalding's correlation by about 15–20% towards the end of the lifetime of the droplet. According to Prakash and Sirignano, the difference could be due to the contribution of the wake region to vaporization which was neglected in their model. The contribution may be important when the droplet is heated. A part of the disagreement could be due to the experimental error in the correlation.

For a comparison of the droplet lifetime for the three fuels, the change of droplet radius with respect to time is plotted on the same graph in Fig. 6.22. The time scale is nondimensionalized with the thermal diffusion time r_{s0}^2/α_l of the liquid droplet. It can be seen from this figure that the variations in the lifetimes for the three fuels are only about 10% although their volatilities are quite different. It is also interesting to note from this figure that the vaporization rates in the initial part of the lifetime are substantially smaller for the less volatile fuels. To understand why the lifetimes of these three fuels are about the same, we need to examine the temperature–time history of these fuel droplets. A plot of the surface temperature variation (at $\theta = 90°$) and vortex-center temperature variation versus dimensionless time is given in Fig. 6.23. The surface temperature rises sharply during the initial heating period, and

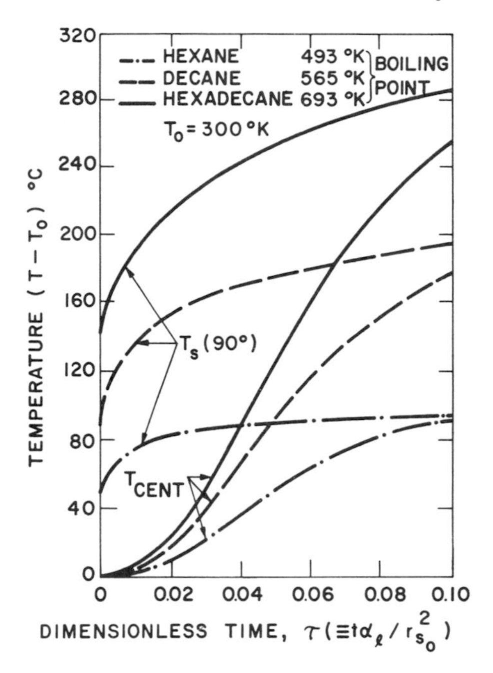

Figure 6.23 Surface and vortex-center temperature variation with respect to time (after Prakash and Sirignano[20]).

then the change becomes smaller at later times. The surface temperature for each of the three fuels is less than the boiling point of the fuel even at the end of the droplet lifetime. This effect is strongest for the least volatile fuel (hexadecane). The temperature at the center of the vortex of each fuel droplet is considerably lower than the droplet surface temperature during the earlier part of the lifetime. At the end of the droplet lifetime, the vortex center temperature is about 20 K lower for hexadecane and about 6 K lower for hexane (the most volatile fuel).

The reason for the small variation in the overall lifetime of a single droplet is that the surface temperature for the less volatile fuel rises quickly to a high value in a very short time, during which the quasisteady thermal boundary layer is established. After the initial establishment of the thermal boundary layer, the surface temperature rises slowly, so that the less volatile fuel is always at a higher temperature at any time, and hence the droplet lifetime is not significantly affected by the differences in volatility.

According to Prakash and Sirignano,[19,20] the internal circulation model lies between two extreme models: (a) a pure conduction model without any circulation, and (b) a rapid-mixing model. The pure conduction model has a zero vortex strength, while the rapid-mixing model has an infinite circulation rate. Their model showed that internal circulation leads to a shorter characteristic heating length than in the pure conduction case. In particular, the distance from the droplet surface to the vortex center is about $\frac{1}{3}$ of the droplet radius.

This implies that the characteristic heating time (which is proportional to the length squared) decreases by about an order of magnitude due to internal circulation. The heating time can be defined as the time required to reach a nearly uniform temperature profile in the droplet. In general, it is expected that the internal circulation via increased heat transfer rates to the droplet interior will yield lower surface temperatures and lower vaporization rates during the initial portion of the droplet lifetime than would be achieved with pure conduction only. The rapid-mixing model should yield still lower vaporization rates during this initial time interval. For an isolated droplet, Prakash and Sirignano concluded that the unsteadiness in droplet vaporization persists for most of the droplet lifetime, especially for the less volatile fuels. The temperature distribution inside the droplet is nonuniform for most of the lifetime; the difference between the surface temperature and the temperature in the interior is higher for the heavier and less volatile fuels.

REFERENCES

1. S. P. Burke and T. E. W. Schumann, "Diffusion Flames," *Indust. Eng. Chem.*, Vol. 20, No. 10, p. 998, 1928.
2. H. C. Hottel and W. R. Hawthorne, *Third International Symposium on Combustion*, p. 255, 1949.
3. D. B. Spalding, *Combustion and Mass Transfer*, 1st ed., Chapters 7, 9, 10, Pergamon Press, New York, 1979.
4. H. Schlichting, *Boundary Layer Theory*, 6th ed., Chapter XI, McGraw-Hill Book Co., New York, 1968.
5. G. M. Faeth, "Current Status of Droplet and Liquid Combustion," *Prog. Energy Combustion Sci.*, Vol. 3, pp. 191–224, 1977.
6. A. M. Kanury, *Introduction to Combustion Phenomena*, Chapter 5, Gordon and Breach Science Publishers, New York, 1977.
7. D. B. Spalding, *Fourth International Symposium on Combustion*, p. 847, 1953.
8. D. B. Spalding, *Some Fundamentals of Combustion*, Chapter 4, Butterworths, London, 1955.
9. P. L. Blackshear, Jr., *An Introduction to Combustion*, Chapter IV, Dept. of Mech. Engrg., University of Minnesota, Minneapolis, 1960.
10. I. Glassman, *Combustion*, Chapter 6, Academic Press, New York, 1977.
11. G. A. E. Godsave, *Fourth International Symposium on Combustion*, p. 818, 1953.
12. M. Summerfield, "Theory of Burning of a Composite Solid Propellant," ARS preprint, 1958.
13. D. B. Spalding, "Theory of Particle Combustion at High Pressure," *ARS J.*, pp. 828–835, November 1959.
14. D. E. Rosner, "On Liquid Droplet Combustion at High Pressures," *AIAA J.*, Vol. 5, No. 1, pp. 163–166, 1967.
15. D. P. Dominicis, *An Experimental Investigation of Near Critical and Supercritical Burning of Bipropellant Droplets*, NASA CR-72399, 1968.
16. A. Chervinsky, *AIAA J.*, Vol. 7, pp. 1815–1817, 1969.
17. G. M. Faeth, D. P. Dominicis, J. F. Tulpinsky, and D. R. Olson, "Supercritical Bipropellant Droplet Combustion," *Twelfth International Symposium on Combustion*, pp. 9–18, 1969.

18. S. Prakash, *Unsteady Theory of Droplet Vaporization with Large Gas and Liquid Reynolds Numbers*, Ph.D. Thesis, Princeton University, 1978.
19. S. Prakash and W. A. Sirignano, "Liquid Fuel Droplet Heating with Internal Circulation," *Internat. J. Heat Mass Transfer*, Vol. 21, pp. 885–895, 1978.
20. S. Prakash and W. A. Sirignano, "Theory of Convective Droplet Vaporization with Unsteady Heat Transfer in the Circulating Liquid Phase," *Internat. J. Heat Mass Transfer*, Vol. 23, No. 3, pp. 253–268, 1980.
21. W. E. Ranz and W. R. Marshall, "Evaporation from Drops," *Chem. Engrg. Prog.*, Vol. 48, No. 3, pp. 141–173, 1952.
22. D. B. Spalding, "Experiments on Burning and Extinction of Liquid Fuel Spheres," *Fuel* (London), Vol. 32, No. 2, pp. 169–185, 1953.
23. R. C. Reid and T. K. Sherwood, *The Properties of Gases and Liquids*, Chapters 3, 8, McGraw-Hill, New York, 1958.
24. S. S. Penner and M. Y. Bahadori, "Laminar Diffusion Flames: Revisiting a Classical Combustion Problem," International Colloquium on Gas Dynamics of Explosions and Reactive Systems, Poitier, France, 1983.
25. S. H. Chung and C. K. Law, "Burke-Schumann Flame with Steamwise and Preferential Diffusion," *Comb. Sci. and Tech.*, Vol. 37, p. 21, 1984.

HOMEWORK

1. Integrate the x-momentum equation (6-43) from $r = 0$ to $r = \infty$, and show that I_u of an axisymmetric jet is an invariant. Discuss the physical meaning of I_u.
2. Show that I_f of a cylindrical jet is an invariant. What is the physical meaning of I_f?
3. Derive Eq. (6-179) by considering the total amount of fuel mass (burned and unburned) to be constant.
4. A turbojet engine has a flight speed of 200 m/sec. If the combustor length is 2 m and if the temperature in the combustor is 1000 K, what is the maximum allowable size of n-pentane droplets if they have to be totally consumed by (a) vaporization alone, (b) vaporization and combustion, before they exit from the combustor?
5. Find (a) the evaporation time; (b) the rate of evaporation per unit area; and (c) the temperature of the surface at $t = 0$, for a 4-mm-diameter water droplet evaporating in dry air at 760 Torr if the ambient temperature is 288 K. Repeat the problem for ambient temperature equal to 1273 K. Use the following data:

$$\Delta H_v = 590 \text{ kcal/kg}$$

$$C_{p,g} = 0.24 \text{ kcal/kg K}$$

$$\rho_l = 1000 \text{ kg/m}^3$$

$$C_{p,l} = 1 \text{ kcal/kg K}$$

at 288 K, $\rho_g = 1.264\ \text{kg/m}^3$ and $\alpha_g = 2.06 \times 10^{-2}\ \text{m}^2/\text{sec}$

at 1273 K, $\rho_g = 0.2707\ \text{kg/m}^3$ and $\alpha_g = 0.2583\ \text{m}^2/\text{sec}$

6. For a laminar jet diffusion flame burning in a still atmosphere, the flame surface can be considered as a source of heat and a source (or sink) of chemical species dividing the flow into two regions containing either fuel or oxidant.
   (a) Specify the governing equations (in cylindrical coordinates) which are needed for the solution of concentration, temperature, and velocity distributions throughout the entire flow field.
   (b) State the fundamental assumptions.
   (c) Give the necessary boundary conditions.
   (d) Suggest a method of simplification in the solution of this problem.

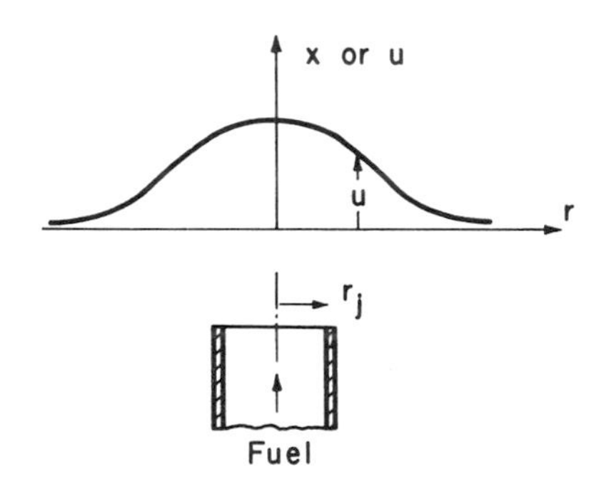

7. Consider a spherical metal particle which burns on the surface and forms a nonvolatile oxide which immediately dissolves upon formation in the metal itself. The surface reaction and dissolving rates are very fast compared to the oxidizer diffusion rate. Derive an expression for the burning rate of this metal.

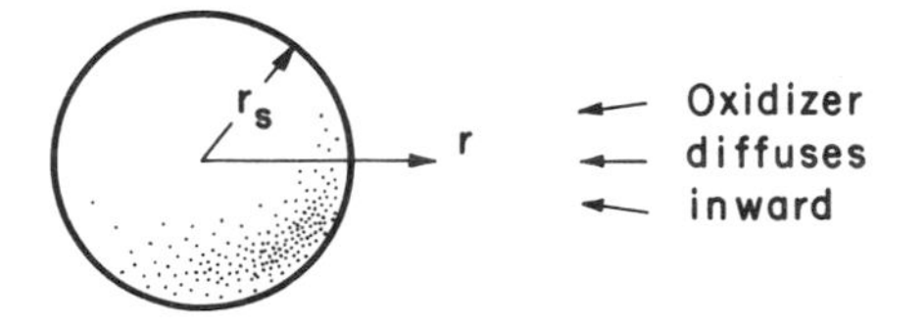

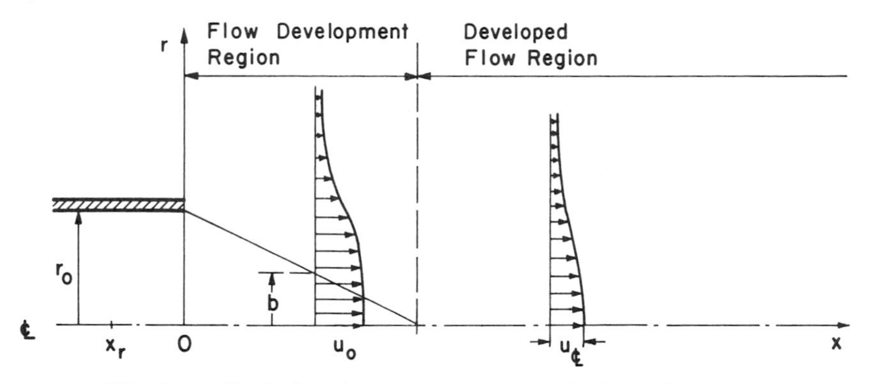

Velocity profiles in flow development region and developed flow region.

PROJECT

The laminar-jet flow field can be divided into a developing region and a developed region (see figure below). In the developing region, a potential core can be assumed to exist. The potential core, having uniform velocity u_0 and local radius b, is bounded by an annular free-shear layer. Make the following assumptions:

a. The surrounding air far from the jet is at rest.
b. Chemical reaction is absent.
c. The flow is steady and axisymmetric.
d. The pressure in the flow field is uniform.
e. The jet issues from a virtual origin x_v inside the cylindrical tube having radius r_0.

Consider the following Schlichting jet velocity profile in the free-shear layer surrounding the potential core:

$$\frac{u}{u_0} = 1 \qquad \text{for} \quad 0 \le R \le B \tag{A}$$

$$\frac{u}{u_0} = \frac{2(\Gamma/\mathrm{Re})^2}{X + X_v}\left[1 + \frac{(\Gamma/\mathrm{Re})^2(R - B)^2}{4(X + X_v)^2}\right] \qquad \text{for} \quad B \le R < \infty \tag{B}$$

where

$$B \equiv \frac{b}{2r_0}$$

$$R \equiv \frac{r}{2r_0}, \qquad X \equiv \frac{x}{\mathrm{Re} \cdot 2r_0}$$

$$\mathrm{Re} \equiv \frac{2u_0 r_0}{\nu}, \qquad X_v \equiv \frac{x_v}{\mathrm{Re} \cdot 2r_0}$$

with x_v the distance between the virtual origin and the jet exit, and

$$\Gamma \equiv \left(\frac{3}{8} \frac{I_u}{\nu} \right)^{1/2}$$

is the jet spread-angle parameter. In the above expression, the centerline velocity of the Schlichting jet has been shifted to the edge of the potential core. In order to insure velocity continuity at $R = B$, we must have

$$2\left(\frac{\Gamma}{\mathrm{Re}} \right)^2 = X + X_v$$

Therefore,

$$\frac{u}{u_0} = \left[1 + \frac{(R - B)^2}{16(\Gamma/\mathrm{Re})^2} \right]^{-2} \qquad \text{for} \quad B \leq R < \infty \tag{C}$$

1. Derive Eq. (B) by rearranging Eq. (6-56) and replacing r with $r - b$ and x with $x + x_v$.
2. Using Eqs. (A) and (C) and the fact that I_u and $\int_0^\infty u^3 r\, dr$ are jet invariants, derive the ordinary differential equations for B and Γ/Re as functions of X.
3. Show that the above two differential equations can be combined and integrated to give

$$k_1 \ln\left[\frac{12\eta^2 + 15\pi\eta + 64}{(512 + 70\pi\eta)^2} \right] + k_2 \tan^{-1}\left[\frac{24\eta + 15\pi}{\sqrt{3072 - 225\pi^2}} \right] + \frac{k_3\eta + k_4}{(12\eta^2 + 15\pi\eta + 64)} + k_5 = X \tag{D}$$

where

$$k_1 = -0.10395, \qquad k_2 = 0.02699$$

$$k_3 = -0.51613, \qquad k_4 = 2.81768$$

$$k_5 = -0.90537,$$

$$\eta \equiv \frac{B}{\Gamma/\mathrm{Re}} \tag{E}$$

4. Plot the flow field (velocity profiles) and potential core boundary using Eqs. (D), (A), (C), and (E).

CHAPTER

7 Turbulent Flames

ADDITIONAL SYMBOLS

Symbols	Description	Dimension
a	Dimensionless age of a gas parcel	—
a_1	Value for complete reaction of parcel	—
d	Flame thickness	L
f_i	Body force per unit volume in the ith direction	F/L^3
k	Kinetic energy of turbulence	L^2/t^2
l_T	Characteristic scale of turbulence	L
$\mathscr{P}$	Probability density function	—
r	Oxidant-to-fuel weight ratio defined in Eq. (7-169) or reaction progress variable used in Fig. 7.24	—
R	Correlation coefficient	—
S	Flame speed	L/t
t	Time	t
u	Instantaneous velocity in x-direction	L/t
v	Instantaneous velocity in y-direction	L/t
w	Instantaneous velocity in z-direction	L/t
$\bar{u}$	Time-average velocity in x-direction	L/t
$\bar{v}$	Time-average velocity in y-direction	L/t
$\bar{w}$	Time-average velocity in z-direction	L/t
u'	Fluctuating velocity component in the x-direction	L/t
v'	Fluctuating velocity component in the y-direction	L/t
w'	Fluctuating velocity component in the z-direction	L/t
K_z	Kovasznay number	—

α_T	Turbulent thermal diffusivity	L^2/t
ν_T	Turbulent eddy viscosity	L^2/t
$\mathscr{D}_T$	Turbulent mass diffusivity	L^2/t
ε	Dissipation rate of turbulence kinetic energy	L^2/t^3
λ_T	Turbulent thermal conductivity	$Q/(LTt)$
Λ_T	Taylor microscale of turbulence	L
η	Kolmogorov microscale of turbulence	L
τ_c	Characteristic chemical reaction time	t
τ_m	Characteristic aerodynamic time	t
τ	Reactedness (used in EBU model)	—
Ω'_k	Vorticity of turbulence	$1/t$

Diacriticals

$\bar{}$	Time-averaged quantity
$\tilde{}$	Mass-weighted average quantity

Superscripts

$'$	Fluctuation quantity in Reynolds averaging
$''$	Fluctuation quantity in Favre averaging
$*$	Lagrangian quantity; critical value

Subscripts

T	Turbulence case
L	Laminar case
b	Burned gas
u	Unburned gas
crit	Critical value
rms	Root-mean-square

1 INTRODUCTION

In preceding chapters, we have discussed some important concepts for laminar flames. For laminar flows, the adjacent layers of fluid slide past one another in a smooth, orderly manner. The only mixing possible is due to molecular diffusion. The velocity, temperature, and concentration profiles measured in laminar flow with a high-sensitivity instrument will be quite smooth.

At very high Reynolds or Grashof numbers the flow becomes turbulent. In turbulent flow, eddies move randomly back and forth and across the adjacent fluid layers. The flow no longer remains smooth and orderly. Typical superimposed contours of instantaneous flame boundaries are shown in Fig. 7.1.

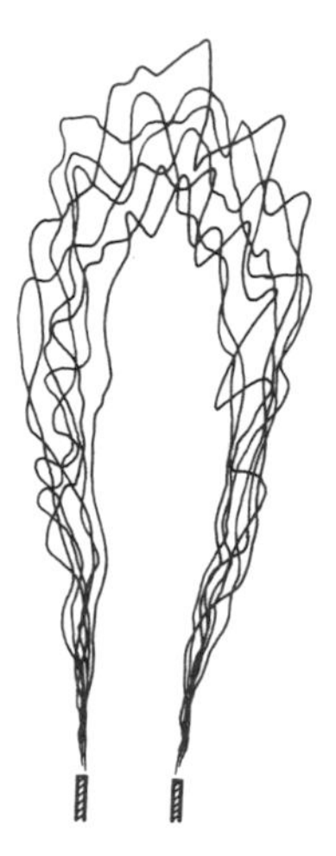

Figure 7.1 Superimposed contours of instantaneous flame boundaries (after Fox and Weinberg[18]).

As described by Tennekes and Lumley,[1] it is very difficult to give a precise definition of turbulence. All one can do is to list some of the characteristics of turbulent flows:

A. Irregularity. All turbulent flows are irregular, or random. This makes a deterministic approach to turbulence problems impossible; one relies on statistical methods.

B. Diffusivity. The diffusivity of turbulence, which causes rapid mixing and increased rates of momentum, heat, and mass transfer, is another important feature of all turbulent flows.

C. Large Reynolds Number. Turbulent flows always occur at high Reynolds numbers. Turbulence often originates as an instability of laminar flows if the Reynolds number becomes too large. The instability is related to the interaction of viscous terms and nonlinear inertia terms in the equation of motion.

D. Three-Dimensional Vorticity Fluctuations. Turbulence is rotational and three-dimensional. It is characterized by high levels of fluctuating vorticity. For this reason, vorticity dynamics plays an essential role in the description of turbulent flows.

E. Dissipation. Turbulent flows are always dissipative. Viscous shear stresses perform deformation work which increases the internal energy of the fluid at the expense of the kinetic energy of the turbulence. Turbulence needs a continuous supply of energy to make up for these viscous losses.

F. Continuum. Turbulence is a continuum phenomenon, governed by the equations of fluid mechanics. Even the smallest scales occurring in a turbulent flow are ordinarily far larger than any molecular length scale.

G. Turbulent Flows Are Flows. Turbulence is not a feature of fluids but of fluid flows. Most of the dynamics of turbulence is the same in all fluids, whether they are liquids or gases, if the Reynolds number of the turbulence is large enough; the major characteristics of turbulent flows are not controlled by the molecular properties of the fluid in which the turbulence occurs.

Even chemically nonreacting turbulent flows are highly challenging due to the above characteristics. When chemical reactions occur, the problems become even more complex, since the turbulent fluid flow is further coupled with chemical kinetics and quite often with phase changes. This is why the study of turbulent reacting flows is one of the most challenging fields of engineering science. It has attracted increased interest in recent years due to its application in a variety of devices, ranging from jet engines to industrial power plants to gas-dynamic lasers. The increase in interest in recent years has led to a series of specialized conferences, workshops, reviews, and so on. Several recent excellent reviews by Bracco,[2] Jones,[3] Libby and Williams,[4] and Jones and Whitelaw[5] have classified various models into different categories, discussed their applications and limitations, indicated the difficulties in understanding turbulent flames, and suggested directions for future approaches, both theoretical and experimental. Some of our discussion in the later portion of this chapter follows that of Jones and Whitelaw.[5]

The turbulent flame, unlike the laminar one, is often accompanied by noise and rapid fluctuations of the flame envelope. For the laminar flame, it is possible to define a flame velocity that, within reasonable limits, is independent of the experimental apparatus. It would be equally desirable to define a propagation velocity for turbulent flames that would be independent of the experimental apparatus and depend only on the fuel–air ratio and some transport properties such as λ, μ and $\mathscr{D}$ for laminar flow. However, this is not possible, because the transport properties of turbulent flame are functions of the flow rather than the fluid. At some stoichiometric ratios the effective thermal diffusivity can be 100 times larger than the molecular thermal diffusivity in the laminar case. Thus, the theoretical concepts for turbulent flames are not so well defined as for laminar flames.

In addition to the characteristics of nonreacting turbulent flows mentioned earlier, some of the flames are described briefly in the following:

1. The flame surface is very complex, and it is difficult to locate the various surfaces that are used to characterize laminar flames.
2. Also due to the enhanced transport properties, the turbulent flame speed is much greater than the laminar flame speed.
3. The height of a turbulent flame is smaller than that of a laminar flame at the same flow rate, fuel–air ratio, and burner size. This is shown by comparison of direct photographs. At a given velocity, the flame height decreases as the intensity of the turbulence increases.
4. Unlike the laminar flame, the reaction zone is usually quite thick.
5. The flame velocity S_T increases as Re increases.

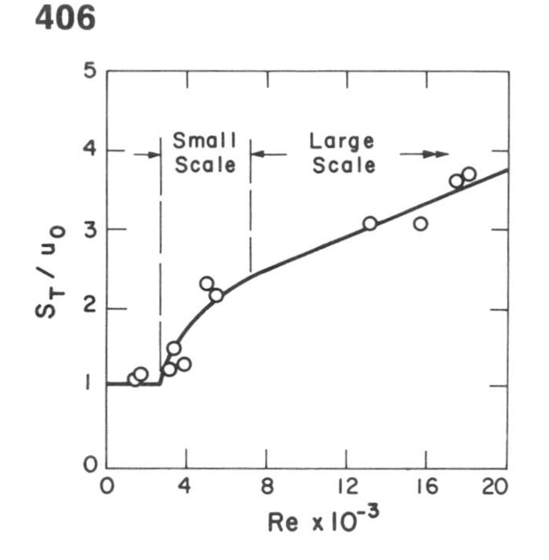

Figure 7.2 Effects of Reynolds number on flame speed (after Damköhler[6]).

The approaches developed in studying turbulent flames can be generally classified as phenomenological approaches and comprehensive approaches. The former methods were developed earlier, in the 1940s and 1950s; the latter methods have been developed more recently, in the 1970s and 1980s. To clarify the historical development of turbulent-flame theories, both approaches are given.

2 PHENOMENOLOGICAL APPROACHES TO TURBULENT FLAMES

2.1 Damköhler's Analysis[6,7] (1940)

Although Mallard and LeChatelier recognized in very early days (1883) that turbulence affected burning velocities, systematic investigations of turbulent flames began in 1940 with Damköhler's classical theoretical and experimental study. Figure 7.2 shows the Bunsen-burner measurements of the flame speed at various Reynolds numbers by Damköhler.[6] He found that the flame speed is (a) independent of Reynolds number when $Re \lesssim 2300$, (b) proportional to the square root of the Reynolds number when $2300 \lesssim Re \lesssim 6000$, and (c) proportional to the Reynolds number when $Re \gtrsim 6{,}000$. Obviously, items (b) and (c) are influenced by turbulence and hence the measured flame speeds depend on geometry and flow.

In the range $2300 \lesssim Re \lesssim 6000$, turbulence is of fine scale; that is, the eddy size and mixing length are much smaller than the flame-front thickness. The effect of these fine-scale eddies is to enhance the intensity of transport processes within the combustion wave. Under these circumstances, transport of heat and species is due to the turbulent diffusivity but not the molecular diffusivity.

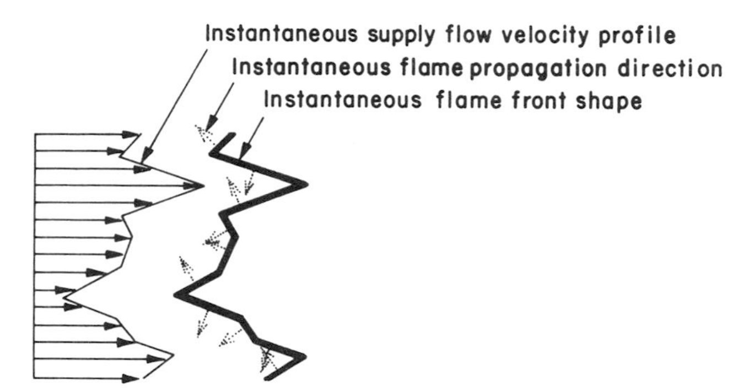

Figure 7.3 An exaggerated view of the turbulent flame front (adapted from Kanury[8]).

When Re $\gtrsim$ 6000, the turbulent eddies are of dimensions comparable with the tube diameter, much larger than the laminar flame-front thickness. The large eddies do not increase the diffusivities as the small eddies do, but they distort the otherwise smooth "laminar" flame front as shown in Fig. 7.3. The influence of these folds in the flame front increases the flame-front area per unit cross section of the tube. As a consequence, the apparent flame speed is increased without any change in the instantaneous local flame structure itself.

Damköhler stated that small-scale turbulence simply increases the transport properties in the wave, and he investigated these changes as a function of Reynolds number in the following manner. For a laminar flame, the flame speed S_L is given by Eq. (5-5). It is reasonable to expect the speed of turbulent flame, S_T, to be

$$S_T \propto \sqrt{\alpha_T \mathrm{RR}} \tag{7-1}$$

Thus

$$\frac{S_T}{S_L} \approx \left(\frac{\alpha_T}{\alpha}\right)^{1/2} \tag{7-2}$$

If both the turbulent Prandtl number ($\equiv \nu_T/\alpha_T$) and the Prandtl number based upon molecular transport properties are approximately equal to one, then Eq. (7-2) becomes

$$\frac{S_T}{S_L} \approx \left(\frac{\nu_T}{\nu}\right)^{1/2} \tag{7-3}$$

For pipe flow, $\nu_T/\nu \approx 0.01\,\mathrm{Re}$. Therefore,

$$\frac{S_T}{S_L} \approx 0.1\,\mathrm{Re}^{1/2} \tag{7-4}$$

This equation indeed predicts the trend of Damkohler's small-scale burning speed measurement. However, Eq. (7-1) has a serious drawback: as $\nu_T \to 0$, $S_T \to 0$ instead of approaching S_L.

In the case of large-scale, low-intensity turbulence, the flame will be wrinkled but the molecular transport properties will remain the same; therefore, the laminar-flame speed will remain constant. Since for constant S_L, the flame area is proportional to the flow velocity, it would be expected that the increase in flame surface area due to turbulence would be proportional to u'_{rms}. Since ν_T is proportional to the product of intensity and mixing length (which can be considered constant), u_{rms} is proportional to ν_T. Also $\nu_T \propto \text{Re}$ for constant ν, so we have

$$\frac{S_T}{S_L} \propto \text{area} \propto \text{fluctuation} \propto \nu_T \propto \text{Re} \tag{7-5}$$

It is not surprising, therefore, that certain experimental results appear to correlate as

$$S_T = A\,\text{Re} + B \tag{7-6}$$

where A and B are constants. This relationship describes Damköhler's large-scale burning speeds quite satisfactorily.

In summary, for given turbulence intensity, sufficiently large-scale turbulence merely wrinkles a premixed laminar flame without significantly modifying its internal structure. Sufficiently small-scale turbulence alters the effective transport coefficients in the flame without significantly wrinkling the flame front.

2.2 Schelkin's Analysis[9,10] (1943)

Schelkin considered the importance of the times (τ) associated with the turbulence. For laminar flames,

$$S_L \propto \sqrt{\frac{\alpha}{\tau_r}} \propto \sqrt{\frac{\lambda}{\rho C_p \tau_r}} \propto \sqrt{\frac{\lambda}{\tau_r}} \tag{7-7}$$

For turbulent flames, he proposed a similar relationship,

$$S_T \propto \sqrt{\frac{\lambda + \lambda_T}{\tau_r}} = \sqrt{\frac{\lambda}{\tau_r}\left(1 + \frac{\lambda_T}{\lambda}\right)} = S_L\sqrt{1 + \frac{\lambda_T}{\lambda}} \tag{7-8}$$

where τ_r is the reaction time.

Schelkin also considered large-scale, low-intensity turbulence. He assumed surfaces are distorted into cones whose base area was proportional to the

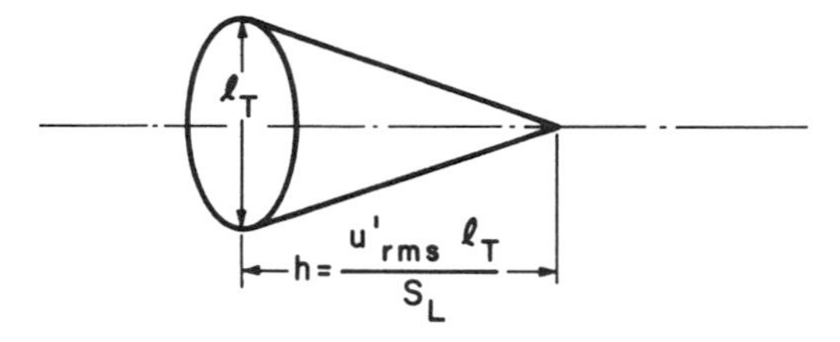

Figure 7.4 A conical flame front formed from distortion of a flat flame front.

square of the average eddy diameter l_T, as shown in Fig. 7.4. The height of the cone is proportional to $u'_{\rm rms}$ and to the time during which an element of the combustion wave is associated with an eddy motion in the direction normal to the wave. This time, then, can be taken equal to l_T/S_L. Schelkin then proposed that the ratio of S_T to S_L equals the ratio of the average cone area to the average cone base. From the geometry,

$$A_C = A_B\left(1 + \frac{4h^2}{l_T^2}\right)^{1/2} \tag{7-9}$$

where A_C is the area of the cone, A_B is the area of the base, and h is the cone height, which can be represented by

$$h = u'_{\rm rms}t = u'_{\rm rms}\frac{l_T}{S_L} \tag{7-10}$$

Therefore,

$$S_T = S_L\left[1 + \left(\frac{2u'_{\rm rms}}{S_L}\right)^2\right]^{1/2} \tag{7-11}$$

For large values of $u'_{\rm rms}/S_L$, the above expression reduces to the one developed by Damköhler, $S_T \propto u'_{\rm rms}$.

2.3 Karlovitz's Analysis[11] (1951)

As mentioned previously, Damköhler proposed that S_T is proportional to $\sqrt{\alpha_T \text{RR}}$. However, there is a defect in this expression: when α_T approaches zero, S_T will go to zero. Actually S_T will be close to S_L when α_T goes to zero. In order to avoid this drawback, Karlovitz et al.[11] (1951) suggested that

$$S_T = S_L + u'_{\rm rms} \tag{7-12}$$

where $u'_{\rm rms}$ is the turbulent intensity defined as the root mean square of the fluctuation velocity, and

$$u'_{\mathrm{rms}} \propto \sqrt{\alpha_T \mathrm{RR}} \tag{7-13}$$

Therefore, we have

$$\frac{S_T}{S_L} = 1 + K_1 \sqrt{\frac{\alpha_T}{\alpha}} \tag{7-14}$$

where K_1 is an empirical constant.

Karlovitz's work was modified and expanded further by some other researchers. For butane–air flames, Wohl et al.[12] found

$$S_T = S_L \left[1 + K_2 U^{\beta} \left(\frac{u'_{\mathrm{rms}}}{U} + 0.01 \right) \right] \tag{7-15}$$

where U is the approach velocity in the burner tube, and K_2 and β depend only on the fuel–air ratio. Bowditch[13] found agreement with the empirical expression

$$S_T = S_L \exp\left(\frac{1.15 u'_{\mathrm{rms}}}{S_L} \right) \tag{7-16}$$

With the same type of apparatus, Kozachenko[14] obtained

$$S_T = 3.9 S_L + u'_{\mathrm{rms}} \tag{7-17}$$

which is very similar to the expression suggested by Karlovitz.

2.4 Summerfield's Analysis[15] (1955)

Summerfield et al. (1955) proposed to consider the time scale associated with the turbulent flame to be different from that for laminar flame. They suggested evaluating the ratio of S_T to S_L as

$$\frac{S_T}{S_L} = \frac{(\nu_T/\tau_T)^{1/2}}{(\nu/\tau_L)^{1/2}} \tag{7-18}$$

The new feature is that the reaction time is affected by the turbulence. The reaction time can be represented by

$$\tau_L = \frac{\delta_L}{S_L}, \qquad \tau_T = \frac{\delta_T}{S_T} \tag{7-19}$$

where δ is the flame thickness. This approach considers the reaction zone to be extended in the turbulent case. Conceptually, it considers S_L to be a measure of the chemical kinetics alone. With the expressions for τ it follows, then, that

$$\frac{S_T^2}{S_L^2} = \frac{\nu_T S_T/\delta_T}{\nu S_L/\delta_L} \tag{7-20}$$

or cross-multiplying,

$$\frac{S_T \delta_T}{\nu_T} = \frac{S_L \delta_L}{\nu} \tag{7-21}$$

Experimental data on laminar flames give

$$\frac{S_L \delta_L}{\nu} \approx 10 \tag{7-22}$$

Thus

$$\frac{S_T \delta_T}{\nu_T} = \frac{S_L \delta_L}{\nu} \approx 10 \tag{7-23}$$

so that S_T may be determined from the turbulent-flame thickness and eddy viscosity.

2.5 Kovasznay's Characteristic-Time Approach

The assumption that the turbulent flame consists of a continuous laminar flame is explicit in many of the earlier qualitative studies of turbulent combustion. The work of Damköhler and Schelkin has had some more recent proponents who, following the same concept of the wrinkled flame, considered the breakup of the flame surface more extensively and thus derived more complicated formulas. However, information obtained on many turbulent flames, particularly those in which small-scale high-intensity turbulence was evident, showed that laminar flamelets do not exist. The failure of this wrinkled-flame concept led others to the concept of a distributed reaction zone. Somewhat later, more precise experiments on how turbulence affects flame radiation led to the proposal that there are a series of possible mechanisms which describe the effect of turbulence on the combustion zone.

One of these postulations, given by John and Mayer[16] (1957), says that the mechanisms can be interpreted in terms of a characteristic chemical time τ_c and a characteristic aerodynamic time τ_m. The chemical-reaction time is defined as

$$\tau_c = \frac{\delta_L}{S_L} \tag{7-24}$$

where δ_L is the thickness of the laminar flame. τ_c increases as one lowers either the pressure or the chemical activity or both. The aerodynamic time τ_m is defined as

$$\tau_m = \frac{\Lambda_T}{u'_{\text{rms}}} \tag{7-25}$$

where Λ_T is the Taylor microscale of turbulence discussed in Section 3.4.

A nondimensional number can be formed from both these times and has been called the Kovasznay number:[10,17]

$$\text{Kz} \equiv \frac{\tau_c}{\tau_m} \tag{7-26}$$

Weak turbulence (u'_{rms} small and $\tau_m \gg \tau_c$) merely wrinkles the flame front. τ_m, in this case, can be regarded as being inversely proportional to the velocity gradient characteristic of the flow approaching the flame front; and τ_c is inversely proportional to the gradient of the laminar flame speed.

Stronger turbulence ($\tau_m \approx \tau_c$) disrupts the laminar flame front. Thereupon, τ_m and τ_c lose their significance as reciprocal velocity gradients. In this case, τ_m is interpreted as the mean lifetime of an eddy and τ_c as the time for chemical reaction in these pocket combustibles.

Still stronger turbulence ($\tau_m \ll \tau_c$) shows its effects by locally diluting and preheating the initial centers of deflagration, and in the limit results in homogeneous reaction mixtures. This limit in this context is sometimes called that of the *continuously stirred reactor*.

3 FUNDAMENTALS OF TURBULENT FLOW

Since turbulent flames are very complicated, it is useful to cover some fundamentals of nonreacting turbulent flow in the beginning and then consider and incorporate the chemical reactions. To facilitate our discussions of turbulent flames, some of the fundamentals of turbulent flow are given in this section.

As the turbulent motion is random and irregular, it has a broad range of length scales. In order to obtain theoretical solutions by solving three-dimensional, time-dependent problems, we would have to have a computer with enormous storage capacity, greatly exceeding those of existing computers. Thus we are forced to restrict ourselves, at the present time, to the consideration of some type of averaged quantities in turbulent flows. There are two different averaging procedures[19] commonly used: (1) conventional time averaging (also called Reynolds averaging) and (2) mass-weighted averaging (also called Favre[20] averaging). The detailed averaging procedures and the relationships between the quantities averaged by these procedures are described in the following subsections.

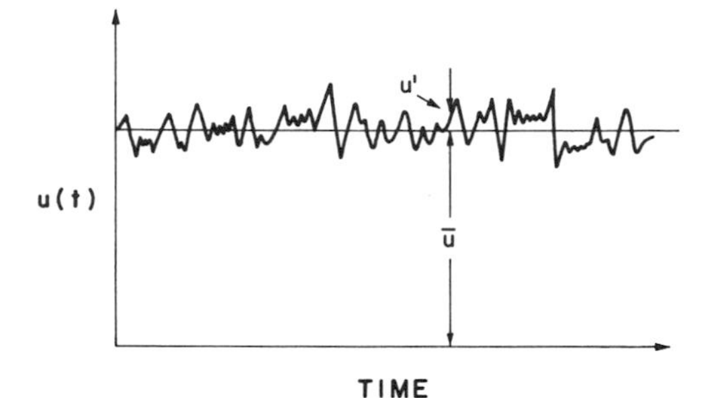

Figure 7.5 Instantaneous turbulent velocity versus time.

3.1 Conventional Time Averaging and Mass-Weighted Averaging

In order to obtain governing conservation equations for turbulent flames, it is convenient to decompose instantaneous quantities into mean and fluctuating quantities as shown in Fig. 7.5. In the conventional time-averaging procedure, the mean quantity $\bar{q}$ is defined by

$$\bar{q} = \lim_{\Delta t \to \infty} \frac{1}{\Delta t} \int_{t_0}^{t_0 + \Delta t} q(t)\, dt \tag{7-27}$$

Clearly, the time average is useful only if it is independent of t_0; then, the averaged quantities are called *statistically stationary*. Physical quantities for stationary mean flows can now be decomposed into two parts, related to the mean motion and the fluctuation, or eddy, motion:

$$u_i(x_i, t) = \overline{u_i(x_i)} + u_i'(x_i, t) \tag{7-28}$$

$$p(x_i, t) = \overline{p(x_i)} + p'(x_i, t) \tag{7-29}$$

$$\rho(x_i, t) = \overline{\rho(x_i)} + \rho'(x_i, t) \tag{7-30}$$

$$h_t(x_i, t) = \overline{h_t(x_i)} + h_t'(x_i, t) \tag{7-31}$$

$$T(x_i, t) = \overline{T(x_i)} + T'(x_i, t) \tag{7-32}$$

where the average of fluctuating quantities, $\overline{q'}(t)$, is zero, that is,

$$\overline{\rho'} = \overline{h_t'} = \overline{h'} = \overline{T'} = 0 \tag{7-33}$$

In addition, if f and g are two dependent variables and if s denotes any one of the independent variables x, y, z, and t, then the Reynolds averaging rule requires that

$$\bar{\bar{f}} = \bar{f} \tag{7-34}$$

$$\overline{f + g} = \bar{f} + \bar{g} \tag{7-35}$$

$$\overline{\bar{f}g} = \bar{f}\bar{g}, \qquad \overline{fg} = \bar{f}\bar{g} + \overline{f'g'} \tag{7-36}$$

$$\overline{\frac{df}{ds}} = \frac{d\bar{f}}{ds} \tag{7-37}$$

$$\overline{\int f\, ds} = \int \bar{f}\, ds \tag{7-38}$$

and the conservation equations can be expressed in terms of the average and fluctuation correlation quantities.

The continuity equation is

$$\frac{\partial \bar{\rho}}{\partial t} + \frac{\partial \bar{\rho}\bar{u}_i}{\partial x_i} + \frac{\partial \overline{\rho' u_i'}}{\partial x_i} = 0 \tag{7-39}$$

The momentum equation for an incompressible fluid is

$$\rho\left[\frac{\partial \bar{u}_i}{\partial t} + \bar{u}_j \frac{\partial \bar{u}_i}{\partial x_j}\right] = -\frac{\partial \bar{p}}{\partial x_i} + \frac{\partial}{\partial x_j}\left[\mu \frac{\partial \bar{u}_i}{\partial x_j} - \rho\overline{u_i' u_j'}\right] + \bar{f}_i \tag{7-40}$$

where $-\rho\overline{u_i' u_j'}$ is called the apparent stress or Reynolds stress.

Expanding and rearranging the above equation, one gets the incompressible equation in the three basic directions:

x-direction:

$$\rho\left[\frac{\partial \bar{u}}{\partial t} + \bar{u}\frac{\partial \bar{u}}{\partial x} + \bar{v}\frac{\partial \bar{u}}{\partial y} + \bar{w}\frac{\partial \bar{u}}{\partial z}\right] = -\frac{\partial \bar{p}}{\partial x} + \mu\nabla^2 \bar{u} + \bar{f}_x - \rho\left[\frac{\partial \overline{u'^2}}{\partial x} + \frac{\partial \overline{u'v'}}{\partial y} + \frac{\partial \overline{u'w'}}{\partial z}\right] \tag{7-41}$$

y-direction:

$$\rho\left[\frac{\partial \bar{v}}{\partial t} + \bar{u}\frac{\partial \bar{v}}{\partial x} + \bar{v}\frac{\partial \bar{v}}{\partial y} + \bar{w}\frac{\partial \bar{v}}{\partial z}\right] = -\frac{\partial \bar{p}}{\partial y} + \mu\nabla^2 \bar{v} + \bar{f}_y - \rho\left[\frac{\partial \overline{u'v'}}{\partial x} + \frac{\partial \overline{v'^2}}{\partial y} + \frac{\partial \overline{v'w'}}{\partial z}\right] \tag{7-42}$$

z-direction:

$$\rho\left[\frac{\partial \bar{w}}{\partial t}+\bar{u}\frac{\partial \bar{w}}{\partial x}+\bar{v}\frac{\partial \bar{w}}{\partial y}+\bar{w}\frac{\partial \bar{w}}{\partial z}\right]=-\frac{\partial \bar{p}}{\partial z}+\mu\nabla^2\bar{w}+\bar{f}_z - \rho\left[\frac{\partial \overline{u'w'}}{\partial x}+\frac{\partial \overline{v'w'}}{\partial y}+\frac{\partial \overline{w'^2}}{\partial z}\right] \quad (7\text{-}43)$$

The components of the stress tensor due to the fluctuation velocity components of the flow are

$$\begin{pmatrix} \sigma_{xx} & \tau_{xy} & \tau_{xz} \\ \tau_{xy} & \sigma_{yy} & \tau_{yz} \\ \tau_{xz} & \tau_{yz} & \sigma_{zz} \end{pmatrix}_T = -\begin{pmatrix} \rho\overline{u'^2} & \rho\overline{u'v'} & \rho\overline{u'w'} \\ \rho\overline{u'v'} & \rho\overline{v'^2} & \rho\overline{v'w'} \\ \rho\overline{u'w'} & \rho\overline{v'w'} & \rho\overline{w'^2} \end{pmatrix} \quad (7\text{-}44)$$

The mean velocities of turbulent flow satisfy the same equations as those of laminar flow except that the laminar stresses must be increased by the additional stresses known as apparent (or Reynolds) stresses. Generally speaking, the apparent stresses far outweigh the viscous components, and consequently the latter may be omitted in many actual cases with a good degree of approximation. Then we have

$$(\tau_{xy})_T = -\rho\overline{u'v'} \equiv \rho\nu_T\frac{\partial \bar{u}}{\partial y} \quad (7\text{-}45)$$

where ν_T is the eddy viscosity. Equation (7-45) can be regarded as the definition of ν_T, which may be evaluated by using some empirical formula or with the use of a turbulent model.

In the immediate neighborhood of a wall, the apparent stresses are small compared with the viscous stresses, and it follows that in every turbulent flow, there exists a very thin laminar sublayer. All turbulent components must vanish at the wall.

Reliable measurements of the fluctuation velocity components have been obtained with the aid of the hot-wire anemometer, the laser Doppler anemometer, and other instruments for flow visualization.

Consider, for example, a mean flow given by $\bar{u}=\bar{u}(y)$, $\bar{v}=\bar{w}=0$ with $d\bar{u}/dy>0$ as shown in Fig. 7.6. We can see that the correlation $\overline{u'v'}$ is different from zero. The Reynolds stress $\tau_{xy}=-\rho\overline{u'v'}$ can be interpreted as the transport of x-momentum of the flow element through a surface normal to the y-axis. A particle which travels upwards as a result of the turbulent fluctuation ($v'>0$) arrives at a layer from a region where a smaller mean velocity $\bar{u}$ prevails. Conversely, a particle which arrives from a higher layer ($v'<0$) gives

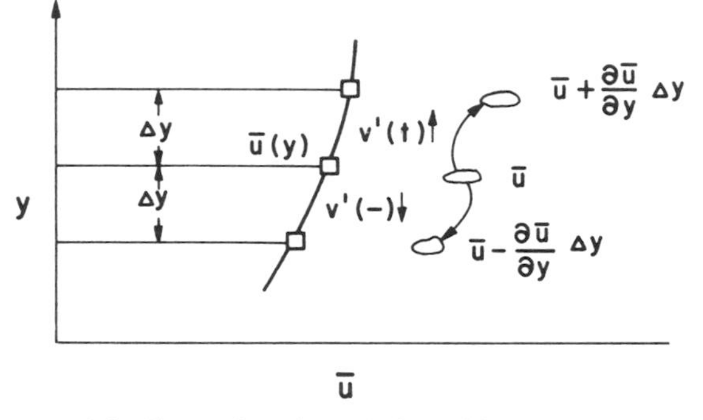

Figure 7.6 The motion of a turbulent eddy at any given time.

rise to a positive u' in it. Therefore a positive v' is generally associated with a negative u', and a negative v' with a positive u'. We may thus expect that the time-averaged $\overline{u'v'}$ is not only different from zero but in fact negative. This is also expressed by stating that there exists a correlation between the longitudinal and transverse fluctuations of velocity at a given point. We can define various correlation coefficients as the ratio of a covariance to the product of standard deviations. The one-point double correlation, $\mathscr{R}_{u'v'}$, is defined as

$$\mathscr{R}_{u'v'} \equiv \frac{\overline{u'v'}}{\sqrt{\overline{u'^2}}\sqrt{\overline{v'^2}}} \tag{7-46}$$

A two-point double correlation can be defined as

$$\mathscr{R}(r) \equiv \frac{\overline{u_1'u_2'}}{\sqrt{\overline{u_1'^2}}\sqrt{\overline{u_2'^2}}} \tag{7-47}$$

A two-point triple correlation can be defined as

$$\mathscr{R}(r) \equiv \frac{\overline{u_1'u_2'v_2'}}{\sqrt{\overline{u_1'^2}}\sqrt{\overline{u_2'^2}}\sqrt{\overline{v_2'^2}}} \tag{7-48}$$

In general, a correlation coefficient is a function of r. The definitions of fluctuating velocity components are shown in Fig. 7.7, and a typical relation

Figure 7.7 Fluctuating velocity components at two different locations.

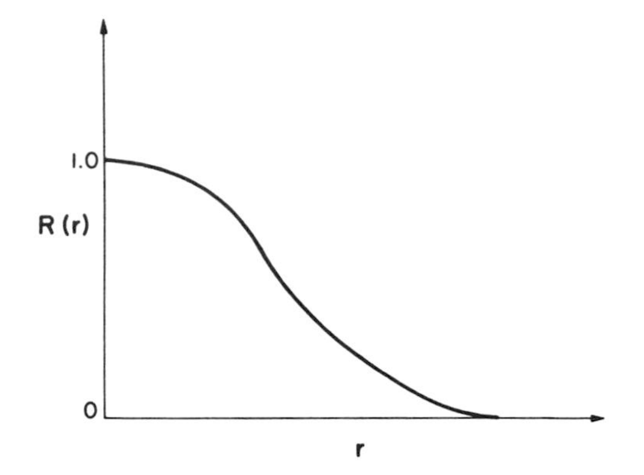

Figure 7.8 A typical r dependence of a correlation function $\mathscr{R}$.

between $\mathscr{R}$ and r is shown in Fig. 7.8. It is interesting to note that

$$\lim_{r \to 0} \mathscr{R}(r) \to 1 \tag{7-49}$$

$$\lim_{r \to \infty} \mathscr{R}(r) \to 0 \tag{7-50}$$

It is necessary to specify a characteristic length which is a measure of the "magnitude of a turbulent eddy," that is, the scale of turbulence. Usually, this can be done experimentally by measuring the correlation coefficient of longitudinal fluctuation u_1' and u_2' at two points whose transverse distance is y:

$$\mathscr{R}(y) \equiv \frac{\overline{u_1' u_2'}}{\sqrt{\overline{u_1'^2}}\sqrt{\overline{u_2'^2}}} \tag{7-51}$$

The results of measurements taken behind a screen of mesh width l_{mesh} in a region of isotropic turbulence are plotted against y in Fig. 7.9.

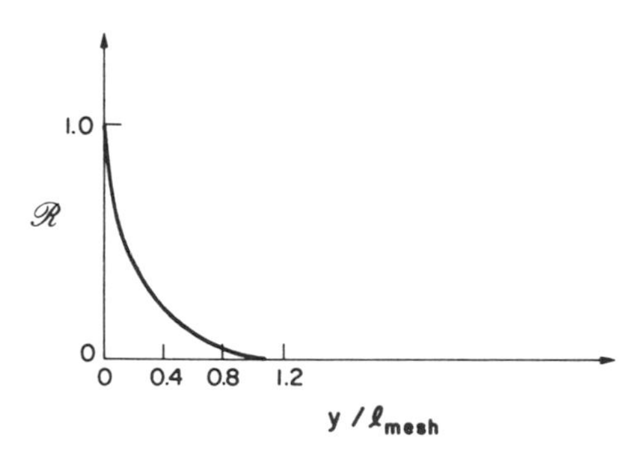

Figure 7.9 A typical relation between $\mathscr{R}$ and y/l_{mesh} in isotropic turbulent flow.

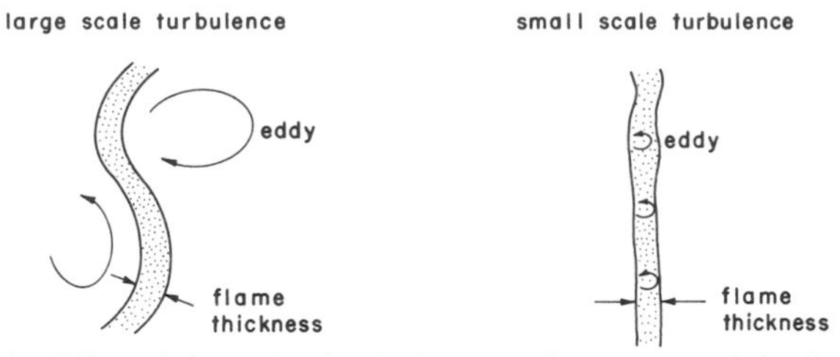

Figure 7.10 Effect of the scale of turbulence on the structure of the flame front.

The correlation coefficient decreases rapidly to zero with increasing y, and the quantity which is the characteristic scale of turbulence is given by

$$l_T = \int_0^\infty \mathscr{R}(y)\, dy \tag{7-52}$$

where l_T is an Eulerian scale. l_T is a measure of the size of the lumps of fluid which move together as a unit and thus describes the size of individual eddies. When l_T is very large there is small influence of turbulence on the flame; only the small-scale eddies will affect the structure of the flame front, as shown in Fig. 7.10.

The Lagrangian correlation coefficient $\mathscr{R}^*(t)$ can be defined as

$$\mathscr{R}^*(t) \equiv \frac{\overline{u'(t)u'(t+\Delta t)}}{\sqrt{\overline{u'(t)^2}}\sqrt{\overline{u'(t+\Delta t)^2}}} \tag{7-53}$$

Here, the Lagrangian characteristic scale of turbulence is given by

$$l_T^* = u'_{\text{rms}} t^*_{\text{mixing}} \tag{7-54}$$

where t^*_{mixing} is defined as follows

$$t^*_{\text{mixing}} \equiv \int_0^\infty \mathscr{R}^*(t)\, dt \tag{7-55}$$

Generally speaking, there are several length scales for turbulent flow. The largest is the dimension of flow field; the smallest is the diffusive scale due to molecular viscosity.

From the above discussion, we see that the integral scale of turbulence can be regarded as the distance an eddy moves through before dissolving and losing its identity. Usually, it is of the same order of magnitude as the size of the eddy itself. In addition to the scale of turbulence, the intensity of the turbulence is another important parameter characterizing turbulent flows. The intensity I is a measure of the "violence" of the eddies, that is, the time-average fluctuation of the velocity in the fluid. It can be defined for isotropic turbulence as

$$I \equiv \frac{\sqrt{\overline{u'^2}}}{\overline{u}} \tag{7-56}$$

or for more general cases as

$$I \equiv \frac{\sqrt{\frac{1}{3}\left(\overline{u'^2} + \overline{v'^2} + \overline{w'^2}\right)}}{\overline{u}} \tag{7-57}$$

The time-averaged conservation equations can be obtained by applying the Reynolds averaging procedure. However, there are several immediate drawbacks: (1) in the continuity equation (7-39), there is an extra term, $(\partial/\partial x_j)(\overline{\rho' u_j'})$, which needs to be modeled; (2) this term implies that there is a mean mass interchange across the mean streamline, which is not consistent with the usual concept of a streamline; (3) most sampling probes measure values that approximate mass-weighted concentrations rather than time-average concentrations; and (4) as we shall see later, the resulting equations are of simpler form when the mass-weighted averaging procedure is used. For these reasons, mass-weighted averaging procedures are highly recommended by Favre,[20,21] Laufer and Ludloff,[22] Bilger,[23,24] and others.

We can define a mass-weighted mean velocity as

$$\boxed{\tilde{u}_i \equiv \frac{\overline{\rho u_i}}{\overline{\rho}}}, \tag{7-58}$$

The velocity may then be written as

$$u_i(x_i, t) \equiv \tilde{u}_i(x_i) + u_i''(x_i, t) \tag{7-59}$$

where $u_i''(x_i, t)$ is the superimposed velocity fluctuation. Multiplying Eq. (7-59) by $\rho(x_i, t)$ and decomposing the velocity u_i into two parts, we have

$$\rho u_i = \rho(\tilde{u}_i + u_i'') = \rho\tilde{u}_i + \rho u_i''$$

By time-averaging the above equation, we get

$$\overline{\rho u_i} = \overline{\rho}\tilde{u}_i + \overline{\rho u_i''}$$

From the definition of $\tilde{u}_i$ given by Eq. (7-58), it follows that

$$\overline{\rho u_i''} = 0 \tag{7-60}$$

Similarly, we can define the static enthalpy, static temperature, and total enthalpy thus:

$$h(x_i, t) = \tilde{h}(x_i) + h''(x_i, t) \tag{7-61}$$

$$T(x_i, t) = \tilde{T}(x_i) + T''(x_i, t) \tag{7-62}$$

$$h_t(x_i, t) = \tilde{h}_t(x_i) + h_t''(x_i, t) \tag{7-63}$$

where

$$\tilde{T} = \frac{\overline{\rho T}}{\bar{\rho}}, \qquad \tilde{h} = \frac{\overline{\rho h}}{\bar{\rho}}, \qquad \tilde{h}_t = \tilde{h} + \tfrac{1}{2}\tilde{u}_i\tilde{u}_i + \frac{1}{2}\frac{\overline{\rho u_i'' u_i''}}{\bar{\rho}}$$

$$h_t'' = h'' + \tilde{u}_i u_i'' + \tfrac{1}{2}u_i''u_i'' - \frac{1}{2}\frac{\overline{\rho u_i'' u_i''}}{\bar{\rho}} \tag{7-64}$$

Also,

$$\overline{\rho T''} = \overline{\rho h''} = \overline{\rho h_t''} = 0 \tag{7-65}$$

The expression for $\tilde{h}_t$ can be obtained by the following procedure. Multiplying both sides of $h_t = h + \frac{1}{2}u_i u_i$ by ρ, and introducing the definition of u_i given by Eq. (7-59) into the resulting expression, we can write

$$\overline{\rho h_t} = \overline{\rho h} + \tfrac{1}{2}\bar{\rho}\tilde{u}_i\tilde{u}_i + \tfrac{1}{2}\overline{\rho u_i'' u_i''} \tag{7-66}$$

Since $\overline{\rho h_t} = \bar{\rho}\tilde{h}_t$ and $\overline{\rho h} = \bar{\rho}\tilde{h}$, Eq. (7-66) can be written as

$$\tilde{h}_t = \tilde{h} + \tfrac{1}{2}\tilde{u}_i\tilde{u}_i + \frac{1}{2}\frac{\overline{\rho u_i'' u_i''}}{\bar{\rho}}$$

Also, one can follow the procedure given below to show the validity of Eq. (7-64)

$$h_t = \tilde{h}_t + h_t'' = \tilde{h} + h'' + \tfrac{1}{2}(\tilde{u}_i + u_i'')^2 = \tilde{h} + h'' + \tfrac{1}{2}\tilde{u}_i\tilde{u}_i + \tilde{u}_i u_i'' + \tfrac{1}{2}u_i''u_i''$$

Substituting the expression for $\tilde{h}_t$ into the above expression, we get

$$h_t'' = h'' + \tilde{u}_i u_i'' + \tfrac{1}{2}u_i''u_i'' - \frac{1}{2}\frac{\overline{\rho u_i'' u_i''}}{\bar{\rho}} \tag{7-67}$$

The relationship given by Eqs. (7-59) through (7-66) are convenient for deducing various quantities through turbulence measurements. In hot-wire anemometry, the quantities measured at low speeds are the fluctuations of ρu_i *and*

T, and those measured at supersonic speeds are the fluctuations of ρu_i and of a quantity that is very close to the total enthalpy. The mean pressure is directly measurable. For this reason, the conventional time average is convenient for p. Furthermore, we shall also use the conventional time averaging procedures for the stress tensor τ_{ij} and heat-flux vector q_j.

3.2 Relation between Conventional Time-Averaged Quantities and Mass-Weighted Averaged Quantities

A relationship between $\bar{u}_i$ and $\tilde{u}_i$ can be established as follows. Using Eq. (7-60), we can write a general form

$$\overline{\rho\phi''} = \overline{(\bar{\rho} + \rho')\,\phi''} = 0$$

where ϕ can be u, h, h_t or T. The above equation can be rearranged to give

$$\overline{\phi''} = -\frac{\overline{\rho'\phi''}}{\bar{\rho}} \tag{7-68}$$

Taking the time average of $\phi = \tilde{\phi} + \phi''$ and rearranging, we get

$$\tilde{\phi} - \bar{\phi} = -\overline{\phi''} \tag{7-69}$$

Hence,

$$\tilde{\phi} - \bar{\phi} = \frac{\overline{\rho'\phi''}}{\bar{\rho}} \tag{7-70}$$

Recalling the two separate averaging schemes, ϕ can be written as $\phi = \bar{\phi} + \phi'$ $= \tilde{\phi} + \phi''$. Multiplying both sides of that expression by ρ and averaging, we get

$$\bar{\rho}\tilde{\phi} + \overline{\rho\phi''} = \bar{\rho}\bar{\phi} + \overline{\rho\phi'}$$

If we recall that $\overline{\rho\phi''} = 0$ and that $\overline{\rho\phi'}$ can be written as $\overline{(\bar{\rho} + \rho')\phi'} = \overline{\rho'\phi'}$, we can combine the above expression and Eq. (7-70) to get

$$\tilde{\phi} - \bar{\phi} = \frac{\overline{\rho'\phi''}}{\bar{\rho}} = \frac{\overline{\rho'\phi'}}{\bar{\rho}} \tag{7-71}$$

From Eq. (7-71), we see that the difference between the two average quantities depends on the density-variable correlation term.

The well-known Navier–Stokes equations of motion for a compressible, viscous, heat-conducting perfect gas may be written in the following form (see

Chapter 3 for details):

Continuity,

$$\frac{\partial \rho}{\partial t} + \frac{\partial}{\partial x_j}(\rho u_j) = 0 \tag{7-72}$$

Momentum (assuming that the body force is negligible),

$$\frac{\partial}{\partial t}(\rho u_i) + \frac{\partial}{\partial x_j}(\rho u_i u_j) = -\frac{\partial p}{\partial x_i} + \frac{\partial \tau_{ij}}{\partial x_j} \tag{7-73}$$

Energy,

$$\frac{\partial}{\partial t}(\rho h_t) + \frac{\partial}{\partial x_j}(\rho u_j h_t) = \frac{\partial p}{\partial t} + \frac{\partial}{\partial x_j}(u_i \tau_{ij} - q_j) \tag{7-74}$$

where the stress tensor τ_{ij}, heat-flux vector q_j, and total enthalpy h_t are given by

$$\tau_{ij} = \left(\mu' - \tfrac{2}{3}\mu\right)\frac{\partial u_k}{\partial x_k}\delta_{ij} + \mu\left(\frac{\partial u_i}{\partial x_j} + \frac{\partial u_j}{\partial x_i}\right) \tag{7-75}$$

$$q_j = -\lambda \frac{\partial T}{\partial x_j} \tag{7-76}$$

$$h_t = h + \tfrac{1}{2}u_i u_i \tag{7-77}$$

3.3 Mass-Weighted Conservation and Transport Equations

3.3.1 Continuity and Momentum Equations

If we substitute the expressions given by Eqs. (7-29, 7-30, and 7-59) into Eqs. (7-72) and (7-73), we obtain

$$\frac{\partial}{\partial t}(\bar{\rho} + \rho') + \frac{\partial}{\partial x_j}(\rho \tilde{u}_j + \rho u_j'') = 0 \tag{7-78}$$

and

$$\frac{\partial}{\partial t}(\rho \tilde{u}_i + \rho u_i'') + \frac{\partial}{\partial x_j}(\rho \tilde{u}_i \tilde{u}_j + \rho u_j'' \tilde{u}_i + \rho u_i'' \tilde{u}_j + \rho u_i'' u_j'')$$

$$= -\frac{\partial \bar{p}}{\partial x_i} - \frac{\partial p'}{\partial x_i} + \frac{\partial \tau_{ij}}{\partial x_j} \tag{7-79}$$

Taking the time average of the terms appearing in these equations, we obtain the mean-continuity and mean-momentum equations for compressible turbulent flow:

$$\boxed{\underbrace{\frac{\partial \bar{\rho}}{\partial t}}_{\text{I}} + \underbrace{\frac{\partial}{\partial x_j}(\bar{\rho}\tilde{u}_j)}_{\text{II}} = 0} \tag{7-80}$$

$$\boxed{\underbrace{\frac{\partial}{\partial t}(\bar{\rho}\tilde{u}_i) + \frac{\partial}{\partial x_j}(\bar{\rho}\tilde{u}_i\tilde{u}_j)}_{\text{I}} = -\underbrace{\frac{\partial \bar{p}}{\partial x_i}}_{\text{II}} + \frac{\partial}{\partial x_j}\left(\underbrace{\bar{\tau}_{ij}}_{\text{III}} - \underbrace{\overline{\rho u_i'' u_j''}}_{\text{IV}}\right)} \tag{7-81}$$

The last term on the right-hand side of Eq. (7-81) represents the turbulent stresses due to turbulent diffusion of momentum.

3.3.2 Energy Equations

If we substitute the expressions given by Eqs. (7-29), (7-30), (7-59), and (7-63) into Eq. (7-74) and use Eq. (7-77), we obtain

$$\frac{\partial}{\partial t}(\rho\tilde{h}_t + \rho h_t'') + \frac{\partial}{\partial x_j}(\rho\tilde{h}_t\tilde{u}_j + \rho h_t''\tilde{u}_j + \rho u_j''\tilde{h}_t + \rho h_t'' u_j'')$$

$$= \frac{\partial}{\partial t}(\bar{p} + p') + \frac{\partial}{\partial x_j}(u_i\tau_{ij} - q_j) \tag{7-82}$$

After subtracting the mechanical energy from the total enthalpy equation, we have

$$\frac{\partial}{\partial t}(\rho\tilde{h} + \rho h'') + \frac{\partial}{\partial x_j}(\rho\tilde{h}\tilde{u}_j + \rho h''\tilde{u}_j + \rho u_j''\tilde{h} + \rho u_j'' h'')$$

$$= \frac{\partial}{\partial t}(\bar{p} + p') + (\tilde{u}_j + u_j'')\frac{\partial}{\partial x_j}(\bar{p} + p') + \tau_{ij}\frac{\partial u_i}{\partial x_j} - \frac{\partial q_j}{\partial x_j} \tag{7-83}$$

Taking the time average of the terms appearing in the equations, we obtain the mean-energy equations in terms of static enthalpy:

$$\underbrace{\frac{\partial}{\partial t}(\bar{\rho}\tilde{h}) + \frac{\partial}{\partial x_j}(\bar{\rho}\tilde{h}\tilde{u}_j)}_{\text{I}} = \underbrace{\frac{\partial \bar{p}}{\partial t} + \tilde{u}_j \frac{\partial \bar{p}}{\partial x_j}}_{\text{II}} + \underbrace{\overline{u_j'' \frac{\partial p}{\partial x_j}}}_{\text{III}} + \frac{\partial}{\partial x_j}\Bigg(\underbrace{-\bar{q}_j}_{\text{IV}} \underbrace{- \overline{\rho h'' u_j''}}_{\text{V}}\Bigg) + \underbrace{\bar{\tau}_{ij}\frac{\partial \tilde{u}_i}{\partial x_j} + \overline{\tau_{ij}\frac{\partial u_i''}{\partial x_j}}}_{\text{VI}} \tag{7-84}$$

The physical meaning of each term can be given as follows:

I:	$\frac{\partial}{\partial t}(\bar{\rho}\tilde{h}) + \frac{\partial}{\partial x_j}(\bar{\rho}\tilde{h}\tilde{u}_j)$	Average rate of change of ρh per unit volume per unit time
II:	$\frac{\partial \bar{p}}{\partial t} + \tilde{u}_j \frac{\partial \bar{p}}{\partial x_j}$	Pressure work due to macroscopic motion
III:	$\overline{u_j'' \frac{\partial p}{\partial x_j}}$	Work due to turbulence
IV:	$-\frac{\partial}{\partial x_j}(\bar{q}_j)$	Transport of heat due to conduction
V:	$\frac{\partial}{\partial x_j}(-\overline{\rho h'' u_j''})$	Turbulent diffusion of ρh
VI:	$\bar{\tau}_{ij}\frac{\partial \tilde{u}_i}{\partial x_j} + \overline{\tau_{ij}\frac{\partial u_i''}{\partial x_j}}$	Dissipation due to molecular friction

3.3.3 Mean-Kinetic-Energy Equation

By considering the scalar product of $\tilde{u}_j$ and the mean momentum equation for $\tilde{u}_i$,

$$\tilde{u}_j\left[\frac{\partial}{\partial t}(\bar{\rho}\tilde{u}_i) + \frac{\partial}{\partial x_k}(\bar{\rho}\tilde{u}_i\tilde{u}_k) = -\frac{\partial \bar{p}}{\partial x_i} + \frac{\partial}{\partial x_k}\left(\bar{\tau}_{ik} - \overline{\rho u_i'' u_k''}\right)\right] \tag{7-85}$$

and the scalar product of $\tilde{u}_i$ and the mean momentum equation for $\tilde{u}_j$,

$$\tilde{u}_i\left[\frac{\partial}{\partial t}(\bar{\rho}\tilde{u}_j) + \frac{\partial}{\partial x_k}(\bar{\rho}\tilde{u}_j\tilde{u}_k) = -\frac{\partial \bar{p}}{\partial x_j} + \frac{\partial}{\partial x_k}\left(\bar{\tau}_{jk} - \overline{\rho u_j'' u_k''}\right)\right] \tag{7-86}$$

and adding these two equations and rearranging, we obtain

$$\frac{\partial}{\partial t}(\bar{\rho}\tilde{u}_i\tilde{u}_j) + \frac{\partial}{\partial x_k}(\bar{\rho}\tilde{u}_i\tilde{u}_j\tilde{u}_k)$$

$$= -\tilde{u}_j\frac{\partial \bar{p}}{\partial x_i} - \tilde{u}_i\frac{\partial \bar{p}}{\partial x_j} + \tilde{u}_j\frac{\partial}{\partial x_k}\left(\bar{\tau}_{ik} - \overline{\rho u_i'' u_k''}\right) + \tilde{u}_i\frac{\partial}{\partial x_k}\left(\bar{\tau}_{jk} - \overline{\rho u_j'' u_k''}\right) \tag{7-87}$$

For $i = j$, Eq. (7-87) becomes

$$\underbrace{\bar{\rho}\frac{D}{Dt}\left(\frac{\tilde{u}_i\tilde{u}_i}{2}\right)}_{\text{I}} = \underbrace{-\tilde{u}_i\frac{\partial \bar{p}}{\partial x_i}}_{\text{II}} + \underbrace{\tilde{u}_i\frac{\partial \bar{\tau}_{ik}}{\partial x_k}}_{\text{III}} \underbrace{- \tilde{u}_i\frac{\partial}{\partial x_k}\left(\overline{\rho u_i'' u_k''}\right)}_{\text{IV}} \tag{7-88}$$

Terms I through IV in the above equation have the following physical meanings:

I: $\bar{\rho}\frac{D}{Dt}\left(\frac{\tilde{u}_i\tilde{u}_i}{2}\right)$ — Rate of change of kinetic energy of mean motion, or gain of kinetic energy of the mean motion by "advection" (not thermal convection)

II: $-\tilde{u}_i\frac{\partial \bar{p}}{\partial x_i}$ — Flow work done by mean pressure forces

II: $\tilde{u}_i\frac{\partial \bar{\tau}_{ik}}{\partial x_k}$ — Work of molecular friction due to mean motion

IV: $-\tilde{u}_i\frac{\partial}{\partial x_k}\overline{\rho u_i'' u_k''}$

$= -\frac{\partial}{\partial x_k}(\tilde{u}_i\overline{\rho u_i'' u_k''})$ — Spatial transport of mean kinetic energy by turbulent fluctuations

$+\overline{\rho u_i'' u_k''}\frac{\partial \tilde{u}_i}{\partial x_k}$ — Production of turbulent energy from mean flow energy

3.3.4 Reynolds-Stress Transport Equations

Let us consider the scalar product of u_j and momentum equation for u_i,

$$u_j\left[\frac{\partial}{\partial t}(\rho u_i) + \frac{\partial}{\partial x_k}(\rho u_i u_k) = -\frac{\partial p}{\partial x_i} + \frac{\partial \tau_{ik}}{\partial x_k}\right] \tag{7-89}$$

and the scalar product of u_i and the momentum equation for u_j,

$$u_i\left[\frac{\partial}{\partial t}(\rho u_j)+\frac{\partial}{\partial x_k}(\rho u_j u_k)=-\frac{\partial p}{\partial x_j}+\frac{\partial \tau_{jk}}{\partial x_k}\right] \tag{7-90}$$

The sum of the two equations is

$$\frac{\partial}{\partial t}(\rho u_i u_j)+\frac{\partial}{\partial x_k}(\rho u_i u_j u_k)=-u_j\frac{\partial p}{\partial x_i}-u_i\frac{\partial p}{\partial x_j}+u_j\frac{\partial \tau_{ik}}{\partial x_k}+u_i\frac{\partial \tau_{jk}}{\partial x_k} \tag{7-91}$$

Using Eq. (7-59), we can write Eq. (7-91) as

$$\frac{\partial}{\partial t}\left[\rho(\tilde{u}_i+u_i'')(\tilde{u}_j+u_j'')\right]+\frac{\partial}{\partial x_k}\left[\rho(\tilde{u}_i+u_i'')(\tilde{u}_j+u_j'')(\tilde{u}_k+u_k'')\right]$$

$$=-(\tilde{u}_j+u_j'')\frac{\partial p}{\partial x_i}-(\tilde{u}_i+u_i'')\frac{\partial p}{\partial x_j}+(\tilde{u}_j+u_j'')\frac{\partial \tau_{ik}}{\partial x_k}+(\tilde{u}_i+u_i'')\frac{\partial \tau_{jk}}{\partial x_k}$$

where $\tau_{ik}=\bar{\tau}_{ik}+\tau_{ik}'$ and $\tau_{jk}=\bar{\tau}_{jk}+\tau_{jk}'$. Taking the time average of the above equation, we obtain

$$\frac{\partial}{\partial t}\left(\bar{\rho}\tilde{u}_i\tilde{u}_j+\overline{\rho u_i''u_j''}\right)$$

$$+\frac{\partial}{\partial x_k}\left(\bar{\rho}\tilde{u}_i\tilde{u}_j\tilde{u}_k+\tilde{u}_k\overline{\rho u_i''u_j''}+\tilde{u}_i\overline{\rho u_j''u_k''}+\tilde{u}_j\overline{\rho u_i''u_k''}+\overline{\rho u_i''u_j''u_k''}\right)$$

$$=-\tilde{u}_j\frac{\partial\bar{p}}{\partial x_i}-\overline{u_j''\frac{\partial p}{\partial x_i}}-\tilde{u}_i\frac{\partial\bar{p}}{\partial x_j}-\overline{u_i''\frac{\partial p}{\partial x_j}}$$

$$+\tilde{u}_j\frac{\partial\bar{\tau}_{ik}}{\partial x_k}+\overline{u_j''\frac{\partial\tau_{ik}}{\partial x_k}}+\tilde{u}_i\frac{\partial\bar{\tau}_{jk}}{\partial x_k}+\overline{u_i''\frac{\partial\tau_{jk}}{\partial x_k}} \tag{7-92}$$

Subtracting Eq. (7-87) from the above equation and rearranging (with the understanding that τ_{jk} contains ρ through μ) we obtain the following Reynolds-stress transport equation:

$$\boxed{\begin{aligned}&\frac{D}{Dt}\left(\overline{\rho u_i''u_j''}\right)+\frac{\partial}{\partial x_k}\left(\overline{\rho u_i''u_j''u_k''}\right)\\&\quad=-\overline{u_j''\frac{\partial p}{\partial x_i}}-\overline{u_i''\frac{\partial p}{\partial x_j}}+\overline{u_j''\frac{\partial\tau_{ik}'}{\partial x_k}}+\overline{u_i''\frac{\partial\tau_{jk}'}{\partial x_k}}\\&\qquad-\overline{\rho u_i''u_k''}\frac{\partial\tilde{u}_j}{\partial x_k}-\overline{\rho u_j''u_k''}\frac{\partial\tilde{u}_i}{\partial x_k}-\overline{\rho u_i''u_j''}\frac{\partial\tilde{u}_k}{\partial x_k}\end{aligned}} \tag{7-93}$$

3.3.5 Turbulence-Kinetic-Energy Equation

For $i = j$, Eq. (7-93) becomes the turbulence-kinetic-energy equation:

$$\underbrace{\frac{D}{Dt}\left(\overline{\tfrac{1}{2}\rho u_i''u_i''}\right)}_{\text{I}} + \underbrace{\frac{\partial}{\partial x_k}\overline{u_k''\left(\tfrac{1}{2}\rho u_i''u_i''\right)}}_{\text{II}} = -\underbrace{\overline{u_i''\frac{\partial p}{\partial x_i}}}_{\text{III}} + \underbrace{\overline{u_i''\frac{\partial \tau_{ik}'}{\partial x_k}}}_{\text{IV}} - \underbrace{\tfrac{1}{2}\,\overline{\rho u_i''u_k''}\frac{\partial \tilde{u}_i}{\partial x_k}}_{\text{V}} \tag{7-94}$$

The physical meanings of the terms of Eq. (7-94) are as follows:

I:	$\frac{D}{Dt}(\overline{\tfrac{1}{2}\rho u_i''u_i''})$	Rate of change of kinetic energy of turbulence
II:	$\frac{\partial}{\partial x_k}[\overline{u_k''(\tfrac{1}{2}\rho u_i''u_i'')}]$	Kinetic energy of fluctuations convected by the fluctuations (i.e., diffusion of fluctuation energy)
III:	$\overline{u_i''\frac{\partial p}{\partial x_i}}$	Work due to turbulence
IV:	$\overline{u_i''\frac{\partial \tau_{ik}'}{\partial x_k}}$	Work of viscous stresses due to fluctuation motion
V:	$-\overline{\rho u_i''u_k''}\frac{\partial \tilde{u}_i}{\partial x_k} \quad -\overline{\rho u_i''u_i''}\frac{\partial \tilde{u}_k}{\partial x_k}$	Product of turbulent stress and mean rate of strain: production of turbulent energy

3.3.6 Species Conservation Equation

The species conservation equation based upon the Favre averaging can be shown to have the following form:

$$\frac{\partial}{\partial t}(\bar{\rho}\tilde{Y}_k) + \frac{\partial}{\partial x_i}(\bar{\rho}\tilde{Y}_k\tilde{u}_i) = \frac{\partial}{\partial x_i}\left[\mathscr{D}\bar{\rho}\frac{\partial \tilde{Y}_k}{\partial x_i} - \overline{\rho Y_k''u_i''}\right] + \frac{\partial}{\partial x_i}\mathscr{D}\overline{\rho\frac{\partial Y_k''}{\partial x_i}} + \bar{\dot{\omega}}_k \tag{7-95}$$

It will be useful for the reader to determine the physical meanings of the terms in the above equation.

3.4 Effect of Vorticity on Turbulent Flames

The vorticity equation can be obtained by taking the curl of the momentum equation. The reason that we are interested in the vorticity equation is the

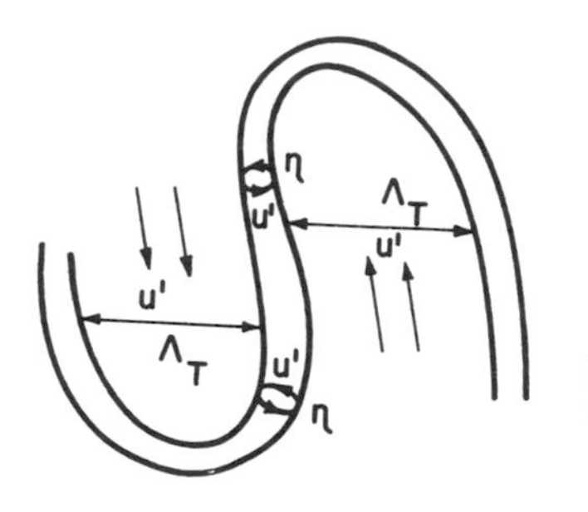

Figure 7.11 Tennekes's model of a region of concentrated vorticity (after Chomiak[26]).

tendency towards dissipation intermittency discovered initially by Batchelor and Townsend.[25] They showed that for high-Reynolds-number turbulence, the turbulent fine structure, which is decisive for the molecular mixing and dissipative process, is concentrated in isolated regions. The total volume of these isolated regions is a very small fraction of the volume of the fluid. As described clearly by Chomiak,[26] the problem was then studied from two points of view: (a) statistical modeling of the energy cascade of turbulent fluctuations and (b) consideration of the hydrodynamic vorticity production mechanism due to stretching of vortex lines. Kuo and Corrsin[27] made an extensive review of works on the vortex-stretching mechanism. They also succeeded[28] in showing experimentally that the severely intermittent small-scale structures are typically ribbons or tubes of activity. Tennekes[29] developed a model for the dissipative regions; he considered that the turbulence fine structure is composed of randomly distributed tubes with diameters on the order of the Kolmogorov microscale η and a spacing on the order of the Taylor microscale Λ_T (see Fig. 7.11).

The Kolmogorov microscale is defined[30] as

$$\eta \equiv \left[\frac{\nu^3}{\varepsilon}\right]^{1/4} \tag{7-96}$$

where ν is the kinematic viscosity and ε is the rate of energy dissipation per unit fluid mass. Mathematically, ε can be defined[30] as

$$\varepsilon \equiv \nu \overline{\frac{\partial u_i'}{\partial x_j}\frac{\partial u_i'}{\partial x_j}} = \nu\overline{\Omega_k'\Omega_k'} \tag{7-97}$$

It is equal to the kinematic viscosity times the mean square of the rate of strain —or times the mean squared vorticity of the turbulence, where the vorticity of turbulence is defined as

$$\Omega_k' \equiv -e_{ijk}\frac{\partial u_i'}{\partial x_j} \tag{7-98}$$

with e_{ijk} the permutation symbol defined by

$$e_{ijk} = \begin{cases} 0 & \text{if any two of } i, j, k \text{ are the same} \\ 1 & \text{if } ijk \text{ is an even permutation of 123} \\ -1 & \text{if } ijk \text{ is an odd permutation of 123} \end{cases} \tag{7-99}$$

The Kolmogorov microscale is the dissipative scale of turbulence; it represents the smallest scale in a turbulent flow. The Taylor microscale, which can be obtained using order-of-magnitude analysis by setting the production term equal to the dissipation term in turbulence kinetic energy equation, is defined as

$$\Lambda_T \equiv \frac{u'_{\text{rms}}}{\sqrt{\overline{\dfrac{\partial u'_1}{\partial x_1} \dfrac{\partial u'_1}{\partial x_1}}}} \tag{7-100}$$

which is much larger than η, as shown in Fig. 7.11. Tennekes's model is consistent with the small-scale dynamics of turbulence; it gives good explanations of the changes of intermittency factor with Reynolds number.

Taking the curl of the momentum equation, the vorticity equation can be obtained (see Thomson[31]):

$$\frac{\partial \mathbf{\Omega}}{\partial t} + \mathbf{v} \cdot \nabla \mathbf{\Omega} = \mathbf{\Omega} \cdot \nabla \mathbf{v} + \nu \nabla^2 \mathbf{\Omega} - \mathbf{\Omega} \nabla \cdot \mathbf{v} - \nabla \frac{1}{\rho} \times \nabla p \tag{7-101}$$

where the vector $\mathbf{\Omega}$ is defined as $\frac{1}{2}\nabla \times \mathbf{v}$. The curl of the velocity is known as the vorticity. The meanings of the terms on the right-hand side of Eq. (7-101) are listed below:

$\mathbf{\Omega} \cdot \nabla \mathbf{v}$ represents the change in vorticity due to stretching of the vortices.[32]

$\nu \nabla^2 \mathbf{\Omega}$ is the viscous-dissipation term.

$\mathbf{\Omega} \nabla \cdot \mathbf{v}$ corresponds to the effect of a change in density for the vortices, which rotate more slowly if their sizes increase.

$\nabla \dfrac{1}{\rho} \times \nabla p$ represents the vorticity generation due to the existence of pressure gradients in a medium of varying density.

If we disregard all the terms on the right-hand side of Eq. (7-101) except the vortex stretching term, then we have

$$\frac{D\mathbf{\Omega}}{Dt} = \mathbf{\Omega} b(t) \tag{7-102}$$

where $b(t)$ is the effective stretching of the fluid element. It represents the component of the gradient of the velocity vector $\mathbf{v}$ along the vorticity vector $\boldsymbol{\Omega}$. It can also be interpreted as the instantaneous rate of extension of the local vortex tubes. From Eq. (7-102), it follows that if the stretching of vortex line is a random function of time, $|\Omega|$ will have a lognormal distribution as time $t \to \infty$, since

$$\ln \frac{|\Omega(t)|}{|\Omega(0)|} = \int_0^t b(t^*)\, dt^* \tag{7-103}$$

This result matches with most statistical studies of the energy cascade.

According to Chomiak,[26] the vortex structures are not successively produced and dissipated. They are permanent and follow the regions of maximum stretching. The velocity of the vortex regions relative to the adjacent fluid is

$$v = \left(\varepsilon \nu \, \mathrm{Re}_{\Lambda_T}\right)^{1/4} \tag{7-104}$$

If the region of intense stretching shifts, the concentrated vortex will diffuse and will be rebuilt in the region of maximum stretching. Therefore, the movement of maximum dissipative regions has the form of "jumping" or "marching" from maximum-vorticity regions to regions of maximum stretching.

In general, regions of concentrated vorticity are the only ones where all the molecular processes occur in the turbulent fluid, that is, viscous dissipation of turbulent energy, molecular mixing, molecular heat exchange, and consequently, chemical reactions. The dissipative regions are like well-stirred reactors with very high effective chemical reaction rates, placed in a background of large regions free of chemical reaction. Based upon the review of Monin and Yaglom,[34] even if the initial vorticity has a very flat spatial distribution, the vorticity generation mechanism in turbulent flow leads to a narrowly peaked spatial distribution after a time which is large compared with large eddy turnover time. Vorticity is generally concentrated in isolated regions whose entire volume is a small fraction of the volume of the fluid. Also, the concentrated-vorticity regions become less and less space-filling as the Reynolds number of the flow increases.

Heat release due to chemical reactions has effects on vorticity. The expansion of the fluid due to heat release reduces the vorticity and destroys the local vortex tubes. If the vorticity generation terms in Eq. (7-101) can not compensate for this effect, the turbulence energy decays. Then combustion is not generating turbulence but damping it and laminarizing the flow. On the other hand, if the heat release in one portion of the system causes the flow field to change so that the vorticity generation terms are larger, then combustion in one region can intensify turbulence in the other regions. Any readers interested in vortex method in turbulent flames are referred to Sec. 8.3 for further discussions.

4 THE APPLICATION OF PROBABILITY DENSITY FUNCTIONS IN TURBULENT FLAMES

During recent years, intensive research activity in the field of turbulent combustion has occurred, due to the demand for higher performance combustors and concern for pollutant emission. The use of probability-density-function (pdf) equations, so common in kinetic theory, statistical mechanics, and quantum field theory, has recently become more popular in the study of turbulent flames.

Reacting turbulent flow is a very complex phenomenon due to continuous fluctuations in density, velocities, temperature, pressure, and concentrations of species. The governing equations take on a very complicated, unclosed form (since there are more unknowns than equations). Thus special formulation and closure methods are needed.

The approaches taken by different investigators can be classified in the following manner:

1. *Moment Methods.* These include correlations and spectral methods.
2. *Probability Density Function (pdf) Methods.* These methods use a statistical description of the turbulent field together with the governing conservation equations.
3. *Mixed Methods.* These methods use the moment formalism for the fluid-mechanical description of the problem and use the pdf method, either explicitly or implicitly, for the scalars, such as the species mass fractions and the enthalpy of the mixture.

There has not been significant development in the use of moment methods in modeling reaction parts of the combustion problem, because of the difficulties in modeling the unclosed species-production-rate term ω_i occurring in the species mass-balance equation

$$\rho \frac{\partial Y_i}{\partial t} + \rho \mathbf{v} \cdot \nabla Y_i - \nabla \cdot \rho \mathscr{D}_i \nabla Y_i = \omega_i \tag{7-105}$$

where

$$\omega_i = -\rho^2 k Y_i Y_j \exp(-T_a/T) \tag{7-106}$$

Donaldson and Varma[118] gave a set of general conservation equations for a turbulent multicomponent reacting system. They found that the production rate and its constituent variables need to be modeled. For a simple three-component reaction of the type $A + B \rightarrow C$, thirteen moments must be modeled, and the task of modeling them correctly is very difficult. For more complex chemical systems, the closure problem is even more complicated.

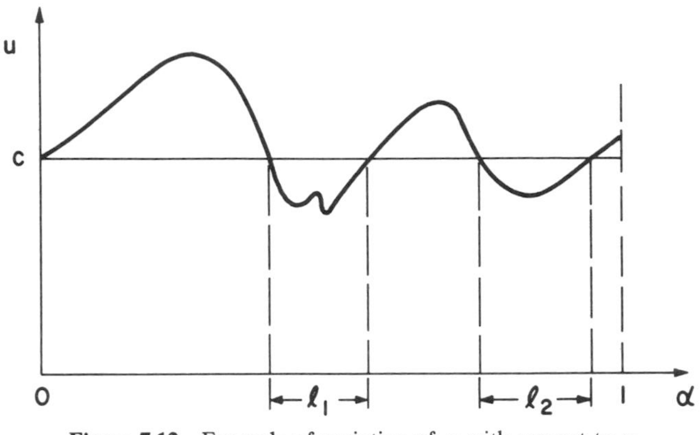

Figure 7.12 Example of variation of u with respect to α.

4.1 Probability Density Function

Before we introduce the pdf, we define probability (Pr) as follows. Let us consider a continuous distribution of a variable u, each of which has been given a serial number α, lying between zero and unity ($\alpha \in [0,1]$). Then the probability that $u < c$ (a fixed constant) is defined as

$$\Pr\{u < c\} = \frac{\text{number of experiments in which } u < c}{\text{total number of experiments}} \tag{7-107}$$

Also,

$$\Pr\{u < c\} = \text{fraction of } [0,1] \text{ on which } u < c$$

If for example the variation u with respect to α is as shown in Fig. 7.12, then corresponding to this profile

$$\Pr\{u < c\} = l_1 + l_2$$

Let us define an indicator function $\phi(\alpha)$ for the set of points on which $u < c$:

$$\phi(\alpha) \equiv \begin{cases} 1 & \text{for} \quad u < c \\ 0 & \text{otherwise} \end{cases} \tag{7-108}$$

We can then write

$$\Pr\{u < c\} = \int_0^1 \phi(\alpha)\, d\alpha \tag{7-109}$$

There are several more terms that must be introduced before defining the pdf. These terms are: measure, Lebesgue integral, and distribution function.

4.1.1 Measure

Measure of a set is a generalization of the concept of length; if a set of points consists of intervals only, then the measure of that set is just the sum of the lengths of the intervals.

For more general sets (let us consider sets lying in [0, 1]), one can consider *coverings* of the set: collections of open (i.e., not including the end points) intervals which contain the set. Among all the possible coverings the sum of the lengths of the intervals will have a least upper bound, which is called the *outer measure* of the set. If the same is done to the *complement* of the set (i.e., all those points in [0, 1] which are not in the set), and the result subtracted from unity, the result is the *inner measure* of the set. If the inner and the outer measure are the same, one says that the set is *measurable*, and the common value is its (Lebesgue) *measure*, denoted for example $\mu(\{\alpha \mid u < c\})$, which is read "the measure of the set of points α at which $u < c$."

We can define probability in terms of the measure:

$$\Pr\{u < c\} = \mu(\{\alpha \mid u < c\}) \tag{7-110}$$

4.1.2 Lebesgue Integral

If function is integrable in the ordinary way (Riemann-integrable), it is also Lebesgue-integrable. Let us divide the function $f(\alpha)$ shown in Fig. 7.13 vertically instead of horizontally by considering a mesh of possible values that the function can take on, say $\{c_k\}$ with $c_{k-1} < c_k < c_{k+1}$. If the function $f(\alpha)$ is a measurable function (i.e., the set of points on which it takes on the values in any mesh interval are measurable), then one can write an approximating sum as

$$\int_a^b f(\alpha)\, d\alpha \approx \sum_k \zeta_k \mu(\{\alpha \mid c_{k-1} \le f < c_k\}) \tag{7-111}$$

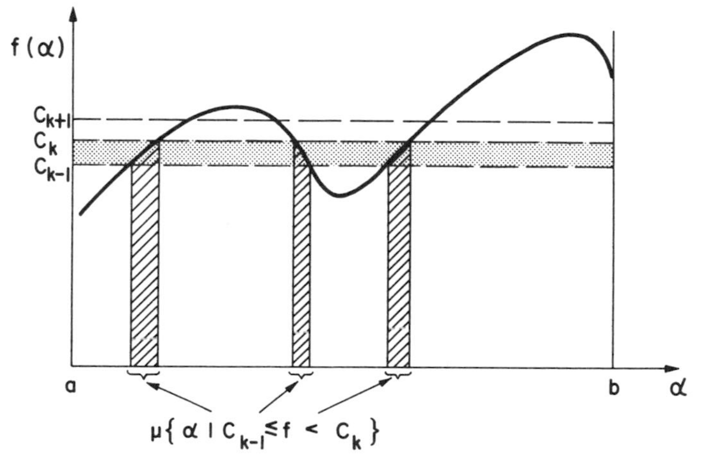

Figure 7.13 Lebesgue integration of a function $f(\alpha)$.

where

$$c_{k-1} \leq \zeta_k < c_k \tag{7-112}$$

As the mesh is drawn finer and finer, the summation may go to a limit. If it does, one says the function is Lebesgue-integrable.

For the indicator function $\phi(\alpha)$ defined earlier, we have

$$\int_0^1 \phi(\alpha)\, d\alpha = \mu(\{\alpha \mid u < c\}) \tag{7-113}$$

4.1.3 Distribution Function

The function defined by Eq. (7-109) or (7-110) can be considered as a function of c, since the value of $\Pr\{u < c\}$ changes with c. The distribution function is defined as

$$F(c) \equiv \Pr\{u < c\} \tag{7.114}$$

Several important properties of the distribution function are given below:

A. $F(c) \geq 0$. (Since it is a probability, or measure of a portion of the interval [0, 1], it cannot be negative.)

B. $F(c_1) \leq F(c_2)$ if $c_1 \leq c_2$. (We have $\Pr\{u < c_2\} = \Pr\{u < c_1\} + \Pr\{c_1 < u < c_2\}$; but $\Pr\{c_1 < u < c_2\}$ is always positive, and therefore $\Pr\{u < c_1\} \leq \Pr\{u < c_2\}$.)

C. $F(c) \leq 1$. [This is apparent from the definition of $F(c)$.]

D. $F(-\infty) = 0$ and $F(\infty) = 1$. (Since the value of u is always bounded everywhere.)

A typical distribution function is shown in the accompanying figure.

4.1.4 Defintion of the PDF

Let us now consider a function $\psi(\alpha)$ defined by

$$\psi(\alpha) = \begin{cases} 1 & \text{for} \quad c < u < c + \Delta c, \text{ where } \Delta c > 0 \\ 0 & \text{otherwise} \end{cases} \tag{7-115}$$

It can be shown by modifying Fig. 7.12 and using the definitions of the

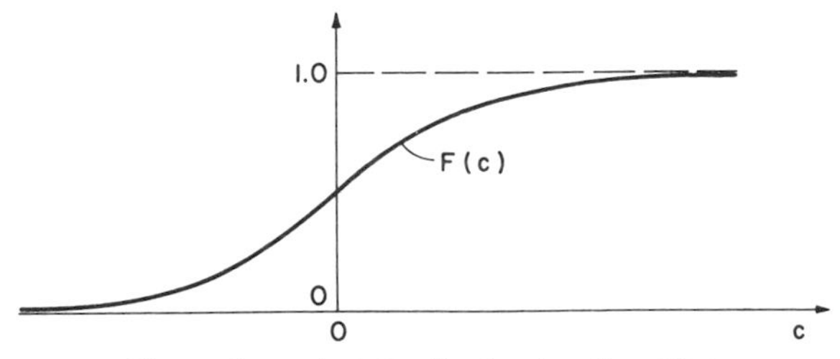

Shape of a typical distribution function $F(c)$.

indicator functions $\phi(\alpha)$ and $\phi^*(\alpha)$ that

$$\psi(\alpha) = \phi^*(\alpha)[1 - \phi(\alpha)] \tag{7-116}$$

where $\phi(\alpha)$ is defined by Eq. (7-108) and $\phi^*(\alpha)$ is defined as

$$\phi^*(\alpha) = \begin{cases} 1 & \text{if } u < c + \Delta c \\ 0 & \text{otherwise} \end{cases} \tag{7-117}$$

The probability for u to satisfy $c \leq u < c + \Delta c$ can be obtained by integrating $\psi(\alpha)$ from 0 to 1:

$$\Pr\{c \leq u < c + \Delta c\} = \int_0^1 \psi(\alpha)\, d\alpha \tag{7-118}$$

We shall always assume that the following limit exists for all values of c, and define it as the probability density function $\mathscr{P}(c)$:

$$\lim_{\Delta c \to 0} \frac{\int_0^1 \psi(\alpha)\, d\alpha}{\Delta c} \equiv \mathscr{P}(c) \tag{7-119}$$

From property B of $F(c)$, we have

$$\int_0^1 \phi^*(\alpha)\, d\alpha \geq \int_0^1 \phi(\alpha)\, d\alpha \tag{7-120}$$

Also, based upon the above inequality, we can write

$$\int_0^1 \phi(\alpha)\phi^*(\alpha)\, d\alpha = \int_0^1 \phi(\alpha)\, d\alpha \tag{7-121}$$

Now we can express the integral of $\psi(\alpha)$ as

$$\int_0^1 \psi(\alpha)\, d\alpha = \int_0^1 [\phi^*(\alpha) - \phi^*(\alpha)\phi(\alpha)]\, d\alpha = \int_0^1 [\phi^*(\alpha) - \phi(\alpha)]\, d\alpha$$

$$= F(c + \Delta c) - F(c) \tag{7-122}$$

From Eqs. (7-119) and (7-122), we have

$$\boxed{\mathscr{P}(c) = \lim_{\Delta c \to 0} \frac{F(c + \Delta c) - F(c)}{\Delta c} = \frac{dF(c)}{dc}} \tag{7-123}$$

The generalized function $\mathscr{P}(c)$ is called the *probability density function* (pdf). Several important properties of the pdf [or $\mathscr{P}(c)$] are given below:

A. $\mathscr{P}(c) \geq 0$. [This implies that the slope of $F(c)$ is always greater than or equal to zero.]

B. $F(b) - F(a) = \int_a^b \mathscr{P}(c)\, dc$. [This is an integrated form of Eq. (7-123).] In terms of probability, we can write

$$\Pr\{a \le u < b\} = \Pr\{u < b\} - \Pr\{u < a\} = F(b) - F(a)$$

Therefore,

$$\Pr\{a \le u < b\} = \int_a^b \mathscr{P}(c)\, dc = F(b) - F(a) \qquad (7\text{-}124)$$

C. If we extend the limits to ∞ and $-\infty$, Eq. (7-124) gives

$$\Pr\{-\infty \le u < \infty\} = \int_{-\infty}^{\infty} \mathscr{P}(c)\, dc = F(\infty) - F(-\infty)$$

Using the property D of the distribution function, we have

$$\int_{-\infty}^{\infty} \mathscr{P}(c)\, dc = 1 \qquad (7\text{-}125)$$

Example 7.1

Consider an ensemble made of the function

$$u = \sin(2\pi\alpha)$$

that is, the experimental values consist of samples taken from a sinusoidal function at arbitrary phase. Derive expressions for the distribution function and probability density function. Plot them as functions of c.

Solution:

Let us first sketch the function $u(\alpha)$ and draw a horizontal line at $u = c$. As we can see,

$$F(c) = \begin{cases} 1 & \text{if} \quad c > +1 \\ a + b + d & \text{if} \quad -1 \le c \le +1 \\ 0 & \text{if} \quad -1 > c \end{cases}$$

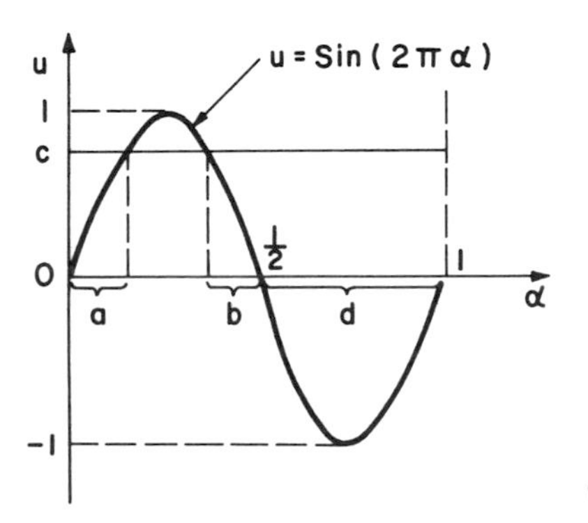

Graph of u with respect to α in Example 7.1.

Since $a = (1/2\pi)\sin^{-1}c = b$ and $d = \frac{1}{2}$, $F(c)$ can be written as

$$F(c) = \begin{cases} 1 & \text{if} \quad c > +1 \\ \dfrac{1}{2} + \dfrac{1}{\pi}\sin^{-1}c & \text{if} \quad -1 \le c \le +1 \\ 0 & \text{if} \quad -1 > c \end{cases}$$

Differentiating $F(c)$ with respect to c, the probability density function $\mathscr{P}(c)$ can be written as

$$\mathscr{P}(c) = \begin{cases} 0 & \text{if} \quad c > 1 \\ \pi^{-1}(1 - c^2)^{-1/2} & \text{if} \quad -1 \le c \le +1 \\ 0 & \text{if} \quad -1 > c \end{cases}$$

$F(c)$and $\mathscr{P}(c)$ are plotted below:

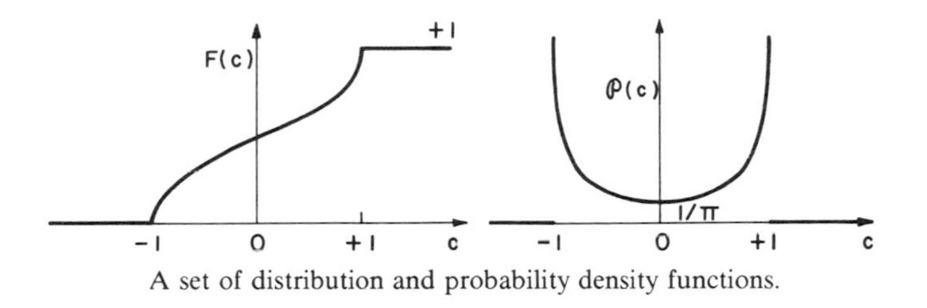

A set of distribution and probability density functions.

The fact that $\mathscr{P}(c)$ is unbounded at $+1$ and -1 is not surprising. This is because $u = \sin(2\pi\alpha)$ is very flat near $+1$ and -1, so that there are, correspondingly, infinitely many values of α at which u is near $+1$ and -1.

Consider the variation of the mass fraction of a certain species Y with respect to time as given by the sinusoidal function

$$Y = \overline{Y} + \tfrac{1}{2}(Y_+ - Y_-)\sin(2\pi\omega t) \tag{7-126}$$

where ω is frequency. Then based upon the above example, we have

$$F(Y) = \begin{cases} 1 & \text{if} \quad Y > Y_+ \\ \dfrac{1}{2} + \dfrac{1}{\pi}\sin^{-1}\left(\dfrac{Y - \overline{Y}}{Y_+ - \overline{Y}}\right) & \text{if} \quad Y_- \le Y \le Y_+ \\ 0 & \text{if} \quad Y_- > Y \end{cases} \tag{7-127}$$

$$\mathscr{P}(Y) = \begin{cases} 0 & \text{if} \quad Y > Y_+ \\ \dfrac{1/\pi}{\sqrt{(Y_+ - Y)(Y - Y_-)}} & \text{if} \quad Y_- \le Y \le Y_+ \\ 0 & \text{if} \quad Y_- > Y \end{cases} \tag{7-128}$$

The plots of $F(Y)$ and $\mathscr{P}(Y)$ are identical to those in Example 7.1 except that the values -1 and $+1$ of C are replaced by Y_- and Y_+ on the horizontal axis of Y.

The expected value (or ensemble average) of any physical quantity q as a function of u can be written as

$$E\{q(u)\} = \int_0^1 q[u(\alpha)]\, d\alpha \tag{7-129}$$

Referring to Fig. 7.13, the integral can be evaluated by Lebesgue integration in the following manner

$$\int_0^1 q[u(\alpha)]\, d\alpha = \lim_{\|\Delta c\| \to 0} \sum_k \zeta_k \mu(\{\alpha \mid c_k \leq u < c_{k+1}\}) \tag{7-130}$$

where

$$\|\Delta c\| = \max_{\substack{k \\ c_k \leq \zeta_k < c_{k+1}}} |c_{k+1} - c_k|$$

In terms of c, the ensemble average of $q(u)$ can be written as

$$E\{q(u)\} = \int_{c=-\infty}^{c=\infty} q[u(c)]\, dF(c) = \int_{-\infty}^{\infty} q[u(c)]\mathscr{P}(c)\, dc \tag{7-131}$$

This equation says that the ensemble average of $q(u)$ can be evaluated easily from the right-hand integral as soon as the pdf is specified. The density-weighted mean value of any scalar quantity q (uniquely related to mixture fraction f, e.g., T and Y_j) can be given, according to Jones and Whitelaw,[5] by

$$\tilde{q} = \int_0^1 q(f)\tilde{\mathscr{P}}(f, x_i)\, df \tag{7-132}$$

and unweighted Reynolds-averaged values by

$$\bar{q} = \bar{\rho}\int_0^1 \frac{q(f)}{\rho(f)}\tilde{\mathscr{P}}(f, x_i)\, df \tag{7-133}$$

where $\tilde{\mathscr{P}}(f, x_i)$ is the density-weighted pdf. The density can be obtained from

$$\bar{\rho} = \left[\int_0^1 \frac{\tilde{\mathscr{P}}(f, x_i)}{\rho(f)}\, df\right]^{-1} \tag{7-134}$$

where x_i represents the coordinate in the direction i.

4.2 Joint Distribution Function and Joint Probability Density Function

We can define the probability that $u < c$ and $v < c'$ are simultaneously satisfied:

$$\Pr\{u < c, v < c'\} = \frac{\text{number of experiments in which } u < c,\ v < c'}{\text{total number of experiments}} \tag{7-135}$$

Recall the indicator function $\phi(\alpha)$ given by Eq. (7-108), and define $\phi'(\alpha)$ in a similar manner:

$$\phi'(\alpha) = \begin{cases} 1, & v < c' \\ 0 & \text{otherwise} \end{cases} \tag{7-136}$$

Then

$$\phi(\alpha)\phi'(\alpha) = \begin{cases} 1, & u < c \text{ and } v < c' \\ 0 & \text{otherwise} \end{cases} \tag{7-137}$$

The measure of this set is obtained by integrating over the ensemble; as before,

$$\Pr\{u < c, v < c'\} = \mu(\{\alpha | u < c, v < c'\}) = \int_0^1 \phi(\alpha)\phi'(\alpha)\, d\alpha \tag{7-138}$$

The joint distribution function $F(c, c')$ is defined by

$$F(c, c') \equiv \Pr\{u < c, v < c'\} \tag{7-139}$$

Several important properties of the joint distribution function are given below:

A. $\lim_{c \to -\infty} F(c, c') = 0$, $\lim_{c' \to -\infty} F(c, c') = 0$. Similarly, $\lim_{c, c' \to \infty} F(c, c') = 1$.
B. $F(c, c') \le F(c, c'')$ if $c' \le c''$; $F(c, c') \le F(c'', c')$ if $c \le c''$.
C. $F(c_2, c_1') - F(c_1, c_1') + F(c_2, c_2') - F(c_2, c_1') \ge 0$ if $c_2 > c_1$ and $c_2' > c_1'$.
D. $F(c, +\infty) = F(c)$ and $F(\infty, c') = F(c')$, where $F(c) \equiv \Pr\{u < c\}$.

The joint pdf $\mathscr{P}_J(c, c')$ can be defined as

$$\mathscr{P}_J(c, c') \equiv \frac{\partial^2}{\partial c\, \partial c'} F(c, c') \tag{7-140}$$

Several important properties of $\mathscr{P}_J(c, c')$ are given below:

A. $\mathscr{P}_J(c, c') \ge 0$
B. $\int_{-\infty}^{\infty}\int_{-\infty}^{\infty} \mathscr{P}_J(c, c')\, dc\, dc' = 1$

C. $\int_{-\infty}^{\infty}\mathscr{P}_J(c, c')\,dc' = \mathscr{P}(c)$, where $\int_a^b \mathscr{P}(c)\,dc = \Pr\{a \le u < b\}$

D. $\int_a^b[\int_d^e \mathscr{P}_J(c, c')\,dc']\,dc = \Pr\{a \le u < b,\ d \le v < e\}$

E. By definition

$$\psi(\alpha) \equiv \begin{cases} 1, & c \le u < c + \Delta c \\ 0 & \text{otherwise} \end{cases}$$

$$\psi'(\alpha) = \begin{cases} 1, & c' \le v < c' + \Delta c' \\ 0 & \text{otherwise} \end{cases}$$

Therefore,

$$\lim_{\substack{\Delta c \to 0 \\ \Delta c' \to 0}} \left(\frac{1}{\Delta c\,\Delta c'}\right) \int_0^1 \psi(\alpha)\psi'(\alpha)\,d\alpha = \mathscr{P}_J(c, c')$$

The ensemble average of physical quantity q can be evaluated by

$$\begin{aligned} E\{q(u, v)\} &= \int_0^1 q[u(\alpha), v(\alpha)]\,d\alpha \\ &= \int_{c=-\infty,\ c'=-\infty}^{c=\infty,\ c'=\infty} q(c, c')\,dF(c, c') \\ &= \int_{-\infty}^{\infty}\int_{-\infty}^{\infty} q(c, c')\mathscr{P}_J(c, c')\,dc\,dc' \end{aligned} \tag{7-141}$$

4.3 Various Forms of Probability Density Function Used in Turbulent Flames

Various forms of pdf have been adopted by different investigators in turbulent-combustion calculations. The commonly used ones are the following:

4.3.1 Rectangular-Wave Variation of Mixture Fraction with Time

Spalding,[35] Gosman and Lockwood,[36] and Khalil et al.[37] used a rectangular-wave variation of f as the pdf for turbulent flames in furnaces:

$$\mathscr{P}(f, x_i) = a\delta(f - f^+) + (1 - a)\delta(f - f^-) \tag{7-142}$$

Thus the pdf at location x_i is constructed out of two δ-functions located at f^+ and f^-, where the parameters a, f^+, and f^-, were determined from the values of $\tilde{f}$ and $\widetilde{f''^2}$. This double-δ-function pdf was shown by Jones[38] to be unsatisfactory in comparison with Kent and Bilger's measurements[39] on turbulent diffusion flames.

4.3.2 Clipped Gaussian distribution

Lockwood and Naguib[40] proposed and used a clipped Gaussian distribution:

$$\mathscr{P}(f, x_i) = \begin{cases} \displaystyle\int_{-\infty}^{0} \frac{1}{\sigma(2\pi)^{1/2}} \exp\left[-\frac{1}{2}\left(\frac{f-\mu}{\sigma}\right)^2\right] df & \text{if } f = 0 \\ \displaystyle\frac{1}{\sigma(2\pi)^{1/2}} \exp\left[-\frac{1}{2}\left(\frac{f-\mu}{\sigma}\right)^2\right] & \text{if } 0 < f < 1 \\ \displaystyle\int_{1}^{\infty} \frac{1}{\sigma(2\pi)^{1/2}} \exp\left[-\frac{1}{2}\left(\frac{f-\mu}{\sigma}\right)^2\right] df & \text{if } f = 1 \end{cases} \tag{7-143}$$

The distribution is represented by the Gaussian function for the range $0 < f < 1$, but the tails of the distribution are represented by δ-functions at $f = 0$ and 1. The most probable value of the distribution, μ, and the variance, σ^2, were determined by the value of $\tilde{f}$ and g. They are defined as

$$\tilde{f} \equiv \int_0^1 f \mathscr{P}(f, x_i)\, df \tag{7-144}$$

$$g \equiv \widetilde{f''^2} = \int_0^1 (f - \tilde{f})^2 \mathscr{P}(f, x_i)\, df \tag{7-145}$$

It should be noted that Lockwood and Naguib used Reynolds averaging in their work in 1975 instead of Favre averaging. Explicit expressions for μ and σ (the standard deviation) cannot be obtained; their values must be obtained iteratively from the implicit equations (7-144) and (7-145). For computational convenience, this is often done once and for all, and the results stored in tabular form. According to Jones and Whitelaw,[5] the values of $\tilde{f}$ and $\widetilde{f''^2}$ are obtained from the following Favre-averaged partial differential equations

$$\rho\tilde{u}_j \frac{\partial \tilde{f}}{\partial x_j} = \frac{\partial}{\partial x_j}\left\{\frac{\mu_t}{\sigma_t}\frac{\partial \tilde{f}}{\partial x_j}\right\} \tag{7-146}$$

$$\rho\tilde{u}_j \frac{\delta \widetilde{f''^2}}{\partial x_j} = \frac{\partial}{\partial x_j}\left[\frac{\mu_t}{\sigma_t}\frac{\delta \widetilde{f''^2}}{\partial x_j}\right] + 2\frac{\mu_t}{\sigma_t}\left(\frac{\partial \tilde{f}}{\partial x_j}\right)^2 - C_D \frac{\bar{\rho}\varepsilon}{k}\widetilde{f''^2} \tag{7-147}$$

where σ_t is the turbulent Prandtl/Schmidt number and the turbulence constant C_D has a value of 2.0. The form of the clipped Gaussian pdf in terms of f is shown in Fig. 7.14.

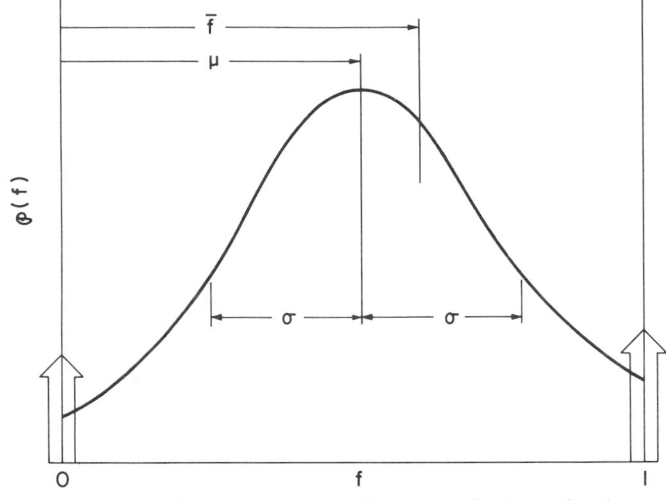

Figure 7.14 Clipped Gaussian pdf in terms of mixture fraction.

4.3.3 *Alternative Formulation of the Clipped Gaussian PDF*

Kent and Bilger[41] considered an alternative formulation of the clipped Gaussian pdf whereby intermittency was utilized. They proposed the following formula:

$$\mathscr{P}(f, x_i) = [1 - I(x_i)]\delta(f) + I(x_i)\mathscr{P}_1(f, x_i) \tag{7-148}$$

where $\mathscr{P}_1(f, x_i)$ is the pdf for turbulent flow, for which a clipped Gaussian distribution was used, and $I(x_i)$ is the intermittency obtained from empirical correlation. Kent and Bilger found that nitric oxide concentrations are particularly sensitive to the form of the pdf used.

4.3.4 β-*Probability Density Function*

The β-pdf was utilized by Richardson et al.,[42] Rhodes et al.,[43] Jones and Priddin,[44] and Jones and McGuirk.[45,46] It can be written, for $0 < f < 1$, as

$$\mathscr{P}(f, x_i) = \frac{f^{a-1}(1-f)^{b-1}}{\int_0^1 f^{a-1}(1-f)^{b-1}\, df} \tag{7-149}$$

where a and b can be determined explicitly from the values of $\tilde{f}$ and $\widetilde{f''^2}$. The integral in the denominator is called Beta function $B(a, b)$, which can be expressed in terms of several Gamma functions as $\Gamma(a)\Gamma(b)/\Gamma(a+b)$, where $\Gamma(a) \equiv (a-1)!$.

Jones's[38] calculation, using various pdfs, showed that the major species and temperature appear to be relatively insensitive to the precise shape of the pdf once the mean and variance are determined, with the provision that it be constructed from continuous functions with Dirac δ-functions at the bounds. There were small differences between the calculated temperature and species mass fraction profiles using the clipped Gaussian and β-pdfs.

4.3.5 Joint PDF for Mixture Fraction and Reaction-Progress Variable

In the consideration of NO_x formation in hydrogen–air flames, Janicka and Kollmann[47] considered the joint pdf for the mixture fraction and the reaction-progress variable r (see Fig. 6.9 for the mixture state as a function of f and r). They assumed that f and r are statistically independent and constructed a pdf using a β-function for f and three Dirac δ-functions for r. They found that the predictions for NO using the joint pdf were in good agreement with measurements.

4.3.6 Student's t-Distribution

Student's t-distribution[48] can sometimes be used to assess experimental uncertainties. The pdf of Student's t-distribution has the following form:

$$\mathscr{P}(t) = \frac{\Gamma\left(\dfrac{n+1}{2}\right)}{\sqrt{n\pi}\,\Gamma\left(\dfrac{n}{2}\right)\left(1 + \dfrac{t^2}{n}\right)^{(n+1)/2}} \tag{7-150}$$

for $-\infty < t < \infty$. In the above equation, n is the number of degrees of freedom, the mean $\mu = 0$, the variance

$$\sigma^2 = \frac{n}{n-2} \qquad \text{for} \quad n > 2 \tag{7-151}$$

and

$$t \equiv \frac{\chi^* \sqrt{n}}{Y} \tag{7-152}$$

where χ^* is an independent random variable which has mean equal to zero and variance equal to σ^2. The ratio Y^2/σ^2 is distributed as χ^2 with n degrees of freedom. The χ^2 (chi-square) distribution can be written as

$$\chi^2 = \sum_{i=1}^{n} Z_i^2 \tag{7-153}$$

where $Z_1, Z_2, \ldots, Z_n$ are normally and independently distributed with mean zero and variance 1.

As n approaches ∞, the Student's t pdf approaches the normal distribution. The formula for the χ^2 distribution is

$$\mathscr{P}(\chi^2) = \frac{(\chi^2)^{(n-2)/2}}{2^{n/2}\Gamma(n/2)} e^{-\chi^2/2} \quad \text{for} \quad 0 < \chi^2 < \infty \tag{7-154}$$

The form of the χ^2-distribution depends upon the value of n.

More detailed discussions of the use of pdfs in connection with turbulent flames are given in subsequent sections, which contain various models developed in recent years. These models are classified into turbulence models and combustion models.

5 TURBULENCE MODELS

Various models have been proposed in recent years to achieve closure of the equations for turbulence in reacting flow systems. They can be classified into the following types.

5.1 Direct Closure of the Equations for the Components of the Stress Tensor

The Reynolds stress tensor may be expressed as the dependent variable of the exact conservation equation[5] in slightly different form than Eq. (7-93):

$$\begin{aligned}
&\bar{\rho}\frac{\partial}{\partial t}\widetilde{u_i''u_j''} + \bar{\rho}\tilde{u}_k\frac{\partial}{\partial x_k}\widetilde{u_i''u_j''} \\
&= -\frac{\partial}{\partial x_k}\left\{\overline{\rho u_i''u_j''u_k''}\right\} - \overline{u_i''}\frac{\partial \bar{p}}{\partial x_j} - \overline{u_j''}\frac{\partial \bar{p}}{\partial x_i} \\
&\quad + \left\{\overline{u_j''\frac{\partial p'}{\partial x_i}} + \overline{u_i''\frac{\partial p'}{\partial x_j}}\right\} \\
&\quad - \bar{\rho}\widetilde{u_i''u_k''}\frac{\partial \tilde{u}_j}{\partial x_k} - \bar{\rho}\widetilde{u_j''u_k''}\frac{\partial \tilde{u}_i}{\partial x_k} - \bar{\rho}\varepsilon_{ij}
\end{aligned} \tag{7-155}$$

In order to close the above equation, the triple-velocity correlation, correlations involving fluctuating pressure, and the viscous dissipation ε_{ij} must all be approximated in terms of lower-order known quantities.

This direct-closure method has previously been proposed in the context of isothermal flows by Launder,[49] Bradshaw,[50] Lumley,[51] and Bradshaw et al.[52] Extension to reacting flows with associated large density variations has been considered by Jones.[38] However, such models are extremely complex, and for recirculating flows there are substantial difficulties in obtaining error-free numerical solutions.[5] Second-order stress closure models must be regarded at the present time to be largely untested in the recirculating flows.

5.2 Two-Equation Models

In 1972, Jones and Launder[53] introduced the k–ε two-equation model for recirculating flows. They assumed a linear relationship between Reynolds stress and rate of strain:

$$\bar{\rho}\widetilde{u_i''u_j''} = \tfrac{2}{3}\delta_{ij}\left(\bar{\rho}k + \mu_t\frac{\partial \tilde{u}_k}{\partial x_k}\right) - \mu_t\left(\frac{\partial \tilde{u}_i}{\partial x_j} + \frac{\partial \tilde{u}_j}{\partial x_i}\right) \tag{7-156}$$

A gradient-diffusion model is adopted for turbulent flux of scalar quantities, namely

$$\bar{\rho}\widetilde{u_j''\phi_\alpha''} = -\frac{\mu_t}{\sigma_t}\frac{\partial \tilde{\phi}_\alpha}{\partial x_j} \tag{7-157}$$

where σ_t is the turbulent Prandtl/Schmidt number for ϕ. The turbulent (or eddy) viscosity is given by

$$\mu_t = C_\mu \bar{\rho}\frac{k^2}{\varepsilon} \tag{7-158}$$

where the turbulence constant $C_\mu = 0.09$. The turbulence kinetic energy k $(\equiv \widetilde{u_i''u_i''}/2)$ can be obtained from its transport equation in the reorganized form of Eq. (7-94):

$$\bar{\rho}\tilde{u}_j\frac{\partial k}{\partial x_j} = \frac{\partial}{\partial x_j}\left[\left(\frac{\mu_t}{\sigma_k} + \mu\right)\frac{\partial k}{\partial x_j}\right] - \bar{\rho}\widetilde{u_i''u_j''}\frac{\partial \tilde{u}_i}{\partial x_j} + \frac{\mu_t}{\bar{\rho}^2}\frac{\partial \bar{\rho}}{\partial x_i}\frac{\partial \bar{p}}{\partial x_i} - \bar{\rho}\varepsilon \tag{7-159}$$

where σ_k is the turbulent Prandtl/Schmidt number for k. The turbulence dissipation rate ε is found from the following ε transport equation:

$$\bar{\rho}\tilde{u}_j \frac{\partial \varepsilon}{\partial x_j} = \frac{\partial}{\partial x_j}\left[\left(\frac{\mu_t}{\sigma_\varepsilon} + \mu\right)\frac{\partial \varepsilon}{\partial x_j}\right]$$

$$- C_{\varepsilon 1}\frac{\varepsilon}{k}\left(\overline{\bar{\rho} u_i'' u_j''}\frac{\partial \tilde{u}_i}{\partial x_j} + \frac{\mu_t}{\bar{\rho}^2}\frac{\partial \bar{\rho}}{\partial x_i}\frac{\partial \bar{p}}{\partial x_i}\right)$$

$$- C_{\varepsilon 2}\bar{\rho}\frac{\varepsilon^2}{k} \tag{7-160}$$

The following set of constants is recommended for turbulence calculations:

$$C_\mu = 0.09, \qquad C_{\varepsilon 1} = 1.44, \qquad C_{\varepsilon 2} = 1.92$$

$$\sigma_k = 1.0, \qquad \sigma_\varepsilon = 1.30, \qquad \sigma_t = 0.7$$

Equations (7-156) through (7-160) can also be written in conventional time-averaged form and, as indicated by Bray,[54] include additional terms which stem from the fluctuating density and which must be modeled.

It is important to point out that the above two-equation model, after neglecting terms involving $(\partial \bar{\rho}/\partial x_i)(\partial \bar{p}/\partial x_i)$, entails the assumptions and approximations which are normally used for constant-density flows. The model assumes that the gradient-diffusion model can be rewritten in the density-weighted form without any explicit account being taken of density fluctuations. These approximations and assumptions have not been and are unlikely to be fully tested in a direct sense,[5] but their validity has been supported in many cases. Work of Launder and Spalding,[55] Jones and McGuirk,[45] and Ribeiro and Whitelaw[56] supports the k–ε two-equation model.

Moss,[57] on the basis of his measurements of concentration and velocity in an open premixed turbulent flame, suggested considering the effects of *counter-gradient diffusion*, that is, turbulent diffusion of a species against its mean gradient. Also, Libby and Bray[58,59] predicted that such behavior would be important in premixed planar flames. Counter-gradient diffusion can be attributed to the preferential influence of the mean pressure gradient on low- and high-density gases which is manifested through the appearance of terms like

$$\overline{\rho' u_j''}\frac{\partial \bar{p}}{\partial x_i} \quad \text{and} \quad \overline{\rho' \phi_\alpha''}\frac{\partial \bar{p}}{\partial x_i}$$

in the Favre-averaged Reynolds-stress and turbulent scalar flux-transport equations. According to Jones and Whitelaw,[5] at the present time it is difficult to assess whether or not countergradient-diffusion effects have significant influence in practical systems.

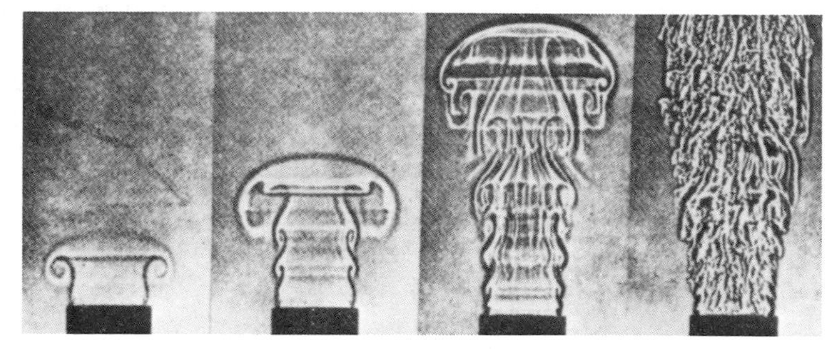

Figure 7.15 Schlieren photographs of a starting free jet, showing separate vortices in the initial stages and highly turbulent flow in the later stages (after J. E. Garside et al.[60]).

6 INTERMITTENCY AND COHERENT STRUCTURES

6.1 Intermittency

Experimental evidence (such as that shown in Figs. 7.15 and 7.16) suggests that the dissipation rate is not evenly distributed over the volume occupied by a turbulent flow. The distribution of the dissipation rate appears to be intermittent, with large dissipation rates occupying a small volume fraction (see Section 7.3.4). Intermittency is a general feature of turbulent flows; it is present whether there are chemical reactions or not. The phenomenon of intermittency is also illustrated in Fig. 7.17, which was photographed by Fernholz[61] and also discussed by Bradshaw[50] and by Libby and Williams.[4] This figure shows the intermittency in a turbulent boundary layer with the turbulent fluid tagged with smoke. As we can see, the boundary between the clear (unsmoked) and smoked regions is highly convoluted.

As mentioned by Libby and Williams,[4] measurements of the properties of the flow at a point involving intermittency result in probability density functions which reflect the intermittency. To understand the significance of intermittency relative to probability density function, we can consider a nonreacting turbulent flow with a foreign species A injected at some upstream locations. Let the mass fraction of the injected species be denoted by $Y_A(\mathbf{x}, t)$,

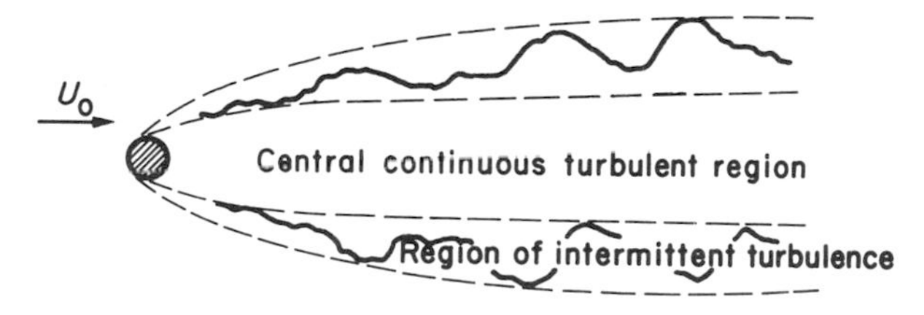

Figure 7.16 Intermittency near the edges of the wake behind a cylinder (after J. O. Hinze[30]).

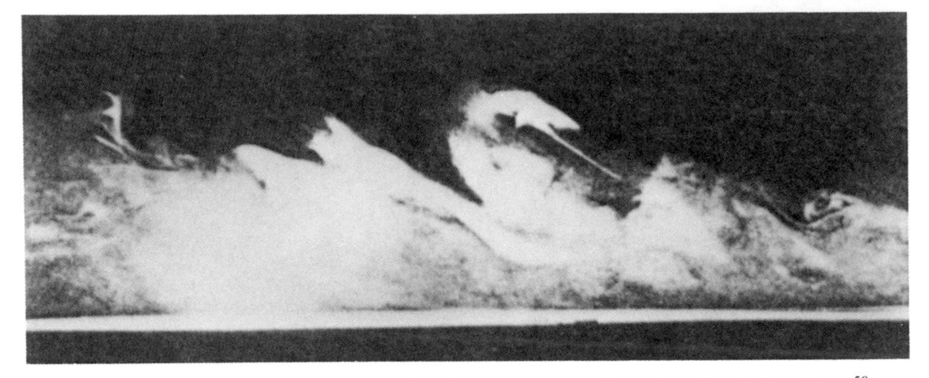

Figure 7.17 Visualization of a boundary layer by means of smoke (after P. Bradshaw[50]).

and imagine that we are able to measure the instantaneous concentration at a fixed spatial point in the flow wherein intermittency occurs. The instantaneous $Y_A(\mathbf{x}, t)$ is shown schematically in Fig. 7.18. We can see that time periods with $Y_A \approx 0$ are interspersed between periods with $Y_A > Y_{Ag}$. The former correspond to intervals when the fixed point is in a region nearly uncontaminated by the foreign species A, whereas the latter correspond to intervals when the point is within the contaminated region of the turbulent flow. Because of the inaccuracies in concentration measurement, it is necessary to identify contaminated regions with $Y_A > Y_{Ag}$, where Y_{Ag} is a suitable chosen level (or gate).

The intermittency function $I(\mathbf{x}, t)$ can be defined according to Fig. 7.18, in which a specified gate is chosen, as

$$I(\mathbf{x}, t) = \begin{cases} 0 & \text{if} \quad Y_A < Y_{Ag} \\ 1 & \text{if} \quad Y_A > Y_{Ag} \end{cases} \tag{7-161}$$

The mean value of the intermittency function, $\bar{I}(\mathbf{x})$, is called the *intermittency*; it represents the fraction of the time that the flow is contaminated at the point in question. The probability density function $\mathscr{P}(Y_A; \mathbf{x})$ is also schematically shown in Fig. 7.18. As indicated in the figure, there is a spike in the neighborhood of $Y_A = Y_{Ag}$.

One can consider that the uncontaminated region represents irrotational flow of uniform composition, density, and temperature. The contaminated region can be considered as the turbulent (rotational) flow region with all species present in various amounts. Then, the multivariable pdf for a single point $\mathbf{x}$ can be decomposed into two separate probability density functions:

$$\begin{aligned} &\mathscr{P}(u_1, u_2, u_3, \rho, T, Y_1, Y_2, \ldots, Y_N; \mathbf{x}) \\ &\quad = [1 - \bar{I}(\mathbf{x})] \mathscr{P}_0(u_1, u_2, u_3, \mathbf{x}) \delta(\rho - \rho_0) \delta(T - T_0) \\ &\qquad \times \delta(Y_1 - Y_{1_0}) \cdots \delta(Y_N - Y_{N_0}) \\ &\qquad + \bar{I}(\mathbf{x}) \mathscr{P}_1(u_1, u_2, u_3, \rho, T, Y_1, Y_2, \ldots, Y_N; \mathbf{x}) \end{aligned} \tag{7-162}$$

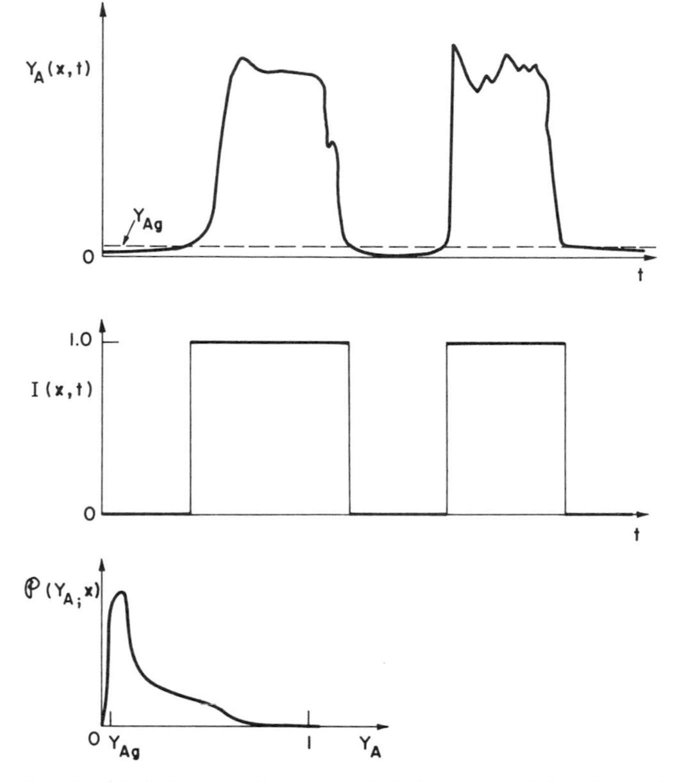

Figure 7.18 Relationship between instantaneous turbulent signal, intermittency function, and probability density function.

where the subscripts in $\mathscr{P}_0$ and $\mathscr{P}_1$ designate conditions in the uncontaminated and contaminated regions, respectively. Both $\mathscr{P}_0$ and $\mathscr{P}_1$ are normalized to unity. The so-called "unconditioned" Reynolds-stress-tensor component $\overline{\rho u_1'' u_2''}$ can be expressed as[4]

$$\overline{\rho u_1'' u_2''} = [1 - \bar{I}(\mathbf{x})]\left(\overline{\rho u_1'' u_2''}\right)_0 + \bar{I}(\mathbf{x})\left(\overline{\rho u_1'' u_2''}\right)_1 \qquad (7\text{-}163)$$

The terms on the right-hand side of Eq. (7-163) are conditioned by the intermittency function.

6.2 Coherent Structures

The term "coherent structures" has been used broadly in connection with turbulent flows, both nonreacting and reacting. Coherent structures represent large, rather well-organized lumps of fluid and display their own dynamic behavior in turbulent flows. Their large size implies a long lifetime. Coherent structures arise in various types of turbulent flows, such as boundary layers,

Figure 7.19 Vortices at boundary of a half jet (after J. O. Hinze;[30] original photograph by G. Flügel).

jets, wakes, and mixing layers. In contrast with our usual understanding of turbulent flows as involving the cascading of large eddies into smaller ones, we must realize that some coherent structures may increase in size as they move downstream. In some cases, the increase is caused by the merging of two smaller structures into a single large one; in other cases, the sizes of coherent structures may stay fairly constant. One obvious manifestation of coherence is intermittency, discussed earlier; Corrsin's[62] discovery of intermittency has been considered as the first evidence of coherent structures. Presently, there has been substantial interest in the study of coherent structures. Some recent work in this area is included in the symposium volumes edited by Fiedler.[63] The genesis of these large lumps of turbulent eddies is being studied extensively. Particular attention has been given to two-dimensional mixing layers and two-dimensional boundary layers. The development of vortices at the boundary of a jet can be seen clearly in Fig. 7.19. These vortices, with axes parallel to the splitting plate, have been considered to be created by an instability mechanism.

According to Libby and Williams,[4] the existence of coherent structures implies that unconditioned averaging, whether conventional or mass-weighted, disregards an important physical phenomenon. Coherent structure can not be ignored, since the large-scale, orderly structures contribute more than half of the Reynolds-stress terms, as evidenced by experimental results. In spite of the fact that these coherent structures are present in turbulent flows and may have significant influence on the overall flow property distribution, there is practically no theoretical work being done before 1976 to describe turbulent flows on the basis of coherent structures. Libby's[64] and Dopazo's[65] work can be regarded as pioneering in this area. The work of Chorin et al.[116,112] and Ghoniem et al.[119,120] on random vortex method belongs to the large scale simulation. Their approach is discussed in Section 8.3. It is believed that more emphasis will be placed in the future on the large-scale structure of reacting flows.

7 TURBULENT REACTING FLOWS WITH NONPREMIXED REACTANTS (TURBULENT DIFFUSION FLAMES)

In many combustion systems, reactants enter in separate streams, oxidizer and fuel being nonpremixed. The resulting turbulent diffusion flames are one of the most challenging topics in engineering sciences; there are many unsolved problems. An excellent review of turbulent flows with nonpremixed reactants is given by Bilger.[66] The heart of the problem in modeling reacting turbulent flows with nonpremixed reactants lies in the handling of the mean-chemical-production term and in the effects of heat release on turbulent structures. There are several basic assumptions associated with various limiting cases, such as fast chemistry for high-temperature environments or fast mixing for high turbulence intensity. The application of these assumptions, the simplifications resulting from them, and various approaches are described in subsequent sections.

7.1 The Conserved-Scalar Approach

Many chemical reactions have high rates and can be considered complete as soon as the reactants are mixed. If the reaction time is negligibly short in comparison with the mixing time, the turbulent flame can be approximated adequately with the *fast-chemistry assumption*. According to Bilger,[66] this assumption implies that the instantaneous molecular-species concentrations and the temperature are functions only of the conserved scalar concentration at that instant. In the limit of fast chemistry, the problems of chemistry–turbulence interactions are drastically simplified, since the molecular-species concentrations are directly related to the conserved scalar, and the statistics of all thermodynamic variables are obtainable from the knowledge of the statistics of that scalar. Therefore, the need to evaluate mean reaction rates is removed. For example, the equilibrium composition, temperature, and density of a gaseous mixture can be determined if the elemental mass fractions of all elements present, the pressure, and the enthalpy are known. A typical choice for the strictly conserved (i.e., zero-source) scalar variable is the mixture fraction f, defined as the mass fraction of fuel both burned and unburned (see Chapter 1).

Even in fast-chemistry cases, we still have to link the means and higher moments of species concentrations and temperature to those of the conserved scalar. These linkages are not simple, since the functions relating them to conserved scalar are nonlinear.

The mixing patterns in a turbulent diffusion flame jet are illustrated in Fig. 7.20. Figure 7.20a shows the instantaneous isopleths for a conserved scalar ξ.

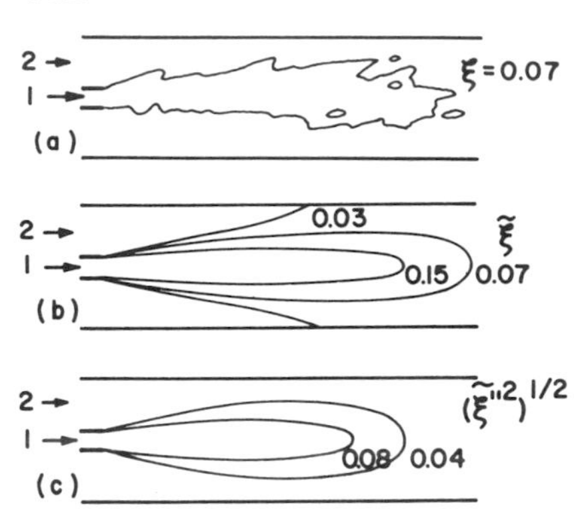

Figure 7.20 Mixing patterns in a typical turbulent reacting flow with nonpremixed reactants entering in streams 1 and 2: (*a*) instantaneous isopleth for a conserved scalar ξ; (*b*) isopleths for the Favre average of ξ; (*c*) contours of root-mean-square fluctuation of ξ (after R. W. Bilger[66]).

Such mixing patterns can be measured experimentally by determining the concentration of an inert species doped in one of the streams. The concentration measurements can be made by laser Raman spectroscopy. Element concentrations can also be obtained by laser spark spectroscopy,[66] in which a very small volume of gas is pulsed with a high-intensity laser, converting it into plasma. The spectroscopic analysis of the light emission gives the concentrations of elements, since each element will show its own atomic and ionic lines with intensity proportional to its concentration.

7.1.1 Conserved Scalars, Mixture Fraction, and the Fast-Chemistry Assumption

There are many scalar variables which are conserved under chemical reactions and hence can be used as the major parameter for describing the mixing in a nonpremixed reacting flow. The mass fraction of element i (Z_i) is one such variable which has no chemical source term in its conservation equation,

$$\frac{\partial}{\partial t}(\rho Z_i) + \frac{\partial}{\partial x_k}(\rho u_k Z_i) = \frac{\partial}{\partial x_k}\left(\rho \mathscr{D} \frac{\partial Z_i}{\partial x_k}\right), \qquad i = 1, 2, \ldots, L \quad (7\text{-}164)$$

where Z_i is defined as

$$Z_i \equiv \sum_{j=1}^{N} \mu_{ij} Y_j, \qquad i = 1, 2, \ldots, L \quad (7\text{-}165)$$

and

$$\mu_{ij} \equiv \frac{\text{number of grams of } i\text{th element in } j\text{th species}}{1 \text{ gram of } j\text{th species}} \quad (7\text{-}166)$$

The chemical source term vanishes in Eq. (7-164), because in chemical reactions elements are conserved. There are $L - 1$ variables for a system involving L elements, since

$$\sum_{i=1}^{L} Z_i = 1 \quad (7\text{-}167)$$

The Favre-averaged element conservation equation[4] can be written as

$$\frac{\partial}{\partial x_k}(\bar{\rho}\tilde{u}_k\tilde{Z}_i) = \frac{\partial}{\partial x_k}\left(\overline{\rho\mathscr{D}\frac{\partial Z_i}{\partial x_k}} - \overline{\rho u_k''Z_i''}\right) \tag{7-168}$$

Solution of the $L - 1$ equations for $\tilde{Z}_i$ gives the mean elemental composition throughout the reacting turbulent flow field, as shown in Fig. 7.20*b*. The molecular-species composition, based upon the fast-chemistry assumption, can be determined from this elemental composition. Even though $\tilde{Z}_i$ is simpler to use than $\tilde{Y}_i$, to facilitate our discussions of several previous studies, we shall use both $\tilde{Z}_i$ and $\tilde{Y}_i$ in following sections.

If the combustion process can be suitably represented by a single one-step reaction, such as

$$\{1 \text{ kg of fuel } (F)\} + \{r \text{ kg of oxidant } (O)\} \rightarrow \{(1+r) \text{ kg of product } (P)\} \tag{7-169}$$

then simpler conserved scalars may be used. Based upon Eq. (7-169), we have

$$\dot{\omega}_F = \frac{1}{r}\dot{\omega}_O = -\frac{1}{r+1}\dot{\omega}_P \tag{7-170}$$

Therefore, the so-called Shvab–Zel'dovich coupling parameters (see Chapter 3)

$$\beta_{FO} \equiv Y_F - \frac{Y_O}{r}$$

$$\beta_{FP} \equiv Y_F + \frac{Y_P}{r+1}$$

$$\beta_{OP} \equiv Y_O + \frac{rY_P}{r+1} \tag{7-171}$$

are conserved scalars. One can form other conserved scalars from the sensible enthalpy. If all the conserved scalars are linearly related, then the solution for one yields solutions for all others, and the choice of the conserved scalar is then arbitrary. Under other circumstances, the choice of conserved scalar will depend upon the magnitude of the turbulence Reynolds number, the complexity of chemical-kinetic mechanisms, and the uniformity of the reactant feeding streams.

It should be pointed out that the chemical source terms in the conservation equation of the above conserved scalars are not zero if the mass diffusivities of the molecular species are not equal. Fortunately, in most turbulent flows at moderate to high Reynolds numbers, such differential diffusion effects are negligible. In other words, the molecular-diffusion term in Eq. (7-168) can be neglected in comparison with the turbulent-diffusion term. For conserved scalar ξ in moderate- to high-Reynolds-number flows, Eq. (7-168) can be reduced to

$$\frac{\partial}{\partial x_k}(\bar{\rho}\tilde{u}_k\tilde{\xi}) = -\frac{\partial}{\partial x_k}(\overline{\rho u_k''\xi''}) \tag{7-172}$$

It is also interesting to note that linear relationships among all conserved scalars exist only when there are two uniform reactant feeding streams. If enthalpies are to be included in the conserved scalar quantities, then the enthalpies of the two feeding streams must be uniform and have the same rate of change with respect to time. The conserved scalars of the Shvab–Zel'dovich coupling parameters are useful only when a single one-step reaction such as Eq. (7-169) is involved. For more complex chemical systems involving many multistep and competitive reactions with high heat release rates, it is more convenient to use conserved scalars based on elements.

The mixture fraction for two feeding streams with equal species mass diffusivities can be defined as

$$f \equiv \frac{Z_i - Z_{i2}}{Z_{i1} - Z_{i2}} = \frac{\beta - \beta_2}{\beta_1 - \beta_2} \tag{7-173}$$

where the subscripts 1 and 2 refer to the uniform composition in the two feeding streams, and β is a Shvab–Zel'dovich coupling parameter defined in Eq. (7-171). Since f is a conserved scalar, it satisfies

$$\frac{\partial}{\partial t}(\rho\xi) + \frac{\partial}{\partial x_k}(\rho u_k \xi) = \frac{\partial}{\partial x_k}\left(\rho \mathscr{D}\frac{\partial \xi}{\partial x_k}\right) \tag{7-174}$$

If ξ is substituted by f (see homework problem 7.11)

Its statistical behavior will be like that of other conserved scalars such as the mass fraction of an inert species. The Favre variance of ξ in a stationary turbulent mean flow, $\widetilde{\xi''^2}$, can be described by the following transport equation:

$$\underbrace{\bar{\rho}\tilde{u}_k\frac{\partial \widetilde{\xi''^2}}{\partial x_k}}_{\substack{\text{transport of}\\ \widetilde{\xi''^2}\text{ by}\\ \text{convection}}} = \underbrace{-2\overline{\rho u_k''\xi''}\frac{\partial \tilde{\xi}}{\partial x_k}}_{\substack{\text{production}\\ \text{by mean}\\ \text{gradient}}} \underbrace{-\frac{\partial}{\partial x_k}\left(\overline{\rho u_k''\xi''^2}\right)}_{\substack{\text{diffusion by}\\ \text{velocity}\\ \text{fluctuations}}} \underbrace{-2\rho\mathscr{D}\overline{\frac{\partial \xi''}{\partial x_k}\frac{\partial \xi''}{\partial x_k}}}_{\substack{\text{dissipation}\\ \text{by molecular}\\ \text{diffusion}}} \tag{7-175}$$

The solution of the above equation yields the results shown in Fig. 7.20c.

With two-stream feeding, there exists a special value of the mixture fraction, f_{st}, which corresponds to the stoichiometric condition. For the one-step single reaction shown by Eq. (7-169), with no oxidant in the fuel feeding stream and no fuel in the oxidant feeding stream, we have

$$f_{\rm st} = \frac{Y_{O2}}{rY_{F1} + Y_{O2}} \tag{7-176}$$

For fuel lean cases with fast chemistry, we have the following relationships for burnt mixtures ($f \leq f_{\rm st}$, $\beta_{FO} \leq 0$):

$$\begin{aligned} Y_F &= 0 \\ Y_O &= r\beta_{FO} = rY^*(f_{\rm st} - f) \\ Y_P &= (r+1)Y^*f(1 - f_{\rm st}) \end{aligned} \tag{7-177}$$

where

$$Y^* \equiv Y_{F1} + \frac{1}{r}Y_{O2} = \frac{Y_{F1}}{1 - f_{\rm st}} \tag{7-178}$$

For fuel-rich cases with fast chemistry ($f \geq f_{\rm st}$, $\beta_{FO} \geq 0$) we have

$$\begin{aligned} Y_F &= Y^*(f - f_{\rm st}) = \beta_{FO} \\ Y_O &= 0 \\ Y_P &= (r+1)Y^*f_{\rm st}(1 - f) \end{aligned} \tag{7-179}$$

The fast-chemistry assumption requires that the forward and backward reactions be fast in comparison with turbulent mixing processes. This implies that the instantaneous chemical composition of the mixture at a given spatial location is at chemical equilibrium, since the local mixture can be considered isolated and to have enough time to react. The burned plane in the plot of f versus species mass fractions for irreversible one-step reactions is shown in Chapter 1. The species concentrations are unique functions of f, since closure is obtained by the use of the equilibrium constant for the reaction. If the reaction is reversible, there will be both fuel and oxidant present near $f_{\rm st}$ due to the back reaction. The plot of species mass fraction versus f will have gradual changes near the stoichiometric point as shown in Fig. 7.21. The mass fraction

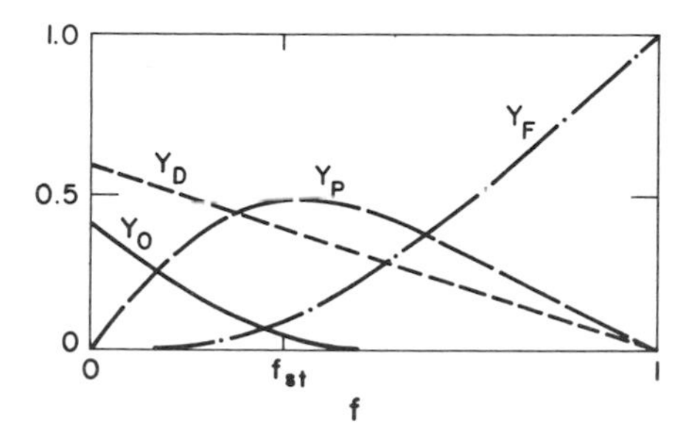

Figure 7.21 Mass fraction of species as a function of mixture fraction in a one-step reversible reaction with fast chemistry.

of species i can be written as

$$Y_i = Y_i^e(f) \tag{7-180}$$

where the superscript e denotes equilibrium. We can also write

$$T = T^e(f) \quad \text{and} \quad \rho = \rho^e(f) \tag{7-181}$$

The mean properties and variances, based upon Favre averaging and the fast-chemistry assumption, can be written as

$$\tilde{Y}_i(\mathbf{x}) = \int_0^1 Y_i^e(f)\tilde{\mathscr{P}}(f;\mathbf{x})\,df \tag{7-182}$$

$$\widetilde{Y_i''^2}(\mathbf{x}) = \int_0^1 [Y_i^e(f) - \tilde{Y}_i(\mathbf{x})]^2\tilde{\mathscr{P}}(f;\mathbf{x})\,df \tag{7-183}$$

$$\tilde{T}(\mathbf{x}) = \int_0^1 T^e(f)\tilde{\mathscr{P}}(f;\mathbf{x})\,df \tag{7-184}$$

$$\widetilde{T''^2}(\mathbf{x}) = \int_0^1 [T^e(f) - T(\mathbf{x})]^2\tilde{\mathscr{P}}(f;\mathbf{x})\,df \tag{7-185}$$

where

$$\tilde{\mathscr{P}}(f;\mathbf{x}) \equiv \frac{1}{\bar{\rho}}\int_0^\infty \rho\mathscr{P}(\rho, f;\mathbf{x})\,d\rho \tag{7-186}$$

is called the Favre pdf. With $\rho = \rho^e(f)$, the above equation becomes

$$\tilde{\mathscr{P}}(f;\mathbf{x}) = \frac{\rho^e(f)}{\bar{\rho}(\mathbf{x})}\mathscr{P}(f;\mathbf{x}) \tag{7-187}$$

For the one-step irreversible reaction represented by Eq. (7-169), it can be shown that (see Homework 9)

$$\begin{aligned}\tilde{Y}_F &= Y^*\int_{f_{st}}^1 (f - f_{st})\tilde{\mathscr{P}}(f;\mathbf{x})\,df \\ &= Y_F^e(\tilde{f}) + \alpha_c Y^*\sqrt{\widetilde{f''^2}}\,J_1\left(\frac{f_{st} - \tilde{f}}{\sqrt{\widetilde{f''^2}}}\right)\end{aligned} \tag{7-188}$$

$$\begin{aligned}\tilde{Y}_0 &= rY^*\int_0^{f_{st}} (f_{st} - f)\tilde{\mathscr{P}}(f;\mathbf{x})\,df \\ &= Y_0^e(\tilde{f}) + \alpha_c rY^*\sqrt{\widetilde{f''^2}}\,J_1\left(\frac{f_{st} - \tilde{f}}{\sqrt{\widetilde{f''^2}}}\right)\end{aligned} \tag{7-189}$$

$$\tilde{Y}_P = (r+1)Y^*\left\{(1-f_{st})\int_0^{f_{st}} f\tilde{\mathscr{P}}(f;\mathbf{x})\,df + f_{st}\int_{f_{st}}^1 (1-f)\tilde{\mathscr{P}}(f;\mathbf{x})\,df\right\}$$

$$= Y_P^e(\tilde{f}) - \alpha_c(r+1)Y^*\sqrt{\widetilde{f''^2}}\,J_1\left(\frac{f_{st}-\tilde{f}}{\sqrt{\widetilde{f''^2}}}\right) \quad (7\text{-}190)$$

where the function J_1 is defined as

$$J_1\left(\frac{f_{st}-\tilde{f}}{\sqrt{\widetilde{f''^2}}}\right) \equiv \int_0^{f_{st}} \frac{f_{st}-f}{\sqrt{\widetilde{f''^2}}}\tilde{\mathscr{P}}(f;\mathbf{x})\,df - H(f_{st}-\tilde{f})\frac{f_{st}-\tilde{f}}{\sqrt{\widetilde{f''^2}}} \quad (7\text{-}191)$$

In the above equations, Y^* appears many times, as defined in Eq. (7-178), and H is the Heaviside function. The parameter α_c is equal to one for one-step irreversible reactions. It serves as a correction factor for reversible and multi-step reactions. J_1 can be considered as the representation of *unmixedness*—a term first suggested by Hawthorne et al.[67] to describe the effect of concentration fluctuations on the mean composition. As pointed out by Bilger,[66] J_1 is nonnegative, and from Eqs. (7-188) through (7-190) one can see that the effect of turbulence is to increase the mean concentration of reactants above the laminar value at the mean mixture fraction. Bilger[66] also reported that J_1 is not too sensitive to the form of the pdf for values of $\tilde{f}$ close to f_{st}. However, for $\tilde{f}$-values differing significantly from f_{st}, the form of the pdf does have some significant influence on the value of J_1.

7.1.2 PDFs of Conserved Scalars

Knowledge of the pdfs of a given conserved scalar quantity is required to obtain the mean and other moments of thermochemical properties. In general, the form of the pdf will depend on the flow conditions and chemical heat release rates. This dependence might seem to severely limit the application of the conserved-scalar approach to turbulent nonpremixed flames. However, the detailed shape of the pdf has been found to have very small effects on the computed mean and variance quantities from Eqs. (7-172) and (7-175).

The pdf of the mixture fraction can be obtained from measurements of the concentration of an inert species or of fine particles in the flow. However, not many measurements have been made in turbulent reacting flows. Pdfs of temperature fluctuations have been measured over a broad range of nonreacting flows, including mixing layers, jets, boundary layers, and wakes. Figure 7.22 shows schematically the type of pdf found in various flows. As we can see, the functional form of the pdf varies from point to point in a given flow and also changes with the type of flow. The broad arrow at $f = 0$ or 1 represents a δ-function. In practice, measured pdfs show these δ-functions smeared into Gaussian-like peaks due to electronic noise and residual temperature fluctua-

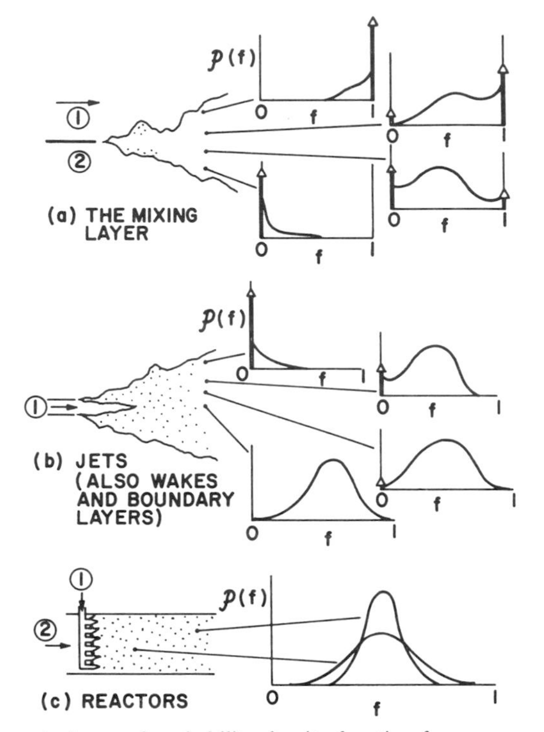

Figure 7.22 Schematic forms of probability density function for a conserved scalar in various types of flow (after R. W. Bilger[66]).

tions in the free stream. In a jet region where intermittency plays an important role, one could use Eq. (7-148), suggested by Kent and Bilger,[41] for the pdf.

For reacting flows with significant heat release, there are very limited data available on the conserved-scalar pdfs. Measurements of conventional and Favre pdfs by Kennedy and Kent[68] in turbulent jet diffusion flames are shown in Fig. 7.23. This figure shows the pdfs on a radial traverse across a turbulent diffusion flame of hydrogen injected into quiescent air at an axial distance of $x/D = 40$. It can be seen from this figure that Favre pdfs emphasize intermittency spikes. For the Favre pdf, Eq. (7-148) becomes

$$\tilde{\mathscr{P}}(f;\mathbf{x}) = [1 - \tilde{I}(\mathbf{x})]\delta(f) + \tilde{I}(\mathbf{x})\tilde{\mathscr{P}}_1(f;\mathbf{x}) \tag{7-192}$$

where the Favre intermittency $\tilde{I}(\mathbf{x})$ is related to the conventional $\bar{I}(\mathbf{x})$ by

$$\bar{\rho}[1 - \tilde{I}(\mathbf{x})] = \rho_0[1 - \bar{I}(\mathbf{x})] \tag{7-193}$$

with ρ_0 as the density of the fluid in the external flow. Kent and Bilger[41] used

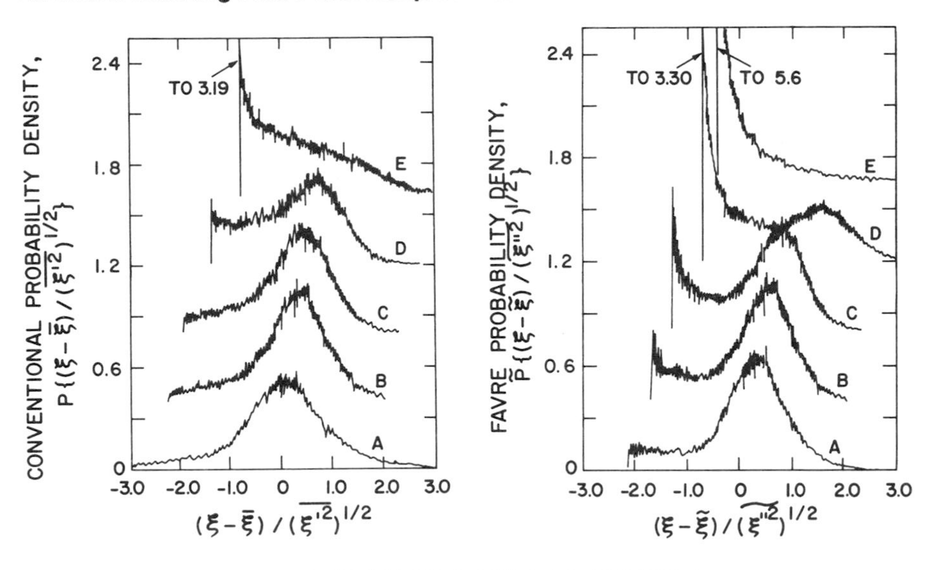

Figure 7.23 Conventional and Favre pdfs measured in a hydrogen–air diffusion flame at 40 jet diameters from the nozzle by Kennedy and Kent.[68] Radial distances in nozzle radii: (A) 0.0, (B) 1.16, (C) 2.33, (D) 4.65, (E) 9.31.

an empirical correlation for $\tilde{I}(x)$,

$$\tilde{I} = \frac{K + 1}{\left(\widetilde{f''^2}/\tilde{f}^2\right) + 1} \tag{7-194}$$

where $K \approx 0.25$. $\tilde{\mathscr{P}}_1$ in Eq. (7-192) was considered to have a Gaussian distribution with mean $\tilde{f}_t$ and variance $\widetilde{f_t''^2}$; these two parameters were evaluated from

$$\tilde{f} = \tilde{I}\tilde{f}_t \tag{7-195}$$

and

$$\widetilde{f_t''^2} = K\tilde{f}_t^2 \tag{7-196}$$

7.1.3 Reaction Rate and Reaction-Zone Structure

The essence of the conserved-scalar approach is to avoid the complex problem of handling the chemical-production term. As a consequence of this approach, a simple formula for the instantaneous reaction rate is obtained. Substituting Eq. (7-180) into the equation for conservation of individual species,

$$\frac{\partial Y_i}{\partial t} + u_k \frac{\partial Y_i}{\partial x_k} - \frac{1}{\rho}\frac{\partial}{\partial x_k}\left(\rho \mathscr{D} \frac{\partial Y_i}{\partial x_k}\right) = \frac{\dot{\omega}_i}{\rho}, \qquad i = 1, 2, \ldots, N \tag{7-197}$$

we have

$$\dot{\omega}_i = \rho \frac{dY_i^e}{df}\left[\frac{\partial f}{\partial t} + u_k \frac{\partial f}{\partial x_k} - \frac{1}{\rho}\frac{\partial}{\partial x_k}\left(\rho\mathscr{D}\frac{\partial f}{\partial x_k}\right)\right]^{0} - \rho\mathscr{D}\frac{\partial f}{\partial x_k}\frac{\partial f}{\partial x_k}\frac{d^2Y_i^e}{df^2}$$

$$= -\rho\mathscr{D}\frac{\partial f}{\partial x_k}\frac{\partial f}{\partial x_k}\frac{d^2Y_i^e}{df^2} \tag{7-198}$$

Taking the mean of Eq. (7-198), we have

$$\bar{\dot{\omega}}_i = -\tfrac{1}{2}\bar{\rho}\int_0^1\left[\int_0^\infty \chi \frac{d^2Y_i^e}{df^2}\tilde{\mathscr{P}}(\chi, f)\,d\chi\right]df \tag{7-199}$$

where

$$\chi \equiv 2\mathscr{D}\frac{\partial f''}{\partial x_k}\frac{\partial f''}{\partial x_k} \approx 2\mathscr{D}\frac{\partial f}{\partial x_k}\frac{\partial f}{\partial x_k} \tag{7-200}$$

for moderate to high turbulence Reynolds number reacting flows. The appropriate turbulence Reynolds number is defined as

$$\mathrm{Re}_T \equiv \frac{\sqrt{2k_{\mathrm{ref}}}\,l_T}{\nu_{\mathrm{ref}}} = \frac{u''_{\mathrm{rms}}l_T}{\nu_{\mathrm{ref}}} \tag{7-201}$$

where k_{ref} is the turbulence kinetic energy at a suitable reference point, and $\sqrt{2k_{\mathrm{ref}}}$ is a representative rms velocity fluctuation. l_T can be characterized as a length beyond which various fluid-mechanical quantities become essentially uncorrelated. In general, l_T is less than the characteristic length scale of the main flow, but of the same order of magnitude. ν_{ref} is the molecular kinematic viscosity at a reference temperature. The function $\tilde{\mathscr{P}}(\chi, f)$ is the joint Favre pdf of χ and f. The mean of χ is the scalar dissipation in Eq. (7-175) if we set $f = \xi$. For the one-step irreversible reaction with fast chemistry in Eq. (7-169), $d^2Y_F^e/df^2$ resembles a δ-function at $f = f_{\mathrm{st}}$ and

$$\bar{\dot{\omega}}_F = -\tfrac{1}{2}\bar{\rho}Y^*\tilde{\chi}(f_{\mathrm{st}})\tilde{\mathscr{P}}(f_{\mathrm{st}}) \tag{7-202}$$

In the limit of fast chemistry, the reaction will be confined to the surface given by $f = f_{\mathrm{st}}$ for systems with one-step irreversible reactions. The flame surface is usually contorted and possibly multiply connected, as shown for the contour in Fig. 7.20*a*.

For finite-rate chemistry, the reaction zone will be broadened around the $f = f_{\mathrm{st}}$ contour. Gibson and Libby[69] estimated this thickness to be

$$\delta_r \sim L_B(Da_T)^{1/3} \tag{7-203}$$

where L_B is the Batchelor length scale[66] defined as

$$L_B = \left(\frac{\nu \mathscr{D}^2}{\varepsilon} \right)^{1/4} \tag{7-204}$$

and Da_T is the turbulent Damköhler number defined as

$$Da_T \equiv \left(\frac{\nu k^2}{\varepsilon} \right)^{1/2} \tag{7-205}$$

where k is the chemical-reaction rate constant.

In flames with complex reaction mechanism and involving significant reverse reactions, the relationship $Y_i \propto f$ becomes that shown in Fig. 7.21 with a finite second derivative of $d^2Y_i^e/df^2$ over a range of f near f_{st}. The reaction rate equation (7-198) indicates that chemical reaction occurs over this range in f, with the reaction zone becoming "equilibrium-broadened" around f_{st}. This broadening[66] is due to the back reactions shifting the equilibrium products on either side of f_{st}. Turbulent flame structure can also be conceived under some circumstances to be an ensemble of stretched diffusion flamelets. The stretching is due to the turbulent strain, which produces steep concentration gradients and thereby increases the rate of diffusion of reactants toward the reaction zone.

7.1.4 Pollutant Formation in Diffusion Flames

It is generally accepted that the formation of nitric oxide can be described by the slow Zel'dovich mechanism[66,10],

$$O + N_2 \overset{k_1}{\rightarrow} NO + N \tag{7-206}$$

$$N + O_2 \overset{k_2}{\rightarrow} NO + O \tag{7-207}$$

An excellent review of kinetics of these reactions was given by Bowman.[70] Nitric oxide is usually present in trace quantities and has negligible influence on reactions having significant amounts of heat release or on the overall flow processes. The formation of NO is several orders of magnitude slower than the main heat-release reactions and is thus kinetically limited. If the reverse reactions of the above equations are negligible and Eq. (7-206) is rate-limiting, we have

$$\dot{\omega}_{NO} = 2\rho^2 k_1 Y_O Y_{N_2} \tag{7-208}$$

The rate constant k_1 is purely a function of temperature. If the oxygen-atom concentration is assumed to be in equilibrium, then Y_O, k_1, ρ, and Y_{N_2} are functions of the mixture fraction only. Then the mean formation rate of NO can be determined from

$$\bar{\dot{\omega}}_{NO}(\mathbf{x}) = \int_0^1 \dot{\omega}_{NO}(f)\tilde{\mathscr{P}}(f;\mathbf{x})\,df \tag{7-209}$$

When the above equation is applied to hydrogen–air flames, it does not reproduce the "rich shift" of the maximum NO concentration observed by Bilger and Beck[71] and gives NO levels substantially different from those measured. This discrepancy is probably due to the strong dependence of the NO formation rate on temperature and mixture fraction. The pdf, $\mathscr{P}(f;\mathbf{x})$, may be a source of this discrepancy. According to Jones and Whitelaw,[5] a more widely accepted explanation is the oxygen-atom equilibrium assumption; the oxygen-atom concentration can be one order of magnitude greater than equilibrium values. In view of the above discrepancy, Janicka and Kollmann[72] used a two-variable approach for the hydrogen–air flame. This is discussed in the next section.

7.2 Two-Variable Approaches

7.2.1 Janicka and Kollmann's Model[72]

To extend the conserved-scalar approach to more complicated multistep chemical reactions, it is possible to use an additional variable, the reaction-progress variable (or reactedness) r, discussed in Chapter 6. As we can see from Fig. 6.9, concentrations are linearly related to the reaction-progress variable and mixture fraction. In their study of hydrogen–air systems, Janicka and Kollmann[72] considered the following simplified system of seven equations:

$$\left.\begin{aligned} H + O_2 &\rightleftharpoons OH + O \\ O + H_2 &\rightleftharpoons OH + H \\ H_2 + OH &\rightleftharpoons H_2O + H \\ 2OH &\rightleftharpoons H_2O + O \end{aligned}\right\} \quad \text{Shuffle reactions} \tag{7-210}$$

$$\left.\begin{aligned} H + OH + M &\rightleftharpoons H_2O + M \\ H + O + M &\rightleftharpoons OH + M \\ H + H + M &\rightleftharpoons H_2 + M \end{aligned}\right\} \quad \text{Three-body recombination reactions} \tag{7-211}$$

The first four reactions (7-210) represent fast "shuffle" reactions, whereas those in Eq. (7-211) represent slow three-body recombination reactions. This system can be expected to give reasonable results at high temperatures under conditions close to stoichiometric. For lower temperatures and lean mixtures, other radicals such as HO_2 become important.[73] To make the system tractable with a

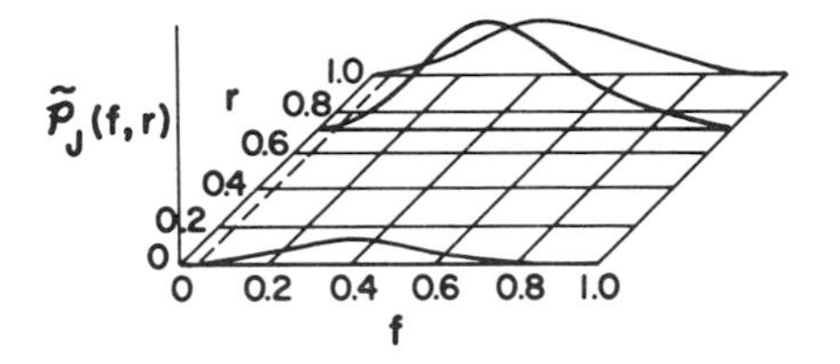

Figure 7.24 Joint probability density function of mixture fraction f and reaction-progress variable r (after Janicka and Kollmann[72]).

reasonable and manageable number of species, Janicka and Kollmann[72] considered the above simplified system to be satisfactory. The joint probability density function, $\mathscr{P}_J$, used by Janicka and Kollmann is shown in Fig. 7.24.

In their model, nine transport equations were solved together. Mean O-radical concentrations were predicted and used for the estimation of nitric oxide formation in the flame. Reasonable agreement between theory and experiments was obtained.

7.2.2 Chemical Production Term Closure for the Second Variable

Some researchers in turbulent-diffusion-flame studies choose to use merely a global kinetic rate based on the mean quantities in the following equation:

$$\bar{\dot{\omega}}_F = -\bar{k}\bar{\rho}^2\bar{Y}_F\bar{Y}_O\exp\left(-\frac{T_a}{\bar{T}}\right) \tag{7-212}$$

This equation is highly unsatisfactory, since it totally ignores the effect of turbulent fluctuations and the important role of mixing, which governs reaction rates. Some researchers, such as Borghi[74] and Hutchinson et al.,[75] applied a correction factor $(1 + F_c)$ to Eq. (7-212), where

$$F_c = F_c\left(\bar{T}, \overline{T'^2}, \bar{Y}_O, \bar{Y}_F, \overline{Y_O'T'}, \overline{Y_F'Y_O'}\right) \tag{7-213}$$

or

$$F_c = F_c\left(\tilde{T}, \widetilde{T''^2}, \tilde{Y}_O, \tilde{Y}_F, \overline{Y_O''T''}, \overline{Y_F''Y_O''}\right)$$

The correction factor can be obtained by the truncation of the following series expansion:[4]

$$-\bar{\dot{\omega}}_F = \widetilde{\rho k_f}\left(\bar{\rho}\tilde{Y}_F\tilde{Y}_O + \overline{\rho Y_F''Y_O''}\right) + \overline{\rho(\rho k_f)''Y_O''}\tilde{Y}_F + \overline{\rho(\rho k_f)''Y_F''}\tilde{Y}_O$$
$$+ \overline{\rho(\rho k_f)''Y_F''Y_O''} \tag{7-214}$$

where k_f is the specific reaction rate of

$$F + O \xrightarrow{k_f} P \tag{7-215}$$

k_f can be expanded in the form

$$k_f = Ae^{-T_a/\tilde{T}}\left\{1 + \overset{\ll 1}{\frac{T_a}{\tilde{T}^2}} T'' + \left[\frac{T_a^2}{2\tilde{T}^4} - \overset{\text{negligible}}{\frac{T_a}{\tilde{T}^3}}\right] T''^2 + \cdots\right\} \tag{7-216}$$

The Favre average of k_f is

$$\widetilde{k_f} = Ae^{-T_a/\tilde{T}}\left\{1 + \left(\frac{T_a^2}{2\tilde{T}^4}\right)\frac{\overline{\rho T''^2}}{\bar{\rho}} + \cdots\right\} \tag{7-217}$$

where $T_a/\tilde{T}^3$ has been neglected in comparison with $T_a^2/2\tilde{T}^4$ on the basis of $T_a/\tilde{T}$ being large.[4] It should be noted that the above series expansion approach is valid only if $T_a T''/\tilde{T}^2 \ll 1$ and the higher-order terms in the expansion are negligible.

The turbulence correlations, such as $\overline{Y_O' T'}$ and $\overline{Y_F' Y_O'}$ (or $\overline{Y_O'' T''}$ and $\overline{Y_F'' Y_O''}$), have to be obtained by second-order closure of their transport equations. This type of method seems to gain little advantage from the use of the conserved scalar. Further discussion of these direct approaches is given in Section 7.7.3.

An alternative way of approaching the chemical closure problem was suggested by Janicka and Kollmann,[72] who used a combined variable $Y_{H_2}^*$ proposed initially by Kaskan and Schott.[76] $Y_{H_2}^*$ is defined as

$$Y_{H_2}^* \equiv Y_{H_2} + \frac{1}{2}\left(\frac{W_{H_2}}{W_{OH}}\right) Y_{OH} + \left(\frac{W_{H_2}}{W_O}\right) Y_O + \frac{3}{2}\left(\frac{W_{H_2}}{W_H}\right) Y_H \tag{7-218}$$

With the introduction of this combined variable, the net rates of reactions in Eq. (7-210) are eliminated. Turbulent diffusion flames with fast reactions in partial equilibrium and slow reactions in nonequilibrium can then be described instantaneously by only two dependent variables, namely f and $Y_{H_2}^*$ (or f and r). In their work, the nonequilibrium O-atom concentrations were obtained directly from $Y_{H_2}^*$ and f, so that the instantaneous NO production and NO concentrations can be calculated using the joint pdf.

According to Bilger,[66] a basic problem with all of these two-variable approaches is that they are not well suited to systems in which the chemistry is reasonably fast and the composition is close to equilibrium. Small errors in the estimation of these variables can lead to large errors in the extent of departure from equilibrium.

7.2.3 Bilger's Perturbation Analysis[24]

In many combustion systems of practical interest, such as flames with nonequilibrium radical contributions to NO formation, the departure from the equilibrium state is then important. A perturbation of the equilibrium state was considered by Bilger.[24,66] He suggested representing the species mass fraction

Y_i by

$$Y_i = Y_i^e(f) + y_i \tag{7-219}$$

where f is a conserved scalar (e.g., mixture fraction) and y_i is the departure of species i from equilibrium. Substituting Eq. (7-219) into the species conservation equation (3-36), we have

$$\dot{\omega}_i = L(Y_i) = L(y_i) + \frac{dY_i^e}{df} \cancelto{0}{L(f)} - \rho \mathscr{D} \frac{\partial f}{\partial x_k} \frac{\partial f}{\partial x_k} \frac{d^2 Y_i^e}{df^2} \tag{7-220}$$

where

$$\begin{aligned} L(y_i) &= \rho \frac{\partial y_i}{\partial t} + \rho u_k \frac{\partial y_i}{\partial x_k} - \frac{\partial}{\partial x_k}\left(\rho \mathscr{D} \frac{\partial y_i}{\partial x_k}\right) \\ &= \dot{\omega}_i + \rho \mathscr{D} \frac{\partial f}{\partial x_k} \frac{\partial f}{\partial x_k} \frac{d^2 Y_i^e}{df^2} \end{aligned} \tag{7-221}$$

Applying Favre averaging to Eq. (7-221) at high Reynolds number, we have

$$\bar{\rho} \tilde{u}_k \frac{\partial \tilde{y}_i}{\partial x_k} + \frac{\partial}{\partial x_k}\left(\widetilde{\bar{\rho} u_k'' y_i''}\right) = \bar{\dot{\omega}}_i + \overline{\frac{1}{2} \rho \chi \frac{d^2 Y_i^e}{df^2}} \tag{7-222}$$

where χ is the instantaneous dissipation $2\mathscr{D}(\partial f / \partial x_k)^2$ defined by Eq. (7-200).

Some physical insight into the source term $\overline{\frac{1}{2} \rho \chi \, d^2 Y_i^e / df^2}$ can be gained by considering the mixing of two equal lumps of fluid, each in chemical equilibrium, and with mixture fractions $f + \delta f$ and $f - \delta f$, respectively. These are shown as points A and B on the $Y_i^e(f)$ curve for one of the constituent species in Fig. 7.25. On mixing without chemical reaction, the composition would be at C, whereas at chemical equilibrium, the composition would be at D. The compositions at A and B can be related to that at D through a Taylor series expansion

$$Y_{iA}^e = Y_{iD}^e + \left(\frac{dY_i^e}{df}\right)_D \delta f + \frac{1}{2}\left(\frac{d^2 Y_i^e}{df^2}\right)_D \delta f^2 + \cdots \tag{7-223}$$

$$Y_{iB}^e = Y_{iD}^e - \left(\frac{dY_i^e}{df}\right)_D \delta f + \frac{1}{2}\left(\frac{d^2 Y_i^e}{df^2}\right)_D \delta f^2 + \cdots \tag{7-224}$$

The deviation of the mixture from the equilibrium composition is given by

$$Y_{iC} - Y_{iD}^e = \frac{1}{2}(Y_{iA}^e + Y_{iB}^e) - Y_{iD}^e = \frac{1}{2}\left(\frac{d^2 Y_i^e}{df^2}\right)_D \delta f^2 + \cdots \tag{7-225}$$

The rate at which this out-of-equilibrium material is produced will depend on

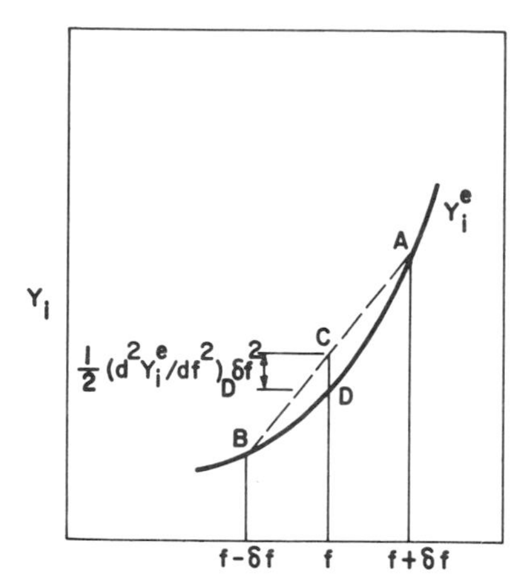

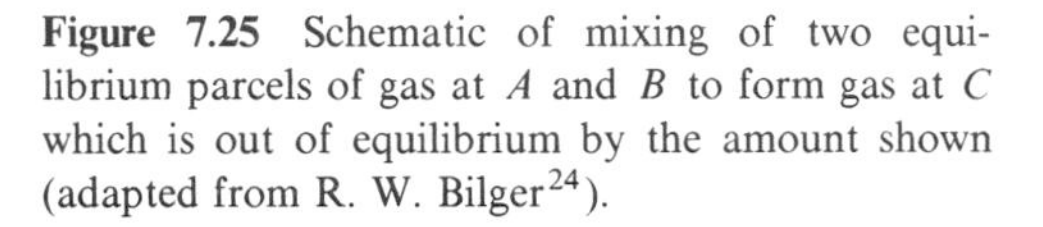

Figure 7.25 Schematic of mixing of two equilibrium parcels of gas at A and B to form gas at C which is out of equilibrium by the amount shown (adapted from R. W. Bilger[24]).

the rate at which mixing of lumps of composition $f \pm \delta f$ occurs. This is clearly related to the fluctuation dissipation, so that the rate of creation of out-of-equilibrium material is plausibly on average equal to $\frac{1}{2}\overline{\rho\chi\, d^2Y_i^e/df^2}$.

It should be noted that the term $\bar{\dot{\omega}}_i$ on the right-hand side of Eq. (7-222) is the chemical sink term for y_i, since it is negative if i is a reactant. In the limit of fast chemistry, $\tilde{y}_i \to 0$ and the chemical sink term $\bar{\dot{\omega}}_i$ becomes everywhere equal to the turbulent microscale mixing source term

$$\overline{\rho\mathscr{D}\frac{\partial f}{\partial x_k}\frac{\partial f}{\partial x_k}\frac{d^2Y_i^e}{df^2}}$$

With $d^2Y_i^e/df^2$ obtained directly from chemical equilibrium (see Fig. 7.26),

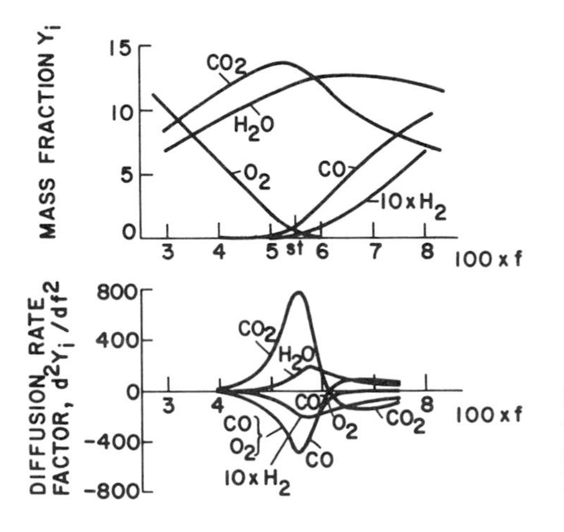

Figure 7.26 Structure and reaction rate within a methane–air diffusion flame for a shifting-equilibrium model (after R. W. Bilger[24]).

the source term becomes

$$\overline{\rho \mathscr{D} \frac{\partial f}{\partial x_k} \frac{\partial f}{\partial x_k} \frac{d^2 Y_i^e}{df^2}} = \frac{1}{2} \bar{\rho} \tilde{\chi} \int_0^1 \frac{d^2 Y_i^e}{df^2} \tilde{\mathscr{P}}(f)\, df \tag{7-226}$$

The term $\bar{\dot{\omega}}_i$ in Eq. (7-222) can be modeled by perturbation closure with a two-variable formulation. Bilger[24,66] suggested the consideration of

$$\dot{\omega}_y = \dot{\omega}_y(f, y) \tag{7-227}$$

where y denotes the departure of a particular mass-fraction function from its equilibrium value. For example, in Janicka and Kollmann's[72] treatment of the hydrogen–air system, we can define

$$y \equiv \left(Y^*_{H_2}\right)^e - Y^*_{H_2} \tag{7-228}$$

where $Y^*_{H_2}$ is defined by Eq. (7-218) and $(Y^*_{H_2})^e$ is the equilibrium value. Bilger[24] showed that, at atmospheric pressure, $\dot{\omega}_y$ can be expressed as

$$\frac{\dot{\omega}_y}{\rho} = -1.6 \times 10^6 y^2 \text{ sec}^{-1} \qquad \text{for} \quad 0.02 \le f \le 0.05 \tag{7-229}$$

For the photochemical-smog system of dilute NO_2, NO, and O_3 in air under strong sunlight, he proposed that the reaction rate is of the form

$$\frac{\dot{\omega}_y}{\rho} = -G_1(f)y + G_2(f)y^2 \approx -G_1(f)y \tag{7-230}$$

He found that linearization is possible over much of the range of interest and that $G_1(f)$ was slowly varying over a wide range of f. It is useful to point out that the above equations (7-229) and (7-230) satisfy the condition $\dot{\omega}_y \to 0$ as $y \to 0$, as required for equilibrium. Bilger believes that the direct dependence of $\dot{\omega}_y$ on y indicates that the perturbation approach may be better than the direct-closure approach for systems of moderately fast chemistry. In the direct-closure approach, the departure from equilibrium is usually arrived at from calculations of $\tilde{Y}_i$ and $\tilde{Y}_i^e$ and of moments of these parameters. For moderately fast chemistry, the departure from equilibrium is small and y is obtained as a small difference between relatively large quantities. Therefore, the perturbation approach is likely to yield a better estimate of $\tilde{y}$.

If one neglects the effect of fluctuation, the average reaction rate can be expressed as

$$\bar{\dot{\omega}}_y \approx \dot{\omega}_y(\tilde{f}, \tilde{y}) \tag{7-231}$$

For the hydrogen–air system, this equation can be written in the same form as Eq. (7-229):

$$\frac{\bar{\dot{\omega}}_y}{\bar{\rho}} \approx -1.6 \times 10^6 \tilde{y}^2 \ \text{sec}^{-1} \tag{7-232}$$

For the one-step reversible photochemical-smog system, we can follow Eq. (7-230) to write

$$\frac{\bar{\dot{\omega}}_y}{\bar{\rho}} \approx -G_1(\tilde{f})\,\tilde{y} \tag{7-233}$$

The magnitude of the truncation errors in the above approximations has yet to be explored. Also, closure by means of using joint pdf of f and y has not been investigated. In general, the two-variable approach is not suitable for complex chemical reactions which are far from equilibrium. Under these conditions, direct-closure approaches should be considered.

7.3 Direct-Closure Approaches

Although the conserved scalar approach is well established and generally preferred, there are certain turbulent flame problems for which it is either inappropriate or impossible to apply. For example, when the turbulence Reynolds number is low and differential diffusion effects are important or in some systems the reactant feeds are not uniform and constant in composition and enthalpy, it is either not possible to define an appropriate conserved scalar, or no unique relationships exist among conserved scalars. In other cases, there are complex chemical reactions far from equilibrium and thus the two-variable approach is not suitable. In such cases, it is necessary to consider the direct closure of chemical production terms.

Two methods have been developed for direct closure of chemical-production terms, namely the moment closure methods and the pdf closure methods. So far, these methods have been developed for relatively simple chemistry and for equal molecular diffusivities. They are discussed below to show how more difficult problems may be solved by these methods.

7.3.1 Moment Closure Methods

Following the initial development of Borghi,[74] for a simple one-step irreversible reaction (7-215), the reaction rate is given by

$$\dot{\omega}_F = -k_f \rho^2 Y_F Y_O \tag{7-234}$$

with

$$k_f = BT^{\alpha} \exp\left(-\frac{T_a}{T}\right) \tag{7-235}$$

where $T_a \equiv E_a/R_u$ [see Eq. (2-5)]. Using the perfect-gas law and employing the conventional Reynolds decomposition, we have

$$\dot{\omega}_F = -\frac{BP^2}{R^2}(\bar{T}+T')^{\alpha-2}\exp\left(-\frac{T_a}{\bar{T}+T'}\right)(\bar{Y}_F + Y_F')(\bar{Y}_O + Y_O') \quad (7\text{-}236)$$

The exponential term may be written as

$$\exp\left(-\frac{T_a}{\bar{T}+T'}\right) = \exp\left(-\frac{T_a}{\bar{T}}\right)\exp\left(\frac{T_a}{\bar{T}}\frac{T'/\bar{T}}{1+T'/\bar{T}}\right)$$

which may be expanded to yield

$$\bar{\dot{\omega}}_F = -\frac{Bp^2}{R^2}\bar{T}^{\alpha-2}\exp\left(-\frac{T_a}{\bar{T}}\right)\bar{Y}_F\bar{Y}_O(1 + F_c) \quad (7\text{-}237)$$

with

$$F_c \equiv \frac{\overline{Y_F'Y_O'}}{\bar{Y}_F\bar{Y}_O} + (P_2 + Q_2 + P_1Q_1)\frac{\overline{T'^2}}{\bar{T}^2} + (P_1 + Q_1)\left(\frac{\overline{T'Y_F'}}{\bar{T}\bar{Y}_F} + \frac{\overline{T'Y_O'}}{\bar{T}\bar{Y}_O}\right)$$

$$+P_1\frac{\overline{T'Y_F'Y_O'}}{\bar{T}\bar{Y}_F\bar{Y}_O} + P_2\left(\frac{\overline{T'^2Y_F'}}{\bar{T}^2\bar{Y}_F} + \frac{\overline{T'^2Y_O'}}{\bar{T}^2\bar{Y}_O}\right) + (P_3 + Q_3)\frac{\overline{T'^3}}{\bar{T}^3} + \cdots \quad (7\text{-}238)$$

where

$$P_n \equiv \sum_{k=1}^{n}(-1)^{n-k}\frac{(n-1)!}{(n-k)![(k-1)!]^2k}\left(\frac{T_a}{\bar{T}}\right)^k \quad (7\text{-}239)$$

$$Q_n \equiv \frac{(\alpha-2)(\alpha-1)\cdots(\alpha+1+n)}{n!} \quad (7\text{-}240)$$

The series (7-238) is convergent only if T_a is not large compared with $\bar{T}$ and the fluctuation levels are small. However, in practice $T_a/\bar{T} \gg 1$ for many combustion problems. This presents a severe restriction on the applicability of Eq. (7-238) to practical problems.

Assuming the series is convergent, the procedure can be followed by considering the transport equations for the second-order correlations $\overline{Y_F'Y_O'}$, $\overline{Y_F'T'}$, $\overline{Y_O'T'}$, $\overline{T'^2}$. Usually, several terms in each of these transport equations are modeled in terms of gradient transport. Closure can be obtained by appropriate treatment of the third- and higher-order correlations which appear in the chemical-product terms. The common procedure is to set these correlations equal to zero.[74] However, this procedure can lead to erroneous results, since these zero values may be inconsistent with the inequalities that exist between moments of various orders.[66]

In general, the method of moment closure appears to be best suited to isothermal or near-isothermal problems or reactions with low activation energies. Usually, the number of partial differential equations that need to be solved simultaneously is large, and there are severe limitations to the validity of the expansion of the exponential function. It should be noted that a development similar to the above Reynolds decomposition can be followed for Favre decomposition and averaging.

7.3.2 PDF Closure Methods

An alternative approach to closing the reaction-rate term is pdf closure. The reaction-rate term can be written as

$$\dot{\omega}_i = \dot{\omega}_i(\rho, T, Y_1, Y_2, \ldots, Y_N) \tag{7-241}$$

At the spatial location $\mathbf{x}$, we can write $\bar{\dot{\omega}}_i$ in terms of a multivariable pdf as follows:

$$\bar{\dot{\omega}}_i = \int_0^1 \cdots \int_0^1 \int_0^1 \int_0^\infty \int_0^\infty \dot{\omega}_i \mathscr{P}(\rho, T, Y_1, Y_2, \ldots, Y_N; \mathbf{x})\, d\rho\, dT\, dY_1 \cdots dY_N \tag{7-242}$$

Specification of the multivariable pdf can be constrained by its moments $\bar{\rho}$, $\tilde{T}$, $\tilde{Y}_i$, $\widetilde{T''^2}$, and so on as defined by equations such as Eq. (7-132). The multivariable pdf as a function of $\mathbf{x}$ is determined by solving Favre-averaged conservation equations and transport equations such as (7-95) and (7-84). The second-order correlation equations involving terms of the type $\overline{\dot{\omega}_i Y_j''}$ can be determined directly as

$$\overline{\dot{\omega}_i Y_j''} = \int_0^1 \cdots \int_0^1 \int_0^1 \int_0^\infty \int_0^\infty (Y_j - \tilde{Y}_j)\dot{\omega}_i \times \mathscr{P}(\rho, T, Y_1, Y_2, \ldots, Y_N; \mathbf{x})\, d\rho\, dT\, dY_1 \cdots dY_N \tag{7-243}$$

Currently, there is no experimental evidence available to guide the construction of this multivariable pdf.

Donaldson and Varma[77] considered a "typical eddy" for the modeling of this pdf. The pdf was represented by seven Dirac δ-functions corresponding to the pure components with two reactants and one product, two half mixtures of reactants and products, and an equally proportioned mixture of each reactant and the product. The name "typical eddy" is due to the physical simulation of this pdf as a series of cells of the compositions passing the field point, each having a duration proportional to the strength of the Dirac δ-function. The need for series expansions for the reaction rate as in Eqs. (7-237) and (7-238) is avoided, since each cell has a unique value of the reaction rate and the density associated with it. However, the composition of the cells in the "typical eddy" approach is purely arbitrary.

This "typical eddy" approach can be regarded as an attempt to construct a joint pdf at fixed locations in composition space. Since there exists a certain arbitrariness in specifying the locations of the Dirac δ-functions, calculated averaged reaction rates are expected to be sensitive to the details of the "typical eddy" treatment, particularly where the chemistry is fast. An alternative method to overcome this difficulty was suggested by Jones;[38] he recommended constructing joint pdfs from continuous functions. However, the presumed shape of a joint pdf constructed from exponential functions suffers from the disadvantage that the number of moment equations increases very rapidly with the number of independent species present. For a system of N species, it requires the solution of $N(N+3)/2$ transport equations for the moments.[5] The approach is thus only practical for very simple kinetic schemes involving few independent scalars.

7.4 Lundgren, Monin, and Novikov's Probabilistic Approach[78-81]

Rather than specify or construct a joint pdf, a potentially more powerful technique is to obtain the joint pdf from its transport equation. Lundgren[78] and Monin[79-80] independently derived the hierarchy of turbulence pdf equations. Similar equations were simultaneously obtained by Novikov[81] for the vorticity field. Their work is in the context of the incompressible Navier–Stokes equations. Lundgren[78] used the "fine-grained density" method, which is simpler and more efficient, to reach the same results. Therefore, we shall follow Lundgren's derivation in our discussion.

The application of a probabilistic approach to turbulent flames was initiated by Dopazo[82-85] and O'Brien.[84] Comprehensive reviews of this statistical method for reacting turbulent flows were given by O'Brien.[86,87]

7.4.1 Lundgren's Development of the PDF Equation

Consider a large ensemble of fluid systems. Each member of the ensemble is an incompressible fluid with mass fraction $Y(\mathbf{x}, t)$ in an infinite space satisfying the conservation equations. The only difference between members of the ensemble is in the initial conditions, which vary from member to member. Distribution functions are a way of determining the ensemble averages in terms of those in the initial state. The mass fraction $Y(\mathbf{x}, t)$ is a scalar field. At a fixed point $\mathbf{x}$ it takes on random values because of turbulent velocity fluctuations, species reactions, and molecular diffusion. The species conservation equation is to be solved subject to the initial condition that $Y(\mathbf{x}, 0) = Y_0$. The solution of this problem gives Y as a function of Y_0. It is assumed that the variation of Y_0 over the ensemble is specified statistically by assigning a probability function. Then the ensemble averages of a function such as $G[Y(\mathbf{x}_1, t), Y(\mathbf{x}_2, t), \ldots]$ can be calculated by multiplying by the probability function and integrating over the Y_0 function space. The fine-grained density $\mathscr{P}_{\text{fg}}$ is defined as the concentration distribution function for one member of the ensemble.

Suppose the function $Y^*(\mathbf{x}, t)$ describes one realization of the random field $Y(\mathbf{x}, t)$ and denote the fine-grained density of this realization by

$$\mathscr{P}_{\text{fg}}(\hat{Y}; \mathbf{x}, t) = \delta[Y^*(\mathbf{x}, t) - \hat{Y}] \tag{7-244}$$

This function is clearly a function of $\hat{Y}$ and $Y^*(\mathbf{x}, t)$, since it depends on the entire set of values of Y^*. It has the following properties. For a fixed value of $\hat{Y} = \hat{Y}_F$ and at any fixed point $(\mathbf{x}, t)$

$$\mathscr{P}_{\text{fg}}(\hat{Y}; \mathbf{x}, t) = \begin{cases} 0 & \text{if} \quad Y^*(\mathbf{x}, t) \neq \hat{Y}_F \\ \infty & \text{if} \quad Y^*(\mathbf{x}, t) = \hat{Y}_F \end{cases} \tag{7-245}$$

In general, the fine-grained density for the random scalar field $Y(\mathbf{x}, t)$ is defined as

$$\mathscr{P}_{\text{fg}}(\hat{Y}; \mathbf{x}, t) \equiv \delta[Y(\mathbf{x}, t) - \hat{Y}] \tag{7-246}$$

It is also true that $\mathscr{P}_{\text{fg}}$ has the required properties of a pdf including normalization, since

$$\int \mathscr{P}_{\text{fg}}(\hat{Y}; \mathbf{x}, t)\, d\hat{Y} = 1 \tag{7-247}$$

by definition of the δ-function. Note also that for the $Y^*(\mathbf{x}, t)$ realization, the nth moment of Y is by direct calculation

$$Y^n(\mathbf{x}, t) = \int \hat{Y}^n \mathscr{P}_{\text{fg}}(\hat{Y}; \mathbf{x}, t)\, d\hat{Y} \tag{7-248}$$

for all n.

The fine-grained density $\mathscr{P}_{\text{fg}}(\hat{Y}; \mathbf{x}, t)$ can be thought of as a device by which each realization of the random field is written in a pdf format. In particular, the ensemble average of $\mathscr{P}_{\text{fg}}(\hat{Y}; \mathbf{x}, t)$ is the customary pdf of the field, denoted by

$$\mathscr{P}(\hat{Y}; \mathbf{x}, t) = \langle \mathscr{P}_{\text{fg}}(Y; \mathbf{x}, t) \rangle \tag{7-249}$$

where $\langle\ \rangle$ denotes ensemble average. $\mathscr{P}(\hat{Y}; \mathbf{x}, t)d\hat{Y}$ is the probability that the mass fraction $Y(\mathbf{x}, t)$ at any point $\mathbf{x}$ at any t is within the limits $\hat{Y} \pm d\hat{Y}$. The explanation for obtaining mostly smooth, continuous pdfs from such a "spiky' field of realizations was first given by Stratonovich.[88]

We now derive the pdf equation for the simplest case with constant density and constant diffusivity. The species conservation equation can be written as

$$\frac{\partial Y}{\partial t} + \mathbf{v} \cdot \nabla Y = \mathscr{D}\nabla^2 Y + \frac{\dot{\omega}}{\rho} \tag{7-250}$$

Differentiating Eq. (7-246) with respect to t, we have

$$\frac{\partial \mathscr{P}_{\text{fg}}}{\partial t} = \frac{\partial \delta}{\partial t} = \frac{\partial \delta}{\partial Y}\frac{\partial Y}{\partial t} = -\frac{\partial \mathscr{P}_{\text{fg}}}{\partial \hat{Y}}\frac{\partial Y}{\partial t} \tag{7-251}$$

Substituting $\partial Y/\partial t$ of Eq. (7-250) into the above equation, we have

$$\frac{\partial \mathscr{P}_{\text{fg}}}{\partial t} = -\frac{\partial \mathscr{P}_{\text{fg}}}{\partial \hat{Y}}\left\{-\mathbf{v} \cdot \nabla Y + \mathscr{D}\nabla^2 Y + \frac{\dot{\omega}}{\rho}\right\}$$

The third term on the right-hand side of the above equation can be written as

$$\mathscr{D}\frac{\partial^2 Y}{\partial x_k\,\partial x_k}\frac{\partial \mathscr{P}_{\text{fg}}}{\partial \hat{Y}} = \mathscr{D}\frac{\partial}{\partial x_k}\left(\frac{\partial Y}{\partial x_k}\frac{\partial \mathscr{P}_{\text{fg}}}{\partial \hat{Y}}\right) - \mathscr{D}\frac{\partial Y}{\partial x_k}\frac{\partial^2 \mathscr{P}_{\text{fg}}}{\partial x_k\,\partial \hat{Y}}$$

Consider a chemical reaction system with 3 major species as unknown and let $Y = Y_\alpha$, wc have

$$\mathscr{D}\nabla^2 Y_\alpha \frac{\partial \mathscr{P}_{\text{fg}}}{\partial \hat{Y}_\alpha} = -\mathscr{D}\nabla^2 \mathscr{P}_{\text{fg}} + \mathscr{D}\frac{\partial Y_\alpha}{\partial x_k}\sum_{\gamma=1}^{3}\frac{\partial^2 \mathscr{P}_{\text{fg}}}{\partial \hat{Y}_\alpha\,\partial \hat{Y}_\gamma}\frac{\partial Y_\gamma}{\partial x_k}$$

Therefore,

$$\mathscr{D}\nabla^2 Y_\alpha \frac{\partial \mathscr{P}_{\text{fg}}}{\partial \hat{Y}_\alpha} = -\mathscr{D}\nabla^2 \mathscr{P}_{\text{fg}} + \mathscr{D}\sum_{\gamma=1}^{3}\left(\nabla Y_\alpha \cdot \nabla Y_\gamma\right)\frac{\partial^2 \mathscr{P}_{\text{fg}}}{\partial \hat{Y}_\alpha\,\partial \hat{Y}_\gamma} \tag{7-252}$$

For the simple reaction mentioned above, we have

$$\frac{\partial \mathscr{P}_{\text{fg}}}{\partial t} = \mathbf{v} \cdot \frac{\partial \mathscr{P}_{\text{fg}}}{\partial \hat{Y}}\nabla Y - \mathscr{D}\sum_{\alpha=1}^{3}\nabla^2 Y_\alpha \frac{\partial \mathscr{P}_{\text{fg}}}{\partial \hat{Y}_\alpha} - \sum_{\alpha=1}^{3}\frac{\partial}{\partial \hat{Y}_\alpha}\left(\frac{\dot{\omega}_\alpha}{\rho}\mathscr{P}_{\text{fg}}\right)$$

Therefore,

$$\frac{\partial \mathscr{P}_{\text{fg}}}{\partial t} + \mathbf{v} \cdot \nabla \mathscr{P}_{\text{fg}} - \mathscr{D}\sum_{\alpha=1}^{3}\nabla^2 \mathscr{P}_{\text{fg}} + \sum_{\alpha=1}^{3}\frac{\partial}{\partial \hat{Y}_\alpha}\left(\frac{\dot{\omega}_\alpha}{\rho}\mathscr{P}_{\text{fg}}\right)$$
$$+\mathscr{D}\sum_{\alpha=1}^{3}\sum_{\gamma=1}^{3}\nabla Y_\alpha \cdot \nabla Y_\gamma \frac{\partial^2 \mathscr{P}_{\text{fg}}}{\partial \hat{Y}_\alpha\,\partial \hat{Y}_\gamma} = 0 \tag{7-253}$$

This is a partial differential equation for the fine-grained density in the sense that for any infinitely differentiable function $\psi(\hat{Y})$, the integral of the product of $\psi(\hat{Y})$ and the terms on the left hand side of Eq. (7-253) with respect to $\hat{Y}$ is equal to zero. Taking the ensemble average of Eq. (7-253), one can show that

$$\frac{\partial \mathscr{P}}{\partial t} + \langle \mathbf{v} \cdot \nabla \mathscr{P}_{\text{fg}} \rangle + \sum_{\alpha=1}^{3} \frac{\partial}{\partial \hat{Y}_\alpha}\left(\frac{\dot{\omega}}{\rho}\mathscr{P}\right)$$

$$= \mathscr{D} \sum_{\alpha=1}^{3} \nabla^2 \mathscr{P} - \mathscr{D} \sum_{\gamma=1}^{3} \sum_{\alpha=1}^{3} \frac{\partial^2}{\partial \hat{Y}_\alpha \partial \hat{Y}_\gamma} \langle \nabla Y_\alpha \cdot \nabla Y_\gamma \mathscr{P}_{\text{fg}} \rangle \quad (7\text{-}254)$$

This is the evolution equation we seek. It demonstrates many of the advantages and disadvantages of the pdf formulation. A more general form of Eq (7-254) can be written as

$$\frac{\partial}{\partial t} \bar{\rho} \tilde{\mathscr{P}} + \nabla \cdot \bar{\rho} \langle \mathbf{v} \mathscr{P}_{\text{fg}} \rangle = -\nabla \cdot \int \left\langle \rho \mathscr{D} \nabla Y \frac{\partial \mathscr{P}_{\text{fg}}}{\partial \hat{Y}} \right\rangle d\hat{\rho} - \frac{\partial}{\partial \hat{Y}} \int \dot{\omega} \langle \mathscr{P}_{\text{fg}} \rangle \, d\hat{\rho} \quad (7\text{-}254\text{a})$$

The dependent variable $\mathscr{P}$ is a function of five independent variables $Y, \mathbf{x}, t$, while in the original formulation the dependent variable Y is a function of only four independent variables $\mathbf{x}, t$. This is a part of the price we have to pay for getting rid of the problem of closing the chemical production rate. Here $\dot{\omega}$ can be considered to be a known parameter, since it has no differential or integral space operators in its definition; and that remains true for all such specifications of chemical kinetics, no matter how complicated. This can be regarded as the major advantage of the pdf method.

A general combustion problem will consist of N reactants, three velocity components, one density, and one enthalpy, with no symmetry in space or time. If we write a single point pdf for such a general case, it will be a function of $N + 9$ variables. However, this will lead to a mammoth computing task. In deriving Eq. (7-254) we have used an externally prescribed advecting velocity field. In effect we have neglected the effect of reaction on the turbulent flow. Recent attempts at using the pdf formulation have used moment methods for the continuity and momentum equations and the pdf method for scalar equations such as those for the enthalpy and species concentration.

7.4.2 Closure Considerations for Turbulence PDF Equations

The unclosed terms in Eq. (7-254) are the advective transport term $\langle \mathbf{v} \cdot \nabla \mathscr{P}_{\text{fg}} \rangle$ and the molecular diffusion term $-\mathscr{D}(\partial/\partial \hat{Y})\langle \nabla^2 Y \mathscr{P}_{\text{fg}} \rangle$. The most popular approach to approximating the advective transport term has been the use of gradient transport hypothesis, that is,

$$\langle \mathbf{v} \cdot \nabla \mathscr{P}_{\text{fg}} \rangle = \mathbf{v} \cdot \nabla \mathscr{P} - \nabla \cdot \mathbf{K} \nabla \mathscr{P} \quad (7\text{-}255)$$

where $\mathbf{K}$ is an eddy diffusivity tensor.

The unclosed molecular-diffusion term has been a peculiarly confusing problem in the study of turbulence. Any closure assuming statistical independence of $\nabla^2 Y$ and $\mathscr{P}_{\text{fg}}$ results in the loss of proper representation of the microscale mixing and yields unphysical results. At moderate Reynolds number, this term is usually negligible compared to the advective transport term. The former can be written in the form

$$-\mathscr{D}\frac{\partial^2}{\partial \hat{Y}^2}\left\langle \frac{\partial Y}{\partial x_i}\frac{\partial Y}{\partial x_i}\mathscr{P}_{\text{fg}}\right\rangle$$

The factor $(\partial Y/\partial x_i)(\partial Y/\partial x_i)$ is easily recognized as the dissipation function of the scalar field $Y(\mathbf{x}, t)$. It is always nonnegative, as is $\mathscr{P}_{\text{fg}}$. At present, considerable effort has been made and much controversy has been generated in attempts to model this term. Readers are referred to Refs. 86 and 87 for more discussions.

7.4.3 Favre-Averaged Single-Point Joint PDF Transport Equation

In most combustion problems, density fluctuation is not negligible. In order to take account of the density fluctuations, use of Favre averaging has become very common. In general the density in a combustion problem is a random variable due to such effects as turbulent compressibility. To incorporate Favre averaging in the pdf formulation we redefine the fine-grained pdf in terms of mass-fraction fluctuations and density fluctuations for one member of the ensemble:

$$\tilde{\mathscr{P}}_{\text{fg}}(\hat{Y}, \hat{\rho}; \mathbf{x}, t) = \delta[Y(\mathbf{x}, t) - \hat{Y}]\delta[\rho(\mathbf{x}, t) - \hat{\rho}] \tag{7-256}$$

The Favre-averaged pdf defined by Bilger[66] can be written as

$$\tilde{\mathscr{P}}(\hat{Y}; \mathbf{x}, t) = \frac{1}{\langle \rho(\mathbf{x}, t)\rangle}\int \hat{\rho}\left\langle \mathscr{P}_{\text{fg}}(\hat{Y}, \hat{\rho}; \mathbf{x}, t)\right\rangle d\hat{\rho} \tag{7-257}$$

With this definition, we have the Favre-averaged single-point joint pdf transport equation[5] for a set of N scalar quantities:

$$\bar{\rho}\frac{\partial \tilde{\mathscr{P}}}{\partial t} + \bar{\rho}\tilde{u}_j\frac{\partial \tilde{\mathscr{P}}}{\partial x_j} - \frac{\partial}{\partial x_j}\left\{\frac{\mu}{\sigma}\frac{\partial \tilde{\mathscr{P}}}{\partial x_j}\right\} + \sum_{k=1}^{N}\frac{\partial}{\partial \hat{\phi}}\{\bar{\rho}\tilde{\mathscr{P}}S(\hat{\phi})\}$$
$$= -\frac{\partial}{\partial x_j}\{\bar{\rho}\langle u_j''\tilde{\mathscr{P}}_{\text{fg}}\rangle\} - \sum_{k=1}^{N}\sum_{l=1}^{N}\frac{\partial^2}{\partial \hat{\phi}_k\,\partial \hat{\phi}_l}\left\langle \frac{\mu}{\sigma}\frac{\partial \phi_k}{\partial x_j}\frac{\partial \phi_l}{\partial x_j}\tilde{\mathscr{P}}_{\text{fg}}\right\rangle \tag{7-258}$$

where

$$\tilde{\mathscr{P}}_{\text{fg}} = \prod_{k=1}^{N}\delta(\hat{\phi}_k - \phi_k) \tag{7-259}$$

is the fine-grained pdf and S is the source term in the transport equation of ϕ.

All terms on the left-hand side of Eq. (7-258) are closed and thus do not require approximation. The terms on the right-hand side of Eq. (7-258) are not well understood, and their modeling appears to be more difficult than for second-order moment closures. It is shown by Pope[89] that a satisfactory modeling of the last molecular-mixing terms purely in terms of local (in $\hat{\phi}$-space) quantities is not possible; an integral formulation is required. Dopazo and O'Brian[84] used the quasi-Gaussian approximation to model the molecular-mixing term. This can be expected to work well for near-Gaussian pdfs, but in general has severe limitations.[5]

Due to the complex nature of pdf transport equation, solutions have been obtained only for a few simple flows. Janicka et al.[90] obtained finite-difference solutions for the pdf of a single strictly conserved scalar for an H_2–air flame. Their calculated mean concentration and temperature profiles were not noticeably better than could be obtained with a presumed pdf. More recently, they have compared finite-difference solutions of a single passive (i.e., strictly conserved) scalar in a round jet and in plane and axisymmetric mixing-layer flows.[5] Excellent agreement between predictions and measurements were obtained for the round jet. Good results were also obtained for mixing layers.

In the case of joint pdf with two or more scalar variables, the finite-difference method is not adequate, due to computer-storage requirements. To overcome this, Pope[91] devised a Monte Carlo method for solving multidimensional pdf transport equations. Only modest computer-storage requirements are claimed for the method, and the computational time increases linearly with the scalar dimensionality of the pdf. Pope has applied this calculation procedure to a passive scalar in a mixing layer and found good agreement with measurements of Batt.[92] Pope[93] also made computations on a one-dimensional premixed turbulent propane–air flame, and the results were in plausible agreement with the data of Robinson.[94] The Monte Carlo solution method thus offers considerable promise for solving joint pdf equations. According to Jones and Whitelaw's recent review,[5] the success or failure of this solution procedure depends to a certain extent on the type of approximations utilized to close the pdf equation, particularly for the molecular-mixing term.

7.5 Spalding's ESCIMO Theory[95-97]

The so-called ESCIMO model developed by Spalding and coworkers[95-97] represents another approach to turbulent combustion problems for both diffusion and premixed flames. The acronym ESCIMO stands for engulfment, stretching, coherence, interdiffusion, and moving observer. These terms represent the following physical concepts and processes (see Fig. 7.27):

Engulfment. Tongues of fluid from the jet thrust into the free surrounding stream; they wrap themselves around the free-stream fluid and draw it into the interior of the jet.

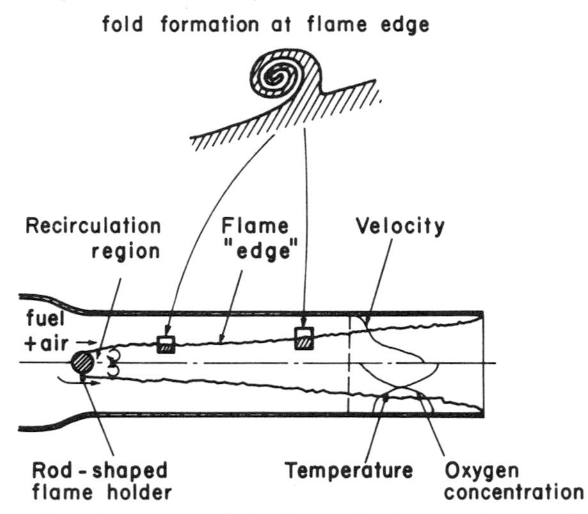

Figure 7.27 The schematic structure of the flame edge in a combustor (adapted from D. B. Spalding[97, 99]).

Stretching. The interleaved elements of jet fluid and free-stream fluid are distorted, so that the thickness of the leaves is continuously reduced.

Coherence. The leaves of jet and free-stream fluids, once interspersed, can never be separated.

Interdiffusion. The molecular-scale processes of interpretation of the two cohering fluids and of their consequent chemical reactions, diffusion, heat conduction, and the creation and destruction of chemical species are all involved. These processes are regarded in ESCIMO as obeying the laws of laminar flow.

Moving observer. The interdiffusion processes are analyzed in a frame of reference which moves with the coherent "block" of fluid (the Lagrangian aspect of ESCIMO).

In the ESCIMO model, the analysis is separated into two parts: (1) a biographic part, in which the details of reaction and molecular diffusion within folds are treated in an essentially one-dimensional manner and (2) a demographic part, which involves the specification and description of the fold distribution. The *biographic* part concerns what happens within the fold between its birth and its death, whereas the *demographic* part concerns the statistics of the "population" of coherent fluid parcels (also called "blocks," "sandwiches"): how many of each kind are "born," where they travel to, to what environmental conditions they are subjected, and how long they "live."

Interested readers are referred to Refs. 95, 96, and 97 for more detailed discussion and formulation. At present stage, the model appears to have been formulated on largely intuitive grounds, and no use is made of the exact conservation equations in the derivation. A complete formulation and application to a problem of practical interest involving moderately fast chemistry has yet to appear.

8 TURBULENT REACTING FLOWS WITH PREMIXED REACTANTS

Perfectly premixed flames occur less frequently in practical combustion systems, and thus have received less attention than diffusion flames. A thorough treatment and discussion of premixed turbulent flames is given by Bray.[98] A brief review of various methods developed for premixed turbulent flames is also discussed by Jones and Whitelaw.[5] In general, most work on premixed flames has assumed that combustion can be characterized by a global single-step reaction of the type shown by Eq. (7-215). The rates of the chemical kinetics are strongly dependent on temperature. In laminar premixed flames, the flame propagation requires thermal conduction and diffusion of radical species from the hot burned region to the unburned region (see Chapter 5). In turbulent premixed flames, these molecular processes are augmented by direct turbulent mixing and by distortion of flame surfaces, leading to an increase in contact area between the burned and unburned. The end result is that the mass consumption rate is greatly enhanced by turbulence and the mean rate of heat release is generally more strongly influenced by the turbulence than by chemical kinetics. Therefore, premixed turbulent combustion can be regarded as primarily a fluid-mechanical problem, except for ignition and flame quenching, in which both chemical kinetics and fluid mechanics are important.

8.1 Spalding's Eddy-Breakup Model

Before we focus our attention to the model, let us consider the combustion phenomenon, as shown in Fig. 7.27. A stream of premixed fuel gas and air flows steadily through a duct. In the center of the duct there is a bluffbody flame to be anchored at the baffle, and the flame spreads obliquely across the duct into the unburned stream, ultimately consuming the reactants. Although the eye perceives a continuous and thick sheet of flame, high-speed photography reveals that the reaction zone is highly convoluted, with many isolated pockets of hot and cold gas which are isolated from the main masses. There are many investigations[101-106] of this phenomenon. The most striking feature of the experimental findings has been that the angle of spread of the flame is almost unaffected by the experimental conditions: the distance from the symmetry plane to the flame edge is equal to about 0.1 times the distance from the baffle, regardless of the mixture ratio, the approach velocity, the initial temperature of the mixture, and the level of free-stream turbulence, i.e., θ is not a function of x, F/O, $\overline{U}$, T_0, u'_{rms}.

Among all the theoretical models describing this phenomenon quantitatively, the most plausible one is Spalding's eddy-breakup (EBU) model,[99-100] which was developed in 1970 and was modified further in 1976. Spalding suggested that turbulent combustion processes are best understood by focusing attention on *coherent bodies of gas*, which are squeezed and stretched during their travel through the flame. This suggestions leads to an expression for the rate of reaction which can be employed, together with a suitable scheme for solving the relevant differential equations, for predicting turbulent-flame phenomenon. Spalding[35,99] proposed that the mean rate of product formation under EBU should be written as

$$\bar{S}_P = C_{\text{EBU}} \frac{\varepsilon}{k} \sqrt{\overline{Y_F'^2}} \tag{7-260}$$

This equation implies that the mean reaction rate is determined solely by the rate of scale reduction via a process of turbulent vortex stretching. The ratio of turbulence dissipation rate ε to turbulence kinetic energy k has dimensions of $1/t$. This equation takes no explicit account of chemical kinetics and presumes that the rate of combustion is entirely turbulent-mixing controlled. This is in agreement with the well-known fact that the local rate of reaction in a turbulent fluid is governed by the local time-average and fluctuating quantities of the gaseous mixture. Spalding's EBU model offers such an expression.

The basic ideas of EBU model are given briefly below.

8.1.1 Formation and Coherence of Gas Parcels

It is assumed that the turbulent entrainment processes which occur at the outer boundary of a flame result in the formation of parcels of gas, having interleaved layers of different condition; thus fuel-rich gas layers may be interspersed with air-rich layers, and hot layers may mingle with cold ones.

The thickness of such a two-part layer, when first formed, is denoted by λ_0. It is assumed that the gas layers formed in this way travel together through the flame, but the layer thickness λ diminishes with time in accordance with the equation

$$\frac{D\lambda}{Dt} = -\lambda R^* \tag{7-261}$$

where R^* is the local stretching rate. In a *parabolic* flow, R^* is given by

$$R^* = \left| \frac{\partial \bar{u}}{\partial y} \right| \tag{7-262}$$

where $\bar{u}$ is the time-average velocity in the mainstream direction and y is the distance in the direction of steepest gradient of $\bar{u}$.

8.1.2 Small-Scale Flame Propagation

For premixed gases, in which a flame can propagate from the hot layer into the cold by heat conduction, by molecular diffusion, and by chemical reaction, it is assumed that a flame front propagates at a velocity S relative to the unburned gas. Consequently, the rate of fuel consumption is given by

$$\frac{DY_F}{Dt} = -\frac{1}{\lambda}S(Y_{F,u} - Y_{F,b}) \tag{7-263}$$

where Y_F is the average mass fraction of unburned fuel in the gas, and $Y_{F,u}$ and $Y_{F,b}$ are the Y_F's of the unburned and burned gases respectively.

8.1.3 Constancy of R*

In real flames, coherent parcels of gas pass through regions of both high and low shear rate, so R^* varies with time. However, in the interest of securing simple algebraic relations, the variations R^* with time will be ignored, as they were by Spalding.[100]

This assumption permits Eq. (7-261) to be integrated, with the result

$$\frac{\lambda}{\lambda_0} = \exp(-R^*t) \tag{7-264}$$

8.1.4 Constancy of S

If the laminar flame front is thin compared with the gas layer in question, the laminar flame speed can be regarded as independent of time until all the unburned fuel is consumed. Then combination of Eqs. (7-263) and (7-264) leads to

$$\dot{r} \equiv \frac{Dr}{Dt} = \frac{S}{\lambda_0}\exp(R^*t) \tag{7-265}$$

where r, the *reactedness*, is defined by

$$r \equiv \frac{Y_{F,u} - Y_F}{Y_{F,u} - Y_{F,b}} \tag{7-266}$$

Integration of Eq. (7-265) yields the variation of reactedness with time:

$$r = \frac{S}{R^*\lambda_0}[\exp(R^*t) - 1] \tag{7-267}$$

and elimination of the exponential between this and Eq. (7-265) leads to a relationship between the reaction rate and reactedness r:

$$\dot{r} = R^* r + \frac{S}{\lambda_0}, \qquad r \leq 1 \tag{7-268}$$

It is interesting to compare this linear relation with those typical of laminar gases,[107] such as

$$\dot{r} = c r^n (1 - r) \tag{7-269}$$

where c and n are constants, the latter being of the order of 10. Evidently the temperature dependence of laminar flames is much greater than that of turbulent ones; the differences in their behavior are in part due to this effect.

8.1.5 The Age Spectrum

Let the *age function* a be defined by

$$a \equiv \exp(R^* t) - 1 \tag{7-270}$$

At any location in the flame, it is to be expected that, as a consequence of turbulent mixing, gas parcels of various ages are present. The age distribution can be represented by a probability density function $\mathscr{P}$. The simplest possible $\mathscr{P}$ function will be presumed, i.e., that of "top-hat" form:

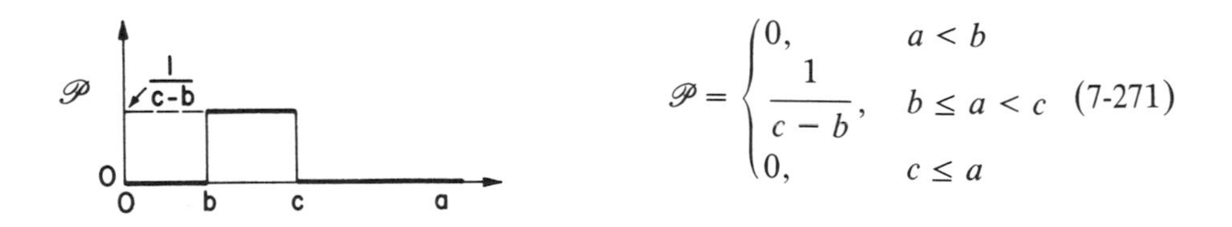

$$\mathscr{P} = \begin{cases} 0, & a < b \\ \dfrac{1}{c - b}, & b \leq a < c \\ 0, & c \leq a \end{cases} \tag{7-271}$$

The probability density function in the form of a top hat.

Later, it will be assumed that the fraction b/c is a constant. Of course, points farthest from the point of origin of eddies will have the largest values of b and c.

8.1.6 Analysis of the Reaction-Rate Relation

With the aid of the above basic ideas and assumptions, we can investigate the relationship between the time-average reactedness and reaction rate further.

Define the time-averaged r and $\dot{r}$ respectively as follows:

$$\bar{r} \equiv \int_0^\infty r\mathscr{P}\,da \tag{7-272}$$

$$\bar{\dot{r}} \equiv \int_0^\infty \dot{r}\mathscr{P}\,da \tag{7-273}$$

Here r and $\dot{r}$ can be expressed as functions of a, by reason of Eqs. (7-267), (7-268), and (7-270), in the following manner:

$$a \le a_1: \qquad r = \frac{S}{R^*\lambda_0} \cdot a$$

$$\dot{r} = \frac{S}{\lambda_0}(1 + a) \tag{7-274}$$

$$a > a_1: \qquad r = 1$$

$$\dot{r} = 0 \tag{7-275}$$

Here the quantity a_1, which is the age of a gas parcel that is just fully burned, is defined by

$$a_1 \equiv \frac{R^*\lambda_0}{S} \tag{7-276}$$

Substitution of these expressions into Eqs. (7-272) and (7-273), and evaluation of the integrals with the aid of the probability density function [Eq. (7-271)], yields the following results:

$$a_1 < b: \qquad \bar{r} = 1$$

$$\bar{\dot{r}} = 0 \tag{7-277}$$

$$b \le a_1 < c: \qquad \bar{r} = 1 - \frac{(a_1 - b)(1 - b/a_1)}{2(c - b)} \tag{7-278}$$

$$\frac{\bar{\dot{r}}}{R^*} = \frac{a_1 - b}{c - b}\left(\frac{1 - b/a_1}{2} + \frac{1}{a_1}\right) \tag{7-279}$$

$$c \le a_1: \qquad \bar{r} = \frac{b + c}{2a_1} \tag{7-280}$$

$$\frac{\bar{\dot{r}}}{R^*} = \frac{b + c}{2a_1} + \frac{1}{a_1} \tag{7-281}$$

Hence, $\bar{\dot{r}}$ can be expressed explicitly in terms of $\bar{r}$, as follows:

$$\bar{r} \le r^*: \qquad \bar{\dot{r}} = R^* \bar{r} + S/\lambda_0 \tag{7-282}$$

$$r^* \le \bar{r} < 1: \qquad \bar{\dot{r}} \approx R^* r^* \frac{1 - \bar{r}}{1 - r^*} \tag{7-283}$$

where $r^* \equiv \frac{1}{2}(1 + b/c)$.

8.1.7 Discussion of the Reaction-Rate Relation

Several points should be noted about the above simplified analysis. Firstly, the "top-hat" pdf is an idealization. However, we may reasonably expect that the spread of "ages" between old and newly formed gas parcels will be such that b/c is of the order of 0.5. Then r^* will be about 0.75. The actual value of r^* should be determined experimentally. Secondly, for common fuel–air mixtures $S/R^*\lambda_0$ has a value of the order of 0.1; hence, S/λ_0 can be omitted from Eq. (7-282) with little loss. This is a confirmation that the chemical-kinetic properties of turbulent burning mixtures have little influence on the burning rate.

Thirdly, the $\bar{\dot{r}}$–$\bar{r}$ relation is "tent-shaped"; the larger the ratio of b/c, the farther the shift of its peak to the right. The smallest possible value of b/c is zero. This would make r^* equal to one-half, and it would make the $\bar{\dot{r}}$–$\bar{r}$ relation symmetrical. However, according to Spalding,[100] the peak is actually expected to lie at $\bar{r} = 0.75$.

It should also be noted that $Y_{F,u}$ and $Y_{F,b}$, appearing in the definition (7-266), vary with the fuel–air ratio. Therefore, for generality, the relations (7-282) and (7-283) must be rewritten in the following form:

$$\frac{Y_{F,u} - Y_F}{Y_{F,u} - Y_{F,b}} \le r^*: \qquad \frac{D}{Dt} Y_F = -(Y_{F,u} - Y_{F,b}) R^* + \frac{S}{\lambda_0} \tag{7-284}$$

$$r^* < \frac{Y_{F,u} - Y_F}{Y_{F,u} - Y_{F,b}} < 1: \qquad \frac{D}{Dt} Y_F = \frac{r^*}{1 - r^*} (Y_F - Y_{F,b}) R^* \tag{7-285}$$

where the overbar is omitted from Y_F but time-average values are intended. These equations are easy to use in practice because they express the local rate of consumption of fuel in terms of local properties (Y_F, $Y_{F,b}$, $Y_{F,u}$, R^*), which are all easy to calculate. The approximations made by Spalding in his derivation have been selected with this end in view. However, the existence and implications of these approximations must not be forgotten.

8.2 Statistical Model of Bray and Moss[108]

As we learned from the previous discussion on premixed turbulent flames, there are two alternative hypotheses regarding the physical structure of a premixed turbulent flame:

A. *Wrinkled-Laminar-Flame Hypothesis.* The effect of turbulence is to stretch and wrinkle the thin laminar flame structure, increasing its area and hence the effective flame speed. Intense turbulence tears the wrinkled laminar flame sheet into pieces, thus forming packets of burned and unburned gas separated from each other by narrow combustion zones. A recent formulation of this concept is the eddy-breakup model of Spalding, discussed in the previous section.[100]

B. Distributed-Reaction Hypothesis. The instantaneous region of combustion is distributed throughout the time average of combustion zone, rather than being confirmed to be narrow, laminar-like flames moving within the time-averaged combustion zone.

It seems likely that, in different circumstances, each of these alternative models may have a region of validity. The purpose of Bray and Moss's work[108] is to create a unified description of the premixed turbulent flame, in order to define these regions of validity and to bridge the gaps between these two hypotheses.

Bray and Moss considered the following assumptions in their model:

1. Only two chemical species are considered, namely, "reactant" (mass fraction $1 - c$) and "product" (mass fraction c), which are defined for a combustible mixture of given stoichiometry, and which may include diluents.
2. A single-step, irreversible chemical reaction, reactant → product, occurs at a rate described by a global reaction-rate expression.
3. Reactant and product species are treated as ideal gases.
4. The specific heat at constant pressure, C_p, of both reactant and product is the same, and is a constant.
5. Thermal-diffusion and pressure-diffusion effects are neglected, while normal binary diffusion is represented through Fick's law.
6. The Lewis number $\rho \mathscr{D} C_p/\lambda$ is equal to unity for both reactant and product.
7. The flow within the flame occurs at a Mach number much less than unity, so that the terms in the energy-balance equation representing effects of pressure changes and viscous dissipation may be neglected.
8. Pressure fluctuations are assumed to be of small intensity, and are neglected.
9. The flow is adiabatic and far upstream of the combustion zone is steady, one-dimensional, and uniform in all its properties.

Bray and Moss took a pdf for the product mass fraction having the general form

$$\mathscr{P}(c) = \alpha\delta(c) + \beta\delta(1 - c) + [H(c) - H(c - 1)]\gamma f(c) \quad (7\text{-}286)$$

where $\delta(c)$ and $H(c)$ are the Dirac δ and Heaviside functions, respectively. The coefficients α, β, and γ are nonnegative functions of position x. The function $f(c)$ is typically a continuous function of c, and satisfies the condition

$$\int_0^1 f(c)\, dc = 1 \quad (7\text{-}287)$$

It should be noted that both $\mathscr{P}$ and f are functions of x. For simplicity in mathematical forms, they are represented here explicitly only as functions of c. It can be shown readily, from the characteristics of a pdf, that the coefficients α, β, and γ must satisfy the following relationship:

$$\alpha + \beta + \gamma = 1 \quad (7\text{-}288)$$

If we identify the δ-functions at $c = 0$ and $c = 1$ with pockets of unburned and all-burned mixture, and $f(c)$ with product distributions in a burning mode, then the coefficients α, β, and γ describe the partitioning between these three possible modes. The time-averaged mean product mass fraction can be written as

$$\bar{c} = \int_0^1 c\mathscr{P}(c)\, dc = \beta + \gamma I_1 \quad (7\text{-}289)$$

where

$$I_1 \equiv \int_0^1 cf(c)\, dc \quad (7\text{-}290)$$

Also, it can be shown that

$$\overline{c'^2} = \bar{c}(1 - \bar{c}) - \gamma(I_1 - I_2) \quad (7\text{-}291)$$

where

$$I_2 \equiv \int_0^1 c^2 f(c)\, dc \quad (7\text{-}292)$$

The thermal equation of state for the mixture is written as

$$\bar{p} = \rho R_u T\left(\frac{1 - c}{W_1} + \frac{c}{W_2}\right) \quad (7\text{-}293)$$

where W_1 and W_2 are the effective molecular weights of reactant and product, respectively. The caloric equation of state is

$$h = C_p T - c\,\Delta H_r \tag{7-294}$$

where ΔH_r is the heat of reaction per unit mass of product formed and h is the specific enthalpy. Using the assumptions listed above, the energy equation can be written as

$$\rho\frac{\partial h}{\partial t} + \rho v_\beta \frac{\partial h}{\partial x_\beta} = \frac{\partial}{\partial x_\beta}\left(\rho\mathscr{D}\frac{\partial h}{\partial x_\beta}\right) \tag{7-295}$$

This equation is also satisfied if h is a constant within the flow, that is, $h = h_0 = c_p T_0$. Here, T_0 is the assumed uniform temperature of the premixed reactants, far upstream of the combustion zone, where $c = 0$. For downstream, where combustion is complete and $c = 1$, the temperature reaches the adiabatic flame temperature given by

$$T_{f\infty} = \frac{h_0 + \Delta H_r}{C_p} \tag{7-296}$$

The equation of state (7-293) after decomposition and time averaging becomes

$$\bar{p} = R\left(\bar{\rho}\bar{T} + \overline{\rho' T'}\right)\left(\frac{1-\bar{c}}{W_1} + \frac{\bar{c}}{W_2}\right)$$
$$+ R\left(\frac{1}{W_2} - \frac{1}{W_1}\right)\left(\bar{\rho}\overline{c'T'} + \bar{T}\overline{\rho' c'} + \overline{\rho' T' c'}\right) \tag{7-297}$$

and the caloric equation of state (7-294) becomes

$$C_p\bar{T} - \bar{c}\Delta H_r = h_0 \tag{7-298}$$

It can be shown that the fluctuating quantities T' and ρ' are related to c' by

$$T' = \frac{\Delta H_r}{C_p}c' \quad \text{and} \quad \rho' = -\bar{\rho}\psi c' \tag{7-299}$$

where

$$\psi \equiv \frac{\Delta H_r}{C_p\bar{T}} + \frac{\dfrac{1}{W_2} - \dfrac{1}{W_1}}{\dfrac{1-\bar{c}}{W_1} + \dfrac{\bar{c}}{W_2}} \tag{7-300}$$

Finally, the time-averaged equation of state is written in the form

$$\bar{\rho} = \bar{\rho}\left(h_0, \bar{p}, \bar{c}, \overline{c'^2}\right) \tag{7-301}$$

Turbulent transport terms have been modeled in an entirely conventional manner through an eddy viscosity ν_T, by analogy with the modeling of the Reynolds-stress term:

$$-\overline{g'v'} = \frac{\nu_T}{\sigma_g}\frac{\partial \bar{g}}{\partial y} \tag{7-302}$$

where g is any property (c', c'^2, T', etc.) and the turbulent Prandtl/Schmidt number σ_g is expected to be of order unity. Modeling of $\overline{\rho'v'}$ in this manner is unnecessary, since from Eq. (7-299)

$$\overline{\rho'v'} = -\bar{\rho}\psi\overline{c'v'} = \bar{\rho}\psi\frac{\nu_T}{\sigma_c}\frac{\partial \bar{c}}{\partial y} \tag{7-303}$$

where σ_c is also a turbulent Prandtl/Schmidt number. ν_T can be expressed as $\nu_T = bk^{1/2}l$, where turbulent length scale l can be obtained either from experiment or a simple proportionality to the shear-layer thickness, k is the turbulence kinetic energy, and b is a constant.

Assuming that the buoyancy effect is negligible in the x-momentum equation, the conservation equations are as follows:

Continuity equation:

$$\frac{\partial}{\partial x}(\bar{\rho}\bar{u}) + \frac{\partial}{\partial y}\left(\bar{\rho}\bar{v} + \bar{\rho}\psi\frac{\nu_T}{\sigma_c}\frac{\partial \bar{c}}{\partial y}\right) = 0 \tag{7-304}$$

x-momentum equation:

$$\bar{\rho}\bar{u}\frac{\partial \bar{u}}{\partial x} + \left(\bar{\rho}\bar{v} + \bar{\rho}\psi\frac{\nu_T}{\sigma_c}\frac{\partial \bar{c}}{\partial y}\right)\frac{\partial \bar{u}}{\partial y} = -\frac{\partial \bar{p}}{\partial x} + \frac{\partial}{\partial y}\left(\bar{\rho}\nu_T\frac{\partial \bar{u}}{\partial y}\right) \tag{7-305}$$

k-equation:

$$\begin{aligned}\bar{\rho}\bar{u}\frac{\partial \bar{k}}{\partial x} &+ \left(\bar{\rho}\bar{v} + \bar{\rho}\psi\frac{\nu_T}{\sigma_c}\frac{\partial \bar{c}}{\partial y}\right)\frac{\partial \bar{k}}{\partial y} \\ &= -a_4\bar{\rho}\bar{k}\left(\frac{\partial \bar{u}}{\partial x} + \frac{\partial \bar{v}}{\partial y}\right) + \bar{\rho}\nu_T\left(\frac{\partial \bar{u}}{\partial y}\right)^2 \\ &\quad + \frac{\partial}{\partial y}\left(\bar{\rho}\frac{\nu_T}{\sigma_k}\frac{\partial \bar{k}}{\partial y}\right) - a_2\frac{\bar{\rho}\bar{k}^{3/2}}{l}\left(1 + \frac{a_3}{\mathrm{Re}_T}\right)\end{aligned} \tag{7-306}$$

where the a_i are constants, Re_T is the turbulence Reynolds number defined as

$$\text{Re}_T = \frac{\rho k^{1/2} l}{\bar{\mu}} \tag{7-307}$$

and σ_k is a turbulent Prandtl/Schmidt number.

Product-species equation:

$$\bar{\rho}\bar{u}\frac{\partial \bar{c}}{\partial x} + \left(\bar{\rho}\bar{v} + \bar{\rho}\psi\frac{\nu_T}{\sigma_c}\frac{\partial \bar{c}}{\partial y}\right)\frac{\partial \bar{c}}{\partial y} = \frac{\partial}{\partial y}\left(\bar{\rho}\frac{\nu_T}{\sigma_c}\frac{\partial \bar{c}}{\partial y}\right) + \omega_{\max}\gamma I_3 \tag{7-308}$$

where

$$I_3 = \int_0^1 \frac{\omega(c)}{\omega_{\max}} f(c)\, dc \tag{7-309}$$

in which $\omega(c)$ represents the instantaneous mass rate of product formation, and $\omega_{\max}$ is its maximum value.

The fraction of time, γ, during which a mixture of reactant and product exists at a point in the flow is unknown, and therefore requires an additional equation. It is shown in Eq. (7-291) that γ is related to $\bar{c}$ and $\overline{c'^2}$. In order to evaluate γ, it is necessary to consider the following balance equation for $\overline{c'^2}$:

$$\bar{\rho}\bar{u}\frac{\partial}{\partial x}\overline{c'^2} + \bar{\rho}\bar{v}\frac{\partial}{\partial y}\overline{c'^2} = -2\bar{\rho}\overline{v'c'}\frac{\partial \bar{c}}{\partial y} - \frac{\partial}{\partial y}\left(\bar{\rho}\overline{c'^2 v'}\right)$$

$$- C_* \bar{\rho} k^{1/2}\overline{c'^2}/l + \overline{2\omega' c'} \tag{7-310}$$

It should be noted that the original term representing the dissipation of composition fluctuations due to turbulent mixing followed by molecular diffusion in Eq. (7-310) has been replaced by

$$2\bar{\rho}\bar{\mathscr{D}}\overline{\frac{\partial c'}{\partial x_\beta}\frac{\partial c'}{\partial x_\beta}} = C_* \bar{\rho} k^{1/2}\overline{c'^2}/l \tag{7-311}$$

following Spalding.[35]

From Eq. (7-310), after some manipulation, Bray and Moss[108] derive a new form of $\overline{c'^2}$ balance equation, which is an effective balance equation given in the following form:

$$\bar{\rho}\tilde{u}\frac{\partial}{\partial x}[\tilde{c}(1-\tilde{c})-(I_1-I_2)\gamma]+\bar{\rho}\tilde{v}\frac{\partial}{\partial y}[\tilde{c}(1-\tilde{c})-(I_1-I_2)\gamma]$$

$$=2\bar{\rho}\frac{\nu_T}{\sigma_c}\left(\frac{\partial\tilde{c}}{\partial y}\right)^2+\frac{\partial}{\partial y}\left\{\bar{\rho}\frac{\nu_T}{\sigma_*}\frac{\partial}{\partial y}[\tilde{c}(1-\tilde{c})-(I_1-I_2)\gamma]\right\}$$

$$-C_*\frac{\bar{\rho}\bar{k}^{1/2}}{l}[\tilde{c}(1-\tilde{c})-(I_1-I_2)\gamma]+2\omega_{\max}\gamma(I_4-\tilde{c}I_3) \quad (7\text{-}312)$$

where

$$I_4=\int_0^1\frac{\omega(c)}{\omega_{\max}}f(c)\,c\,dc \quad (7\text{-}313)$$

The balance equations for $\overline{c'^2}$ and k both involve extensive approximation, some of which can only be justified on grounds of expediency.

The premixed-turbulent-flame model developed by Bray and Moss has eight major dependent variables: $\bar{p}$, $\bar{\rho}$, $\overline{T}$, $\tilde{c}$, $\tilde{u}$, $\tilde{v}$, k, and γ. Seven equations have thus far been specified: state [Eqs. (7-301) and (7-298)], continuity [Eq. (7-304)], x-momentum [Eq. (7-305)], turbulence kinetic energy [Eq. (7-306)], product species [Eq. (7-308)], and species fluctuation [Eq. (7-312)]. The set of differential and algebraic equations presented above provides a complete, closed description of a premixed turbulent flame if the pressure can be specified by

$$\bar{p}=p_e(x) \quad (7\text{-}314)$$

where $p_e(x)$ is the pressure outside the shear layer, obtained in a separate calculation. The function $f(c)$ which appears via the integrals I_1, I_2, I_3, and I_4 must also be supplied as an empirical input to the model.

The pdf $\mathscr{P}(c)$ may be visualized as depending on the dimensionless ratios l_{lam}/l and $S_{\text{lam}}/u'_{\text{rms}}$, where l_{lam} and S_{lam} are respectively measures of the undisturbed laminar flame thickness and flame speed in the premixed combustible mixture at temperature T_0, while l and u'_{rms} are respectively a characteristic length scale and velocity of the turbulence. Several alternative limiting cases have been distinguished clearly by Bray and Moss. Through their classification, some physical insight may be obtained into the approximate form of $\mathscr{P}(c)$, as set out below.

Case I. Small-scale turbulence in a thick reaction zone ($l_{lam}/l \gg 1$).

(a) $S_{lam}/u'_{rms} \gg 1$. In this case, the relative turbulent intensity is very low, so the turbulent perturbation to the relatively thick laminar flame is small, and $\mathscr{P}(c)$ consists of a narrow spike centered on $c = \bar{c}$. This spike may be expected to have an almost Gaussian shape. The instantaneous reaction is distributed throughout the time-average reaction zone.

(b) $S_{lam}/u'_{rms} \sim 1$. As the relative turbulent intensity increases, the flame structure is modified by turbulent transport and the spike in $\mathscr{P}(c)$ centered near $c = \bar{c}$ is expected to be broadened. Then δ-functions may appear in the pdf at $c = 0, 1$, representing packets of material entrained into the flame zone, but not yet mixed on a molecular scale.

Case II. Large-scale turbulence interacting with thin reaction zones ($l_{lam}/l \ll 1$).

(a) $S_{lam}/u'_{rms} > 1$. This case of relatively weak, but large-scale turbulence corresponds to the classical wrinkled laminar flame discussed by Glassman.[10] Here the instantaneous reaction occurs only within the thin laminar flame, which is wrinkled and distorted by the turbulence, so as to move over the whole of the time-average heat-release zone. The pdf $\mathscr{P}(c)$ contains δ-functions at $c = 0, 1$, representing unburned and burned material, respectively. It also contains continuous structure, arising from periodic passage of the wrinkled laminar flame past the measuring station. If the turbulence is sufficiently weak, this structure will be predictable from the undisturbed laminar flame.

(b) $S_{lam}/u'_{rms} < 1$. As the relative intensity of the turbulence is increased, the thin wrinkled laminar flame front is expected to be lacerated by the turbulent motion; packets of burned or unburned material will then be formed by thin flame regions (see, for example, Howe and Shipman[109]). These packets will appear as δ-functions at $c = 0, 1$ in $\mathscr{P}(c)$. If the thin flame regions retain a laminar-like structure, the continuous part of $\mathscr{P}(c)$ will remain similar to that of case II(a) above.

(c) $S_{lam}/u'_{rms} \ll 1$. In the limit of high-intensity turbulence, it may be expected that laminar-like combustion will be severely distorted and randomly quenched by intense shear in the fluctuating turbulent velocity field, giving a more uniformly distributed reaction field. It is not clear what form $\mathscr{P}(c)$ will take in this case. However, the nearly random nature of the velocity and hence of the quenching of the reaction suggests $\mathscr{P}(c)$ may perhaps again resemble a Gaussian function.

The form (7-286) proposed for $\mathscr{P}(c)$ by Bray and Moss is quite general. It readily admits of interpretation in terms of two models of turbulent combus-

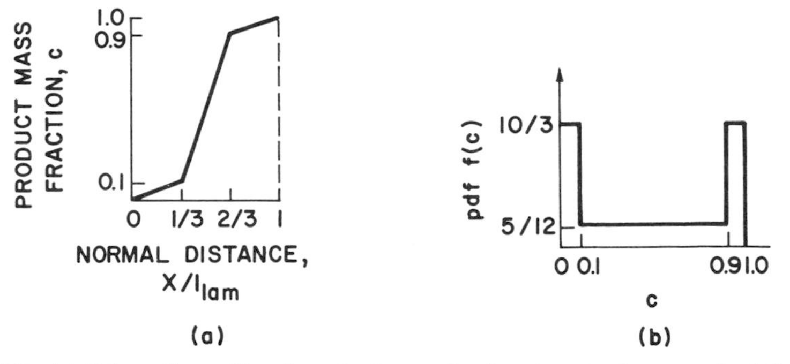

Figure 7.28 (a) Schematic profile of product mass fraction in a laminar premixed flame. (b) Reaction-product probability density function (after Bray and Moss[108]).

tion, which have been identified above: that of the wrinkled laminar flame [Cases II(a) and (b)], and that of distributed reaction [Cases I(a) and (b), and possibly II(c)].

For wrinkled laminar flames with fast chemistry in which the laminar-flame thickness l_{lam} is small compared with the turbulence scales, Bray and Moss envisage that molecular processes predominate in the region of burning. They choose therefore to model such combustion within the turbulent flame brush by identifying alternate packets of unburned mixture and product species, separated by burning interfaces which possess structure similar to the laminar flame. A typical example of the wrinkled laminar flame is

$$c = c(x/l_{\text{lam}}) \tag{7-315}$$

where $c(x/l_{\text{lam}})$ is increasing monotonically from $c(0) = 0$ to $c(1) = 1$. Sampling at arbitrarily chosen points (x/l_{lam}) through the flame reveals that

$$f(c) = \left[\frac{dc}{d(x/l_{\text{lam}})}\right]^{-1} \tag{7-316}$$

A schematic profile $c(x/l_{\text{lam}})$ is shown in Fig. 7.28*a*, which demonstrates characteristics of premixed laminar flames. The associated pdf $f(c)$ is shown in Fig. 7.28*b*. Since we shall be concerned wholly with integral relationships involving $f(c)$, the fine detail neglected by this fairly rudimentary profile is unimportant.

It was emphasized by Bray and Moss that the term "packet" is not necessarily intended to mean a discernible volume bounded by a laminar flame. A single flame surface flapping back and forth past an observation station would create a similar impression and be similarly described.

A small-γ approximation (corresponding to large Damköhler and Reynolds numbers) to the full set of model equations has also been developed by Bray and Moss.[108] An explicit equation has been derived for the time-average reaction rate

$$\overline{S_p} = C_R \frac{\varepsilon}{k} \overline{c'^2} \propto \frac{\varepsilon}{k} \overline{Y_F'^2} \tag{7-317}$$

where C_R is a constant dependent on the continuous part $f(c)$ of the pdf. By comparing Eq. (7-317) with (7-260), we can see that the result of Bray and Moss is very similar to the eddy-breakup result of Spalding, except that the mean-square fluctuation in Eq. (7-317) is replaced by the root-mean-square fluctuation in Eq. (7-260).

The burning-mode part of the pdf $f(c)$ can be determined from a laminar-flamelet description though results are found by Bray[110] to be fairly insensitive to the shape chosen. For small and intermediate Damköhler numbers, the shape of the burning-mode part of the pdf will be greater in importance, though its form can still be determined from a laminar-flamelet model. The model, to be realistic, must be extended to include more complex reaction mechanisms. Champion, Bray, and Moss[111] adopted this latter approach to treat a one-dimensional premixed propane–air flame. They considered two sequential stages in the combustion mechanism. In the first stage (the delay zone), propane was taken to combine with oxygen to form hydrogen and carbon monoxide via a single global reaction. The second stage of the reaction involved the oxidation of hydrogen and carbon monoxide (the combustion zone). Fast bimolecular reactions for achieving partial equilibrium are invoked to simplify the reaction mechanism in the combustion zone.

For the delay zone, calculations were presented using two methods: one suggested by Borghi[74] whereby the averaged species mass consumption rate is evaluated by an averaged Taylor series expansion truncated at second order, the other based upon an assumed pdf for the reaction-progress variable. The pdf used comprised a Dirac δ-function at the equilibrium temperature plus a continuous contribution function. Small differences between the two methods were noted, and the results appear to be insensitive to the pdf shape. The combustion zone is a region with large fluctuations in temperature. Results showed a large and plausible influence of temperature fluctuations on the mean temperature in the combustion zone. However, no comparison with experiment was presented, for lack of suitable measurements.

The Bray–Moss model for premixed turbulent flames has been extended by Libby and Bray.[58,59] They used the concept of laminar flamelets to derive models for turbulent transport and dissipative processes in one-dimensional planar flames. They employed a joint pdf $\mathscr{P}(u, r)$ which is a function of the velocity component normal to the average position of the turbulent flame front and the reaction-progress variable. The fast-chemistry assumption was invoked, and the joint pdf was separated into two conditioned pdfs corresponding to reactant and product nodes. The Libby–Bray model can be specified in terms of three quantities, the mean velocity $\tilde{u}$ and the fluxes $\widetilde{\bar{\rho} u'' Y_i''}$ and $\widetilde{\bar{\rho} u''^2}$, the values of which are obtained from the solution of their respective transport equations. The closure of these equations was conducted through the use of laminar flamelet model for dissipative terms. From Jones and Whitelaw's point of view,[5] the neglect of quantities involving fluctuating pressure, pressure redistribution, and pressure scrambling terms is unacceptable. A major advantage of Libby and Bray's approach is that recourse to gradient-diffusion arguments is completely avoided. The extension of the Libby–Bray model to multidimensional flows is likely to require considerable effort, and the procedure is not straightforward, since the joint pdf of the velocity vector and the reaction-progress variable must be specified. Its dimensionality must be substantially increased and the model becomes very complicated.

8.3 Chorin, Ghoniem, and Oppenheim's Random-Vortex Method[112]

The influence of the exothermic process of combustion on the turbulent flow field, where it occurs, has been studied by Chorin et al.[112] through a numerical analysis based on the *random-vortex method* (RVM). The algorithm developed traces the action of elementary turbulent eddies and their cumulative effects without imposing any restriction on their motion. The technique has been applied to flow in a combustion tunnel where the flame is stabilized by a back-facing step.

Traditionally, numerical analysis of turbulent flow has been based on some form of finite-difference treatment of averaged (mass-averaged or Favre-averaged) Navier–Stokes equations, along with an adequate set of closure relations to correlate the turbulent-flow parameters. An averaging method inherently deprives the equations of essential information about the mechanism of turbulence, necessitating the introduction of some sort of artificial closure equations. Also the finite-differencing technique introduces numerical diffusion (in the form of viscosity), leading to smoothing of local perturbations when high-shear regions appear in the field. The random-vortex method provides a solution to the Navier–Stokes equations, introducing effects of turbulent diffusion and combustion, without any averaging. It is grid-free and hence is unaffected by numerical viscosity effects of finite differencing. The RVM models the turbulent eddies by vortex elements, and it keeps track of the position and strength of all the vortex elements in the flow field.

The basic assumptions used in their model are:

1. The mean flow is essentially two-dimensional.
2. The flow field consists of two incompressible media: the unburned mixture and burned gases.
3. The flame is treated as constant-pressure deflagration wave propagating locally at the prescribed normal burning velocity of 12 cm/sec.
4. The effect of combustion is manifested entirely by an increase in the specific volume of the transformation of one component into other.

5. All three-dimensional effects, compressibility effects, kinetic effects, and thermal effects are neglected.

The governing equations from the above idealizations are:

$$\nabla \cdot \mathbf{V} = \varepsilon^{\diamond}(\boldsymbol{r}_f) \tag{7-318}$$

$$\frac{D\mathbf{V}}{Dt} = \mathrm{Re}^{-1}\nabla^2\mathbf{V} - \nabla p \tag{7-319}$$

where $\mathbf{V} = (u, v)$ is the velocity vector normalized by the inlet velocity u_∞, $\varepsilon^{\diamond}$ is the corresponding local rate of expansion, $\boldsymbol{r} = (x, y)$ is the position vector normalized by a reference length L, the time t is normalized by L/u_∞, and pressure normalized by ρu_∞^2, ρ denoting the reference density of the medium. The subscript f refers to the flame front.

The local rate of expansion $\varepsilon^{\diamond}$ varies throughout the field, and its distribution is determined by the location of the flame front $\boldsymbol{r}_f$, governed by the flame propagation equation:

$$\frac{D\boldsymbol{r}_f}{Dt} = S_L \mathbf{n}_f \tag{7-320}$$

where S_L is the normal burning velocity and $\mathbf{n}_f$ is the unit vector normal to the flame surface. The boundary conditions for Eqs. (7-318) and (7-319) are

$$\mathbf{V} = \begin{cases} 0 & \text{along all solid boundaries} \\ (1,0) & \text{at the inlet} \end{cases} \tag{7-321}$$

The solution procedure is based on the principle of fractional steps,[113,114] according to which the governing equations are split into a sum of elementary components and the solution is determined by treating these components in succession. The vorticity is defined by

$$\boldsymbol{\xi} = \nabla \times \mathbf{V} \tag{7-322}$$

The vortex transport equation for the two-dimensional flow field is

$$\frac{D\boldsymbol{\xi}}{Dt} = \frac{1}{\mathrm{Re}}\nabla^2\boldsymbol{\xi} \tag{7-323}$$

Equations (7-318) and (7-322) are used to determine the velocity field. Equation (7-323) is employed to update the vorticity field. $\varepsilon^{\diamond}(x, y)$ is determined from the flame-propagation algorithm by solving Eq. (7-320).

The velocity vector $\mathbf{V}$ is decomposed into a divergence-free vector field $\mathbf{V}_\xi$ and a curl-free field $\mathbf{V}_\varepsilon$:

$$\mathbf{V} \equiv \mathbf{V}_\xi + \mathbf{V}_\varepsilon \tag{7-324}$$

By following the Hodge decomposition theorem,[115,116] the governing equations for $\mathbf{V}_\xi$ and $\mathbf{V}_\varepsilon$ are then obtained immediately, after substituting Eq. (7-324) into Eqs. (7-318) and (7-322):

$$\nabla \cdot \mathbf{V}_\xi = 0 \tag{7-325}$$

$$\nabla \times \mathbf{V}_\xi = \boldsymbol{\xi} \tag{7-326}$$

and

$$\nabla \cdot \mathbf{V}_\varepsilon = \varepsilon^\diamond \tag{7-327}$$

$$\nabla \times \mathbf{V}_\varepsilon = 0 \tag{7-328}$$

Both $\mathbf{V}_\xi$ and $\mathbf{V}_\varepsilon$ are required to satisfy the zero-normal-velocity boundary condition independently, namely

$$\mathbf{V}_\xi \cdot \mathbf{n} = 0 \tag{7-329}$$

$$\mathbf{V}_\varepsilon \cdot \mathbf{n} = 0 \tag{7-330}$$

where $\mathbf{n}$ is the unit vector normal to the walls. However, only the total velocity $\mathbf{V}$ is required to satisfy the nonslip condition

$$\mathbf{V} \cdot \mathbf{S} = 0 \tag{7-331}$$

where $\mathbf{S}$ is the unit vector tangent to the walls.

The solution algorithm consists of two loops, one for handling *vortex dynamics*, and the other for *flame propagation*. These are linked together to yield the total velocity field. The detailed solution procedure is shown in Fig. 7.29, which was given by Ghoniem et al. in Ref. 119.

The RVM was used to obtain a solution for turbulent flow with combustion in a tunnel behind a step. A propane–air mixture with $\phi = 0.5$ was used, with the specific volume ratio equal to 4.25. The laminar flame speed was taken as 12 cm/sec, and $u_\infty = 6$ m/sec, corresponding to Re = 10,000. The computer output depicts sequentially the variation of vortex and velocity fields and the flame front. The solution showed reasonable development of the flow field and growth of large-scale eddies. Chorin et al.[112] thus considered the RVM to be capable of reproducing the essential features of turbulent combustion and providing a clarification of the essential mechanism of turbulent combustion. Ng and Ghoniem[120] applied RVM to study the vortex development in a confined shear layer. A part of their numerical solutions showing the growth of the vorticity field is given in Fig. 7.30.

The RVM is still a relatively new method. Its capability of predicting quantitative stochastic turbulent-flow parameters has not yet been fully established. However, the preliminary qualitative results are very encouraging. The

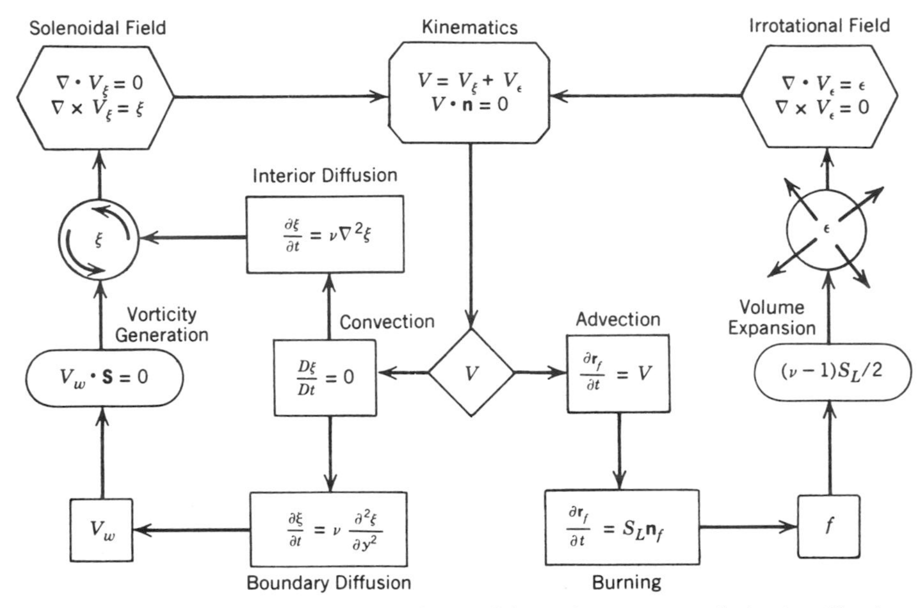

Figure 7.29 Block diagram of the algorithm used in random vortex method (after Ghoniem, Chen, and Oppenheim[119]).

main advantage of the method is that it is unencumbered by numerical diffusion. Thus, the instabilities in the flow field, which arise as a characteristic feature of turbulence, can be sustained without distorting their effects. The following features of turbulent combustion were displayed in the RVM solution:

1. Fluid-mechanical processes of formation of large-scale turbulent flow structure.
2. A rationale for the role played by the intrinsic instability of the flow system in stabilizing the flow.
3. The fluid-mechanical process of ignition in the turbulent flow of premixed gases.
4. The mechanism of exothermic processes in turbulent combustion.
5. Detailed features of entrainment and mixing in the turbulent flow.

Since this method has been tested for a limited situation only, all the successes mentioned above need to be verified. Moreover, the test case was a simple one, and so the suitability of the method for accurately depicting the

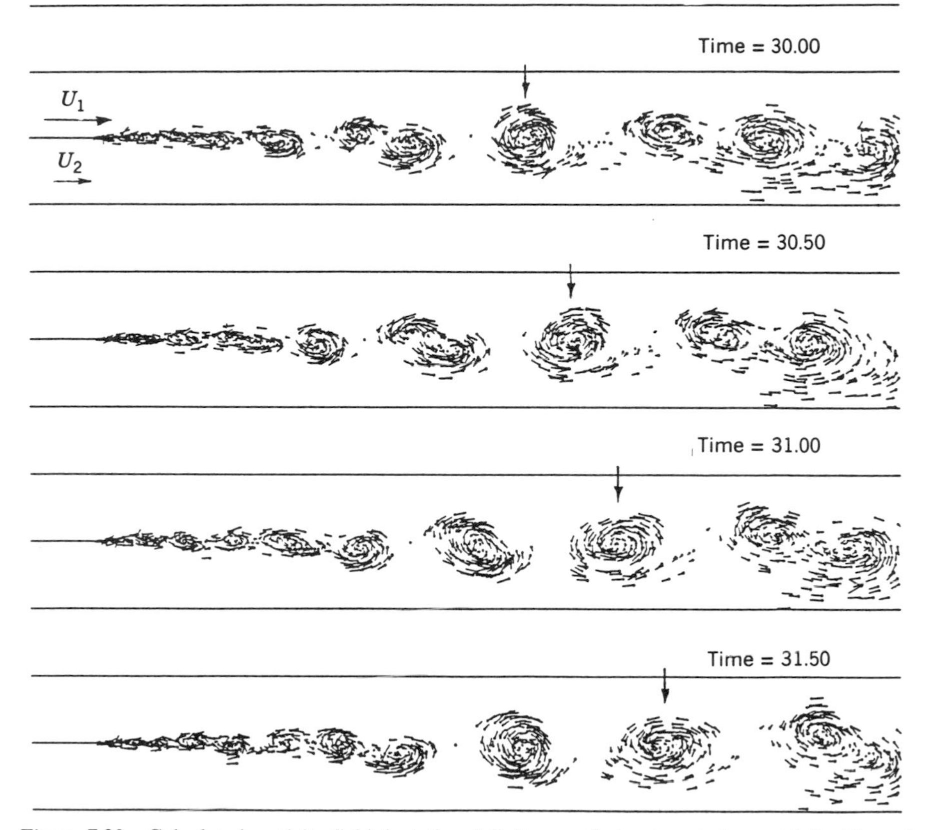

Figure 7.30 Calculated vorticity field depicting definite growth due to entrainment (after Ng and Ghoniem[120]).

flow field in complex reacting flows is still an open question. Its complexity in three-dimensional situations may render it unsuitable due to a large number of vortex blobs and sheets involved in modeling the flow. The vortex-stretching mechanism which is inherent in three-dimensional turbulent flows has not been studied by means of the RVM. Hence, much further research has to be carried out to determine the usefulness of the method.

9 CHARACTERISTIC SCALE OF WRINKLES IN TURBULENT PREMIXED FLAMES

The effect of the turbulence scale of unburned gas on the scale of wrinkles of premixed turbulent flames was directly observed experimentally by Yoshida and Tsuji[117] through Schlieren photography. The characteristic scale of the

TABLE 7.1 Turbulence-producing Grids, Turbulence Parameters, and Length Scale of Wrinkles[a]

Grid no.	$\overline{U}$ (m/sec)	u' (m/sec)	l[b] (mm)	l_f[c] (mm)
G-1[d]	4.0	0.12	2.46	—
	8.0	0.15	1.41	4.14–5.59
G-4[e]	4.0	0.36	1.54	4.67–5.48
	8.0	0.45	1.79	5.23–5.75

[a]After Yoshida and Tsuji.[117]
[b]Integral scale of turbulence.
[c]Length scale of wrinkles.
[d]40-mesh wire gauze.
[e]1-mm-thick perforated plate with mesh size 3.0 mm and hole diameter 1.5 mm.

wrinkles was found to be much larger than that of the unburned-gas turbulence, which was produced by a grid or wire gauze. Throughout their study, the extensive use of the thermocouple in studies of turbulent premixed flames was made; the integral length scale of the temperature fluctuations was measured, which corresponds to the mean diameter of unburned and burned gas lumps. This technique was applied to determine the time scale of wrinkles of the flame front, from which the length scale was deduced.

Experiments were conducted using an open cylindrical burner with 10-mm inner diameter, surrounded by eight small city-gas pilot flames. A lean propane–air mixture was used to minimize the catalytic effect of the thermocouple. The equivalence ratio of the unburned gas was fixed at 0.68, whereas the mean velocity $\overline{U}$ was varied from 4 to 8 m/sec. In their experiments, several wire meshes and perforated plates were used to generate isotropic turbulence, whose characteristics differed. In Ref. 117, the results using two typical grids are reported. The details of each grid and turbulence parameters are presented in Table 7.1. The turbulence characteristics of the grids were determined by a hot-wire anemometer and checked by a laser Doppler velocimeter.

The temperature measurements were made with an uncoated 50-μm Pt–Pt : 13% Rh thermocouple, and time-resolved temperature signals were obtained up to 3 kHz. The thermocouple time constant was determined by careful consideration of the pdfs of the temperature fluctuation.

9.1 Schlieren Photographs

Schlieren photography has been very useful in recognizing the existence of wrinkled laminar flames. Figure 7.31 shows the Schlieren photographs taken during the above experimental study. As can be seen from this figure, for the case of $\overline{U} = 8$ m/sec the G-1 flame appears to have a continuous wavy flame front. On the other hand, for the G-4 flame, increasing the turbulence intensity causes a more distorted flame front and decreases the flame length, which

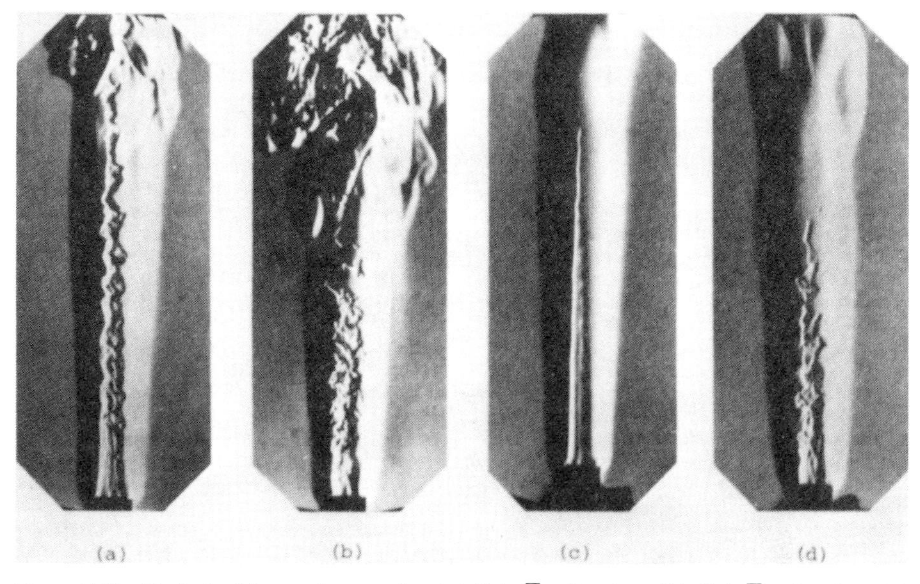

Figure 7.31 Typical schlieren photographs: (*a*) G-1, $\overline{U} = 8$ m/sec, (*b*) G-4, $\overline{U} = 8$ m/sec, (*c*) G-1, $\overline{U} = 4$ m/sec, (*d*) G-4, $\overline{U} = 4$ m/sec (after Yoshida and Tsuji[117]).

corresponds to an increase in the turbulent burning velocity. When $\overline{U}$ is reduced to 4 m/sec, the wrinkles are still observed in the G-4 flame, but they disappear in the G-1 flame, which has a nearly laminar appearance (Fig. 7.31*c*) even though the unburned gas is turbulent.

9.2 Observations on the Structure of Wrinkled Laminar Flames

The thermal structure of wrinkled laminar flames was studied using a small-scale turbulent flame with high velocity and a fine-wire thermocouple. Figure 7.32 shows the mean and fluctuating temperature distributions for the G-1 flame with $\overline{U} = 8$ m/sec.

In the upstream region of the turbulent flame (small x/D), no heat release is observed near the flame axis. Moving radially outward, one can see that abrupt increase in both mean and fluctuating temperatures occurs on the unburned side of the turbulent-flame zone. The fluctuating temperature reaches a maximum near the center of the flame zone, and then decreases on the burned side, whereas the mean temperature increases monotonically in the flame zone and then reaches a temperature plateau. In the downstream region of the turbulent flame (large x/D), the mean and fluctuating temperatures are high even on the flame axis, due to the closure of the flame zone at the top.

The maximum fluctuating temperature is nearly constant, independent of the downstream location. This is one of the characteristic features of wrinkled laminar flames. Figure 7.33 shows a typical pdf taken at $x/D = 10$ for the same flame.

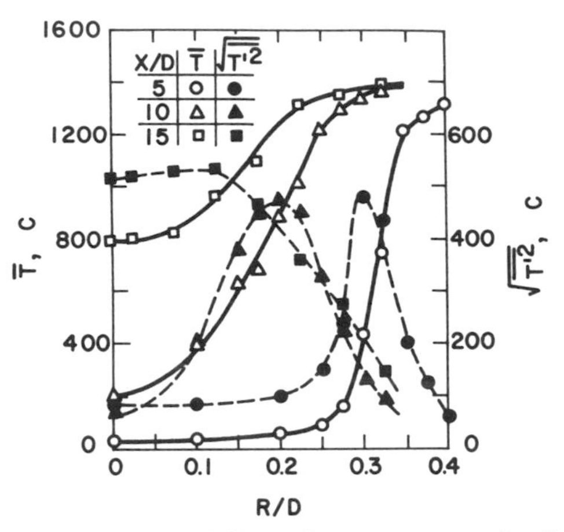

Figure 7.32 Distributions of mean and fluctuating temperatures for G-1, $\overline{U} = 8$ m/sec (after Yoshida and Tsuji[117]).

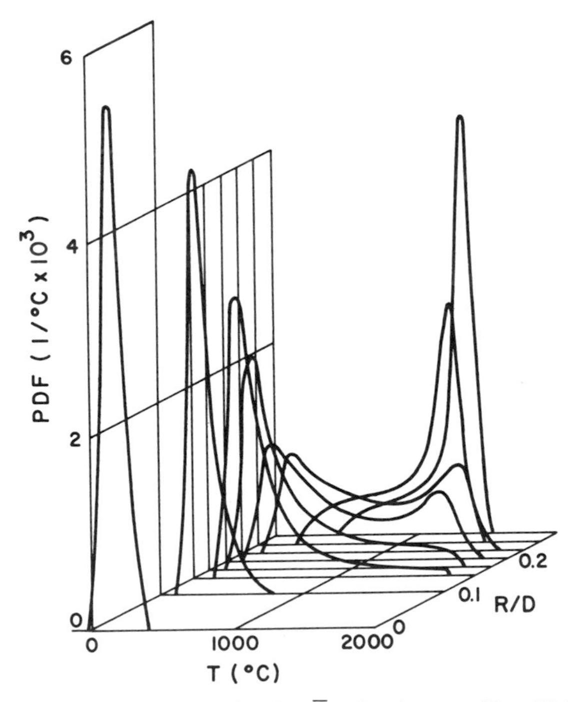

Figure 7.33 Probability density functions for G-1, $\overline{U} = 8$ m/sec at $x/D = 10$ (after Yoshida and Tsuji[117]).

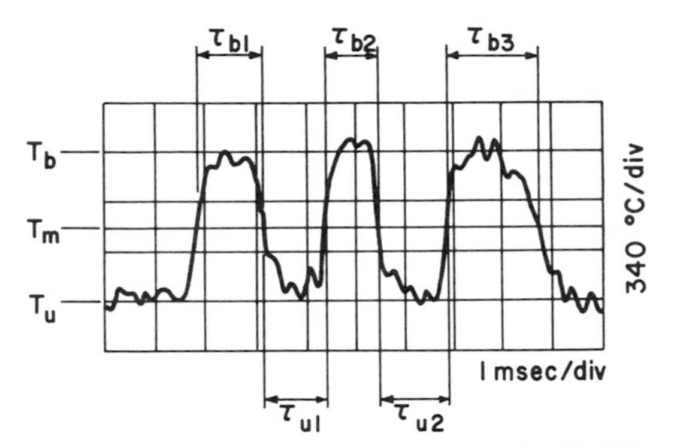

Figure 7.34 Typical time record of temperature signal (after Yoshida and Tsuji[117]).

From unburned to burned side, the probability that the thermocouple is exposed to the unburned gas decreases, whereas that of burned gas increases. At the center of the flame zone, the probability of the unburned gas is equal to that of burned gas.

9.3 Measurements of Scales of Unburned and Burned Gas Lumps

Figure 7.34 shows the time records of temperature signal at the center of the flame zone. In this figure, the high and low signal levels correspond to the burned and unburned gas lumps respectively. With regard to the temperature measurements, it can be assumed that the temperature corresponds to a burned gas lump if it is higher than the algebraic mean of high and low levels $[T_m = (T_b + T_u)/2]$, and to an unburned gas lump if it is lower than T_m.

Thus the mean period during which the thermocouple is exposed to unburned gas or burned gas can be determined as follows:

$$\tau_u = \lim_{n \to \infty} \frac{1}{n} \sum \tau_{un} \tag{7-332}$$

$$\tau_b = \lim_{n \to \infty} \frac{1}{n} \sum \tau_{bn} \tag{7-333}$$

In this study, n was taken to be 150–300, depending on the signal condition. The time scale of a disturbed laminar G-1 flame with $\overline{U} = 4$ m/sec (Fig. 7.31c) could not be measured, since the time record of the temperature signal did not include a clear enough sequence of pulses.

Figure 7.35 shows the time scales of unburned and burned gas lumps for G-1 and $\overline{U} = 8$ m/sec. From unburned to burned side, the time scale of the

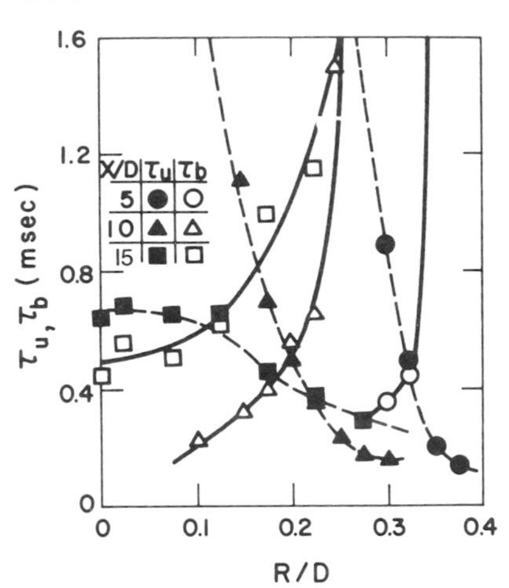

Figure 7.35 Time scales of unburned and burned gas for G-1, $\overline{U} = 8$ m/sec (after Yoshida and Tsuji[117]).

burned gas lumps increases rapidly from near zero to infinite in the turbulent flame zone, whereas that of unburned gas lumps decreases from infinite to zero. Near the tip of turbulent flame zone, the flame closure makes both time scales of unburned and burned gas lumps finite even on the axis. Similar trends were observed for the other two cases.

The time scale of wrinkles is determined at the point where the scales of unburned and burned gas lumps are equal. At this point (i.e., at the center of turbulent flame zone) the fluctuating temperature is maximum.

Figure 7.36 shows distributions of time scales of wrinkles of flame front. For $\overline{U} = 8$ m/sec, the time scale falls in the range from 0.45 to 0.62 msec. Even though the G-1 flame is much longer than the G-4 flame, in the downstream region of the turbulent flame the time scale of wrinkles for both flames is about 0.6 msec. Although the characteristics of turbulence generated by G-1 and G-4 grids were different as shown in Table 7.1, the time scale is found to be

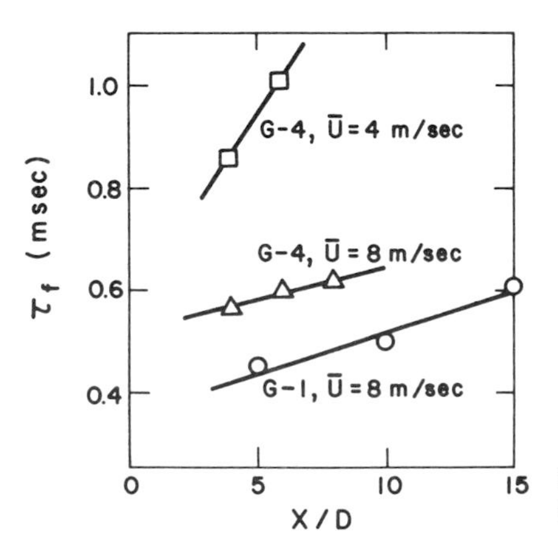

Figure 7.36 Time scales of wrinkles of the flame front (after Yoshida and Tsuji[117]).

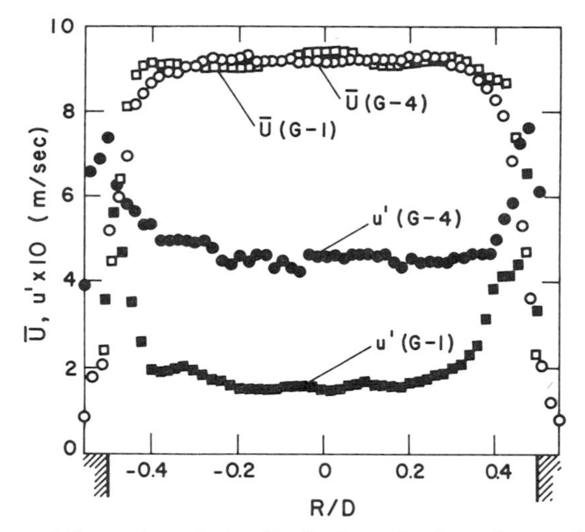

Figure 7.37 Mean and fluctuating velocity distributions of unburned gas at burner exit. $\overline{U} = 8$ m/sec.

unaffected by the upstream turbulence. On the other hand, for $\overline{U} = 4$ m/sec, it falls into the range from 0.86 to 1.01 msec, and is roughly twice that for $\overline{U} = 8$ m/sec.

From Fig. 7.36, it is clear that the time scale of wrinkles tends to increase slightly with the downstream distance in all three cases. This means that the wrinkles of the laminar flame front are not only transported with the mean flow velocity but also propagate by themselves.

9.4 Length Scale of Wrinkles

After obtaining the time scale of wrinkles, they were defined and converted into the length scale by using the local uniform flow velocity in the middle of the turbulent flame.

Figure 7.37 shows the radial distributions of mean and fluctuating velocities at the burner exit. The velocities are uniform except for the pipe-wall boundary layer, where the mean velocity decreases whereas the fluctuating velocity increases due to the wall shear stress.

For $\overline{U} = 8$ m/sec, the length scale determined falls in the range from 4.14 to 5.75 mm for both grids; for $\overline{U} = 4$ m/sec, from 4.67 to 5.48 mm.

Several conclusions can be gathered from Yoshida and Tsuji's experimental investigation:

1. The length scale of wrinkles is around 5 mm, almost independent of the upstream turbulence and the unburned-gas velocity.
2. The schlieren photographs show that the flame length decreases with increase in the fluctuating velocity, resulting in an increase of the

turbulent burning velocity. For the wrinkled laminar flame, the turbulent burning velocity is determined by the flame area, which therefore increases with the fluctuating velocity.

3. Periodicity of the temperature signal is found. This fact suggests that the wrinkles which appear upstream are transported downstream along the flame front.
4. The wavelength derived from the length scale grows longer along the flame front, with a growth rate of the order of the laminar burning velocity.

REFERENCES

1. H. Tennekes and J. L. Lumley, *A First Course in Turbulence*, MIT Press, Cambridge, Mass., 1972.
2. F. V. Bracco, "Reacting Turbulent Flows," *Combustion Sci. and Technol.*, Vol. 13, Special Issue, 1976.
3. W. P. Jones, "Models for Turbulent Flows with Variable Density," VKI Lecture Series 1979-2, *Prediction Methods for Turbulent Flows*, Hemisphere Publ. Corp., New York, 1980.
4. P. A. Libby and F. A. Williams, "Turbulent Reacting Flows," *Topics in Applied Physics*, Vol. 44, Springer-Verlag, New York, 1980.
5. W. P. Jones and J. H. Whitelaw, "Calculation Methods for Reacting Turbulent Flows: A Review," *Combustion and Flame*, Vol. 48, pp. 1–26, 1982.
6. G. Damköhler, *Z. Electrochem.*, Vol. 46, p. 601, 1940.
7. G. Damköhler, NACA Tech. Memo. No. 1112, 1947.
8. A. M. Kanury, *Introduction to Combustion Phenomena*, Gordon and Breach, New York, September 1975.
9. K. I. Schelkin, NACA Tech. Memo. No. 1110, 1947.
10. I. Glassman, *Combustion*, Academic Press, New York, 1977.
11. B. Karlovitz, D. W. Denniston, Jr., and F. E. Wells, *J. Chem. Phys.*, Vol. 19, p. 541, 1951.
12. K. Wohl, L. Shore, H. von Rosenberg, and C. W. Weil, *4th Symposium on Combustion*, pp. 620–635, 1953.
13. F. W. Bowdith, *4th Symposium on Combustion*, pp. 674–681, 1953.
14. L. S. Kozachenko, *Izvest. Akad Nauk SSSR Otd. Tekh. Nauk*, *Energetika i Avtomatika*, No. 2, p. 21, 1959; English translation, *ARS J.*, Vol. 29, p. 761, 1959.
15. M. Summerfield, S. H. Reiter, V. Kebely, and R. W. Mascolo, *Jet Propul.*, Vol. 25, p. 377, 1955.
16. R. R. John and E. Mayer, Arde Assoc. Tech. Note 4555-5, 1957.
17. L. S. G. Kovasznay, *Jet Propul.*, Vol. 26, p. 485, 1956.
18. M. D. Fox and F. J. Weinberg, *Proc. R. Soc. London A*, Vol. 268, pp. 222–239, 1962.
19. T. Cebeci and A. M. O. Smith, *Analysis of Turbulent Boundary Layers*, Academic Press, New York, 1974.
20. A. Favre, "Equations des gaz turbulents compressibles," *J. Mechanique*, Vol. 4, p. 361, 1965.
21. A. Favre, "Statistical Equations of Turbulent Gases," *Problems of Hydrodynamics and Continuum Mechanics* (Sedov 60th birthday volume), SIAM, Philadelphia, 1969.

22. J. Laufer and K. G. Ludloff, *Conservation Equations in a Compressible Turbulent Fluid and a Numerical Scheme for their Solution*, McDonnell-Douglas Paper WD 1355, 1970.

23. R. W. Bilger, "Turbulent Jet Diffusion Flames," *Progr. Energy Comb. Sci.*, Vol. 1, p. 87, 1976.

24. R. W. Bilger, "Perturbation Analysis of Turbulent, Non-Premixed Combustion," *Comb. Sci. Tech.*, Vol. 22, p. 251, 1980.

25. G. K. Batchelor and A. A. Townsend, *Proc. R. Soc. London A*, Vol. 199, p. 238, 1949.

26. J. Chomiak, "Basic Considerations in the Turbulent Flame Propagation in Premixed Gases," *Prog. Energy Combustion Sci.*, Vol. 5, pp. 207–221, 1979.

27. A. Y. Kuo and S. J. Corrsin, *J. Fluid Mech.*, Vol. 50, p. 285, 1971.

28. A. Y. Kuo and S. J. Corrsin, *J. Fluid Mech.*, Vol. 56, p. 447, 1972.

29. H. Tennekes, *Phys. Fluids*, Vol. 11, p. 669, 1968.

30. J. O. Hinze, *Turbulence*, 2nd ed., Chapter 2, McGraw-Hill, New York, 1975.

31. P. A. Thomson, *Compressible Fluid Dynamics*, McGraw-Hill, New York, 1972.

32. G. K. Batchelor, *Introduction to Fluid Dynamics*, Cambridge University Press, Cambridge, England, 1967.

33. J. Chomiak, "Problems of Turbulent Combustion," presented at VIIth International Symposium on Combustion Processes, Jablonna, 1981.

34. A. S. Monin and A. M. Yaglom, "Statistical Fluid Mechanics," *Mechanics of Turbulence*, Vol. 2, Chapter 25, MIT Press, Cambridge, Mass., 1975.

35. D. B. Spalding, "Concentration Fluctuations in a Round Turbulent Free Jet," *Chem. Eng. Sci.*, Vol. 26, p. 95, 1971.

36. A. D. Gosman and F. C. Lockwood, "Prediction of the Influence of Turbulent Fluctuations on Flow and Heat Transfer in Furnaces," *Proc. Heat Mass Transfer Seminar, Int. Summer School, Trogir, Yugoslavia*, Vol. 4, p. 215, 1973.

37. E. E. Khalil, D. B. Spalding, and J. H. Whitelaw, "The Calculation of Local Flow Properties in Two-Dimensional Furnaces," *Int. J. Heat Mass Transfer*, Vol. 18, p. 775, 1975.

38. W. P. Jones, "Models for Turbulent Flows with Variable Density," VKI Lecture Series 1979-2, *Prediction Methods for Turbulent Flows* (W. Kollmann, Ed.), Hemisphere Publ. Corp., New York, 1980.

39. J. H. Kent and R. W. Bilger, *Turbulent Diffusion Flames*, Report 7N F-37, University of Sydney; see also *14th Symposium in Combustion*, p. 615, 1973.

40. F. C. Lockwood and A. S. Naguib, "The Prediction of the Fluctuations in the Properties of Free, Round-Jet, Turbulent, Diffusion Flame," *Combustion and Flame*, Vol. 24, pp. 109–124, 1975.

41. J. H. Kent and R. W. Bilger, "The Prediction of Turbulent Diffusion Flame Fields and Nitric Oxide Formation," *16th Symposium (Int.) on Combustion*, pp. 1643–1656, 1977.

42. J. M. Richardson, H. C. Howard, and R. W. Smith, "The Relation between Sampling-Tube Measurements and Concentration Fluctuations in a Turbulent Gas Jet," *IV Symposium (Int.) on Combustion*, p. 814, 1953.

43. R. P. Rhodes, P. T. Harsha, and C. E. Peters, "Turbulent Kinetic Energy and Analysis of Hydrogen–Air Diffusion Flames," *Acta Astronaut.*, Vol. 1, p. 443, 1974.

44. W. P. Jones and C. Priddin, "Prediction of the Flow Field and Local Gas Composition in Gas Turbine Combustors," *XVIIth Symposium (Int.) on Combustion*, p. 399, 1978.

45. W. P. Jones and J. McGuirk, "Computation of a Round Turbulent Jet Discharging into a Confined Cross Flow," *Turbulent Shear Flows*, Vol. 2, p. 233, 1980.

46. W. P. Jones and J. McGuirk, "A Comparison of Two Droplets Models for Gas Turbine Combustion Chamber Flows," *Proc. 5th ISABE*, Bangalore, India, 1981.

47. J. Janicka and W. Kollmann, "A Prediction Model for Turbulent Diffusion Flames Including NO Formation," *AGARD Proc.*, No. 275, 1980.

48. M. R. Speigel, *Probability and Statistics*, Schaum's Outline Series in Mathematics, pp. 116–117, McGraw-Hill, New York, 1975.

49. B. E. Launder, "Progress in the Modeling of Turbulent Transport," *Lecture Series 76*, von Karman Institute for Fluid Dynamics, Belgium, 1975.

50. P. Bradshaw, *Turbulence*, Topics in Applied Physics, Vol. 12, Springer-Verlag, New York, 1976.

51. J. L. Lumley, "Computational Modeling of Turbulent Flows," *Advances in Applied Mechanics*, No. 18, Academic Press, New York, 1978.

52. P. Bradshaw, T. Cebeci, and J. H. Whitelaw, *Calculation Methods for Turbulent Flows*, Academic Press, New York, 1981.

53. W. P. Jones and B. E. Launder, "The Prediction of Laminarisation with a Two-Equation Turbulence Model," *Int. J. Heat Mass Transfer*, Vol. 15, p. 301, 1972.

54. K. N. C. Bray, *Equations of Turbulent Combustion I. Fundamental Equations of Reacting Turbulent Flow*, Univ. of Southhampton AASU Rept. No. 330, 1973.

55. B. E. Launder and D. B. Spalding, "The Numerical Computation of Turbulent Flows," *Computer Methods Appl. Mech. Engrg.*, Vol. 3, p. 239, 1974.

56. M. M. Ribeiro and J. H. Whitelaw, "Coaxial Jets With and Without Swirl," *J. Fluid Mech.*, Vol. 96, p. 769, 1980.

57. J. B. Moss, "Simultaneous Measurements of Concentration and Velocity in an Open Premixed Turbulent Flame," *Combustion Sci. and Technol.*, Vol. 22, p. 115, 1980.

58. P. A. Libby and K. N. C. Bray, *Counter Gradient Diffusion in Premixed Turbulent Flames*, AIAA Paper No. 80-0013, 1980.

59. P. A. Libby and K. N. C. Bray, "Implications of the Laminar Flamelet Model in Premixed Turbulent Combustion," *Combust. Flame*, Vol. 39, p. 33, 1980.

60. J. E. Garside, A. R. Hall, and D. T. A. Townend, *Nature*, Vol. 152, p. 748, 1943.

61. H. H. Fernholz, "External Flows," Ref. 50.

62. S. Corrsin, *Investigation of Flow in an Axially Symmetrical Heated Jet of Air*, ACR 3L23, NACA Wartime Report W-94, 1943.

63. H. Fiedler (Ed.), *Structure and Mechanisms of Turbulence, I, II*, Lecture Notes in Physics, Vols. 75, 76, Springer-Verlag, Berlin, Heidelberg, New York, 1978.

64. P. A. Libby, *Phys. Fluids*, Vol. 19, pp. 494–501, 1976.

65. C. Dopazo, *J. Fluid Mech.*, Vol. 81, pp. 433–438, 1977.

66. R. W. Bilger, "Turbulent Flows with Non-premixed Reactants," *Turbulent Reacting Flows* (P. A. Libby and F. A. Williams, Eds.), Chapter 3, pp. 65–114, Topics in Applied Physics, Springer-Verlag, New York, 1980.

67. W. R. Hawthorne, D. S. Weddell, and H. C. Hottel, "Mixing and Combustion in Turbulent Gas Jets," *Third Symposium on Combustion, Flame and Explosion Phenomena*, pp. 266–288, 1949.

68. I. M. Kennedy and J. H. Kent, "Measurements of a Conserved Scalar in Turbulent Jet Diffusion Flames," *Seventeenth International Symposium on Combustion*, pp. 279–287, 1979.

69. C. H. Gibson and P. A. Libby, *Combustion Sci. and Technol.*, Vol. 6, p. 29, 1972.

70. C. T. Bowman, *Prog. Energy Combust. Sci.*, Vol. 1, p. 33, 1975.

71. R. W. Bilger and R. E. Beck, "Further Experiments in Turbulent Jet Diffusion Flames," *15th Symposium (International) on Combustion*, p. 541, 1975.

72. J. Janicka and W. Kollmann, "A Two-Variables Formalism for the Treatment of Chemical Reactions in Turbulent H_2-Air Diffusion Flames," *Seventeenth Symposium (International) on Combustion*, pp. 421–430, 1979.

73. J. H. Ay and M. Sichel, "Theoretical Analysis of NO Formation Near the Primary Reaction Zone in Methane Combustion," *Combustion and Flame*, Vol. 26, pp. 1–15, 1976.
74. R. Borghi, "Chemical Calculations in Turbulent Flows—Application to CO Containing Turbojet Flame," *Adv. Geophys.*, Vol. 18B, p. 349, 1974.
75. P. Hutchinson, E. E. Khalil, and J. H. Whitelaw, *Turbulent Combustion* (L. A. Kennedy, Ed.), Progress in Astronautics and Aeronautics, Vol. 58, p. 211, 1978.
76. W. E. Kaskan and G. L. Schott, *Combustion and Flame*, Vol. 6, p. 73, 1962.
77. C. du P. Donaldson and A. K. Varma, *Combustion Sci. and Technol.*, Vol. 13, p. 55, 1976.
78. T. S. Lundgren, "Distribution Functions in Statistical Theory of Turbulence," *Phys. Fluids*, Vol. 10, pp. 969–975, 1967.
79. A. S. Monin, "Equations of Turbulent Motion," *Prikl. Mat. Mekh.*, Vol. 31, pp. 1057–1067, 1967.
80. A. S. Monin, "Equations for Finite Dimensional Probability Distributions of a Field of Turbulence," *Dokl. Akad. Nauk SSSR*, Vol. 177, pp. 1036–1038, 1967.
81. E. A. Novikov, "Kinetic Equations for a Vortex Field," *Dokl. Akad. Nauk SSSR*, Vol. 117, pp. 299–301, 1967.
82. C. Dopazo, *Non-isothermal Turbulent Reactive Flows*; *Stochastic Approaches*, Ph.D. Dissertation, State University of New York at Stony Brook, 1973.
83. C. Dopazo, "Probability Density Function Approach for a Turbulent Axisymmetric Heated Jet: Centerline Evolution," *Phys. Fluids*, Vol. 18, pp. 397–404, 1975.
84. C. Dopazo and E. E. O'Brien, "An Approach to the Autoignition of a Turbulent Mixture," *Acta Astronaut.*, Vol. 1, No. 9/10, pp. 1239–1266, 1974.
85. C. Dopazo, "A Probabilistic Approach to Turbulent Flame Theory," *Acta Astronaut.*, Vol. 3, pp. 853–878, 1976.
86. E. E. O'Brien, "Statistical Methods in Reacting Turbulent Flows," AIAA Paper 80-0137, presented at AIAA 18th Aerospace Sciences Meeting, January 14–16, 1980.
87. E. E. O'Brien, "The Probability Density Function (pdf) Approach to Reacting Turbulent Flows," *Turbulent Reacting Flows*, (P. A. Libby and F. A. Williams, Eds.), Chapter 5, pp. 185–218, Topics in Applied Physics, Springer-Verlag, New York, 1980.
88. R. L. Stratonovich, Doctoral Dissertation, Moscow University, 1965.
89. S. B. Pope, "The Probability Approach to the Modeling of Turbulent Reacting Flows," *Combustion and Flame*, Vol. 27, p. 299, 1976.
90. J. Janicka, W. Kolbe, and W. Kollmann, "The Solution of a pdf Transport Equation for Turbulent Diffusion Flames," *Proc. 26th Heat Transfer and Fluid Mechanics Institute*, p. 296, 1978.
91. S. B. Pope, "A Monte Carlo Method for the pdf Equations of Turbulent Reactive Flow," *Combustion Sci. and Technol.*, Vol. 25, p. 159, 1981.
92. R. G. Batt, "Turbulent Mixing of Passive and Chemically Reacting Species in a Low Speed Shear Layer," *J. Fluid Mech.*, Vol. 82, p. 53, 1977.
93. S. B. Pope, "Monte Carlo Calculations of Premixed Turbulent Flames," presented at 18th Symposium on Combustion, 1980.
94. F. Robinson, *Pollutant Formation in Turbulent Flames*, Ph.D. Thesis, Northwestern University, 1974.
95. D. B. Spalding, "A General Theory of Turbulent Combustion," AIAA Paper 77-141, presented at AIAA 15th Aerospace Meeting, Los Angeles, January 1977.
96. D. B. Spalding, "The Influence of Laminar Transport and Chemical Kinetics on the Time-Mean Reaction Rate in Turbulent Flow," *17th Symposium (Int.) on Combustion*, p. 431, 1978.

97. A. S. C. Ma, M. A. Noseir, and D. B. Spalding, "An Application of the ESCIMO Theory of Turbulent Combustion," AIAA Paper No. 80-0014, presented at AIAA 18th Aerospace Sciences Meeting, January 14–16, 1980.

98. K. N. C. Bray, "Turbulent Flows with Premixed Reactants," *Turbulent Reacting Flows* edited by (P. A. Libby and F. A. Williams, Eds.), Chapter 4, pp. 115–183, Topics in Applied Physics, Springer-Verlag, New York, 1980.

99. D. B. Spalding, "Mixing and Chemical Reaction in Steady, Confined Turbulent Flames," *13th Symposium on Combustion*, p. 643, 1970.

100. D. B. Spalding, "Development of the Eddy-Breakup Model of Turbulent Combustion," *16th Symposium on Combustion*, p. 1657, 1976.

101. G. C. Williams, H. C. Hottell, and A. C. Scurlock, *Third Symposium on Combustion, Flame and Explosion Phenomena*, p. 21, Williams and Wilkins, Baltimore, 1949.

102. F. H. Wright and E. E. Zukoski, *Eighth Symposium (International) on Combustion*, p. 933, Williams and Wilkins, Baltimore, 1962.

103. V. P. Solntsev, *Stabilization of a Flame and Development of the Process of Combustion in a Turbulent Stream* (G. M. Gorbunov, Ed.), p. 75, Oborongiz, Moscow, 1961.

104. N. M. Howe, C. W. Shipman, and A. Vranos, *Ninth Symposium (International) on Combustion*, p. 36, Academic Press, New York, 1963.

105. E. C. Wilkerson and J. B. Fenn, *Fourth Symposium (International) on Combustion*, p. 749, Williams and Wilkins, Baltimore, 1953.

106. R. J. Petrein, J. P. Longwell, and M. A. Weiss, *Jet Propulsion*, Vol. 26, p. 81, 1956.

107. D. B. Spalding, *Combustion and Flame*, Vol. 1, No. 3, pp. 287–295, 1957.

108. K. N. C. Bray and J. B. Moss, "A Unified Statistical Model of the Premixed Turbulent Flame," *Acta Astronaut.*, Vol. 4, pp. 291–319, 1977.

109. N. M. Howe and C. W. Shipman, "A Tentative Model for Rates of Combustion in Confined Turbulent Flames," *10th International Symposium on Combustion*, p. 1139, 1965.

110. K. N. C. Bray, "The Interaction Between Turbulence and Combustion," *17th Symposium (Int.) on Combustion*, p. 233, 1978.

111. M. Champion, K. N. C. Bray, and J. B. Moss, "The Turbulent Combustion of a Propane–Air Mixture," *Acta Astronaut.*, 5, p. 1063, 1978.

112. A. J. Chorin, A. F. Ghoniem, and A. K. Oppenheim, "Numerical Modeling of Turbulent Flow in Premixed Combustion," presented at 18th Symposium (Int.) on Combustion, 1981.

113. S. Lie and F. Engel, *Theorie der Transformationsgruppen*, 3 vols., Teubner, Leipzig, 1880.

114. A. A. Samarski, "An Efficient Method for Multi-dimensional Problems in an Arbitrary Domain," *Vycisl. Mat. i Mat. Fiz.*, Vol. 2, p. 787, 1962.

115. G. K. Batchelor, *An Introduction to Fluid Mechanics*, Cambridge University Press, London, 1967.

116. A. J. Chorin and J. E. Marsden, *A Mathematical Introduction of Fluid Mechanics*, Springer-Verlag, Berlin, vii + 205 pp., 1979.

117. A. Yoshida and H. Tsuji, "Characteristic Scale of Wrinkles in Turbulent Premixed Flames," *19th Symposium (Int.) on Combustion*, pp. 403–411, 1982.

118. C. du P. Donaldson and A. K. Varma, "Remarks on Construction of a Second-Order Closure Description of Turbulent Reacting Flows," *Combustion Sci. Technol.*, Vol. 13, pp. 55–78, 1976.

119. A. F. Ghoniem, D. Y. Chen, and A. K. Oppenheim, "Formation and Inflammation of a Turbulent Jet," to be published in *AIAA Journal*, 1985.

120. K. K. Ng and A. F. Ghoniem, "Numerical Simulation of a Confined Shear Layer," presented at the 10th International Colloquium on Dynamics of Explosions and Reactive System, Berkeley, California, August 4–9, 1985.

HOMEWORK

1. Use Reynolds's averaging procedure to derive the x-momentum equation for turbulent compressible reacting fluid flow in the rectangular Cartesian coordinate system. Also derive the momentum equation in the Favre-averaged form.

2. Derive the vorticity equation (7-101), and discuss the differences of the vorticity equation between two-dimensional and three-dimensional flow conditions.

3. If the variation of the mass fraction of a certain species, Y, with respect to time is given by the sinusoidal function on the left-hand side of Eq. (7-126), show that the distribution function $F(Y)$ and probability density function $\mathscr{P}(Y)$ must follow the forms given by Eqs. (7-127) and (7-128), respectively.

4. Use the β-pdf defined in Eq. (7-149) and the definitions of $\tilde{f}$ and $\widetilde{f''^2}$ given by Eqs. (7-144) and (7-145) to solve the two coefficients a and b in terms of $\tilde{f}$ and $\widetilde{f''^2}$.

5. Convert Eq. (7-94) into the final form of the turbulence kinetic-energy equation (7-159). List all the required assumptions.

6. Derive the turbulence dissipation rate equation (7-160) from the instantaneous and the mean momentum equation. List all necessary assumptions in order to obtain the exact form of Eq. (7-160).

7. Derive the transport equation (7-175) for $\widetilde{\xi''^2}$ from Eq. (7-174).

8. Show that for the one-step single reaction given by Eq. (7-167) with no oxidant in the fuel feeding stream and no fuel in the oxidant feeding stream, the stoichiometric mixture fraction can be written in the form given by Eq. (7-176).

9. Show that for the one-step irreversible reaction of Eq. (7-169), the Favre-averaged mass fractions of fuel, oxidizer, and product can be written as Eqs. (7-188), (7-189), and (7-190).

10 (a) Consider a low speed flow situation and derive the scalar flux equation for $\overline{u_2'\theta'}$ by Reynolds averaging procedure, where θ is a scalar. The density fluctuations can be mostly ignored except that the body force fluctuation due to density fluctuation cannot be ignored and one can use $B_i' = \rho' g_i$ in the ith direction and

$$\frac{\rho'}{\rho} = -\alpha_c \frac{\theta'}{\theta}$$

10 (b) Derive the scalar flux equation for $u_i''\theta''$ by Favre averaging procedure. In this particular case, please consider the chemically reacting flow to be compressible.

11 Derive the Favre Average Form of mixture fraction $\tilde{f}$ equation

PROJECT NO. 1

Consider the mixing and combustion of two turbulent streams of premixed gases (streams A and B have the same F/O ratio but different velocities and temperatures) in a constant-area duct. List the major governing equations you would use. State the boundary conditions, method of closure, and basic assumptions.

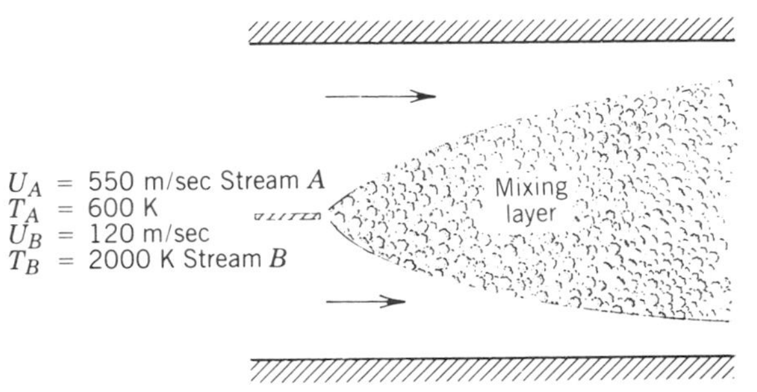

PROJECT NO. 2

Consider a solid-fuel ramjet (SFRJ) having two combustor geometries shown in the following drawings:

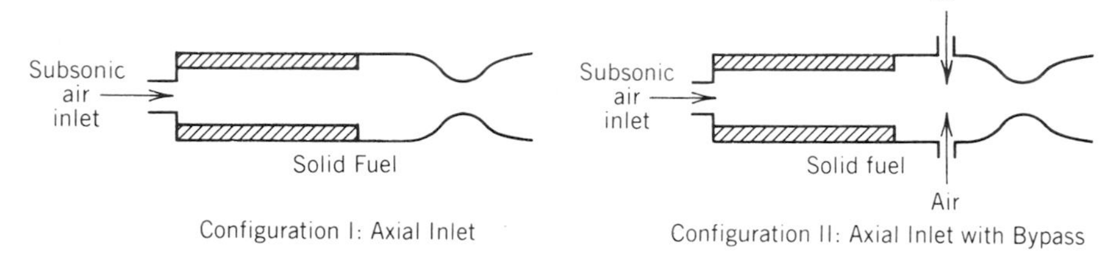

The solid-fuel grain provides a part of the walls for the combustion chamber. A sudden expansion at the air inlet (axial, with or without bypass) can be used to provide flame stabilization. A turbulent boundary layer develops and includes a diffusion-controlled flame between the fuel-rich zone near the wall and the oxygen-rich central core. Due to the diffusion flame, heat is transferred by convection to the solid surface, causing vaporization of the fuel. The fuel regression rate can be represented by the following Arrhenius form

$$\dot{r} = A \exp\left(- \frac{E}{R_u T_w} \right)$$

The fuel pyrolysis and hydrocarbon combustion of the fuel vapor with air can be approximated by a four-step process as follows:

$$C_xH_y \rightarrow C_xH_{y-2} + H_2$$

$$C_xH_{y-2} + \frac{x}{2}O_2 \rightarrow x\,CO + \frac{y-2}{2}H_2$$

$$CO + \tfrac{1}{2}O_2 \rightarrow CO_2$$

$$H_2 + \tfrac{1}{2}O_2 \rightarrow H_2O$$

(a) Formulate the three-dimensional model for a quasisteady, subsonic flow including finite-rate chemical kinetics.
(b) List all your assumptions. (Note: To simplify this problem, radiation heat transfer is assumed to be negligible.)
(c) Sketch your anticipated axial distributions of u, Y_{O_2}, $Y_{C_xH_y}$, and T along the centerline of configuration I.

PROJECT NO. 3

Demonstrate the use of either conserved scalar approach or two-variable approach by presenting an example in turbulent diffusion flame. Please give (a) a problem statement (including the source of reference), (b) procedures used in the treatment, (c) major results obtained, and (d) your comments of merits and drawbacks of the method which you select to describe.

PROJECT NO. 4

To answer the question "whether or not a laminar flame can exist in a premixed turbulent flow," Klimov[a] and Williams[b,c] showed, from solutions of the laminar flame equations in an imposed shear flow, that a propagating laminar flame may exist only if the following stretch factor K is less than a critical value of the order of unity

$$K \equiv \frac{\delta_L}{S_L} \frac{1}{A} \frac{dA}{dt} \tag{1}$$

where δ_L, S_L, and A represent the thickness, speed, and elemental areas of laminar flame, respectively. It is also known that the percentage of elemental area variation can be approximated by

$$\frac{1}{A} \frac{dA}{dt} \cong \frac{u'}{\Lambda_T} \tag{2}$$

where u' and Λ_T represents the turbulence intensity and Taylor microscale.

(i) Use the above equations and any other necessary relationships to show that the stretch factor K can be expressed as

$$K \cong \left(\frac{\delta_L}{\eta}\right)^2 \tag{3}$$

[a] A. M. Klimov, *Zh Prikl. Mekh Tekh Fiz.*, Vol. 3, 49–58 (1963).
[b] F. A. Williams, "A Review of Some Theoretical Considerations of Turbulent Flame Structure," in *Analytical and Numerical Methods for Investigation of Flow Fields with Chemical Reactions, Especially Related to Combustion* (AGARD Conf. Proc. 164, NATO, Paris 1975), pp. II-1 to II-25.
[c] F. A. Williams, *Combustion and Flame*, Vol. 26, 269–270 (1976).

where η is the Kolmogorov microscale. By setting K equal to 1, the following criterion for laminar flame to exist in a premixed turbulent flow can be obtained.

$$\delta_L \leq \eta \tag{4}$$

This is called *Klimov–Williams criterion*[d]; when it is satisfied, wrinkled laminar flames may occur.

(ii) Construct a dimensionless intensity (u'/S_L) versus Reynolds number map to show different regimes of turbulent premixed flames. Sketch lines corresponding to constant values (e.g. 0.1, 1, 10, 100, 1000) of η/δ_L, Da, and l/δ_L. Please note the following definitions of various Reynolds numbers.

$$\mathrm{Re}_l \equiv \frac{u'l}{\nu}, \qquad \mathrm{Re}_{\Lambda_T} \equiv \frac{u'\Lambda_T}{\nu}, \qquad \mathrm{Re}_\eta \equiv \frac{u'\eta}{\nu}$$

They are related by

$$\mathrm{Re}_\eta^4 \cong \mathrm{Re}_{\Lambda_T}^2 \cong \mathrm{Re}_l$$

Also indicate the special line which represents the Klimov–Williams criterion.

(iii) Discuss physical meanings for each regime.

[d] K. N. C. Bray, "Turbulent Flows with Premixed Reactants," Chapter 4 of *Turbulent Reacting Flow*, edited by P. A. Libby and F. A. Williams, Springer-Verlag, Berlin/Heidelberg (1980), pp. 115–183.

CHAPTER

8 Combustion in Two-Phase Flow Systems

ADDITIONAL SYMBOLS

Symbols	Description	Dimension
B	Transfer number	—
C_i	Parameters in turbulence model $i = \mu, \varepsilon 1, \varepsilon 2, g1, g2$	—
$\mathscr{D}$	Diffusivity	L^2/t
d_p or D	Particle diameter	L
E	Total energy per unit mass	Q/M
e_m	Internal energy per unit mass of mixture	Q/M
f	Mixture fraction	—
F	Body force	F
F_{Em}	Rate of energy generation per unit volume of mixture	Q/L^3t
$F^{(qp)}$	Time constant of momentum transfer	$1/t$
g	Square of mixture-fraction fluctuations	—
g_a	Acceleration of gravity	L/t^2
$G^{(qp)}$	Time constant for energy transfer	$1/t$
h_c	Convective heat-transfer coefficient	Q/L^2Tt
h_v	Heat of vaporization	Q/M
$J_i^{(q)}$	Barycentric diffusion motion of component q	M/L^2t
k	Kinetic energy of turbulence	L^2/t^2
$K^{(q)}$	Rate of q generation by reaction in gaseous phase	M/L^3t
$K_p^{(s)}$	Rate of s generation by breakup or coalescence of other particle species	M/L^3t
$\bar{v}^0$	Weighted mean velocity in r-direction	L/t

μ_t	Turbulent viscosity	L^2/t
$\mu_m^{(q)}$	Viscosity of q in mixture	L^2/t
σ_ϕ	Turbulent Prendtl/Schmidt number ($\phi = \rho, k, g, f$)	—
δ	Boundary-layer thickness	L
τ_p	Particle relaxation time	t
ϕ	Generic property (scalar) or void fraction of a two-phase mixture	—
ϵ	Emissivity	—
ε	Dissipation rate of turbulence kinetic energy	L^2/t^3
ε_i	Mass-flux fraction of ith species	—
ϕ_s	Volume fraction (solid) of species s	—
Γ	Total rate of generation to particulate phase per unit volume	M/L^3t
$\Gamma^{(q)}$	Rate of generation of q per unit volume	M/L^3t
$(\Delta m)_{ji}$	Deformation tensor	$1/t$
θ_m	Dilatation	$1/t$

Diacritical

—	Time-averaged mean

Superscripts

$'$	Turbulent fluctuating component based upon time averaging
(q)	Particular component of mixture
(p)	Components of mixture other than q
$(s), (r)$	Additional phases among particles

Subscripts

∞	Ambient condition
f	Liquid fuel
p	Particle or particle phase
s	Particle or droplet surface
r_s	Droplet radius
m	Mixture

There are two major topics covered in this chapter: spray combustion of liquid fuel droplets, and combustion of solid particles in multiphase reacting flow systems. Both of these topics have broad industrial applications in power production, jet and rocket propulsion, material processing, and pollution control. Spray combustion is covered in Sections 1–9. Before discussing solid-phase combustion systems, a continuum model for multiphase systems is introduced in Section 10. The combustion of solid particles in fluidized beds is presented in Section 11.

1 INTRODUCTION TO SPRAY COMBUSTION

Because of convenience in transport, flexibility in storage, and availability in liquid phase, a significant portion of the current total energy demand has been met by the combustion of liquid fuels injected as spray into the combustion chamber. For efficient combustion to occur, intimate mixing of fuel and air is a necessity; therefore, the study of the mixing process in an evaporating spray is important and constitutes an essential part of spray-combustion studies. In some special cases, mixing can be separated from combustion, but most often combustion of the spray proceeds concurrently with mixing, which makes the physicochemical process more closely coupled.

Spray combustion occurs in liquid-fuel rocket engines, gas turbines, diesel engines, industrial furnaces, and so on. In view of these varied applications, the establishment of predictive models for spray combustion processes is im-

TABLE 8.1 Reviews of Spray Combustion Processes (after Faeth[1])

Topic	References
Injection Processes	
Injector configurations, spray breakup, drop-size correlation and distributions, initial spread rates	3, 6, 7, 10, 14, 16, 18, 19, 20, 21, 22, 23, 24, 25, 29, 30, 31, 32, 36
Single-Drop Processes	
Transient effects, ignition, evaporation and combustion, convection effects, drag, extinction	3, 7, 8, 9, 10, 12, 13, 14, 15, 17, 18, 19, 26, 27, 30, 35
Spray Processes	
Profiles of mean and turbulent quantities, spread rates, models	2, 3, 4, 5, 6, 7, 8, 10, 12, 13, 14, 17, 18, 19, 27, 28, 30, 32, 35, 37, 38
Pollutants in Sprays	
Formation of HC, NO_x, CO, and soot	2, 4, 5, 6, 9, 17, 34, 37
Experimental Techniques	
Spray size distributions, velocities, and temperatures; droplet burning.	3, 7, 8, 11, 14, 18

portant in order to reduce the cost of development by trial-and-error methods. Numerous studies of spray combustion and associated processes have helped designers to establish criteria to design efficient and stable combustors, determine rates of heat transfer to combustion-chamber surfaces, and examine the formation of pollutants such as soot, unburned hydrocarbons, NO_x, and CO. For example, it is known that NO_x formation is strongly dependent on temperature, and significant reductions in NO_x emissions can be achieved by reducing maximum flame temperatures. Also, it has been found from various investigations that conditions most favorable for soot formation occur when fuel-rich zones have strong temperature gradients. The basic method of reducing soot formation in spray combustion is not only to reduce temperature gradients in fuel-rich zones but also to reduce the size of zones where strong temperature gradients and fuel-rich concentration can arise. In order to achieve these goals, the flow properties in combusting sprays must be determined.

A realistic analytical model of a combusting spray must involve consideration of diverse phenomena such as the hydrodynamic characteristics of injection and spray formation, the transport characteristics of individual droplets, the turbulent two-phase flow of a spray, chemical reactions in a turbulent environment, and interactions of radiation with flame chemistry and turbulence. Some of the earlier reviews[1-44] which have touched upon these aspects were summarized (see Table 8.1) by Faeth,[1] who has made significant contributions to the study of spray combustion. Therefore, a substantial portion of the discussion of spray combustion covered in this chapter has been drawn from several excellent review papers by Faeth[1,46,47] and many papers published by him and his coworkers. Recently developed group-combustion theories of droplet clouds and fuel spray by Chiu and his coworkers[49-54] are also included in the discussion. Finally, to end the part on spray combustion, a list of problems for future research in this area is given.

2 SPRAY-COMBUSTION SYSTEMS

Sprays are burned in various ways, and each poses different problems for the development of a reliable spray model. Several cases which are typical of the range of configurations encountered in practice are summarized by Faeth[1] as shown in Table 8.2.

In the prevaporizing system (the first case in Table 8.2), the spray is injected into a heated air stream. The drops are almost completely evaporated before reaching the flame. Typical examples of such a configuration are afterburners and the carburetors of spark ignition engines. One-dimensional models can generally provide useful results in this case, except near the injector or when only a limited number of injectors are employed. The two-phase portion of the flow is usually noncombusting.

In liquid-fueled rocket engines (the second case in Table 8.2), both fuel and oxidizer are injected from one end, providing a more or less premixed

TABLE 8.2 Spray Combustion Systems[a]

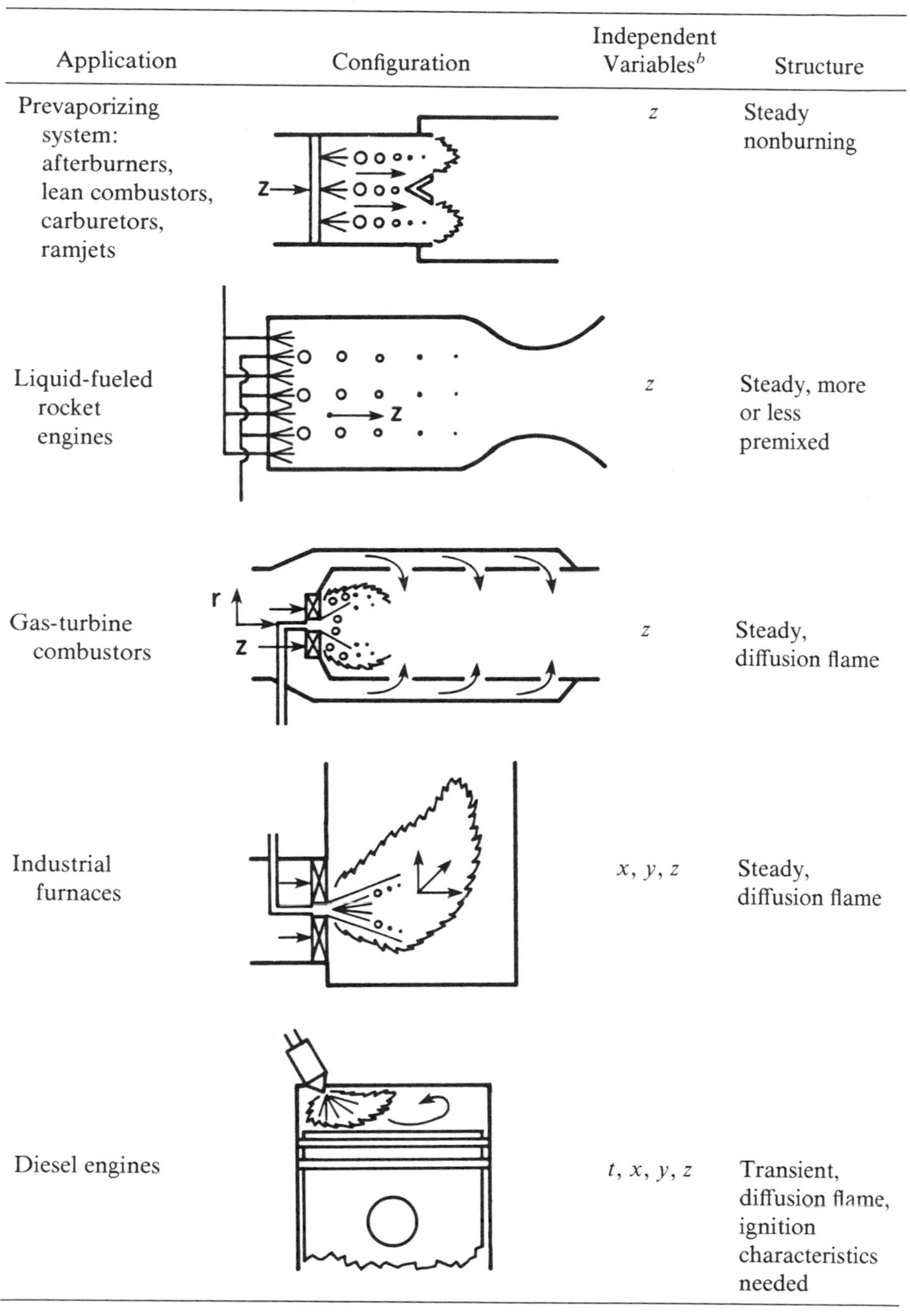

Application	Configuration	Independent Variables[b]	Structure
Prevaporizing system: afterburners, lean combustors, carburetors, ramjets		z	Steady nonburning
Liquid-fueled rocket engines		z	Steady, more or less premixed
Gas-turbine combustors		z	Steady, diffusion flame
Industrial furnaces		x, y, z	Steady, diffusion flame
Diesel engines		t, x, y, z	Transient, diffusion flame, ignition characteristics needed

[a]After Faeth.[1]

[b]Simplest realistic approximation; all systems are axisymmetric or three-dimensional near the injector.

combustion system. In many designs, one-dimensional flow dominates most of the flow field, yielding relatively simple models for performance predictions. Near the injector face, and when only a few injectors are used, mixing effects are important and more complex models must be considered.

The gas-turbine combustor can be divided into three zones: a primary zone where liquid is injected into an air stream to form a nearly stoichiometric mixture of reactants in a two-phase flow, a secondary zone where combustion is completed, and a dilution zone where the combustion products are mixed with air to reduce the temperature of the flow to levels acceptable for expansion through the turbine. Since the fuel and air are not extensively premixed before combustion, the flame has the characteristics of a diffusion flame in which mixing of fuel and oxygen strongly influences the rate of reaction. One-dimensional models are not suitable for this configuration, although lumped-parameter models have been used. Industrial furnaces are qualitatively similar to gas turbines, although unsymmetric configurations are more common and buoyancy effects can be important due to the large physical size of the injector and lower gas velocities (refer to Chapter 6 for reasons for using large injectors in industrial furnaces).

It is generally accepted that diesel engines (the last case in Table 8.2) represent the most difficult modeling problem. The process is primarily a diffusion flame; however, it is transient, and fuel impingement on surfaces can be important. The flow is three-dimensional, and prediction of ignition characteristics is necessary, since the combustion process is intermittent. It is important to note that for the cases illustrated in Table 8.2, the designation "premixed" or "diffusion flame" should be interpreted as only a general indication of the dominant behavior. For two-phase combusting flows, portions of the flow may be burning in a very different manner than other portions, depending upon the effectiveness of the mixing process. This is because the mixing and combustion processes are closely coupled in two-phase reacting flows.

3 FUEL ATOMIZATION

3.1 Injector Types

The performance of a spray combustion system has a crucial bearing on the design of the injector. An injector may be evaluated according to the distribution of drop sizes it produces, the angle of the spray, and the nature of the spray pattern [i.e., whether the spray pattern completely fills the outermost boundaries of the spray (full cone) or has a region relatively free of drops, along the axis of the injector (hollow cone)]. However, the flow conditions and properties of the gas within the combustion chamber also influence the spray pattern.

According to Faeth,[1] injectors may be classified into two major categories as follows:

1. Pressure-atomizing injectors: only the liquid passes through such injectors, and atomization is achieved by virtue of the pressure drop, as the name suggests.
2. Twin-fluid injectors: here atomization of the liquid is aided by a flow of high-velocity gas through injector passages.

TABLE 8.3 Various Types of Injector Systems[a]

Type	Configuration	Structure	Application
	Pressure-Atomizing Injectors		
Plain orifice	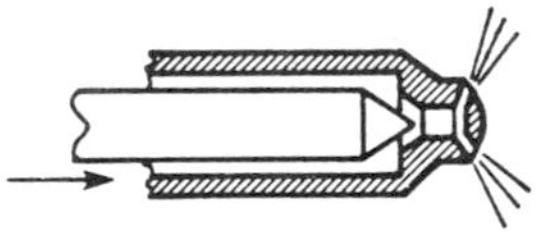	Hollow cone	Diesel engines
Pintle nozzle	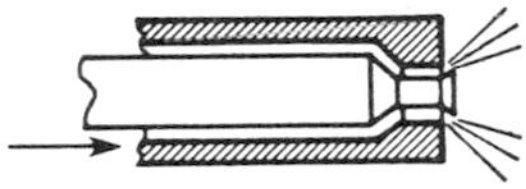	Full cone or multiple cones	Diesel engines, gas turbines
Swirl nozzle (spill type) return	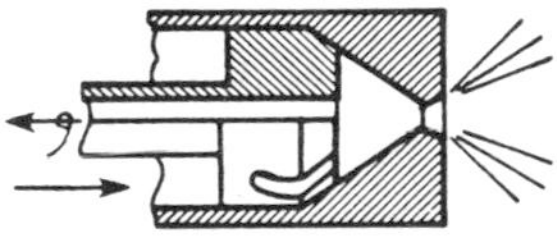	Hollow cone	Furnaces, gas turbines
Impinging jet	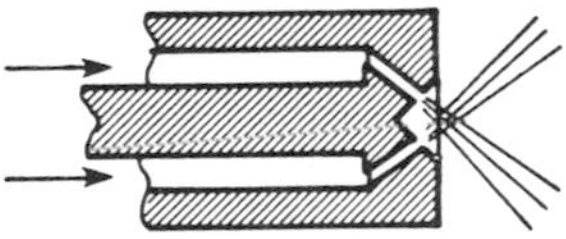	Fan spray	Rocket engines
	Twin-Fluid Injectors		
Internal Mixing	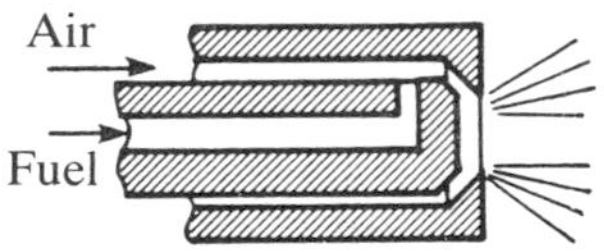 Air, Fuel	Full or hollow cone	Furnaces, gas turbines
External mixing	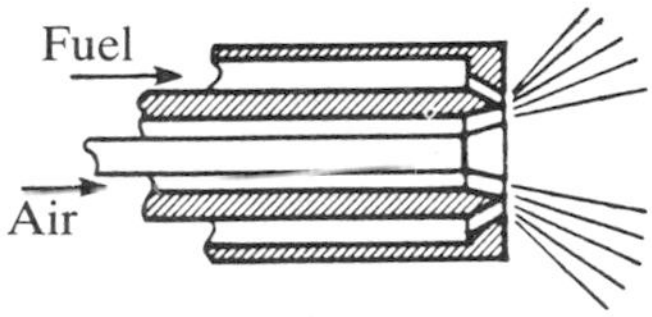 Fuel, Air	Hollow cone	Furnaces, gas turbines

[a]After Faeth.[1]

Although twin-fluid injectors involve additional complexity, they can achieve much finer atomization than pressure-atomizing injectors, particularly at low fuel flow rates during off-design operation. Some typical injector types used in combustion systems are listed in Table 8.3.

3.2 Atomization Characteristics

Theoretical modeling and numerical evaluation of sprays require information on the distribution of droplet sizes and velocities produced by the injector. The spray formation process, however, complicates this specification, since it involves complicated processes such as breakup of the primary jet, secondary droplet breakup, and collisions between drops. There are several other difficulties associated with the specification of the required atomization characteristics. First, spray characterization is normally done under cold conditions; drop vaporization and the influence of combustion gases on the breakup process itself can alter these characteristics.[40] Second, complete distributions of size and velocity as a function of position in the dilute portion of the spray are rarely available.[41] Third, measurement of two-phase flow conditions near the injector exit is extremely difficult. Because of these difficulties, average spray characteristics at some downstream location from the injector exit are used in most cases.

4 SPRAY STATISTICS

4.1 Particle Characterization

According to Williams,[13] an understanding of the mechanism of spray combustion requires a knowledge of (a) the burning mechanism of individual particles, (b) the statistical methods for describing groups of particles, and (c) the manner in which these groups modify the behavior of the gas in flow systems.

The shape of liquid droplets may be considered to be spherical when the following two conditions are met:

1. The droplet collision, agglomeration and micro-explosion effects are small.
2. The Weber number is low: $We \ll 20$.

Condition 1 requires that the volume occupied by the condensed phase be much less than the total spatial volume. This means that the spray must be diluted so that the droplets collide with each other so seldom that collision-induced oscillations are viscously damped to a negligible amplitude for most droplets. This condition is usually valid for hydrocarbon spray combustion systems. The degree of deformation of a liquid droplet caused by the slip velocity between the droplet and gas depends upon the ratio of the dynamic force to the surface-tension force, which is represented by the Weber number We, defined as

$$\mathrm{We} \equiv \frac{2r\rho_g|v_p - v_g|^2}{\sigma_s} = \frac{\text{dynamic force}}{\text{surface-tension force}} \tag{8-1}$$

When We $\ll 20$, droplets are nearly spherical. As We increases the droplets deform and eventually break up at We ≈ 20. When the shape of a droplet is very close to spherical, the droplet size can then be adequately specified by a single parameter r, the radius of the particle.

4.2 Distribution Function

The distribution function defined below,

$$f_j(r, \mathbf{x}, \mathbf{v}, t)\, dr\, d\mathbf{x}\, d\mathbf{v}$$

gives a statistical description of the spray, and the above product represents the probable number of particles of chemical composition j in the radius range dr about r located in the spatial range $d\mathbf{x}$ about $\mathbf{x}$ with velocities in the range $d\mathbf{v}$ about $\mathbf{v}$ at time t. Here $d\mathbf{x}$ and $d\mathbf{v}$ (shown in Fig. 8.1) are the three-dimensional elements of physical space and velocity space, respectively. The dimension of $f_j(r, \mathbf{x}, \mathbf{v}, t)$ is (Number of particles)/$[LL^3(L/t)^3]$.

If the velocity dependence of the distribution function is not of primary interest, one can define another distribution function G_j such that

$$G_j \equiv \int f_j\, d\mathbf{v}, \qquad j = 1, 2, \ldots, M \tag{8-2}$$

Then G_j represents the number of droplets of the jth component per unit volume per unit range of radius. Similarly, one can integrate over the physical

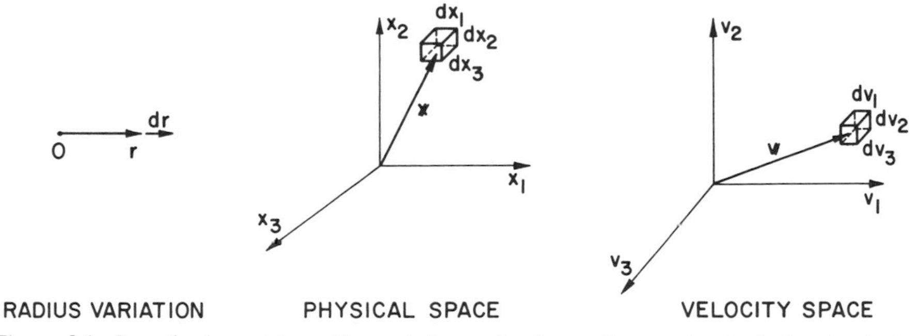

Figure 8.1 Length element in radius variation and volume elements in physical and velocity spaces.

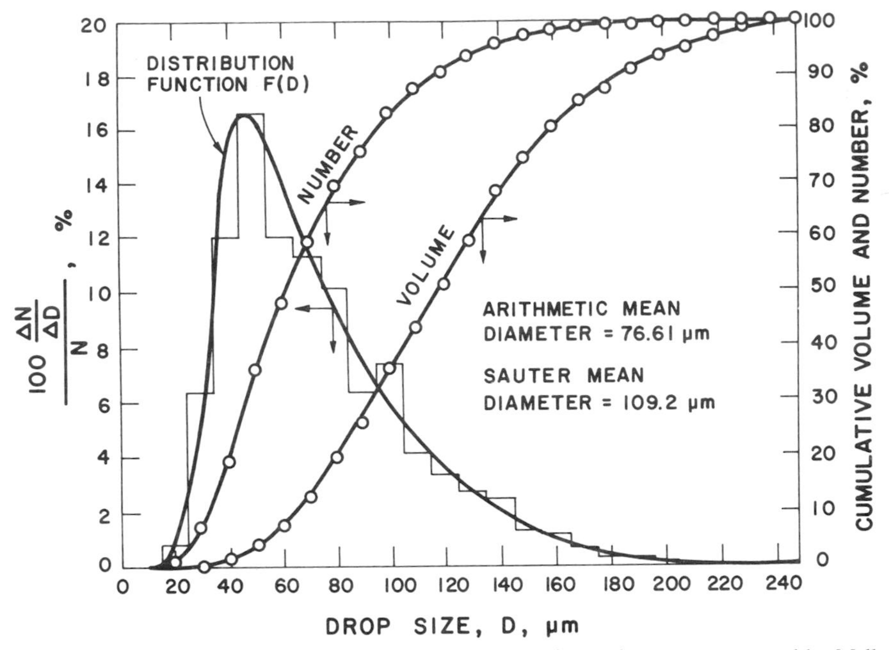

Figure 8.2 Drop-size distribution curves for a water spray in an air stream, measured by Mellor et al.[41] (adapted from Ref. 1).

space to define a distribution function F_j for droplet size (radius or diameter). If only one type of liquid (a single component) is present, the subscript j can be ignored. Therefore, the simplest drop-size distribution function can be written $F(D)$, which represents the fraction of particles per unit diameter range about D. Drop-size distribution curves obtained by Mellor et al.[41,1] for a water spray in an air stream are shown in Fig. 8.2.

Overall spray characteristics are represented by distribution curves which can be given in terms of the cumulative percentage of droplet number, surface area, or volume as a function of drop diameter. Figure 8.2, besides showing the distribution function $F(D)$, illustrates several of these methods of spray description for water injection through a swirl-type pressure atomizer into a moving air stream.[41]

In many mass-transfer and flow processes it is desirable to work only with average diameters instead of the complete drop-size distribution. Therefore, the droplet size is frequently represented by a mean or median diameter. However, there are at least half a dozen to choose from. A general expression for a mean diameter D_{jk} in terms of the distribution function dN/dD can be written as

$$\left(D_{jk}\right)^{j-k}\int_{D_{\min}}^{D_{\max}} D^k \frac{dN}{dD}\,dD = \int_{D_{\min}}^{D_{\max}} D^j \frac{dN}{dD}\,dD \tag{8-3}$$

where j and k are integers. A more general form of the mean of the droplet

TABLE 8.4 Popular Mean Diameters and Their Fields of Application

k	j	Order $k+j$	Name	Field of Application
0	1	1	Linear	Comparisons, evaporation
0	2	2	Surface	Absorption
0	3	3	Volume	Hydrology
1	2	3	Surface diameter	Adsorption
1	3	4	Volume diameter	Evaporation, molecular diffusion
2	3	5	Sauter	Combustion, mass transfer, and efficiency studies
3	4	7	De Brouckere	Combustion equilibrium

diameter raised to a power $j-k$ can be calculated from

$$(D_{jk})^{j-k} \equiv \frac{\int_0^\infty D^j F(D)\,dD}{\int_0^\infty D^k F(D)\,dD} \tag{8-4}$$

which yields D_{10} as the average (arithmetic) droplet diameter, D_{20} as the diameter of a droplet whose surface area times the total number of droplets equals the total area of the spray, and D_{30} as the volume-average diameter. The Sauter mean diameter (SMD), D_{32}, is the diameter of a droplet whose ratio of volume to surface area is equal to that of the entire spray. It is commonly used to represent the size of an equivalent monodisperse spray for approximate analysis of evaporation and combustion. Median droplet diameters are also used to describe a spray by dividing it into two equal portions by number, length, area, etc.[20,1] A summary of the popular mean diameters and their fields of application is given in Table 8.4.

An accurate knowledge of the drop-size distribution is a prerequisite for fundamental analysis of the transport of mass or heat or the separation of phases in spray combustion systems. Numerous distribution functions have been proposed to correlate drop size in a spray. General features of four important size distribution functions are given below. Among them, the Rosin–Rammler[43] and Nukiyama–Tanasawa[44] distributions are the best known.

4.2.1 Logarithmic Probability Distribution Function

Drop diameters can be graded in size ranges exponentially rather than linearly, by defining

$$y \equiv \ln \frac{D}{D_{30}} \tag{8-5}$$

Then, the volume distribution equation can be written as

$$\frac{dV}{dy} = \phi(y) \tag{8-6}$$

When sets of spray data are plotted on this basis, it is found that they ordinarily give a fairly symmetrical distribution about a certain value of y which is close to a single maximum of the curve. A good guess for $\phi(y)$ based upon statistical analysis is the *normal distribution function* $(\delta/\sqrt{\pi})e^{-\delta^2 y^2}$. Then we have

$$\frac{dV}{dy} = \frac{\delta}{\sqrt{\pi}} e^{-\delta^2 y^2} \tag{8-7}$$

The general expression for the mean diameter of this distribution can be written as

$$D_{jk} = D_{30} e^{(k+j-6)/4\delta^2} \tag{8-8}$$

where δ is the distribution factor and has to be fitted to the data by trial and error.

4.2.2 Rosin–Rammler Distribution Function[43]

The volume distribution equation is of the form

$$\frac{dV}{dD} = \frac{\delta D^{\delta-1}}{D_{\text{ref}}^{\delta}} e^{-(D/D_{\text{ref}})\delta} \tag{8-9}$$

where D_{ref} is a characteristics size $(= D_{30})$ and δ is the distribution factor. The expression for the mean diameter is

$$(D_{jk})^{j-k} = (D_{\text{ref}})^{j-k} \frac{\Gamma\left(\dfrac{j-3}{\delta} + 1\right)}{\Gamma\left(\dfrac{k-3}{\delta} + 1\right)} \tag{8-10}$$

4.2.3 Nukiyama–Tanasawa Distribution Function[44]

Nukiyama and Tanasawa,[44] from their experiments with air atomization, obtained an empirical expression for drop-size distributions. The volume distribution is of the form

$$\frac{dV}{dD} = \frac{b^6/\delta}{\Gamma(3/\delta)} D^5 e^{-bD^\delta} \tag{8-11}$$

where Γ is the gamma function, δ is the distribution factor, and b is the size parameter, which has a dimension of $D^{-\delta}$. The equation for the mean diameter is

$$(D_{jk})^{j-k} = b^{-(j-k)/\delta} \frac{\Gamma\left(\frac{j+3}{\delta}\right)}{\Gamma\left(\frac{k+3}{\delta}\right)} \tag{8-12}$$

4.2.4 Upper-Limit Distribution Function of Mugele and Evans[42]

It is quite obvious that both the Rosin–Rammler and the Nukiyama–Tanasawa distribution functions have the following generic form:

$$F(D) = aD^p \exp(-bD^n) \tag{8-13}$$

where a, p, b, and n are empirical parameters. Distributions of this type place no restriction on the size of the largest drop that can exist in a spray. Therefore, extrapolation of distributions given in Eq. (8-13) can yield values of D_{32} greater than any observed droplet diameter in the spray in some cases, and the use of this equation can also cause significant underestimation of spray evaporation rates. In fact, all three distribution functions considered so far fail to place an upper limit on the drop diameter. Mugele and Evans[42] have proposed a distribution function which imposes a limit.

Basically, it is a modification of the logarithmic probability distribution function. The dimensionless parameter y is defined as

$$y \equiv \ln \frac{aD^s}{D_m^s - D^s} \tag{8-14}$$

or

$$y \equiv \ln \frac{D(D_m - D_{\text{vmd}})}{D_{\text{vmd}}(D_m - D)} \tag{8-15}$$

where a is a dimensionless constant, s is a positive integer (e.g., 1, 2, or 3), D^s represents a simple function of the drop size (e.g., diameter, surface, or volume), and D_m represents the maximum drop size. So $D_m^s - D^s$ is a measure of the "size deficiency" or reduction from maximum drop size. D_{vmd} in Eq. (8-15) represents the volume median diameter. The expression for the Sauter mean diameter (SMD, D_{32}) can be calculated from

$$D_{32} = \frac{D_m}{1 + ae^{-4\delta^2}} \tag{8-16}$$

The volume distribution can be calculated from

$$\frac{dV(D)}{dD} = \frac{\delta}{\sqrt{\pi}} \exp(-\delta^2 y^2) \tag{8-17}$$

where δ is the drop-size parameter.

In the study of Mugele and Evans,[42] all four distribution functions were critically compared with experimental data. The following general conclusions were reached:

a. The logarithmic probability distribution function is so far the best one for calculating mean diameters. It predicts equality of means of the same order (e.g., $D_{30} = D_{21}$, $D_{40} = D_{31}$), and this agrees well with experimental data.
b. The Nukiyama–Tanasawa distribution function sometimes gives a completely wrong trend for the volume distribution when the parameters are calculated from numerical distribution data. Uncertainty in δ leads to uncertainty in the mean diameter.
c. Unreasonable values of mean diameters sometimes result from the Rosin–Rammler distribution function.
d. The ULDF predicts mean diameters as well as the logarithmic pdf and agrees quite well with experimental data. The ULDF may be capable of eventual interpretation in terms of fundamental mechanisms.

Aside from the distribution function, there are numerous correlations for various average and median drop-size parameters. These are specific for various injector designs, but generally have the form[44,1]

$$\frac{D_{jk}}{d_0} = f_{jk}\left(\frac{L}{d_0}, \phi, \frac{d_0 \rho v}{\mu}, \frac{\mu v}{\sigma_s}, \frac{\rho_\infty}{\rho}, \frac{\mu_\infty}{\mu}, \frac{v_\infty}{v}\right) \tag{8-18}$$

4.3 Transport Equation of the Distribution Function

Following Williams,[39,13] an equation governing the time rate of change of the distribution function f_j mentioned in Section 4.2 may be derived phenomenologically by using reasoning analogous to that employed in the kinetic theory of gases. Let us define the following quantities:

$R_j \equiv (dr/dt)_j$, the rate of change of the size r of a particle of kind j at $(r, \mathbf{x}, \mathbf{v}, t)$

$\mathbf{F}_j \equiv (d\mathbf{v}/dt)_j$, the force per unit mass on a particle of kind j at $(r, \mathbf{x}, \mathbf{v}, t)$

$\hat{Q}_j$, the rate of increase of f_j with time due to particle formation (from smaller particles) or destruction (from larger particles), as in nucleation or liquid breakup

Γ_j, the rate of increase of the distribution function caused by collisions with other particles (these collisions must occur sufficiently seldom that the aerodynamic contributions to Γ_j are separable from those in F_j)

$-\nabla_x \cdot (\mathbf{V}f_j)$, the increase of f_j due to the motion of particles into and out of the spatial element $d\mathbf{x}$ by virtue of their velocity $\mathbf{V}$

$-\nabla_v \cdot (\mathbf{F}_j f_j)$, the increase of f_j in the velocity element $d\mathbf{v}$ because of the acceleration $\mathbf{F}_j$

Adding the changes in f_j, we have

$$\frac{\partial f_j}{\partial t} = -\frac{\partial}{\partial r}(R_j f_j) - \nabla_x \cdot (\mathbf{V}f_j) - \nabla_v \cdot (\mathbf{F}_j f_j) + \hat{Q}_j + \Gamma_j \qquad (8\text{-}19)$$

for $j = 1, 2, \ldots, M$, where M is the total number of different kinds of particles classified according to their chemical composition. Equation (8-19) is called the transport equation of the distribution function, or the spray equation.[39,13]

Under many circumstances the intensity of burning is comparatively low in the vicinity of the atomizer and most of the combustion takes place in regions where particle interactions and sources are of no more than secondary importance. Since we focus our attention on the burning process, we may then legitimately neglect $\hat{Q}_j$ and Γ_j. If a steady process is considered, then $\partial f_j/\partial t = 0$. Hence for many combustion problems, the steady-state equation of motion of a statistical ensemble of particles of type j is

$$\frac{\partial}{\partial r}(R_j f_j) + \nabla_x \cdot (\mathbf{V}f_j) + \nabla_v \cdot (\mathbf{F}_j f_j) = 0 \qquad (8\text{-}20)$$

This equation is coupled with the hydrodynamical equations of motion of the gas through the variables R_j and $\mathbf{F}_j$, which depend upon the local properties of the fluid. However, if the spray is sufficiently dilute, then the statistical fluctuations in the gas properties induced by the spray can be neglected and use of accurate average fluid-flow variables can be made. Then the motion of the gas can be described by the ordinary fluid-mechanical equations with appropriate boundary conditions determined by average spray properties.

4.4 Simplified Spray Combustion Model for Liquid-Fuel Rocket Engines

As mentioned in Section 2, the spray combustion processes occurring in liquid-fuel rocket engines can be treated with a steady or quasisteady approximation. Also, the reactant mixtures can be considered as premixed. The general flow characteristics are quasi-one-dimensional. Under these idealized conditions, simplified spray combustion models have been developed to study the combustion efficiency of liquid-fuel rocket-engine combustors with variable cross-sectional area. Probert[55] initially investigated this problem, and later the model was refined by Williams,[12,56] Tanasawa,[57,58] and Tesima.[58] By neglecting the velocity dependence of the distribution function, Eq. (8-20) reduces to

$$\frac{\partial}{\partial r}\left(\bar{R}_j G_j\right) + \nabla_x \cdot \left(\bar{\mathbf{v}}_j G_j\right) = 0, \qquad j = 1, 2, \ldots, M \tag{8-21}$$

where G_j is given in Eq. (8-2) and the bar denotes an average over all velocities, that is,

$$\bar{R}_j = \frac{1}{G_j} \int R_j f_j \, d\mathbf{v}, \qquad j = 1, 2, \ldots, M \tag{8-22}$$

and

$$\bar{\mathbf{v}}_j = \frac{1}{G_j} \int \mathbf{v} f_j \, d\mathbf{v}, \qquad j = 1, 2, \ldots, M \tag{8-23}$$

Equation (8-21) can be rewritten in the form of Eq. (8-24) to take account of the variation of the cross-sectional area $A(x)$ along the combustor:

$$\frac{\partial}{\partial r}\left(\bar{R}_j G_j\right) + \frac{1}{A} \frac{\partial}{\partial x}\left(A \bar{v}_j G_j\right) = 0, \qquad j = 1, 2, \ldots, M \tag{8-24}$$

where $\bar{v}_j$ is the x-component of $\bar{v}_j$, and the quantities $\bar{R}_j$, G_j, and $\bar{v}_j$ have been assumed to be essentially independent of the spatial coordinates normal to x. These quantities may be regarded as averages over the cross-sectional area.

It is known from the droplet vaporization correlation that the dependence of $\bar{R}_j$ on droplet size may be expressed by the relation

$$\bar{R}_j = -\frac{\chi_j}{r^{\alpha_j}}, \qquad j = 1, 2, \ldots, M \tag{8-25}$$

where $0 \le \alpha_j \le 1$ and χ_j represents a positive constant independent of r. It is useful to note that $\alpha_j = 1$ corresponds to the d^2 evaporation law mentioned in Chapter 6.

In view of the dependence of $\bar{v}_j$ on droplet size, one can generally expect that the velocity distributions may differ for particles of different sizes. However, when the droplets are very small, they can be totally entrained by the gas. Under this circumstance, $\bar{v}_j$ is independent of r. Using these approximations or assumptions, Eq. (8-24) can be integrated. The solution of the distribution function G_j by integrating Eq. (8-24) was given by Williams,[13] and the expression obtained is

$$G_j = \frac{\left(A_0 \bar{v}_{j,0}\right)}{A \bar{v}_j} \left(\frac{r}{\gamma_j}\right)^{\alpha_j} G_{j,0}(\gamma_j), \qquad j = 1, 2, \ldots, M \tag{8-26}$$

where

$$\gamma_j \equiv \left[r^{\alpha_j+1} + (\alpha_j + 1) \int_0^x \frac{x_j}{\bar{v}_j} \, dx \right]^{1/(\alpha_j+1)}, \qquad j = 1, 2, \ldots, M \quad (8\text{-}27)$$

The subscript 0 in Eq. (8-26) designates parameters at the station $x = 0$.

The most interesting parameter in the study of liquid-fuel rocket engines is the combustion efficiency for a chamber of length L from the size distribution at position L. Let Q_j denote the heat of reaction (heat released) per unit mass of material evaporated from a droplet of kind j, and let $\rho_{1,j}$ be the density of the liquid droplets of kind j. The mass per unit volume of the spray of kind j is therefore

$$\int_0^\infty \tfrac{4}{3}\pi r^3 \rho_{1,j} G_j \, dr$$

and the corresponding mass flow rate is $A\bar{v}_j$ times this integral term. The total amount of heat released per second by the jth spray fuel between the injector and position L is

$$Q_j \left(A_0 \bar{v}_{j,0} \int_0^\infty \tfrac{4}{3}\pi r^3 \rho_{1,j} G_{j,0} \, dr - A\bar{v}_j \int_0^\infty \tfrac{4}{3}\pi r^3 \rho_{1,j} G_j \, dr \right)$$

Since the maximum possible heat release from burning all the droplets is

$$\sum_{j=1}^{M} Q_j A_0 \bar{v}_{j,0} \int_0^\infty \tfrac{4}{3}\pi r^3 \rho_{1,j} G_{j,0} \, dr$$

the combustion efficiency at $x = L$ can be calculated from

$$\eta_c = \frac{\displaystyle\sum_{j=1}^{M} Q_j \left(A_0 \bar{v}_{j,0} \int_0^\infty \tfrac{4}{3}\pi r^3 \rho_{1,j} G_{j,0} \, dr - A\bar{v}_j \int_0^\infty \tfrac{4}{3}\pi r^3 \rho_{1,j} G_j \, dr \right)}{\displaystyle\sum_{j=1}^{M} Q_j A_0 \bar{v}_{j,0} \int_0^\infty \tfrac{4}{3}\pi r^3 \rho_{1,j} G_{j,0} \, dr} \quad (8\text{-}28)$$

Using Eq. (8-26), the combustion efficiency can be written as

$$\eta_c = 1 - \frac{\displaystyle\sum_{j=1}^{M} (Q_j \rho_{1,j} \bar{v}_{j,0}) \int_0^\infty r^3 \left(\frac{r}{\gamma_j} \right)^{\alpha_j} G_{j,0}(\gamma_j) \, dr}{\displaystyle\sum_{j=1}^{M} (Q_j \rho_{1,j} \bar{v}_{j,0}) \int_0^\infty r^3 G_{j,0}(r) \, dr} \quad (8\text{-}29)$$

γ_j in the above equation can be calculated from Eq. (8-27) by setting the upper limit of the integral on the right-hand side of Eq. (8-27) equal to L. Equation

(8-29), together with Eq. (8-27), gives the combustion efficiency for an arbitrary initial droplet size distribution $G_{j,0}(r)$.

5 SPRAY COMBUSTION CHARACTERISTICS

Although combustion of liquid fuel droplets in a spray is inherently governed by the diffusion of fuel vapor and oxidizer species, both premixed- and diffusion-flame theories have been applied to spray combustion problems. Premixed-flame theories can sometimes be applied to spray combustion in liquid rocket engines, the stabilization of spray flames on flame holders, and the development of deflagration and detonation waves in spray fields.[1]

Figure 8.3 shows a schematic comparison of temperature and reaction-rate distributions in premixed gaseous flames and a spray combustion flame in a one-dimensional model. In both cases, a simple one-step chemical reaction

$$\text{fuel} + O_2 \rightarrow \text{products}$$

was assumed. The distribution of concentration and temperature profiles in the

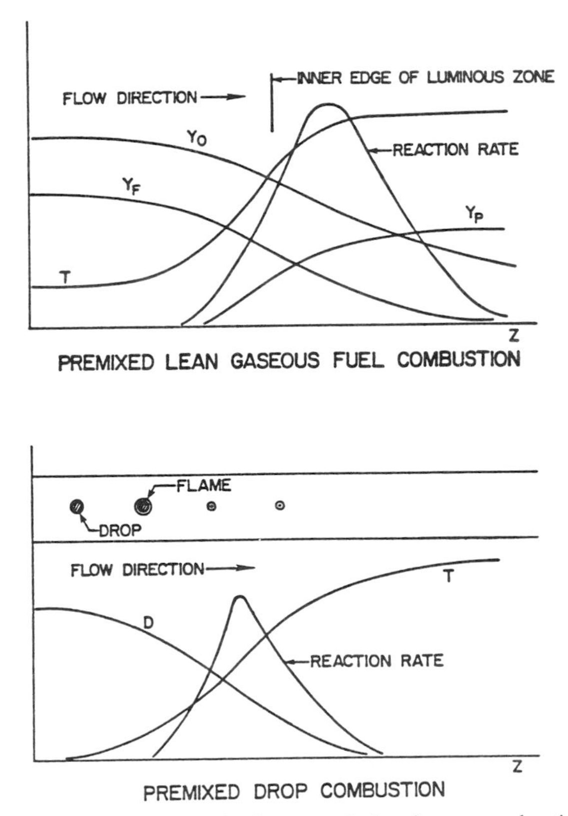

Figure 8.3 Schematic diagram of premixed gaseous fuel and spray combustion (after Faeth[1]).

premixed gaseous flame is shown on the upper portion of this figure. The treatment in Chapter 5 leads to distribution curves of this form. In the combustion of fuel droplets, one must consider the size and volatility of the droplet. For very small droplets of a high-volatility fuel, droplet evaporation may be completed in the heatup process, so that the flame structure is only mildly influenced by the effect of two-phase flow. Burgoyne and Cohen[59] conducted a classic experiment using monodisperse tetralin sprays, showing that droplet sizes less than 10 μm resulted in flame speeds that are not appreciably different from those in pure gas. They also showed that the flame speed decreases as the initial drop size increases. This is mainly due to the increased time required to evaporate the fuel. For larger drop sizes and nonvolatile fuels, the fuel evaporation rate decreases significantly and a region of combustion surrounds each droplet. The end of the flame zone then nearly corresponds to the disappearance of the droplet. A simplified one-dimensional model of this process has been developed and presented by Williams.[13] General agreement between his theoretical results and the experimental data of Burgoyne and Cohen[59] was obtained. However, one must be extremely cautious in using the theory of premixed flames to treat spray combustion problems.

Diffusion-flame theories can definitely be applied to many spray combustion problems, since the fuel vapor evaporated from the droplet surface has to mix with the ambient oxidizer before chemical reaction can occur. The background material on diffusion flames surrounding a single droplet as well as in a jet has been covered extensively in Chapter 6. Readers are encouraged to become familiar with Chapter 6 before proceeding with the following discussions.

In the study of spray diffusion flames, Mizutani et al.[60] obtained radial profiles of the temperature, velocity, and droplet mass flux at various axial stations of a spray flame stabilized in a heated air stream (see Fig. 8.4). In their experimental setup, pressure-atomized injection was used with a moderate swirl in the fluid stream, so that a full-cone spray was obtained. The two-phase flow was sampled by a specially designed sampling probe, and the liquid fraction was determined by separation and weighing. Although full-cone spray was established, the spray had some hollow-cone characteristics: near the injector, the maximum droplet mass flux occurred at some distance away from the axis of the spray. The flame front, determined from the appearance of blue light, is located about 10 mm from the injector exit, and the hot air entrained by the spray provides a premixed region which stabilized the flame. Near the centerline, the temperature levels are low and the droplet concentrations are relatively high; hence the droplets are evaporating, and no significant combustion occurs in this cool central region. However, in the outer regions of the spray, fuel droplets are actively involved in the combustion process. The zone bounded by the maximum temperature was found to be slightly smaller than the spray cone. They concluded that spray flames were stabilized in a high-temperature stream, principally by the flame propagation mechanism. The flame front therefore approaches the injector exit far more closely than would be expected from the ignition-delay data.

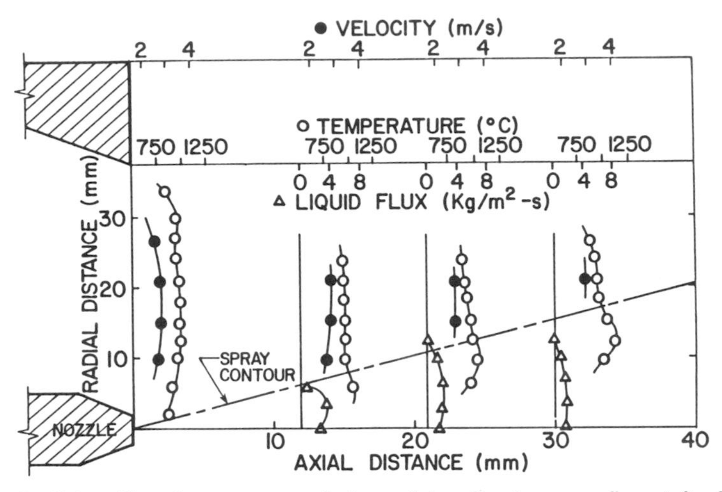

Figure 8.4 Radial profiles of temperature, velocity, and drop flux in a spray flame (after Mizutani et al.[60]).

Khalil and Whitelaw[61] studied spray combustion of kerosene fuel and employed several injectors to achieve sprays with different drop sizes ($4.5 < \text{SMD} < 100$ μm). Measurements were made of droplet velocity, velocity fluctuations, and number density using a laser Doppler velocimeter. Values of mean temperatures were measured with a Pt : 40% Rh–Pt : 20% Rh fine-wire thermocouple of bead diameter 180 μm. Values of the rms temperature fluctuation were determined with a thermocouple of bead diameter 40 μm and Pt : 30% Rh and Pt : 60% Rh wires. Contours of isovelocity, isoturbulence, and isotherms for one of their hollow-cone spray flames with SMD = 45 μm are shown in Fig. 8.5. The outline of the reaction zone is indicated by the velocity and temperature distributions. It roughly corresponds to the cone angle of the injector. Along the axis, temperatures and velocities are relatively low near the injector, suggesting that portions of the spray are evaporating in a relatively cool environment, with a major reaction zone along the periphery of the spray. This observation is similar to those of Mizutani et al.[60] With larger droplet sizes and cone angles, the cool central region expanded both axially and radially, suggesting relatively strong influences of the injector on the combustion process.

They also made an attempt to examine the effect of two-phase (liquid–gas) flow by studying the corresponding gas flame with mass velocity and cone angle similar to the spray. The assumptions of fast chemical reaction and a clipped Gaussian probability distribution of scalar fluctuations were used in their theoretical calculations based upon a two-equation turbulence model. Figure 8.6 gives the comparison of the centerline velocity and temperature distributions of the spray flame, and the corresponding gaseous simulation. This indicates that the gaseous simulation attains a maximum temperature before the spray flame. The maximum centerline temperature of the gaseous

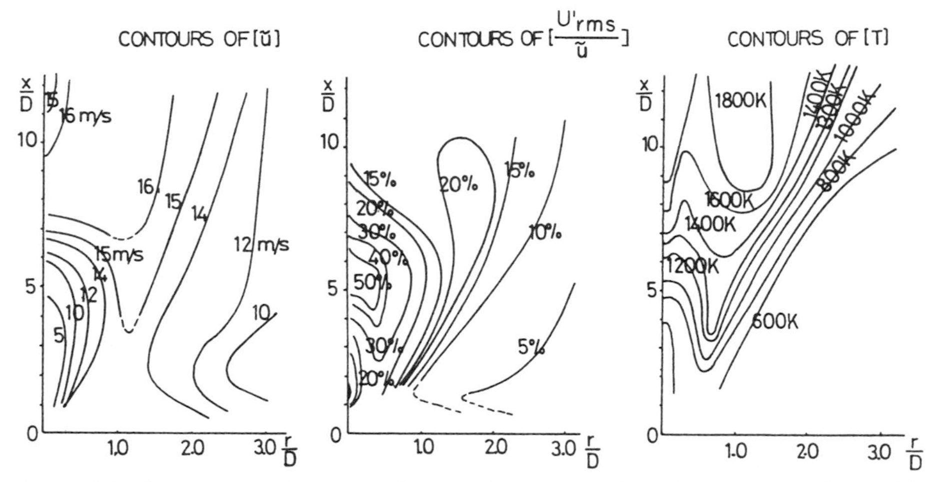

Figure 8.5 Contours of velocity, turbulence, and temperature for a kerosene spray flame with SMD = 45 μm (after Khalil and Whitelaw[61]).

simulation is less than that of the spray flame, but as can be seen from Fig. 8.7, the maximum temperatures in two different spray flames were achieved at some distance from the centerline. This situation stems from the characteristic of hollow-cone spray flow caused by the angle of the spray and the corresponding gaseous simulation. Figure 8.7 indicates that the radial profiles of spray flames can not be accurately simulated by corresponding gaseous flames. Even with fine sprays (SMD = 45 μm), two-phase effects must be considered in modeling.

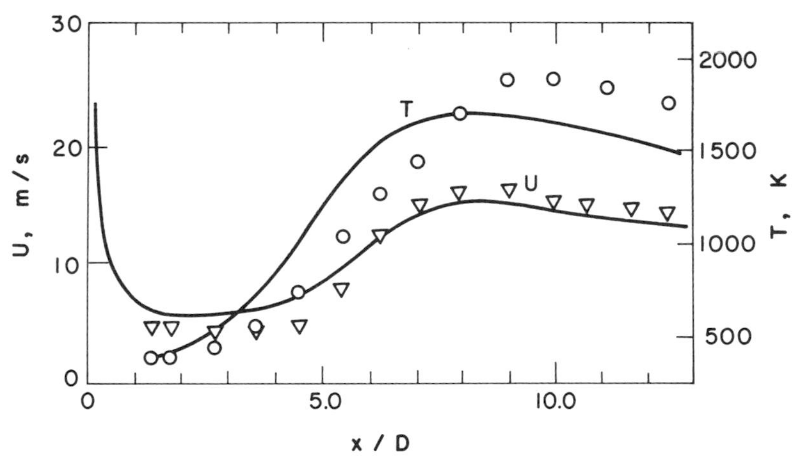

Figure 8.6 Centerline distribution of velocity and temperature in a kerosene spray flame with SMD = 45 μm: curves, simulated gas flame; ▽, measured values of U; ○, measured values of T (after Khalil and Whitelaw[61]).

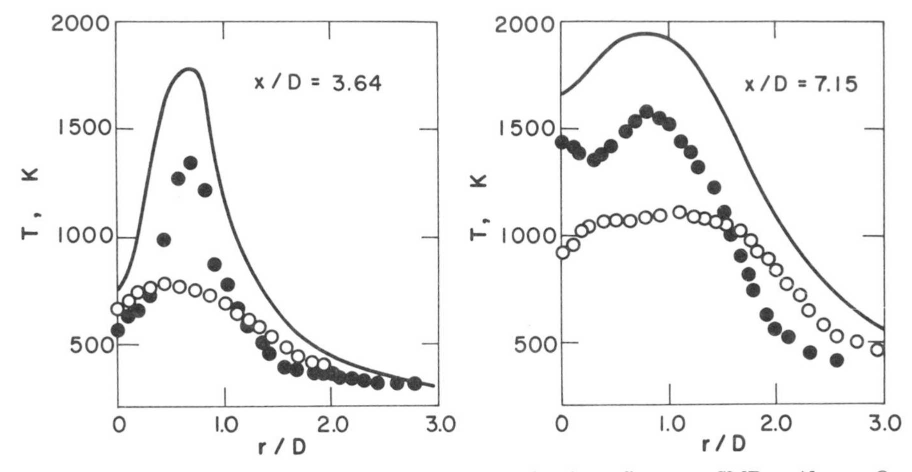

Figure 8.7 Radial profiles of temperature: curves, simulated gas flame; ●, SMD = 45 μm; ○, SMD = 100 μm (after Khalil and Whitelaw[61]).

Onuma and Ogasawara[62] conducted experiments on spray combustion flames of axial jets of kerosene under low-turbulence conditions. Droplet and temperature distributions, flow velocity, and gas composition were measured in the flame of an air-atomizing burner. They found that the region where the droplets exist is limited to a small area above the burner nozzle. From the correlation between various distributions measured, they concluded that most of the droplets in the flame do not burn individually with envelope flames, but that fuel vapor from the droplets forms a cloud and burns like a gaseous diffusion flame. Under the same conditions (using the same apparatus, and changing only the fuel from liquid kerosene to gaseous propane), the spray combustion flame was found to be very similar to the turbulent gas diffusion flame in structure. Figures 8.8 and 8.9 show various distributions along the flame axis obtained for these flames. It is quite obvious that there is a qualitative resemblance between the two cases.

However, there was a slight difference. In the gas diffusion flame, chemical reactions occurred slightly downstream as compared with the spray combustion flame. This is due to the higher initial flow velocity of propane than of kerosene. In the spray combustion flame, the temperature drop downstream is slightly steeper than in the gas diffusion flame. This difference is probably caused by the higher emissivity and radiation cooling rate for the spray combustion flame. Despite these differences, two flames are still very similar in various property distributions. This similarity provides experimental support for group combustion theories developed by Chiu and his coworkers.[49-54]

Onuma et al.[63] extended their earlier investigation in similar experiments to determine (1) whether the similarity between the spray combustion flame and the turbulent gas diffusion flame also exists in nitric oxide formation, and (2) whether their earlier conclusions on the similarity can be applied to the flame

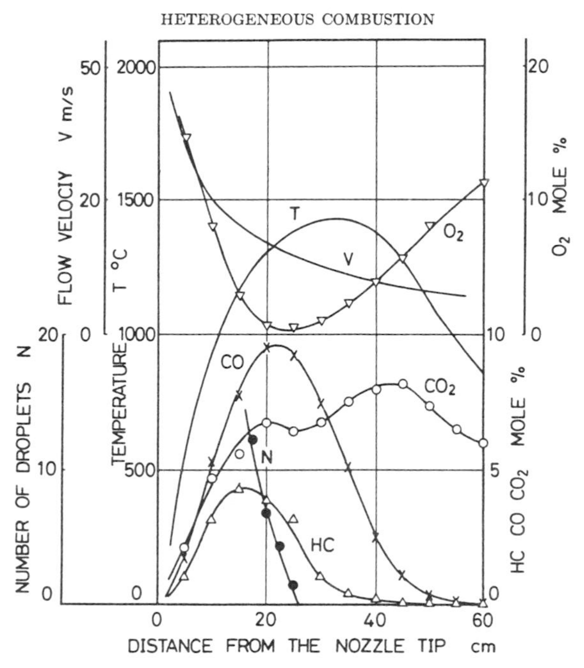

Figure 8.8 Various distributions along the flame axis in the spray combustion flame (after Onuma and Ogasawara[62]).

of a low-volatility heavy oil. They observed that the radial profile of NO concentration in a kerosene flame exhibited two peaks symmetrical with respect to the flame axis. These peaks coincide approximately with the peaks of the temperature profile, and with the positions where the local equivalence ratio is unity. This tendency is the same as in turbulent gas diffusion flames, and so it was concluded that the spray combustion flame is also similar to the turbulent gas diffusion flame in the NO formation process. A heavy-oil flame was also experimentally compared with a kerosene flame under the same conditions. The results showed that the shapes and measured profiles of various quantities were almost the same for both flames. This confirms that the heavy-oil flame does not differ significantly in structure from the kerosene flame, even though the heavy oil had somewhat larger drop sizes. Styles and Chigier[64] have also conducted measurements in a kerosene spray flame using an air-blast atomizer. Their photographs revealed an initial dense spray region with no combustion, surrounded by a gaseous diffusion flame. General agreement with the results of Onuma et al.[62,63] were obtained.

An overall picture of the spray combustion process which emerges from these studies was presented by Faeth[1] as shown schematically in Fig. 8.10. This figure illustrates the relative locations of the cold core region, reaction zone, spray boundary, and jet boundary of a coaxial spray diffusion flame. It is generally understood that the spray leaving the injector is highly nonuniform,

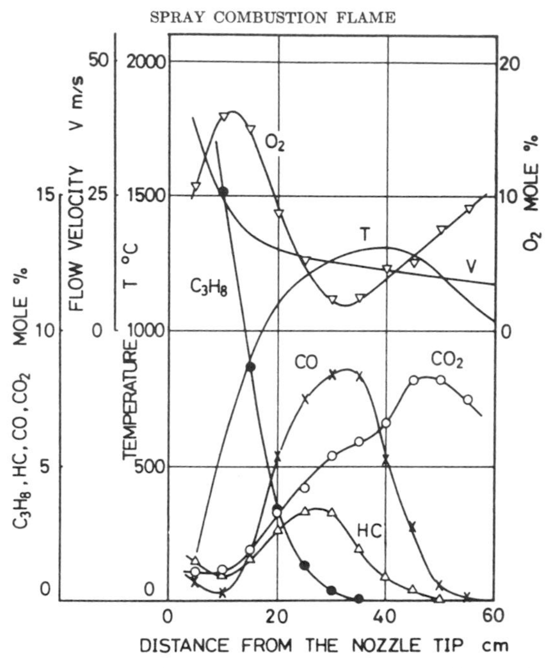

Figure 8.9 Various distributions along the flame axis in the gas diffusion flame (after Onuma and Ogasawara[62]).

with smaller droplets at the periphery and larger droplets near the centerline. Near the injector exit, the velocity difference (slip velocity) between the droplets and the surrounding gas is greatest, and momentum of the liquid is transferred to the gas over an extended axial distance. The small droplets around the periphery of the spray exchange momentum with the gas and cause the spray jet to entrain surrounding gas. These small droplets also evaporate rapidly to provide fuel vapor which is consumed near the outer portion of the turbulent diffusion flame. In general, beyond the maximum reaction zone, small droplets tend to follow the flow and may be largely confined to fuel-rich eddies as shown in the insert of Fig. 8.10. These small droplets may evaporate significantly before high concentrations of oxygen appear in their immediate surroundings. Large droplets, however, can travel through a considerable distance before evaporation and combustion processes begin; this is due to the relatively high inertia of a large droplet.

With increasing distance from the injector exit, the centerline temperature increases due to combustion of the spray, causing more rapid evaporation of large droplets close to the centerline. According to Faeth,[1] measurements from various experiments show that the disappearance of droplets is closely correlated with positions of maximum temperature, in both the radial and axial directions. A number of investigations have been done to study these phenomena under more complex conditions akin to those in gas-turbine combustors

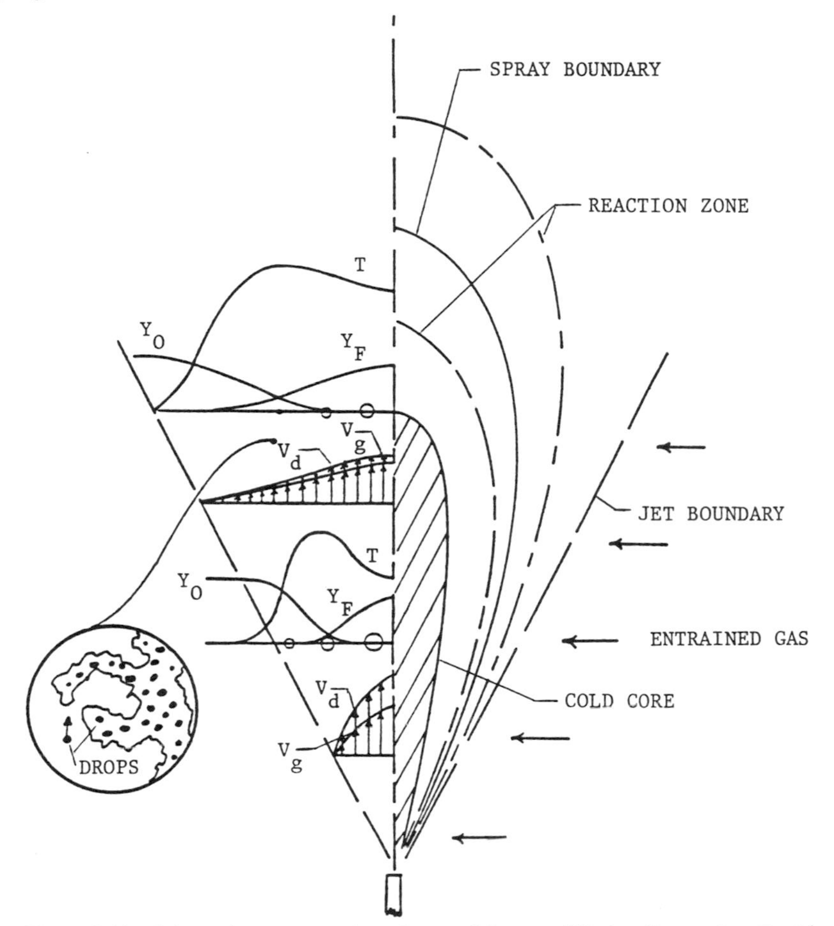

Figure 8.10 Schematic representation of a coaxial spray diffusion flame (after Faeth[1]).

and furnaces. The conclusion is that the bulk of the fuel evaporation occurs in relatively cool, low-oxygen-concentration regions. However, this conclusion may be modified in some cases where poor atomization in conjunction with large gas velocities yields regions where evaporation of large drops occurs in an oxidizing environment.

Faeth concluded in his review paper[1] that the present understanding of spray combustion in various practical systems shows that although in some cases spray combustion can be modeled by ignoring the details of spray evaporation and treating the system in the same manner as a gaseous diffusion flame, in many circumstances such simplifications are unwarranted and the turbulent two-phase flow must be considered. There is a need for improved injector characterization methods, more information on droplet transport characteristics in turbulent flow, and continued development of more complete two-phase turbulent models. In particular, the following aspects need further consideration:

A. Although one-dimensional models are acceptable methods for the design of liquid rocket engines, information is lacking on three-dimensional effects near the injector, liquid jet breakup processes, transient effects, the sensitivity of model predictions to parameters, and injector characteristics in hot recirculating flows.

B. Additional work is necessary to firmly establish the limitations of the local-homogeneity assumption (to be discussed later on).

C. A consistent method of estimating drop characteristics in two-phase flows is needed. Systematic studies are required to resolve all the transport effects.

D. Criteria for drop ignition and the presence of stable envelope flames need more attention. The effect of convection on burning drops is also of substantial importance and needs to be studied further.

The following section describes various spray models developed and recent advances in spray combustion processes.

6 CLASSIFICATION OF MODELS DEVELOPED FOR SPRAY COMBUSTION PROCESSES

In view of the two-phase nature of the turbulent reaction processes involved in spray combustion, there are numerous difficulties one can encounter in the development of predictive models. Models with different basic assumptions and levels of sophistication have been proposed, ranging from simple correlations to complex turbulent two-phase reacting models. In the following, classes and representative samples of such models are described.

6.1 Correlations

Numerous investigators have summarized experimental results by simple power-law correlations,[65,66,1] yielding expressions for the percentage of fuel evaporation as a function of pressure, temperature, air velocity, injector characteristics, and distance from the injector. Empirical correlations have also been developed for the rate of spread of the spray.[66] For diesel engines, empirical correlations have been developed to relate the rate of injection and the rate of heat release. Such correlations are limited to the specific engine and injector considered in the study. For the purpose of air-pollution control, a number of correlations for NO emissions from specific engines have been developed. Some success was obtained for those correlations by considering the characteristics of drop evaporation, fuel-and-air mixing rate, and so on, in order to define characteristic times which suggest appropriate forms of correlation for specific operating conditions.

6.2 Droplet Ballistic Models

In this type of model, the ambient gas temperature and velocity are assumed to be constant, and the effect of entrainment and cooling of the ambient gases by the spray is ignored. The spray characteristic is solely determined by processes associated with individual droplets. A more detailed description of this type of model is given in Section 4.3. As can be seen clearly from the definition of the distribution function f_j and the transport equation of f_j, polydisperse injected sprays are conveniently handled using Eq. (8-19) derived by Williams.[39,13] An example of this type of model is the pioneering analysis of Probert[55] mentioned in Section 4.4, in which the evolution of the size distribution for a drop was considered under the assumption of no relative motion between the ambient gas and fuel droplets. The effects of convection and drag for monodisperse injected sprays have been studied by Natarajan and Ghosh.[67] Westbrook[68] recently developed a numerical solution technique for the spray equation (8-19) and applied it to thin sprays injected into a combustion chamber. The effects of independent variations of a large number of system parameters were studied. These parameters include initial spray dispersion, amount and type of gas swirl, gas density, injection timing, chamber geometry, initial droplet size distribution, injection velocity, drag coefficient, vaporization rate, injector aperture size, droplet specific gravity, and direction of injection. The calculations were made for three-dimensional and transient conditions, neglecting the source terms Q_j and Γ_j in Eq. (8-19). While this is an impressive calculation, neglecting the effect of the spray in modifying gas temperatures and velocities limits the application of the results to general spray combustion problems.[1] Bracco[69] used this type of model to study the spray combustion of ethanol drops in oxygen in a rocket motor of constant cross-sectional area. His model adopts a Nukiyama–Tanasawa distribution function of initial drop radii, a Stokes drag equation, and either a modified Priem–Heidmann[70] or a modified Spalding[71] drop vaporization-rate equation. Bracco found that his calculated results reproduced accurately the steady-state data on the engine tested. The Priem–Heidmann and Spalding vaporization-rate equations, which were suggested for single fuel droplets vaporizing and burning in infinite oxidizing media, were found to overestimate the vaporization rate of the droplets within the spray.

6.3 One-Dimensional Models

In one-dimensional models, interactions between liquid and gas phases are considered; however, the complexities of droplet diffusion in turbulent gas streams are neglected. As mentioned before, for liquid-fuel rocket engines with fine sprays, one-dimensional spray combustion models are broadly accepted.

In their studies of spray drying processes, Dickinson and Marshall[72] used a one-dimensional model to determine the effect of the spray size distribution function. They assumed that the effects of drag and droplet evaporation on gas velocity are negligible and the droplet temperature was constant. In a later study, Law[73] improved this analysis by allowing for gas velocity and temperature changes due to evaporation.

In a simulation of droplet evaporation in a Wankel engine, Bracco[74] solved the transient, one-dimensional spray equation together with the gas-phase energy equation, assuming no spatial pressure gradients. In his analysis, convective heating of droplets during evaporation was considered. However, the effect of transient heating of droplets was neglected. None of the above models have been critically evaluated by comparison with experimental data. In order to study the combustion efficiency of a variety of rocket engines, Priem and Heidmann[70] developed a one-dimensional model assuming constant gas temperature. Droplet heatup and drag were considered, but burning around individual droplets was neglected. Although droplet shattering was not considered in the model, the condition for droplet breakup was found when a critical Weber number We_c was exceeded. Mador and Roberts[75] applied the stream-tube approach to the analysis of gas-turbine combustors. In their analysis, exchange of mass and heat by turbulent mixing is allowed. Although the model gave some encouraging results in the prediction of exhaust emissions, the validity of the model for predicting other characteristics of spray combustion needs to be examined.

According to Faeth's detailed review,[1] one-dimensional models are comprehensive but still involve a substantial amount of empiricism. Experience with these models has led to optimized selection of physical properties and empirical parameters. These models can be helpful in design, but this in no way implies that one-dimensional models are satisfactory in all respects for predicting spray combustion processes.

6.4 Stirred-Reactor Models

In order to assist in the design of certain combustion chambers for the evaporation and burning of fuel droplets, some engineers have employed simplified analyses based upon the concept of stirred reactors. In this approach, the treatment of recirculating flow patterns and many detailed two-phase reacting-flow phenomena are significantly simplified by considering the so-called "well-stirred" or "plug-flow" reactors. Swithenbank et al.[76] developed a model of this type based upon interconnected and partially stirred reactors, whose individual performance was computed utilizing energy-balance principles. Their model was developed to predict performance variables such as blowoff stability limits, combustion efficiency, combustion intensity, and overall pressure loss. Turbulence levels within the combustor were also calculated to determine the noise output, ignition condition, and heat transfer rates.

Satisfactory agreement was obtained between the predicted and measured overall stability loops from a high-intensity gas-turbine-type combustor.[76] Courtney[77] applied a stirred-reactor analysis to study rocket-engine combustion processes. Munz and Eisenklam[78] modeled combustion processes in a high-intensity spray combustion chamber in terms of the flow configuration of stirred and plug-flow reactor elements. The evaporation of a polysize spray followed by homogeneous pyrolysis and complex chemical reactions was calculated in order to determine the trends in NO emission.

Although various researchers and designers have used this type of model as a tool to guide their combustion-chamber design, the concept and application of stirred reactors have severe limitations. The diffusion processes of droplets and gaseous species are strongly coupled with the combustion processes. Also, the fuel vapors from the droplet surfaces have to mix well with the oxidizer before significant chemical reaction can occur. Therefore, the application of this type of model must be carefully examined before it is used.

6.5 Locally-Homogeneous-Flow Models

In an earlier work, Thring and Newby[79] recommended that the length of a turbulent spray flame could be estimated by assuming locally homogeneous flow (LHF), neglecting the slip effect between the condensed and gas phases. Under this assumption, the two phases are assumed to be in dynamic and thermodynamic equilibrium, that is, at each point in the flow, they have the same velocity and temperature and are in phase equilibrium. The spray is essentially equivalent to a gas jet having the same momentum and stoichiometric conditions. It is useful to note that the LHF condition may be regarded as the limiting case of a spray consisting of infinitely small droplets. Avery and Faeth[80] applied the LHF model to study the combustion of halogen gases injected into a molten alkali-metal bath. The integral method was employed for the jet, using a variable-density entrainment parameter. Since bubbles quickly assume local liquid velocities and temperatures, the LHF model was successful in this case, and the flame lengths in this heterogeneous bubbly flow situation exhibited the same characteristics as homogeneous gas-jet flames. Velocities and temperature levels were also predicted satisfactorily in the region downstream from the flame.

In their study of evaporation of liquid sprays at supercritical conditions, Newman and Brzustowski[81] employed the locally homogeneous flow assumption in the analysis. Since liquid and gas densities are similar under supercritical conditions, the LHF assumption is reasonable. Atomization, mean velocities, and temperatures were predicted quite well. Khalil and Whitelaw[61] used a second-order turbulence closure and LHF approximation in studying kerosene spray flames. As discussed in Section 5, they found their solutions not completely satisfactory even for small spray droplets (SMD $\approx$ 45 μm). More extensive discussion of LHF models is given in Section 7.

6.6 Separated-Flow (Two-Phase-Flow) Models

This type of model is the most logical approach in studying spray combustion, since effects of finite rates of transport between the two phases are included in the analysis. A systematic development of turbulence models for two-phase flows with the consideration of concentration, temperature, and velocity gradients has just begun in the last few years. The development of separated-flow (SF) models is of great importance, since LHF models are severely limited to the condition when the drops in the spray are extremely small. Various SF models have been proposed; some have been solved, others are being solved. More extensive discussion of SF models is given in Section 8, following the detailed discussion of LHF models in the next section.

7 LOCALLY-HOMOGENEOUS-FLOW MODELS

The basic premise of locally-homogeneous-flow (LHF) models is that rates of transport between the phases are fast compared to the rate of development of the flow field as a whole. This approximation requires that at each point in the flow, all phases have the same velocity and are in thermodynamic equilibrium. The LHF requirement is most easily met by a gas phase dispersed in a continuous liquid phase, due to the relatively low inertia and thermal capacity of bubbles, as in the experiments of Avery and Faeth.[80] For sprays (dispersed liquid phase and continuous gaseous phase), the LHF approximation is most appropriate when the drop sizes are small, the densities of the phases are almost the same, and the rate of development of the process itself is slow.[47]

The justification of LHF models for some spray combustion processes is given by the observations made by Onuma et al.,[62,63] Styles and Chigier,[64] and Komiyama et al.[82] They showed that there exist striking similarities between the structure of flames fueled with gases and with well-atomized sprays having maximum drop number densities for $10 < \mathrm{SMD} < 20\ \mu\mathrm{m}$.

There are several advantages of LHF models. First, they require minimum information concerning injector characteristics, since initial drop size and velocity distributions play no role in the computations. It is well known that injector properties are difficult to obtain: the liquid jet breakup and droplet formation processes in the dense spray are still far from being understood. The second advantage of LHF models is the saving in computer time. Computations for sprays are nearly identical to those for single-phase flows. A third advantage is that LHF models require far fewer empirical constants than SF models.

According to Faeth,[47] LHF models provide a reasonable first estimate of the extent and character of a spray process. They are useful in giving a lower bound on the size of the spray. LHF models can also provide an indication of potential process improvements by improving atomization, prior to testing.

7.1 Classification of LHF Models

There are various kinds of LHF models of spray combustion; some are formulated by integral approaches (e.g., those of Thring and Newby,[79] Newman and Brzustowski,[81] Shearer and Faeth[83]), and some employ higher-order turbulence models calibrated using measurements in combusting and noncombusting gas flows (e.g., Khalil and Whitelaw;[61] Khalil;[84] Shearer, Tamura, and Faeth;[85] Mao et al.[86,87]). A summary of LHF models of spray has been compiled by Faeth[47] and is given in Table 8.5.

According to Reynolds,[88] besides integral models, there are two major classes of turbulence models: full-field modeling (FFM) and large-eddy simulation (LES).[47] The FFM method uses partial differential equations to describe the change in certain averaged quantities. These average quantities or variables can be grouped into two categories:

a. Mean flow properties, such as the velocity, mixture fraction, and temperature
b. Turbulence parameters, such as the turbulence kinetic energy k, dissipation rate ε, square of the mixture-fraction fluctuation g, and turbulent stress components

As described in Chapter 7, models must be constructed for various terms appearing in the governing equations for turbulence quantities, due to the requirement of turbulence closure. When these models are constructed, contributions from all scales of turbulent motion must be considered.[47] This presents some difficulties, since large-scale aspects of turbulent flows are frequently anisotropic, while processes at small scales approach isotropy.

The LES approach involves completing calculations of the time-dependent, three-dimensional structure of the turbulent flow with specified initial conditions to reflect randomness. In this approach, modeling is only necessary for turbulence scales smaller than the computational grid spacing. It is well known that the small scale of turbulence is nearly universal in character and can be modeled more reliably than situations where the full range of turbulence scales must be modeled.[88] As clearly noted by Faeth,[47] "While it appears that LES approaches have the great potential to remove many of the uncertainties of turbulence modeling, this approach is not so well developed as full-field models." At the present time, LES solutions generally require substantial computation times for practical problems. With projected computer development, routine use of LES models should be possible within a decade.[88] For the time being, spray models have generally employed the FFM approach.

It is useful to point out that either Reynolds (time) averaging or Favre (mass-weighted) averaging can be used for FFM spray models. The latter is even more suitable for compressible flows. Detailed discussions for these

TABLE 8.5 Summary of Locally-Homogeneous-Flow Models of Sprays[a]

Date	Reference	Flow configuration[b]	Model[c]	Experiment[d]	Assessment
1953	Thring and Newby[79]	Axisymmetric, boundary-layer combustion	Integral, EQ, parabolic	Combustion of steam-atomized spray in air, 0.1 MPa	Qualitative agreement for flame length
1971	Newman and Brzustowski[81]	Axisymmetric, boundary-layer evaporation	Integral, EQ, parabolic	Evaporation of pressure-atomized liquid near its critical point, 6–9 MPa	Good agreement for spray boundary
1977	Shearer and Faeth[83]	Axisymmetric, boundary-layer combustion	Integral, EQ, parabolic	Combustion of pressure-atomized spray, no swirl, still air, SMD = 30 μm (est.), 0.1–9 MPa	Spray and flame boundaries underestimated by 30–50% poor estimation of flow width
1977	Khalil and Whitelaw[61]	Axisymmetric, swirling flow with recirculation and combustion	k–ε–g, EQP, elliptic	Combustion of a pressure-atomized spray, swirl, SMD = 45, 100 μm, 0.1 MPa	Rate of development of process overestimated
1978	Khalil[84]	Axisymmetric, swirling flow with recirculation and combustion	k–ε–g, MEBU, elliptic	Combustion of a pressure-atomized spray, swirl, SMD = 45, 100 μm, 0.1 MPa	Improved prediction in some cases
1979	Shearer et al.[85]	Axisymmetric, boundary-layer evaporation	k–ε–g, EQP, parabolic	Evaporation of air-atomized spray, no swirl, in still air, SMD = 29 μm, 0.1 MPa	Rate of development of process overestimated, predicted mean velocities and mixture fraction 20–40% below measurements near the injector
1980	Mao et al.[86]	Axisymmetric, boundary-layer combustion	k–ε–g, EQP, parabolic	Combustion of air-atomized spray, no swirl, in still air, SMD = 35 μm, 0.1 MPa	Rate of development of process overestimated, flame length underestimated by 20%
1980	Mao et al.[87]	Axisymmetric, boundary-layer combustion	k–ε–g, EQP, parabolic	Combustion of pressure-atomized spray, no swirl, in still air, SMD = 30 μm (est.), 3–9 MPa	Rate of development of process overestimated, spray lengths underestimated by 20%

[a] After Faeth.[47]
[b] All cases shown here are steady.
[c] EQ implies local thermal equilibrium, EBU implies use of eddy-breakup model to estimate fuel concentration in conjunction with local thermal equilibrium, MEBU is the same as EBU except for an empirical modification for drop combustion, EQP implies use of local thermal equilibrium with pdf for mean properties.
[d] Pressures are those of the spray environment.

averaging procedures are given in Chapter 7. In general, when Reynolds averaging is used, the governing equations can only reach the Favre-averaged form when a number of terms involving density fluctuations, which may not be negligible, are ignored. The solution of Reynolds-averaged equations then corresponds to a Favre-averaged solution, without computation of mass averages by solution of equations involving correlations of density and other variables.[88,47] Since this is only one of many approximations in turbulent-flow models, Reynolds averaging has generally been adopted for sprays due to its computational convenience, even though either method could be applied.[47] Therefore, the following discussions will be limited to the Reynolds-averaged model equations.

7.2 Mathematical Formulation of LHF Models

There are several major components of the theoretical model, including basic assumptions, equation of state, conservation equations, turbulent transport equations, boundary conditions, and physical input correlations and constants.

7.2.1 Basic Assumptions

The following assumptions are usually made for LHF models, to yield a simple form of the theoretical formulation:

1. Transport coefficients of all species and heat are the same ($\mathscr{D}_t = \alpha_t = \nu_t$).
2. The combustion process is adiabatic.
3. Radiation, viscous dissipation, and kinetic energy are negligible.
4. Molecular rates of reaction are infinitely fast, so local thermodynamic equilibrium is maintained.
5. The mean flow is steady and axisymmetric.
6. Due to the computational convenience and extensive validation of the k–ε two-equation model, the following isotropic-turbulent-viscosity expression is assumed to be valid:

$$\mu_t = C_\mu \bar{\rho} \frac{k^2}{\varepsilon} \tag{8-30}$$

The assumption that species diffusion coefficients are the same implies equality of both the molecular and the turbulent components of the diffusivities. Local thermodynamic equilibrium, when combined with the assumption of equal species diffusion coefficients, implies that the local state of the mixture is completely specified by the pressure, the velocity, the mixture fraction, and the temperature or total enthalpy. The equal diffusion coefficient assumption, when coupled with the assumption that radiation, viscous dissipation, and kinetic energy can be neglected, implies that the local state of the mixture can be completely specified by the pressure and mixture fraction alone.

7.2.2 Equation of State

The relationships between mixture enthalpy, composition, temperature, density, and mixture fraction can be regarded as equations of state. Each type of spray considered requires a separate equation of state.

Under low enough pressures, all gases may be assumed to be ideal gases.

Considering N species in the flow, the mass fraction for the ith species in the mixture can be written as

$$Y_i = Y_{i0} f + Y_{i\infty}(1 - f), \qquad i = 1, \ldots, N \tag{8-31}$$

where subscripts 0 and ∞ designate the injector-exit and far-field conditions, respectively. Since each species may exist in gaseous and/or liquid state, we have

$$Y_i = Y_{fi} + Y_{gi} \tag{8-32}$$

The enthalpy of the mixture can be expressed as

$$h = h_0 f + h_\infty (1 - f) \tag{8-33}$$

In terms of mass fractions of all species in their possible states, the enthalpy can also be written as

$$h = \sum_N \left(Y_{fi} h_{fi} + Y_{gi} h_{gi} \right), \qquad i = 1, \ldots, N \tag{8-34}$$

When the mixture temperature used is at the appropriate value, these two enthalpy expressions should yield equal values. The density of the mixture can be given as

$$\rho = \sum_N \left(Y_{fi} v_{fi} + Y_{gi} v_{gi} \right)^{-1}, \qquad i = 1, \ldots, N \tag{8-35}$$

where v_{fi} and v_{gi} are the partial specific volumes of species i in the liquid and gas phases.

For a specified mixture composition, given the relationships between the enthalpy and density of each species, the temperature, and the partial pressure of each species, Eqs. (8-31) through (8-35) are sufficient to describe the composition, temperature, and density of the mixture. The relative mass fractions of the gas and liquid phases of the ith species can be obtained from Eq. (8-32) and the requirement that the chemical potential of each species must be the same in both phases.

Example 8.1

Specify the necessary equation-of-state relationships for a pure air jet injected into a stagnant bath of water at 298 K and 101 kPa at the plane of injection.

Solution:

An equation of state for a two-phase air-water system can be formulated as follows. The spray can be considered isothermal and the air can be assumed to behave as an ideal gas. The effect of water vapor can be considered negligible. Also, one can assume that no air is dissolved in the water. Under the above assumptions, the mass fraction of air and water can be found from Eq. (8-31) as follows:

$$Y_A = f \tag{1}$$

$$Y_W = 1 - f \tag{2}$$

The density of the two-phase mixture in the jet can be obtained from Eq. (8-35)

$$\rho = \left(\frac{f}{\rho_A} + \frac{1-f}{\rho_W} \right)^{-1} \tag{3}$$

where the density of both the water and the air can be assumed to be constant. The air density can be computed assuming that it is an ideal gas. The plot of

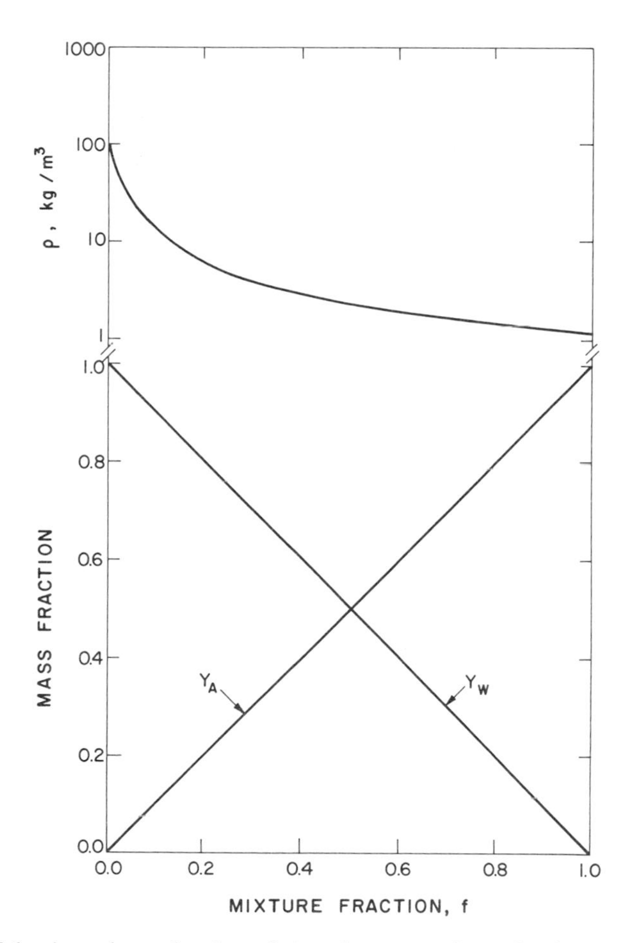

Dependence of density and mass fractions of air and water on mixture fractions (Equation of state for the air–water system).[83]

Y_A, Y_W, and ρ versus mixture fraction f is shown in the accompanying figure.[83] It is interesting to note that the mass fractions of the air and water vary linearly with the mixture fraction. However, the variation of density with mixture fraction is very nonlinear.

Example 8.2

Specify the necessary equation-of-state relationships for air-atomized injection of Freon-11 as an evaporating spray into a stagnant air at 298 K and 101 kPa.

Solution:

The mass fraction of Freon-11 can be determined from Eq. (8-31):

$$Y_F = Y_{F0} f \tag{1}$$

where Y_{F0} is known from the injector exit condition. The mass fraction of air, which is the complementary part, is obviously

$$Y_A = 1 - Y_F \tag{2}$$

The enthalpy of the two-phase mixture can be determined from Eq. (8-33):

$$h = h_0 f + h_{A\infty}(1 - f) \tag{3}$$

The enthalpy of air is given by

$$h_A = C_{pA}(T - T_R) \tag{4}$$

where the reference temperature T_R is 298 K. The enthalpy of Freon-11 in the liquid and gas phases can be expressed as

$$h_{Ff} = C_{p,Ff}(T - T_R) \tag{5}$$

$$h_{Fg} = C_{p,Fg}(T - T_R) + h_{F,fgR} \tag{6}$$

where $h_{F,fgR}$ is the heat of vaporization of Freon-11 at the reference temperature T_R.

The enthalpy at the injector exit is the sum of the contributions due to Freon-11 and the injected air:

$$h_0 = Y_{A0} h_{A0} + Y_{F0} h_{F0} \tag{7}$$

The gas phase consists of both Freon-11 vapor and air. Assuming ideal-gas behavior, the total pressure is equal to the sum of the partial pressures of the individual components:

$$p = \sum p_i = p_A + p_{Fg} \tag{8}$$

When the liquid is present in the spray and is at equilibrium with the vapor, the partial pressure of Freon-11 in the gas phase must equal the vapor pressure. The vapor pressure of Freon-11 can be correlated by the following expression:

$$\log_{10} p_{Fg} = A' - \frac{B'}{T} \tag{9}$$

where $A' = 6.7828$ and $B' = 1416.1$ if T is in K and p is kPa. Then, since mole fractions in the gas phase are proportional to the partial pressures, the mass fraction of Freon-11 in the gas phase is

$$Y_{Fg} = \frac{Y_A p_{Fg} W_F}{p_A W_A (1 - p_{Fg}/p_A)} \tag{10}$$

Then

$$Y_{Ff} = Y_F - Y_{Fg} \tag{11}$$

The enthalpy of the mixture can also be determined from Eq. (8-34):

$$h = Y_A h_A + Y_{Ff} h_{Ff} + Y_{Fg} h_{Fg} \tag{12}$$

As long as the mixture is saturated, the mixture temperature and composition can be calculated at any mixture fraction using Eqs. (1)–(12). A temperature is estimated and the partial pressure of Freon vapor is determined from Eq. (9). The mass fractions and mixture enthalpy are then calculated from Eqs. (10)–(12). The enthalpy computed in this manner is compared with the enthalpy determined directly from Eq. (3). An interval-halving procedure can be employed to adjust the temperature until the two calculated values of the mixture enthalpy agree.

Once all the liquid has vaporized, $Y_{Ff} = 0$ and the mass fractions of air and Freon-11 vapor can be determined directly from

$$Y_{Fg} = Y_{F0} f \tag{13}$$

$$Y_A = 1 - Y_{Fg} \tag{14}$$

The mixture temperature can then be determined from Eqs. (4), (3), (6), and (12).

The mixture density can be determined from

$$\rho = \left(\frac{Y_{Ff}}{\rho_{Ff}} + \frac{Y_{Fg}}{\rho_{Fg}} + \frac{Y_A}{\rho_A} \right)^{-1} \tag{15}$$

The density of the Freon-11 liquid can be taken as a linear function of temperature. The air and vapor densities can be obtained from the ideal-gas expression at the temperature and total pressure of the mixture. The equation-

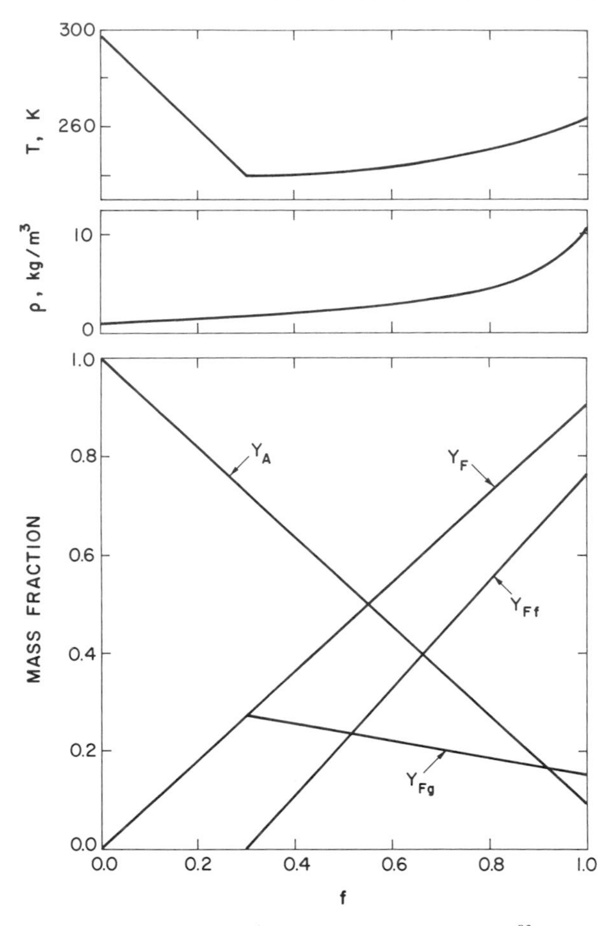

Equation of state for the evaporating Freon-11 spray.[83]

of-state relationships for the evaporating Freon-11 spray in an air-atomized jet into a stagnant air is shown in the accompanying figure (after Shearer and Faeth[83]). The total mass fractions of air and Freon-11 are linear. The presence of liquid, however, causes nonlinear behavior of mixture temperature and density. The physical properties in the computation are

$$W_F = 137.37 \text{ kg/kg-mole}, \qquad C_{p,Fg} = 0.674 \text{ kJ/kg K},$$

$$C_{p,Ff} = 0.879 \text{ kJ/kg K}, \qquad h_{F,fgR} = 171.17 \text{ kJ/kg},$$

$$\rho_{Ff} = 2143.7 - 2.235T \text{ kg/m}^3$$

7.2.3 Conservation Equations

With the equation of state defined for the two-phase mixture, the subsequent LHF model formulation reduces to the conservation equations, which are the same as for single-phase flows. For the case of a spray directed into an infinite stagnant medium, the flow can be assumed axisymmetric and the boundary-layer approximations are applicable. The following conservation equations are limited to cases of steady-state mean flow, with axisymmetric geometries.

Continuity Equation

The generalized form of the continuity equation is

$$\frac{\partial}{\partial x_i}\left(\bar{\rho}\bar{u}_i + \overline{\rho' u_i'}\right) = 0 \tag{8-36}$$

Now, considering an order-of-magnitude analysis for boundary layers where the thickness is very small compared to the axial distance, we assume

$$x, \bar{u}, \bar{\rho} \sim O(1)$$

$$r \sim O(\delta) \tag{8-37}$$

where the symbol O designates the order of magnitude.

The generalized continuity equation (8-36) can be expressed in cylindrical coordinates for the two-dimensional axisymmetric flow as:

$$\overbrace{\frac{\partial}{\partial x}(\bar{\rho}\bar{u})}^{(a)} + \overbrace{\frac{\partial}{\partial x}\left(\overline{\rho' u'}\right)}^{(b)} + \overbrace{\frac{1}{r}\frac{\partial}{\partial r}(r\bar{\rho}\bar{v})}^{(c)} + \overbrace{\frac{1}{r}\frac{\partial}{\partial r}\left(r\overline{\rho' v'}\right)}^{(d)} = 0 \tag{8-38}$$

The Boussinesq approximation can be employed to model the density-velocity correlations appearing in terms (b) and (d):

$$-\bar{\rho}\overline{\rho' u'} = \frac{\mu_t}{\sigma_\rho}\frac{\partial\bar{\rho}}{\partial x} \tag{8-39}$$

$$-\bar{\rho}\overline{\rho' v'} = \frac{\mu_t}{\sigma_\rho}\frac{\partial\bar{\rho}}{\partial r} \tag{8-40}$$

where μ_t is the "turbulent viscosity" and σ_ρ is the turbulent Prandtl/Schmidt number. Substituting Eqs. (8-39) and (8-40) into Eq. (8-38), applying the order-of-magnitude analysis using Eq. (8-37), and setting $\sigma_\rho = 1$, we have

$$\underbrace{\overbrace{\frac{\partial}{\partial x}(\bar{\rho}\bar{u})}^{(a)}}_{1} - \underbrace{\overbrace{\frac{\partial}{\partial x}\left(\frac{\mu_t}{\bar{\rho}}\frac{\partial \bar{\rho}}{\partial x}\right)}^{(b)}}_{\mu_t} + \underbrace{\overbrace{\frac{1}{r}\frac{\partial}{\partial r}(r\bar{\rho}\bar{v})}^{(c)}}_{\bar{v}/\delta} - \underbrace{\overbrace{\frac{1}{r}\frac{\partial}{\partial r}\left(r\frac{\mu_t}{\bar{\rho}}\frac{\partial \bar{\rho}}{\partial r}\right)}^{(d)}}_{\mu_t/\delta^2} = 0 \quad (8\text{-}41)$$

where the order of magnitude of each term is indicated. In order that Eq. (8-41) be nontrivial, terms (c) and (d) must be retained. This implies that

$$\bar{v} \sim 0(\delta) \quad (8\text{-}42)$$

$$\mu_t \sim 0(\delta^2) \quad (8\text{-}43)$$

Therefore term (b) can be neglected with respect to the other terms and Eq. (8-41) reduces to

$$\frac{\partial}{\partial x}(\bar{\rho}\bar{u}) + \frac{1}{r}\frac{\partial}{\partial r}(r\bar{\rho}\bar{v}^\circ) = 0 \quad (8\text{-}44)$$

where

$$\bar{\rho}\bar{v}^\circ = \bar{\rho}\bar{v} + \overline{\rho' v'} \quad (8\text{-}45)$$

The combination of mean and fluctuating terms given by Eq. (8-45) is a convenient formulation for the remainder of the analysis.

Momentum Equation

The time-averaged momentum equations valid for nonuniform property flow in a generalized form may be expressed as (see Hinze[89] for detailed derivation)

$$\overbrace{\left[\overline{\rho u_j} + \left(\overline{\rho' u_j'}\right)\right]\frac{\partial \bar{u}_i}{\partial x_j}}^{(a)} = (\bar{\rho}_\infty - \bar{\rho})g_{ai} - \frac{\partial p}{\partial x_i} - \frac{\partial}{\partial x_j}\left[\overbrace{\bar{\rho}\overline{u_i' u_j'}}^{(b)} + \overbrace{\overline{\rho' u_i' u_j'}}^{(c)} + \overbrace{\bar{u}_j \overline{\rho' u_j'}}^{(d)}\right]$$

$$+ \frac{\partial}{\partial x_j}\left[-\overbrace{\tfrac{2}{3}\bar{\mu}\frac{\partial \bar{u}_i}{\partial x_j}\delta_{ij}}^{(e)} + \overbrace{\bar{\mu}\left(\frac{\partial \bar{u}_i}{\partial x_j} + \frac{\partial \bar{u}_j}{\partial x_i}\right)}^{(f)} - \overbrace{\tfrac{2}{3}\overline{\mu'\frac{\partial u_i'}{\partial x_j}}\delta_{ij}}^{(g)} + \overbrace{\overline{\mu'\left(\frac{\partial u_i'}{\partial x_j} + \frac{\partial u_j'}{\partial x_i}\right)}}^{(h)}\right]$$

(8-46)

All of the terms involving the molecular viscosity can be neglected for a turbulent free-jet flow. For a two-dimensional case in cylindrical coordinates the momentum equation in the axial direction can be expressed as

$$\underbrace{\bar{\rho}\bar{u}\frac{\partial \bar{u}}{\partial x}}_{(a)} + \underbrace{\bar{\rho}\bar{v}\frac{\partial \bar{u}}{\partial r}}_{(b)} = \underbrace{g_{ax}(\bar{\rho}_\infty - \bar{\rho})}_{(c)} - \underbrace{\frac{\partial}{\partial x}(\bar{\rho}\overline{u'u'})}_{(d)}$$

$$- \underbrace{\frac{1}{r}\frac{\partial}{\partial r}(r\bar{\rho}\overline{u'v'})}_{(e)} - \underbrace{\frac{\partial}{\partial x}(\bar{u}\overline{\rho'u'})}_{(f)} - \underbrace{\frac{1}{r}\frac{\partial}{\partial r}(r\bar{v}\overline{\rho'v'})}_{(g)}$$

$$- \underbrace{\frac{\partial}{\partial x}(\overline{\rho'u'u'})}_{(h)} - \underbrace{\frac{1}{r}\frac{\partial}{\partial r}(r\overline{\rho'u'v'})}_{(i)} - \underbrace{\overline{\rho'u'}\frac{\partial \bar{u}}{\partial x}}_{(j)} - \underbrace{\overline{\rho'v'}\frac{\partial \bar{u}}{\partial r}}_{(k)} \quad (8\text{-}47)$$

As before, the Boussinesq approximation [Eqs. (8-39) and (8-40)] can be used to model terms (f), (g), (j), and (k). Gosman et al.[90] suggest the following expressions for the Reynolds-stress terms:

$$\bar{\rho}\overline{u'u'} = -2\mu_t\frac{\partial \bar{u}}{\partial x} + \frac{2}{3}\frac{\mu_t}{\bar{\rho}}\bar{u}\frac{\partial \bar{\rho}}{\partial x} \quad (8\text{-}48)$$

$$\bar{\rho}\overline{u'v'} = \mu_t\left(\frac{\partial \bar{u}}{\partial r} + \frac{\partial \bar{v}}{\partial x}\right) \quad (8\text{-}49)$$

Only the first term on the right-hand side of Eq. (8-48) is significant, so to simplify the formulation we consider it alone. Also, it is obvious that the first term on the right-hand side of Eq. (8-49) is greater than the second, and hence the latter can be dropped. The triple correlation terms (h) and (i) in Eq. (8-47) may not be negligible; however, they are neglected in this analysis for lack of suitable correlating expressions.

From this discussion Eq. (8-47) may be written with the order of magnitude of the terms indicated as follows:

$$\underbrace{\overbrace{\bar{\rho}\bar{u}\frac{\partial \bar{u}}{\partial x}}^{(a)}}_{1} + \underbrace{\overbrace{\bar{\rho}\bar{v}\frac{\partial \bar{u}}{\partial r}}^{(b)}}_{1} = \underbrace{\overbrace{g_{ax}(\bar{\rho}_\infty - \bar{\rho})}^{(c)}}_{1} + \underbrace{\overbrace{2\frac{\partial}{\partial x}\left(\mu_t\frac{\partial \bar{u}}{\partial x}\right)}^{(d)}}_{\delta^2}$$

$$+ \underbrace{\overbrace{\frac{1}{r}\frac{\partial}{\partial r}\left(r\mu_t\frac{\partial \bar{u}}{\partial r}\right)}^{(e)}}_{1} + \underbrace{\overbrace{\frac{\partial}{\partial x}\left(\bar{u}\frac{\mu_t}{\sigma_\rho}\frac{\partial \bar{\rho}}{\partial x}\right)}^{(f)}}_{\delta^2} + \underbrace{\overbrace{\frac{1}{r}\frac{\partial}{\partial r}\left(r\bar{v}\frac{\mu_t}{\sigma_\rho}\frac{\partial \bar{\rho}}{\partial r}\right)}^{(g)}}_{\delta}$$

$$+ \underbrace{\overbrace{\frac{\mu_t}{\sigma_\rho}\frac{\partial \bar{\rho}}{\partial x}\frac{\partial \bar{u}}{\partial x}}^{(j)}}_{\delta^2} + \underbrace{\overbrace{\frac{\mu_t}{\sigma_\rho}\frac{\partial \bar{\rho}}{\partial r}\frac{\partial \bar{u}}{\partial r}}^{(k)}}_{1} \quad (8\text{-}50)$$

Due to their higher order, terms (d), (f), (g), and (j) may be eliminated, and the equation is reduced to the form

$$\bar{\rho}\bar{u}\frac{\partial \bar{u}}{\partial x} + \bar{\rho}\bar{v}^{\circ}\frac{\partial \bar{u}}{\partial r} - \frac{1}{r}\frac{\partial}{\partial r}\left(r\mu_t\frac{\partial \bar{u}}{\partial r}\right) = g_{ax}(\bar{\rho}_{\infty} - \bar{\rho}) \tag{8-51}$$

Conservation of Mixture Fraction

The mixture fraction f was defined in Chapter 1 so that it has a zero value in a pure oxidant stream and unity in a pure fuel stream. The balance equation for f may be expressed as[90]

$$\overbrace{\left(\bar{\rho}\bar{u}_j + \overline{\rho' u_j'}\right)\frac{\partial \bar{f}}{\partial x_j}}^{(a)} = -\frac{\partial}{\partial x_j}\left(\overbrace{\bar{\rho}\overline{u_j' f'}}^{(b)} + \overbrace{\overline{\rho' u_j' f'}}^{(c)} + \overbrace{\bar{u}_j\overline{\rho' f'}}^{(d)}\right) + \overbrace{\frac{\partial}{\partial x_j}\left(\rho\mathscr{D}_f\frac{\partial f}{\partial x_j}\right)}^{(e)} \tag{8-52}$$

Neglecting term (e) and putting the equation in cylindrical coordinates for the two-dimensional axisymmetric case gives

$$\overbrace{\bar{\rho}\bar{u}\frac{\partial \bar{f}}{\partial x}}^{(a)} + \overbrace{\bar{\rho}\bar{v}\frac{\partial \bar{f}}{\partial r}}^{(b)} = -\overbrace{\frac{\partial}{\partial x}\left(\bar{\rho}\overline{u'f'}\right)}^{(c)} - \overbrace{\frac{1}{r}\frac{\partial}{\partial r}\left(r\bar{\rho}\overline{v'f'}\right)}^{(d)}$$

$$-\overbrace{\frac{\partial}{\partial x}\left(\bar{u}\overline{\rho' f'}\right)}^{(e)} - \overbrace{\frac{1}{r}\frac{\partial}{\partial r}\left(r\bar{v}\overline{\rho' f'}\right)}^{(f)} - \overbrace{\frac{\partial}{\partial x}\left(\overline{\rho' u' f'}\right)}^{(g)}$$

$$-\overbrace{\frac{1}{r}\frac{\partial}{\partial r}\left(r\overline{\rho' v' f'}\right)}^{(h)} - \overbrace{\overline{\rho' u'}\frac{\partial \bar{f}}{\partial x}}^{(i)} - \overbrace{\overline{\rho' v'}\frac{\partial \bar{f}}{\partial r}}^{(j)} \tag{8-53}$$

The triple correlation terms (g) and (h) are again neglected as in the case of the momentum equation.

It has been shown by Gosman et al.[90] and other investigators that the term $\overline{\rho' f'}$ can be approximated by the following integral:

$$\overline{\rho' f'} = \int_0^1 \left(\frac{d\rho}{df}\right)_{\bar{f}} (f - \bar{f})^2 \mathscr{P}(f)\, df \tag{8-54}$$

where $\mathscr{P}(f)$ is the pdf, which is unknown. They hence chose to ignore this term, as is done in the following analysis. This assumption is usually acceptable. However, for the case of an air jet injected into water, a large value of $(d\rho/df)_{\bar{f}}$ may appear at the edge of the jet (where $\bar{f} \approx 0$), and $\overline{\rho' f'}$ is not negligible. For the sake of simplicity, this term is disregarded.

The Boussinesq approximation for any scalar quantity ϕ can be written as

$$\bar{\rho}\overline{\phi' u'} = \frac{-\mu_t}{\sigma_\phi}\frac{\partial\bar{\phi}}{\partial x} \tag{8-55}$$

and

$$\bar{\rho}\overline{\phi' v'} = \frac{-\mu_t}{\sigma_\phi}\frac{\partial\bar{\phi}}{\partial r} \tag{8-56}$$

which can be used to model terms (c) and (d) of Eq. (8-53), with $\phi = f$. Equations (8-39) and (8-40) can be used to model terms (i) and (j). Therefore, Eq. (8-53) can be rewritten as follows:

$$\underbrace{\overbrace{\bar{\rho}\bar{u}\frac{\partial\bar{f}}{\partial x}}^{(a)}}_{1} + \underbrace{\overbrace{\bar{\rho}\bar{v}\frac{\partial\bar{f}}{\partial r}}^{(b)}}_{1} = \underbrace{\overbrace{\frac{\partial}{\partial x}\left(\frac{\mu_t}{\sigma_f}\frac{\partial\bar{f}}{\partial x}\right)}^{(c)}}_{\delta^2} + \underbrace{\overbrace{\frac{1}{r}\frac{\partial}{\partial r}\left(r\frac{\mu_t}{\sigma_f}\frac{\partial\bar{f}}{\partial r}\right)}^{(d)}}_{1}$$

$$+ \underbrace{\overbrace{\frac{\mu_t}{\sigma_\rho}\frac{\partial\bar{\rho}}{\partial x}\frac{\partial\bar{f}}{\partial x}}^{(i)}}_{\delta^2} + \underbrace{\overbrace{\frac{\mu_t}{\sigma_\rho}\frac{\partial\bar{\rho}}{\partial r}\frac{\partial\bar{f}}{\partial r}}^{(j)}}_{1} \tag{8-57}$$

Neglecting higher-order terms and combining terms (b) and (j), we can express Eq. (8-57) in the form

$$\bar{\rho}\bar{u}\frac{\partial\bar{f}}{\partial x} + \bar{\rho}\bar{v}^\circ\frac{\partial\bar{f}}{\partial r} - \frac{1}{r}\frac{\partial}{\partial r}\left(r\frac{\mu_t}{\sigma_f}\frac{\partial\bar{f}}{\partial r}\right) = 0 \tag{8-58}$$

7.2.4 Turbulent Transport Equations

In order to achieve turbulence closure, several turbulent transport equations must be considered.

Turbulent Kinetic Energy

The generalized form of the turbulent-kinetic-energy equation can be treated in a similar manner to obtain the final form as follows:

$$\bar{\rho}\bar{u}\frac{\partial k}{\partial x} + \bar{\rho}\bar{v}^{\circ}\frac{\partial k}{\partial r} - \frac{1}{r}\frac{\partial}{\partial r}\left(r\frac{\mu_t}{\sigma_k}\frac{\partial k}{\partial r}\right) = \mu_t\left(\frac{\partial \bar{u}}{\partial r}\right)^2 - \bar{\rho}\varepsilon \tag{8-59}$$

Equations for the Dissipation Rate (ε) *and the Square of the Mixture-Fraction Fluctuation (g)*

These conservation equations have been presented in the following fashion by Faeth, after employing a treatment similar to that used for the previously discussed conservation equations:

$$\bar{\rho}\bar{u}\frac{\partial \varepsilon}{\partial x} + \bar{\rho}\bar{v}^{\circ}\frac{\partial \varepsilon}{\partial r} - \frac{1}{r}\frac{\partial}{\partial r}\left(r\frac{\mu_t}{\sigma_\varepsilon}\frac{\partial \varepsilon}{\partial r}\right) = C_{\varepsilon 1}\frac{\varepsilon}{k}\mu_t\left(\frac{\partial \bar{u}}{\partial r}\right)^2 - C_{\varepsilon 2}\bar{\rho}\frac{\varepsilon^2}{k} \tag{8-60}$$

$$\bar{\rho}\bar{u}\frac{\partial g}{\partial x} + \bar{\rho}\bar{v}^{\circ}\frac{\partial g}{\partial r} - \frac{1}{r}\frac{\partial}{\partial r}\left(r\frac{\mu_t}{\sigma_g}\frac{\partial g}{\partial r}\right) = C_{g1}\mu_t\left(\frac{\partial \bar{f}}{\partial r}\right)^2 - C_{g2}\bar{\rho}\frac{\varepsilon g}{k} \tag{8-61}$$

where $C_{\varepsilon 1}$, $C_{\varepsilon 2}$, C_g, and C_{g2} are all constants of the turbulence model.

It can be seen that the final forms of all these conservation equations are similar, and so we can define an operator $D(\phi)$ as follows, where $\phi = f, u, k, \varepsilon$, or g:

$$D(\phi) = \bar{\rho}\bar{u}\frac{\partial \phi}{\partial x} + \bar{\rho}\bar{v}^{\circ}\frac{\partial \phi}{\partial r} - \frac{1}{r}\frac{\partial}{\partial r}\left(r\frac{\mu_t}{\sigma_\phi}\frac{\partial \phi}{\partial r}\right) \tag{8-62}$$

Using this operator the conservation equations can now be represented by

$$\frac{\partial \bar{\rho}\bar{u}}{\partial x} + \frac{1}{r}\frac{\partial}{\partial r}(r\bar{\rho}\bar{v}^{\circ}) = 0 \tag{8-63}$$

$$D(\bar{u}) = g_{ax}(\rho_\infty - \bar{\rho}) \tag{8-64}$$

$$D(\bar{f}) = 0 \tag{8-65}$$

$$D(k) = \mu_t\left(\frac{\partial \bar{u}}{\partial r}\right)^2 - \bar{\rho}\varepsilon \tag{8-66}$$

$$D(\varepsilon) = C_{\varepsilon 1}\frac{\varepsilon}{k}\mu_t\left(\frac{\partial \bar{u}}{\partial r}\right)^2 - C_{\varepsilon 2}\bar{\rho}\frac{\varepsilon^2}{k} \tag{8-67}$$

$$D(g) = C_{g1}\mu_t\left(\frac{\partial \bar{f}}{\partial r}\right)^2 - C_{g2}\rho\frac{\varepsilon g}{k} \tag{8-68}$$

The turbulence constants are assigned the following values:

$$C_\mu = 0.09, \quad C_{\varepsilon 1} = 1.44, \quad C_{g1} = 2.8, \quad \sigma_k = 1.0,$$

$$\sigma_\varepsilon = 1.3, \quad \sigma_{\tilde{f}} = 0.7, \quad \sigma_g = 0.7, \quad \sigma_{\overline{Y}_i} = 0.7.$$

For constant-density flows, $C_{\varepsilon 2} = C_{g2} = 1.89$. For variable-density flows, $C_{\varepsilon 2} = C_{g2} = 1.84$.

7.2.5 Boundary Conditions

The boundary conditions for these equations are

$$r = 0, \qquad \frac{\partial \phi}{\partial r} = 0 \tag{8-69}$$

and

$$r = \infty, \qquad \phi = 0 \tag{8-70}$$

Besides these, several other conditions must be specified to give:

1. The mass flow rate of the material injected
2. Thermodynamic state and thrust at the injector
3. Distributions of k and ε at the injector exit

7.2.6 Solution Procedures

Following Faeth,[47] the governing equations for nonswirling flows under steady and axisymmetric conditions are summarized in Table 8.6. The governing equations in this case can be written in the following general divergence form:

$$\frac{\partial}{\partial x}(\bar{\rho}\bar{u}\phi) + \frac{1}{r}\frac{\partial}{\partial r}(r\bar{\rho}\bar{v}^\circ\phi) = \frac{1}{r}\frac{\partial}{\partial r}\left(r\frac{\mu_t}{\sigma_\phi}\frac{\partial \phi}{\partial r}\right) + S_\phi \tag{8-71}$$

The variable H is the total enthalpy of the mixture including sensible, chemical, and kinetic energies:

$$H = \sum Y_i h_i + \tfrac{1}{2}(u^2 + v^2 + w^2) \tag{8-72}$$

where

$$h_i = \Delta h^\circ_{f,i} + \int_{T_{\text{ref}}}^{T} C_{p,i}\, dT \tag{8-73}$$

the source term S_{rad} represents the radiation contribution to the increase of H. The term R_f in the source term of $\overline{Y}_i$ represents the rate of reaction, which will

TABLE 8.6 Source Terms in Eq. (8-71)[a]

ϕ	S_ϕ
1	0
$\bar{u}$[b]	$\pm a(\rho_\infty - \bar{\rho})$
k	$\mu_t\left(\frac{\partial \bar{u}}{\partial r}\right)^2 - \bar{\rho}\varepsilon$
ε	$\frac{\varepsilon}{k}\left(C_{\varepsilon 1}\mu_t\left(\frac{\partial \bar{u}}{\partial r}\right)^2 - C_{\varepsilon 2}\bar{\rho}\varepsilon\right)$
$\bar{f}$	0
$\overline{H}$[c]	S_{rad}
g	$C_{g1}\mu_t\left(\frac{\partial \bar{f}}{\partial r}\right)^2 - C_{g2}\bar{\rho}\frac{\varepsilon g}{k}$
$\overline{Y}_i$	$C_i R_f$

[a]After Faeth.[47]
[b]The positive sign is used for $S_{\bar{u}}$ for vertical upward flow.
[c]Dilation and shear work terms have been ignored in the equation for $\overline{H}$.

be discussed later. A more general table similar to Table 8.6 was compiled by Faeth;[47] it includes momentum equations in r and θ directions.

In many practical situations, some further simplifications can be made in the solution of the set of governing equations listed in Table 8.6. Three separate cases of interest are defined in Table 8.7. In case 1, local chemical equilibrium with no heat losses is assumed.[61, 85–87] The quantities solved for in the conservation and transport equations are

$$\bar{u}, \bar{v}, \bar{w}, \bar{f}, k, \varepsilon, g$$

Other scalar properties such as $\bar{\rho}$, $\overline{T}$, $\overline{Y}_i$ are found by a stochastic procedure involving the selection of a general form for the pdf of the mixture fraction. Since the specification of the pdf requires information on g, this approach is frequently called the k–ε–g procedure. As mentioned in Chapter 7, the method was initially suggested by Spalding[91, 92] and was subsequently developed and applied to flames by Lockwood and Naguib,[93] Bilger,[94] and many others. Once $\bar{f}$, g, k, and ε are found from the governing equations, the mean value of any scalar property θ can be determined from the integral

$$\bar{\theta} = \int_0^1 \theta(f)\mathscr{P}(f)\,df \tag{8-74}$$

For a given system pressure, $\theta(f)$ represents the known state relationship described in Section 7.2.2. The pdfs employed to date are most often characterized by two parameters, generally associated with the most probable value

TABLE 8.7 LHF-Model Approximations[a]

Additional assumptions	Case 1	Case 2	Case 3
Adiabatic flow	✓	✓	
Negligible radiation	✓	✓	
Low Mach number	✓	✓	
Local chemical equilibrium	✓		✓
	Transport Equations Solved		
Mean quantities	$\bar{u}, \bar{v}, \bar{w}, \bar{f}$	$\bar{u}, \bar{v}, \bar{w}, \bar{f}, \overline{Y}_f$	$\bar{u}, \bar{v}, \bar{w}, \bar{f}, \overline{H}$, radiation transport equation
Turbulent quantities	k, ε, g	k, ε, g or g_{Y_F}	k, ε, g

[a] After Faeth.[47]

(μ) and the variance (σ) of the distribution. As described in Chapter 7, these parameters are often solved implicitly from

$$\bar{f} = \int_0^1 f\mathscr{P}(f)\, df \tag{8-75}$$

and

$$g = \int_0^1 (f - \bar{f})^2 \mathscr{P}(f)\, df \tag{8-76}$$

For case 2 of Table 8.7, the assumption of local chemical equilibrium is relaxed. Because of this relaxation, the state of the mixture can no longer be fixed solely by the mixture fraction f. One additional scalar transport equation must be solved. This equation is usually selected to be the partial differential equation for $\overline{Y}_i$, where i represents an important species involved in the chemical reaction. In order to solve for $\overline{Y}_i$, the rate of reaction of species i must be specified. Spalding's eddy-breakup model of turbulent reaction, described in Chapter 7, was developed precisely to give the reaction rate of fuel species. The reaction rate according to Spalding[95] can be written as

$$R_f = -C_R \bar{\rho} \sqrt{g_{\overline{Y}_F}}\, \frac{\varepsilon}{k} \tag{8-77}$$

where C_R is a constant having a value on the order of unity and $g_{\overline{Y}_F}$ is the square of the fuel mixture-fraction fluctuation.

Along the same line of thought, Magnussen and Hjertager[96] proposed the following expression for R_f, which has been applied to various studies of premixed combustion in furnaces:

$$R_f = \text{Min}\left\{ A\bar{\rho}\overline{Y}_F,\ A\bar{\rho}\frac{\overline{Y}_{O_2}}{(O/F)_{st}},\ A'\frac{\bar{\rho}\overline{Y}_P}{1 + (O/F)_{st}} \right\} \frac{\varepsilon}{k} \tag{8-78}$$

where $A \approx 4$, $A' \approx 2$, and Min represents the minimum value among the group of parameters in the parenthesis. Other expressions for R_f have been proposed by various researchers, including Khalil,[84] Gosman et al.,[97] Borghi,[98] Bray and Moss,[99] and others.

After the solution of the governing equations has been obtained, mean values of various scalar quantities such as $\overline{T}, \bar{\rho}, \overline{Y}_i$ can be computed by stochastic averaging procedures. For chemically nonequilibrium processes (finite-rate chemistry), a parameter describing the degree of reactedness is necessary in addition to the mixture fraction for the construction of a formula to obtain the average values of the above scalar quantities. Lockwood[100] defined the reactedness parameter $\imath$ as

$$\imath \equiv \frac{Y_i - Y_{i,u}}{Y_{i,b} - Y_{i,u}}, \qquad i = F, O, \text{ or } P \tag{8-79}$$

where u and b designate completely unburned and burned conditions. The reactedness parameter was introduced in Chapter 6 (see Fig. 6.9). For a single-step forward chemical reaction, $\imath$ is independent of the major species chosen. It can however be extended to more complicated chemical reactions. Local mean properties depend upon the fluctuations of both f and $\imath$. In order to simplify the model, the f and $\imath$ fluctuations are assumed to be uncorrelated ($\overline{f'\imath'} = 0$). The time-averaged value of a fluid property $\theta(f, \imath)$ at any spatial point is then given by

$$\bar{\theta} = \int_0^1 \int_0^1 \theta(f, \imath) \mathscr{P}(f) \mathscr{P}(\imath) \, df \, d\imath \tag{8-80}$$

where $\mathscr{P}(f)$ is found from $\bar{f}$ and g, and $\mathscr{P}(\imath)$ is determined from $\bar{\imath}$ as follows:

$$\mathscr{P}(\imath) = (1 - \bar{\imath})\delta(\imath) + \bar{\imath}\delta(1 - \imath) \tag{8-81}$$

where δ represents the Dirac δ-function. It should be noted that the above computational procedure to obtain mean scalar properties involves the basic assumption of negligible effect of fluctuations of f and $\imath$. When property variations are not linear in f and $\imath$, this simplification is questionable.

In the third case of LHF models, local chemical equilibrium is assumed but radiation transfer is not negligible. In order to consider radiative heat losses, the equation of radiation transfer must also be solved to yield the source term S_{rad} in the total-enthalpy equation (see Table 8.6). As pointed out by Faeth,[47] a number of approximations must be used in the solution of the radiative-transfer equation since radiative transport involves complicated integrodifferential equations. Readers are referred to the work of Bilger,[94] Gosman et al.,[97] and Elghobashi and Pun.[101]

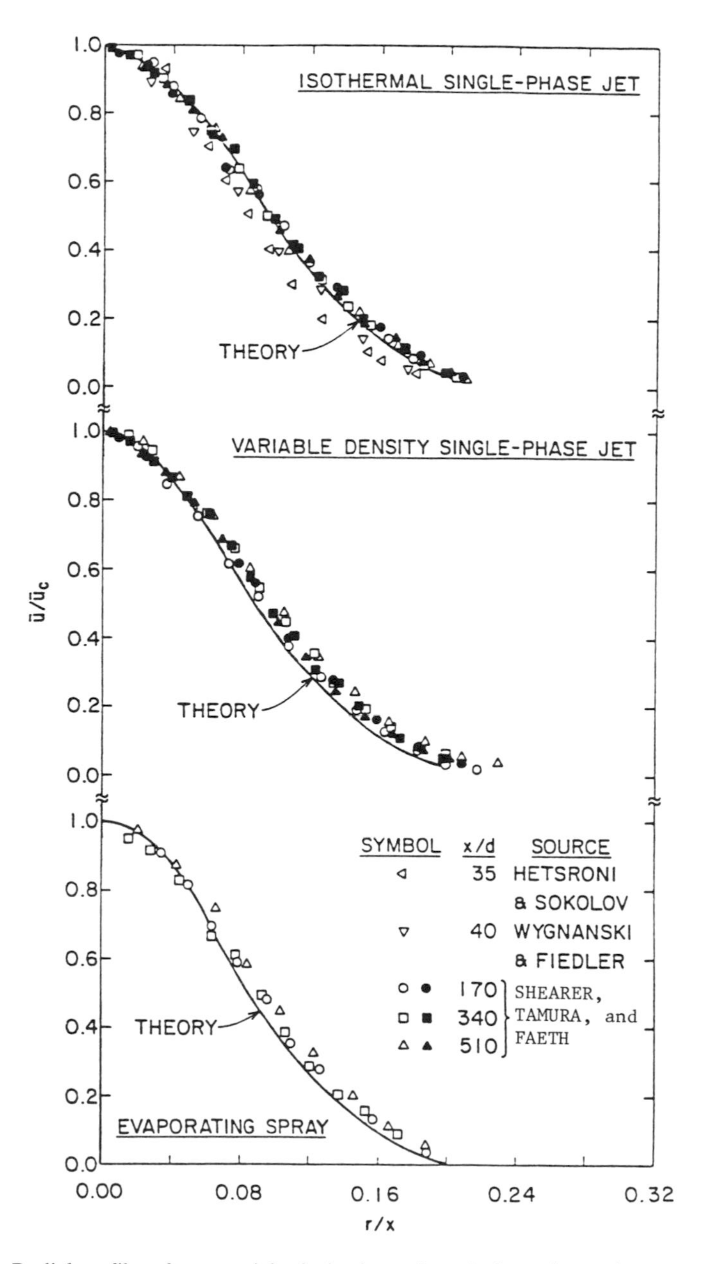

Figure 8.11 Radial profiles of mean axial velocity for various single- and two-phase noncombusting jets (after Faeth[47]).

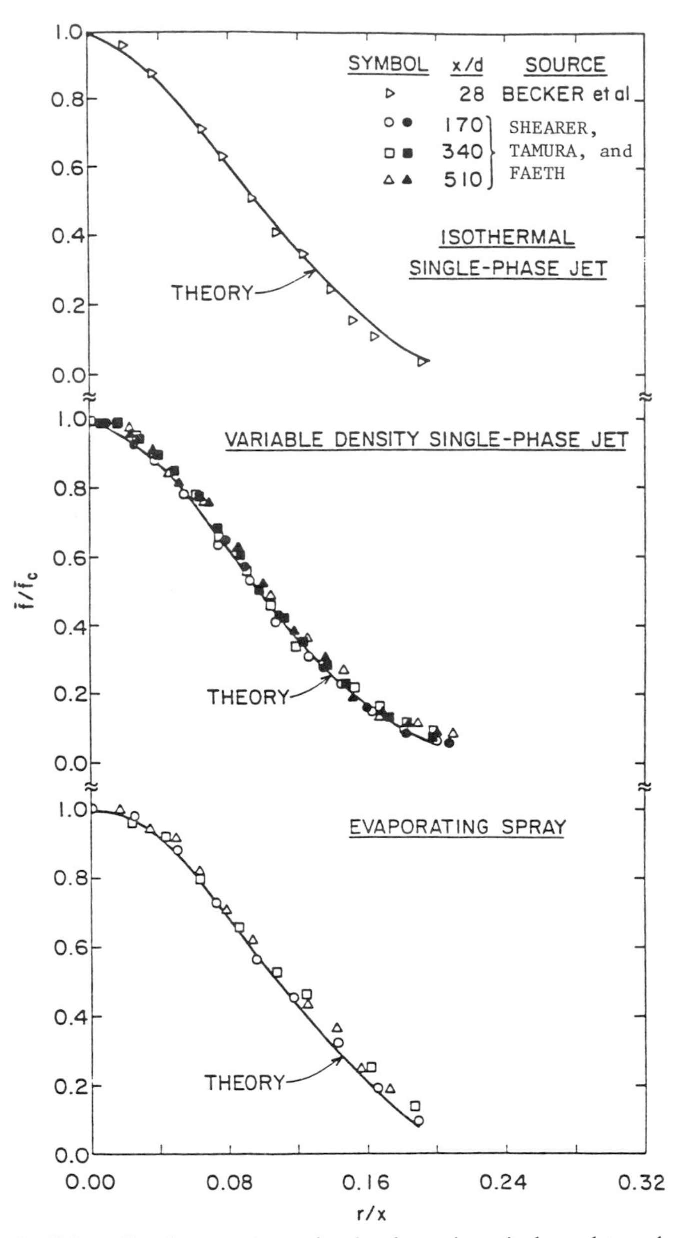

Figure 8.12 Radial profile of mean mixture fraction for various single- and two-phase noncombusting jets (after Faeth[47]).

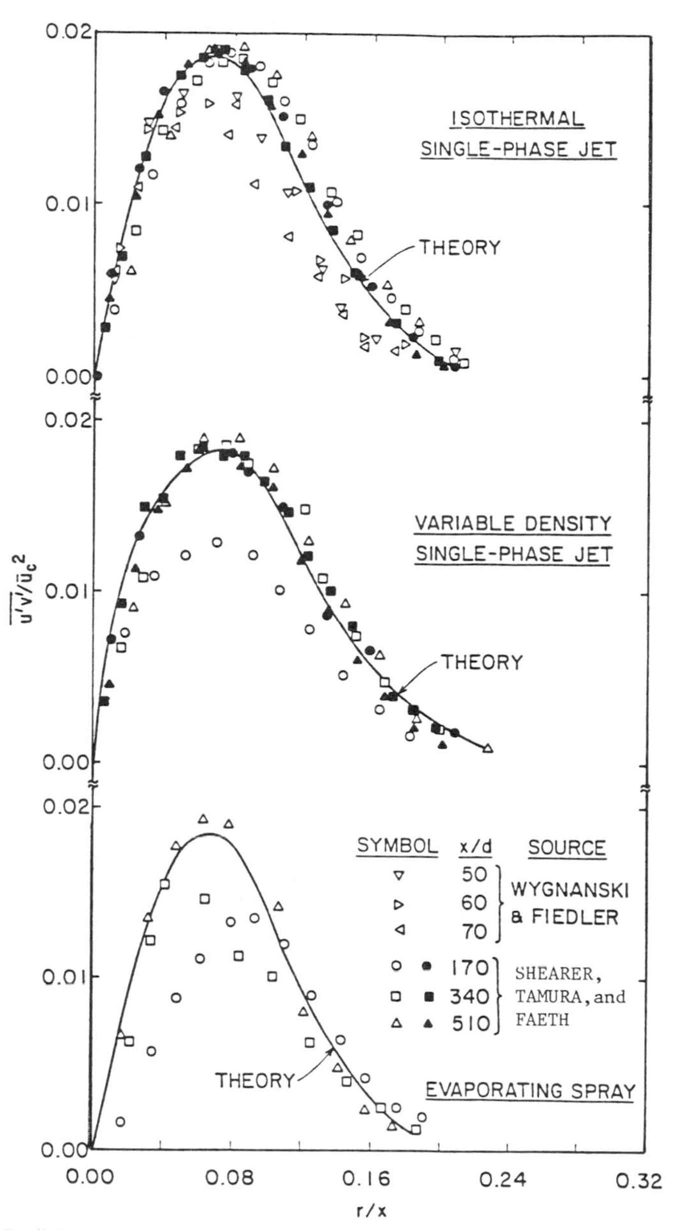

Figure 8.13 Radial profile of Reynolds stress for various single- and two-phase noncombusting jets (after Faeth[47]).

7.2.7 Comparison of LHF-Model Predictions with Experimental Data

LHF models have been compared with several combusting and noncombusting two-phase flows. For noncombusting flows, the model of Shearer, Tamura, and Faeth[85] was compared with a variety of single- and two-phase flows as shown in the following list:

A. *Constant-Density Single-Phase Jet.* Comparison was made with data of Shearer, Tamura, and Faeth,[85] Wygnanski and Fiedler,[102] Hetsroni and Sokolov,[103] and Becker, Hottel, and Williams.[104]

B. *Variable-Density Single-Phase Jet.* Comparison was made with data of Shearer, Tamura, and Faeth[85] and Corrsin and Uberoi.[105]

C. *Air Jet into Water.* Comparison was made with data of Tross.[106]

D. *Evaporating Freon-11 Spray in Air.* Comparison was made with data of Shearer, Tamura, and Faeth.[85]

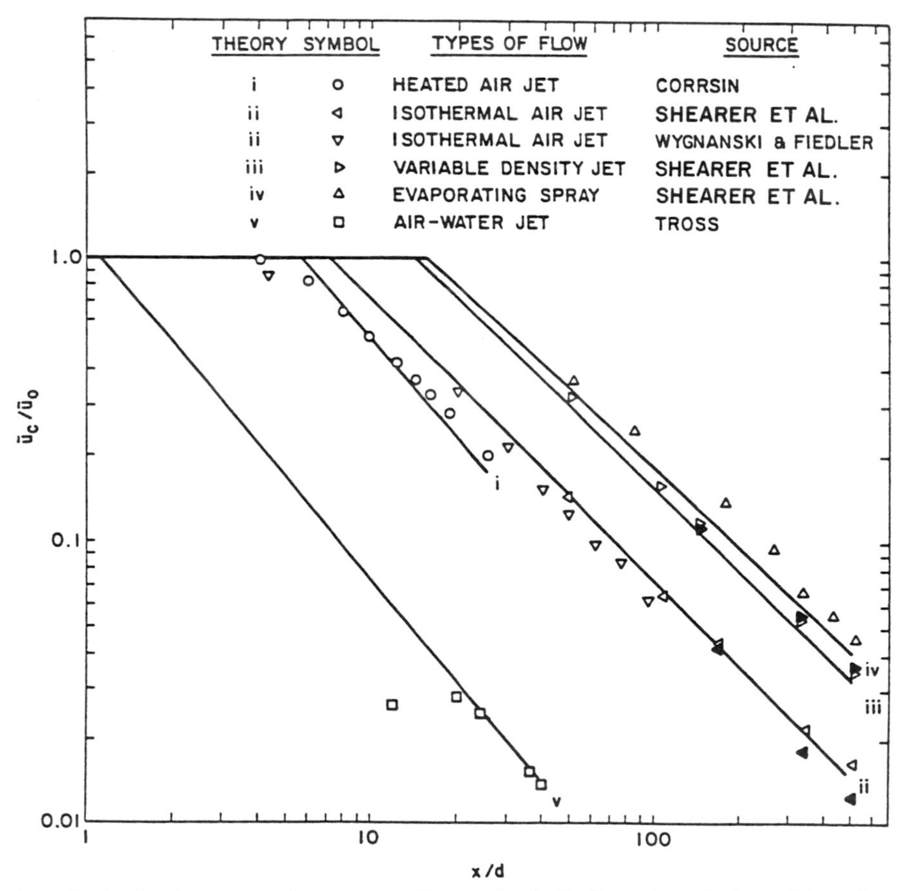

Figure 8.14 Axial variation of mean centerline axial velocity for various single- and two-phase noncombusting jets (after Shearer et al.[85]).

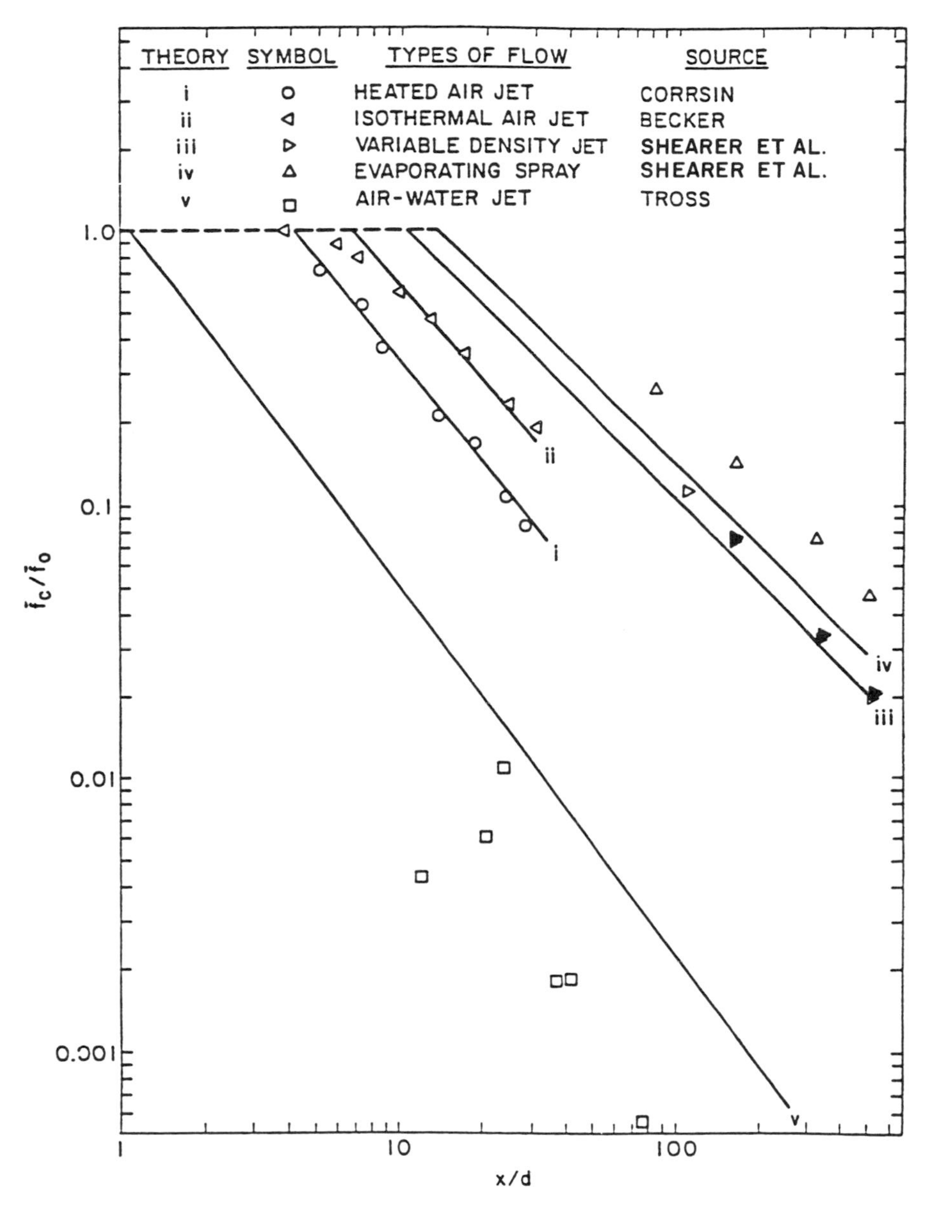

Figure 8.15 Axial variation of mean centerline mixture fraction for various single- and two-phase noncombusting jets (after Shearer et al.[85]).

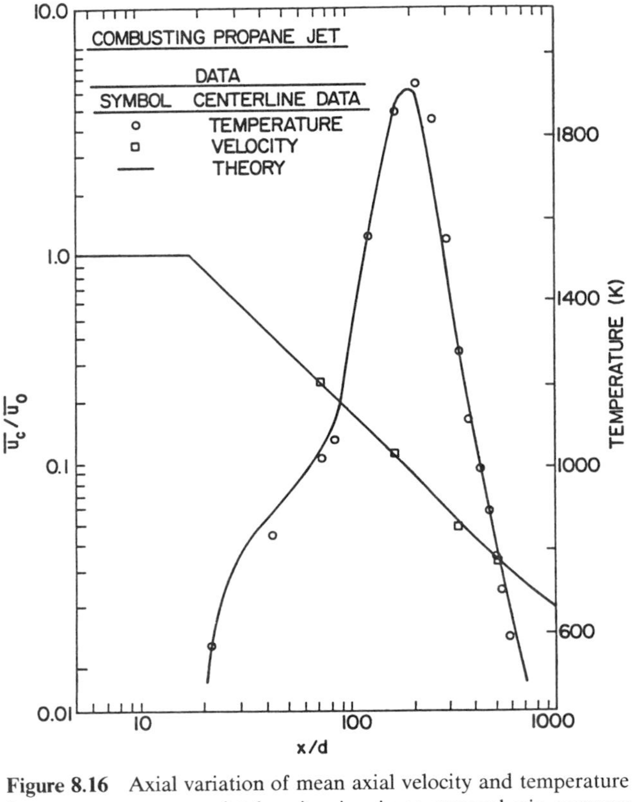

Figure 8.16 Axial variation of mean axial velocity and temperature for an n-propane gas jet burning in air at atmospheric pressure (from Mao et al.[86]).

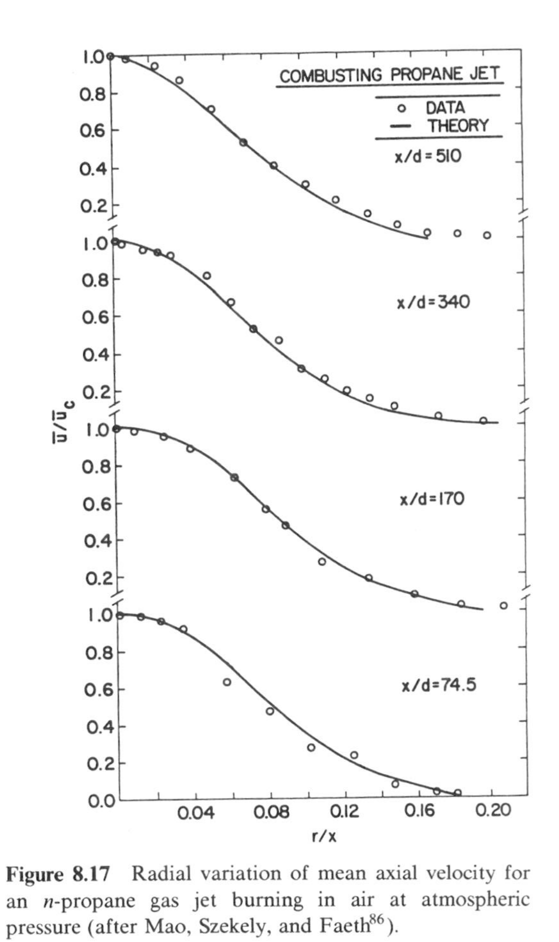

Figure 8.17 Radial variation of mean axial velocity for an n-propane gas jet burning in air at atmospheric pressure (after Mao, Szekely, and Faeth[86]).

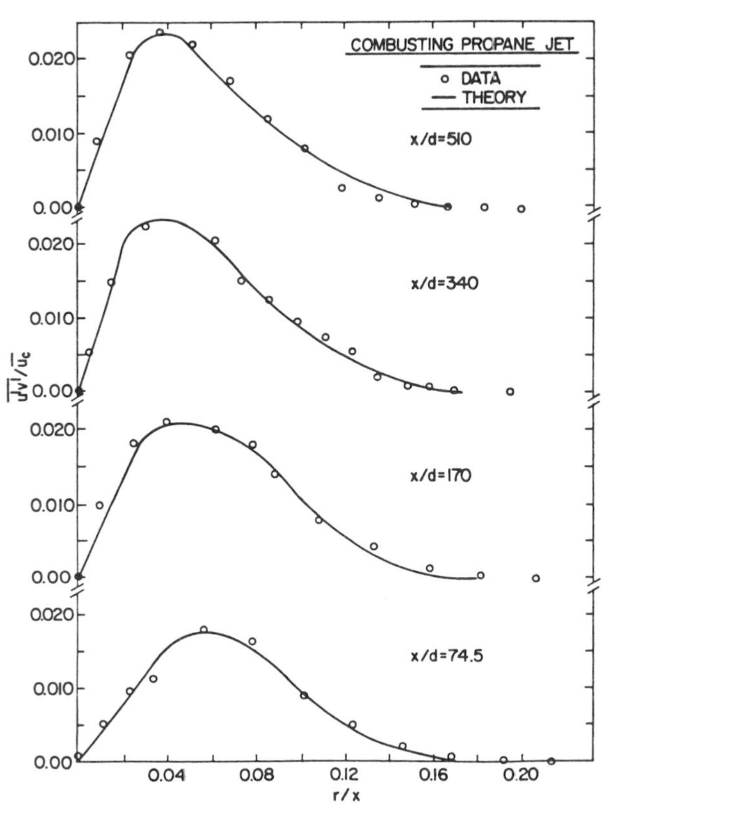

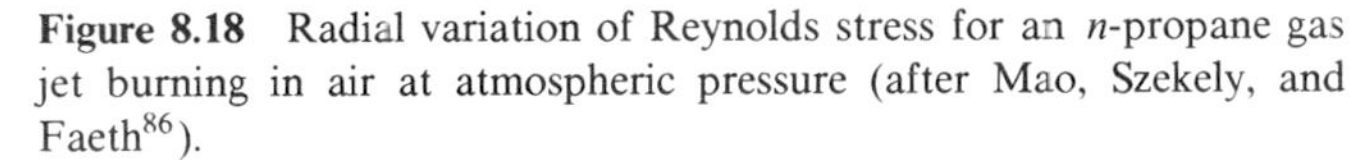

Figure 8.18 Radial variation of Reynolds stress for an n-propane gas jet burning in air at atmospheric pressure (after Mao, Szekely, and Faeth[86]).

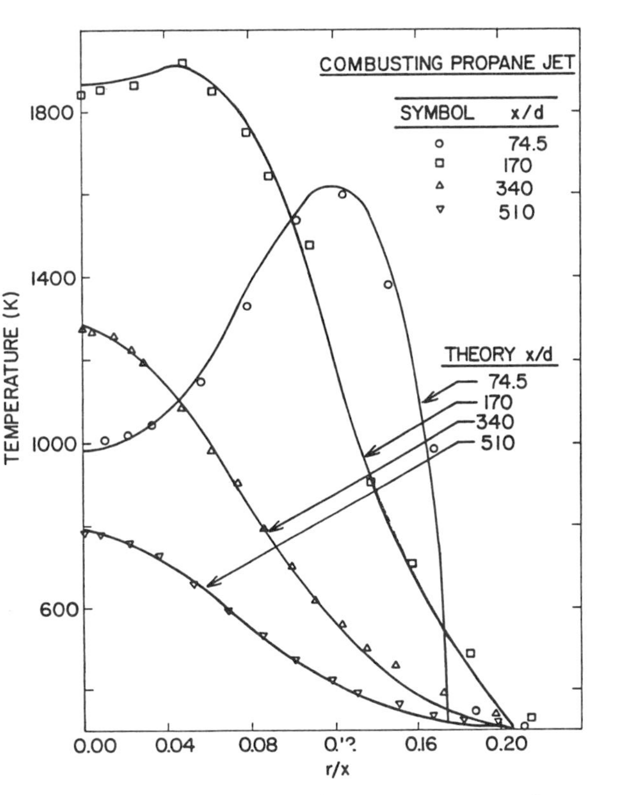

Figure 8.19 Radial variation of mean temperature for an n-propane gas jet burning in air at atmospheric pressure (after Mao, Szekely, and Faeth[86]).

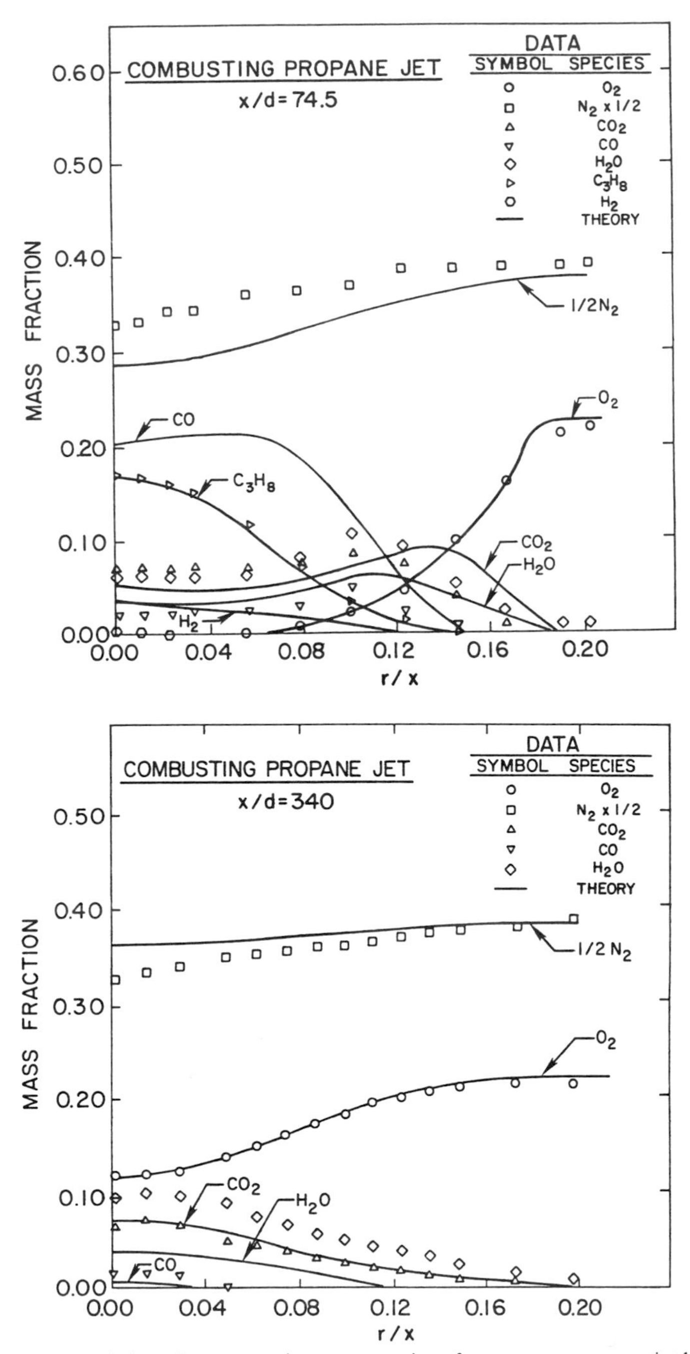

Figure 8.20 Radial variation of mean species concentrations for an *n*-propane gas jet burning in air at atmospheric pressure (from Mao, Szekely and Faeth[86]): (*a*) $x/d = 74.5$, (*b*) $x/d = 340$, (*c*) $x/d = 510$.

It was found[47] that the predicted radial profiles of $\bar{u}/\bar{u}_c$, $\bar{f}/\bar{f}_c$, and $\overline{u'v'}/\bar{u}_c^2$ are all in reasonably good agreement with data (see Figs. 8.11–8.13). The model predicts similarity in these coordinates over the range of data; therefore, only a single theoretical curve is shown in each case. As anticipated, the experimental data and theoretically predicted results on mean axial velocities and mixture fraction along the centerline decrease as the axial distance increases. The comparisons of $\bar{u}_c/\bar{u}_0$ and $\bar{f}_c/\bar{f}_0$ versus x/d between theory and experiments are shown in Figs. 8.14 and 8.15. The data shown in these figures cover a range of ratios of initial jet fluid density to ambient fluid density from 0.0012 to 6.88. As can be seen from these figures, the predicted centerline velocities are 10–20% lower than the measurements, while the predicted centerline mixture fractions are 40% below the measurements.

For combusting sprays, the model of Mao, Szekely, and Faeth[86] was used to predict various combusting sprays listed below:

a. n-Propane gas jet burning in air at 1 atm
b. Air-atomized liquid n-pentane spray burning in air at 1 atm
c. Pressure-atomized liquid n-pentane spray burning in air at pressures of 3–9 MPa

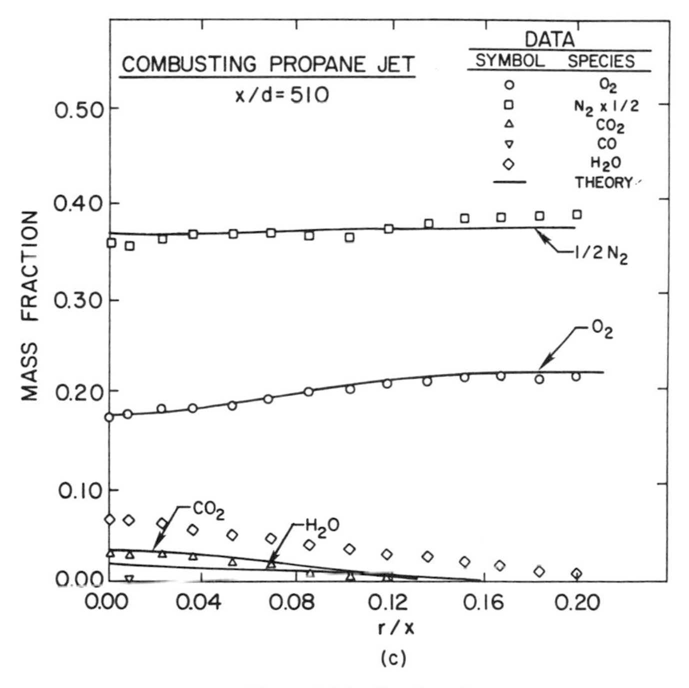

Figure 8.20 Continued.

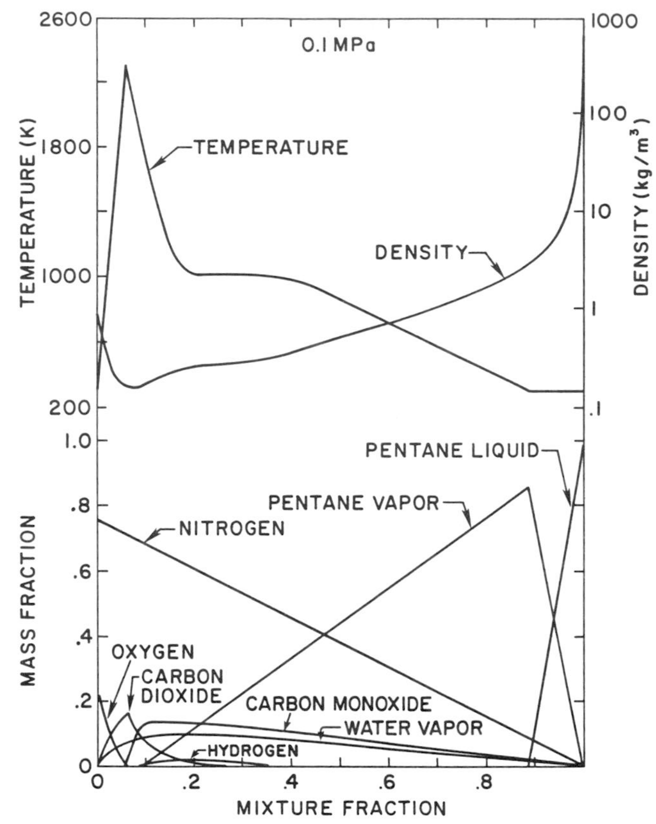

Figure 8.21 Scalar properties as functions of mixture fraction for *n*-pentane spray burning in air at atmospheric pressure (from Mao et al.[86]).

As shown in Fig. 8.16, the agreement between predicted and measured mean axial velocities and temperatures along the centerline is excellent. Predictions of both velocity and Reynolds stress are in good agreement with measurements (see Figs. 8.17 and 8.18). It is interesting to note that due to large density variation in combusting sprays, radial profiles of velocity and Reynolds stress at various axial stations are *not similar*. Predicted radial profiles of $\overline{T}$ at various axial locations are in good agreement with experimental data as shown in Fig. 8.19. As shown in Fig. 8.20*a*–*c*, the predicted species concentration distributions are in reasonable agreement with experimental data. The scalar properties as functions of mixture fraction for *n*-pentane spray burning in air at atmospheric pressure are shown in Fig. 8.21.

Comparison of theoretical results with the data on air-atomized *n*-pentane (liquid) spray burning in air obtained by Mao et al.[86] are given in Figs. 8.22–8.25. Results indicate that:

a. The spray develops more slowly than predicted.
b. The calculated $\overline{T}_{max}$ points are closer to the injector than measured.

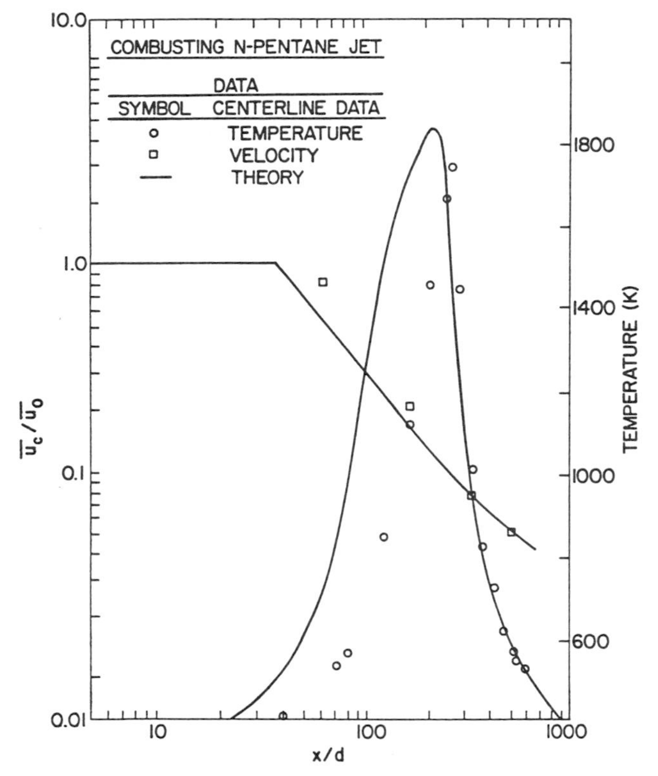

Figure 8.22 Axial variation of mean axial velocity and temperature for an *n*-pentane liquid spray burning in air at atmospheric pressure (from Mao, Szekely, and Faeth[86]).

c. For $x/d > 300$, good agreement was obtained, since it is beyond the region where drops are present.

d. Good agreement is obtained between predictions and measurements for radial profiles of $\bar{u}/\bar{u}_c$ and $\overline{u'v'}/\bar{u}_c^2$.

e. There are large discrepancies between predictions and measurements of the mean temperature near the injector, where the overestimation of the rate of development of the flow is most pronounced.

A comparison of predicted and measured spray boundaries of pressure-atomized liquid *n*-pentane spray burning in air at $p = 3$, 6, and 9 MPa is shown in Fig. 8.26. Both predictions and measurements indicate that the extent of the spray boundary is reduced as the pressure increases. However, the theory overestimates the magnitude of the reduction. Predicted spray lengths are 10–20% less than the measurements.

Khalil and Whitelaw[61] compared their predictions with measurements on an open combusting liquid kerosene spray with swirl. The results, given in Figs.

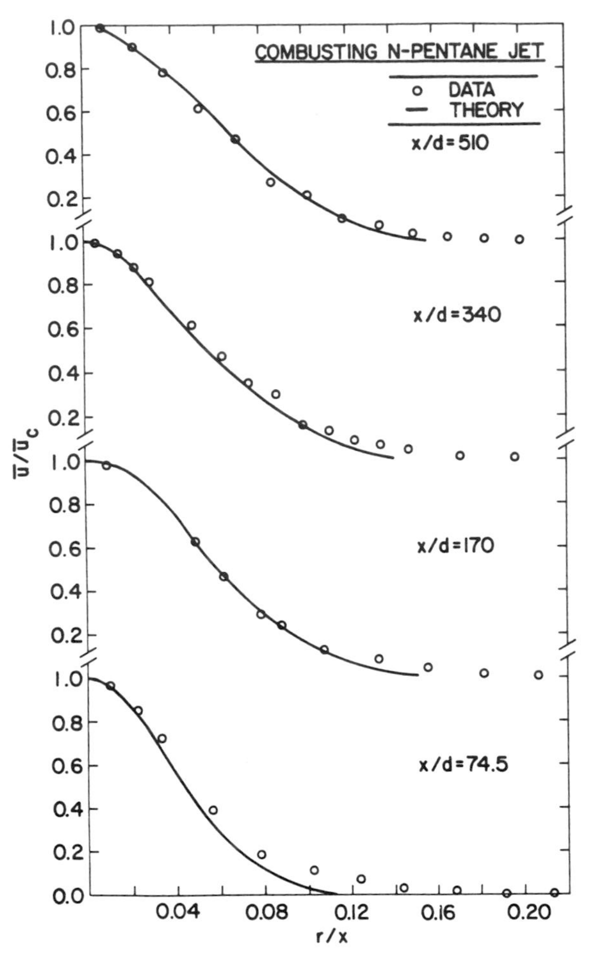

Figure 8.23 Radial variation of mean axial velocity for a spray of liquid n-pentane burning in air at atmospheric pressure (from Mao et al.[86]).

8.6 and 8.7, show that:

a. The $\bar{u}$ distributions along the centerline of the spray show good agreement between the prediction and measurements.

b. The maximum-temperature location along the centerline is closer to the injector than measured. This indicates the overestimation by the LHF model of the development of the flow.

c. The results for the spray flame having an SMD of 45 μm are much better approximated by the theory than the results for the flame having a larger SMD of 100 μm.

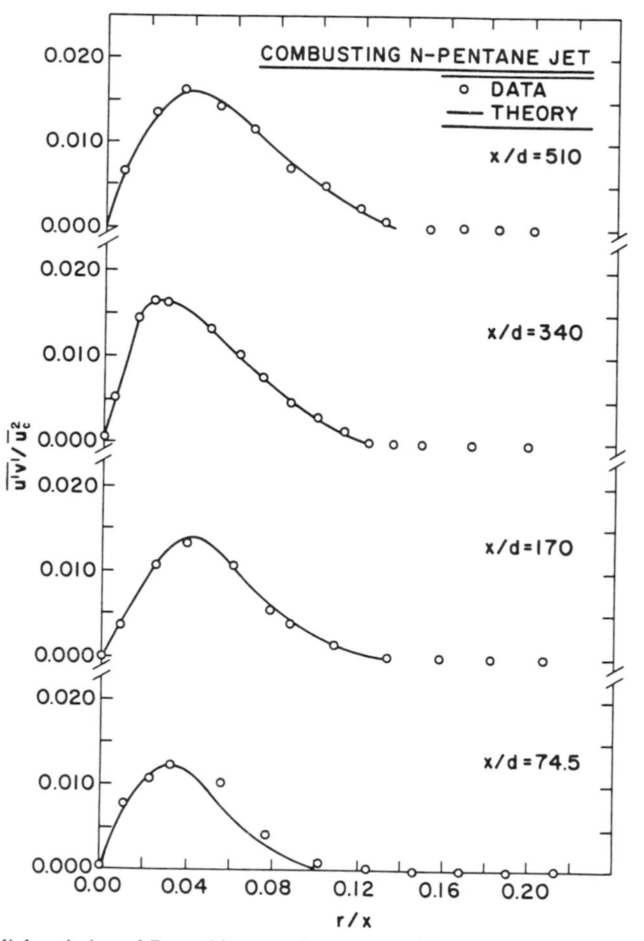

Figure 8.24 Radial variation of Reynolds stress for a spray of liquid *n*-pentane burning in air at atmospheric pressure (from Mao et al.[86]).

In a later study, Khalil[84] was able to achieve better predictions for a portion of the data by using an eddy-breakup (EBU) model. He introduced additional empirical parameters to account for heterogeneous combustion of drops. The usefulness of the extension is limited, since the approach cannot truly account for the effect of relative velocity between the drops and gas.

8 SEPARATED-FLOW (TWO PHASE-FLOW) MODELS

As mentioned in Section 6.6, this type of model is the most logical one for the simulation of spray combustion problems, as such models specifically treat the finite rates of exchange of mass, momentum, and energy between the liquid

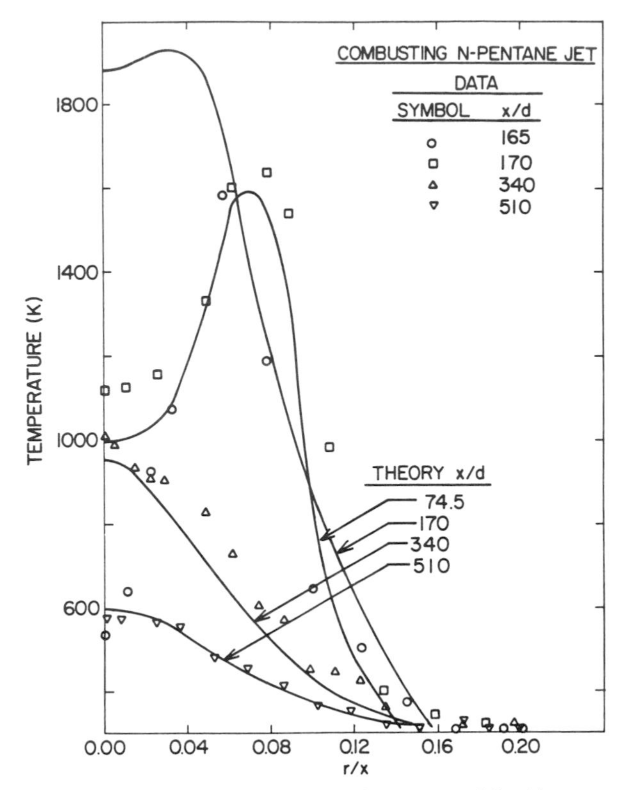

Figure 8.25 Radial variation of mean temperature for a spray of liquid n-pentane burning in air at atmospheric pressure (from Mao et al.[86]).

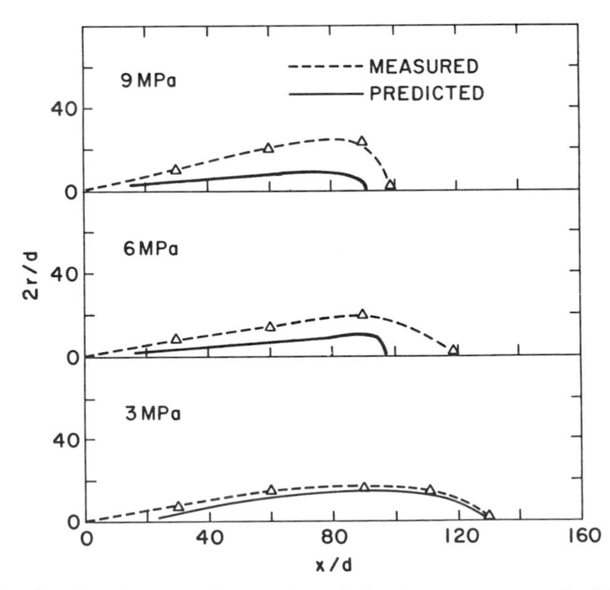

Figure 8.26 Predicted and measured spray boundaries for pressure-atomized liquid n-pentane sprays burning in high-pressure air (from Mao, Wakamatsu, and Faeth[87]).

and gas phases. However, due to limitations on computer storage and cost of computation, researchers developing separated-flow (SF) models have made no attempt to accurately model the details of the flow field around individual drops. Therefore, the exchange processes between phases must be modeled independently. Usually a set of empirical correlations for droplet drag and heat and mass transfer is employed. In general, there are three different approaches in SF analyses for evaporating and combusting sprays. They are briefly introduced below:

A. Particle-Source-in-Cell Model (PSICM), or Discrete-Droplet Model (DDM). Finite numbers of groups of particles are used to represent the entire spray. The motion and transport of representative samples of discrete drops are tracked through the flow field using a Lagrangian formulation, while a Eulerian formulation is used to solve the governing equations for the gas phase. The effect of droplets on the gas phase is taken into account by introducing appropriate source terms in the gas-phase conservation equations.

B. Continuous Droplet Model (CDM). As discussed in Section 4.2, the distribution function $f_j(r, \mathbf{x}, \mathbf{v}, t)$ can be used to evaluate the statistical distributions of the drop temperature, concentration, and so on. The transport equation (8-19) for f_j is solved along with the gas conservation equations to provide all properties of the spray. Similarly to the DDM, the governing equations for the gas phase must also include appropriate source terms.

C. Continuum-Formulation Model (CFM). The motion of both drops and gas are treated as though they were interpenetrating continua. A continuum formulation of the conservation equations for both phases is used to model spray combustion and evaporation problems. In this approach, the governing equations for the two phases are similar; however, there are many difficulties in describing the droplet heatup process, the turbulent stresses, and the turbulent dispersion of droplets.

Each of these approaches is described in the following sections.

8.1 Particle-Source-In-Cell Model (Discrete-Droplet Model)

In this approach, the entire spray is divided into many representative samples of discrete drops whose motion and transport through the flow field are found using a Lagrangian formulation in determining the drop life history, while a Eulerian formulation is used to solve the governing equations for the gas phase. The representative works are those of Crowe[107,108] and his coworkers[109,110] on aircraft gas-turbine combustors; Alpert and Mathews[111] on commercial sprinkler systems for fire safety; El Banhawy and Whitelaw[112] on modeling spray combustion in furnaces; Gosman and Johns[113] and Butler et al.[114] on reciprocating direct-injection stratified-charge (DISC) engines; Bruce et al.,[115] Mongia and Smith,[116] and Switherbank and coworkers[117,118] on

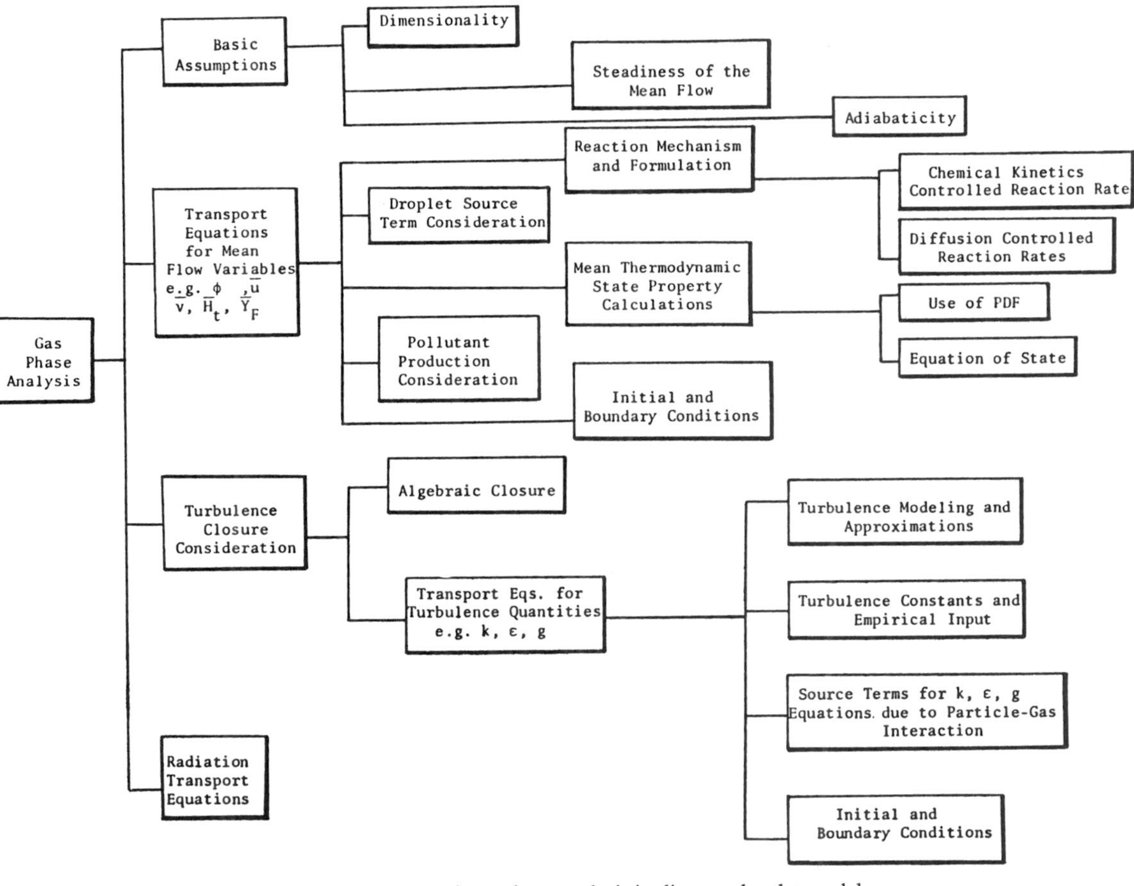

Figure 8.27 Structure of gas-phase analysis in discrete-droplet models.

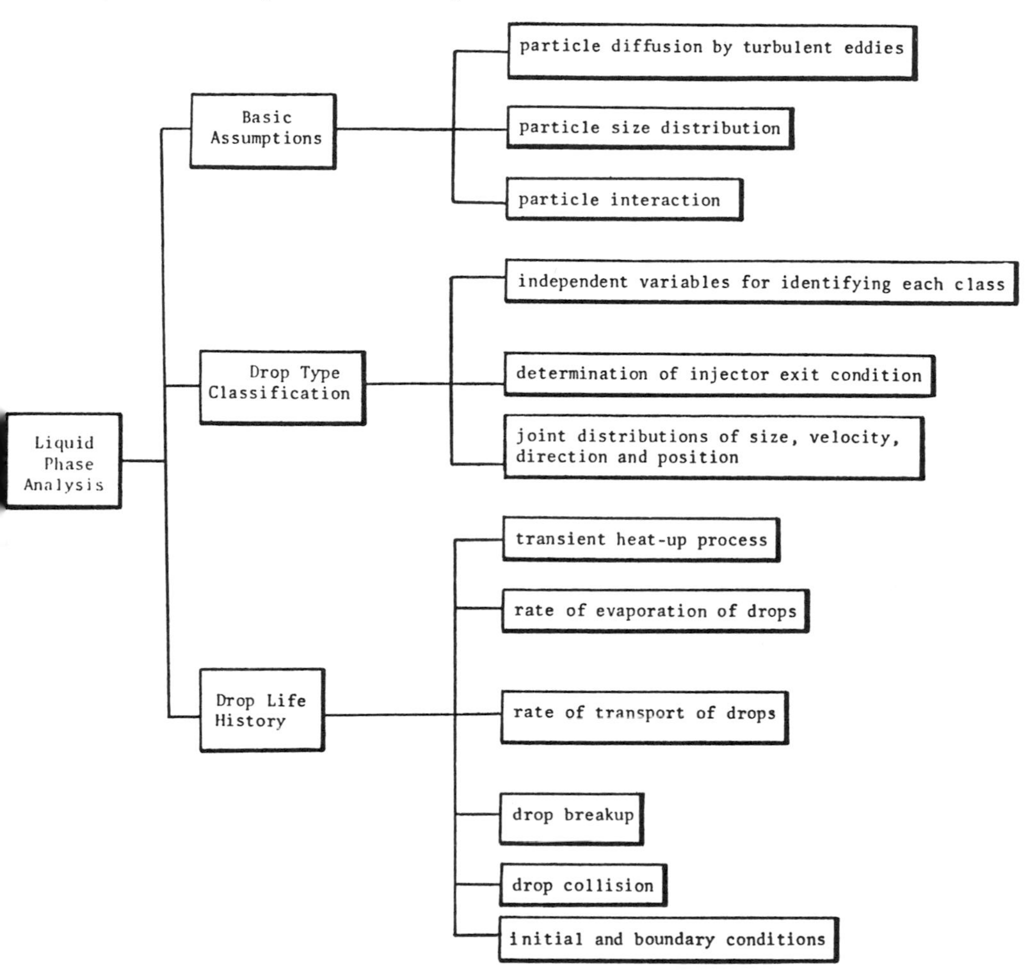

Figure 8.28 Structure of liquid-phase analysis in discrete-droplet models.

gas-turbine combustors; Gosman and coworkers[119,120] on spray combustion in cylindrical furnaces; Solomon[121] on evaporating sprays; Shuen[122] and co-workers[123] on dilute particle-laden turbulent gas jets; and Anderson et al.[124] on spray evaporation in premixed prevaporizing passages. A detailed review of these investigations was given by Faeth.[47]

A comprehensive theoretical model should include both gas-phase analysis and liquid-droplet-phase analysis. To provide readers with an overall picture of the structure of the DDM, Figs. 8.27 and 8.28 show the organization chart for gas-phase and liquid-phase analysis, respectively. The basic elements required in the gas-phase analysis, shown in Fig. 8.27, include basic assumptions, transport equations for mean-flow variables, turbulence closure consideration, and radiative-transport equations. For each of these elements, subelements and their components are also described. Figure 8.28 describes the basic elements required for the liquid-phase analysis: basic assumptions, drop-type classification, and drop-life history. Subelements are also shown as branches to the main elements.

Depending upon the consideration of the effect of turbulent fluctuations on particle motion and the method of treatment of the velocity differences (slip) between the phases, discrete-droplet models (DDMs) are further subdivided into deterministic separated-flow (DSF) models and stochastic separated-flow (SSF) models.

8.1.1 Deterministic Separated-Flow Models

In DSF models, the slip and finite interphase transport rates are considered, but effects of drop dispersion by turbulence and effects of turbulence on interphase transport rates are ignored. Droplets are assumed to interact only with the mean gas motion. As mentioned by Shuen et al.,[123] in the DSF formulation, particles follow deterministic trajectories found by solving their Lagrangian equations of motion. Spray models of this type usually employ the standard drag coefficient for spheres and ignore virtual mass and Bassett forces. These approximations are appropriate for high void fractions and high liquid-to-gas density ratio.

It is generally assumed that the spray is dilute in these models. This implies that although particles interact with the gas phase, they do not interact with each other. Therefore, droplet collisions are ignored, and empirical correlations determined for single drops in an infinite medium are used to estimate interphase transport rates. The volume fraction for the liquid phase is assumed to be negligibly small in most models. Due to experimental limitations, most measurements of drop size and velocity distributions have generally been made at some distance from the injector. This so-called initial station is usually in the dilute portion of the spray. Thus the practical application of models is limited to the dilute region of the spray. Both experimental and theoretical studies of the dense spray region are definitely needed for the future to understand the jet breakup process as well as to accurately specify the flow conditions at a selected station near the injector exit.

In the DSF models, the liquid-droplet flow at a station downstream of the injector exit is divided into a finite number of drop classes. Each drop class is assigned an initial diameter, velocity, direction, temperature, concentration, position, and time of injection. The life history of each class is then computed throughout the flow field. The drop position of a given group is calculated from

$$\mathbf{x}_p = \mathbf{x}_{p0} + \int_0^t \mathbf{v}_p \, dt \tag{8-82}$$

where $\mathbf{x}_{p0}$ is the initial location of the drop. The instantaneous velocity is determined by solution of the equation of motion

$$m\frac{d\mathbf{v}_p}{dt} = -\frac{\pi}{8} d_p^2 \rho C_D |\mathbf{v}_p - \mathbf{v}|(\mathbf{v}_p - \mathbf{v}) + \mathbf{F}_g \tag{8-83}$$

where $\mathbf{F}_g$ is the body force due to gravitation. In many cases, this force will have negligible effect on the motion of the representative droplet. The instantaneous mass of the drop is calculated from

$$m = \frac{\pi}{6} \rho_{\mathrm{bf}} d_p^3 \tag{8-84}$$

where ρ_{bf} is determined from the equation of state for the bulk liquid temperature and composition.

The standard drag coefficient for solid spheres is usually employed in calculations on dilute sprays.[47] An expression proposed by Putnam[125] can be given as

$$C_D = \begin{cases} \dfrac{24}{\mathrm{Re}_{d_p}}\left[1 + \dfrac{\mathrm{Re}_{d_p}^{2/3}}{6}\right] & \text{for} \quad \mathrm{Re}_{d_p} < 1000 \\ 0.44 & \text{for} \quad \mathrm{Re}_{d_p} > 1000 \end{cases} \tag{8-85}$$

where the Reynolds number is defined as

$$\mathrm{Re}_{d_p} = \frac{d_p|\mathbf{v}_p - \mathbf{v}|}{\nu} \tag{8-86}$$

According to Faeth,[47] the following expressions of Dickerson and Schuman[126] with a broader Reynolds-number range yield similar results to Eq. (8.85):

$$C_D = \begin{cases} 27/\mathrm{Re}_{d_p}^{0.84} & \text{for} \quad \mathrm{Re}_{d_p} < 80 \\ 0.271\,\mathrm{Re}_{d_p}^{0.217} & \text{for} \quad 80 \le \mathrm{Re}_{d_p} \le 10^4 \end{cases} \tag{8-87}$$

A useful formula for the multiplicative correction for heat and mass transfer, proposed by Faeth and Lazar,[127] is

$$\frac{h_c}{(h_c)_{\mathrm{Re}_{d_p}=0}} = 1 + \frac{0.278\,\mathrm{Re}_{d_p}^{1/2}\mathrm{Pr}^{1/3}}{1 + 1.232/\left[\mathrm{Re}_{d_p}\mathrm{Pr}^{4/3}\right]^{1/2}} \tag{8-88}$$

$$\frac{\dot{m}}{(\dot{m})_{\mathrm{Re}_{d_p}=0}} = 1 + \frac{0.278\,\mathrm{Re}_{d_p}^{1/2}\mathrm{Sc}^{1/3}}{1 + 1.232/\left[\mathrm{Re}_{d_p}\mathrm{Sc}^{4/3}\right]^{1/2}} \tag{8-89}$$

where $(h_c)_{\mathrm{Re}_{d_p}=0}$ and $(\dot{m})_{\mathrm{Re}_{d_p}=0}$ are found from

$$\frac{(h_c)_{\mathrm{Re}_{d_p}=0}\,d_p}{\lambda} = \frac{\dot{m}C_p/\pi d_p\lambda}{\exp(\dot{m}C_p/2\pi d_p\lambda) - 1} \tag{8-90}$$

and

$$(\dot{m})_{\mathrm{Re}_{d_p}=0} = 2\pi d_p \rho \mathscr{D} \ln\left(\frac{\overline{Y}_{i\infty} - \varepsilon_i}{Y_{isg} - \varepsilon_i}\right), \qquad i = 1, 2, \ldots, N \tag{8-91}$$

where Y_{isg} represents the mass fraction of species i on the gas-phase side of the droplet surface, and ε_i represents the mass-flux fraction of species i. It is defined according to Williams[13] as

$$\varepsilon_i \equiv \frac{\dot{m}_i}{\dot{m}} \qquad \text{for} \quad i = 1, 2, \ldots, N \tag{8-92}$$

From the above definition

$$\sum_{i=1}^{N} \varepsilon_i = 1 \tag{8-93}$$

It is useful to note that the values of ε_i can be greater or less than zero, depending upon whether the species i is evaporating or condensing at the liquid surface. Specification of the values of the ε_i depends on the liquid transport model adopted (to be discussed later). Equation (8.91) is obtained from the conservation of species

$$\frac{d}{dr}\left[r^2\left(\rho v Y_i - \rho \mathscr{D} \frac{dY_i}{dr}\right)\right] = 0, \qquad i = 1, 2, \ldots, N \tag{8-94}$$

and Eq. (8-90) is obtained from the energy equation

$$\frac{d}{dr}\left\{r^2\left[\rho v C_p (T - T_r) - \lambda \frac{dT}{dr}\right]\right\} = 0 \tag{8-95}$$

Detailed steps are given in Ref. 47.

8.1.2 Stochastic Separated-Flow Models

Although the deterministic separated-flow (DSF) model described in the above section considers the interphase slip between particles and the continuous phase, the effects of turbulent fluctuations on particle motion are ignored. Several stochastic separated-flow (SSF) models[119,121–123,128,129] have been developed to treat both slip and the effects of turbulent fluctuations. Yuu et al.[129] employed empirical correlations of turbulence intensities and length scales for their calculations of particle dispersions in jets. Gosman and Ioannides[119] took a more comprehensive approach and used a k–ε model for predicting both flow properties and dispersion. Shuen et al.[123,128] followed the approach of Gosman and Ioannides[119] and modified their method for evaluating the turbulent-eddy lifetime. Shuen et al.[128,123] also made detailed comparisons of theoretical results with many sets of particle-laden jet data obtained by Yuu et al.,[129] Laats and Frishman,[130,131] and Levy and Lockwood.[132] Shuen et al.[128,123] found that the solutions of SSF models agree very closely with experimental data on particle dispersion in turbulent round jets. A discussion of their comparisons is given in Section 8.2 after the presentation of their SSF model.

In the consideration of governing equations for the continuous phase, several basic assumptions can be introduced under certain flow conditions. For example, when the particle mass loading in the jet is sufficiently small ($< 0.5\%$), the particles can be assumed to have negligible effect on mean and turbulent gas-phase properties. For low Mach numbers (< 0.3) at the exit of the injector, density variations, the kinetic energy of the mean flow, and viscous dissipation can be neglected with small error. Also, for high jet Reynolds numbers ($\sim 10^4$), molecular transport can be ignored in comparison with turbulent transport. Under these assumptions, the governing equations for the continuous phase are identical to the DSF model. The transport equations for k and ε are:

$$\frac{\partial}{\partial x}(\bar{\rho}\bar{u}k) + \frac{1}{r}\frac{\partial}{\partial r}(r\bar{\rho}\bar{v}^{\circ}k) = \frac{1}{r}\frac{\partial}{\partial r}\left(r\frac{\mu_t}{\sigma_k}\frac{\partial k}{\partial r}\right) + \mu_t\left(\frac{\partial \bar{u}}{\partial r}\right)^2 - \bar{\rho}\varepsilon + \underline{\overline{uS_{pu}} - \bar{u}\bar{S}_{pu}} \tag{8-96}$$

$$\frac{\partial}{\partial x}(\bar{\rho}\bar{u}\varepsilon) + \frac{1}{r}\frac{\partial}{\partial r}(r\bar{\rho}\bar{v}^{\circ}\varepsilon) = \frac{1}{r}\frac{\partial}{\partial r}\left(r\frac{\mu_t}{\sigma_\varepsilon}\frac{\partial \varepsilon}{\partial r}\right) + C_{\varepsilon 1}\mu_t\frac{\varepsilon}{k}\left(\frac{\partial \bar{u}}{\partial r}\right)^2 - C_{\varepsilon 2}\bar{\rho}\frac{\varepsilon^2}{k} - \underline{2C_{\varepsilon 3}\mu_t\frac{\varepsilon}{k}\frac{\partial \bar{S}_{pu}}{\partial r}} \tag{8-97}$$

where $\bar{S}_{pu}$ is the particle source term in the streamwise momentum equation. For the momentum equation of the continuous phase in the ith direction, we have

$$\frac{\partial}{\partial x_j}(\bar{u}_i\bar{u}_j) = \nu\frac{\partial^2 \bar{u}_i}{\partial x_j\,\partial x_j} + \frac{\bar{S}_{pu_i}}{\bar{\rho}} \tag{8-98}$$

For the streamwise direction, $i = j = 1$; therefore, no subscript for $\bar{S}_{pu}$ is needed in Eqs. (8-96) and (8-97). The underlined terms in these two equations are introduced by the interphase transport.[122] Although these terms are of interest, they were ignored in the theoretical calculations and their comparison[128,123] with experimental data on dilute sprays.

In order to determine particle trajectories, the general equation of particle motion for a spherical particle can be written as follows:[47]

$$\underbrace{\frac{\pi}{6}d_p^3\rho_p\frac{d\mathbf{u}_p}{dt_p}}_{\substack{\text{inertial}\\ \text{force of}\\ \text{sphere}}} = \underbrace{\frac{\pi}{8}d_p^2\rho C_D|\mathbf{u}-\mathbf{u}_p|(\mathbf{u}-\mathbf{u}_p)}_{\substack{\text{drag force on sphere}\\ \text{including skin friction}\\ \text{and form drag}}} \underbrace{-\frac{\pi}{6}d_p^3\frac{\partial p}{\partial r}n_r}_{\substack{\text{force on}\\ \text{sphere due}\\ \text{to static}\\ \text{pressure}\\ \text{gradient}}} + \underbrace{\frac{\pi}{12}d_p^3\rho C_I\frac{d}{dt_p}(\mathbf{u}-\mathbf{u}_p)}_{\substack{\text{force on sphere due to}\\ \text{inertia of adjacent fluid being}\\ \text{displaced by its motion}\\ \text{(virtual-mass term)}}}$$

$$+ \underbrace{\tfrac{3}{2}d_p^2(\pi\rho\mu)^{1/2}C_B\int_{t_{p0}}^{t_p}\frac{(d/d\xi)(\mathbf{u}-\mathbf{u}_p)}{(t_p-\xi)^{1/2}}\,d\xi}_{\substack{\text{Bassett force to}\\ \text{account for effects of}\\ \text{deviation of}\\ \text{flow from steady flow}\\ \text{pattern around sphere}}} + \underbrace{\mathbf{F}_e}_{\substack{\text{external or}\\ \text{body-force}\\ \text{term, e.g.,}\\ \text{gravity}}} \tag{8-99}$$

where $\mathbf{n}_r$ is a unit vector in the positive radial direction and the time derivative is taken following the particle motion

$$\frac{d}{dt_p} = \frac{\partial}{\partial t} + u_p \frac{\partial}{\partial r} \tag{8-100}$$

In the simulations of dilute sprays with $\rho_p/\rho > 200$, Shuen et al.[128,123] considered the virtual mass, Bassett forces, Magnus forces, and so on, negligible. Gravitational forces were also neglected except in some special cases. The simplified equation of motion then became

$$\frac{du_{pi}}{dt} = \left(\frac{3\rho C_D}{4d_p\rho_p}\right)(u_i - u_{pi})|\mathbf{u} - \mathbf{u}_p| + g_i, \qquad i = 1,3 \tag{8-101}$$

The position of a particle was determined from

$$\frac{dx_{pi}}{dt} = u_{pi}, \qquad i = 1,3 \tag{8-102}$$

The above two equations were integrated using a second-order Runge–Kutta algorithm. The trajectories of at least 1000 particles were computed and averaged to obtain dispersion properties.

Following the method of Gosman and Ioannides,[119] the motion of the particles was traced as they interacted with a succession of turbulent eddies, each of which was assumed to have constant flow properties (see Fig. 8.29). A particle was assumed to interact with an eddy for a time taken as the smaller of either the eddy lifetime t_e, or the transit time t_t required for the particle to traverse the eddy. These times were estimated by assuming that the characteris-

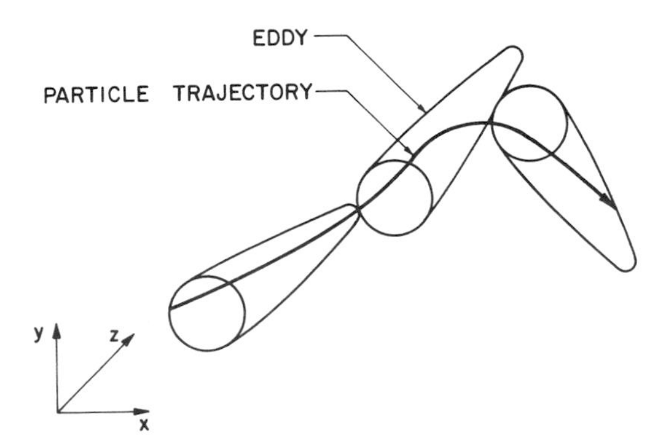

Figure 8.29 Sketch of the particle trajectory model (from Shuen, Chen, and Faeth[47]).

tic size of an eddy is the dissipation length scale

$$L_e = \frac{C_\mu^{3/4} k^{3/2}}{\varepsilon} \tag{8-103}$$

Gosman and Ioannides[119] computed the eddy lifetime as

$$t_e = \frac{L_e}{|\mathbf{u}'|} \tag{8-104}$$

However, Shuen et al.[128] found better agreement with measurements by employing

$$t_e = \frac{L_e}{\left(\frac{2}{3}k\right)^{1/2}} \tag{8-105}$$

The transit time of a particle was found using the linearized equation of motion for a particle in a uniform flow,

$$t_t = -\tau_r \ln\left(1 - \frac{L_e}{\tau_r|\mathbf{u} - \mathbf{u}_p|}\right) \tag{8-106}$$

where $\mathbf{u} - \mathbf{u}_p$ is the velocity at the start of the interaction and τ_r is particle

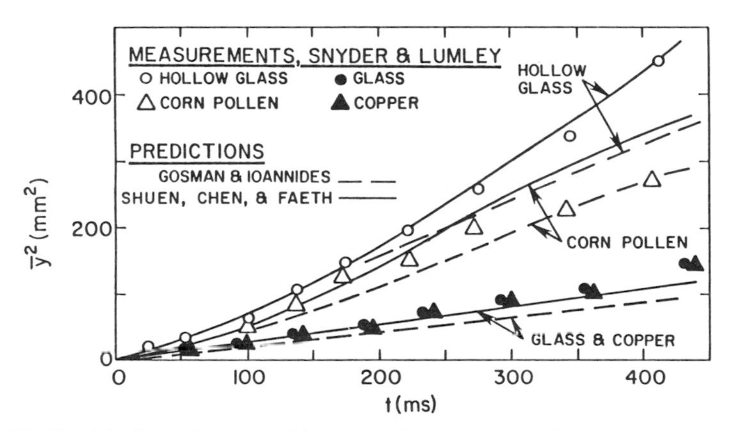

Figure 8.30 Particle dispersion in a grid-generated turbulent duct flow (from Shuen, Chen, and Faeth[128]).

relaxation time defined as

$$\tau_r \equiv \frac{4\rho_p d_p}{3\rho C_D |\mathbf{u} - \mathbf{u}_p|} \tag{8-107}$$

When $L_e > \tau_r|\mathbf{u} - \mathbf{u}_p|$, Eq. (8-106) has no solution. This can be interpreted as the eddy having captured a particle, so that the interaction time becomes t_e.[119,47] The stochastic method generally requires an estimate of the mean and turbulent properties of the continuous phase. Particle trajectories are then calculated using random sampling to determine the instantaneous properties of the continuous phase, similarly to a random-walk calculation. Mean dispersion rates are obtained by averaging over a statistically significant number of particle trajectories.

8.2 Discussion of Results from SSF and DSF Models

The stochastic-model solution of Shuen et al.[128] was compared with the measurements of Snyder and Lumley.[133] In their experiments, the dispersion of solid particles in a uniform turbulent flow downstream of a grid was measured. The comparison between predictions and measurements is illustrated in Fig.

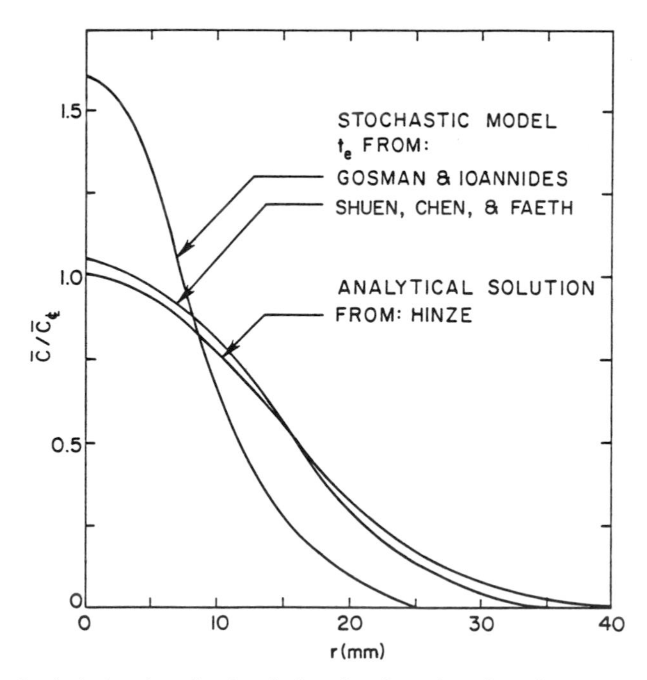

Figure 8.31 Analytical and stochastic solutions for dispersion of small particles in homogeneous isotropic flow with long diffusion times (from Shuen, Chen, and Faeth[128]).

8.30. Both models (Gosman and Ioannides[119] and Shuen et al.[128]) yield the same results for heavier particles, since the interaction times are then largely controlled by particle transit times, which are computed in the same manner.

The radial variation of particle concentration downstream of the point source in homogeneous isotropic flow is shown in Fig. 8.31. In this figure, the analytical solution of Hinze[134] is based upon the diffusion of "marked" fluid particles introduced at a constant rate from a point source into a homogeneous isotropic flow. The predictions of the model of Gosman and Ioannides and that of Shuen et al. are compared with Hinze's solution. It is quite obvious that the predictions of Shuen et al. agree well with the analytical expression, while the result of Gosman and Ioannides tends to underestimate the rate of dispersion.

For a particle-laden round jet, Yuu et al.[129] injected an air jet containing nearly monodisperse particles (with $d_p = 20\ \mu\text{m}$ and $\rho_p = 2000\ \text{kg/m}^3$) into

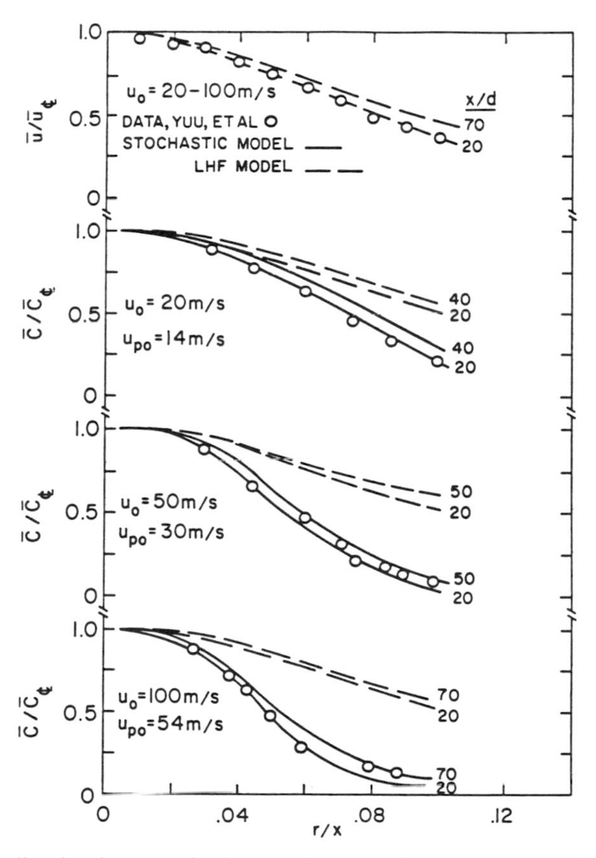

Figure 8.32 Predicted and measured axial gas velocities and particle concentrations in a dust-laden air jet with particle diameter 20 μm and particle density 2000 kg/m^3 (from Shuen, Chen, and Faeth[128]).

still air. Their measured profiles of mean and gas velocities and particle concentrations are given in Fig. 8.32. The predictions of stochastic dispersion model of Shuen et al.[128] and the LHF model are also shown in the figure. As is evident, the stochastic dispersion model yields quite good agreement with the particle-concentration measurements. The predictions of the LHF model overestimate the dispersion of particles. Faeth[47] also indicated that the DSF model underestimates the radial spread of the particles. This underestimation was due to neglect of particle dispersion by turbulence. Particles can move in the radial

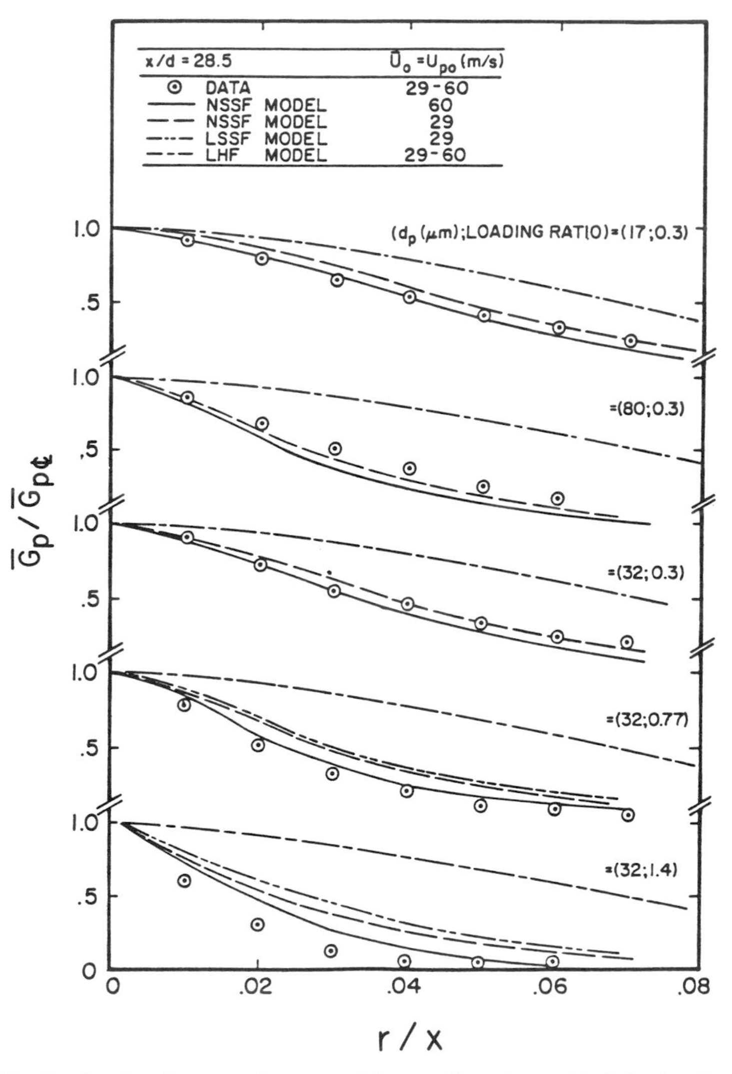

Figure 8.33 Predicted and measured mean particle mass fluxes in particle-laden jets (from Shuen, Chen, and Faeth[123]).

direction only on account of the drag forces exerted by the radial component of velocity.

Results predicted by the SSF model of Shuen et al.[123] and data on mean particle mass fluxes measured by Laats and Frishman[130,131] are shown in Fig. 8.33. As can be seen from this figure, the LHF model overestimates rates of particle dispersion in all cases. Two types of SSF models are used for predictions: a linearized method (LSSF model[128]) for determining transient times from L_e and t_e, and a nonlinear method (NSSF model) in which particles interact with an eddy as long as their residence times and displace-

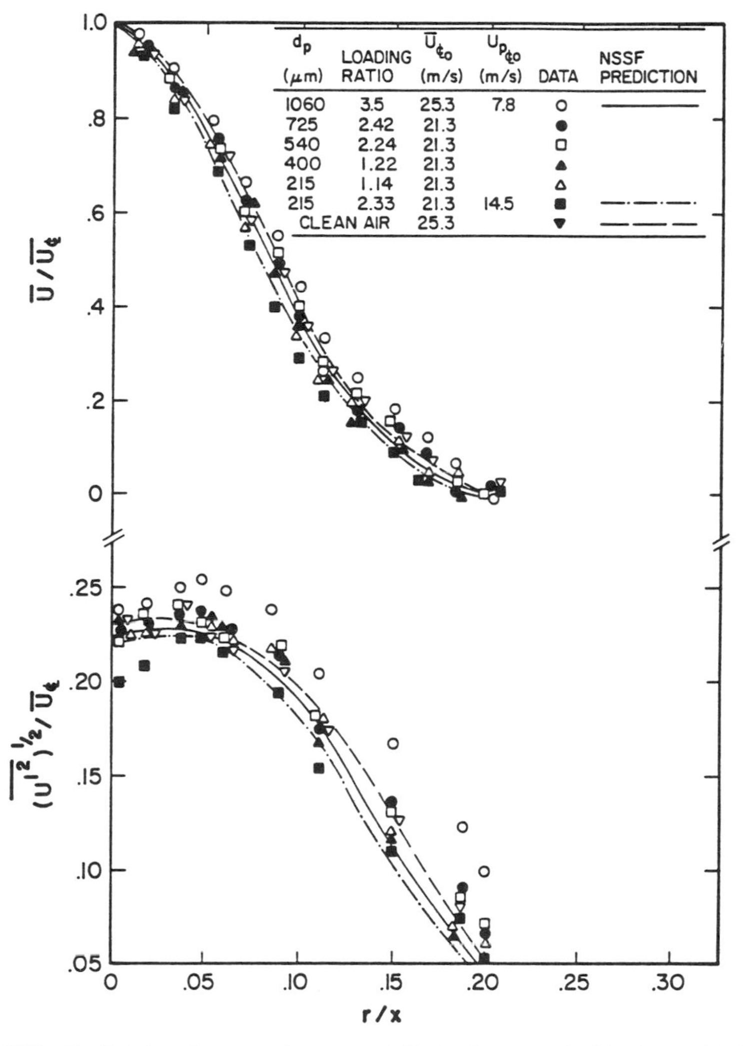

Figure 8.34 Predicted and measured mean and fluctuating gas velocities in particle-laden jets (from Shuen, Chen, and Faeth[123]).

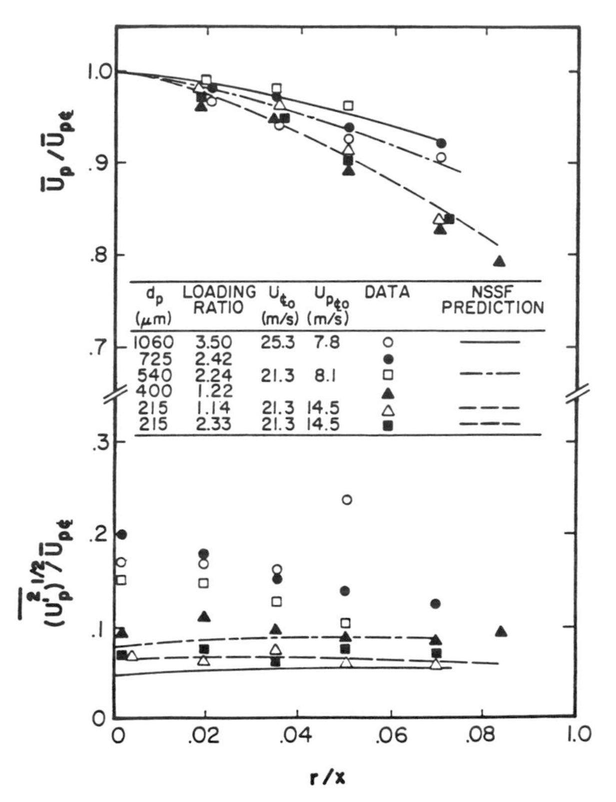

Figure 8.35 Predicted and measured mean and fluctuating particle velocities in particle-laden jets (from Shuen, Chen, and Faeth[123]).

ments are less than t_e and L_e. The SSF models yield reasonably good predictions. As the loading ratio increases, the SSF models overestimate the particle spread; this is probably due to the fact that damping of turbulent fluctuations by particle drag was ignored in Shuen's calculations.[123]

Comparison of mean and fluctuating gas and particle velocities is illustrated in Figs. 8.34 and 8.35 for the data of Levy and Lockwood.[132] Only the predictions of the NSSF model of Shuen et al. are shown; the LSSF predictions were nearly the same, while the LHF predictions overestimated the particle spread rates as before. The results in Fig. 8.34 show that particle size and loading have only a small effect on predicted and measured gas properties, since high loading corresponds to large particles that exchange relatively little momentum with the gas between the jet exit and the measuring station for these data.[123]

In Fig. 8.35, the predictions of mean particle velocities are reasonably good; however, the predictions of velocity fluctuations of large particles are underestimated. This was attributed to an artifact of the particle injection system: the momentum exchange of large particles is too small for the flow to induce the high fluctuation levels shown in Fig. 8.35. For small particles, the velocity-fluctuation predictions are much closer to the measured values, since small particles interact with the flow to a greater degree, so that effects of the injector system are damped out.[123]

In general, it can be concluded that the SSF model provides very encouraging predictions of the structure of particle-laden jets. In contrast, the DSF model was unsatisfactory over the entire data base.[123] The LHF model was satisfactory only for the smallest particles ($\approx$ 2-μm diameter).

9 GROUP-COMBUSTION MODELS OF CHIU

In very dilute sprays with low-volatility fuels, individual drop burning is highly possible. The formulation and solution for a single fuel drop are given in Chapter 6, and combustion models for dilute sprays are given in the earlier sections of this chapter. In this section, the concept of group combustion of drop clouds is introduced. This concept is especially useful for dense spray combustion in gas-turbine engines, industrial furnaces, and diesel engines.

A series of group-combustion models was developed by Chiu and his coworkers[49-54] for studying the combustion behavior of particle clouds of liquid-fuel sprays. Their models were partly based upon the experimental observations of Onuma and Ogasawara[62] that there is structural similarity between two different flames: one burning liquid kerosene and the other a gaseous fuel. It was suggested by Onuma and Ogasawara that the rate-determining step in spray combustion of kerosene was mixing between fuel vapor and air, rather than the vaporization of drops as conventionally thought. (It should be noted, however, that for fuels of lower volatility and in larger droplets, drop vaporization will still be one of the controlling factors in spray combustion.) The second body of experimental evidence for group-combustion models is due to Chigier[135] and his coworkers. In their studies of pressure-jet- and air-blast-atomized spray flames, they observed regions of low oxygen concentration and low temperature within the spray. The existence of these regions causes the displacement of flame toward the outer boundaries of the spray. According to the group-combustion theory of Chiu,[49-54] the collective behavior of drops in liquid sprays forms a fuel-rich mixture in the core region of the spray. This mixture is nonflammable at the spray core, due to insufficient air penetration. The radial transport of gaseous fuel by convection and diffusion leads to the formation of flammable mixtures at some distance from the centerline of the spray. These flammable mixtures burn as gaseous diffusion flames.

As drops move beyond the dense core region of the spray, the separation distances between drops increase and the drop size is reduced, so that the air concentration increases. Under these conditions, some liquid drops may burn individually with flames inside the spray boundary, while other drops may burn as groups. In general, the core region consists of drops vaporizing in atmospheres of low oxygen concentration, while the outer region may contain drops burning with multidrop flames.[48] In group-combustion models, the collective behavior of drops is accounted for by a simultaneous analysis of an inner heterogeneous region and an outer homogeneous gas-phase region. No detailed formulation of the group combustion of liquid droplets is given here. Interested readers should consult the original papers by Chiu and his associates.[49,51–53,136]

9.1 Group-Combustion Numbers

Spray-combustion models are classified according to group-combustion numbers. The group-combustion number of the second kind, G, is defined as ratio of the total heat-transfer rate between two phases to the rate of heat of vaporization associated with the diffusing fuel vapor[136], i.e.,

$$G \equiv \frac{\text{rate of heat exchange between two phases}}{\text{rate of heat of vaporization}} \tag{8-108}$$

$$G \equiv \frac{\mathscr{D}_\infty}{R_\infty U_\infty} G_1 \tag{8-109}$$

where the group-combustion number of the first kind, G_1, is defined as

$$G_1 \equiv \frac{2\pi\lambda_l r_{l0} n_0 R_\infty^2 T_0}{\rho_\infty \mathscr{D}_\infty L}\left[2 + 0.6\,\mathrm{Re}_{d_{l0}}^{1/2}\mathrm{Pr}^{1/3}\right] \tag{8-110}$$

where $\mathscr{D}_\infty$ and ρ_∞ are the mass diffusivity and density of the gas in the undisturbed environment. R_∞ designates the radius of the boundary of the two-phase zone. The number density of the drops at the reference condition, n_0, is given by

$$n_0 \equiv \frac{N}{\frac{4}{3}\pi R_\infty^3} \tag{8-111}$$

where N is the total number of droplets present in the cloud. In Eq. (8-110), L represents the latent heat of vaporization per unit mass, and r_{l0} designates the reference radius of the droplet. After substituting Eq. (8-111) into Eq. (8-110), we have

$$G_1 = \tfrac{3}{2}N\left(\frac{r_{l0}}{R_\infty}\right)\frac{\lambda_l T_0}{\rho_\infty \mathscr{D}_\infty L}\left[2 + 0.6\,\mathrm{Re}_{d_{l0}}^{1/2}\mathrm{Pr}^{1/3}\right] \tag{8-112}$$

In a later paper, Chiu and Liu[49] redefined the dimensionless parameter G_1 as

$$G_1 \equiv \frac{4\pi\lambda_l r_{l0} n_0 R_\infty^2}{\rho_\infty \mathscr{D}_\infty C_p}\left[1 + 0.276\,\mathrm{Re}_{d_{l0}}^{1/2}\mathrm{Sc}^{1/3}\right] \tag{8-113}$$

This new dimensionless number of the first kind gives G the following physical meaning:

$$G \equiv \frac{\text{rate of heat exchange between two phases}}{\text{rate of energy transport by convection}} \tag{8-114}$$

Using the new definition of G_1 and the expression for n_0, one can show that

$$G_1 = 3\left(1 + 0.276\,\mathrm{Re}_{d_{l0}}^{1/2}\mathrm{Sc}^{1/3}\right)\mathrm{Le}\,N\left(\frac{r_{l0}}{R_\infty}\right) \tag{8-115}$$

or, alternatively,

$$G_1 = 3\left(1 + 0.276\,\mathrm{Re}_{d_{l0}}^{1/2}\mathrm{Sc}^{1/3}\right)\mathrm{Le}\,N^{2/3}\left(\frac{r_{l0}}{d_i}\right) \tag{8-116}$$

where Le is the Lewis number and d_i is the interdroplet separation. The dimensionless parameter G or G_1 represents the degree of interaction between the two phases and serves to differentiate strong and weak interactions.

9.2 Modes of Group Burning in Spray Flames

As shown in Fig. 8.36, four group combustion modes of a droplet cloud are possible.[48, 52] In sprays where $G > 10^2$, external sheath burning occurs, which consists of an inner nonvaporizing droplet cloud surrounded by a vaporizing droplet layer with the flame at a "standoff" distance from the spray boundary. High-G sprays usually have large group-burning rates and low core temperatures. For marginally high-G sprays ($G > 1$), external group combustion prevails. The spray zone consists of an inner vaporizing cloud with a standoff diffusion flame from the boundary of the droplets. For $10^{-2} < G < 1$, the mode of combustion is internal group combustion. In this mode, the main flame locates within the spray boundary, while individual drop burning occurs in the outer regions of the spray. For very low values of G ($< 10^{-2}$), the mode becomes individual droplet combustion.

A schematic diagram of liquid-fuel spray group combustion is shown in Fig. 8.37. In this figure, Chiu and Croke[51] subdivided the spray flame into many zones: potential core, external group-combustion zone with evaporating droplets, turbulent envelope diffusion flame at spray core boundary, multidroplet combustion zone with internal group-combustion behavior, and turbulent brush flame.

Chiu and Croke[51] used their group-combustion theory to conduct predictive calculations for a $C_{10}H_{14}$ spray flame. Their predicted temperature and concentration profiles indicate that the flame is stabilized near the spray boundary. They also reported that a relative minimum in temperature profile occurs on the axis of the spray, where the fuel vapor concentration has a

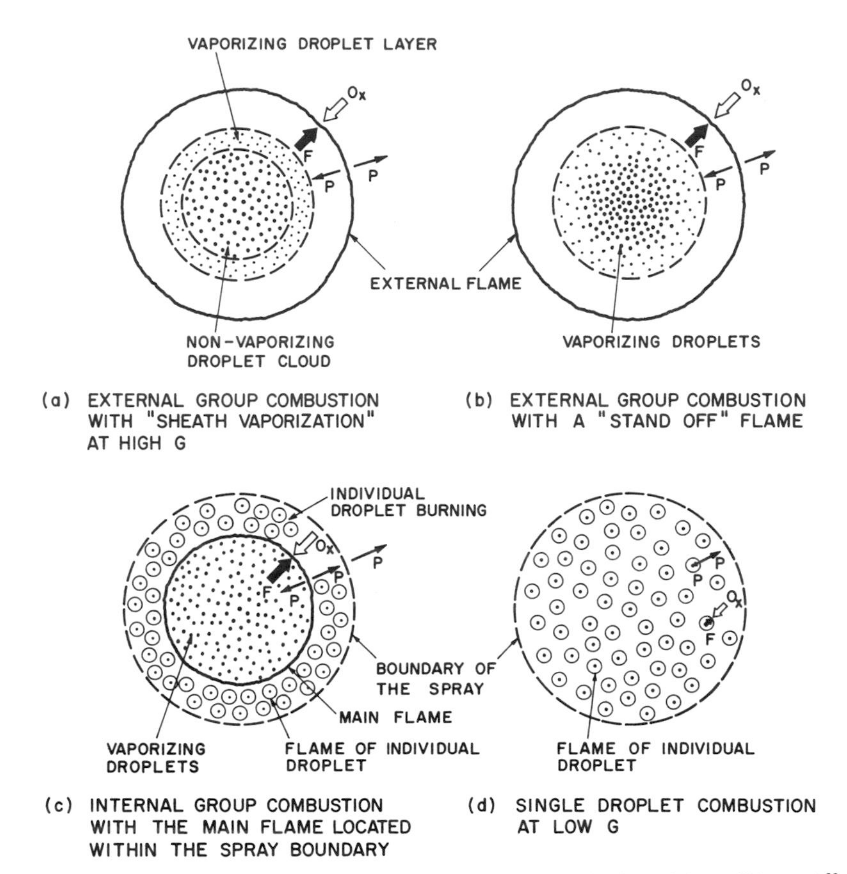

Figure 8.36 Four group combustion modes of a droplet cloud (adapted from Chiu et al.[52]).

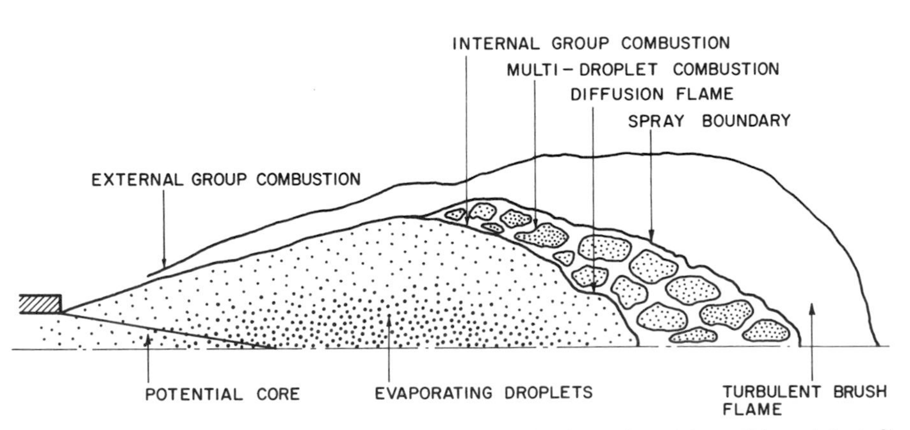

Figure 8.37 Schematic of liquid-fuel spray group combustion (adapted from Chiu and Croke[51]).

maximum. Sprays with smaller G have a smaller flame standoff distance and thus have a higher average gas temperature in the spray core. Higher temperature in the fuel rich core, in turn, increases the rate of pyrolysis of heavier hydrocarbons into lighter hydrocarbons, leading to soot and particulate formation by petroleum coking.[48]

Besides the group-combustion number G, the fuel–air mass-density ratio is also important in determining the relative location of the envelope flame to the spray boundary. The fuel–air mass-density ratio can be defined as

$$\beta \equiv \frac{\frac{4}{3}\pi r_{l0}^3 n_0 \rho_l}{\rho_\infty} \tag{8-117}$$

where ρ_l is the density of the liquid droplet. At a fixed G, an increase in β results in a shift of the flame toward the spray boundary, which is primarily due to reduction in the average group burning rate. According to the calculations of Chiu and Croke,[51] for $G \sim 10^{-2}$, $B \sim 0.1$, the flame is stabilized near the spray boundary. For $G \approx 0.5$, β must be increased to 100 for the flame to approach the spray boundary. They found that in order to have external combustion the value of G must be greater than a critical group-combustion number, G_c, which varies with β in the following manner:

β	G_c
1	9×10^{-3}
10	6×10^{-2}
100	7×10^{-1}

Following the explanation given by Chigier,[48] high-β sprays are characterized by high jet speeds and larger drops. This leads to shorter residence time for a fixed spatial distance, and poor vaporization characteristics with resultant lower group burning rates. For a fixed β, high-G sprays have a higher group burning rate and are characterized by relatively dense small drops which are readily vaporized. High-G sprays have larger flame radii and hence a larger flame area and thus a higher group burning rate.

Chiu and Croke[51] found, in their studies of a cloud of $C_{10}H_{14}$ drops, that the group envelope flame was stabilized on the boundary of the drop cloud for $G = 1.36$. As G decreases, the envelope flame penetrates into the drop cloud and divides the cloud into two zones, a strongly interacting zone located inside the group envelope flame (see Fig. 8.36c) and a weakly interacting zone established between the envelope flame and the boundary of the cloud. In the strongly interacting zone, the drops vaporize and the vapor produced is consumed at the group envelope. Drops in a weakly interacting zone burn with an envelope flame surrounding each drop. Their calculated group combustion modes are shown in Fig. 8.38, where N, the total number of drops, is plotted as

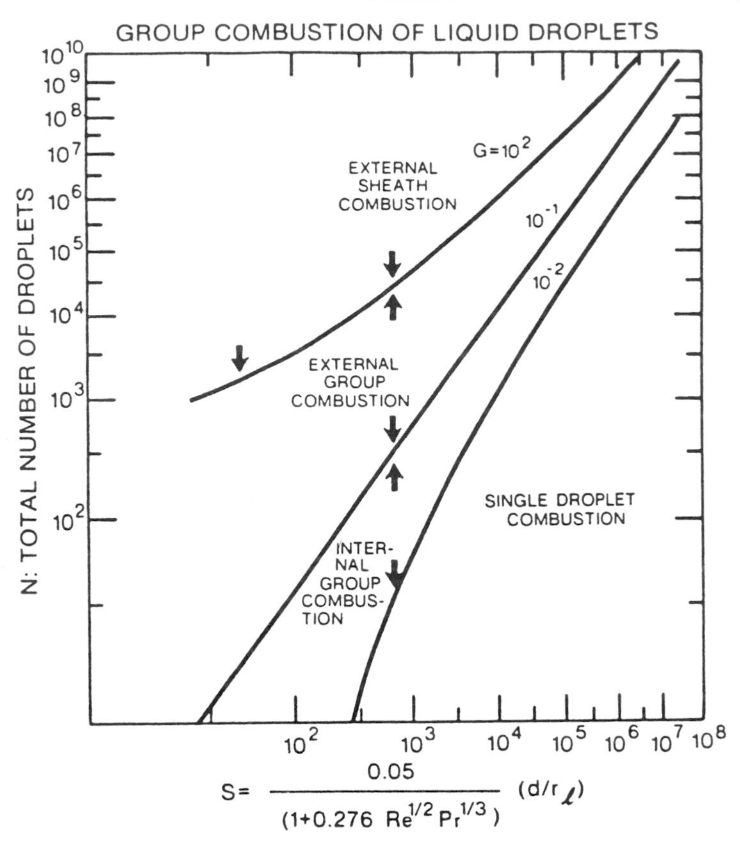

Figure 8.38 Group combustion modes for droplet clouds.[48]

a function of the nondimensional separation distance

$$s = 0.05 \frac{d_i/r_l}{1 + 0.276\,\mathrm{Re}_{d_{l0}}^{1/2}\mathrm{Pr}^{1/3}} \tag{8-118}$$

Zones for four different modes of group combustion are also shown on the figure.

Although the group combustion models developed so far include many important aspects of spray combustion, there are still numerous improvements which could be incorporated into the model: finite thickness of the envelope flame, radiative heat-transfer effects, particle dispersion by turbulence, droplet synergetic phenomena, spray vaporization, and so on. Some of these are being actively studied by Chiu and coworkers.

10 CONTINUUM-MECHANICS APPROACH FOR MULTIPHASE SYSTEMS

For single-phase fluids, the continuum-mechanics approach has achieved great success due to its adequate simplifications of the general transport equations (Boltzmann transport equations) of kinetic theory, as long as the fluid flow satisfies the continuum assumption. The continuum-mechanics approach has been extended by many researchers[137] in their formulation of a two-phase system with spherical particles of one size. As indicated by Soo,[137] any realistic general formulation of multiphase (gas–solid, gas–liquid, or other combinations of a particulate phase and a fluid phase) flow must account for the distribution of the size of particles. Also, appropriate consideration must be given to the interactions among various components in the multiphase system (mixture of solid, liquid, and gas phases).

In this section, the approach of Soo[137] is followed. Both the mixture and its components are treated as true continua. The flow of a mixture is visualized as that of a number of interacting systems occupying the same space; each may have its own streamlines. The motion of the center of mass of the mixture is tracked in the formulation, and the individual motions of components are treated in terms of diffusion through the mixture.

The density of a mixture, ρ_m, is given by

$$\rho_m = \sum_q \rho^{(q)} \tag{8-119}$$

where $\rho^{(q)}$ is the density of component q of the mixture occupying the volume of the mixture of ν components ($q = 1, 2, \ldots, \nu$). The ith component of velocity of the mixture, U_{mi}, is given by

$$\rho_m U_{mi} = \sum_q \rho^{(q)} U_i^{(q)} \tag{8-120}$$

Taking the general case including chemical reaction, the continuity equation of component q of the mixture can be expressed as

$$\frac{\partial \rho^{(q)}}{\partial t} + \frac{\partial}{\partial x_j}\left(\rho^{(q)} U_i^{(q)}\right) = \Gamma^{(q)} \tag{8-121}$$

where t is the time, x_j is the jth component of the space coordinate, and $\Gamma^{(q)}$ is the rate of generation of q per unit volume. Since $\Sigma_q \Gamma^{(q)} = 0$ over the whole mixture, we get

$$\frac{\partial \rho_m}{\partial t} + \frac{\partial}{\partial x_j}(\rho_m U_{mi}) = 0 \tag{8-122}$$

For Newtonian fluids, with negligible thermal diffusion, the momentum and energy equation are as follows:

$$\rho_m \frac{dU_{mi}}{dt_m} = \frac{\partial}{\partial x_j}\left[-p_m\delta_{ji} + \mu_m(\Delta_m)_{ji} + \mu_{m2}\Theta_m\delta_{ji}\right] + \rho\mathscr{F}_{mi} \tag{8-123}$$

$$\rho_m \frac{dE_m}{dt_m} = \frac{\partial}{\partial x_j} U_{mj}\left[-p_m\delta_{ji} + \mu_m(\Delta_m)_{ji} + \mu_{m2}\Theta_m\delta_{ji}\right]$$

$$+ \frac{\partial}{\partial x_j}\frac{\lambda_m \partial T_m}{\partial x_j} + \mathscr{F}_{Em} \tag{8-124}$$

Here

$$\frac{d}{dt_m} = \frac{\partial}{\partial t} + U_{mj}\frac{\partial}{\partial x_j} \tag{8-125}$$

The deformation tensor and the dilatation are

$$(\Delta_m)_{ji} = \frac{\partial U_{mi}}{\partial x_j} + \frac{\partial U_{mj}}{\partial x_i} \tag{8-126}$$

$$\Theta_m = \tfrac{1}{2}(\Delta_m)_{kk} = \frac{\partial U_{mk}}{\partial x_k} \tag{8-127}$$

and p_m is the overall static pressure of the mixture, μ_m the viscosity:

$$\mu_{m2} = \mu' - \tfrac{2}{3}\mu_m \tag{8-128}$$

where μ' is the bulk viscosity. The total energy is

$$E_m = e_m + \frac{U_m^2}{2} \tag{8-129}$$

where e_m is the internal energy per unit mass of the mixture. T_m is the static temperature, $\mathscr{F}_{mi}$ is the ith component of the body force acting on the mixture, and $\mathscr{F}_{Em}$ is the rate of energy generation per unit volume. The cumulative nature of energy gives

$$\rho_m E_m = \sum_q \rho^{(q)} E^{(q)} \tag{8-130}$$

$$\rho_m C_{vm} = \sum_q \rho^{(q)} C_v^{(q)} \tag{8-131}$$

The static temperatures T_m and $T^{(q)}$ are related according to

$$T_m = \left(\rho_m C_{pm}\right)^{-1}\left[\sum_q \rho^{(q)} C_p^{(q)} T^{(q)} + \frac{1}{2}\sum_{(q)} \rho^{(q)} U_i^{(q)}\left(U_i^{(q)} - U_{mi}\right)\right]$$

$$= \left(\rho_m C_{pm}\right)^{-1}\sum_q \left[\rho^{(q)} C_p^{(q)} T^{(q)} + \tfrac{1}{2} U_i^{(q)} J_i^{(q)}\right], \tag{8-132}$$

where $J_i^{(q)} \equiv \rho^{(q)}(U_i^{(q)} - U_{mi})$ is the diffusion flow component of q with respect to that of the center of mass of the mixture (barycentric motion). Obviously $\Sigma_q J^{(q)} = 0$ for the mixture.

Equations (8-123) and (8-124) can be reduced to those of the components, as can be shown by use of Eqs. (8-119), (8-120), and (8-130):

$$\rho_m \frac{dU_{mi}}{dt_m} = \sum_q \left[\rho^{(q)} \frac{dU_i^{(q)}}{dt^{(q)}} + \left(U_i^{(q)} - U_{mi}\right)\Gamma^{(q)} - \frac{\partial}{\partial x_j}\rho^{(q)}\left(U_i^{(q)} - U_{mi}\right)\left(U_j^{(q)} - U_{mj}\right)\right] \tag{8-133}$$

$$\rho_m \frac{dE_m}{dt_m} = \sum_q \left[\rho^{(q)} \frac{dE^{(q)}}{dt^{(q)}} + \Gamma^{(q)} E^{(q)} - \frac{\partial}{\partial x_j}\rho^{(q)}\left(U_j^{(q)} - U_{mj}\right)E^{(q)}\right] \tag{8-134}$$

showing the contributions of the motion about the center of mass of the system, where

$$\frac{d}{dt^{(q)}} = \frac{\partial}{\partial t} + U_j^{(q)}\frac{\partial}{\partial x_j} \tag{8-135}$$

that is, at each point in the space coordinates each component in the mixture can have its own streamlines. Moreover,

$$(\Delta_m)_{ji} = \sum_{(q)}\left[\frac{\rho^{(q)}}{\rho_m}\Delta_{ji}^{(q)} + \rho_m^{-1}\left(U_i^{(q)} - U_{mi}\right)\frac{\partial \rho^{(q)}}{\partial x_j} + \rho_m^{-1}\left(U_j^{(q)} - U_{mj}\right)\frac{\partial \rho^{(q)}}{\partial x_i}\right]$$

$$= \rho_m^{-1}\sum_{(q)}\left[\rho^{(q)}(\Delta_m)_{ji} + \frac{\partial J_i^{(q)}}{\partial x_j} + \frac{\partial J_j^{(q)}}{\partial x_i}\right] \tag{8-136}$$

and

$$\frac{\partial T}{\partial x_j} = \sum_{(q)}\left[\frac{\rho^{(q)} C_p^{(q)}}{\rho_m C_{pm}}\frac{\partial T^{(q)}}{\partial x_j} + T^{(q)}\frac{\partial}{\partial x_j}\frac{\rho^{(q)} C_p^{(q)}}{\rho_m C_{pm}} + \frac{\partial}{\partial x_j}\frac{U_j^{(q)} J_j^{(q)}}{2\rho_m C_{pm}}\right] \tag{8-137}$$

Substitution of Eqs. (8-133) to (8-137) into Eqs. (8-123) and (8-124) and rearranging gives, for component q, the following equations of momentum:

$$\frac{\rho^{(q)} dU_i^{(q)}}{dt^{(q)}} = \frac{\partial}{\partial x_j}\left[-p^{(q)}\delta_{ji} + \mu_m^{(q)}\Delta_{ji}^{(q)} + \mu_{m2}^{(q)}\Theta^{(q)}\delta_{ji}\right]$$

$$+\rho^{(q)}\mathscr{F}_i^{(q)} - \left(U_i^{(q)} - U_{mi}\right)\Gamma^{(q)}$$

$$+\rho^{(q)}\sum_{(p)} F^{(qp)}\left(U_i^{(p)} - U_i^{(q)}\right) \tag{8-138}$$

$F^{(qp)}$ is the time constant for momentum transfer, such that,

$$\sum_q \sum_p \rho^{(q)} F^{(qp)}\left[U^{(p)} - U^{(q)}\right] = 0 \tag{8-139}$$

For the energy,

$$\frac{\rho^{(q)} dE^{(q)}}{dt^{(q)}} = \frac{\partial}{\partial x_j} U_j^{(q)}\left[-p^{(q)}\delta_{ji} + \mu_m^{(q)}\Delta_{ji}^{(q)} + \mu_{m2}^{(q)}\Theta^{(q)}\delta_{ji}\right]$$

$$-\Gamma^{(q)}E^{(q)} + \frac{\partial}{\partial x_j}\left[\frac{\lambda_m^{(q)}\partial T^{(q)}}{\partial x_j}\right] + \mathscr{F}_E^{(q)}$$

$$+C_p^{(q)}\rho^{(q)}\sum_{(p)} G^{(qp)}\left(T^{(p)} - T^{(q)}\right) \tag{8-140}$$

The $G^{(qp)}$'s are the time constants of energy transfer, and

$$\sum_q \sum_p C_p^{(q)}\rho^{(q)} G^{(qp)}\left[T^{(p)} - T^{(q)}\right] = 0 \tag{8-141}$$

for the overall mixture.

The viscosity $\mu_m^{(q)}$ of component q in the mixture is defined by

$$\mu_m^{(q)}\Delta_{ji}^{(q)} = \mu_m \frac{\rho^{(q)}}{\rho_m}\Delta_{ji}^{(q)} + \frac{\mu_m}{\rho_m}\left[\left(U_i^{(q)} - U_{mi}\right)\frac{\partial \rho^{(q)}}{\partial x_j} + \left(U_j^{(q)} - U_{mj}\right)\frac{\partial \rho^{(q)}}{\partial x_i}\right]$$

$$+\rho^{(q)}\left(U_i^{(q)} - U_{mi}\right)\left(U_j^{(q)} - U_{mj}\right)$$

$$= \mu_m \frac{\rho^{(q)}}{\rho_m}\left(\Delta_m\right)_{ji} + \frac{\mu_m}{\rho_m}\left[\frac{\partial J_i^{(q)}}{\partial x_j} + \frac{\partial J_j^{(q)}}{\partial x_i}\right] + J_i^{(q)}\left(U_j^{(q)} - U_{mj}\right) \tag{8-142}$$

and

$$\mu_{m2}^{(q)}\Theta^{(q)} = \mu_{m2}\frac{\rho^{(q)}}{\rho_m}\Theta^{(q)} + \frac{\mu_{m2}}{\rho_m}\left(U_k^{(q)} - U_{mk}\right)\frac{\partial\rho^{(q)}}{\partial x_k}$$

$$= \mu_{m2}\frac{\rho^{(q)}}{\rho_m}\Theta_m + \frac{\mu_{m2}}{\rho_m}\frac{\partial J_k^{(q)}}{\partial x_k} \tag{8-143}$$

The heat flux to component q includes

$$\frac{\lambda_m^{(q)}\partial T^{(q)}}{\partial x_j} = \lambda_m\frac{\rho^{(q)}C_p^{(q)}}{\rho_m C_{pm}}\frac{\partial T^{(q)}}{\partial x_j} + \lambda_m T^{(q)}\frac{\partial}{\partial x_j}\frac{\rho^{(q)}C_p^{(q)}}{\rho_m C_{pm}}$$

$$+\lambda_m\frac{\partial}{\partial x_j}\left[\frac{\rho^{(q)}U_i^{(q)}\left(U_i^{(q)} - U_{mi}\right)}{\rho_m C_{pm}}\right] + \rho^{(q)}\left(U_j^{(q)} - U_{mj}\right)E^{(q)}$$

$$= \lambda_m\left\{\frac{\rho^{(q)}C_p^{(q)}}{\rho_m C_{pm}}\frac{\partial T^{(q)}}{\partial x_j} + T^{(q)}\frac{\partial}{\partial x_j}\frac{\rho^{(q)}C_p^{(q)}}{\rho_m C_{pm}} + \frac{\partial}{\partial x_j}\left(\frac{U_i^{(q)}J_i^{(q)}}{\rho_m C_{pm}}\right)\right\}$$

$$+J_j^{(q)}E^{(q)} \tag{8-144}$$

The above component equations (8-121), (8-128), and (8-140) treat the components of a mixture as interacting continua. They do not incorporate the specific behavior of particles (e.g., fuel drops) suspended in a medium (e.g., oxidant).

In the case of a reactive suspension, consisting of a fluid phase and a particulate phase, the fluid phase may consist of a reacting gaseous mixture or a reacting solution, and the particulate phase may consist of solid particles or liquid droplets. The particulate phase is identified by the subscript (p), and additional phases among particles by superscripts (s) and (r); the unsubscripted quantities denote the fluid phase, and different molecular species in the fluid phase are identified by superscripts (q) and p).

The continuity equation in the fluid phase becomes

$$\frac{\partial\rho^{(q)}}{\partial t} + \frac{\partial}{\partial x_i}\left[\rho^{(q)}\left(U_i + V_i^{(q)}\right)\right] = -\Gamma^{(q)} + K^{(q)} \tag{8-145}$$

where $V_i^{(q)}$ is the ith component of the diffusion velocity of species q, $\Gamma^{(q)}$ is the rate of generation (mass per unit volume and time) of species q going into the particulate phase of condensed phase, and $K^{(q)}$ is the rate at which species q is generated by reaction in gaseous phase. The total rate of generation to the particulate phase is given by

$$\sum_q \Gamma^{(q)} = \Gamma \tag{8-146}$$

and for the net conversion in the vapor phase

$$\sum_q K^{(q)} = 0, \qquad \sum \mathbf{J}^{(q)} = 0 \tag{8-147}$$

Therefore, for the fluid (liquid or gaseous) phase, assumed to have only one mass velocity, the overall continuity equation is

$$\frac{\partial \rho}{\partial t} + \frac{\partial}{\partial x_i} \rho U_i = -\Gamma \tag{8-148}$$

with

$$\rho = \bar{\rho}\left(1 - \sum_s \phi^{(s)}\right) \tag{8-149}$$

as before, where $\phi^{(s)}$ is the volume fraction of species s in the condensed (solid) phase.

For the particulate phases, the continuity equation is

$$\frac{\partial \rho_p^{(s)}}{\partial t} + \frac{\partial}{\partial x_i}\left[\rho_p^{(s)} U_{pi}^{(s)}\right] = \Gamma^{(s)} + K_p^{(s)} \tag{8-150}$$

where $U_{pi}^{(s)}$ includes ith component of the diffusion velocity of species s. $K_p^{(s)}$ is the generation rate of species s by breakup or coalescence of other particulate species. Therefore, in the general sense of the multiphase system, transition from species s to species r through change in particle size by evaporation or condensation is in itself a phase change. Thus

$$\sum_s \Gamma^{(s)} = \Gamma = \sum_q \Gamma^{(q)} \tag{8-151}$$

The latter equality follows by Eq. (8-145) with

$$\sum_{(s)} K_p^{(s)} = 0 \tag{8-152}$$

since the mass of the particulate phase is not altered by $K_p^{(s)}$.

Similarly, the equations of conservation for momentum and energy can be formulated for both the fluid phase and particulate phase, completing the continuum model.

The limitations and drawbacks of the continuum-mechanics approach to multiple-phase problems are noted and discussed in the next section.

11 COMBUSTION OF SOLID PARTICLES IN FLUIDIZED BEDS

11.1 Introduction

Hot gases flowing through an aggregate of combustible solid particles can produce high rates of energy release because of the large burning surface area per unit volume of the solid particles available for ignition and combustion. The subject of fluidized-bed combustion, therefore, has drawn considerable interest from researchers in recent years. In particular, combustion of granular solid propellants in gun propulsion systems has been studied by many investigators[138,142,145] because of its potential for producing high thrusts within extremely short time intervals.

When high-temperature gases enter a packed bed of solid propellant grains, the sequence of events can be described as follows: propellant grains are heated by hot gases; some of the heated particles are ignited and produce more hot product gases to ignite more propellant grains; the high pressure gradients generated along the granular bed cause compaction of the bed; the increased pressure gradient also causes the ignition front to accelerate; at a later stage, the projectile starts to move and the granular bed expands and becomes fluidized. Obviously, the physical process of granular propellant combustion is very involved and requires simplifying assumptions for solution. Several theoretical models have been proposed for this purpose. These can be classified into four categories: (1) two-phase fluid-dynamic models, (2) formal averaging models, (3) continuum-mechanics models, and (4) statistical models.

Of the abovementioned models, only the first two will be discussed in this section. The continuum-mechanics methods[137] and the associated governing equations were given in Section 10 and will not be repeated here. It should be noted that the continuum-mechanics method for this problem has several inconveniences: (1) the coexistence of solid and gas at any spatial location must be assumed, but is not very realistic, especially for granular propellants with large particles; (2) the conservation equations for solid and gas phases are obtained by splitting the equations for the mixture, a procedure that involves some arbitrariness; and (3) when steep temperature gradients exist in the solid phase, the energy equation for the solution phase, obtained by splitting the mixture energy equation, cannot describe the particle surface temperature or the subsurface temperature gradient, so the flame spreading rate and the heat-transfer conditions at the solid–gas interface cannot be described adequately.

Statistical methods[139–141] have an inherent problem in that there is not enough information available to evaluate the net contribution of the microscopic fluctuation terms unless one has at one's disposal the pdfs and reliable statistical data. Therefore, one faces difficulties in closing the problem.

The two-phase fluid-dynamics[142] formulation is quite straightforward and very easy to understand. This will become clear in the following section, where a model developed by Kuo et al.[142] is introduced. For separated-flow models, a formal averaging method was developed by Gough and Zwartz.[145] In this model, one must initially consider the balance of fluxes from the microscopic point of view. The involved integrals in the conservation equations are simplified by proper definitions of averaged quantities. And finally, after much calculation, one obtains a set of governing equations essentially the same as those obtained from the two-phase fluid-dynamics model, with very small differences.

11.2 Two-Phase Fluid-Dynamics Model

In order to determine the transient gas-dynamical behavior of hot-gas penetration, flame propagation, chamber pressurization, and combustion processes in the fluidized bed, the mass, momentum, and energy equations for the gas phase, and the mass and momentum equations for the solid phase, are derived and expressed in a quasi-one-dimensional form. They are approached by considering the balance of fluxes over a control volume small enough to give the desired spatial distributions in the complete system but large enough to contain many solid particles, so that the averaged particle velocity and fractional porosity are meaningful.

The control volume for the gas phase is the portion of void volume occupied by it in a small elementary volume ($A\,\Delta x$); the remaining portion, occupied by the particles, is taken as the control volume for the particle phase. Figure 8.39 shows these control volumes.

The fractional porosity ϕ is defined as

$$\phi \equiv 1 - \frac{\tilde{n} m_p}{\rho_p} = \frac{\text{void volume}}{\text{total volume}} \tag{8-153}$$

where:

$\tilde{n}$ is the number density (number of particles per unit spatial volume).
m_p is the mass of each particle (M).
ρ_p is the density of the particles (M/L^3).

For a system of spherical particles of uniform size, we have

$$\phi = 1 - \tfrac{4}{3}\pi r_p^3 \tilde{n} \tag{8-154}$$

or

$$\tilde{n} = \frac{3(1-\phi)}{4\pi r_p^3}$$

The specific wetted surface area A_s can be expressed as follows:

$$A_s = S_B \tilde{n} = 4\pi r_p^2 \frac{3(1-\phi)}{4\pi r_p^3} = \frac{3(1-\phi)}{r_p} \tag{8-155}$$

where S_B is the burning surface of a spherical particle.

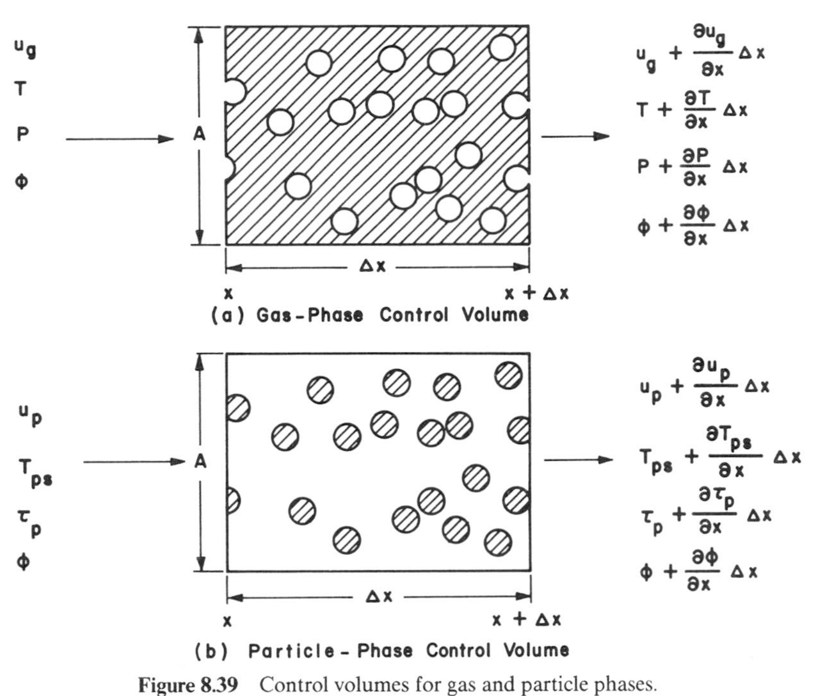

Figure 8.39 Control volumes for gas and particle phases.

11.2.1 Gas-Phase Mass Equation

Consider a duct of constant area A. The average flow cross-sectional area can be represented by ϕA. The mass flux convected into the control volume is (Fig. 8.40)

$$\cancel{\rho A \phi u_g} - \left[\cancel{\rho A \phi u_g} + \frac{\partial(\rho A \phi u_g)}{\partial x} \Delta x \right] = -\frac{\partial(\rho \phi u_g)}{\partial x} A \, \Delta x$$

The rate of gaseous mass addition due to gasification of solid particles is

$$A_s A \, \Delta x \, \rho_p r_b$$

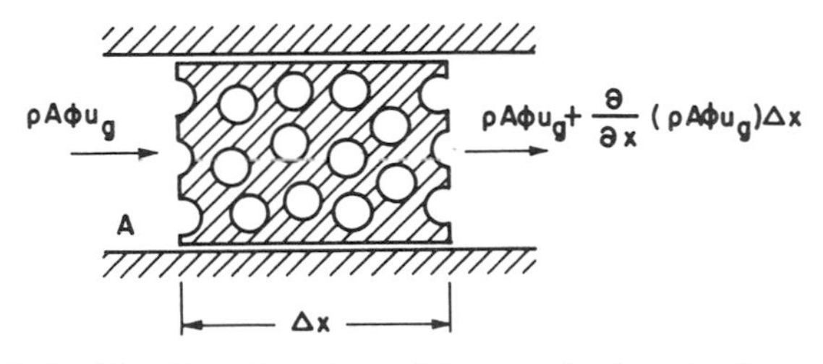

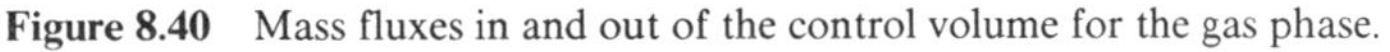
Figure 8.40 Mass fluxes in and out of the control volume for the gas phase.

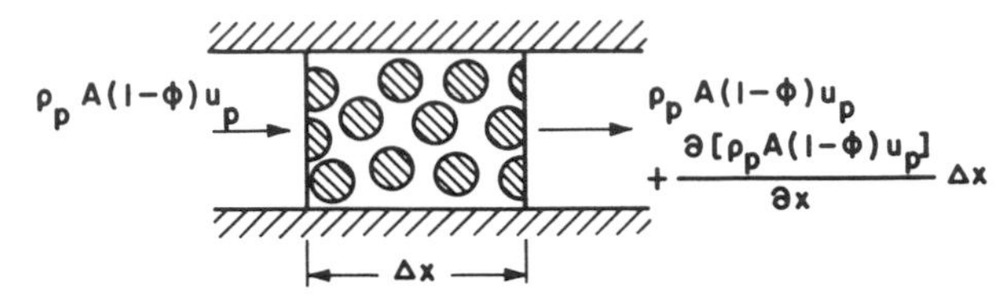

Figure 8.41 Mass fluxes in and out of the particle-phase control volume.

The rate of gaseous mass accumulation in the control volume occupied by gases is

$$\frac{\partial}{\partial t}(\rho \phi A\,\Delta x)$$

The mass balance therefore becomes

$$\frac{\partial(\rho\phi)}{\partial t}(A\,\Delta x) = -\frac{\partial(\rho\phi u_g)}{\partial x}(A\,\Delta x) + A_s \rho_p r_b (A\,\Delta x)$$

$$\therefore \quad \frac{\partial(\rho\phi)}{\partial t} + \frac{\partial(\rho\phi u_g)}{\partial x} = A_s \rho_p r_b \tag{8-156}$$

Note that when there is no particle in the spatial volume $A\,\Delta x$, then $\phi = 1$ and $A_s = 0$, so that the above equation reduces to the conventional continuity equation for unsteady, one-dimensional, constant-area flow.

11.2.2 Solid-Phase Mass Equation

From Fig. 8.41, we see that the rate of particle mass convection into the control volume occupied by particles is

$$\rho_p A(1-\phi)u_p - \left\{\rho_p A(1-\phi)u_p + \frac{\partial\left[\rho_p A(1-\phi)u_p\right]}{\partial x}\right\}\Delta x$$

$$= \frac{\partial\left[\rho_p(1-\phi)u_p\right]}{\partial x} A\,\Delta x$$

The rate of particle mass reduction due to gasification of solid particles is

$$-A_s A\,\Delta x\,\rho_p r_b$$

The rate of particle mass accumulation in the control volume occupied by particles is

$$\frac{\partial}{\partial t}\left[\rho_p(1-\phi)A\,\Delta x\right]$$

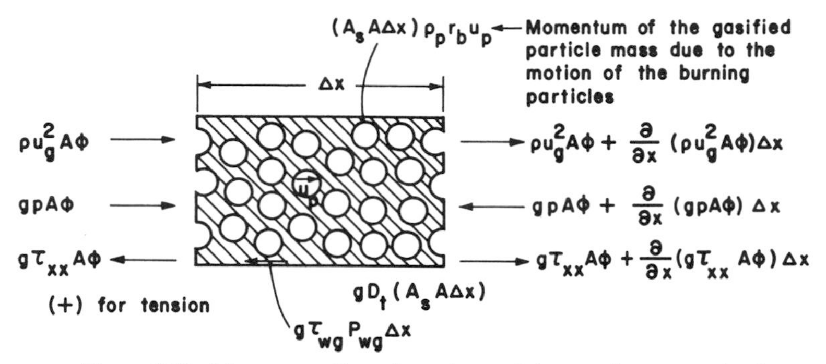

Figure 8.42 Momentum fluxes in and out of the gas-phase control volume.

The mass balance can be written as

$$\frac{\partial\left[\rho_p(1-\phi)\right]}{\partial t}+\frac{\partial\left[\rho_p(1-\phi)u_p\right]}{\partial x}=-A_s\rho_p r_b \tag{8-157}$$

Note that the sign of the source term in Eq. (8-157) is opposite to that in Eq. (8-156). When Eq. (8-156) and Eq. (8-157) are added, the source term vanishes, implying that the mass of the gas–solid mixture is conserved.

11.2.3 Gas-Phase Momentum Equation

The derivation of the gas-phase momentum equation (Fig. 8.42) is based upon the following balance:

rate of increase of momentum

= net rate of momentum flux into control volume

+ summation of forces acting on the control volume

The rate of increase of momentum is

$$\frac{\partial}{\partial t}\left(\rho\phi A\,\Delta x\,u_g\right)=\frac{\partial\left(\rho\phi u_g\right)}{\partial t}A\,\Delta x$$

The net rate of momentum flux into the control volume is

$$\rho u_g^2 A\phi-\left[\rho u_g^2 A\phi+\frac{\partial}{\partial x}\left(\rho u_g^2 A\phi\right)\Delta x\right]+A_s A\,\Delta x\,\rho_p r_b u_p$$

$$=\left[-\frac{\partial}{\partial x}\left(\rho u_g^2\phi\right)+A_s\rho_p r_b u_p\right]A\,\Delta x$$

The summation of forces acting on the control volume is

$$gpA\phi - \left[gpA\phi + \frac{\partial}{\partial x}(gpA\phi)\,\Delta x\right] - gD_tA_sA\,\Delta x$$

$$-g\tau_{xx}A\phi + \left[g\tau_{xx}A\phi + \frac{\partial}{\partial x}(g\tau_{xx}A\phi)\,\Delta x\right] - g\tau_{wg}P_{wg}\,\Delta x$$

$$= g\left[-\frac{\partial}{\partial x}(p\phi) - D_tA_s + \frac{\partial}{\partial x}(\tau_{xx}\phi) - \frac{\tau_{wg}P_{wg}}{A}\right]A\,\Delta x$$

where P_{wg} is the perimeter between the wall and gas phase.

$$\therefore \quad \frac{\partial(\rho\phi u_g)}{\partial t} = \left[-\frac{\partial}{\partial x}\left(\rho u_g^2\phi\right) + A_s\rho_p r_b u_p\right]$$

$$+\left[-g\frac{\partial}{\partial x}(p\phi) - gD_tA_s + g\frac{\partial}{\partial x}(\tau_{xx}\phi) - g\frac{\tau_{wg}P_{wg}}{A}\right]$$

or

$$\frac{\partial(\rho\phi u_g)}{\partial t} + \frac{\partial\left(\rho\phi u_g^2\right)}{\partial x} = -g\frac{\partial}{\partial x}(p\phi) + A_s\rho_p r_b u_p - gA_sD_t$$

$$+g\frac{\partial}{\partial x}(\tau_{xx}\phi) - g\frac{\tau_{wg}P_{wg}}{A} \qquad (8\text{-}158)$$

The total drag force between the gas and particle phases can be represented by the sum of D_v and D_p, where D_v represents the drag due to the relative velocity between gas and particles, and D_p represents the drag due to the porosity gradient. The total drag can be written as

$$D_t \equiv D_v + D_p = D_v - \frac{p}{A_s}\frac{\partial\phi}{\partial x}$$

After substitution of D_t into Eq. (8-158), we have

$$\frac{\partial(\rho\phi u_g)}{\partial t} + \frac{\partial\left(\rho\phi u_g^2\right)}{\partial x} = -g\phi\frac{\partial p}{\partial x} + A_s\rho_p r_b u_p - gA_sD_v$$

$$+g\frac{\partial}{\partial x}(\tau_{xx}\phi) - g\frac{\tau_{wg}P_{wg}}{A} \qquad (8\text{-}159)$$

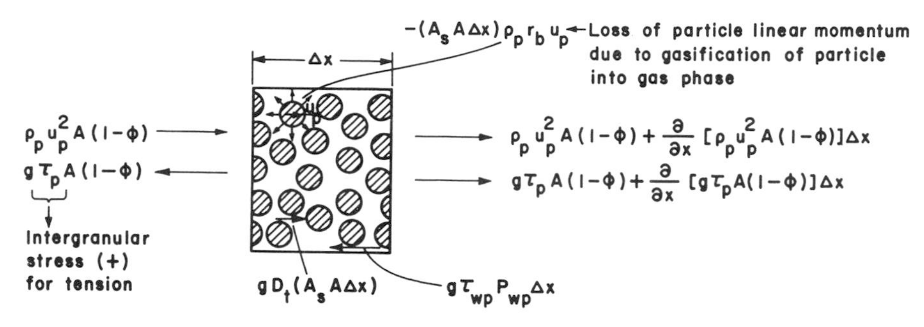

Figure 8.43 Momentum fluxes in and out of the particle-phase control volume.

11.2.4. Solid-Phase Momentum Equation

The derivation of the solid-phase momentum equation is based upon the following balance:

rate of increase of particle momentum

= net rate of momentum flux into control volume

+ summation of forces acting on particle control volume

The momentum fluxes in and out of the particle-phase control volume are shown in Fig. 8.43. The rate of increase of particle momentum is

$$\frac{\partial}{\partial t}\left[\rho_p(1-\phi)A\,\Delta x\,u_p\right] = \frac{\partial\left[\rho_p(1-\phi)u_p\right]}{\partial t}A\,\Delta x$$

The net rate of momentum flux into the particle control volume is

$$\rho_p u_p^2 A(1-\phi) - \left\{\rho_p u_p^2 A(1-\phi) + \frac{\partial}{\partial x}\left[\rho_p u_p^2 A(1-\phi)\right]\Delta x\right\}$$

$$-(A_s A\,\Delta x)\rho_p r_b u_p$$

$$= \left\{-\frac{\partial}{\partial x}\left[\rho_p u_p^2(1-\phi)\right] - A_s \rho_p r_b u_p\right\}A\,\Delta x$$

The summation of forces acting on the particle control volume is

$$-g\tau_p A(1-\phi) + \left\{g\tau_p A(1-\phi) + \frac{\partial}{\partial x}\left[g\tau_p A(1-\phi)\right]\Delta x\right\}$$

$$+gD_t A_s A\,\Delta x - g\tau_{wp}P_{wp}\,\Delta x$$

$$= g\left\{\frac{\partial}{\partial x}\left[\tau_p(1-\phi)\right] + A_s D_t - \frac{\tau_{wp}P_{wp}}{A}\right\}A\,\Delta x$$

where P_{wp} is the perimeter between wall and particles phase.

$$\therefore \quad \frac{\partial\left[\rho_p(1-\phi)u_p\right]}{\partial t} = \left\{-\frac{\partial\left[\rho_p u_p^2(1-\phi)\right]}{\partial x} - A_s\rho_p r_b u_p\right\} + \left\{g\frac{\partial\left[\tau_p(1-\phi)\right]}{\partial x} + gA_sD_t - g\frac{\tau_{wp}P_{wp}}{A}\right\}$$

or

$$\frac{\partial\left[\rho_p(1-\phi)u_p\right]}{\partial t} + \frac{\partial\left[\rho_p(1-\phi)u_p^2\right]}{\partial x} = g\frac{\partial\left[\tau_p(1-\phi)\right]}{\partial x} - A_s\rho_p r_b u_p + gA_sD_t - g\frac{\tau_{wp}P_{wp}}{A} \tag{8-160}$$

Note that:

1. τ_p is equivalent to $\tau_{xx} - p$ in the gas phase.
2. $1 - \phi$ in the solid-phase momentum equation is equivalent to ϕ in the gas-phase momentum equation.
3. The momentum equation for the solid–gas mixture can be obtained by adding Eq. (8-158) and Eq. (8-160):

$$\frac{\partial\left[\phi\rho u_g\right]}{\partial t} + \frac{\partial\left[(1-\phi)\rho_p u_p\right]}{\partial t} + \frac{\partial\left[\phi\rho u_g^2\right]}{\partial x} + \frac{\partial\left[(1-\phi)\rho_p u_p^2\right]}{\partial x} - g\frac{\partial}{\partial x}\left[\phi(\tau_{xx}-p)\right] - g\frac{\partial}{\partial x}\left[(1-\phi)\tau_p\right] = -g\frac{\tau_{wg}P_{wg}}{A} - g\frac{\tau_{wp}P_{wp}}{A} \tag{8-161}$$

When the shear stress between the gas and tube wall (τ_{wg}) and that between the particle and tube wall (τ_{wp}) are zero, the above equation reduces to

$$\frac{\partial\left[\phi\rho u_g + (1-\phi)\rho_p u_p\right]}{\partial t} + \frac{\partial\left[\phi\rho u_g^2 + (1-\phi)\rho_p u_p^2\right]}{\partial x} = g\frac{\partial}{\partial x}\left[\phi(\tau_{xx}-p) + (1-\phi)\tau_p\right] \tag{8-162}$$

The above equation can be interpreted as

rate of change of momentum of the gas–solid mixture

+ net momentum flux out of spatial volume $A\,\Delta x$

= total force due to normal stresses in x-direction

4. When $\phi = 1$, Eq. (8-161) becomes

$$\frac{\partial \rho u_g}{\partial t} + \frac{\partial \rho u_g^2}{\partial x} = -g\frac{\partial p}{\partial x} + g\frac{\partial \tau_{xx}}{\partial x} - g\frac{\tau_{wg} P_{wg}}{A} \tag{8-163}$$

which is the conventional momentum equation for one-dimensional, unsteady, single-phase flow through a constant-area duct.

5. When the flow is steady and inviscid, the above equation reduces to

$$\frac{d}{dx}\left[\rho u_g^2 + gp\right] = 0$$

or

$$\rho u_g^2 + gp = \text{constant} \tag{8-164}$$

which implies that the stream thrust is unchanged in a steady inviscid flow.

11.2.5 Gas-Phase Energy Equation

The first law of thermodynamics says

$$\underbrace{dE}_{\substack{\text{increase}\\ \text{in stored}\\ \text{energy of}\\ \text{the system}}} = \underbrace{\delta Q}_{\substack{\text{energy}\\ \text{input to}\\ \text{system}}} + \underbrace{\delta W}_{\substack{\text{work done}\\ \text{on system}}}$$

Let

$$E = \text{total stored energy} = e + \frac{u_g^2}{2gJ}$$

The rate of change of the total stored energy can then be written as

$$\frac{dE}{dt} = \frac{\delta Q}{\delta t} + \frac{\delta W}{\delta t}$$

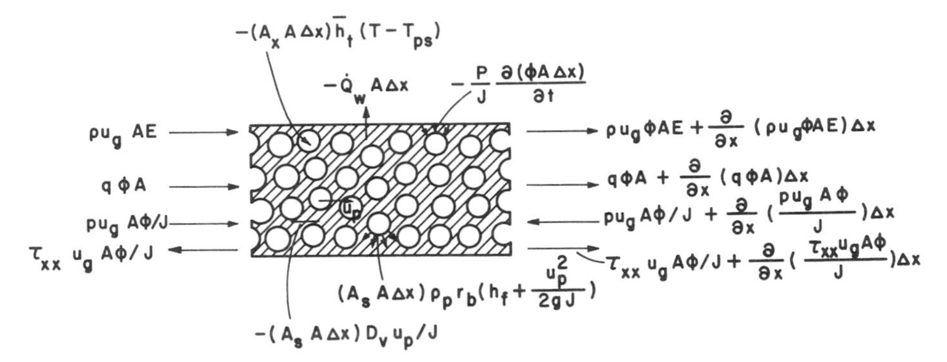

Figure 8.44 Energy fluxes in and out of the gas-phase control volume.

From Fig. 8.44 we see that the rate of increase of total stored energy is

$$\frac{\partial}{\partial t}[\rho E\phi A\,\Delta x] = \frac{\partial(\rho\phi E)}{\partial t}A\,\Delta x$$

The energy flux convected into the control volume is

$$\rho u_g\phi AE - \left[\rho u_g\phi AE + \frac{\partial}{\partial x}(\rho u_g\phi AE)\,\Delta x\right] = \frac{-\partial(\rho u_g\phi E)}{\partial x}A\,\Delta x$$

The heat conducted into the control volume is

$$q\phi A - \left[q\phi A + \frac{\partial}{\partial x}(q\phi A)\,\Delta x\right] = -\frac{\partial(q\phi)}{\partial x}A\,\Delta x$$

The rate of heat loss to solid particles is

$$-A_s\bar{h}_t(T - T_{ps})A\,\Delta x$$

The rate of heat loss to the tube wall is

$$-\dot{Q}_W A\,\Delta x$$

The rate of energy input due to particle burning is

$$(A_s A\,\Delta x)\rho_p r_b\left(h_f + \frac{u_p^2}{2gJ}\right)$$

The rate of pressure work on the end surfaces of the control volume is

$$\cancel{\frac{p}{J}u_g A\phi} - \left[\cancel{\frac{p}{J}u_g A\phi} + \frac{\partial}{\partial x}\left(\frac{p}{J}u_g A\phi\right)\Delta x\right] = -\frac{1}{J}\left[\frac{\partial}{\partial x}(pu_g\phi)\right]A\,\Delta x$$

The rate of work done by viscous normal stresses is

$$-\cancel{\frac{\tau_{xx}u_g A\phi}{J}} + \cancel{\frac{\tau_{xx}u_g A\phi}{J}} + \frac{\partial}{\partial x}\left[\frac{\tau_{xx}u_g A\phi}{J}\right]\Delta x = \frac{1}{J}\left[\frac{\partial}{\partial x}(\tau_{xx}u_g\phi)\right]A\,\Delta x$$

The rate of work done by the drag force acting on the gas phase due to the solid particles is

$$-(A_s A\,\Delta x)\frac{D_v}{J}u_p = -\frac{A_s D_v u_p}{J}A\,\Delta x$$

The rate of pressure work for the dilatation of the gaseous control volume is

$$-\frac{p}{J}\frac{\partial(\phi A\,\Delta x)}{\partial t} = -\frac{p}{J}\frac{\partial\phi}{\partial t}A\,\Delta x$$

Substituting the above terms into the first law of thermodynamics, we have

$$\frac{\partial(\rho\phi E)}{\partial t} + \frac{\partial(\rho u_g\phi E)}{\partial x} + \frac{1}{J}\frac{\partial(pu_g\phi)}{\partial x}$$

$$= A_s\rho_p r_b\left(h_f + \frac{u_p^2}{2gJ}\right) - A_s\bar{h}_t(T - T_{ps}) - \frac{A_s D_v u_p}{J}$$

$$-\frac{p}{J}\frac{\partial\phi}{\partial t} - \frac{\partial(q\phi)}{\partial x} + \frac{1}{J}\frac{\partial(\tau_{xx}u_g\phi)}{\partial x} - \dot{Q}_W \tag{8-165}$$

11.2.6 Solid-Phase Heat-Conduction Equation

The heat equation written in terms of the Lagrangian time derivative and the spherical coordinates of the particle (Fig. 8.45) is

$$\left(\frac{DT_p}{Dt}\right)_p = \frac{\alpha_p}{r}\frac{\partial^2(rT_p)}{\partial r^2} \tag{8-166}$$

with initial condition

$$T_p(o, r) = T_0 \tag{8-167}$$

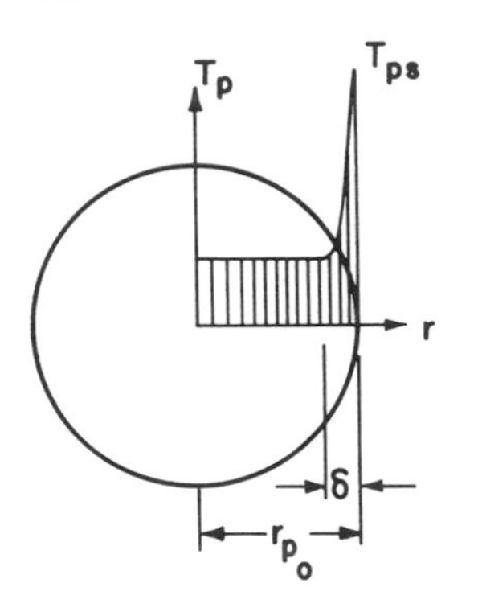

Figure 8.45 Temperature profile inside a spherical particle.

and boundary conditions

$$\frac{\partial T_p}{\partial r}(t, o) = 0 \tag{8-168}$$

$$\frac{\partial T_p}{\partial r}(t, r_{po}) = \frac{\bar{h}_t(t)}{\lambda_p}\left[T(t) - T_{ps}(t)\right] \tag{8-169}$$

where

$$\bar{h}_t(t) \equiv \bar{h}_c(t) + \bar{h}_r(t) = \bar{h}_c(t) + \epsilon_p \sigma \left[T(t) + T_{ps}(t)\right]\left[T^2(t) + T_{ps}^2(t)\right] \tag{8-170}$$

ϵ_p is the average emissivity of the particle and σ is the Stefan–Boltzmann's constant. The heatup equation and its initial and boundary conditions can be recast by an integral method to yield a first-order ordinary differential equation describing the increase of propellant surface temperature with respect to time:

$$\left(\frac{DT_{ps}}{Dt}\right)_p = \frac{\dfrac{12\alpha_p}{\delta r_{po}}\left[T_{ps} - T_o + \dfrac{r_{po}\bar{h}_t}{\lambda_p}(T - T_{ps})\right] + \delta\left[\dfrac{\bar{h}_t}{\lambda_p}\left(\dfrac{DT}{Dt}\right)_p + \dfrac{T - T_{ps}}{\lambda_p}\left(\dfrac{D\bar{h}_t}{Dt}\right)_p\right]}{\left[\dfrac{6r_{po} - \delta}{r_{po}} + \dfrac{\bar{h}_t\delta}{\lambda_p}\right]} \tag{8-171}$$

where δ is the thermal-wave penetration depth in a spherical particle. As shown by Kuo et al.,[147] it can be calculated from

$$\delta(t) = \frac{3r_{po}\left[T_{ps}(t) - T_o\right]}{\left[T_{ps}(t) - T_o\right] + \left[r_{po}\bar{h}_t(t)/\lambda_p\right]\left[T(t) - T_{ps}(t)\right]} \tag{8-172}$$

11.2.7 Equation of State

For the gas-phase the Noble–Abel dense gas law can be used:

$$p\left(\frac{1}{\rho} - b\right) = RT \tag{8-173}$$

The solid particles can be assumed to be incompressible, that is,

$$\rho_p = \text{constant} \tag{8-174}$$

In order to close the problem, the following four correlations are needed:

1. constitutive law for intergranular stress, τ_p;
2. drag correlation for D_v;
3. convective-heat-transfer ($\bar{h}_c$) correlation; and
4. a burning-rate (r_b) law for the solid particle.

Depending upon the particle geometry and flow situations, suitable correlations can be obtained from the literature. It is obvious that a closed-form analytical solution of the set of equations derived above is not possible. Hence, they must be solved using numerical techniques.

11.2.8 Boundary Conditions for Combustion Processes in a Mobile Granular Propellant Bed

In the modeling of transient one-dimensional two-phase combustion of granular propellants in gun interior ballistics, the boundary conditions at both the breech end and the base of the projectile must be specified adequately. It was stressed by Chen et al.[148] that the form and the total number of the boundary conditions required depend upon the relative velocities of the gases and solid particles with respect to the solid boundary and the condition of fluidization. In the study of a simulated gun system, the flow properties at the boundary are obtained by considering (1) the local balances of mass, momentum, and energy over a small control volume adjacent to the boundary, (2) the compatibility relationships along the characteristic lines, and (3) the number of algebraic relationships according to the instantaneous flow conditions. These considerations are necessary to provide the extraneous boundary conditions required for solving the partial differential equations with a second-order numerical scheme. Numerical results were compared and found in agreement with test-firing data. To limit the scope of presentation in this chapter, detailed mathematical formulation of the boundary conditions is not included. Interested readers are referred to the papers cited by Chen et al.[148]

A novel approach to the theoretical formulation of two-phase interpenetrating media was developed by Kuo et al.[150] In this approach a comprehensive theoretical model for predicting the flame-spreading, combustion, and grain deformation phenomena of long unslotted stick propellants was presented. The formulation is based upon a combined Eulerian–Lagrangian approach to simulate special characteristics of the two-phase combustion processes in a cartridge loaded with a bundle of sticks. Interested readers are referred to the original paper for detailed discussions.

11.3 Formal Averaging Method

As the name suggests, in this method a formal mathematical definition of local mean variables is used to derive the macroscopic two-phase conservation equations from the point or microscopic conservation equations. This method was first introduced by Anderson and Jackson.[143] Later, Whitaker[144] and then Gough and Zwarts[145] used the same method, though in a slightly different form.

In this method, a weighting function is defined to obtain the average values of properties. Following Anderson and Jackson,[143] we shall define a weighting function which depends on spatial coordinates only. Gough and Zwarts[145] have defined a weighting function which depends on both the spatial coordinates and time. The proof of the basic mathematical relationships used by them, which relate the average of a spatial derivative of a property to the spatial derivative of the mean property, is given in Ref. 146.

11.3.1 Definitions

The following definition and approximations are used for the derivation of the governing equations using the formal averaging method.

1. Let $g(r)$ be a function, defined for $r > 0$ and $t > 0$, with the following properties:

   a. $g(r) \geq 0$ for all r, and g decreases monotonically with increasing r.
   b. $g(r)$ possess derivatives $g^n(r)$ of all orders for each value of r.
   c. $\int_{V_\infty} g^n(r)\, dV$ exists for all values of n, where r denotes the distance from a point in three-dimensional space, and V_∞ represents the whole space. The weighting function is normalized so that

$$\int_{V_\infty} g(r)\, dV = 1 \tag{8-175}$$

2. The local mean void fraction ϕ at a time t and location $\mathbf{x}$ is defined as

$$\phi(\mathbf{x}, t) = \int_{V_{f\infty}(t)} g(\mathbf{x} - \mathbf{y})\, dV_y \tag{8-176}$$

where $V_{f\infty}(t)$ is the volume occupied by the fluid at time t, and dV_y is an elemental volume in the neighborhood of point $\mathbf{y}$.

3. The local mean value of the gas property $\psi(x, t)$ is defined by

$$\langle \psi(\mathbf{x}, t) \rangle = \frac{1}{\phi(\mathbf{x}, t)} \int_{V_{f\infty}(t)} g(\mathbf{x} - \mathbf{y}) \psi(\mathbf{y}, t) \, dV_y \tag{8-177}$$

4. The spatial derivatives of the mean quantities can be expressed as

$$\begin{aligned}
\frac{\partial}{\partial x_i}[\phi(\mathbf{x}, t)\langle \psi(\mathbf{x}, t)\rangle] &\equiv \frac{\partial}{\partial x_i}\left[\int_{V_{f\infty}(t)} g(\mathbf{x} - \mathbf{y}) \psi(\mathbf{y}, t) \, dV_y\right] \\
&= \int_{V_{f\infty}(t)} \frac{\partial}{\partial x_i}[g(\mathbf{x} - \mathbf{y}) \psi(\mathbf{y}, t)] \, dV_y \\
&= \int_{V_{f\infty}(t)} \psi(\mathbf{y}, t) \frac{\partial}{\partial x_i} g(\mathbf{x} - \mathbf{y}) \, dV_y \\
&= -\int_{V_{f\infty}(t)} \psi(\mathbf{y}, t) \frac{\partial}{\partial y_i} g(\mathbf{x} - \mathbf{y}) \, dV_y \\
&= -\int_{V_{f\infty}(t)} \frac{\partial}{\partial y_i}[\psi(\mathbf{y}, t) g(\mathbf{x} - \mathbf{y})] \, dV_y \\
&\quad + \int_{V_{f\infty}(t)} g(\mathbf{x} - \mathbf{y}) \frac{\partial}{\partial y_i} \psi(\mathbf{y}, t) \, dV_y
\end{aligned} \tag{8-178}$$

The first term on the right-hand side of Eq. (8-178) can be changed into a surface integral using Green's theorem:

$$\int_{V_{f\infty}(t)} \frac{\partial}{\partial y_i}[\psi(\mathbf{y}, t) g(\mathbf{x} - \mathbf{y})] \, dV_y = \int_{s_f(t)} \psi(\mathbf{y}, t) g(\mathbf{x} - \mathbf{y}) n_i \, ds \tag{8-179}$$

where $s_f(t)$ is the surface bounding the fluid phase at time t. Also, $s_f(t) = s_{f\infty}(t) + s_p(t)$, where $s_{f\infty}(t)$ is the surface bounding the whole system, and $s_p(t)$ is the fluid–solid interface surface. Therefore, the right-hand side of Eq. (8-179) can be written as

$$\int_{s_{f\infty}(t)} \psi(\mathbf{y}, t) g(\mathbf{x} - \mathbf{y}) n_i \, ds - \int_{s_p(t)} \psi(\mathbf{y}, t) g(\mathbf{x} - \mathbf{y}) n_i \, ds \tag{8-180}$$

[In the second term of (8-180), n_i drawn outward from the propellant surface is taken as positive; hence the negative sign.] Now, provided the shortest distance from point $\mathbf{x}$ to the surface $s_{f\infty}$ is considerably larger than the distance r_0, over which the weighting function has nonzero value, the first term of Eq. (8-180) will be considerably smaller than the second term. Neglecting this term and using Eq. (8-179) and Eq. (8-180), we can write Eq. (8-178) as follows:

$$\int_{V_{f\infty}(t)} g(\mathbf{x}-\mathbf{y}) \frac{\partial}{\partial y_i} \psi(\mathbf{y}, t)\, dV_y = \frac{\partial}{\partial x_i}[\phi(\mathbf{x}, t)\langle \psi(\mathbf{x}, t)\rangle] - \int_{s_p(t)} \psi(\mathbf{y}, t) g(\mathbf{x}-\mathbf{y}) n_i\, ds \quad (8\text{-}181)$$

5. The time derivative of the mean quantities can be expressed as follows:

$$\frac{\partial}{\partial t}[\phi(\mathbf{x}, t)\langle \psi(\mathbf{x}, t)\rangle] \equiv \frac{\partial}{\partial t}\left[\int_{V_{f\infty}(t)} g(\mathbf{x}-\mathbf{y}) \psi(y, t)\, dV_y\right] \quad (8\text{-}182)$$

Using Leibnitz's rule, the right-hand side of Eq. (8-182) can be written as

$$\int_{V_{f\infty}(t)} \frac{\partial}{\partial t}[g(\mathbf{x}-\mathbf{y}) \psi(\mathbf{y}, t)]\, dV_y + \int_{s_f(t)} \psi(\mathbf{y}, t) g(\mathbf{x}-\mathbf{y}) v_i n_i\, ds \quad (8\text{-}183)$$

The second term of Eq. (8-183) can be approximated for integration over the surface $s_p(t)$ as explained before. Also note that the integrand of the first term in Eq. (8-183) is equal to

$$g(\mathbf{x}-\mathbf{y}) \frac{\partial}{\partial t} \psi(\mathbf{y}, t) \quad (8\text{-}184)$$

Using Eqs. (8-183) and (8-184), we can express Eq. (8-182) as

$$\int_{V_{f\infty}(t)} g(\mathbf{x}-\mathbf{y}) \frac{\partial}{\partial t} \psi(\mathbf{y}, t)\, dV_y = \frac{\partial}{\partial t}[\phi(\mathbf{x}, t)\langle \psi(\mathbf{x}, t)\rangle] + \int_{s_p(t)} \psi(\mathbf{y}, t) g(\mathbf{x}-\mathbf{y}) v_i n_i\, ds \quad (8\text{-}185)$$

6. $\int_{s_p} g\, ds$ can be considered as the average interface surface area per unit volume, and accordingly

$$\int_{s_p} g\, ds = \frac{(1-\phi) s_p}{V_p} \quad (8\text{-}186)$$

7. The surface average of a gas-phase property ψ is defined as

$$\langle \psi \rangle^s \equiv \frac{\int_{s_p} g\psi \, ds}{\int_{s_p} g \, dA} \tag{8-187}$$

In general, the surface-average values will be quite different from the overall average values.

8. The fluctuating properties ψ' are introduced as follows:

$$\psi = \langle \psi \rangle + \psi' \tag{8-188}$$

If it is assumed that $\langle\langle \psi \rangle\rangle = \langle \psi \rangle$, one will get

$$\langle \psi' \rangle = 0 \tag{8-189}$$

9. Finally,

$$\langle \varepsilon \psi \rangle = \langle \varepsilon \rangle \langle \psi \rangle + \langle \varepsilon' \psi' \rangle \tag{8-190}$$

However, throughout the analysis $\langle \varepsilon' \psi' \rangle$ is assumed to be zero.

11.3.2 Continuity Equation

In the following, the abovementioned definitions and approximations will be used to derive the macroscopic gas-phase continuity equation from the continuity equation of the fluid at a point, namely

$$\frac{\partial \rho}{\partial t} + \frac{\partial}{\partial y_i}(\rho u_i) = 0 \tag{8-191}$$

Multiplying this equation by $g(\mathbf{x} - \mathbf{y})$ and integrating over the fluid volume, we get

$$\int_{V_{f\infty}(t)} \left[\frac{\partial \rho}{\partial t} + \frac{\partial}{\partial y_i}(\rho u_i) \right] g(\mathbf{x} - \mathbf{y}) \, dV_y = 0 \tag{8-192}$$

Using Eq. (8-185), the first term of this equation can be written as

$$\frac{\partial}{\partial t}[\phi(x, t)\langle \rho(\mathbf{x}, t) \rangle] + \int_{s_p(t)} \rho(\mathbf{y}, t) g(\mathbf{x} - \mathbf{y}) v_i n_i \, ds \tag{8-193}$$

Using Eq. (8-181), the second term of Eq. (8-192) can be written as

$$\frac{\partial}{\partial x_i}[\phi\langle \rho u_i\rangle] - \int_{s_p(t)} \rho u_i g n_i \, ds \tag{8-194}$$

From Eq. (8-190) we know

$$\langle \rho u_i\rangle \approx \langle \rho\rangle\langle u_i\rangle \tag{8-195}$$

Using Eqs. (8-192), (8-193), (8-194), and (8-195), we get

$$\frac{\partial}{\partial t}[\phi\langle \rho\rangle] + \frac{\partial}{\partial x_i}[\phi\langle \rho\rangle\langle u_i\rangle] = \int_{s_p} g\rho(u_i - v_i)n_i \, ds \tag{8-196}$$

Now at the particle surface we can use the boundary condition

$$\rho(\mathbf{u} - \mathbf{v}) \cdot \mathbf{n} = \rho_p(\mathbf{u}_p - \mathbf{v}) \cdot \mathbf{n} \tag{8-197}$$

Also the interface velocity $\mathbf{v}$ can be expressed in terms of surface regression rate $\dot{d}$:

$$\mathbf{v} = \mathbf{u}_p + \dot{d}\mathbf{n} \tag{8-198}$$

Using Eqs. (8-197) and (8-198), the right-hand side of Eq. (8-196) becomes

$$\int_{s_p} g\rho(u_i - v_i)n_i \, ds = \int_{s_p} \dot{d} g \, ds \tag{8-199}$$

Using the definitions (8-187) and (8-186), we then get

$$\frac{\partial}{\partial t}[\phi\langle \rho\rangle] + \frac{\partial}{\partial x_i}[\phi\langle \rho\rangle\langle u_i\rangle] = \frac{(1-\phi)s_p}{V_p}\langle \dot{d}\rangle^s \tag{8-200}$$

which is exactly the same as Eq. (8-156) derived using the fluid-dynamics model. In an analogous manner the solid-phase continuity equation can be derived; it is

$$\frac{\partial}{\partial t}\left[(1-\phi)\langle \rho_p\rangle\right] + \frac{\partial}{\partial x_i}\left[(1-\phi)\langle \rho_p\rangle\langle u_{pi}\rangle\right] = -(1-\phi)\frac{s_p}{V_p}\langle \dot{d}\rangle^s \tag{8-201}$$

11.3.3 Momentum Equation

The point momentum equation for the gas phase (neglecting body forces) is

$$\frac{\partial}{\partial t}(\rho u_i) + \frac{\partial}{\partial y_i}(\rho u_i u_j) = \frac{\partial}{\partial y_j}\sigma_{ji} \tag{8-202}$$

Multiplying (8-202) by $g(\mathbf{x} - \mathbf{y})$ and integrating over the fluid volume, we get

$$\int_{V_{f\infty}(t)} \left[\frac{\partial}{\partial t}(\rho u_i) + \frac{\partial}{\partial y_i}(\rho u_i u_j - \sigma_{ji}) \right] dV_y = 0 \tag{8-203}$$

that is,

$$\frac{\partial}{\partial t}[\phi \langle \rho u_i \rangle] + \int_{s_p(t)} \rho u_i g v_j n_j \, ds + \frac{\partial}{\partial x_i}\left[\phi(\langle \rho u_i u_j \rangle - \langle \sigma_{ji} \rangle)\right]$$

$$-\int_{s_p(t)} (\rho u_i u_j - \sigma_{ji}) g n_j \, ds = 0 \tag{8-204}$$

Now, if we make another approximation that

$$\langle \varepsilon \psi n \rangle \approx \langle \varepsilon \rangle \langle \psi \rangle \langle n \rangle \tag{8-205}$$

we get

$$\frac{\partial}{\partial t}(\phi \langle \rho \rangle \langle u_i \rangle) + \frac{\partial}{\partial x_j}\left[\phi(\langle \rho \rangle \langle u_i \rangle \langle u_j \rangle - \langle \sigma_{ij} \rangle)\right]$$

$$= -\int_{s_p(t)} g\left[\sigma_{ji} + \rho u_i (u_j - v_j)\right] n_j \, ds \tag{8-206}$$

Using the boundary condition

$$\left[\sigma_{ji} + \rho u_i (u_j - v_j)\right] n_j = \left[\sigma_{pji} + \rho_p u_{pi} (u_{pj} - v_j)\right] n_j \tag{8-207}$$

and

$$(\sigma_{pji} + \sigma_{ji}) n_j \approx n_i \, \Delta p \tag{8-208}$$

as well as Eq. (8-198), the right-hand side can also be written as

$$\left(\sigma_{pji} + \rho_p u_{pi} \dot{d} n_j\right) n_j \tag{8-209}$$

so that we have

$$\int_{s_p(t)} \sigma_{pji} n_j \, ds = \int_{s_p(t)} g \, \Delta p \, n_i \, ds + \int_{s_p(t)} \left[\langle \sigma_{ji} \rangle + \sigma'_{ji}\right] g n_j \, ds$$

$$= \int_{s_p(t)} g \, \Delta p \, n_i \, ds + \int_{s_p(t)} \sigma'_{ji} g n_j \, ds + \langle \sigma_{ji} \rangle \int_{s_p(t)} g n_j \, ds$$

$$= \int_{s_p(t)} g \, \Delta p \, n_i \, ds + \int_{s_p(t)} \sigma'_{ji} g n_j \, ds + \langle \sigma_{ji} \rangle \int_{V_{f\infty}} \frac{\partial g}{\partial y_j} \, dV_y$$

$$= \int_{s_p(t)} g \, \Delta p \, n_i \, ds + \int_{s_p(t)} \sigma'_{ji} g n_j \, ds + \langle \sigma_{ij} \rangle \frac{\partial \phi}{\partial x_j} \tag{8-210}$$

Making the assumption that $\Delta p \approx 0$ and defining

$$\int_{s_p(t)} \sigma'_{ji} g n_j \, ds \equiv (1-\phi)\frac{s_p}{V_p}\langle F \rangle^s \tag{8-211}$$

we get [use Eqs. (8-186), (8-187), and (8-206) through (8-211)]

$$\frac{\partial}{\partial t}(\phi\langle\rho\rangle\langle u_i\rangle) + \frac{\partial}{\partial x_j}(\phi\langle\rho\rangle\langle u_i\rangle\langle u_j\rangle) - \phi\frac{\partial}{\partial x_j}\sigma_{ji}$$

$$= -(1-\phi)\frac{s_p}{V_p}\langle F\rangle^s + (1-\phi)\frac{\langle\rho_p\rangle s_p}{V_p}\langle u_p\rangle\langle\dot{d}\rangle^s \tag{8-212}$$

where $\langle F \rangle^s$ can be interpreted[145] as the interphase drag per unit area. For the fluid, $\sigma_{ji} \approx -p\delta_{ji}$, and (8-212) becomes

$$\frac{\partial}{\partial t}(\phi\langle\rho\rangle\langle u_i\rangle) + \frac{\partial}{\partial x_j}(\phi\langle\rho\rangle\langle u_i\rangle\langle u_j\rangle)$$

$$= -\phi\frac{\partial p}{\partial x_i} - (1-\phi)\frac{s_p}{V_p}\langle F\rangle^s + (1-\phi)\frac{\langle\rho_p\rangle s_p}{V_p}\langle u_p\rangle\langle\dot{d}\rangle^s \tag{8-213}$$

which is very similar to (8-159). The other conservation equations can be similarly derived. To limit the scope of this presentation, we omit the detailed formulation of Gough's model for interior ballistic simulation, as well as his calculated results. Interested readers should refer to his publications or to the book edited by Krier and Summerfield.[149]

REFERENCES

1. G. M. Faeth, *Progress in Energy and Combustion Science*, Vol. 3, Pergamon Press, Oxford, New York, 1977.
2. *Exploring Energy Choices*, Preliminary Report, Energy Policy Project, Ford Foundation, Washington, 1974.
3. N. A. Chigier, *Progress in Energy and Combustion Science*, Vol. 2, Pergamon Press, Oxford, New York, 1977.
4. L. S. Caretto, *Progress in Energy and Combustion Science*, Vol. 1, Pergamon Press, Oxford, New York, pp. 47–71, 1976.
5. A. M. Mellor, *Progress in Energy and Combustion Science*, Vol. 1, Pergamon Press, Oxford, New York, pp. 111–133, 1976.
6. N. A. Henein, *Progress in Energy and Combustion Science*, Vol. 1, Pergamon Press, Oxford, New York, pp. 165–207, 1976.

7. A. Williams, *Combustion and Flame*, Vol. 21, Pergamon Press, Oxford, New York, pp. 1–31, 1973.
8. A. Williams, *Oxidation and Combustion Rev.*, Vol. 3, pp. 1–45, 1968.
9. A. Williams, *Progress in Energy and Combustion Science*, Vol. 2, Pergamon Press, Oxford, New York, pp. 167–179, 1976.
10. A. B. Hedley, A. S. M. Nuruzzaman, and G. F. Martin, *J. Inst. Fuel*, Vol. 71, pp. 38–54, 1971.
11. N. A. Chigier, "Instrumentation Techniques for Studying Heterogeneous Combustors," presented at Spring Technical Meeting, Central States Section of the Combustion Institute, Cleveland, March 1977.
12. F. A. Williams, *Eighth Symposium (International) on Combustion*, pp. 50–69, Williams and Wilkins, Baltimore, 1962.
13. F. A. Williams, *Combustion Theory*, Addison-Wesley, Reading, Mass., 1965.
14. D. T. Harrje and F. H. Reardon (Eds.), *Liquid Propellant Rocket Combustion Instability*, pp. 37–104, NASA SP-194, Washington, 1962.
15. W. A. Sirigano and C. K. Law, "A Review of Transient Heating and Liquid-Phase Mass Diffusion in Fuel Droplet Vaporization," presented at Symposium on Evaporation–Combustion of Fuel Droplets, San Francisco, August 29–September 3, 1976.
16. R. D. Reitz and F. V. Bracco, "Breakup Regimes of a Single Liquid Jet," presented at Fall Technical Meeting, Eastern Section of the Combustion Institute, Philadelphia, November 1976.
17. J. Odgers, *Fifteenth Symposium (International) on Combustion*, pp. 1321–1338, The Combustion Institute, Pittsburgh, 1975.
18. A. A. Putnam et al., *Injection and Combustion of Liquid Fuels*, WADC Technical Report 56-334, Wright Air Development Center, Wright-Patterson Air Force Base, 1957.
19. F. B. Cramer and P. D. Baker, *Combustion Processes in a Bipropellant Liquid Rocket Engine —a Critical Review*, JPL Report 900-2, Jet Propulsion Laboratory, Pasadena, Calif., 1967.
20. R. W. Tate, "Sprays," *Encyclopedia of Chemical Technology*, Vol. 18, pp. 634–654, 1969.
21. W. E. Ranz, *On Sprays and Spraying*, Dept. of Engrg. Res. Bulletin 65, The Pennsylvania State University, University Park, Pa., 1956.
22. K. J. de Juhasz, *Spray Literature Abstracts*, Vol. 1, ASME, New York, 1959; Vol. 2, ASME New York, 1964; Vol. 3, ASME, New York, 1966; Vol. 4, ASME, New York, 1969.
23. E. Giffen and A. Muraszew, *The Atomization of Liquid Fuels*, Chapman and Hall, London, 1953.
24. C. J. Orr, *Particulate Technology*, Macmillan, New York, 1966.
25. V. G. Levich, *Physicochemical Hydrodynamics*, Prentice-Hall, Englewood Cliffs, N.J., 1962.
26. S. S. Penner, *Chemistry Problems in Jet Propulsion*, pp. 276–296, Macmillan, New York, 1957.
27. S. Lambiris, L. P. Combs, and R. S. Levine, "Stable Combustion Processes in Liquid Propellant Rocket Engines," *Combustion and Propulsion, Fifth AGARD Colloquium: High Temperature Phenomena*, pp. 596–634, Macmillan, 1963.
28. D. B. Spalding, *Some Fundamentals of Combustion Processes*, Gas Turbine Series, Vol. 2, Academic Press, New York, 1955.
29. D. Altman, J. M. Carter, S. S. Penner, and M. Summerfield, *Liquid Propellant Rockets*, Princeton Aeronautical Paperbacks, Princeton University Press, Princeton, N.J. 1960.
30. M. Barrere et al., *Rocket Propulsion*, Elsevier Publishing Co., New York, 1960.
31. J. A. Browing, *Advances in Chemistry Series*, No. 20, pp. 135–154, American Chemical Society, Washington, 1958; J. M. Pilcher and R. E. Thomas, *Ibid.*, pp. 155–165.

32. J. D. Lewis, "Studies of Atomization and Injection Processes in Liquid Propellant Rocket Engines," *Combustion and Propulsion, Fifth AGARD Colloquium: High Temperature Phenomena*, pp. 141–169, Macmillan, New York, 1963.
33. H. Krier and C. L. Foo, *Oxidation and Combustion Rev.*, Vol. 6, pp. 111–143, 1973.
34. A. H. Lefebvre, *Fifteenth Symposium (International) on Combustion*, pp. 1169–1180, The Combustion Institute, Pittsburgh, 1975.
35. S. Kumagai, *Sixth Symposium (International) on Combustion*, pp. 668–674, Reinhold, New York, 1957.
36. R. P. Fraser, *Sixth Symposium (International) on Combustion*, pp. 687–701, Reinhold, New York, 1957.
37. L. S. Caretto, *Fourteenth Symposium (International) on Combustion*, pp. 803–817, The Combustion Institute, Pittsburgh, 1973.
38. A. H. Lefebvre, *Tenth Symposium (International) on Combustion*, pp. 1129–1137, The Combustion Institute, Pittsburgh, 1964.
39. F. A. Williams, *Combustion and Flame*, Vol. 3, pp. 215–228, 1959.
40. N. Dombrowski, W. Horne, and A. Williams, *Combustion Sci. and Technol.*, Vol. 9, pp. 247–254, 1974.
41. R. Mellor, N. A. Chigier, and J. M. Beer, *Pressure Jet Spray in Air Stream*, ASME Paper 70-GT-101, 1970.
42. R. Mugele and H. D. Evans, *Ind. Engrg. Chem.*, Vol. 43, pp. 1317–1324, 1951.
43. P. Rosin and E. Rammler, *J. Inst. Fuel*, Vol. 7, pp. 29–36, 1933.
44. S. Nukiyama and Y. Tanasawa, *Trans. SME Japan*, Vol. 6, pp. 5–7, 1940.
45. R. A. Mugele, *AIChE J.*, Vol. 6, pp. 3–8, 1960.
46. G. M. Faeth, "Spray Combustion Model—A Review," presented at 17th Aerospace Sciences Meeting, 1979.
47. G. M. Faeth, "Evaporation and Combustion of Sprays," *Progress in Energy and Combustion Science*, Vol. 9, pp. 1–76, Pergamon Press, New York, 1983.
48. N. Chigier, "Group Combustion Models and Laser Diagnostic Methods in Sprays: A Review," *Combustion and Flame*, Vol. 51, pp. 127–139, 1983.
49. H. H. Chiu and T. M. Liu, "Group Combustion of Liquid Droplets," *Combustion Sci. and Technol.*, Vol. 17, pp. 127–142, 1977.
50. H. H. Chiu, R. K. Ahluwalia, B. Koh, and E. J. Croke, "Spray Group Combustion," AIAA paper 78–75, presented at AIAA 16th Aerospace Sciences Meeting, Huntsville, Ala., January 16–18, 1978.
51. H. H. Chiu and E. J. Croke, *Group Combustion of Liquid Fuel Sprays*, Energy Technology Lab Report 81–2, Univ. of Illinois at Chicago, 1981.
52. H. H. Chiu, H. Y. Kim, and E. J. Croke, "Internal Group Combustion of Liquid Droplets," *19th Symposium (International) on Combustion*, pp. 971–980, The Combustion Institute, Pittsburgh, 1982.
53. H. Y. Kim and H. H. Chiu, "Group Combustion of Liquid Fuel Sprays," AIAA paper No. 83-0150, presented at AIAA 21st Aerospace Sciences Meeting, Reno, Nevada, January 10–13, 1983.
54. H. H. Chiu and X. Q. Zhou, *Turbulent Spray Group Vaporization and Combustion—Part I*, Energy Technology Lab Report, University of Illinois at Chicago, 1983.
55. R. P. Probert, *Phil. Mag.*, Vol. 37, p. 94, 1946.
56. F. A. Williams, *Phys. Fluids*, Vol. 1, p. 541, 1958.
57. Y. Tanasawa, *Tech. Rept. Tohoku Univ.*, Vol. 18, 195 pp., 1954.
58. Y. Tanasawa and T. Tesima, *Bull. JSME*, Vol. 1, p. 36, 1958.

59. J. H. Burgoyne and L. Cohen, *Proc. Roy. Soc. London A*, Vol. 225, pp. 375–392, 1954.

60. Y. Mizutani, G. Yasuma, and M. Katsuki, "Stabilization of Spray Flames in a High-Temperature Stream," *Sixteenth Symposium (International) on Combustion*, pp. 631–638, The Combustion Institute, Pittsburgh, 1976.

61. E. E. Khalil and J. H. Whitelaw, "Aerodynamic and Thermodynamic Characteristics of Kerosene-Spray Flames," *Sixteenth Symposium (International) on Combustion*, pp. 569–576, The Combustion Institute, Pittsburgh, 1976.

62. Y. Onuma and M. Ogasawara, "Studies of the Structure of a Spray Combustion Flame," *Fifteenth Symposium (International) on Combustion*, pp. 453–465, The Combustion Institute, Pittsburgh, 1974.

63. Y. Onuma, M. Ogasawara, and T. Inoue, "Further Experiments on the Structure of a Spray Combustion Flame," *Sixteenth Symposium (International) on Combustion*, pp. 561–567, The Combustion Institute, Pittsburgh, 1976.

64. A. C. Styles and N. A. Chigier, "Combustion of Air Blast Atomized Spray Flames," *Sixteenth Symposium (International) on Combustion*, pp. 619–630, The Combustion Institute, Pittsburgh, 1976.

65. K. V. L. Rao and A. H. Lefebvre, "Evaporation Characteristics of Kerosine Sprays Injected into a Flowing Air Stream," *Combustion and Flame*, Vol. 26, pp. 303–309, 1976.

66. W. Bahr, *Evaporation and Spreading of Iso-octane Sprays in High Velocity Air Streams*, NACA RM E 53I14, Washington, Nov. 1953.

67. R. Natarajan and A. K. Ghosh, *Fuel*, Vol. 54, pp. 153–161, 1975.

68. C. K. Westbrook, "Three Dimensional Numerical Modeling of Liquid Fuel Sprays," *Sixteenth Symposium (International) on Combustion*, pp. 1517–1526, The Combustion Institute, Pittsburgh, 1976.

69. F. V. Bracco, "Applications of Steady-State Spray Equations to Combustion Modeling," *AIAA J.*, Vol. 12, No. 11, pp. 1534–1540, 1974.

70. R. J. Priem and M. F. Heidmann, *Propellant Vaporization as a Design Criterion for Rocket-Engine Combustion Chambers*, NASA TR R-67, 1960.

71. D. B. Spalding, "The Combustion of Liquid Fuels," *Fourth Symposium (International) on Combustion*, pp. 847–864, The Combustion Institute, Pittsburgh, 1953.

72. D. R. Dickinson and W. R. Marshall, Jr., *AIChE J.*, Vol. 14, pp. 541–552, 1968.

73. C. K. Law, *Int. J. Heat and Mass Transfer*, Vol. 18, pp. 1285–1292, 1975.

74. F. V. Bracco, *Combust. Sci. and Technol.*, Vol. 8, pp. 69–84, 1973.

75. R. J. Mador and R. Roberts, *A Pollutant Emissions Model for Gas Turbine Combustors*, AIAA Paper No. 74-1113, 1974.

76. J. Swithenbank, I. Poli, and M. W. Vincent, "Combustion Design Fundamentals," *Fourteenth Symposium (International) on Combustion*, pp. 627–638, The Combustion Institute, Pittsburgh, 1973.

77. W. G. Courtney, *ARS J.*, Vol. 30, p. 356, 1960.

78. N. Munz and P. Eisenklam, "The Modelling of a High Intensity Spray Combustion Chamber," *Sixteenth Symposium (International) on Combustion*, pp. 593–604, The Combustion Institute, Pittsburgh, 1976.

79. M. W. Thring and M. P. Newby, "Combustion Length of Enclosed Turbulent Jet Flames," *Fourth Symposium (International) on Combustion*, pp. 789–796, The Combustion Institute, Pittsburgh, 1953.

80. J. F. Avery and G. M. Faeth, "Combustion of Submerged Gaseous Oxidizer Jet in a Liquid Fuel," *Fifteenth Symposium (International) on Combustion*, pp. 501–512, The Combustion Institute, Pittsburgh, 1974.

81. J. A. Newman and T. A. Brzustowski, *AIAA J.*, Vol. 9, pp. 1595–1602, 1971.

82. K. Komiyama, R. C. Flagan, and J. B. Heywood, "The Influence of Droplet Evaporation on Fuel–Air Mixing Rate in a Burner," *Sixteenth Symposium (International) on Combustion*, pp. 549–560, The Combustion Institute, Pittsburgh, 1976.
83. A. J. Shearer and G. M. Faeth, *Combustion of Liquid Sprays at High Pressures*, NASA CR-135210, 1977.
84. E. E. Khalil, *A Simplified Approach for the Calculation of Free and Confined Spray Flames*, AIAA Paper No. 78-029, 1978.
85. A. J. Shearer, H. Tamura, and G. M. Faeth, *J. Energy*, Vol. 3, p. 271, 1979.
86. C. P. Mao, G. A. Szekely, Jr., and G. M. Faeth, *J. Energy*, Vol. 4, p. 78, 1980.
87. C. P. Mao, Y. Wakamatsu, and G. M. Faeth, *Eighteenth Symposium (International) on Combustion*, pp. 337–347, The Combustion Institute, Pittsburgh, 1981.
88. W. C. Reynolds, *Combustion Modeling in Reciprocating Engines*, (J. N. Mattavi and C. A. Amann, Eds.), pp. 41–68, Plenum Press, New York, 1980.
89. J. O. Hinze, *Turbulence*, McGraw-Hill Book Company, New York, 1975.
90. A. D. Gosman, F. C. Lockwood, and S. A. Syod, "Prediction of a Horizontal Free Turbulent Diffusion Flame," *Sixteenth Symposium (International) on Combustion*, pp. 1543–1555, The Combustion Institute, Pittsburgh, 1976.
91. D. B. Spalding, *Chem. Engrg. Sci.*, Vol. 26, p. 95, 1971.
92. D. B. Spalding, *Combust. Sci. Technol.*, Vol. 13, p. 3, 1976.
93. F. C. Lockwood and A. S. Naguib, *Combustion and Flame*, Vol. 24, p. 109, 1975.
94. R. W. Bilger, *Prog. Energy Combust. Sci.*, Vol. 1, p. 87, 1976.
95. D. B. Spalding, *Thirteenth Symposium (International) on Combustion*, pp. 649–657, The Combustion Institute, Pittsburgh, 1971.
96. B. F. Magnussen and B. W. Hjertager, "On Mathematical Modeling of Turbulent Combustion with Special Emphasis on Soot Formation and Combustion," *Sixteenth Symposium (International) on Combustion*, pp. 719–729, The Combustion Institute, Pittsburgh, 1976.
97. A. D. Gosman, F. C. Lockwood, and A. P. Salooja, "The Prediction of Cylindrical Furnaces Gaseous Fueled with Premixed and Diffusion Burners," *Seventeenth Symposium (International) on Combustion*, pp. 747–760, The Combustion Institute, Pittsburgh, 1979.
98. R. Borghi, *Adv. Geophys.*, Vol. 18B, p. 349, 1974.
99. K. N. C. Bray and J. B. Moss, *Acta Astronautica*, Vol. 4, p. 291, 1977.
100. F. C. Lockwood, "The Modeling of Turbulent Premixed and Diffusion Combustion in the Computation of Engineering Flows," *Combustion and Flame*, Vol. 29, pp. 111–122, 1977.
101. S. E. Elghobashi and W. M. Pun, "A Theoretical and Experimental Study of Turbulent Diffusion Flames in Cylindrical Furnaces," *Fifteenth Symposium (International) on Combustion*, pp. 1353–1365, The Combustion Institute, Pittsburgh, 1974.
102. I. Wygnanski and H. E. Fiedler, *J. Fluid Mech.*, Vol. 38, p. 577, 1969.
103. G. Hetsroni and M. Sokolov, *J. Appl. Mech.*, Vol. 38, p. 314, 1971.
104. H. A. Becker, W. C. Hottel, and G. C. Williams, *J. Fluid Mech.*, Vol. 30, p. 285, 1967.
105. S. Corrsin and M. S. Uberoi, *Further Experiments on Flow and Heat Transfer in a Heated Turbulent Air Jet*, NACA Report No. 988, Washington, 1950.
106. S. R. Tross, *Characteristics of a Submerged Two-Phase Free Jet*, M.S. Thesis, The Pennsylvania State University, 1974.
107. C. T. Crowe, "A Computational Model for the Gas-Droplet Flow in the Vicinity of an Atomizer," Paper No. 74-25, Western States Section, The Combustion Institute, 1974.
108. C. T. Crowe, "A Numerical Model for the Gas Droplet Flow Field Near an Atomizer," presented at First International Conference on Liquid Atomization and Spray Systems, Tokyo, August 1978.

109. C. T. Crowe, M. P. Sharma, and D. E. Stock, *J. Fluids Engrg.*, Vol. 99, p. 325, 1977.

110. J. T. Jurewicz, D. T. Stock, and C. T. Crowe, "The Effect of Turbulent Diffusion on Gas Particle Flow in an Electric Field," *First Symposium on Turbulent Shear Flows*, pp. 12.27–12.33, University Park, Pa., April 1977.

111. R. L. Alpert and M. K. Mathews, *Calculation of Large-Scale Flow Fields Induced by Droplet Sprays*, Technical Report No. FMRCJ.1.OEOJ4.BU, Factory Mutual Research Corp., Norwood, Mass., 1979.

112. Y. El Banhawy and J. H. Whitelaw, *AIAA J.*, Vol. 18, p. 1503, 1980.

113. A. D. Gosman and R. J. R. Johns, *Computer Analysis of Fuel-Air Mixing in Direct-Injection Engines*, SAE Paper No. 800091, 1980.

114. T. D. Butler, L. D. Cloutman, J. K. Dukowicz, J. D. Ramshaw, and R. B. Krieger, *Combustion Modeling in Reciprocating Engines*, (J. N. Mattari and C. A. Amann, Eds.), pp. 231–264, Plenum Press, New York, 1980.

115. T. W. Bruce, H. C. Mongia, and R. S. Reynolds, *Combustor Design Criteria Validation*, Vols. I–III, USARTL-TR-78-55(A, B, C), Fort Eustis, Va., 1979.

116. H. C. Mongia, and K. Smith, *An Empirical/Analytical Design Methodology for Gas Turbine Combustors*, AIAA Paper No. 78-998, 1978.

117. J. Swithenbank, A. Turan, and P. G. Felton, *Gas Turbine Combustor Design Problems* (A. H. Lefebvre, Ed.), pp. 249–314, Hemisphere Publishing, Washington, 1980.

118. F. Boyson, W. H. Ayers, J. Swithenbank, and Z. Pan, *Three-Dimensional Model of Spray Combustion in Gas Turbine Combustors*, AIAA Paper No. 81-0324, 1981.

119. A. D. Gosman, and E. Ioannides, *Aspects of Computer Simulation of Liquid-Fueled Combustors*, AIAA Paper No. 81-0323, 1981.

120. A. D. Gosman, E. Ioannides, D. A. Lever, and K. A. Cliffe, *A Comparison of Continuum and Discrete Droplet Finite-Difference Models Used in the Calculation of Spray Combustion in Swirling Turbulent Flows*, AERE Harwell Report TP865, 1980.

121. A. S. P. Solomon, *A Theoretical and Experimental Investigation of Turbulent Sprays*, Ph.D. Thesis, The Pennsylvania State University, May 1984.

122. J. S. Shuen, *A Theoretical and Experimental Investigation of Dilute Particle-Laden Turbulent Gas Jets*, Ph.D. Thesis, The Pennsylvania State University, May 1984.

123. J. S. Shuen, L. D. Chen, and G. M. Faeth, "Predictions of the Structure of Turbulent Particle-Laden Round Jets," *AIAA J.*, Vol. 21, No. 11, pp. 1483–1484, 1983.

124. O. L. Anderson, L. M. Chiappetta, D. E. Edwards, and J. B. McVey, *Analytical Modeling of Operating Characteristics of Premixing–Prevaporizing Fuel–Air Mixing Passages*, Report No. UTRC 80-102, Vols. I, II, United Technologies Research Center, East Hartford, Conn., 1980.

125. A. Putnam, *ARS J.*, Vol. 31, p. 1467, 1961.

126. R. A. Dickerson and M. D. Schuman, *J. Spacecraft*, p. 99, 1960.

127. G. M. Faeth, and R. R. Lazar, *AIAA J.*, Vol. 9, p. 2165, 1971.

128. J. S. Shuen, L. D. Chen, and G. M. Faeth, "Evaluation of a Stochastic Model of Particle Dispersion in a Turbulent Round Jet," *AIChE J.*, Vol. 29, No. 1, pp. 167–170, 1983.

129. S. Yuu, N. Yasukouchi, Y. Hirosawa, and T. Jotaki, "Particle Turbulent Diffusion in a Dust Laden Round Jet," *AIChE J.*, Vol. 24, p. 509, 1978.

130. M. K. Laats and F. A. Frishman, "Assumptions Used in Calculating the Two-Phase Jet," *Fluid Dynamics*, Vol. 5, pp. 333–338, March–April 1970.

131. M. K. Laats and F. A. Frishman, "Scattering of an Inert Admixture of Different Grain Size in a Two-Phase Axisymmetric Jet," *Heat Transfer—Soviet Res.*, Vol. 2, pp. 7–12, Nov. 1970.

132. Y. Levy and F. C. Lockwood, "Velocity Measurements in a Particle Laden Turbulent Free Jet," *Combustion and Flame*, Vol. 40, pp. 333–339, March 1981.

133. W. H. Snyder and J. L. Lumley, *J. Fluid Mech.*, Vol. 48, pp. 41–47, 1971.
134. J. O. Hinze, *Turbulence*, 2nd ed., McGraw-Hill, New York, 1975.
135. N. Chigier, *Energy Combustion and Environment*, Chapter 7, McGraw-Hill, New York, 1981.
136. T. Suzuki and H. H. Chiu, "Multi-Droplet Combustion of Liquid Propellants," *Proceedings of the Ninth International Symposium on Space Technology and Science*, pp. 145–154, 1971.
137. S. L. Soo, *Fluid Dynamics of Multiphase Systems*, Blaisdell Publishing Co., 1967.
138. H. Krier, S. Rajan, and W. F. Van Tassel, "Flame Spreading and Combustion in Packed Beds of Propellant Grains," *AIAA J.*, Vol. 14, No. 3, pp. 301–309, March 1976.
139. Yu A. Buyevich, "Statistical Hydromechanics of Disperse Systems—1. Physical Background and General Equations," *J. Fluid Mech.*, Vol. 49, Part 3, pp. 498–507, 1971.
140. Yu A. Buyevich, "Statistical Hydromechanics of Disperse Systems—2. Solution of the Kinetic Equation for Suspended Particles," *J. Fluid Mech.*, Vol. 52, Part 2, pp. 345–355, 1972.
141. Yu A. Buyevich, "Statistical Hydromechanics of Disperse Systems—3. Pseudo-turbulent Structure of Homogeneous Suspensions," *J. Fluid Mech.*, Vol. 56, Part 2, pp. 313–336, 1972.
142. K. K. Kuo, J. H. Koo, T. R. Davis, and G. R. Coates, "Transient Combustion in Gas-Permeable Propellants," *Acta Astronautica*, Vol. 3, pp. 573–591, 1976.
143. T. B. Anderson and T. Jackson, "A Fluid Mechanical Description of Fluidized Beds," *I and EC Fundamentals*, Vol. 6, No. 4, pp. 527–539, 1967.
144. S. Whitaker, "The Transport Equations for Multi-Phase Systems," *Chem. Engrg. Sci.*, Vol. 28, pp. 139–147, 1973.
145. P. S. Gough and F. J. Zwarts, "Modeling Heterogeneous Two-Phase Reacting Flow," *AIAA J.*, Vol. 17, No. 1, pp. 17–25, 1979.
146. W. G. Gray and P. C. Y. Lee, "On the Theorems for Local Volume Averaging of Multiphase Systems," *Int. J. Multiphase Flow*, Vol. 3, pp. 333–340, 1977.
147. K. K. Kuo, R. Vichnevetsky, and M. Summerfield, "Theory of Flame Front Propagation in Porous Propellant Charges under Confinement," *AIAA J.*, Vol. 11, pp. 444–451, April 1973.
148. D. Y. Chen, V. Yang, and K. K. Kuo, "Boundary Condition Specification for Mobile Granular Propellant Bed Combustion Processes," *AIAA J.*, Vol. 19, No. 11, pp. 1429–1437, November 1981.
149. H. Krier and M. Summerfield, *Interior Ballistics of Guns*, Progress in Astronautics and Aeronautics, Vol. 66, pp. 176–196, 1979.
150. K. K. Kuo, K. C. Hsieh and M. M. Athavale, "Modeling of Combustion Processes of Stick Propellants via Combined Eulerian-Lagrangian Approach" *Proceedings of the 8th International Symposium on Ballistics*, pp. I-55–I-67, Oct. 23–25, 1984.

HOMEWORK

1. In the estimation of the rate of reduction of diameter of a droplet traveling within a dilute spray flame, one can usually make the following assumptions:

(a) The temperature variation and oxygen concentration are low within the spray.

(b) No interaction takes place between droplets.

(c) Le = Sc = Pr = 1.

(d) The initial velocity of the droplets, U_{d0}, is equal to the discharge velocity of the fuel.

(e) The initial temperature of the fuel is close to the boiling point.

Using the above assumptions and the d^2 evaporation law, develop two ordinary differential equations describing the rates of change of droplet velocity and diameter.

2. Derive the transport equation (8-96) for the turbulence kinetic energy k of a continuous phase which contains numerous dispersed fuel droplets.

3. Derive the transport equation (8-97) for the turbulence dissipation rate ε of a continuous phase which contains numerous dispersed fuel droplets.

4. Consider an isotropic turbulent flow of a continuous liquid which entrains many dispersed immiscible liquid droplets. Let the droplet diameter of the dispersed liquid be d, the pressure inside the droplet be P_d, and the pressure of the continuous fluid be p.

(a) Show that the static equilibrium condition of the above mixture is

$$P_d - p = \frac{4\sigma}{d}$$

where σ is the surface tension of the liquid droplet residing in the continuous liquid.

(b) Consider the droplet breakup process under the condition that inertial forces dominate over viscous forces. What is the relationship between the dynamic pressure fluctuation (caused by turbulence), surface tension, and the maximum diameter of a stable droplet?

5. For the same problem stated above, except that the droplet breakup is caused by viscous forces rather than inertial forces, what is the relationship between fluctuation velocity, surface tension, and the viscosity of the two liquids?

CHAPTER

9 Chemically Reacting Boundary-Layer Flows

ADDITIONAL SYMBOLS

Symbols	Description	Dimension
Da_i	Damköhler number for species i involved in surface reaction	—
Da_{gi}	Damköhler number for species i involved in gas-phase reaction	—
$\mathscr{D}^{\mathrm{Th}}$	Thermal diffusion coefficient	M/Lt
$\overline{D}$	A reference diffusion coefficient	L^2/t
F	Dimensionless stream function defined by Eq. (9-69)	—
L	Length of the plate	L
l	Dimensionless product of ρ and μ defined in Eq. (9-72)	—
ℓ_m	Mixing length	L
$\mathscr{D}_v$	Van Driest's damping coefficient	—
R_h	Surface roughness height	L
u_*	Friction velocity	L/t
V_T	Turbulent velocity	L/t
$\mathscr{Y}_k$	Mass fraction of element k defined by Eq. (9-146)	—
x^*	Dimensionless axial distance along the body ($x^* \equiv x/L$)	—
Y_i^*	Mass fraction of ith species defined in Eq. (9-72)	—

τ_{gi}	Characteristic time for gas-phase chemical reaction	t
ξ	Transformed axial distances defined by Eq. (9-65)	M^2/L^2t^2
η	Transformed similarity coordinate defined by Eq. (9-66)	—
ζ	Similarity variable defined in Eq. (9-99)	—
δ	Boundary layer thickness	L
ψ	Stream function	M/Lt
ω_i	Rate of production of species i per unit volume	M/L^3t
ω'_a	Rate of production of species a per unit area	M/L^2t
δ^*	Displacement thickness of boundary-layer flow	L

Subscripts

dc	Diffusion-controlled
E	Equilibrium
e	Edge of boundary layer
f	Frozen
se	Sensible
t	Total (stagnation)
w	Wall

Superscripts

$'$	Ordinary differentiation with respect to ζ

1 INTRODUCTION

Chemical reactions in boundary-layer flows have interested engineers and scientists for ages. A boundary layer can be roughly characterized as a flow region of a moving fluid in which there is a single predominant direction of flow, and in which shear stresses, heat fluxes, and mass-diffusion fluxes are significant only in directions normal to the predominant direction of flow. It should be noted that boundary-layer flow does not necessarily require the presence of a solid wall. Two adjacent jet streams with different momentum, species concentration, or enthalpy can provide significant shear stresses, diffusional mass fluxes, or heat fluxes in the direction normal to the predominant direction of flow. Of course, flow over solid surfaces usually produces a shear layer due to nonslip conditions at the fluid–solid interface. This chapter deals with chemical reactions in boundary-layer flows; these reactions can either occur in the gas phase or at the interface.

Reacting boundary layers can either be laminar or turbulent. This chapter deals with both cases; however, we shall devote more effort to describing turbulent boundary layers, since most reacting boundary layers occurring in engineering practice are turbulent. Also, there is much more to discuss in regard to the relevant transport properties of turbulent flow than those of laminar flow.

1.1 Applications of Reacting Boundary-Layer Flows

Classically, reacting boundary-layer flows have received much attention in connection with the combustion of solid and liquid fuels. The greatest surge of interest in the chemically reacting flows adjacent to solid or liquid surfaces has occurred during the last 25 years with the advent of the hypervelocity vehicles in the aerospace technology. Due to the development of digital computers with large storage for scientific applications, significant advances have been made in this area, especially for various propulsion systems. Many areas requiring the knowledge and solution of reacting boundary-layer flows are listed below:

- ☐ Thermal protection of reentry vehicles by ablative heat shields
- ☐ Catalytic reactors in chemical industries
- ☐ Metal-surface erosion by high-temperature shear flows
- ☐ Erosive burning of solid propellants in high-performance rocket motors
- ☐ Recession of graphitic nozzle material in high-temperature, high-pressure environments
- ☐ Solid lubricants in chemically corrosive environments
- ☐ Fire on vertical walls
- ☐ Flame stabilization by plates and rods in premixed flowing gases
- ☐ Burning of fuel-rich solid propellants in air-augmented propulsive devices
- ☐ Combustion of a pool of liquid fuel under forced and natural convection

1.2 High-Temperature Experimental Facilities Used in Investigation

These include:

- ☐ Electric-arc-heated gases
- ☐ Shock tubes or tunnels
- ☐ Ballistic compressors
- ☐ Rocket motors for hot-gas generation
- ☐ Wind tunnels

1.3 Theoretical Approaches and Boundary-Layer Flow Classifications

Most of the problems associated with chemical reactions described in Section 1.1 can be analyzed within the framework of a boundary-layer theory:

A. Equilibrium Case. Chemical reactions, whether occurring in the gas phase or on the surface, are considered to take place infinitely fast, so that the

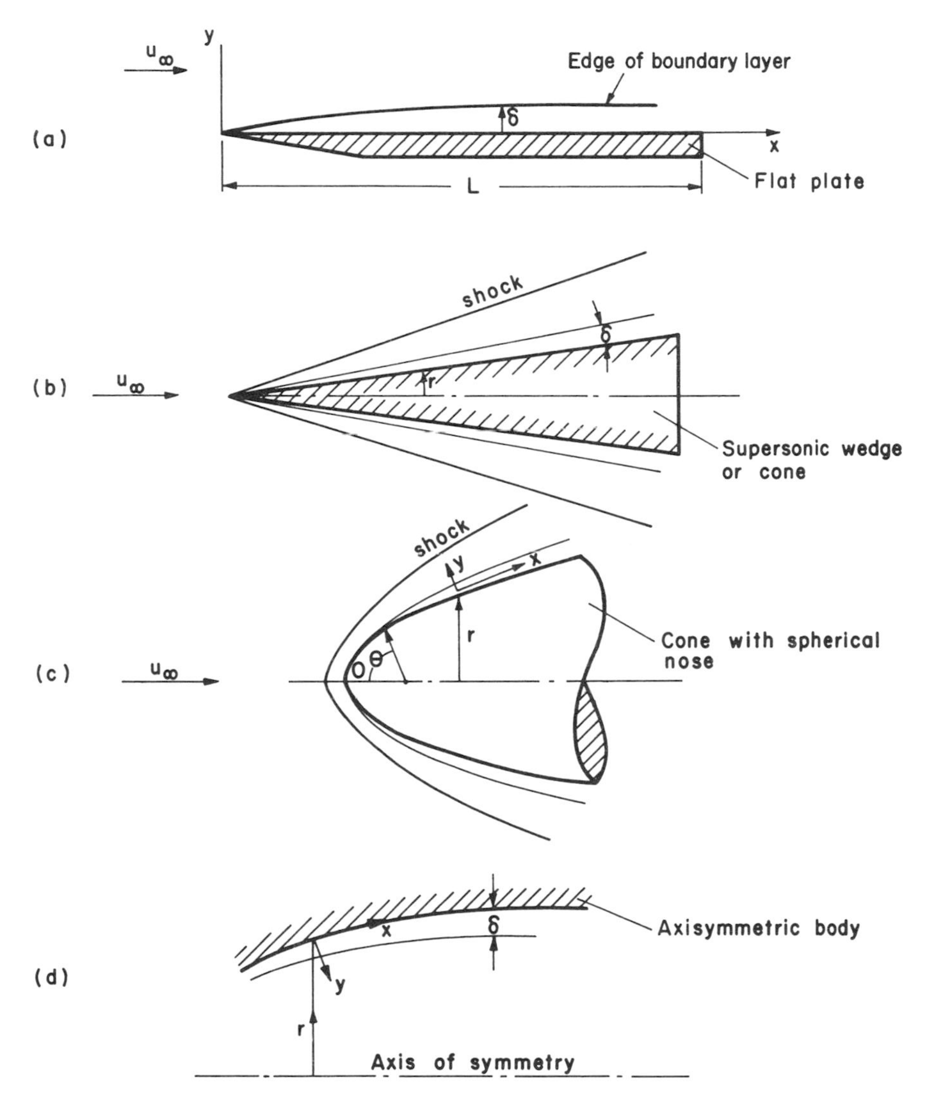

Figure 9.1 Coordinate system of boundary-layer flows over various geometric contours: (a) flat plate, (b) supersonic wedge or cone, (c) cone with spherical nose, (d) axisymmetric body.

chemical rate kinetics is no longer a factor. For this extreme limit, the solution of the problem can usually be obtained in terms of the existing solutions of the conventional boundary-layer equations, and the problem is more or less reduced to algebraically reevaluating the proper energy driving potential. The work of Lees[4] (1958) and Dorrance[5] (1962) falls into this category. Also, when chemical reactions are considered to occur infinitely fast, the governing equations are essentially the same as those for chemically inert boundary layers, and the chemical reactions are governed by the characteristics of the boundary layer alone.

B. Nonequilibrium Case. Chemical reactions take place at a finite rate. The most interesting features of the chemically reacting boundary layers are due to the coupling of the boundary-layer characteristics with the finite-rate chemical reactions.

Reacting laminar boundary layers are treated thoroughly by Chung[6] (1965), Blottner et al.[25] (1971), and Evans[9] (1975). Reacting turbulent boundary layers have been studied by Anderson and Lewis[8] (1971), Evans[9] (1975), Patankar and Spalding[7] (1970), Razdan and Kuo[10] (1979), and many others.

In terms of the solution method, one can classify work into integral methods and numerical partial-differential-equation methods. In terms of the site of chemical reaction, one can classify reactions into gas-phase and heterogeneous reactions. In terms of the geometry of solid bodies, we have two-dimensional, axisymmetric, and three-dimensional cases. A two-dimensional problem case can be regarded as a special case of an axisymmetric problem with the radius of curvature approaching infinity. The coordinate system for different geometries is specified in Fig. 9.1.

Various types of reacting boundary layer can be distinguished according to the terminology in Table 9.1.

TABLE 9.1 Types of Chemically Reacting Boundary-Layer Flows

Parameter	Types
Magnitude of reaction rate	Equilibrium or nonequilibrium
Site of chemical reaction	Gas-phase, heterogeneous, or both
Turbulence level	Laminar or turbulent
Geometric contour	Planar (two-dimensional), axisymmetric, or three-dimensional
Steadiness of boundary-layer flow	Steady or unsteady
Presence of particles	Single-phase or multiphase
Free-stream Mach number	Subsonic, transonic, supersonic, or hypersonic
Exothermicity of reaction	Exothermic or endothermic
Mixing condition of reactants	Premixed or diffusion

1.4 Historical Survey

There are voluminous works on chemically reacting boundary-layer flows; a long survey would be required to cover all of them. The brief discussion presented here is meant to provide the essence of many of the theories, to indicate major developments in the field, and also to show the main lines along which the theories are divided.

A. von Karman and Millan[1] (1953) used boundary-layer concepts in treating the cooling region of a premixed laminar flame near a cold wall.

B. Marble and Adamson[2] (1954) developed a comprehensive analysis of chemical reactions in boundary layers for studying ignition of premixed combustibles in a laminar zone between cold combustible and hot inert gases (see Fig. 9.2).

C. Emmons[3] (1956) studied the burning of solid fuel (in the shape of a flat plate) in an oxidizing stream. The velocity, temperature, and concentration profiles are shown in Fig. 9.3.

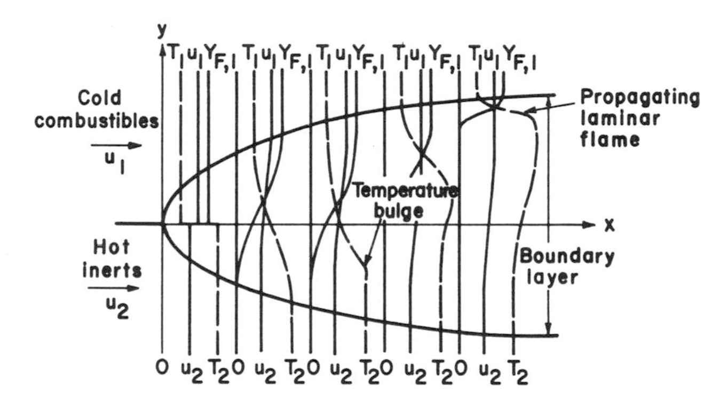

Figure 9.2 Property distributions in reacting shear flows (after Marble and Adamson[2]).

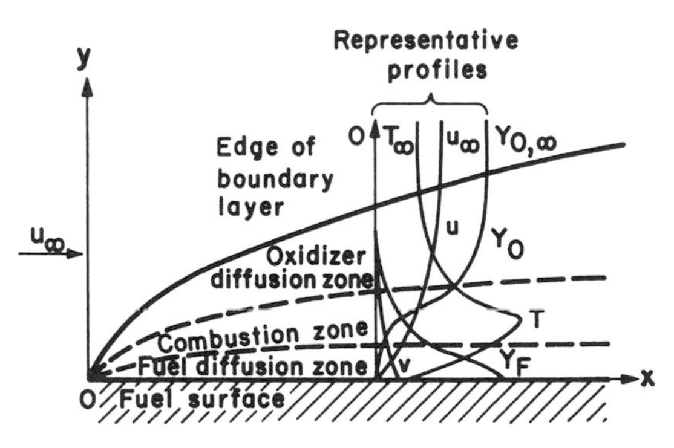

Figure 9.3 Flow-property distributions in the boundary layer of a burning solid fuel (after Emmons[3]).

D. Lees[4] (1958) showed that the spirit of the Shvab–Zel'dovich formulation can easily be retained in reacting boundary-layer equations. He concluded that under the assumption of Lewis number equal to unity, the convective heat transfer in a laminar boundary layer is independent of the model of the chemical reaction. Reaction rates play a secondary role in determining the species mass fractions at the surface, and the heat released in chemical reactions appears as an additional "enthalpy potential."

E. Dorrance[5] (1962) addressed the problem of viscous hypersonic flows in the development of space-exploration vehicles and long-range missiles. He considered both laminar and turbulent boundary layers, and also took into account the effects of dissociation of the gases in the boundary layer, reactions other than dissociation, effects of mass transfer, effects of a melting surface, and interactions between the boundary-layer gases and the solid surfaces. However, the treatment of chemical reactions is limited to equilibrium conditions.

F. Chung[6] (1965) wrote a comprehensive article on chemically reacting nonequilibrium boundary layers and emphasized the important differences between nonequilibrium and equilibrium cases. According to Chung, only in the nonequilibrium cases, where the chemical reaction takes place at a finite rate, does the true behavior of a chemically reacting boundary layer become manifest. His monograph discusses only reacting laminar boundary layers, since at that time there were very few published works on turbulent nonequilibrium boundary layers.

G. Patankar and Spalding[7] (1970) developed a very general numerical scheme using streamline coordinates for solving reacting boundary-layer flow problems. They used the k–ε two-equation method for obtaining turbulence closure. They solved two resulting turbulence-transport equations together with conservation equations for boundary-layer flows.

H. Anderson and Lewis[8] (1971) considered turbulent reacting boundary-layer flows by using two eddy-viscosity models. Chemical equilibrium was assumed for the reacting gas mixture.

I. Evans[9] (1975) developed a computer code based on the boundary-layer integral matrix procedure which has been used by many engineers for rocket engine performance prediction and evaluation calculations.

J. Razdan and Kuo[10] (1979) combined the eddy-breakup concept of Spalding[11] and the k–ε two-equation solution procedure of Patankar and Spalding[7] for their prediction of erosive burning of composite solid propellants, and obtained good agreement with erosive-burning data under various flow conditions.

K. Wu et al.[12] (1982) extended the analysis of Razdan and Kuo[10] and developed a comprehensive aerothermochemical model of erosive burning of double-base propellants. The conservation equations were Favre-averaged, and finite-rate chemical kinetics were considered for simulating reactions in various flame zones. Predicted burning-rate results for a wide range of cross-flow situations compared well with the experimental data of Burick and Osborn.[13]

L. Kuo and Keswani[14] (1985) studied the recession of graphitic nozzles due to high-temperature product gases by considering both mass diffusion and chemical kinetics. The recession process was found to be limited by diffusion of oxidizing species such as H_2O and CO_2 to the nozzle surface.

2 GOVERNING EQUATIONS FOR TWO-DIMENSIONAL REACTING BOUNDARY-LAYER FLOWS

The governing partial differential equations for chemically reacting boundary-layer flow are deduced from the conservation equations for multicomponent reacting systems given in Chapter 3. The following order-of-magnitude analysis is used to simplify the generalized momentum, energy, and species equations to the form commonly used for two-dimensional laminar reacting boundary-layer flows.

Let us consider a steady two-dimensional laminar flow over a flat plate. The boundary layer formed on top of the plate is shown in Fig. 9.1*a*. The thickness of the boundary layer, δ, is considered to be much smaller than the length of the plate, L; that is, $\delta/L \ll 1$. The usual boundary-layer approximation[15,16] states that the orders of magnitude of the gradients are

$$\frac{\partial u}{\partial x} \sim \frac{u}{L} \tag{9-1}$$

and

$$\frac{\partial u}{\partial y} \sim \frac{u}{\delta} \tag{9-2}$$

The continuity equation for a two-dimensional steady boundary layer can be written as

$$\frac{\partial(\rho u)}{\partial x} + \frac{\partial(\rho v)}{\partial y} = 0 \tag{9-3}$$

Based upon Eqs. (9-1) and (9-3), we have

$$\frac{\partial v}{\partial y} \sim \frac{u}{L} \tag{9-4}$$

and

$$v \sim \frac{u\delta}{L} \tag{9-5}$$

Using Eqs. (9-1) through (9-5) and setting $\partial/\partial t = 0$, $B_i = 0$, and $\mu' = 0$, the x-momentum equation (3-58), after setting i in the x-direction, can be written as

$$\underbrace{u\frac{\partial u}{\partial x}}_{\left(\frac{u^2}{L}\right)} + \underbrace{v\frac{\partial u}{\partial y}}_{\left(\frac{u\delta}{L}\right)\left(\frac{u}{\delta}\right)} = \underbrace{-\frac{1}{\rho}\frac{\partial p}{\partial x}}_{\left(\frac{1}{\rho}\right)\left(\frac{\rho u^2}{L}\right)}$$

$$+ \underbrace{\frac{1}{\rho}\left\{ \cancel{-\frac{\partial}{\partial x}\left[\tfrac{2}{3}\mu\left(\frac{\partial u}{\partial x} + \frac{\partial v}{\partial y}\right)\right]} + \overbrace{\cancel{\frac{\partial}{\partial x}\left(2\mu\frac{\partial u}{\partial x}\right)}}^{\text{negligible}} + \frac{\partial}{\partial y}\left[\mu\left(\overbrace{\cancel{\frac{\partial v}{\partial x}}}^{\text{negligible}}\right.\right.\right.}_{\left(\frac{u}{\rho u L}\right)\left(\frac{u^2}{L}\right)} + \underbrace{\left.\left.\left.\frac{\partial u}{\partial y}\right)\right]\right\}}_{\left(\frac{\mu}{\rho u L}\right)\left(\frac{u^2 L}{\delta^2}\right)} \tag{9-6}$$

The orders of magnitude of the terms in Eq. (9-6) are indicated beneath each term. Since $\delta/L \ll 1$, terms in the second group on the right-hand side of Eq. (9-6) are much smaller when compared with the last term and hence, can be neglected.

The y-momentum equation can be written as

$$\underbrace{u\frac{\partial v}{\partial x}}_{\left(u\frac{u\delta}{L^2}\right)} + \underbrace{v\frac{\partial v}{\partial y}}_{\left(\frac{u\delta}{L}\frac{u}{L}\right)} = -\frac{1}{\rho}\frac{\partial p}{\partial y}$$

$$+ \underbrace{\frac{1}{\rho}\left\{ -\frac{\partial}{\partial y}\left[\tfrac{2}{3}\mu\left(\frac{\partial u}{\partial x} + \frac{\partial v}{\partial y}\right)\right] + \frac{\partial}{\partial y}\left(2\mu\frac{\partial v}{\partial y}\right) + \frac{\partial}{\partial x}\left[\mu\left(\frac{\partial u}{\partial y}\right.\right.\right.}_{\left(\frac{\mu}{\rho L u}\right)\left(\frac{u^2}{\delta}\right)} + \underbrace{\left.\left.\left.\overbrace{\cancel{\frac{\partial v}{\partial x}}}^{\text{negligible}}\right)\right]\right\}}_{\left(\frac{\mu}{\rho L u}\right)\left(\frac{u^2\delta}{L^2}\right)} \tag{9-7}$$

If the Reynolds number $\rho u L/\mu$ is sufficiently large, namely,

$$\text{Re} \sim \left(\frac{L}{\delta}\right)^2$$

then the transverse pressure-gradient term, $(1/\rho)\partial p/\partial y$, can at most be of the same order of magnitude as the other terms in Eq. (9-7), which are of the order

$u^2\delta/L^2$. Therefore, the $\partial p/\partial y$ term can be considered negligible. This is why, in the Prandtl boundary-layer equations, the x- and y-momentum equations are simplified to

$$u\frac{\partial u}{\partial x} + v\frac{\partial u}{\partial y} = -\frac{1}{\rho}\frac{\partial p}{\partial x} + \frac{1}{\rho}\frac{\partial}{\partial y}\left(\mu\frac{\partial u}{\partial y}\right) \tag{9-8}$$

and

$$\frac{\partial p}{\partial y} = 0 \tag{9-9}$$

The continuity equation for ith species from Eq. (A) of Table 3.2 can be written as

$$\underbrace{\rho u\frac{\partial Y_i}{\partial x}}_{(\rho u\frac{Y_i}{L})} + \underbrace{\rho v\frac{\partial Y_i}{\partial y}}_{(\rho\frac{u\delta}{L}\frac{Y_i}{\delta})} - \underbrace{\overset{\text{negligible}}{\frac{\partial}{\partial x}\left(\rho\mathscr{D}_i\frac{\partial Y_i}{\partial x}\right)}}_{(\frac{\mathscr{D}}{Lu})(\frac{\rho u Y_i}{L})} - \underbrace{\frac{\partial}{\partial y}\left(\rho\mathscr{D}_i\frac{\partial Y_i}{\partial y}\right)}_{(\frac{\mathscr{D}}{Lu})(\frac{\rho u L Y_i}{\delta^2})} = \omega_i \tag{9-10}$$

It is interesting to note that if we suppose the convective terms to be of the same order of magnitude as the diffusive term, we have

$$\frac{u\delta^2}{\mathscr{D}L} \sim 1 \tag{9-11}$$

On replacing L^2/δ^2 by Re_L, Eq. (9-11) becomes

$$\mathrm{Sc} \equiv \frac{\mu}{\rho\mathscr{D}} \sim 1 \tag{9-12}$$

For most gases, the Schmidt number Sc is in fact approximately equal to 1. The rate of generation of species i, W_i, is therefore of order $\rho u Y_i/L$. The simplified continuity equation for the ith species can be written as

$$\rho\left(u\frac{\partial Y_i}{\partial x} + v\frac{\partial Y_i}{\partial y}\right) - \frac{\partial}{\partial y}\left(\rho\mathscr{D}_i\frac{\partial Y_i}{\partial y}\right) = \omega_i \tag{9-13}$$

or

$$\rho\left(u\frac{\partial Y_i}{\partial x} + v\frac{\partial Y_i}{\partial y}\right) + \frac{\partial}{\partial y}(\rho Y_i V_i) = \omega_i \tag{9-14}$$

where V_i, the mass diffusion velocity in the y-direction, is defined by Fick's law as

$$V_i \equiv -\frac{\mathscr{D}_i}{Y_i}\frac{\partial Y_i}{\partial y} \tag{9-15}$$

Neglecting the body-force terms and radiation energy input, the energy equation written in terms of enthalpy can be reduced from Eq. (3-78) to the following form:

$$\underbrace{\rho\left(u\frac{\partial h}{\partial x}\right.}_{\left(\rho u\frac{h}{L}\right)} + \underbrace{\left.v\frac{\partial h}{\partial y}\right)}_{\left(\rho\frac{u\delta}{L}\right)\left(\frac{h}{\delta}\right)} - \left(\underbrace{u\frac{\partial p}{\partial x}}_{\left(u\frac{\rho u^2}{L}\right)} + \overbrace{\cancel{v\frac{\partial p}{\partial y}}}^{\text{negligible}}\right) = -\left(\overbrace{\cancel{\frac{\partial q_x}{\partial x}}}^{\text{negligible}} + \underbrace{\frac{\partial q_y}{\partial y}}_{\left(\frac{\lambda h}{\delta^2 C_p}\right)}\right)$$

$$+\mu\left\{\underbrace{2\left[\left(\frac{\partial u}{\partial x}\right)^2 + \overbrace{\cancel{\left(\frac{\partial v}{\partial y}\right)^2}}^{\text{negligible}}\right]}_{\left(\frac{u}{L}\right)^2} + \underbrace{\left[\overbrace{\cancel{\frac{\partial v}{\partial x}}}^{\text{negligible}} + \frac{\partial u}{\partial y}\right]^2}_{\left(\cancel{\frac{\delta u}{L^2}} + \frac{u}{\delta}\right)^2} - \underbrace{\frac{2}{3}\left[\frac{\partial u}{\partial x} + \overbrace{\cancel{\frac{\partial v}{\partial y}}}^{\text{negligible}}\right]^2}_{\left(\frac{u}{L}\right)^2}\right\} \tag{9-16}$$

After simplification, the energy equation can be written as

$$\rho\left(u\frac{\partial h}{\partial x} + v\frac{\partial h}{\partial y}\right) = u\frac{\partial p}{\partial x} - \frac{\partial q_y}{\partial y} + \mu\left(\frac{\partial u}{\partial y}\right)^2 \tag{9-17}$$

where the enthalpy h is defined to include the chemical energy:

$$h = \sum_{i=1}^{N} Y_i h_i \tag{9-18}$$

and

$$h_i = \int_{T^\circ}^{T} C_{p,i}\, dT + \Delta h_{fi}^\circ \tag{9-19}$$

The heat flux q_y is mainly due to the heat conduction and the energy carried by the interdiffusing species. Thus, after neglecting the Dufour effect,

$$q_y = -\lambda\frac{\partial T}{\partial y} + \rho\sum_{i=1}^{N}(Y_i V_i h_i) \tag{9-20}$$

Using Fick's law of diffusion, we have

$$q_y = -\lambda \frac{\partial T}{\partial y} - \rho \sum_{i=1}^{N} \left(h_i \mathscr{D}_i \frac{\partial Y_i}{\partial y} \right) \tag{9-21}$$

Substituting Eq. (9-21) into (9-17), we have

$$\rho \left(u \frac{\partial h}{\partial x} + v \frac{\partial h}{\partial y} \right) = u \frac{\partial p}{\partial x} + \frac{\partial}{\partial y} \left(\lambda \frac{\partial T}{\partial y} + \rho \sum_{i=1}^{N} h_i \mathscr{D}_i \frac{\partial Y_i}{\partial y} \right) + \mu \left(\frac{\partial u}{\partial y} \right)^2 \tag{9-22}$$

With a little manipulation, the energy equation can be given in terms of the total enthalpy $h_t \equiv (h + u^2/2)$ as

$$\rho \left(u \frac{\partial h_t}{\partial x} + v \frac{\partial h_t}{\partial y} \right)$$

$$= \frac{\partial}{\partial y} \left\{ \frac{\mu}{\text{Pr}} \left[\frac{\partial h_t}{\partial y} + (\text{Pr} - 1) \frac{\partial (u/2)^2}{\partial y} + \sum_{i=1}^{N} \left(\text{Le}_i^{-1} - 1 \right) h_i \frac{\partial Y_i}{\partial y} \right] \right\} \tag{9-23}$$

where the Lewis number and Prandtl number are defined as

$$\text{Le}_i = \frac{\lambda}{\rho \mathscr{D}_i C_p}, \qquad \text{Pr} = \frac{C_p \mu}{\lambda} \tag{9-24}$$

An alternative form of the energy equation can be given in terms of temperature. From Eq. (9-22) and the relationship (see Problem 1)

$$dh = C_p \, dT + \sum_{i=1}^{N} h_i \, dY_i \tag{9-25}$$

we have

$$\rho C_p \left(u \frac{\partial T}{\partial x} + v \frac{\partial T}{\partial y} \right) = u \frac{\partial p}{\partial x} + \frac{\partial}{\partial y} \left(\lambda \frac{\partial T}{\partial y} \right)$$

$$+ \sum_{i=1}^{N} \rho \mathscr{D}_i \left(\frac{\partial Y_i}{\partial y} \right) \left(\frac{\partial h_i}{\partial y} \right) + \mu \left(\frac{\partial u}{\partial y} \right)^2 - \sum_{i=1}^{N} h_i \omega_i \tag{9-26}$$

When analyzing a problem of surface reaction with frozen gas-phase reactions, it is convenient to use the frozen (sensible) total enthalpy $h_{t\,\text{se}}$ defined by

$$h_{t\,\text{se}} = \sum_{i=1}^{N} Y_i h_{i\,\text{se}} + \frac{u^2}{2} \tag{9-27}$$

where

$$h_{i\,\mathrm{se}} \equiv \int_{T^\circ}^{T} C_{p,i}\, dT \tag{9-28}$$

A frozen-total-energy equation is identical to Eq. (9-23) except that h_t and h_i are replaced by $h_{t\,\mathrm{se}}$ and $h_{i\,\mathrm{se}}$, respectively.

The equation of state for multicomponent systems based upon the ideal-gas assumption can be written as

$$p = \left(\sum_{i=1}^{N} \frac{Y_i}{W_i} \right) \rho R_u T \tag{9-29}$$

Now, basically, the boundary-layer problem is to solve for the 5 + N variables ρ, u, v, h, T, and the Y_i's, by the use of the same number of equations: Eqs. (9-3), (9-8), (9-13), (9-22), (9-18), and (9-29). The energy equation (9-22) can be replaced either by Eq. (9-23) or (9-26). Also, one of the species continuity equations can be replaced by

$$\sum_{i=1}^{N} Y_i = 1 \tag{9-30}$$

It should be noted that even though pressure appears many times in the above set of equations, it is not considered as an unknown in the boundary-layer solution; it is usually found from the potential-flow equations.

3 BOUNDARY CONDITIONS

The boundary conditions for the continuity and momentum equations are straightforward: at $y = 0$

$$u = 0$$

$$v = v_{w+} \tag{9-31}$$

and at $y = \infty$ (or δ)

$$u = U_e \tag{9-32}$$

where the subscripts w and e refer to the wall and the edge of boundary layer, respectively.

Now consider the boundary conditions for the energy equations (9-22), (9-23), and (9-26). The enthalpies and the temperature are usually considered to be known at the boundary-layer edge. Thus, at $y = \infty$ (or δ),

$$h = h_e = \sum_{i=1}^{N} Y_{ie} \int_{T}^{T_e} C_{p,i}\, dT + \sum_{i=1}^{N} Y_{ie} \Delta h_{fi}^{\circ}$$

$$= h_{se\,e} + \sum_{i=1}^{N} Y_{ie} \Delta h_{fi}^{\circ} \tag{9-33}$$

$$h_t = h_{te} = h_e + \frac{U_e^2}{2} \tag{9-34}$$

$$T = T_e \tag{9-35}$$

At the wall $y = 0^+$,

$$h = h_w = \sum_{i=1}^{N} Y_{iw+} \int_{T^\circ}^{T_w} C_{p,i}\, dT + \sum_{i=1}^{N} Y_{iw+} \Delta h_{fi}^{\circ}$$

$$= h_{se\,w+} + \sum_{i=1}^{N} Y_{iw+} \Delta h_{fi}^{\circ} \tag{9-36}$$

$$h_t = h_{tw} = h_w \tag{9-37}$$

$$T = T_w \tag{9-38}$$

Equation (9-36) indicates that the value of enthalpy cannot be specified unless the wall mass fractions Y_{iw} are known. Usually Y_{iw} are not known *a priori*, and therefore the wall boundary conditions (9-22) and (9-23) are coupled to the solutions of the species conservation equations. The wall temperature may or may not be known. Even if the wall temperature is given, the use of Eq. (9-26) instead of Eq. (9-22) or (9-23) does not necessarily expedite the solution, because Eq. (9-26) itself is more strongly coupled to the species conservation equation than the other two energy equations.

Now considering the boundary condition for the species conservation equation (9-13), we have, at $y = \infty$,

$$Y = Y_{ie} \tag{9-39}$$

At the wall (gas–solid or gas–liquid interface), component i is transported from the gas to the solid by diffusion at the rate $(\rho \mathscr{D}_i\, \partial Y_i / \partial y)_w$ (see Fig. 9.4). At the same time, the component i is transported away from the interface by the normal bulk motion of the fluid at the rate $(\rho v)_w (Y_{iw})_+$ in the gas and toward the interface at the rate $(\rho v)_w (Y_{iw})_-$ in the solid (which may be porous) or in the liquid. The net rate of production of component i per unit surface

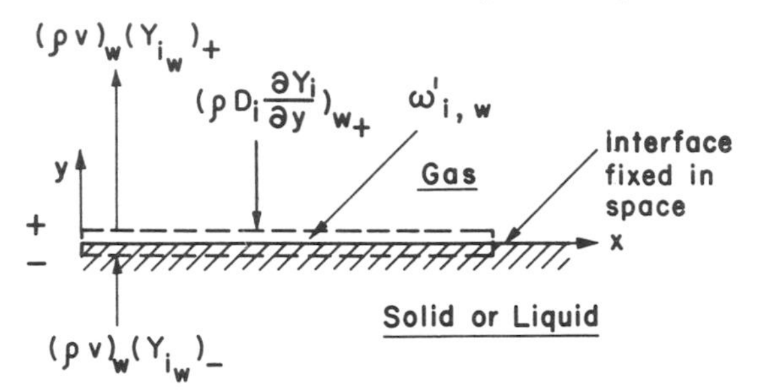

Figure 9.4 Mass balance of species i at the interface.

area by the surface reaction is represented by ω'_{iw}. Then the mass-flux balance of the ith species at the interface is given by

$$\omega'_{i,w} = -\left(\rho \mathscr{D}_i \frac{\partial Y_i}{\partial y}\right)_{w+} + (\rho v)_w\left[(Y_{iw})_+ - (Y_{iw})_-\right] \tag{9-40}$$

where Y_{iw} is zero when the ith component is not included in the injected, ablated, or pyrolyzed gas mixture. The surface reaction rate ω'_{iw}, together with the gas-phase reaction rate ω_i, will be discussed in a subsequent section.

At the interface, the energy fluxes from the boundary layer to the solid or liquid by conduction and radiation are $(\lambda\, \partial T/\partial y)_{w+}$ and $q_{\text{rad}+}$ (see Fig. 9.5). The energies transported away from the interface by mass diffusion and upward bulk motion of the gas are $(\rho \Sigma_{i=1}^N h_i Y_i V_i)_{w+}$ and $(\rho v)_w (h + v^2/2)_{w+}$. At the same time, energy is transferred away from the interface by conduction and radiation, represented by $(\lambda\, \partial T/\partial y)_{w-}$ and $q_{\text{rad}-}$, respectively. Supposing that the interface is located at a fixed position in space (this requires the feeding of solid or liquid material upward at the rate of surface regression), the energy flux transported to the interface is $(\rho v)_w (h + v^2/2)_{w-}$. In case mass diffusion is present below the interface, the energy flux thereby transported to the interface can be represented by $(\rho \Sigma_{i=1}^N h_i Y_i V_i)_{w-}$.

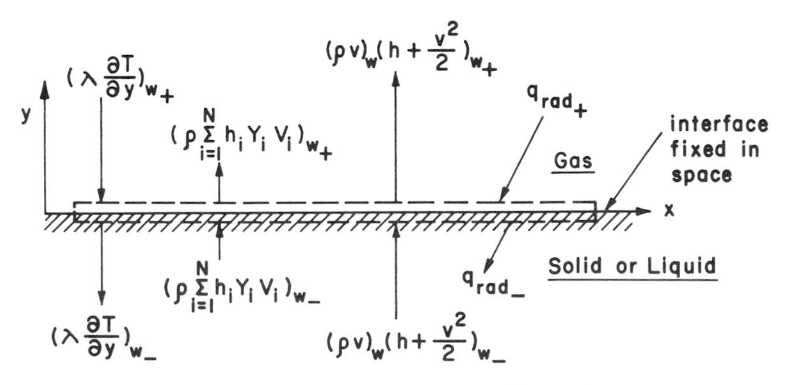

Figure 9.5 Energy-flux balance at the interface.

The energy-flux balance at the interface can be written as

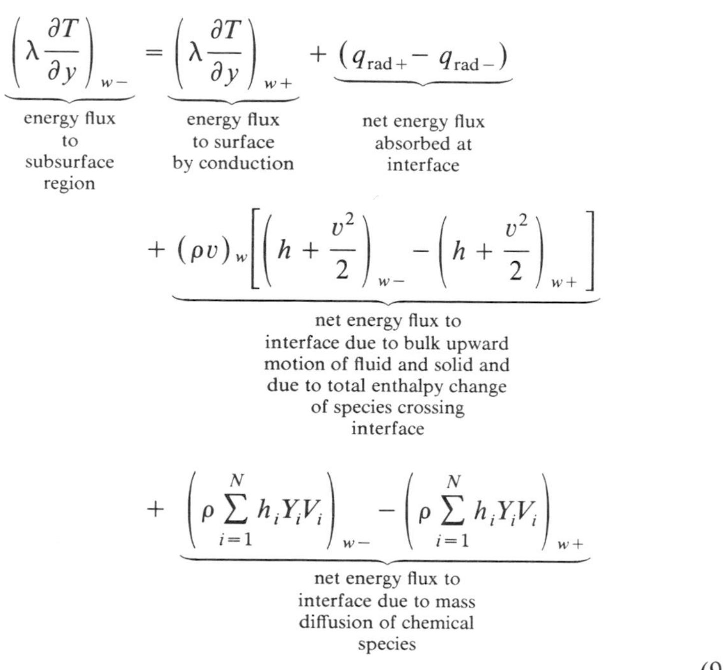

(9-41)

It can be shown that, after neglecting the kinetic-energy terms within the third term in the right-hand side of Eq. (9-41), the energy-flux balance can be written as

$$\left(\lambda\frac{\partial T}{\partial y}\right)_{w-} = \left(\lambda\frac{\partial T}{\partial y}\right)_{w+} + (q_{\text{rad}+} - q_{\text{rad}-}) + \sum_{i=1}^{N}[\rho Y_i(v+V_i)]_{w-}\Delta h_{fi-}$$
$$- \sum_{i=1}^{N}[\rho Y_i(v+V_i)]_{w+}\Delta h^{\circ}_{fi+} + \sum_{i=1}^{N}[\rho Y_i(v+V_i)]_{w-}\int_{T^\circ}^{T} C_{p,i-}\,dT$$
$$- \sum_{i=1}^{N}[\rho Y_i(v+V_i)]_{w+}\int_{T^\circ}^{T} C_{p,i+}\,dT \qquad (9\text{-}42)$$

4 CHEMICAL KINETICS

Chemical kinetics, which provides the general method of examining the mechanism of reaction, is a very complex and extensive subject. It is not intended here, therefore, to give it a comprehensive treatment. We shall only consider those portions of chemical kinetics necessary to continue our study of reacting boundary layers. The reader is referred to Chapter 2 and other literature for a more comprehensive study.

Chemical reactions can be classified into two types: *homogeneous* and *heterogeneous* reactions. The former occur in a homogeneous phase such as a completely gaseous system. The latter occur preferentially at an interface such as the surface of a solid or a liquid.

4.1 Homogeneous Chemical Reactions

According to the law of mass action discussed in Chapter 2, the rate of production of a particular component i from the chemical reaction

$$\sum_{i=1}^{N} \nu_i' M_i \to \sum_{i=1}^{N} \nu_i'' M_i \tag{9-43}$$

is proportional to the product of the concentrations of the reacting chemical species with each concentration raised to a power equal to the corresponding stoichiometric coefficient. Thus,

$$\frac{dC_{M_i}}{dt} = (\nu_i'' - \nu_i')k \prod_{j=1}^{N} \left(C_{M_j}\right)^{\nu_j'} \tag{9-44}$$

The quantity $\nu_i'' - \nu_i'$ is included in the equation because component i may appear in both sides of Eq. (9-43). The specific reaction-rate coefficient k is independent of the concentration and depends only on the temperature. In general, k is given by the Arrhenius equation

$$k = A(T)\exp\left[-\frac{E_a}{R_u T}\right] \tag{9-45}$$

where E_a is the activation energy for the reaction and A was first considered by Arrhenius as an empirical constant to correlate experimental data. But in many actual cases, A is a function of T, since it includes the effects of the steric factor and the collision frequency, which are temperature-dependent. (Refer to Chapter 2 for more detailed discussions.)

For a more general chemical reaction which can proceed both forward and backward, the stoichiometric equation can be written as

$$\sum_{i=1}^{N} \nu_i' M_i \underset{k_b}{\overset{k_f}{\rightleftharpoons}} \sum_{i=1}^{N} \nu_i'' M_i \tag{9-46}$$

where k_f and k_b represent the forward and the backward rate coefficients. The law of mass action then gives, for the net production rate of component i,

$$\frac{dC_{M_i}}{dt} = (\nu_i'' - \nu_i')\left[k_f \prod_{j=1}^{N} (C_{M_j})^{\nu_j'} - k_b \prod_{j=1}^{N} (C_{M_j})^{\nu_j''}\right] \tag{9-47}$$

Writing Eq. (2-25) in a slightly different form or setting the left-hand side of Eq. (9-47) equal to zero, we have

$$K_c \equiv \frac{k_f}{k_b} = \frac{\prod_{j=1}^{N} (C_{M_j,E})^{\nu_j''}}{\prod_{j=1}^{N} (C_{M_j,E})^{\nu_j'}} \tag{9-48}$$

where the subscript E denotes the equilibrium state. Now Eq. (9-48) defines the equilibrium constant K_c based on the concentrations of species at equilibrium. After replacing the backward specific rate coefficient k_b by k_f/K_c, Eq. (9-47) becomes

$$\frac{dC_{M_i}}{dt} = (\nu_i'' - \nu_i')k_f\left[\prod_{j=1}^{N} (C_{M_j})^{\nu_j'} - \frac{1}{K_c}\prod_{j=1}^{N} (C_{M_j})^{\nu_j''}\right] \tag{9-49}$$

Numerous studies have been conducted on boundary-layer flows with dissociation and recombination reactions, due to their importance in hypersonic flows. Let us first concentrate on the dissociation–recombination processes in diatomic gases (such as O_2, N_2). The following equation can be used to represent such processes:

$$A_1 + A_1 + X \underset{k_b}{\overset{k_f}{\rightleftharpoons}} A_2 + X \tag{9-50}$$

where A_1 and A_2 designate the atomic and molecular species, respectively. X denotes the gas particles, which act as a catalyst (third body). In a binary mixture of atoms and molecules, X may be either the atoms or the molecules. The recombination, therefore, is a third-order reaction, whereas the dissociation is a second-order reaction. From the law of mass action, we have

$$\frac{dC_{A_1}}{dt} = -2k_f\left[(C_{A_1})^2(X) - \frac{1}{K_c}(C_{A_2})(X)\right] \tag{9-51}$$

where

$$K_c = \frac{k_f}{k_b} = \frac{(C_{A_2,E})}{(C_{A_1,E})^2} = Z_E(T)\exp\left[\frac{-\Delta G^\circ}{R_u T}\right] \tag{9-52}$$

TABLE 9.2 Several Major Neutral-Gas Reaction Rates for Air Boundary Layers[6,17]

Reaction	Catalyst X	k_f $(\mathrm{cm^6/mole^2\ sec})^a$
(1) $O + O + X \xrightarrow{k_f} O_2 + X$	O	$1.125 \times 10^{20}\ T^{-3/2}$
	O_2	$0.4 \times 10^{20}\ T^{-3/2}$
	N_2	$3.095 \times 10^{15}\ T^{-1/2}$
	N, NO	$1.51 \times 10^{15}\ T^{-1/2}$
(2) $N + N + X \xrightarrow{k_f} N_2 + X$	N	$1.18 \times 10^{21}\ T^{-3/2}$
	N_2	$1.38 \times 10^{16}\ T^{-1/2}$
	O_2, O, NO	$0.545 \times 10^{16}\ T^{-1/2}$
(3) $N + O + X \xrightarrow{k_f} NO + X$	NO	$1 \times 10^{21}\ T^{-3/2}$
	O, N, O_2, N_2	$0.51 \times 10^{20}\ T^{-3/2}$
(4) $NO + N \xrightarrow{k_f} O + N_2$	—	$0.81 \times 10^{13}\ \mathrm{cm^3/mole\ sec}$

$^a T$ is in K.

It is known that three-body recombination processes of the type given by Eq. (9-50) require negligible activation energies. Therefore,

$$k_f(T) = A(T) \propto \frac{1}{T^{\omega}}, \qquad \text{where} \quad 1 \le \omega \le 2 \tag{9-53}$$

Some major reaction rates for air boundary layers reported by Lin and Teare[17,6] are given in Table 9.2.

Practically no dissociation takes place in air at temperatures below about 2000 K. The oxygen begins to dissociate at about 2000 K. The nitrogen begins to dissociate at temperatures between 3500 and 4500 K, depending on the pressure. The degrees of dissociation of O_2 and N_2 can be estimated by the magnitudes of $Y_{A_1}^2/(1 - Y_{A_1}^2)$. According to Chung,[6] these quantities can be expressed as

$$\left(\frac{Y_{\mathrm{O}}^2}{1 - Y_{\mathrm{O}}^2}\right)_E = \frac{1}{4p(\mathrm{atm})} \exp\left[16.2 - \frac{61000}{T(\mathrm{K})}\right] \tag{9-54}$$

$$\left(\frac{Y_{\mathrm{N}}^2}{1 - Y_{\mathrm{N}}^2}\right)_E = \frac{1}{4p(\mathrm{atm})} \exp\left[16.2 - \frac{116000}{T(\mathrm{K})}\right] \tag{9-55}$$

4.2 Heterogeneous Chemical Reactions

Heterogeneous reactions occur at a gas–solid, gas–liquid, or liquid–solid interface. For simplicity, gas–solid interface reactions are discussed here. The detailed theoretical treatment of the surface chemical kinetics is even more complex than for homogeneous reactions; we shall only consider here the simple phenomenological description of the reaction laws.

There are basically two types of surface reactions that are of interest: those in which the surface acts as a catalyst for the gas reactions, and those in which the surface participates in the chemical reaction. The catalytic surface recombination of atoms is a good example of the former category, whereas surface combustion is a good example of the latter.

Consider the following general surface reaction:

$$A + \text{wall} \underset{k_{wb}}{\overset{k_{wf}}{\rightleftharpoons}} B + \text{wall} \tag{9-56}$$

The wall may either act only as a catalyst or react with A and B. Here k_{wf} and k_{wb} are the specific rate coefficients for the forward and backward surface reaction, respectively. Phenomenologically, the rate of production of the reactant per unit area, $\omega'_{A,w}$, may be expressed by

$$\omega'_{A,w} = -k_{wf}(C_A)^{n_f} + k_{wb}(C_B)^{n_b} \tag{9-57}$$

where n_f and n_b are the orders of the forward and backward reactions, respectively. The units of $\omega'_{A,w}$ and the k_w's depend on the units employed for the concentrations of the reactants and products.

For a reacting boundary layer in which surface reactions take place at much higher rates than the rate at which the gaseous reactants are diffused to or from the surface, the surface condition will approach an equilibrium state, since a sufficient amount of time is allowed for the reactions to occur. At equilibrium, Eq. (9-57) gives

$$\frac{k_{wb}}{k_{wf}} = \frac{(C_A)_E^{n_f}}{(C_B)_E^{n_b}} \tag{9-58}$$

and Eq. (9-57) can be written as

$$\omega'_{A,w} = -k_{wf}\left[(C_A)^{n_f} - \frac{(C_A)_E^{n_f}}{(C_B)_E^{n_b}}(C_B)^{n_b}\right] \tag{9-59}$$

The reaction rates, such as those given by Eq. (9-59), are equated to the right-hand side of Eq. (9-40) to obtain the wall concentrations of reactants and products.

Let us first consider the catalytic recombination of dissociated diatomic gases. The surface reactions associated with such recombination are usually known to be first-order reactions. Thus, if we let A and B of Eq. (9-56) represent atoms A_1 and molecules A_2, respectively, then

$$A_1 + \text{wall} \underset{k_{wb}}{\overset{k_{wf}}{\rightleftharpoons}} A_2 + \text{wall} \tag{9-56a}$$

and the rate equation (9-57) becomes

$$\omega'_{A_1,w} = -k_{wf}\left[(C_{A_1})^{n_f} - \frac{(C_{A_1})_E^{n_f}}{(C_{A_2})_E^{n_b}} (C_{A_2})^{n_b} \right] \tag{9-60}$$

The equilibrium ratio $(C_{A_1})_E/(C_{A_2})_E$ is extremely small for oxygen and nitrogen atoms recombining on a surface whose temperature is below about 2000 K. [See Eqs. (9-54) and (9-55).] Therefore, the second term of Eq. (9-60), which is related to the rate of dissociation of recombined molecules, can usually be neglected for the recombination of oxygen and nitrogen atoms. For most of the surface reactions prominent in the catalytic reactors of chemical industries, the equilibrium ratio $(C_A)_E^{n_f}/(C_B)_E^{n_b}$ is usually also of negligible magnitude. The reaction rate of Eq. (9-59) can therefore be commonly written as

$$\omega'_{A,w} = -k_{wf}(C_A)^{n_f} \tag{9-61}$$

The specific reaction rate coefficient k_{wf} depends on the surface material as well as on the gaseous reactant. It is also a quite strong function of the surface temperature. Values of k_{wf} have been obtained experimentally for recombination of various gases on different types of surfaces.

For the case of surface combustion, the gaseous reactant A will usually represent air or oxygen. The backward reaction can usually be neglected. Since, in most experimental work, the rate of change of solid fuel rather than that of the gaseous oxygen is measured, we express the consumption rate of fuel, from Eq. (9-60), as

$$-\omega'_{F,w} = k_w(C_A)^n \tag{9-62}$$

The specific rate coefficient k_w is expressible in the following Arrhenius form:

$$k_w = K_0 \exp\left[-\frac{E_{aw}}{R_u T} \right] \tag{9-63}$$

where the frequency factor K_0 is either a constant or a function of temperature, and E_{aw} is the activation energy for the surface reaction. For the combustion of carbon, Scala[18] showed that most of the reported kinetic data are bounded by the following two sets of kinetic constants:

"fast" reaction:

$$K_0 = 3.29 \times 10^9 \text{ kg/m}^2 \text{ atm}^n \text{ sec}$$

$$E_{aw} = 44 \text{ kcal/mole}$$

"slow" reaction:

$$K_0 = 2.18 \times 10^5 \text{ kg/m}^2 \text{ atm}^n \text{ sec}$$

$$E_{aw} = 42.3 \text{ kcal/mole}$$

the units of which are chosen so that the partial pressure of oxygen is to be used in Eq. (9-61) instead of the mole concentration (C_A). Walker et al.[19] have found reaction orders n between zero and one.

5 LAMINAR BOUNDARY-LAYER FLOWS WITH SURFACE REACTIONS

5.1 Governing Equations and Boundary Conditions

The governing equations for laminar boundary-layer flows with surface chemical reactions are those derived in Section 2 with $\omega_i = 0$, since the gas-phase reaction is considered to be frozen. Specifically, the equations of immediate concern for two-dimensional cases are the continuity equation (9-3), momentum equation (9-8), species conservation equation (9-13), and energy equation (9-23).

The boundary conditions are given by Eqs. (9-31) to (9-41). With the aid of Eq. (9-59), we can now rewrite the boundary condition (9-40) at $y = 0$ in a more specific manner:

$$\left(\rho \mathscr{D}_i \frac{\partial Y_i}{\partial y}\right)_{w+} - (\rho v)_w[(Y_{iw})_+ - (Y_{iw})_-]$$

$$= W_i k_{wf}\left[\left(\frac{\rho}{W_i}Y_i\right)^{n_f} - \frac{\left(\frac{\rho}{W_i}Y_i\right)_E^{n_f}}{\left(\frac{\rho}{W_i}Y_i\right)_E^{n_b}}\left(\frac{\rho}{W_i}Y_i\right)^{n_b}\right] \tag{9-64}$$

where W_i is multiplied on the right-hand side in order to match the units on the two sides of the equation.

With $\omega_i = 0$, the governing equations and the boundary conditions, except the boundary condition (9-64), are essentially the same as those for the classical boundary layers, which have been fully analyzed elsewhere. Basically then, the only new analytical problem at hand is to solve the species-conservation equation with the wall boundary condition given by Eq. (9-64).

5.2 Transformation to (ξ, η) Coordinates

Following the customary practice for boundary-layer analysis, we shall first investigate the possibility of obtaining a self-similar solution. For this purpose, we have to transform the governing equations and boundary conditions from (x, y) coordinates to (ξ, η) coordinates. Following the Levy–Lees transformation described by Lees[20] and Fay and Riddell,[21] we define

$$\xi \equiv \int_0^x \rho_e \mu_e U_e \, dx \tag{9-65}$$

$$\eta \equiv \frac{U_e}{\sqrt{2\xi}} \int_0^y \rho \, dy \tag{9-66}$$

The continuity equation (9-3) is automatically satisfied by introducing the stream function ψ, which is defined by the usual relations

$$\rho u = \frac{\partial \psi}{\partial y} \tag{9-67}$$

$$\rho v = -\frac{\partial \psi}{\partial x} \tag{9-68}$$

Define a nondimensional stream function F as

$$F(\xi, \eta) \equiv \frac{\psi}{\sqrt{2\xi}} \tag{9-69}$$

We then have

$$\frac{\partial F}{\partial \eta} \equiv F_\eta = \frac{u}{U_e} \tag{9-70}$$

and

$$\rho v = -\left[\left(\sqrt{2\xi}\, F_\xi + \frac{1}{\sqrt{2\xi}} F\right)\frac{\partial \xi}{\partial x} + \sqrt{2\xi}\, F_\eta \frac{\partial \eta}{\partial x}\right] \tag{9-71}$$

Also, we define

$$H \equiv \frac{h_t}{h_{te}}, \qquad l \equiv \frac{\rho\mu}{\rho_e \mu_e}, \qquad Y_i{}^* \equiv \frac{Y_i}{Y_{ie}} \tag{9-72}$$

Now, Eqs. (9-8), (9-13), and (9-23) are respectively transformed from (x, y) coordinates to (ξ, η) coordinates for constant Pr and Sc ($\equiv \nu/\mathscr{D}$) as follows:

Momentum:

$$(lF_{\eta\eta})_\eta + FF_{\eta\eta} + 2\left(\frac{\xi}{U_e}\frac{dU_e}{\xi}\right)\left[\frac{\rho_e}{\rho} - (F_\eta)^2\right] = 2\xi(F_\eta F_{\eta\xi} - F_\xi F_{\eta\eta}) \tag{9-73}$$

Conservation of ith species:

$$\frac{1}{\mathrm{Sc}}\left(lY^*_{i,\eta}\right)_\eta + FY^*_{i,\eta} = 2\xi\left(F_\eta Y^*_{i,\xi} - F_\xi Y^*_{i,\eta}\right) \tag{9-74}$$

Total energy:

$$\frac{1}{\mathrm{Pr}}\left(lH_\eta\right)_\eta + FH_\eta + \frac{1}{\mathrm{Pr}}\left\{\frac{U_e^2}{h_{te}}\left[(\mathrm{Pr}-1)lF_\eta F_{\eta\eta}\right]_\eta + \left[\sum_{i=1}^{N}\left(\mathrm{Le}_i^{-1}-1\right)l\frac{Y_{ie}}{h_{te}}Y^*_{i,\eta}\right]_\eta\right\} = 2\xi\left(F_\eta H_\xi - F_\xi H_\eta\right) \tag{9-75}$$

For the frozen boundary layers of present interest, it is more convenient to employ an energy equation based on the frozen total enthalpy $h_{t\,\mathrm{se}}$ rather than Eq. (9-75). Defining

$$H_{\mathrm{se}} \equiv \frac{h_{t\,\mathrm{se}}}{\left(h_{t\,\mathrm{se}}\right)_e} \tag{9-76}$$

the frozen-total-energy equation takes exactly the same form as Eq. (9-75):

$$\frac{1}{\mathrm{Pr}}\left(lH_{\mathrm{se},\eta}\right) + FH_{\mathrm{se},\eta} + \frac{1}{\mathrm{Pr}}\left\{\frac{U_e^2}{\left(h_{t\,\mathrm{se}}\right)_e}\left[(\mathrm{Pr}-1)lF_\eta F_{\eta\eta}\right]_\eta + \left[\sum_{i=1}^{N}\left(\mathrm{Le}_i^{-1}-1\right)l\frac{Y_{ie}h_{i\,\mathrm{se}}}{\left(h_{t\,\mathrm{se}}\right)_e}Y^*_{i,\eta}\right]_\eta\right\} = 2\xi\left(F_\eta H_{\mathrm{se},\xi} - F_\xi H_{\mathrm{se},\eta}\right) \tag{9-77}$$

The boundary conditions given by Eqs. (9-31) to (9-39) and Eq. (9-64) are transformed as follows:

At $\eta = 0$:

$$F_\eta = 0 \tag{9-78}$$

$$F + 2\xi F_\xi = -\frac{\sqrt{2\xi}}{\rho_e\mu_e U_e}(\rho v)_w \tag{9-79}$$

$$H = H_w(\xi) = \sum_{i=1}^{N} Y^*_{iw}\frac{Y_{ie}\int_{T^\circ}^{T_w} C_{p,i}\,dT}{h_{te}} + \sum_{i=1}^{N} Y^*_{iw}\frac{Y_{ie}\Delta h^\circ_{fi}}{h_{te}} \tag{9-80}$$

or

$$H_{se} = H_{se\,w} = \frac{1}{(h_{t\,se})_e} \sum_{i=1}^{N} \left(Y_{iw}^{*} Y_{ie} \int_{T^\circ}^{T_w} C_{p_i}\, dT \right) \tag{9-81}$$

$$Y_{i,\eta}^{*} + \left(\frac{\mathrm{Sc}}{l_w} \right) (F + 2\xi F_\xi) [(Y_{iw}^{*})_+ - (Y_{iw}^{*})_-]$$

$$= \mathrm{Da}_i(\xi) \left[\left(\frac{Y_i^*}{W_i} \right)^{n_f} - \frac{(\rho_w Y_i^*/W_i)_E^{n_f}}{(\rho_w Y_i^*/W_i)_E^{n_b}} \rho_w^{n_b - n_f} \left(\frac{Y_i^*}{W_i} \right)^{n_b} \right] \tag{9-82}$$

where

$$\mathrm{Da}_i(\xi) = \frac{\mathrm{Sc}\sqrt{2\xi}\,(Y_{ie})^{n_f - 1} (k_{wf} \rho_w^{n_f}) W_i}{l_w \rho_e \mu_e U_e} \tag{9-83a}$$

$$\mathrm{Da}_i(\xi) = \frac{(\rho_{w0} Y_{ie})^{n_f} W_i k_{wf} \left(\dfrac{\rho_w}{\rho_{w0}} \right)^{n_f}}{\dfrac{l_w}{\mathrm{Sc}} Y_{ie} \sqrt{\dfrac{(\rho_e \mu_e)_0 U_\infty}{2L}}} \left[\frac{\sqrt{\displaystyle\int_0^{x^*} \frac{\rho_e \mu_e}{(\rho_e \mu_e)_0} \frac{U_e}{U_\infty}\, dx^*}}{\dfrac{\rho_e \mu_e}{(\rho_e \mu_e)_0} \dfrac{U_e}{U_\infty}} \right] \tag{9-83b}$$

where $x^* = x/L$ (L is the characteristic length of the body), and hence

$$\mathrm{Da}_i(\xi) = \frac{(\rho_{w0} Y_{ie})^{n_f} k_{wf} W_i (\rho_w/\rho_{w0})^{n_f}}{(l_w/\mathrm{Sc}) Y_{ie} \mathscr{F}(x^*) \sqrt{(\rho_e \mu_e)_0 U_\infty/(2L)}} \tag{9-83c}$$

where

$$\mathscr{F}(x^*) \equiv \frac{\dfrac{\rho_e \mu_e}{(\rho_e \mu_e)_0} \dfrac{U_e}{U_\infty}}{\sqrt{\displaystyle\int_0^{x^*} \frac{\rho_e \mu_e}{(\rho_e \mu_e)_0} \frac{U_e}{U_\infty}\, dx^*}} \tag{9-84}$$

At $\eta \to \infty$:

$$F_\eta = 1 \tag{9-85}$$

$$H = 1 \tag{9-86}$$

$$H_{se} = 1 \tag{9-87}$$

$$Y_i^* = 1 \tag{9-88}$$

5.3 Conditions for Decoupling of Governing Equations and Self-Similar Solutions

It is seen that the momentum equation and the frozen-total-energy equation are coupled to the species conservation equation only through l, ρ_e/ρ, and the fourth term of Eq. (9-75). For most engineering purposes, the dependence of l and ρ_e/ρ on Y_i can be neglected in comparison with their dependence on temperature. Moreover, l may be often assumed constant, and the entire third term of the momentum equation is negligible for many cases of practical interest. The fourth term of Eq. (9-77) vanishes when either Le_i is equal to one or all $C_{p,i}$ are equal. Again, for many engineering problems, Le_i is close to unity and the $C_{p,i}$ are similar. This term can therefore be neglected. The boundary condition (9-81) may also be written as

$$H_{\mathrm{se}\,w} = \frac{C_{p,w}T_w}{(h_{t\,\mathrm{se}})_e} \tag{9-89}$$

where $C_{p,w}$ is the specific heat of the gas mixture at the wall and is assumed to be known.

Now we see that the momentum and frozen-total-energy equations, together with their boundary conditions, are those of chemically *inert* boundary layers and, for all practical purposes, are decoupled from the species conservation equation. Under this circumstance, they can be solved separately without detailed knowledge of the species distribution. Furthermore, for certain types of boundary-layer flows, the partial differential equations for momentum and frozen total energy can be reduced to ordinary differential equations with η as the sole independent variable. These types of boundary-layer flows are called *self-similar flows*, and have been explored extensively in studies of inert boundary-layer flows. In general, the types of flows amenable to self-similar analysis are the flows associated with flat plates, subsonic wedges, supersonic cones and wedges, and hypersonic blunt bodies. The momentum equation (9-73) for these flows can be reduced to the following Blasius equation:

$$(lF'')' + FF'' = 0 \tag{9-90}$$

where the ($'$) denotes ordinary differentiation with respect to η. The self-similar energy equation reduces from Eq. (9-77) to

$$\frac{1}{\mathrm{Pr}}(lH_{\mathrm{se}}')' + FH_{\mathrm{se}}' + \frac{1}{\mathrm{Pr}}\left[\frac{U_e^2}{(h_{t\,\mathrm{se}})_e}(\mathrm{Pr}-1)lF'F''\right]' = 0 \tag{9-91}$$

For the self-similarity of the above flows not to be destroyed by surface mass transfer $(\rho v)_w$, the mass transfer must be distributed along the surface in such a manner that the right-hand side of Eq. (9-79) is constant. For this case, Eq.

(9-79) becomes

$$F(0) = \frac{\sqrt{2\xi}}{\rho_e \mu_e U_e}(\rho v)_w = \text{constant} \tag{9-92}$$

5.4 Damköhler Number for Surface Reactions

Let us now study the physical meaning of the function $\text{Da}_i(\xi)$ defined by Eq. (9-83), since this function plays an important role in the subsequent analysis of boundary layers with surface reactions.

Suppose that a uniform diffusion potential for component i, $(Y_{ie} - Y_{iw+})$, is applied across the boundary layer along a body. It is well known for such cases that the right-hand side of Eq. (9-74) vanishes and Y_i^* can be evaluated as a function of η only. The rate of diffusion of component i at the surface can be expressed as

$$\left(\rho \mathscr{D}_i \frac{\partial Y_i}{\partial y}\right)_{w+} = \left(\rho \mathscr{D}_i Y_{ie} Y_{i,\eta}^*\right)_{w+} \frac{U_e \rho_{w+}}{\sqrt{2\xi}}$$

$$= Y_{iw+}^{*\prime} \frac{l_w}{\text{Sc}} Y_{ie} \sqrt{\frac{(\rho_e \mu_e)_e U_\infty}{2L}}\, \mathscr{F}(x^*) \tag{9-93}$$

For a given $Y_{ie} - Y_{iw+}$ and l_w, the quantity Y_{iw+}^* is constant. The rest of the right-hand side of Eq. (9-93) therefore represents the characteristic diffusion rate of component i for unit surface area and for unit concentration difference across the boundary layer. Now returning to Eq. (9-83c), it is seen that the denominator of the equation defining $\text{Da}_i(\xi)$ is exactly this characteristic diffusion rate. The numerator of Eq. (9-83c), $W_i k_{wf}(\rho_w Y_{ie})^{n_f}$, on the other hand, represents the characteristic surface reaction rate for unit surface area. The function $\text{Da}_i(\xi)$, therefore, denotes the ratio of the characteristic surface reaction rate to the characteristic diffusion rate for the ith species. One can also consider $\text{Da}_i(\xi)$ as the ratio of the characteristic diffusion time (τ_{diff}) to the characteristic surface reaction time (τ_{ch}). Thus, we may write

$$\text{Da}_i(\xi) = \left(\frac{\text{characteristic surface reaction rate}}{\text{characteristic diffusion rate}}\right)_i$$

$$= \left(\frac{\tau_{\text{diff}}}{\tau_{\text{ch}}}\right)_i \tag{9-94}$$

This characteristic time ratio is called the *surface Damköhler number* for component i.

Now we are in a position to clearly define the two extreme cases of "frozen" and "equilibrium" surface reactions. Let us first consider the case in which

$Da_i \to 0$. It is seen that the right-hand side of Eq. (9-94) is zero, and therefore the effect of surface reaction on the species boundary layer is practically nil. This case is defined as the flow with *frozen* surface reaction. It should be noted here that a frozen surface reaction does not necessarily imply that $k_{wf} \to 0$. It only implies that the reaction rate is much smaller than the diffusion rate, though the absolute value of the reaction rate itself may be quite large.

Next, consider the other extreme case of $Da_i \to \infty$. The left-hand side of Eq. (9-82) is finite. Therefore, the quantity in the bracket on the right-hand side of the equation must approach zero as $Da_i \to \infty$. This means that $(Y_i^*)_w \to (Y_{iE}^*)_w$. This condition is defined as the flow with *equilibrium* surface reactions. Again, one should note that the equilibrium limit does not necessarily imply that $k_{wf} \to \infty$, but rather that the reaction rate is much greater than the diffusion rate. For these cases, the wall boundary condition (9-82) can be replaced by the *a priori* known value $(Y_i^*)_w = (Y_{iE}^*)_w$.

As pointed out clearly by Chung,[6] the full boundary condition (9-82) must be used for the general cases of finite Da_i. When Da_i is finite, $(Y_i^*)_w \neq (Y_{iE}^*)_w$ and the gas layer at the surface is in a state of chemical nonequilibrium. These general cases are therefore defined as flows with *nonequilibrium* surface reactions.

5.5 Surface Combustion of Graphite Near the Stagnation Region

Surface combustion, instead of gas-phase combustion, prevails for graphite when the surface temperature is below about 2750–4450 K, depending on the particular grade of graphite. Surface combustion, therefore, is prevalent in many engineering applications, including ablative heat shields for space vehicles. Let us now discuss the surface combustion rate of graphite in stagnation boundary-layer flows.

The surface reaction of the present problem can be represented by the following relation:

$$\nu'_{O_2} O_{2(g)} + \nu'_C C_{(s)} \to \nu''_{CO_2} CO_{2(g)} + \nu''_{CO} CO_{(g)} \tag{9-95}$$

For a given supply rate of oxygen, the relative proportions of CO and CO_2 produced by the combustion at the surface are not known *a priori*. The combustion rate of carbon varies by up to a factor of two, depending on the ratio ν''_{CO_2}/ν''_{CO}.

According to Eqs. (9-62) and (9-63), the mass rate of carbon combustion can be written as

$$\omega'_{C,w} = -K_0 \exp\left[-\frac{E_{aw}}{R_u T_w}\right]\left[p\left(\frac{W}{W_{O_2}}\right)_w\right]^n (Y_{O_2})_e^n (Y_{O_2}^*)_w^n \tag{9-96}$$

It should be noted that the partial pressure of O_2 was used, instead of the molar concentration of O_2 in Eq. (9-62), in the derivation of Eq. (9-96). The mass rate of oxygen consumption, $\omega'_{O_2,w}$, is related to $\omega'_{C,w}$ from Eq. (9-95) by

$$\omega'_{O_2,w} = \frac{\nu'_{O_2} w_{O_2}}{\nu'_C W_C} \omega'_{C,w} \tag{9-97}$$

In the following, we determine the ratio ν''_{CO_2}/ν''_{CO}. The carbon combustion rate, $\omega'_{C,w}$, will then be obtained from Eq. (9-97).

For simplicity, we shall confine the analysis to the boundary layers where $\rho_e/\rho_w \ll 1$. It is well known[6,20] that for such boundary layers, Eq. (9-90) is applicable as the momentum equation, and also the third term of the energy equation (9-91) is negligible. Moreover, for most engineering purposes, one may consider l to be constant and equal to

$$l = \left[\left(\frac{\rho_w \mu_w}{\rho_e \mu_e} \right)_0 \right]^{0.2} \tag{9-98}$$

For constant l, we can redefine the similarity variable and the stream function as

$$\zeta \equiv \frac{\eta}{\sqrt{l}} = \frac{U_e}{\sqrt{2\xi l}} \int_0^y \rho \, dy \tag{9-99}$$

and

$$f(\xi, \zeta) = \frac{F(\xi, \eta)}{\sqrt{l}} = \frac{\psi}{\sqrt{2\xi l}} \tag{9-100}$$

The momentum equation (9-90), the frozen-total-energy equation (9-91), and the species conservation equation (9-74) become, respectively,

$$f''' + f'' = 0 \tag{9-101}$$

$$\frac{1}{\Pr} H''_{se} + f H'_{se} = 0 \tag{9-102}$$

$$\frac{1}{\mathrm{Sc}} Y^{*\prime\prime}_{O_2} + f Y^{*\prime}_{O_2} = 0 \tag{9-103}$$

where the prime denotes ordinary differentiation with respect to ζ. Near the stagnation region, where all properties along the edge of boundary layer remain constant and

$$U_e = \beta x \tag{9-104}$$

the boundary conditions given in Eqs. (9-78) to (9-88) become:

At $\zeta = 0$:

$$f' = 0 \tag{9-105}$$

$$f = f_w = -\frac{(\rho v)_w}{\sqrt{(\rho_e \mu_e)_0 \beta l}} \tag{9-106}$$

$$H_{se} = H_{se\,w} = (C_{p,w} T_w)/(h_{t\,se})_e \tag{9-107}$$

$$Y^{*\prime}_{O_2 w} + Scf_w Y^*_{O_2 w} = \text{Da}\left[\frac{\nu'_{O_2} W_{O_2}}{\nu'_C W_C}\right](Y^*_{O_2})^n_w \tag{9-108}$$

At $\zeta \to \infty$:

$$f' = 1 \tag{9-109}$$

$$H_{se} = 1 \tag{9-110}$$

$$Y^*_{O_2} = 1 \tag{9-111}$$

In the above, the surface Damköhler number for constant l is defined more conveniently as

$$\text{Da} \equiv \frac{K_0 \exp[-E_{aw}/R_u T_w)]\,[p(W/W_{O_2})_w]^n (Y_{O_2})_e^{n-1}}{\sqrt{(\rho_e \mu_e)_0}\,\beta l} \tag{9-112}$$

Substituting f of Eq. (9-101) into Eq. (9-103) and integrating, we obtain

$$Y^*_{O_2}(\zeta) - Y^*_{O_2 w} = Y^{*\prime}_{O_2 w} \int_0^{\zeta} \left[\frac{f''(\xi)}{f''(0)}\right]^{\text{Sc}} d\zeta \tag{9-113}$$

Upon applying the boundary condition (9-111) and using approximations (see Ref. 6 for details), Eq. (9-113) becomes

$$1 - Y^*_{O_2 w} = Y^{*\prime}_{O_2 w}\left(\frac{1}{A_1 + A_2 f_w}\right) \tag{9-114}$$

where $A_1 = 0.372$ and $A_2 = 0.266$ for Sc = 0.51 (Le = 1.4 when Pr = 0.72). By combining Eq. (9-106) with (9-96), we have

$$f_w = -\text{Da}(Y_{O_2 e})(Y^*_{O_2})^n \tag{9-115}$$

Now we have three algebraic equations [Eqs. (9-114), (9-108), and (9-115)] with four unknowns [$Y^*_{O_2w}$, $Y^{*\prime}_{O_2w}$, f_w, and $\nu'_{O_2}W'_{O_2}/\nu'_C W_C$]. Eliminating $Y^{*\prime}_{O_2w}$ and f_w from these three equations, we obtain

$$1 - Y^*_{O_2w} = \frac{\mathrm{Sc}\left[\left(\nu'_{O_2}W_{O_2}/\nu'_C W_C\right) + Y_{O_2e}Y^*_{O_2w}\right]}{\left[A_1/\mathrm{Da}\left(Y^*_{O_2w}\right)^n\right] - A_2 Y_{O_2e}} \tag{9-116}$$

This equation still has two unknowns. We next postulate that the carbon surface is an ideal catalyst in promoting the following chemical equilibrium of the gas at the interface:

$$2\,CO_2 \rightleftharpoons 2\,CO + O_2 \tag{9-117}$$

The equilibrium constant K_p for the above reaction is derived as

$$K_p = \frac{(p_{CO})^2(p_{O_2})}{(p_{CO_2})^2} = \exp\left(20.926 - \frac{68.224 \times 10^3}{T(\mathrm{K})}\right) \tag{9-118}$$

Then, the ratio of CO to CO_2 partial pressures is obtained from the above equilibrium relationships as

$$\frac{p_{CO}}{p_{CO_2}} = \sqrt{\frac{K_p}{p_{O_2}}} = \sqrt{\frac{K_p}{Y_{O_2e}\,p\left(W/W_{O_2}\right)_w}}\left(Y^*_{O_2w}\right)^{-1/2} \tag{9-119}$$

Now from Eqs. (9-95), (9-118), and (9-119), the ratio $\nu'_{O_2}W_{O_2}/\nu'_C W_C$ becomes

$$\frac{\nu'_{O_2}W_{O_2}}{\nu'_C W_C} = \frac{\nu''_{CO_2} + \frac{1}{2}\nu''_{CO}}{\nu''_{CO_2} + \nu''_{CO}}\frac{W_{O_2}}{W_C} = \frac{2 + p_{CO}/p_{CO_2}}{2 + 2p_{CO}/p_{CO_2}}\frac{W_{O_2}}{W_C}$$

$$= \frac{2\sqrt{Y^*_{O_2w}} + \sqrt{\dfrac{K_p}{Y_{O_2e}\,p\left(W/W_{O_2}\right)_w}}}{2\sqrt{Y^*_{O_2w}} + 2\sqrt{\dfrac{K_p}{Y_{O_2e}\,p\left(W/W_{O_2}\right)_w}}}\frac{W_{O_2}}{W_C} \tag{9-120}$$

It can be seen from the above equation that as $\mathrm{Da} \to \infty$, and hence $Y^*_{O_2w} \to 0$,

$$\lim_{Y^*_{O_2w} \to 0}\left(\frac{\nu'_{O_2}}{\nu'_C}\right) = \frac{1}{2} \tag{9-121}$$

In this limit, Eq. (9-95) becomes

$$O_{2(g)} + 2C_{(s)} \rightarrow 2CO_{(g)} \tag{9-122}$$

showing that CO is the sole product of combustion. Now Eq. (9-116) can be solved for $Y^*_{O_2w}$ with the aid of Eq. (9-120). The combustion rate of carbon is then obtained from Eq. (9-96).

The maximum diffusion-controlled combustion rate, $(\omega'_{C,w})_{dc}$, wherein $Y^*_{O_2w} \to 0$, may be obtained rather directly from Eq. (9-114) with the aid of Eq. (9-121) as

$$(\omega'_{C,w})_{dc} = -2\frac{W_C}{W_{O_2}}\rho_w \mathscr{D}_w \left(\frac{\partial Y_{O_2}}{\partial y}\right)_w$$

$$= \sqrt{(\rho_e \mu_e)_0 \beta l}\,\frac{A_1 Y_{O_2e}}{\mathrm{Sc}\left(\frac{1}{2}W_{O_2}/W_C\right) + A_2 Y_{O_2e}} \tag{9-123}$$

The Damköhler number Da for the surface combustion of graphite is dependent most strongly on the surface temperature through the exponential function. A few typical combustion rates[22-24] computed from the solution procedure described here are shown in Fig. 9.6. The combustion rates are normalized by the diffusion-controlled rate of Eq. (9-123) and are given as a function of the surface temperature. Also shown are the experimental data of Welsh[23] and Diaconis et al.[24] from their respective arc-tunnel tests.

The results of the theory show the following interesting phenomena. With the initial rise in surface temperature and hence Damköhler number, the surface reaction enters into the nonequilibrium regime. This regime is seen to be traversed rather rapidly, and the temperature rise of about 200 K is seen to

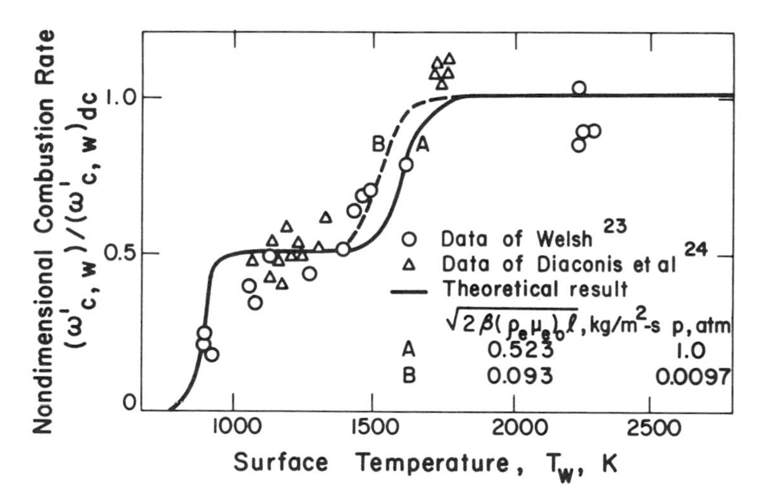

Figure 9.6 Surface combustion rate of graphite as a function of surface temperature.

bring the abscissa to the first plateau with $(\omega'_{C,w})/(\omega'_{C,w})_{dc} = \frac{1}{2}$. On this plateau, $Y^*_{O_2 w}$ is small compared to one, and the combustion is practically diffusion-controlled with CO_2 as the predominant combustion product. With continuous increases in surface temperature, a point is reached where the CO begins to appear among the products. The carbon combustion rate increases with the ratio of CO to CO_2 until the maximum combustion rate is reached, where CO is the sole product.

Summarizing, there seem to be two transitions in carbon combustion as Da is varied. The first is the change of surface chemical condition from the frozen to the equilibrium regime, and the second is the change of the equilibrium combustion product from one composed predominantly of CO_2 to one composed predominantly of CO. The two transitions may occur in two distinct T_w regimes or may occur simultaneously, depending on the values of parameters such as E_{aw} and K_0.

The agreement between theory and experimental data (see Fig. 9.6) is seen to be satisfactory as a whole. Moreover, the trend of experimental data seems generally to support the discussion above. It should be noted that in the analysis presented, the oxygen at the surface is assumed to be in molecular form and l is assumed to be constant. Both assumptions are quite reasonable, since the surface temperature is below about 2000 K and the carbon surface is highly catalytic for atomic recombination. Also, the constant value of l as given by Eq. (9-98) should in fact be satisfactory for the carbon combustion problem.[4]

6 LAMINAR BOUNDARY-LAYER FLOWS WITH GAS-PHASE REACTIONS

Generally, there are two classes of gas-phase reactions in boundary-layer flows: (a) the exothermic combustion of pyrolyzed fuels in the gas stream, and (b) the dissociation–recombination processes occurring in the gas phase. The surface reactions may or may not be important. In some cases they can be considered frozen, and in other cases they may be treated as equilibrium or finite-rate.

6.1 Governing Equations and Coordinate Transformation

The governing equations were derived in Section 2. The continuity equation, momentum equation, and species equations are given by Eqs. (9-3), (9-8), and (9-13), respectively. With them, we may employ any one of the energy equations (9-22), (9-23), or (9-26). The surface heat-transfer rate is given by Eq. (9-42). The expression for the reaction rate term ω_i was derived in Section 4.1. Similarly to the procedure given in Section 5.2, the governing equations can be transformed into (ξ, η) coordinates. The transformed momentum equation is identical to Eq. (9-73). For uniform Y_{ie}, the species equation (9-13) and the energy equation (9-26) are transformed, respectively, to

$$\frac{1}{\mathrm{Sc}}\left(lY^*_{i,\eta}\right)_\eta + FY_{i,\eta} + \frac{2}{Y_{ie}}\frac{L}{U_\infty}\frac{\rho_e\mu_e}{(\rho_e\mu_e)_0}\frac{1}{[\mathscr{F}(x^*)]^2}\frac{\omega_i}{\rho}$$

$$= 2\xi\left(F_\eta Y^*_{i,\xi} - F_\xi Y^*_{i,\eta}\right) \tag{9-124}$$

$$\frac{1}{\mathrm{Pr}}\left[\left(\frac{C_p}{C_{p,e}}\right)l\theta_\eta\right]_\eta + \left(\frac{C_p}{C_{p,e}}\right)F\theta_\eta - \sum_{i=1}^{N}\left[2\frac{L}{U_\infty}\frac{\rho_e\mu_e}{(\rho_e\mu_e)_0}\frac{1}{[\mathscr{F}(x^*)]^2}\frac{h_i}{C_{p,e}T_e}\frac{\omega_i}{\rho}\right]$$

$$+l\left(\frac{U_e^2}{C_{p,e}T_e}\right)(F_{\eta\eta})^2 + \sum_{i=1}^{N}\left(\frac{C_{p,i}}{C_{p,e}}\right)Y_{ie}\frac{l}{\mathrm{Pr\,Le}_i}Y^*_{i,\eta}\theta_\eta$$

$$= F_\eta\left[2\left(\frac{C_p}{C_{p,e}}\right)\theta\left(\frac{d\ln T_e}{d\ln\xi}\right) + \frac{2U_e^2}{C_{p,e}T_e}\frac{\rho_e}{\rho}\left(\frac{d\ln U_e}{d\ln\xi}\right)\right]$$

$$+2\left(\frac{C_p}{C_{p,e}}\right)\xi\left(F_\eta\theta_\xi - F_\xi\theta_\eta\right) \tag{9-125}$$

where

$$\theta \equiv \frac{T}{T_e} \tag{9-126}$$

If l is constant, it can be made to disappear from the governing equations by replacing η and F by ζ and f, respectively, as defined by Eqs. (9-99) and (9-100). The resulting equations in ζ and f are exactly the same as the governing equations given above except that l is missing. The boundary conditions are the same as those developed in the previous section. The right-hand side of the boundary condition (9-82) vanishes when the surface reaction is frozen.

6.2 Damköhler Number for Gas-Phase Reactions

Consider the chemical-reaction factor ω_i/ρ. The dimension of this factor is a reciprocal of time, and we may write

$$\frac{\omega_i}{\rho} = \frac{1}{\tau_{gi}}\omega_i^*(\theta, Y_i^*, p) \tag{9-127}$$

where ω_i^* is the dimensionless rate of production of species i and τ_{gi} is a characteristic time for the gas-phase chemical reaction. The species conservation and energy equations (9-124) and (9-125) now clearly show that the chemical behavior of a reacting boundary layer is largely influenced by the parameter

$$\mathrm{Da}_{gi} \equiv 2\frac{L}{U_\infty}\frac{\rho_e\mu_e}{(\rho_e\mu_e)_0}\frac{1}{[\mathscr{F}(x^*)]^2}\frac{1}{\tau_{gi}} \tag{9-128}$$

Let us investigate the physical meaning of Da_{gi} defined above. The variables governing the characteristic diffusion rate of the ith species are given by Eq. (9-93). The variables which govern the boundary-layer thickness δ are derived from Eq. (9-66) as

$$\delta \propto \sqrt{\frac{2(\rho_e\mu_e)_0 L}{U_\infty}}\frac{1}{\mathscr{F}(x^*)}\frac{\rho_e\mu_e}{(\rho_e\mu_e)_0}\frac{1}{\rho} \tag{9-129}$$

It can be shown that the coefficient in front of $1/\tau_{gi}$ denotes the characteristic diffusion time of the ith species, which is proportional to $\delta^2/\mathscr{D}_i$. Therefore,

$$\mathrm{Da}_{gi} = \left(\frac{\tau_{\mathrm{diff}}}{\tau_g}\right)_i \tag{9-130}$$

In analogy to the surface Damköhler number, we shall call Da_{gi} the *gas-phase Damköhler number*.

As was the case with the surface reaction, $\mathrm{Da}_{gi} \to 0$ implies that the effect of the gas-phase reaction on the boundary-layer flow is negligible. For $\mathrm{Da}_{gi} \to 0$, the chemical-reaction terms in the governing equations vanish and the problem becomes that of the chemically inert boundary layer. The boundary layer is then said to be chemically *frozen*. A frozen boundary layer does not necessarily imply that $1/\tau_{gi} = 0$. It only implies that $\tau_{gi} \gg \tau_{\mathrm{diff}_i}$.

At the other extreme, $\mathrm{Da}_{gi} \to \infty$, ω_i must approach zero because each of the terms in Eq. (9-124) is finite. This means that the forward and the backward reaction rates entering ω_i must be equal at each point in the boundary layer. Thus, there must prevail a state of local chemical equilibrium throughout the boundary layer. Such boundary layers are commonly called *equilibrium* boundary layers. For an equilibrium boundary layer, then, the algebraic equation $\omega_i = 0$ may replace the differential equation (9-124), and the problem is greatly simplified.

Consider the following combustion process in the gas phase between the gasified fuel species and oxygen in the boundary layer to form products:

$$\text{fuel} + \text{oxygen} \xrightarrow{k_f} \text{product} \tag{9-131}$$

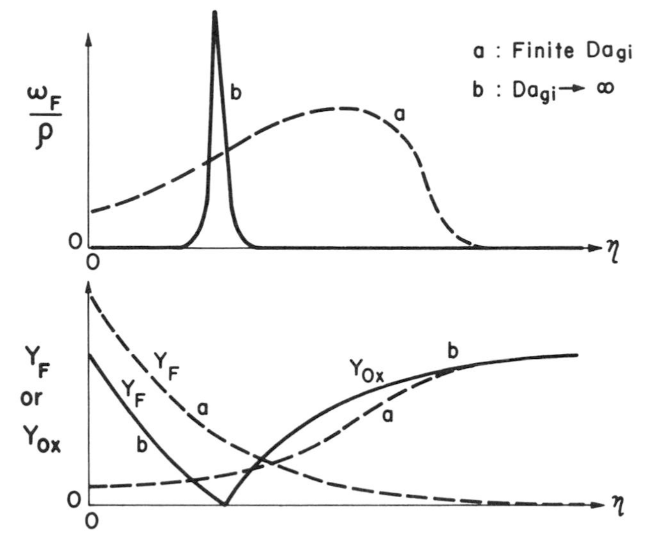

Figure 9.7 Distributions of the rate of reaction and concentrations of a diffusion flame in laminar boundary-layer flow.

The consumption rate of fuel can be written as

$$\frac{\omega_F}{\rho} = -\rho \frac{k_f}{W_{O_x}} (Y_{O_x})(Y_F) \tag{9-132}$$

For smaller values of Da_{gi}, the combustion will take place slowly throughout the boundary layer. The rate of chemical reaction is controlled by the temperature distribution across the boundary layer, and the maximum rate will occur somewhere in the outer portion of the boundary layer where the temperature is maximum (see Fig. 9.7).

As Da_{gi} is increased, most of the injected fuel burns before the boundary-layer edge is reached, and at the same time, most of the oxygen convected and diffused into the boundary layer becomes depleted by reaction before it reaches the wall. Thus, the significant combustion zone becomes narrowed, since combustion requires the coexistence of the fuel and oxygen. With Da_{gi} increased to a sufficiently large value, a condition is reached wherein no fuel and oxygen which enter the boundary layer from the wall and the boundary-layer edge, respectively, survive the journey across the combustion zone. In the limit of $\mathrm{Da}_{gi} \to \infty$, the reaction zone becomes so thin that we may consider it a mathematical discontinuity in the concentration and temperature gradients: it finally becomes a *flame sheet*.

6.3 Extension to Axisymmetric Cases

In the above discussion, the analyses presented have been made rather general so that they will be useful for many problems of reacting laminar boundary-layer flows. The two-dimensional formulation for planar geometry can be extended readily to axisymmetric cases by considering conservation equations in axisymmetric coordinates and using

$$\eta \equiv \frac{r^m U_e}{\sqrt{2\xi}} \int_0^y \rho \, dy \tag{9-133}$$

$$\xi \equiv \int_0^x \rho_e \mu_e u_e r^{2m} \, dx \tag{9-134}$$

to replace Eqs. (9-65) and (9-66), respectively, in Lees's transformation. The stream function ψ can be defined by

$$\rho u r^m = \frac{\partial \psi}{\partial y} \tag{9-135}$$

$$\rho v r^m = -\frac{\partial \psi}{\partial x} \tag{9-136}$$

to automatically satisfy the continuity equation for the axisymmetric case. When one sets m equal to 0, the above equations reduce immediately to the planar case.

7. TURBULENT BOUNDARY-LAYER FLOWS WITH CHEMICAL REACTIONS

7.1 Introduction

Turbulent boundary-layer flow, even of chemically inert fluids, is one of the most challenging subject areas of engineering science. When chemical reactions occur in boundary-layer flows, as is the case in many combustion systems, the challenge takes on a new dimension: the complicated subject of turbulent shear flows is joined with another subject of thermal science, aerothermochemistry, which combines fluid dynamics, heat transfer, thermodynamics, and chemistry. This section is intended to provide readers with background material to unravel the essentials of these complexities and interactions and to present some recent theoretical approaches that have been developed in the field.

In Chapter 7, we have already discussed the basic definition of turbulence, characteristics of turbulent flames, and some general approaches to turbulent flame problems. In this chapter, we shall concentrate on the studies of reacting turbulent boundary layers, introducing two recent methods. One is based on the boundary-layer integral matrix procedure (BLIMP) developed by Evans.[25] The other is based on the method of Patankar and Spalding[7] using a two-equation k–ε or multiequation model for turbulence closure. The former has been used widely in boundary-layer calculations for predicting the performance of various propulsive devices. The latter has been adopted by many researchers in various calculations on reacting turbulent boundary layers.

7.2 Boundary-Layer Integral Matrix Procedure of Evans[25]

The boundary-layer integral matrix procedure (BLIMP) deals with the non-self-similar, chemically reacting, laminar or turbulent boundary layer in ablating, transpiration-cooled, or nonablating internal-flow configurations. The flow can be planar or axisymmetric. The analysis considers either local thermodynamic equilibrium or frozen composition. A wide variety of surface boundary conditions are included, ranging from assigned wall temperature and mass injection rates to surface equilibrium. The model is solved by a numerical solution procedure called an integral matrix approach, which is equivalent to a higher-order finite-difference approach (using spline fits).

7.2.1 General Conservation Equations

In the analysis, the usual turbulent-flow technique of breaking the species, velocity, and enthalpy fields into mean and fluctuating components, time (Reynolds) averaging, and making appropriate order of magnitude approximations is used. The species mass-balance equation can be written as

$$\frac{\partial}{\partial s}\left(\bar{\rho}\bar{u}\overline{Y}_i r^m\right) + \frac{\partial}{\partial y}\left(\bar{\rho}\bar{v}\overline{Y}_i r^m\right) = \frac{\partial}{\partial y}\left[\left(\bar{\rho}\mathscr{D}_{Ti}\frac{\partial \overline{Y}_i}{\partial y} - \bar{J}_i\right) r^m\right] + \bar{\omega}_i r^m \quad (9\text{-}137)$$

where s and y are the streamwise and normal coordinates, respectively, r is the metric coefficient for streamline spreading in three-dimensional flow (radius from the body centerline to the point of interest in a meridian plane for axisymmetric flow), and m is zero for a flat plate and unity for a body of revolution. The radius r is treated as a function of y, and is equal to r_0 on the surface. The relationship between r, r_0, and y is

$$r(s, y) = r_0(s) - y\cos\theta \quad (9\text{-}138)$$

The coordinate system is shown in Fig. 9.8.

The individual-species turbulent eddy diffusivity for mass transport is defined in terms of the correlation of the fluctuating components of concentration and normal velocity, that is,

$$\bar{\rho}\mathscr{D}_{Ti} = -\frac{\overline{(\rho v)'Y_i'}}{\partial \overline{Y}_i/\partial y} \quad (9\text{-}139)$$

In this equation, the turbulent transport term is expressed in Boussinesq form. The triple correlations have been neglected on the basis of order-of-magnitude arguments.

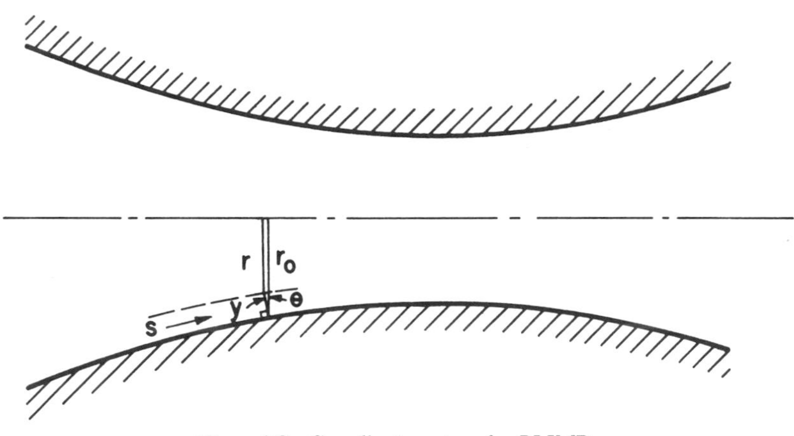

Figure 9.8 Coordinate system for BLIMP.

The molecular diffusion rate $\bar{J}_i$ of the ith species can be expressed, in general, as

$$\bar{J}_i = \frac{\bar{\rho}}{W^2} \sum_{j \neq i} W_i W_j \mathscr{D}_{ij} \frac{\partial X_j}{\partial y} - \mathscr{D}_i^{\mathrm{Th}} \frac{\partial}{\partial y} \ln \bar{T} \tag{9-140}$$

where $\mathscr{D}_{ij}$ is the multicomponent diffusion coefficient of species i into j, $\mathscr{D}_i^{\mathrm{Th}}$ is the multicomponent thermal diffusion coefficient of species i [due to the Soret effect, see Eq. (9-154)], W is the local gas-mixture molecular weight, and W_i is the molecular weight of species i. The Stefan–Maxwell[26] relations may also be used to express J_i:

$$\frac{\partial X_i}{\partial y} = \sum_j \frac{X_i X_j}{\bar{\rho} \mathscr{D}_{ij}} \frac{\bar{J}_j + \mathscr{D}_j^{\mathrm{Th}}\, \partial \ln \bar{T}/\partial y}{\bar{Y}_j} - \frac{\bar{J}_i + \mathscr{D}_i^{\mathrm{Th}}\, \partial \ln \bar{T}/\partial y}{\bar{Y}_i} \tag{9-141}$$

Both of these expressions are complicated in that the multicomponent diffusion coefficients are difficult to evaluate, and the Stefan–Maxwell relations provide only implicit expressions for the $\bar{J}_i$. For the special case when all diffusion coefficients can be assumed equal and thermal diffusion due to the Soret effect can be ignored, Fick's law results:

$$\bar{J}_i = -\bar{\rho} \mathscr{D} \frac{\partial \bar{Y}_i}{\partial y} \tag{9-142}$$

When Eq. (9-137) is summed over all species, the global continuity equation results:

$$\frac{\partial(\bar{\rho}\bar{u} r^m)}{\partial s} + \frac{\partial(\bar{\rho}\bar{v} r^m)}{\partial y} = 0 \tag{9-143}$$

Combining Eqs. (9-143) and (9-137), one obtains the species conservation equation

$$\bar{\rho}\bar{u}\frac{\partial \bar{Y}_i}{\partial s} + \bar{\rho}\bar{v}\frac{\partial \bar{Y}_i}{\partial y} = \frac{1}{r^m}\frac{\partial}{\partial y}\left[r^m\left(\bar{\rho}\mathscr{D}_{Ti}\frac{\partial \bar{Y}_i}{\partial y} - \bar{J}_i\right)\right] + \bar{\omega}_i \qquad (9\text{-}144)$$

In the Evans[25] approach, the conservation of elements is used instead of the conservation of species. The term "element" here refers to those atoms or groupings of atoms which, according to equilibrium relations, are conserved. The "elemental" approach results in significantly fewer simultaneous equations than the conservation-of-species approach, and the equating of all $\mathscr{D}_{Ti}$ gives sufficiently accurate solutions for most types of problems. Reference 27 discusses the merits of this approach in more detail. Defining α_{ki} as the mass fraction of "element" k in species i, multiplying the species equation (9-144) by α_{ki}, and summing over all species results in the following conservation-of-"element" equations:

$$\bar{\rho}\bar{u}\frac{\partial \bar{\mathscr{Y}}_k}{\partial s} + \bar{\rho}\bar{v}\frac{\partial \bar{\mathscr{Y}}_k}{\partial y} = \frac{1}{r^m}\frac{\partial}{\partial y}\left[r^m\left(\bar{\rho}\mathscr{D}_T\frac{\partial \bar{\mathscr{Y}}_k}{\partial y} - \bar{J}_k\right)\right] + \sum_{i=1}^{N}\alpha_{ki}\bar{\omega}_i \quad (9\text{-}145)$$

where $\bar{\mathscr{Y}}_k$ is the mass fraction of "element" k in the system defined by

$$\bar{\mathscr{Y}}_k = \sum_{i=1}^{N}\alpha_{ki}\bar{Y}_i \qquad (9\text{-}146)$$

It has also been assumed that all $\mathscr{D}_i^T = \mathscr{D}^T$. The term $\Sigma\alpha_{ki}\bar{\omega}_i$ in Eq. (9-145) is the production of "element" k, which, for equilibrium chemistry, is set equal to zero. For nonequilibrium chemistry, the production term can be nonzero. By retaining the production term in this conservation equation, the same formulation can be used for both equilibrium and nonequilibrium cases.

The streamwise momentum equation can be written as

$$\bar{\rho}\bar{u}\frac{\partial \bar{u}}{\partial s} + \bar{\rho}\bar{v}\frac{\partial \bar{u}}{\partial y} = \frac{1}{r^m}\frac{\partial}{\partial y}\left[\bar{\rho}r^m(\nu + \nu_T)\frac{\partial \bar{u}}{\partial y}\right] - \frac{\partial \bar{p}}{\partial s} \qquad (9\text{-}147)$$

where ν_T represents the eddy viscosity and is defined in terms of the Reynolds stresses of turbulent flow by

$$\bar{\rho}\nu_T = -\frac{\overline{(\rho v)'u'}}{\partial \bar{u}/\partial y} \qquad (9\text{-}148)$$

The transverse direction momentum equation can be ignored when longitudinal curvature effects are small.

The energy equation for a general chemically reacting boundary layer is

$$\bar{\rho}\bar{u}\frac{\partial \bar{h}_t}{\partial s} + \bar{\rho}\bar{v}\frac{\partial \bar{h}_t}{\partial y} = \frac{1}{r^m}\frac{\partial}{\partial y}\Bigg[\bar{\rho}r^m(\nu_T + \nu)\frac{\partial(\bar{u}^2/2)}{\partial y} + r^m(\lambda + \bar{\rho}\alpha_T\bar{C}_p)\frac{\partial \bar{T}}{\partial y} + r^m\sum_{i=1}^{N}\left(\rho\mathscr{D}_T\frac{\partial \bar{Y}_i}{\partial y} - \bar{J}_i\right)\bar{h}_i - \frac{r^m R_u \bar{T}}{\bar{\rho}}\sum_i\sum_j\frac{\bar{X}_j\mathscr{D}^{\text{Th}}}{W_i\mathscr{D}_{ij}}\left(\frac{\bar{J}_i}{\bar{Y}_i} - \frac{\bar{J}_j}{\bar{Y}_j}\right) + r^m\bar{q}_{\text{rad}}\Bigg] \tag{9-149}$$

where $\bar{h}_t$ is the total enthalpy

$$\bar{h}_t = \bar{h} + \frac{\bar{u}^2}{2} \tag{9-150}$$

and the static enthalpy $\bar{h}$ is defined by Eqs. (9-18) and (9-19). $\bar{C}_p$ is the frozen specific heat of the gaseous mixture defined as

$$\bar{C}_p = \sum_i \bar{Y}_i C_{p,i} \tag{9-151}$$

The turbulent enthalpy-transport coefficient α_T is defined by

$$\bar{\rho}\alpha_T = \frac{\sum_{i=1}^{N}\bar{Y}_i\overline{(\rho v)'h_i'}}{\sum_{i=1}^{N}\bar{Y}_i\partial\bar{h}_i/\partial y} \tag{9-152}$$

7.2.2 Molecular Transport Properties

In the energy equation, as in the species conservation equations, it is necessary to evaluate the molecular diffusion flux $\bar{J}_i$. As discussed earlier, the general expression for these terms is difficult to work with; therefore, an approximate technique for multicomponent diffusion has been developed. Following the bifurcation approximation introduced by Bird[28] and discussed in detail by Bartlett et al.,[29] one can obtain explicit solution of the Stefan–Maxwell relations [Eq. (9-141)] for $\bar{J}_i$ in terms of gradients and properties of species i and of the system as a whole. In this bifurcation approximation, the binary diffusion coefficient $\mathscr{D}_{ij}$ is replaced by

$$\mathscr{D}_{ij} = \frac{\bar{D}(\bar{T}, \bar{p})}{\mathscr{G}_i\mathscr{G}_j} \tag{9-153}$$

where $\overline{D}$ is a reference diffusion coefficient and $\mathscr{g}_i$ is a diffusion factor for species i. The $\mathscr{g}_i$-values are determined for a given system by a least-squares curve fit to actual diffusion data. The accuracy of the approximation was found to be very good; according to Bartlett,[29] it is within 5% for most cases. The multicomponent thermal diffusion coefficients $\mathscr{D}_i^{\mathrm{Th}}$ can be approximated by

$$\mathscr{D}_i^{\mathrm{Th}} = \frac{C_t \bar{\rho} \overline{D} \mu_2}{\mu_1 W} (\overline{Z}_i - \overline{Y}_i) \tag{9-154}$$

which represents a generalization of a correlation of binary-diffusion data. Various parameters introduced in Eq. (9-154) are defined in the following:

$$\overline{Z}_i = \frac{W_i \overline{X}_i}{\mathscr{g}_i \mu_2}, \qquad \overline{\mathscr{Z}}_k = \sum_{i=1}^{N} \alpha_{ki} \overline{Z}_i$$

$$\mu_1 = \sum_{j=1}^{N} \overline{X}_j \mathscr{g}_j, \qquad \mu_2 = \sum_{j=1}^{N} \frac{W_j \overline{X}_j}{\mathscr{g}_j}$$

$$\mu_3 = \sum_{i=1}^{N} \frac{\overline{Z}_i}{W_i}, \qquad \mu_4 = \ln\left(\mu_2 \overline{T}^{C_t}\right)$$

$$C_t = -0.5, \qquad \overline{\mathscr{C}}_p = \sum_{i=1}^{N} \overline{Z}_i C_{p,i}$$

$$\bar{\mathscr{h}} = \sum_{i=1}^{N} \overline{Z}_i \bar{h}_i \tag{9-155}$$

The "elemental" mass flux due to molecular diffusion $\bar{J}_k$ in Eq. (9-145) can be expressed as

$$\bar{J}_k = -\frac{\bar{\rho} \overline{D} \bar{\mu}_2}{\mu_1 W} \left[\frac{\partial \overline{\mathscr{Z}}_k}{\partial y} + (\overline{\mathscr{Z}}_k - \overline{\mathscr{Y}}_k) \frac{\partial \mu_4}{\partial y} \right] \tag{9-156}$$

The diffusive energy-flux terms in Eq. (9-149) can be expressed as

$$\begin{aligned} \bar{q}_a = -\Bigg\{ & \bar{\rho}(\nu_T + \nu) \frac{\partial(\bar{u}^2/2)}{\partial y} + \left(\lambda + \rho \alpha_T \overline{C}_p\right) \frac{\partial \overline{T}}{\partial y} + \bar{\rho} \mathscr{D}_T \left(\frac{\partial \bar{h}}{\partial y} - \overline{C}_p \frac{\partial \overline{T}}{\partial y} \right) \\ & + \frac{\bar{\rho} \overline{D} \mu_2}{\mu_1 W} \Bigg[\frac{\partial \bar{\mathscr{h}}}{\partial y} - \left(\overline{\mathscr{C}}_p + \frac{C_t^2 R_u}{\mu_1 \mu_2} \right) \frac{\partial \overline{T}}{\partial y} + C_t R_u \overline{T} \frac{\partial \mu_3}{\partial y} \\ & + (\bar{\mathscr{h}} - \bar{h} + C_t R_u \overline{T} \mu_3) \frac{\partial \mu_4}{\partial y} \Bigg] \Bigg\} \end{aligned} \tag{9-157}$$

The energy equation can be expressed as

$$\bar{\rho}\bar{u}\frac{\partial \bar{h}_t}{\partial s} + \bar{\rho}\bar{v}\frac{\partial \bar{h}_t}{\partial y} = \frac{1}{r^m}\frac{\partial}{\partial y}\left[r^m(-\bar{q}_a + \bar{q}_{\text{rad}})\right] \tag{9-158}$$

It is interesting to note that if all diffusion coefficients are assumed equal, then $\mu_3 = 1/W$, $\bar{\mathscr{C}}_p = \bar{C}_p$, and $\bar{\hbar} = \bar{h}$. When thermal diffusion is negligible, $C_t = 0$ and $\mu_4 = \ln \mu_2$.

The reference diffusion coefficient $\bar{D}$ in Eq. (9-153) can be evaluated by

$$\bar{D} = \frac{1.719 \times 10^{-5}}{\bar{p}}\bar{T}^{1.659} \tag{9-159}$$

where $\bar{D}$ is in cm^2/sec, $\bar{T}$ is in kelvin, and $\bar{p}$ is in atmospheres. The diffusion factor $\mathscr{g}_i$ for species i in Eq. (9-153) can be calculated from

$$\mathscr{g}_i = \left(\frac{W_i}{26.7}\right)^{0.489} \tag{9-160}$$

It has also been reported[25] that self-diffusion can be better represented if a different correlation is used. Accordingly, $\mathscr{D}_{ii}$ is given by

$$\mathscr{D}_{ii} = \frac{\bar{D}}{G_i^2} \tag{9-161}$$

where

$$G_i = \left(\frac{W_i}{24.3}\right)^{0.454} \tag{9-162}$$

For most gas systems, these correlations are within 5% of more exact values for temperatures on the order of 3000 K.

The viscosity of the mixture can be derived rigorously from first-order kinetic theory.[26] The Buddenberg–Wilke mixture formula can be written as

$$\mu_{\text{mix}} = \sum_{i=1}^{N} \frac{\bar{X}_i\mu_i}{\bar{X}_i + 1.385\left[R_u\bar{T}\mu_i/(\bar{p}W_i)\right]\sum_{j=i}^{N} X_j/\mathscr{D}_{ij}} \tag{9-163}$$

where μ_i is the viscosity of the species i. The μ_i may be expressed in terms of the self-diffusion coefficients $\mathscr{D}_{ii}$ as

$$\mu_i = \frac{5}{6A_{ii}^*}\frac{\bar{p}W_i}{R_u\bar{T}}\mathscr{D}_{ii} \tag{9-164}$$

where A_{ii}^* is a ratio of collision integrals based on a Lennard–Jones intermolecular potential. In BLIMP, A_{ii}^* is set equal to a constant value of 1.13. Substituting Eqs. (9-153), (9-161), and (9-164) into Eq. (9-163), an expression for the viscosity of the multicomponent mixture is obtained:

$$\mu_{\text{mix}} = \frac{\bar{\rho}\bar{D}}{\mu_1} \sum_{i=1}^{N} \left[\frac{\dfrac{\bar{X}_i W_i}{\mathscr{g}_i W}}{1.385 + \dfrac{\bar{X}_i \mathscr{g}_i}{\mu_1}\left(\dfrac{6A_{ii}^* G_i^2}{5\mathscr{g}_i^2} - 1.385 \right)} \right] \tag{9-165}$$

The thermal conductivity in a polyatomic gas mixture may be represented, according to Hirschfelder,[30] by

$$\lambda_{\text{mix}} = \lambda_{\text{mono-mix}} + \lambda_{\text{int}} \tag{9-166}$$

where $\lambda_{\text{mono-mix}}$ is the thermal conductivity of the mixture computed by neglecting all internal degrees of freedom, and λ_{int} is the contribution due to the internal degrees of freedom of the molecules. A simplified expression for $\lambda_{\text{mono-mix}}$ can be derived from the first-order kinetic theory.[26] It takes the form[29]

$$\lambda_{\text{mono-mix}} = \sum_{i=1}^{N} \left[\frac{\bar{X}_i \lambda_{i\,\text{mono}}}{\bar{X}_i + 1.475 \dfrac{R_u \bar{T}\,\mu_i}{\bar{\rho} W_i} \displaystyle\sum_{\substack{j=1 \\ j \neq i}}^{N} \frac{\bar{X}_j}{\mathscr{D}_{ij}}} \right] \tag{9-167}$$

where $\lambda_{i\,\text{mono}}$ is the thermal conductivity of the pure species i neglecting all internal degrees of freedom of the molecule. It may be expressed in terms of the μ_i as

$$\lambda_{i\,\text{mono}} = \frac{15}{4} \frac{R_u}{W_i} \mu_i \tag{9-168}$$

The contribution of internal degrees of freedom to the thermal conductivity may be expressed[29] as

$$\lambda_{\text{int}} = \sum_{i=1}^{N} \frac{\bar{\rho}\bar{X}_i (W_i/W)\left(C_{p,i} - \frac{5}{2} R_u/W_i\right)}{\displaystyle\sum_{j=1}^{N} \bar{X}_j/\mathscr{D}_{ij}} \tag{9-169}$$

By combining Eqs. (9-153) and (9-161) and Eqs. (9-166) through (9-169), the mixture thermal conductivity may be written as

$$\lambda_{\text{mix}} = \frac{\bar{\rho}\bar{D}}{\mu_1}\left\{\sum_{i=1}^{N}\left[\frac{\frac{15}{4}(\bar{X}_i/\mathscr{g}_i)R_u/W}{1.475 + (\bar{X}_i\mathscr{g}_i/\mu_1)\left(\dfrac{6A_{ii}^*G_i^2}{5\mathscr{g}_i^2} - 1.475\right)}\right] + \frac{\mu_2}{W}\left[\bar{\mathscr{C}}_p - \tfrac{5}{2}R_u\mu_3\right]\right\} \tag{9-170}$$

where μ_1, μ_2, μ_3, and $\bar{\mathscr{C}}_p$ are given by Eq. (9-155). Thus Eq. (9-170) can be used to calculate the mixture conductivity in reacting boundary layers.

The Prandtl and Schmidt numbers based on molecular transport properties are defined as

$$\text{Pr} \equiv \frac{\mu}{\lambda_{\text{mix}}} C_{p,\,\text{frozen}} \tag{9-171}$$

$$\text{Sc} \equiv \frac{\mu_1 \mu_{\text{mix}} W}{\mu_2 \bar{\rho}\bar{D}} \tag{9-172}$$

in the BLIMP calculations.

7.2.3 Turbulent Transport Properties

In the conservation equations developed previously, the concepts of eddy viscosity (ν_T), eddy mass diffusivity ($\mathscr{D}_T$), and turbulent thermal diffusivity (α_T) were used to express the correlations of fluctuating velocity, species, and enthalpy in terms of mean flow quantities. This is only one of several possible techniques of closing the set of equations. In general, it is more desirable to describe the turbulent fluctuations in a more complete manner, such as with the turbulent kinetic-energy equation, turbulent dissipation rate equation, and/or local turbulent constitutive equation. Some of these closure methods are discussed in Chapter 7 as well as in Section 7.3 of this chapter. In the BLIMP analysis, the Boussinesq description of turbulence is followed exclusively, since that method has proved to be very useful, particularly for some complex reacting flows.

There is a wide range of latitude possible even within the eddy-viscosity framework of turbulence, particularly in applying classical incompressible models to compressible flows. There are three turbulence models considered in the BLIMP analysis: those of Kendall et al.,[31] Cebeci and Smith,[32-34] and Beckwith and Bushnell.[35] We shall discuss some general features of these models before going into each one separately.

Boussinesq's eddy viscosity is adapted to writing the Reynolds stresses as

$$-\overline{(\rho v)'u'} = \bar{\rho}\nu_T \frac{\partial \bar{u}}{\partial y} \tag{9-173}$$

and a similar relation [a simplified form of Eq. (9-152)] is used to define the turbulent thermal diffusivity α_T:

$$-\overline{(\rho v)'T'} = \bar{\rho}\alpha_T \frac{\partial \bar{T}}{\partial y} \tag{9-174}$$

All three turbulence models considered in the BLIMP analysis employ the Prandtl mixing-length hypothesis in which it is assumed that

$$\nu_T = \ell_m V_T \tag{9-175}$$

where ℓ_m is the mixing length and V_T is the turbulent velocity. The differences between the three models come about through the formulation of ℓ_m and V_T. Kendall and Cebeci treat the boundary layer as a composite layer consisting of inner and outer regions. In the inner, or wall, region, the turbulent velocity is written as

$$V_T = \ell_m \left| \frac{d\bar{u}}{dy} \right| \qquad \text{(wall region)} \tag{9-176}$$

and the mixing length is assumed to be proportional to the distance from the wall. In the outer, or wake, region, the boundary layer is assumed to behave similarly to free turbulent shear flow with

$$V_T = \bar{U}_e \qquad \text{(wake region)} \tag{9-177}$$

and $\ell_m = c\delta^*$ where c is a constant and δ^* is the displacement (velocity-defect) thickness. Thus,

$$\nu_T = C\delta^* U_e \tag{9-178}$$

where

$$\delta^* = \int_0^\infty \left(1 - \frac{u}{U_e}\right) dy \tag{9-179}$$

Bushnell and Beckwith, however, treat the boundary layer as a single layer and apply Eq. (9-175) throughout by introducing the intermittency concept in the definition of ℓ_m.

The most fundamental differences in the models arise, however, from the manner in which the mixing-length expression is obtained. The Cebeci–Smith and Bushnell–Beckwith expressions originate from Prandtl's proposal that in the region of the development of turbulence

$$\frac{d\ell_m}{dy} = \kappa \tag{9-180}$$

which has a solution in the form of

$$\ell_m = \kappa y \tag{9-181}$$

The models are arrived at by significant modifications to this solution to account for the effects of variable properties, pressure gradient, Reynolds number, and so on. It is important to note that these modifications were made to the solution and not to the basic proposition as expressed by Eq. (9-180). The model, due to Kendall et al.,[31] on the other hand, follows from modifications to the basic proposition to account for the effects of variable properties. It has been observed that differences in the models become more pronounced as the degree of property variation increases.[25]

The turbulent transport of scalar quantities is treated the same way as momentum, by introducing the concepts of turbulent thermal diffusivity α_T and turbulent mass diffusivity $\mathscr{D}_T$. That is,

$$\alpha_T = \frac{\nu_T}{\mathrm{Pr}_T} \quad \Rightarrow \quad \frac{-\overline{(\rho v)'h'}}{\bar{\rho}\,\partial\bar{h}/\partial y} = \frac{-\overline{(\rho v)'u'}}{\mathrm{Pr}_T(\partial\bar{u}/\partial y)\bar{\rho}} \tag{9-182}$$

Turbulent Prandtl and Schmidt numbers are defined as

$$\mathrm{Pr}_T \equiv \frac{\nu_T}{\alpha_T}, \qquad \mathrm{Sc}_T \equiv \frac{\nu_T}{\mathscr{D}_T} \tag{9-183}$$

Cebeci and Smith proposed an expression for Pr_T as a function of the distance from the wall, but in the Kendall and Bushnell–Beckwith models, Pr_T is assumed to be a constant. The turbulent Schmidt number is also taken to be constant in all models.

Kendall Model

This model employs the two-layer concept of the turbulent boundary layer. The wall law is based on the following three assumptions:

A. $\lim_{y\to 0} \ell_m = 0$.

B. $\lim_{y\to 0} d\ell_m/dy = 0$.

C. The rate of increase of the mixing length with y is proportional to the difference between the value postulated by Prandtl (κy) and its actual value:

$$\frac{d\ell_m}{dy} \propto (\kappa y - \ell_m) \tag{9-184}$$

The proportionality factor in this relationship is assumed to be dependent on the local shear stress and local kinematic viscosity:

$$\frac{d\ell_m}{dy} = (\kappa y - \ell_m)\frac{\sqrt{\bar{\tau}/\bar{\rho}}}{y_a^+ \nu} \tag{9-185}$$

The values of the constants κ and y_a^+ recommended in this model are 0.44 and 11.823, respectively. These values have been obtained by matching the predictions with experimental data in incompressible turbulent boundary layers with and without blowing.[31] Physically, κ can be considered as a measure of the growth of the mixing length with respect to vertical distance from the surface, and y_a^+ as a measure of the thickness of the laminar sublayer.

The validity of the model for flows with wall blowing and streamwise pressure gradient is argued on the basis of using the local flow properties, such as local shear, in the model. For compressible flow, Eq. (9-185) is modified as follows:

$$\frac{d\bar{\rho}\ell_m}{dy} = \left[\kappa \int_0^y \bar{\rho}\, dy - \bar{\rho}\ell_m\right]\frac{\sqrt{\bar{\tau}/\bar{\rho}}}{y_a^+ \nu} \tag{9-186}$$

where, instead of describing the length scale of a turbulent eddy, the mass of the eddy, represented by $\bar{\rho}\ell_m$, is related to the mass available between the wall and the point of interest. The constants κ and y_a^+, however, are left at their incompressible values. The above integrodifferential equation is solved numerically to obtain the local value of the mixing length ℓ_m.

In the wake region, it is assumed that the eddy viscosity is a constant and is given by Clauser's expression[36] where $c = 0.018$. The wall (inner) and the wake (outer) regions are matched by the following procedure: the ν_T expression for the wall region is used until it exceeds the wake value, at which point the wake value of ν_T is used for the remainder of thc boundary-layer thickness. This value is linearly damped in the outer portions of the boundary layer so that a value of zero occurs at the boundary layer edge.

Cebeci–Smith Model

As mentioned previously, a two-layer model is also used by Cebeci and Smith. In the wall (inner) region, the Van Driest[37] form of mixing length is used:

$$\ell_m = \kappa y\left[1 - \exp\left(-\frac{y^+}{A^+}\right)\right] \tag{9-187}$$

where

$$y^+ \equiv \frac{y\sqrt{\bar{\tau}_w/\bar{\rho}}}{\nu} \tag{9-188}$$

Van Driest suggested constant values of 0.4 and 26 for κ and A^+, respectively. (These have essentially the same meaning as κ and y_a^+ in Kendall's model.) In the Cebeci–Smith model, however, these constants are replaced by functions accounting for pressure gradient and blowing. Compressibility effects are also accounted for by using local values for μ and $\bar{\rho}$.

For flows with pressure gradient and mass transfer, Cebeci and Smith replaced the wall shear stress in the damping parameter by τ_s, which they obtained from the simplified form of the momentum equation in the sublayer:[32]

$$\frac{d\tau_s}{dy} - \frac{v_w}{\nu_w}\tau_s = \frac{d\bar{p}}{dx} \tag{9-189}$$

The solution of this equation at $y^+ = 11.8$ results in

$$A^+ = A\left\{-\frac{P^+}{V_w^+}\left[\exp(11.8V_w^+) - 1\right] + \exp(11.8V_w^+)\right\}^{-1/2} \tag{9-190}$$

where

$$P^+ \equiv \left(-\frac{d\bar{p}}{dx}\right)\frac{\nu}{\rho_w u_\tau^3}, \qquad V_w^+ \equiv \frac{v_w}{u_\tau}, \qquad u_\tau \equiv \sqrt{\frac{\tau_w}{\bar{\rho}}} \tag{9-191}$$

at $A = 26$.

Following Van Driest's approach to arrive at the mixing-length formulation with a damping factor in the inner region, Cebeci and Smith derived the following expression for the turbulent thermal diffusivity:

$$\alpha_T = k_h \kappa y^2 \left[1 - \exp\left(-\frac{y^+}{A^+}\right)\right]\left[1 - \exp\left(-\frac{y^+\sqrt{\mathrm{Pr}}}{B^+}\right)\right]\left|\frac{\partial \bar{u}}{\partial y}\right| \tag{9-192}$$

where

$$B^+ = B\left\{\frac{P^+}{V_w^+}\left[\exp(11.8V_w^+) - 1\right] + \exp(11.8V_w^+)\right\}^{-1/2} \tag{9-193}$$

and $k_h = 0.44$, $B = 34$. The turbulent Prandtl number Pr_T defined in Eq. (9-182) can be expressed in the following manner after the substitution of Eqs. (9-175), (9-176), (9-187), and (9-192):

$$\mathrm{Pr}_T = \frac{\kappa\left[1 - \exp(-y^+/A^+)\right]}{k_h\left[1 - \exp(-y^+\sqrt{\mathrm{Pr}}/B^+)\right]} \tag{9-194}$$

Although Eqs. (9-187) and (9-192) are valid only in the inner region of the boundary layer, Cebeci showed that Eq. (9-194) agrees satisfactorily with experimental data of Simpson et al.,[38] Whitten et al.,[39] Johnson,[40] and Ludweig[41] throughout the boundary layer.

In the wake region, Cebeci and Smith used the Clauser expression for the eddy viscosity, Eq. (9-178) with $c = 0.0168$. This expression is damped in the same way as in the Kendall model.

Bushnell–Beckwith Model

The Bushnell–Beckwith model is a single-layer model in which the mixing length reduces to the Van Driest form near the wall and is modified in the outer region by an intermittency factor γ_I.[42] The mixing-length expression is

$$\frac{\ell_m}{\delta} = k_B\left[1 - \exp\left(-\frac{y^+}{A^+}\right)\right] f\left(\frac{y}{\delta}\right)\sqrt{\gamma_I} \tag{9-195}$$

where

$$\gamma_I = \frac{1 - \operatorname{erf}[5(y/\delta - 0.78)]}{\lambda} \tag{9-196}$$

and

$$f\left(\frac{y}{\delta}\right) = \tanh\left(\frac{\kappa}{k_B}\frac{y}{\delta}\right) \tag{9-197}$$

and the constants are $\kappa = 0.4$, $k_B = 0.08$, $A^+ = 26$. The boundary-layer thickness δ appearing in Eqs. (9-195) through (9-197) is defined as the distance normal to the wall where the velocity ratio $u/U_e = 0.995$.

Bushnell and Beckwith calculated flows with blowing and pressure gradient, and compared their predictions with experimental data of Peterson et al.[43] and Jeromin.[44] They reported[45] that in the application of the model to flows with wall blowing, the effect of blowing could be accounted for only when the wall damping factor of Van Driest, A^+, was made an experimentally based function of the blowing rate. The turbulent Prandtl number is taken to be constant (equal to 0.9) in this model.

7.2.4 Equation of State

The perfect-gas behavior of each species assumed in the BLIMP analysis of Evans means that

$$\bar{p}_i = n_i R_u \bar{T} \tag{9-198}$$

The mixture of gases is treated as an ideal gaseous solution. Basically, this means that the mixture equation of state can be written as

$$\bar{p} = \frac{\bar{\rho} R_u \bar{T}}{W} \tag{9-199}$$

and the mixture thermodynamic state variables can be expressed as

$$\bar{f} = \sum \overline{X}_i \bar{f}_i = \frac{1}{\bar{p}} \sum \bar{p}_i \bar{f}_i \tag{9-200}$$

where f is a property in molar terms, and f_i is evaluated for species i at the mixture temperature.

7.2.5 Integral Matrix Solution Procedure

The solution of the boundary-layer equations described in Section 7.2.1 needs a set of boundary conditions which are similar to those given in Section 5.1 and therefore are not repeated. The integral matrix method has been used for the solution of chemically reacting, nonsimilar, coupled, nonlinear boundary-layer equations. A detailed discussion of the integral matrix procedure is given in Refs. 25 and 46, to which the reader may refer for more details. This section will review only the highlights of the method.

In the integral matrix solution of the boundary-layer equations, the following procedures are used:

A. Coordinate Transformation. The governing equations are transformed from the physical plane (s, y) to a new coordinate space (ξ, η) with the Levy–Lees transformation described in Section 5.2. A stretching parameter, as a function of ξ, is employed to stretch the coordinate η so that the boundary layer remains at constant thickness in the final η.

B. Taylor Series Expansions. The primary dependent variables and their derivatives with respect to η are related by Taylor series expansions such that these dependent variables are represented by connected quadratics or cubics. Then, the series are truncated to reflect the proper polynomial representation. A nodal network is defined through the boundary layer, and the Taylor series expansions are assumed valid between each set of nodes, with an additional requirement of continuous first and second derivatives for a spline fit at each node.

C. Integration Across the Boundary-Layer "Strips." The conservation equations are integrated across a "strip" (between nodal points). The primary reason for this integration is to simplify the η-derivative terms in the energy and species conservation equations, since it is not convenient to express the complex $\bar{q}_a$ and $\bar{J}_k$ terms in derivative forms [see Eqs. (9-157) and (9-156)]. The turbulent mixing length is calculated at each node and used in the next iteration.

D. Newton–Raphson Iteration for a Solution. The linear Taylor series expansions together with linear boundary conditions form a very sparse matrix which has to be inverted only once for a given problem. The nonlinear boundary-layer equations and nonlinear boundary conditions are then solved by driving the errors to zero using Newton–Raphson iteration.

7.2.6 Limitations of the BLIMP Analysis

In view of the assumptions employed in the analysis, BLIMP has several limitations which are listed below:

1. The boundary-layer flow must be steady or quasisteady.
2. The perfect-gas law is assumed for all species.
3. No gas-phase kinetics are included.
4. Surface-roughness effects are not modeled.
5. Even though adverse pressure gradients are allowed, separation is not modeled.
6. Turbulent eddy diffusivities are forced to zero at the edge of the boundary layer.
7. Laminarization of turbulent flows is not modeled.
8. Body forces are not included in the formulation.

7.3 Marching-Integration Procedure of Patankar and Spalding[7]

This section outlines the theoretical model and method of solution for a reacting turbulent boundary layer developed by Patankar and Spalding.[7] Their method can be used to predict flow profiles in boundary layers under various conditions, including shear flows that are incompressible or compressible, laminar or turbulent, reacting or nonreacting, and planar or axisymmetric, and those with blowing or suction.

The theoretical model comprises the conservation equations of mass, momentum, species, enthalpy, turbulent kinetic energy, and turbulent dissipation, and the equation of state. To demonstrate the use of time- and mass-weighted averaging equations, the instantaneous governing partial differential equations are Favre-averaged and a two-equation (k–ε) turbulence closure model is employed. The set of governing equations can be solved numerically using the finite-difference procedure with nonuniform grid spacing to achieve good resolution near the wall, where the gradients are steep.

7.3.1 Description of the Physical Model

The physical model considered in the theoretical analysis consists of flow of gases (a) over a flat plate or (b) inside a cylinder. In both cases, the surface could be subjected to blowing or suction or could be a burning surface. For planar boundary-layer flows, the streamwise pressure gradient can be specified as zero, positive (adverse), or negative (favorable). For the axisymmetric case, the local streamwise pressure gradient cannot be specified *a priori*; instead, it must be determined from the solution for the coupled potential and viscous flow regimes.

7.3.2 Conservation Equations for the Viscous Region

To formulate the theoretical model, a set of general conservation equations for a reacting compressible fluid flow is required. The boundary layer can be assumed quasisteady, planar, or axisymmetric, and chemically reacting. Since the Favre (or mass-weighted) averaging is preferred over the Reynolds (or conventional time) averaging for compressible flows, instantaneous equations were Favre-averaged[47] to arrive at the mean governing equations. (Refer to Section 3 of Chapter 7 for averaging procedures.)

Major assumptions made in the following analysis are:

1. The averaged flow properties are steady.
2. Body forces are absent.
3. There is no radiative heat transfer.
4. The Lewis number is unity.
5. Fick's law of diffusion is valid.

A second-order two-equation k–ε turbulence model is used to achieve closure of the turbulent-flow problem. Using these assumptions and following an order-of-magnitude analysis, conservation equations for mass, momentum, species, enthalpy, turbulent kinetic energy, and turbulent dissipation and the equation of state are obtained as follows:

$$\frac{\partial}{\partial x}(r^m\bar{\rho}\tilde{u}) + \frac{\partial}{\partial y}(r^m\bar{\rho}\tilde{v}) = 0 \tag{9-201}$$

$$\bar{\rho}\tilde{u}\frac{\partial \tilde{u}}{\partial x} + \bar{\rho}\tilde{v}\frac{\partial \tilde{u}}{\partial y} = \frac{1}{r^m}\frac{\partial}{\partial y}\left[r^m\mu_{\text{eff}}\frac{\partial \tilde{u}}{\partial y}\right] - \frac{d\bar{p}}{dx} \tag{9-202}$$

$$\bar{\rho}\tilde{u}\frac{\partial \tilde{Y}_i}{\partial y} + \bar{\rho}\tilde{v}\frac{\partial \tilde{Y}_i}{\partial y} = \frac{1}{r^m}\frac{\partial}{\partial y}\left[r^m\left(\frac{\mu}{\text{Sc}}\right)_{\text{eff}}\frac{\partial \tilde{Y}_i}{\partial y}\right] + \bar{\dot{\omega}}_i \tag{9-203}$$

$$\bar{\rho}\tilde{u}\frac{\partial \tilde{H}}{\partial x} + \bar{\rho}\tilde{v}\frac{\partial \tilde{H}}{\partial y} = \frac{1}{r^m}\frac{\partial}{\partial y}\left\{r^m\left[\left(\frac{\mu}{\text{Pr}}\right)_{\text{eff}}\frac{\partial \tilde{H}}{\partial y} + \left\{\mu_{\text{eff}} - \left(\frac{\mu}{\text{Pr}}\right)_{\text{eff}}\right\}\frac{\partial(\tilde{u}^2/2)}{\partial y}\right]\right\} \tag{9-204}$$

$$\bar{\rho}\tilde{u}\frac{\partial k}{\partial x} + \bar{\rho}\tilde{v}\frac{\partial k}{\partial y} = \frac{1}{r^m}\frac{\partial}{\partial y}\left[r^m\left(\mu + \frac{\mu_t}{C_1}\right)\frac{\partial k}{\partial y}\right] + \mu_t\left(\frac{\partial \tilde{u}}{\partial y}\right)^2 - \bar{\rho}\varepsilon \tag{9-205}$$

$$\bar{\rho}\tilde{u}\frac{\partial \varepsilon}{\partial x} + \bar{\rho}\tilde{v}\frac{\partial \varepsilon}{\partial y} = \frac{1}{r^m}\frac{\partial}{\partial y}\left[r^m\left(\mu + \frac{\mu_t}{C_2}\right)\frac{\partial \varepsilon}{\partial y}\right] + C_3\mu_t\left(\frac{\partial \tilde{u}}{\partial y}\right)^2\frac{\varepsilon}{k} - C_4\bar{\rho}\frac{\varepsilon^2}{k} \tag{9-206}$$

$$\bar{p} = \frac{\bar{\rho}R_u\tilde{T}}{W} \tag{9-207}$$

where $m = 0$ and r is replaced by y for planar flows, and $m = 1$ for axisymmetric flows.

The turbulent viscosity μ_t is expressed in terms of k and ε as suggested in Refs. 48–52:

$$\mu_t = C_\mu \bar{\rho} \frac{k^2}{\varepsilon} \tag{9-208}$$

In arriving at Eqs. (9-202) to (9-206), the following Boussinesq relationships have been used:

$$-\overline{\rho u'' v''} = \mu_t \frac{\partial \tilde{u}}{\partial y} \tag{9-209}$$

$$-\overline{\rho v'' Y_i''} = \frac{\mu_t}{\text{Sc}_t} \frac{\partial \tilde{Y}_i}{\partial y} \tag{9-210}$$

$$-\overline{\rho v'' h_i''} = \frac{\mu_t}{\text{Pr}_t} \frac{\partial \tilde{h}_i}{\partial y} \tag{9-211}$$

7.3.3 Modeling of the Gas-Phase Chemical Reactions

To solve the set of equations (9-201) to (9-207) with appropriate boundary conditions, the term $\bar{\dot{\omega}}_i$ must be known for chemically reacting boundary-layer flows. In this section, the turbulent reacting boundary-layer flow over a solid fuel or propellant is considered. A single-step forward chemical reaction is assumed for the global reaction of gaseous reactants pyrolyzed from the solid surface:

$$\nu_F F + \nu_O O \rightarrow \nu_P P \tag{9-212}$$

where O and F represent the oxidizer- and fuel-rich gases respectively, and P represents the product gases. If an expression involving the Arrhenius law is used in the modeling of $\bar{\dot{\omega}}_i$, time-averaging of the resulting exponential expression presents a difficult problem, as described in Chapter 7, and thus an alternative approach is preferable.

One such approach, first proposed by Spalding,[11] is the eddy-breakup (EBU) model. The use of this model for diffusion flames has been discussed by Lockwood.[53] Following a treatment based on these concepts, an expression for $\dot{\omega}_F$ can be found:[10]

$$\bar{\dot{\omega}}_F = -C_\omega \bar{\rho} \sqrt{k} \left| \frac{\partial \tilde{Y}_F}{\partial r} \right| \tag{9-213}$$

where C_ω is a constant.

It may be noted that the species conservation equations can be combined to yield the conserved scalar $\tilde{Y}_{OF} \equiv [\tilde{Y}_O - (\nu_O W_O / \nu_F W_F) \tilde{Y}_F]$. This procedure eliminates the nonlinear source term in the equation for $\tilde{Y}_{OF}$. The species equation for $\tilde{Y}_F$ can be solved together with $\tilde{Y}_{OF}$. No separate conservation equation is needed for $\tilde{Y}_P$, which can be deduced as $1 - \tilde{Y}_O - \tilde{Y}_F$.

7.3.4 Governing Equations for the Inviscid Region

In the potential-flow region for the developing flow, the following sets of equations are considered:

A. For planar flow, at the free stream,

$$\tilde{u} = U_\infty, \qquad \tilde{T} = T_\infty, \tag{9-214}$$

$$\tilde{Y}_F = \tilde{Y}_O = 0 \qquad \left(\text{or} \quad \frac{\partial \tilde{Y}_F}{\partial y} = \frac{\partial \tilde{Y}_O}{\partial y} = 0\right) \tag{9-215}$$

$$\frac{\partial k}{\partial y} = \frac{\partial \varepsilon}{\partial y} = 0 \tag{9-216}$$

B. For axisymmetric flow, the momentum and energy equations in the core are

$$\rho_c U_c \frac{dU_c}{dx} = -\frac{d\bar{p}}{dx} \tag{9-217}$$

$$T_c = \frac{T_{tc}}{1 + \frac{1}{2}(\gamma - 1)M_c^2} \tag{9-218}$$

The centerline velocity U_c can be calculated from Eq. (9-217), and the axial pressure gradient can be obtained from the overall momentum balance, using the following equations written in terms of the bulk properties ρ_b and U_b:

$$-\frac{d\bar{p}}{dx} = \frac{1}{A}\left[2\pi R \tau_w + \frac{d}{dx}\left(\rho_b A U_b^2\right)\right] \tag{9-219}$$

$$\frac{d}{dx}(\rho_b U_b A) = 2\pi R \rho_s r_b \tag{9-220}$$

The pressure gradient expressed by Eq. (9-219) includes the effect of change in flow area, and the change in bulk flow variables along the x-direction. Equation (9-220) is obtained from the overall mass balance inside the cylinder with instantaneous inner radius R and burning rate r_b.

7.3.5 Boundary Conditions

At the solid–gas interface, mass and energy balance lead to the following boundary conditions:

$$(\bar{\rho}\tilde{v}\tilde{Y}_i)_g - \rho_s r_b Y_{i,s} - \left(\bar{\rho}\mathscr{D}\frac{\partial \tilde{Y}_i}{\partial y}\right)_g = 0 \tag{9-221}$$

$$\lambda \frac{\partial \tilde{T}}{\partial y}\bigg|_g = \lambda_s \frac{\partial T_s}{\partial y}\bigg|_s + \rho_s r_b\left[(C_p - C_s)(T_s - T_{s,\mathrm{ref}}) + Q_{s,\mathrm{ref}}\right] \tag{9-222}$$

where $Q_{s,\mathrm{ref}}$ is defined as the net surface heat release (negative for exothermic reactions) at a reference temperature $T_{s,\mathrm{ref}}$. The net heat flux to the burning solid surface can be obtained by integrating the heat conduction equation in the solid phase,

$$\lambda_s \frac{\partial T_s}{\partial y}\bigg|_s = (T_s - T_{si})\rho_s C_s r_b \tag{9-223}$$

The burning rate of the solid can be expressed as a function of surface temperature through the use of the Arrhenius law of surface pyrolysis:

$$r_b = A_s \exp\left(-\frac{E_{a,s}}{R_u T_s}\right) \tag{9-224}$$

Other boundary conditions are:

At the surface,

$$\tilde{u} = 0, \qquad \tilde{T} = T_s, \qquad \tilde{v} = \frac{\rho_s r_b}{\bar{\rho}_g} \tag{9-225}$$

At the edge of the boundary layer,

$$\tilde{u} = U_\infty, \quad \tilde{T} = T_\infty, \quad \tilde{Y}_F = \tilde{Y}_{F\infty}, \quad \tilde{Y}_O = \tilde{Y}_{O\infty}$$

and

$$\frac{\partial k}{\partial y} = \frac{\partial \varepsilon}{\partial y} = 0 \tag{9-226}$$

where U_∞ and T_∞ are specified or obtained from potential flow analysis.

7.3.6 Near-Wall Treatment of k and ε

In the near-wall region where the Reynolds number is low, it is generally recognized that the models of various turbulence correlations for the k and ε equations are not applicable. In the work of Kuo and his coworkers,[10, 12, 14, 54] the k and ε equations are applied to a region from a small distance above the wall to the edge of the boundary layer. Hence, one must also apply the boundary conditions for these two equations away from the wall rather than at the wall. This avoids the low-turbulent-Reynolds-number region near the wall where the closure models in k and ε are not valid. At the edge of this region, however, it is reasonable to assume that the production and dissipation terms of the k-equation [the last two terms in Eq. (9-205)] are dominant, and can therefore be equated. This yields

$$\varepsilon = \frac{\mu_t}{\bar{\rho}}\left(\frac{\partial \tilde{u}}{\partial y}\right)^2 \tag{9-227}$$

The turbulent viscosity μ_t close to the wall is calculated using a modification of Van Driest's formula:[37, 55]

$$\mu_t = \bar{\rho}\left[\kappa \mathscr{D}_v (y + \Delta y)\right]^2 \left|\frac{\partial \tilde{u}}{\partial y}\right| \tag{9-228}$$

where the damping coefficient $\mathscr{D}_v$ is altered to include the effect of surface roughness as suggested by Cebeci and Chang,[56] and the effect of local shear stress as suggested by Baker and Launder,[55]

$$\mathscr{D}_v = 1 - \exp\left[-\frac{(y + \Delta y)\bar{\rho} u_*}{A^+ \mu} \frac{\tau}{\tau_w}\right] \tag{9-229}$$

where

$$\Delta y = 0.9 \frac{\mu}{\bar{\rho} u_*}\left[\sqrt{R_h^+} - R_h^+ \exp\left(-\frac{R_h^+}{6}\right)\right] \tag{9-230}$$

and

$$R_h^+ = \frac{\bar{\rho} u_* R_h}{\mu} \tag{9-231}$$

By eliminating μ_t from Eqs. (9-208), (9-228), and (9-229), the following expressions for k and ε are obtained:

$$k = \frac{\left[\kappa \mathscr{D}_v (y + \Delta y)\right]^2}{\sqrt{C_\mu}} \left(\frac{\partial \tilde{u}}{\partial y} \right)^2 \tag{9-232}$$

$$\varepsilon = \left[\kappa \mathscr{D}_v (y + \Delta y)\right]^2 \left| \frac{\partial \tilde{u}}{\partial y} \right|^3 \tag{9-233}$$

The near-wall treatment of the k and ε equations presented here is similar to that of Chambers and Wilcox.[57] Detailed justification and validation of the near-wall analysis is given in Ref. 54.

7.3.7 Constants Used in Turbulence Modeling

To test the universality of a set of constants frequently used in turbulence modeling, Arora, Kuo, and Razdan[54] used the near-wall treatment and the abovementioned turbulence closure procedure to obtain numerical solutions for comparison with experimental data. The tests made were stringent, since the data were collected from various researchers for planar and axisymmetric, incompressible and compressible, laminar and turbulent, reacting and nonreacting, subsonic and supersonic flows with and without blowing. Values of the constants used in turbulence modeling, together with their sources, are listed in Table 9.3. These constants were originally obtained for Reynolds-averaged equations and had been successfully tested before in an erosive-burning study of solid propellants.[10] It was tacitly assumed here that the constants remain unchanged for Favre-averaged equations. The comparison of theoretical results using these constants and broad-based experimental data is given in Section 7.3.9.

TABLE 9.3 Constants Used in Turbulence Modeling

Constant	Value	References
C_1	1.0	48–50
C_2	1.3	48–50
C_3	1.57	50
C_4	2.0	49
C_μ	0.09	48–50
C_ρ	1.0	50
C_ω	0.18	52

7.3.8 Coordinate Transformation and Solution Procedure of Patankar and Spalding[7]

The system of coupled, nonlinear, simultaneous partial differential equations (9-201) through (9-206) is parabolic in nature and, due to its complexity, must be solved numerically. Various techniques are available to solve such a system, including the one proposed by Patankar and Spalding,[7] which has been used successfully by numerous researchers.[10,12,14,49–52] This method offers computational economy, particularly for flows in which the field of interest grows rapidly in thickness, as is the case with wall boundary layers. Patankar and Spalding[7] introduced a transformation of coordinates in which the grid points always fit the boundary-layer region even while the thickness of the region is changing.

The following coordinate transformation is introduced for the governing equations:

$$x = x, \qquad \omega = \frac{\psi - \psi_I}{\psi_E - \psi_I} \tag{9-234}$$

The stream function ψ is defined by

$$\frac{\partial \psi}{\partial y} = r\bar{\rho}\tilde{u}$$

$$\frac{\partial \psi}{\partial x} = -r\bar{\rho}\tilde{v} \tag{9-235}$$

with

$$\frac{\partial \psi_I}{\partial x} = -r_I m_I, \qquad \frac{\partial \psi_E}{\partial x} = -r_E m_E \tag{9-236}$$

Subscripts I and E represent the internal and external boundaries of the flow field, respectively (see Fig. 9.9). ψ_I and ψ_E are the stream functions at I and E boundaries of the flow.

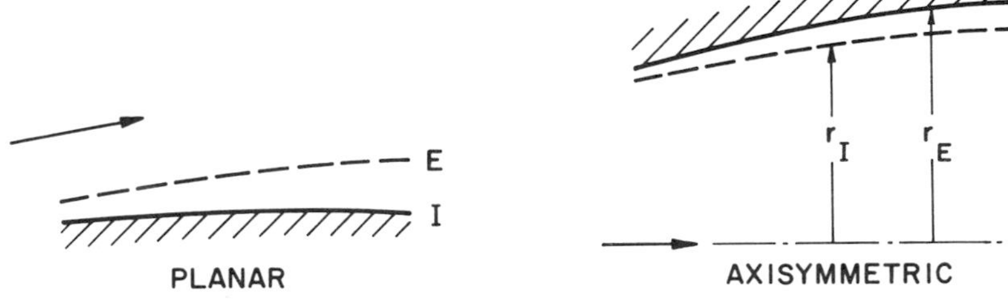

Figure 9.9 E and I boundaries for planar and axisymmetric geometries.

TABLE 9.4 Coefficients c^* and d for Various Conservation Equations

Variable	c^*	d
$\tilde{u}$	$\mu_{\text{eff}} = \mu + \mu_t$	$-\dfrac{1}{\bar{\rho}\tilde{u}}\dfrac{d\bar{p}}{dx}$
$\tilde{Y}_k$	$\left(\dfrac{\mu}{\text{Sc}}\right)_{\text{eff}} = \dfrac{\mu}{\text{Sc}} + \dfrac{\mu_t}{\text{Sc}_t}$	$\dfrac{\bar{\dot{\omega}}_i}{\bar{\rho}\tilde{u}}$
H	$\left(\dfrac{\mu}{\text{Pr}}\right)_{\text{eff}} = \dfrac{\mu}{\text{Pr}} + \dfrac{\mu_t}{\text{Pr}_t}$	$\dfrac{1}{\bar{\rho}\tilde{u}}\left\{\dfrac{\partial}{\partial y}\left[\mu_{\text{eff}} - \left(\dfrac{\mu}{\text{Pr}}\right)_{\text{eff}}\right]\dfrac{\partial(\tilde{u}^2/2)}{\partial y}\right\}$
k	$\mu + \dfrac{\mu_t}{C_1}$	$\dfrac{1}{\bar{\rho}\tilde{u}}\left[\mu_t\left(\dfrac{\partial \tilde{u}}{\partial y}\right)^2 - \bar{\rho}\varepsilon\right]$
ε	$\mu + \dfrac{\mu_t}{C_2}$	$\dfrac{1}{\bar{\rho}\tilde{u}}\left[C_3\mu_t\left(\dfrac{\partial \tilde{u}}{\partial y}\right)^2\dfrac{\varepsilon}{k} - C_4\bar{\rho}\dfrac{\varepsilon^2}{k}\right]$

With the transformations defined by Eqs. (9-234) through (9-236), all governing equations can be transformed into the following general form:

$$\frac{\partial \phi}{\partial x} + (a + b\omega)\frac{\partial \phi}{\partial \omega} = \frac{\partial}{\partial \omega}\left(c\frac{\partial \phi}{\partial \omega}\right) + d \tag{9-237}$$

where

ϕ is any dependent variable,

$$a \equiv \frac{r_I m_I}{\psi_E - \psi_I}$$

$$b \equiv \frac{r_E m_E - r_I m_I}{\psi_E - \psi_I}$$

$$c \equiv \frac{r^2 \bar{\rho}\tilde{u}}{(\psi_E - \psi_I)^2} c^*$$

c^* and d for each equation are listed in Table 9.4.

To form the finite-difference equations for Eq. (9-237), the boundary layer is divided into N strip from $\omega = 0$ to $\omega = 1$. The variation of ϕ between the grid points is assumed to be linear, except in the half interval near the boundaries at $\omega = 0$ and $\omega = 1$ (see Fig. 9.10). The finite-difference equations are formed by integrating Eq. (9-237) in a small control volume of the flow field near a grid point (i, j), where i and j represent the grid locations in the x and ω directions respectively, as shown in Fig. 9.10. In this way, the finite-difference

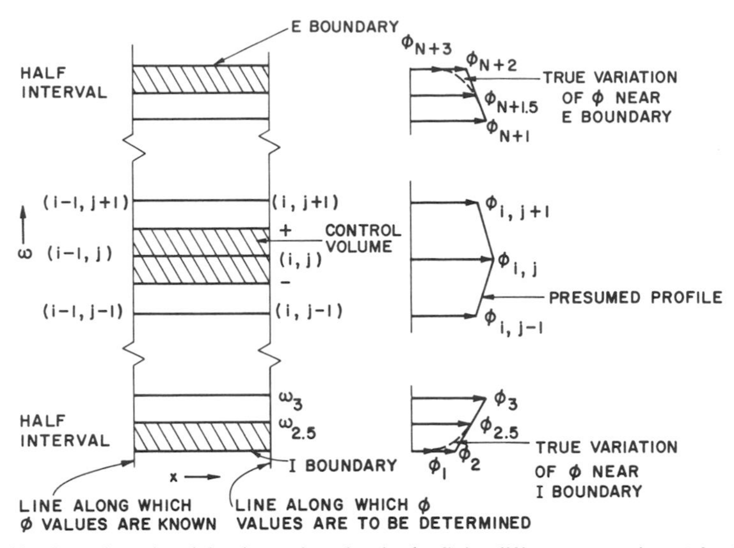

Figure 9.10 Location of nodal points referred to in the finite-difference equations (after Patankar and Spalding[7]).

equations are obtained; these relate $\phi_{i,j}$ to $\phi_{i,j-1}$ and $\phi_{i,j+1}$. At the boundaries, two points are identified, corresponding to the true value of ϕ (ϕ_1 and ϕ_{N+3} at I and E boundaries, respectively), and a false or *slip* value of ϕ (ϕ_2 and ϕ_{N+2} at I and E boundaries, respectively). The slip value is the one which would be obtained if ϕ were assumed to be linear near the boundaries. The integration of Eq. (9-237) over the half interval near the boundaries gives the finite-difference equations expressing the slip value of ϕ in terms of the true value and the value of ϕ at the grid point next closest to the boundary. For example, at the I boundary ϕ_2 is related to ϕ_1 and ϕ_3 through a finite-difference equation. The function of the slip value ϕ_2 is to orient the ϕ_2–ϕ_3 line so that a better representation is obtained for the region from $\omega_{2.5}$ (corresponding to $\phi_{2.5}$ in Fig. 9.10) to ω_3 than that given by the original ϕ_1–ϕ_3 line. The details of the coordinate transformation and integration procedure to obtain the finite-difference equations are given in Ref. 7.

The dependent variables vary steeply in the region close to the wall. Therefore, to obtain the fluxes of mass, momentum, and energy at the wall with reasonable accuracy, exact solutions of the conservation equations for these variables are obtained under the Couette-flow assumption.[7] This implies that the local x-wise convection of the dependent variables is assumed to be negligible. This is a good assumption, particularly if the region near the wall is specified as very small where the velocity $\tilde{u}$ is small. With this assumption, the partial differential equations (9-202) through (9-204) reduce to ordinary differential equations, which can be integrated within the Couette-flow region. Further, it is assumed that within the Couette-flow region the changes in fluid flow properties (μ, μ_{eff}, Pr, etc.) are negligible. Details of the Couette-flow analysis are given in Ref. 7.

7.3.9 Comparison of Theoretical Results with Experimental Data

A comparison between the predictions of the above model and some of the available experimental data is given in this section. Although the main interest is in turbulent reacting flows, the boundary-layer predictions were made for both laminar and turbulent flows. In order to check the validity of the numerical solution and the range of applicability of the computer program, several comparisons were made for both laminar and turbulent boundary-layer flows.[54] Some of these comparisons are discussed below.

Figure 9.11 compares the calculated values with Sparrow and Yu's[58] solution for a laminar boundary layer with blowing. The predicted temperature distributions for two different blowing rates were in excellent agreement with their theoretical results.

A comparison of calculated incompressible turbulent boundary-layer profiles with Klebanoff's[42] data is shown in Fig. 9.12. The data were collected in a boundary layer of a flat plate at zero incidence. As depicted by this figure, all three predicted profiles—velocity, turbulence kinetic energy, and Reynolds

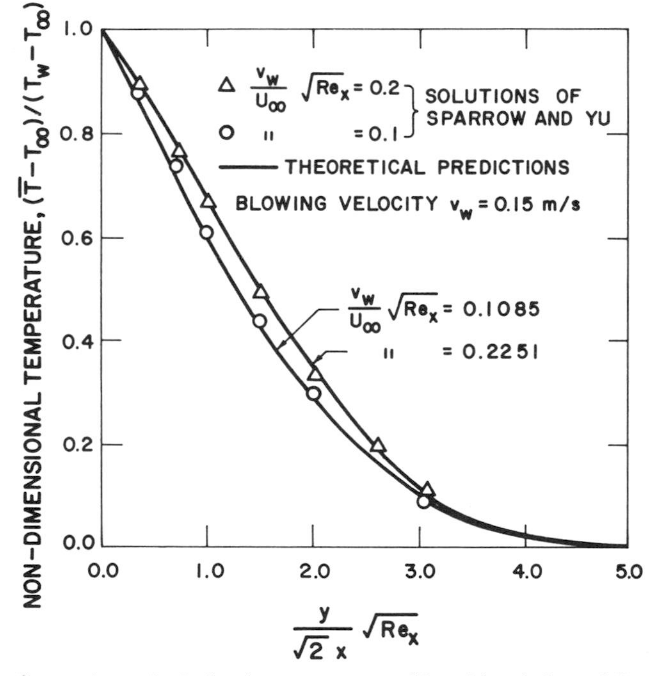

Figure 9.11 Comparison of calculated temperature profiles with solutions of Sparrow and Yu[58] for a laminar boundary layer with blowing (after Arora, Kuo, and Razdan[54]).

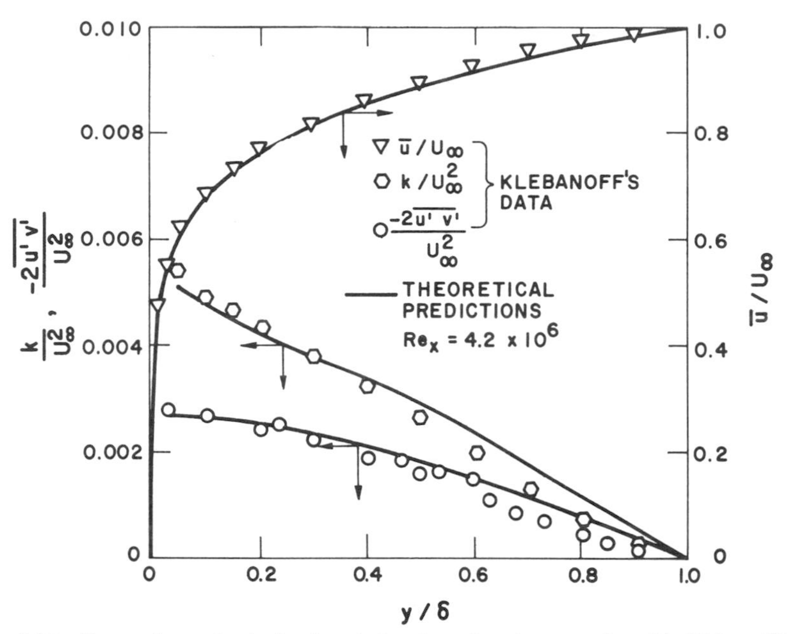

Figure 9.12 Comparison of calculated turbulent-boundary-layer results with Klebanoff's measurements[42] (after Arora, Kuo, and Razdan[54]).

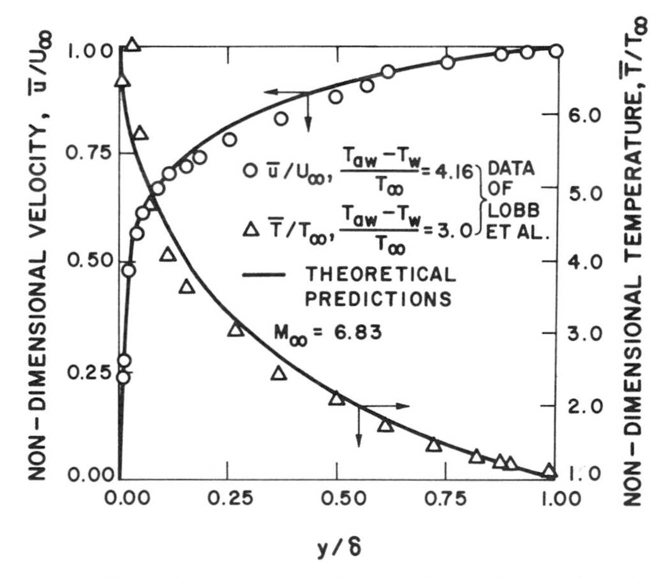

Figure 9.13 Comparison of calculated velocity and temperature profiles in a compressible turbulent boundary layer with data of Lobb et al.[59] (after Arora, Kuo, and Razdan[54]).

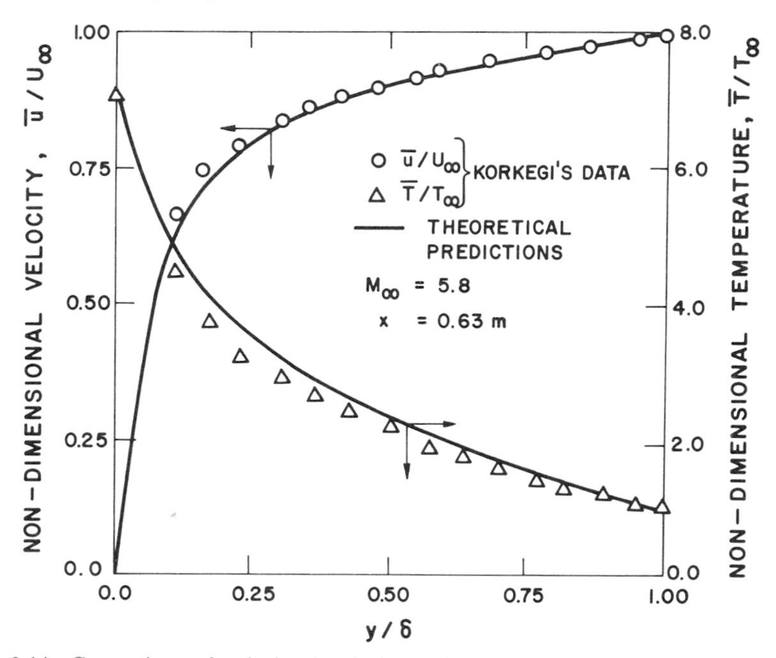

Figure 9.14 Comparison of calculated velocity and temperature profiles in a compressible turbulent boundary layer with Korkegi's data[60] (after Arora, Kuo, and Razdan[54]).

stresses—agree well with the experimental results, both in the trends and in the quantitative values. However, there is a slight discrepancy in the calculated turbulence kinetic energy and Reynolds stresses near the edge of the boundary layer, though this could be considered within the range of experimental error.

The experimental data of Lobb et al.[59] on compressible turbulent boundary layers with heat transfer were compared with the numerical results obtained through the computer simulation. Another set of boundary-layer data obtained by Korkegi[60] from hypersonic wind-tunnel experiments with a nominal Mach number of 5.8 was compared similarly. These comparisons are shown in Figs. 9.13 and 9.14, respectively. Reasonable agreement was found for both velocity and temperature profiles in both sets of data.

Laderman and Demetriades[61,62] used a two-dimensional wind tunnel to make measurements in the boundary layer of a flat plate in a supersonic flow with zero pressure gradient. Data were collected for both adiabatic and isothermal walls and are plotted along with the predicted values in Fig. 9.15. The predicted nondimensionalized velocity and shear-stress profiles were found to be almost independent of the wall temperature in the range considered. Sandborn[63] represented the results of various investigations by a narrow band denoting the variation of τ/τ_w with y/δ, which appears to be valid for Mach numbers ranging from zero to hypersonic speeds. This band is also plotted as the region between the two dashed curves in Fig. 9.15 for comparison. The

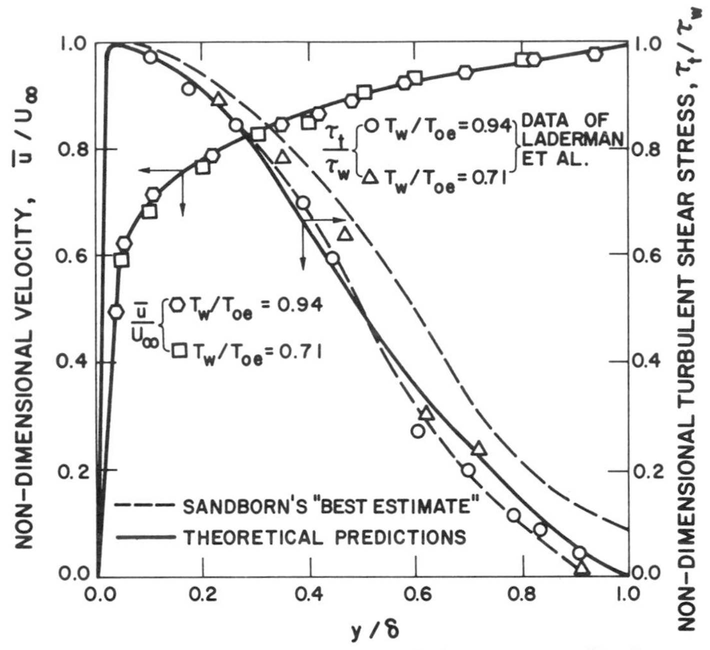

Figure 9.15 Comparison of calculated velocity and shear-stress profiles in a compressible turbulent boundary layer with data of Laderman and Demetriades[61,62] (after Arora, Kuo, and Razdan[54]).

predictions of the numerical model follow the data of Laderman and Demetriades, as well as Sandborn's "best estimate," quite satisfactorily.

Yanta and Lee[64] used a laser-Doppler velocimeter (LDV) to obtain experimental data in a wind tunnel for a flow of Mach number 2.9 under zero pressure gradient. A comparison of the predictions of the program with their turbulent shear stress and kinematic eddy-viscosity profiles is shown in Fig. 9.16; good agreement was found for both profiles. The predictions also compare well with Sandborn's "best estimate," which is reproduced in this figure.

Razdan and Kuo[65] and Marklund and Lake[66] have obtained data on the burning of composite solid propellants under a cross-flow of gases at various values of pressure and velocity. These data are compared with the predictions in Figs. 9.17 and 9.18. Comparison indicates that agreement between the theoretical and experimental results is very close for the conditions considered.

Laufer[67] obtained a detailed set of measurements of turbulence quantities in a fully developed pipe flow. The value of U_{max} for the set used here was 118 m/sec. Profiles of the velocity, turbulent viscosity, turbulent shear stresses, and turbulent kinetic energy are plotted in Figs. 9.19 and 9.20. Figure 9.19 also includes turbulent-viscosity data obtained by Nunner.[68] Both sets are compared with the numerical predictions of the program. In general, agreement is

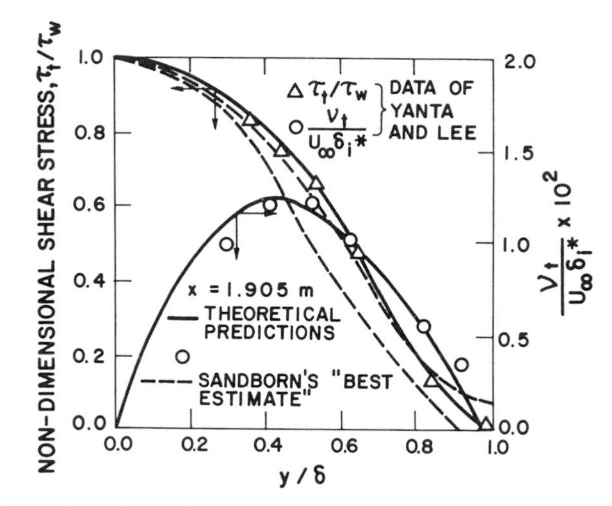

Figure 9.16 Comparison of calculated shear-stress and eddy-viscosity profiles in a compressible turbulent boundary layer with data of Yanta and Lee[64] (after Arora, Kuo, and Razdan[54]).

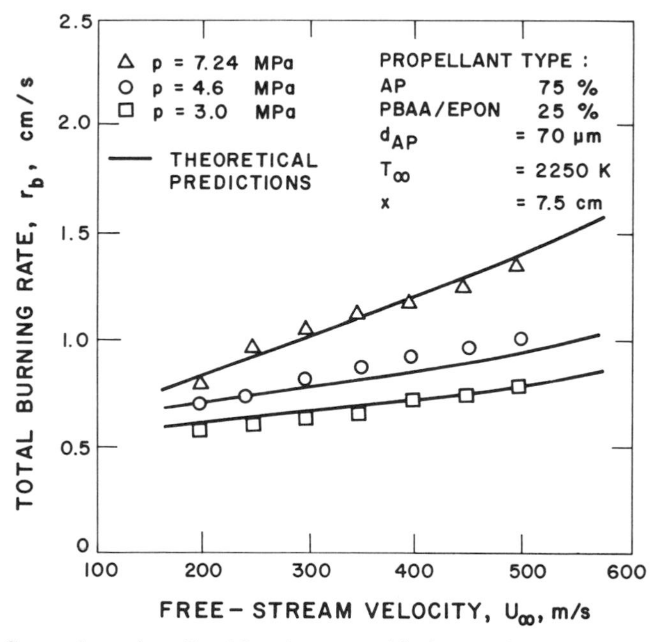

Figure 9.17 Comparison of predicted burning rates with the experimental data of Razdan and Kuo[65] at various pressures and free-stream velocities (after Razdan and Kuo[65]).

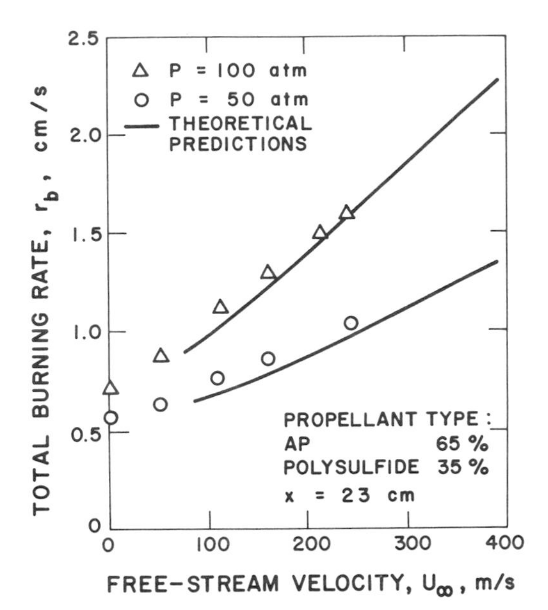

Figure 9.18 Comparison of predicted burning rates with the experimental data of Marklund and Lake[66] (after Razdan and Kuo[65]).

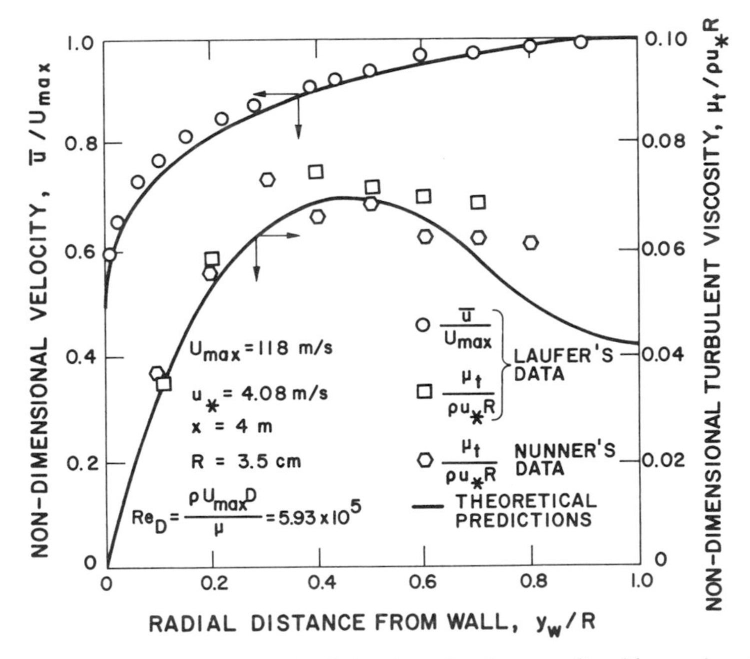

Figure 9.19 Comparison of calculated turbulent-boundary-layer results with experimental measurements of Laufer[67] and Nunner[68] (after Arora, Kuo, and Razdan[54]).

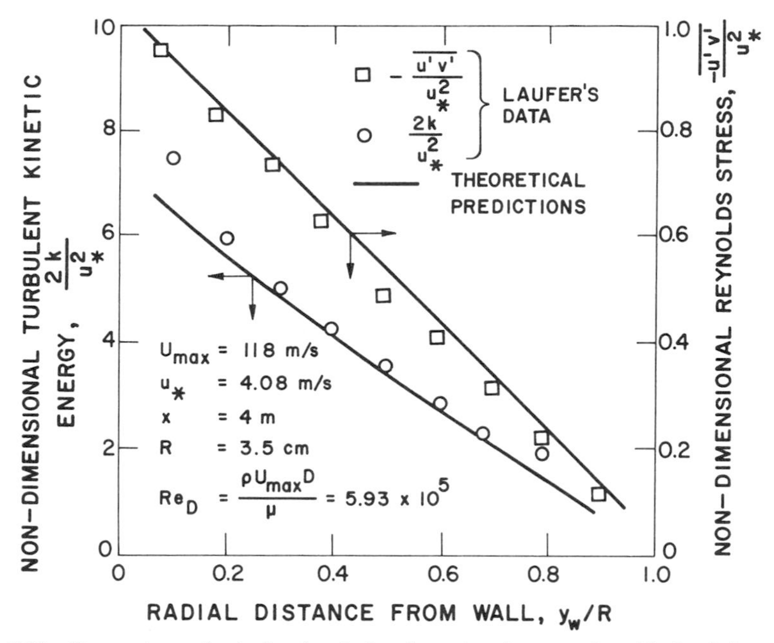

Figure 9.20 Comparison of calculated turbulent-boundary-layer results with Laufer's measurements[67] (after Arora, Kuo, and Razdan[54]).

good, except for the turbulent viscosity, which is underpredicted at y_w/R greater than 0.7. The value of the turbulent kinetic energy near the wall is also slightly underpredicted.

On the basis of the reasonable agreement between predicted profiles of flow variables, distributions of turbulence quantities, and experimental data reported by many investigators, the k–ε model with the near-wall treatment used by Arora et al.[54] appears to be adequate for boundary-layer calculations. Also, the set of turbulence constants used in their computations can be considered universal, since overall results indicate that the set is applicable to a wide range of flow situations. The peak in the turbulent kinetic-energy distribution appearing near the wall was usually underpredicted by this method. However, this does not appear to affect the prediction for mean-flow-property distributions. An improvement in the expression for the Van Driest damping coefficient may remedy this discrepancy.

7.4 Metal Erosion by Hot Reactive Gases

The enhanced erosion of metals subjected to high temperature (2500–3000 K), high-pressure (350 MN/m^2) gases was studied by Gany et al.[69,70] The situation considered involves high-Reynolds-number (up to 5×10^6) flows through a metal disk with an orifice for very short duration (2 msec). (See Fig. 9.21.)

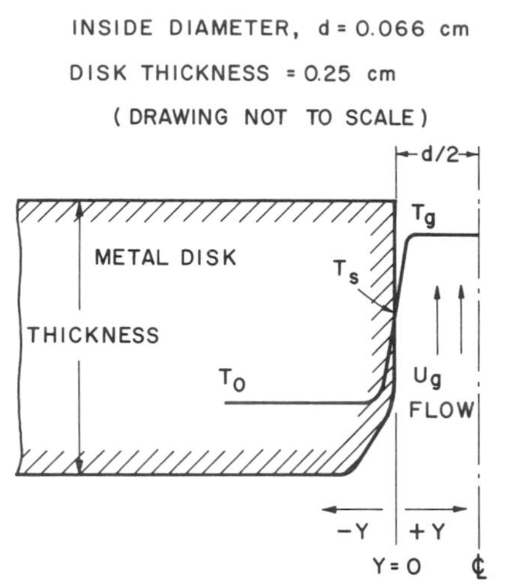

Figure 9.21 Configuration and nomenclature of test specimen (after Gany et al.[69]).

The regression rates for metals subjected to these conditions were found to be much higher than for the specimen subjected to stagnant heating conditions.

Erosion is a result of both chemical attack and removal of molten metal by cross flow. Under high-temperature and high-speed flow conditions, the oxide scales and molten metal are wiped off very fast and fresh surface is exposed to the gases, whereas in stagnant heating, the relative importance of the melting process and the heterogeneous surface reactions is determined by the relative magnitudes of the melting and oxidation temperatures. In the flow situation rapid removal of the oxide and molten metal may prevent the metal surface from reaching the melting point, and hence the heterogeneous oxidation reactions at the surface play an important role in determining the erosion rates of the material regardless of the relative magnitudes of the melting point of the metal and the oxide.

Experimentally, Gany et al.[69] used a ballistic compressor[71] to generate high-temperature, high-pressure gases to flow through nozzle plates. As the pressure inside the ballistic compressor increases, the volumetric flow rate and temperature of the gases flowing over the metal surface increases. Due to rapid rise in the convective film coefficient and the free-stream temperature, the temperature of the metallic nozzle wall increases very rapidly. During this period, the convective heating rate q_g accelerates rapidly (see Fig. 9.22), since both the free-stream temperature and the convective heat-transfer coefficient are increasing. For gases which contain a highly reactive oxidizer, heating due to surface reaction (q_r) becomes prominent above a threshold temperature. For cases in which these reactions are very exothermic, the heating due to surface reaction may be much greater than the nonreactive convective heating. As the metal and oxidizing species react, metallic oxides are formed and may be carried away by the shear forces of the high-speed flow. This is the first mode of material loss or erosion. As the surface temperature continues to increase,

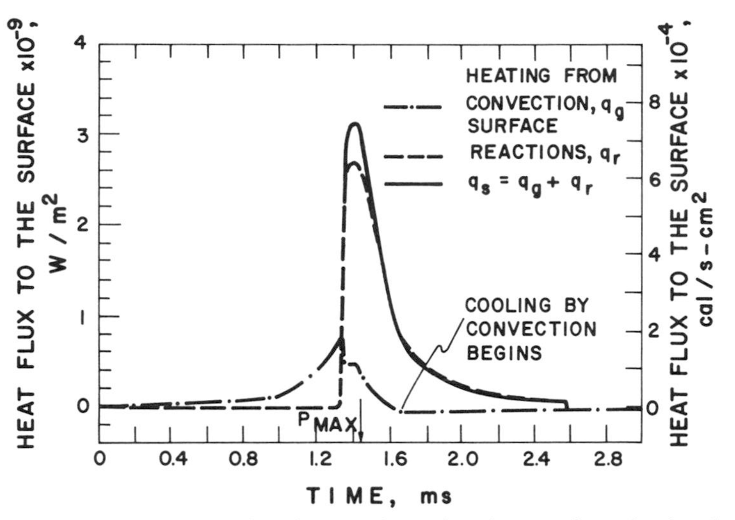

Figure 9.22 Heating rates due to forced convection and surface reactions, showing that surface reaction can be the primary source of heat transfer to the surface (after Gany et al.[69]).

the melting point is reached. Then the metal itself becomes mobile and is carried away by the flow. This is the second mode of material loss.

After peak pressure is achieved (when the piston in the ballistic compressor starts to reverse its direction), the gas temperature begins to decrease rapidly as shown in Fig. 9.23. Accordingly, q_g decreases rapidly. For situations in which surface reactions are prominent, the gas temperature T_g will fall below the surface temperature T_s. As the pressure decreases, the surface temperature falls below the melting point and no further melting occurs.

In the theoretical formulation, a one-dimensional transient solid-phase heat-conduction equation with variable thermal properties was considered:

$$\rho C_p\left(\frac{\partial T}{\partial t} + r_b\frac{\partial T}{\partial y}\right) = \frac{\partial}{\partial y}\left(\lambda\frac{\partial T}{\partial y}\right) \tag{9-238}$$

where r_b represents the linear erosion rate of the metal. The initial condition was

$$t = 0, \qquad T = T_0 \tag{9-239}$$

The boundary condition at the surface ($y = 0$) was specified as

$$\lambda\left(\frac{\partial T}{\partial y}\right)_{y=0} = \begin{cases} q_s & \text{when} \quad T < T_m \\ q_s - L\rho_m r_b & \text{when} \quad T = T_m \end{cases} \tag{9-240}$$

where L is the latent heat of fusion and q_s consists of the heat flux by

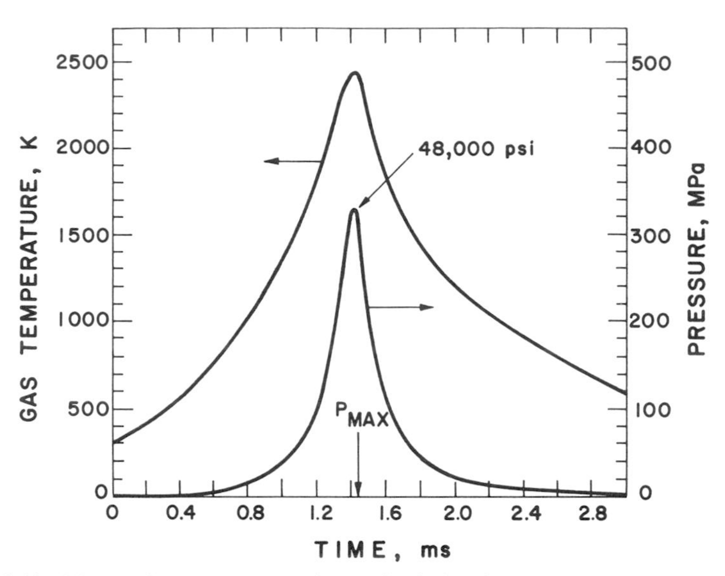

Figure 9.23 Measured pressure versus time and calculated temperature versus time (after Gany et al.[69])

convection (q_g) and the heat flux from the surface chemical reation (q_r):

$$q_s = q_g + q_r \tag{9-241}$$

The boundary condition in the bulk was given as

$$y \to \infty, \qquad T = T_0 \tag{9-242}$$

The convective heat flux was determined from

$$q_g = h_c(T_g - T_s) \tag{9-243}$$

and the convective heat-transfer coefficient h_c corresponds to the Nusselt number in Colburn's equation[72]

$$\mathrm{Nu} = 0.023\,\mathrm{Re}^{0.8}\,\mathrm{Pr}^{0.33} = \frac{h_c d}{\lambda_g} \tag{9-244}$$

where d is the port hydraulic diameter, and the dimensionless numbers are based on the free-stream gas properties. T_g in Eq. (9-243) is the recovery temperature of the gas, and T_s is the surface temperature of the condensed phase.

The surface reaction processes were modeled on the assumptions of heterogeneous surface reactions and chemical equilibrium between the metal and the oxidizer at the gas–metal interface. Thus, the entire process is controlled by the diffusion rate of the oxidizer through the boundary layer to the wall. The equilibrium partial pressure and the mole fraction of the oxidizer at the wall were related to the surface temperature through the appropriate chemical reaction.

When the reactive gas is oxygen, the chemical reaction with the metal (represented by the symbol Me) has the following form:

$$\frac{2x}{y}\mathrm{Me} + \mathrm{O}_2 \rightleftharpoons \frac{2}{y}\mathrm{Me}_x\mathrm{O}_y \tag{9-245}$$

and the expression for the equilibrium constant K_p is

$$K_p = \frac{1}{p_{\mathrm{O}_2}} \tag{9-246}$$

when both the metal and the oxide are in the condensed phase. The tendency for the reaction to occur is expressed by the Gibbs free energy ΔG, which is temperature-dependent. The equilibrium constant is related to ΔG through Eq. (1-165):

$$\Delta G = -R_u T \ln K_p \tag{9-247}$$

Gany et al.[69] found that for reactions of Fe, Al, Ti, or Mo with oxygen, the values of ΔG are highly negative, and hence the values of K_p are very high in the temperature range up to the melting point. These high values indicate that the reaction is basically unidirectional in the temperature range of their investigation and that the equilibrium oxygen partial pressure at the surface is comparatively low.

The mass-transfer coefficient h_m was obtained from the Sherwood number, using

$$\mathrm{Sh} = 0.023\,\mathrm{Re}^{0.8}\,\mathrm{Sc}^{0.33} = \frac{h_m d}{\mathscr{D}} \tag{9-248}$$

The Sherwood number in mass transfer is analogous to the Nusselt number in heat transfer. The molar diffusivity of the oxygen, $\mathscr{D}$, has a complex dependence on molecular interaction and collision functions. Gany et al.[69] used the following formula:

$$\mathscr{D} = 3.85 \times 10^{-4}\frac{T^{1.5}}{pF_c} \tag{9-249}$$

with $\mathscr{D}$ in m^2/sec, T in K, and p in N/m^2. The collision function F_c was approximated by

$$F_c = 1.76 - 0.332\log_{10}T \tag{9-250}$$

The mass flux of the oxidizer to the wall (moles per unit area and time) was evaluated from

$$J_{\mathrm{Ox}} = h_m C_{\mathrm{Ox},\infty}(Y_{\mathrm{Ox},\infty} - Y_{\mathrm{Ox},s}) \tag{9-251}$$

where $C_{\mathrm{Ox},\infty}$ is the molar concentration of oxidizer in the free stream. Chemical equilibrium (or near-equilibrium) conditions at the wall indicate that the heat flux to the wall due to surface chemical reaction, q_r, is directly dependent on the oxidizer mass flux to the wall:

$$q_r = J_{\mathrm{Ox}}(-\Delta H_r)\nu W_m \tag{9-252}$$

where ν is the stoichiometric ratio of metal to oxygen, W_m is the molecular weight of the metal, and ΔH_r is the enthalpy of reaction in joules per kilogram of metal.

The metal regression rate due to metal oxidation during the chemical reaction can be calculated from

$$r_b = \frac{q_r}{-\Delta H_r \cdot \rho_m} \tag{9-253}$$

where ρ_m is the density of the metal. For most situations when melting occurs, the actual mass loss is primarily the result of melting and not of consumption by surface reactions. Due to the high shearing forces of the flow, the melt is rapidly removed by flowing along the surface; thus, the thermal resistance attributable to the melt layer is small. Once the surface reaches the melting temperature, the temperature does not increase further (Fig. 9.24), and the excess of heat from the convective heat flux and chemical reaction causes

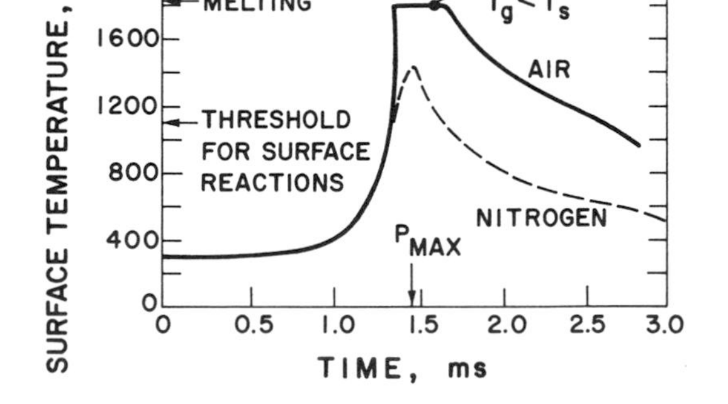

Figure 9.24 Surface temperature versus time, illustrating the onset of surface reactions and melting. Comparison of the results for air and N_2 illustrates the importance of considering the reaction of the AISI 4340 steel with oxygen. (After Gany et al.[69])

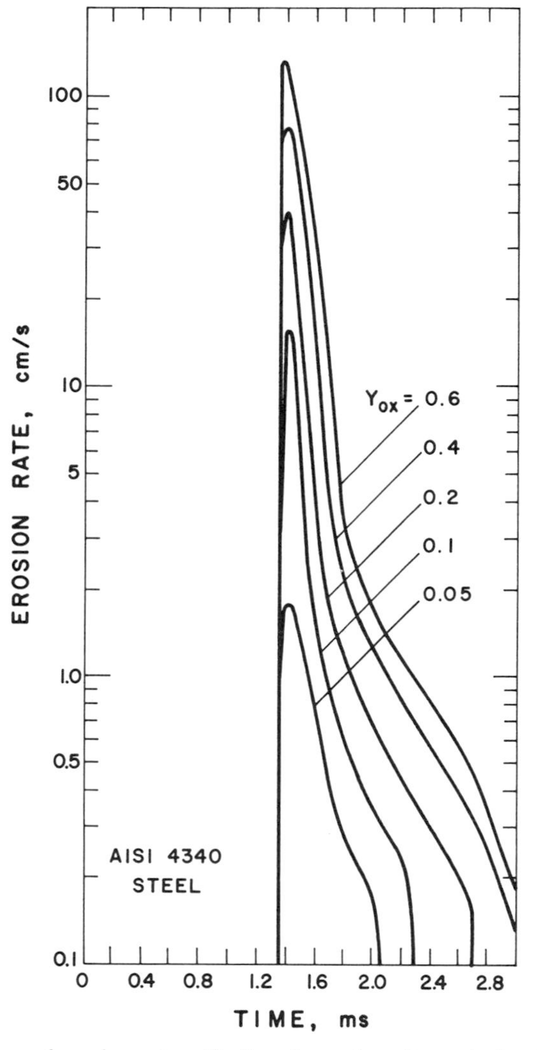

Figure 9.25 Variation of erosion rate with time for various O_2 mole fractions. The rapidly increasing erosion rates with increasing O_2 concentration is attributed to the exothermic surface reactions between Fe and O_2. (After Gany et al.[69])

melting of the metal surface. The erosion rate in this case is

$$r_b = \frac{1}{L\rho_m}\left[q_g + q_r - \lambda\left(\frac{\partial T}{\partial y}\right)_{y=0}\right] \tag{9-254}$$

Figure 9.25 shows the dependence of the erosion rate of AISI 4340 steel on oxygen concentration. As can be seen from this figure, higher oxidizer concentration in the gases leads to significantly enhanced rates of erosion. Accord-

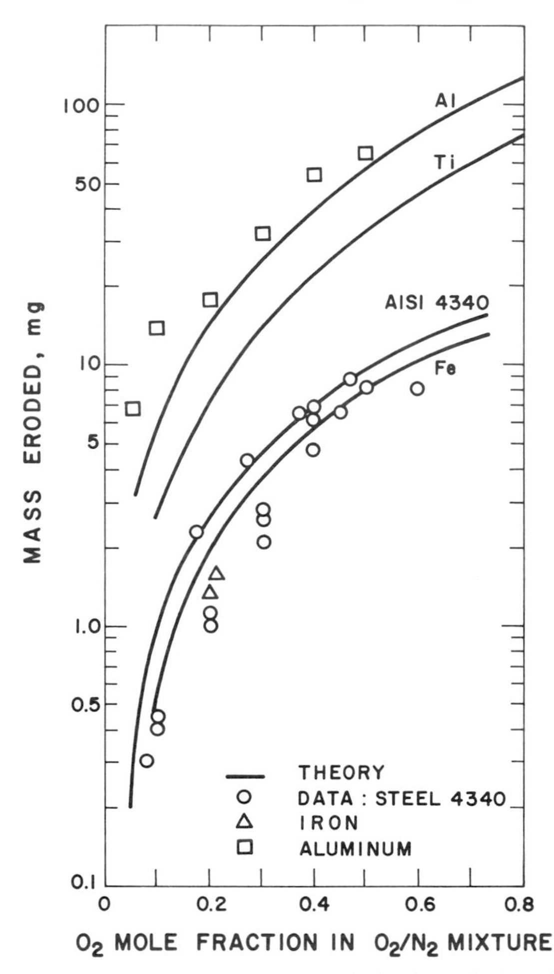

Figure 9.26 Comparison of the experimental data and the theoretical results for erosion of various metals shows that the theoretical model describes the erosion of steel as well as the considerably higher erosion rates of aluminum (after Gany et al.[69]).

ing to Gany et al.,[69] this is mainly attributed to the exothermic surface reactions between Fe and O_2, which greatly augment the surface heat flux and enhance the melting process. Their theoretical predictions of the total erosion are compared with the experimental data for several metals in Fig. 9.26. Good agreement was achieved both for aluminum, which is very reactive with oxygen, and for steel and iron, which are only moderately reactive. The high erosion of aluminum compared to that of steel is due to its high heat of reaction and low melting temperature (see Table 9.5). Titanium, whose melting point is high compared to that of aluminum, exhibits lower erosion, yet higher than that of steel because of its low thermal conductivity. No closely relatable experimental data are available for titanium; however, evidence of high erosion (about one-half of that of aluminum) was reported by Plett et al.[73]

TABLE 9.5 Thermophysical Properties of Some Metals and Alloys[a]

Metal or Alloy	Atomic Weight W_m (kg/kg-mole)	Density ρ (g/cm^3)	Melting Temp. T_{melt} (K)	Latent Heat of Fusion L (kJ/kg)	Enthalpy of Reaction with Oxygen, ΔH_r (10^3 kJ/kg)	To Form
Iron	55.847	7.87	1810	272	−4.73	FeO
Steel AISI 4340	—	7.86	1800	274	−4.73	FeO
Steel AISI 304	—	8.02	1700	274	−4.73	FeO
Aluminum	26.98	2.70	933	389	−31.00	Al_2O_3
Titanium	47.90	4.50	1953	419	−10.93	TiO
					−19.95	TiO_2
Molybdenum	95.94	10.24	2883	253	−5.80	MoO_2

[a]After Gany et al.[69]

As one can see from Fig. 9.26 and Table 9.5, the erosion process is affected by several parameters. With respect to metal properties, low thermal conductivity, melting point, and latent heat of fusion, as well as high heat of reaction and specific heat, all tend to increase erosion. However, the interplay among the various factors is complex. This complexity, according to Gany et al.,[69] is due partially to the strong temperature dependence of some of these parameters. In their study, the surface reactions controlled by gas-phase diffusion explain and correlate the erosion rates in oxygen–metal systems. Erosion is the result of both surface reaction and melt removal by the flowing gases. The surface reactions contribute in two ways to the mass removal: (1) below the melting point the mass loss is due to oxidation and subsequent removal of the oxide, and (2) the oxidation is exothermic and thus augments the heat-transfer rate. The heat transfer produced by the surface reactions under high oxidizer concentrations exceeds the forced convective heating and thus greatly enhances the melting.

In a more recent work, Gany and Caveny[70] refined the theoretical model by considering the flow of a liquid metal layer under high shear force acting on the gas–liquid interface. They further concluded that heterogeneous surface reactions rather than homogeneous gas-phase reactions are established even for metals that would burn in the vapor phase in low flow conditions. Gas-phase reactions which require evaporation of the metal cannot explain the high regression rates observed in experiments. Surface reactions between metals and highly reactive oxidizing gases are gas-phase diffusion-controlled. Below the melting point of the metal, mass loss is due to direct chemical attack. If the metal surface achieves the melting temperature, melting and melt removal can be the main cause of erosion.

Reasonable agreement between calculated results from the theoretical models developed by Gany et al.[69,70] and the experimental data indicates that the most important processes have been taken into account. More accurate predictions may be obtained with further consideration of the flow-field structure of the gas phase, the radiative heat transfer to and from the gas–liquid interface, and empirical correlations which are specially suited to the developing (entrance) region rather than to fully developed turbulent flows in a tube.

7.5 Aerothermochemical Erosion of Graphite Nozzles of Rocket Motors

Carbon–carbon (C/C) composites and other graphitic materials have found increasing use in the manufacture of nozzles for solid-propellant rocket motors. However, some complex problems associated with the use of C/C materials have yet to be solved. One of these problems is the recession of the C/C nozzle during motor operation. The performance of a rocket motor is lowered by the increase in the throat area and the nozzle surface roughness caused by the recession of the C/C composite material. This reduction in performance must be estimated and incorporated into the motor design. One way to determine performance reduction is to conduct test firings of a full-scale motor; however, this requires considerable time and expense. A more efficient and economical way is to accurately model the various aspects of the recession process and predict the recession rate. This section describes the comprehensive model developed by Kuo and Keswani,[14] their major results, and a comparison with experimental data.

During the last few years, the recession behavior of C/C materials has been the subject of many investigations. Geisler[74] measured the recession rate of C/C composites and bulk graphites as a function of aluminum content of the propellant. Geisler found that the recession rate was proportional to the concentration of H_2O, OH, and CO_2 and inversely proportional to the H_2 concentration. Geisler also concluded that chemical attack was the only significant mechanism for throat recession. Klager[75] conducted experimental firings which demonstrated that the primary cause of graphite removal is chemical attack by gases such as H_2O and CO_2. He reported that the chamber temperature showed no correlation with the recession rate, while the chamber pressure strongly influenced the recession rate.

The objective of Kuo and Keswani's study was to predict the recession rate of C/C nozzles and its dependence on the composition of the solid propellant, the motor chamber pressure, the bulk density of the C/C material, and the surface roughness of the nozzle. Also, it is important to determine whether the process of nozzle recession is controlled by diffusion or by chemical kinetics.

As the propellant in the rocket motor burns, the C/C composite nozzle is exposed to the hot combustion gases, which form a turbulent boundary layer over the nozzle surface. Reactants like H_2O and CO_2 diffuse across the boundary layer to the nozzle surface, where heterogeneous reactions occur, causing the surface to recede. The recession rate of the carbon is influenced both by chemical kinetics and diffusion. If the kinetic rates are very high, then the recession rate will be determined solely by the diffusion process. Let this recession rate of carbon be denoted by $\dot{r}_{C,d}$. The other limiting case is that of high diffusion rates and low kinetics rates, in which the recession rate is solely determined by the chemical kinetics. Let this recession rate be denoted by $\dot{r}_{C,ch}$.

In order to account for both chemical kinetics and diffusion, the following conventionally used formula was employed in the analysis:

$$\dot{r}_C = \left[\frac{1}{\dot{r}_{C,d}} + \frac{1}{\dot{r}_{C,ch}}\right]^{-1} \tag{9-255}$$

It can be seen from Eq. (9-255) that if either of $\dot{r}_{C,d}$ or $\dot{r}_{C,ch}$ is much larger than the other, the smaller one will be rate-controlling.

The diffusional recession rate $\dot{r}_{C,ch}$ depends on the kinetics of the heterogeneous reactions, the concentrations of the various reactants, and the surface temperature of the nozzle. The surface temperature was determined by the transient heat-conduction response of the C/C nozzle and the heat transfer from the hot gases to the nozzle.

The physical model considered in the theoretical analysis consisted of an axisymmetric C/C nozzle insert through which high-temperature gases expand from low subsonic to high supersonic flow condition. The flow in the core region was considered to be one-dimensional, compressible, inviscid, and steady. The flow near the nozzle surface was considered to be axisymmetric, viscous, and quasisteady. The heat-conduction process inside the C/C material was considered to be axisymmetric and unsteady. The nozzle material properties were treated as functions of the local temperature.

The formulation of the theoretical model involves the general conservation equations for the gas phase and the solid phase. The gas-phase conservation equations consist of equations for the viscous boundary-layer flow region and the inviscid core region; the solid-phase conservation equation is the transient heat-conduction equation for the C/C composite.

The conservation equations for the turbulent, compressible, reacting boundary layer are the same as those described in Section 7.3.2. One of the major assumptions made in the conservation equations for the boundary-layer flow region was the negligible effect of gas-phase reactions on the recession process. This assumption was justified as follows. Firstly, the concentration of O_2 in the propellant product gases is very small. This prevents the oxidation reaction of H_2 and CO, which are products of heterogeneous reactions, to form H_2O and CO_2. Therefore, the gas-phase reactions can hardly alter the concentration profiles of reactants like H_2O and CO_2. Secondly, the mass generation and consumption rates of species concentrations in the core region are not influenced by the heterogeneous reactions at the nozzle surface.

For the case of diffusion-controlled recession (where the chemical reaction rate is extremely high), the concentration of reactant i at the gas–solid interface is vanishingly small:

$$(\tilde{Y}_i)_g = 0 \tag{9-256}$$

This permits the solution of the species equations and the calculation of $\dot{\omega}_i$ using the following mass-flux balance condition at the solid–gas interface:

$$(\bar{\rho}\tilde{v}\tilde{Y}_i)_g - \left(\bar{\rho}\mathscr{D}\frac{\partial Y_i}{\partial y}\right)_g = \dot{\omega}_i \tag{9-257}$$

If the reaction between a reactant B_i and carbon can be represented as

$$\nu_{C,i}C + \nu_i B_i \to \sum_j \nu_j P_j \tag{9-258}$$

where j represents the jth species in the combustion product. Then it can be shown readily that the recession rate of carbon can be expressed as

$$\dot{r}_{C,d} = \frac{1}{\rho_C}\sum_i \dot{\omega}_i \frac{\nu_{C,i}W_C}{\nu_i W_{B_i}} \tag{9-259}$$

The combustion products from the rocket motor are composed of a variety of gases such as H_2, CO, H_2O, HCl, and CO_2 among others. Klager[75] has shown that the most reactive gases are those which react with carbon to form carbon monoxide. Gases such as H_2O, CO_2, O_2, O, and NO react with carbon to form CO. However, O_2, O, and NO are present in very low concentrations in typical propellant combustion products. Thus, H_2O and CO_2 are considered to be the only major reactive species, and they are assumed to react with carbon as follows:

$$C_{(s)} + H_2O_{(g)} \to CO_{(g)} + H_{2(g)} \tag{9-260}$$

$$C_{(s)} + CO_{2(g)} \to 2CO_{(g)} \tag{9-261}$$

According to Libby and Blake,[76] the reaction rates for the above reactions at high temperatures are approximately equal. They provide the following expression for the reaction rate of CO_2 and H_2O with carbon:

$$\dot{r}_{C,i} = \frac{A_s p_i}{\rho_C}\exp\left(-\frac{E_{a,s}}{R_u T_s}\right) \tag{9-262}$$

where the activation energy $E_{a,s}$ and the preexponential factor A_s are 41.9 kcal/mole and 2470 kg/(m^2/sec/atm), respectively. Using Eq. (9-262), the recession rate of carbon in the limit of chemical-kinetics control can be

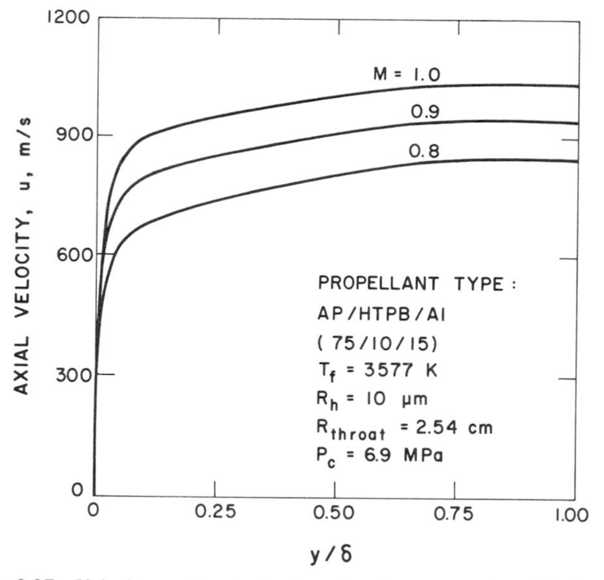

Figure 9.27 Velocity profiles in the boundary layer at various Mach numbers.

expressed as

$$\dot{r}_{C,\,ch} = \sum_i \dot{r}_{C,\,i} \tag{9-263}$$

The near-wall treatment of k and ε used in the analysis is the same as that described in Section 7.3.6. The model was solved by the solution procedure described in Section 7.3.8. The solution of the gas-phase equations permits the calculation of the recession rate for the diffusion-control limit, $\dot{r}_{C,d}$. The calculation of the recession rate for the chemical-kinetics control limit, $\dot{r}_{C,ch}$, required the solution of the transient heat-conduction equation to determine the surface temperature. Sample calculations of the heat-conduction equation show that the nozzle surface temperature quickly rises to 2800 K, after which it remains nearly constant. Thus, the gas-phase equations are solved for a nozzle surface temperature of 2800 K. This permits the calculation of $\dot{r}_{C,d}$, which is nearly independent of changes in the nozzle surface temperature. The solution of the gas-phase equations determines the magnitude of the heat-transfer coefficient. This heat-transfer coefficient is used in the boundary condition of the transient conduction equation. The solution of the conduction equation provides the surface temperature needed for the calculation of $\dot{r}_{C,ch}$ according to Eq. (9-263). Finally, the recession rate is calculated using Eq. (9-255).

Figure 9.27 shows the computed velocity profiles at various Mach numbers. These profiles indicate higher velocity gradient near the wall for higher Mach number. The temperature and turbulent kinetic energy profiles of the nozzle

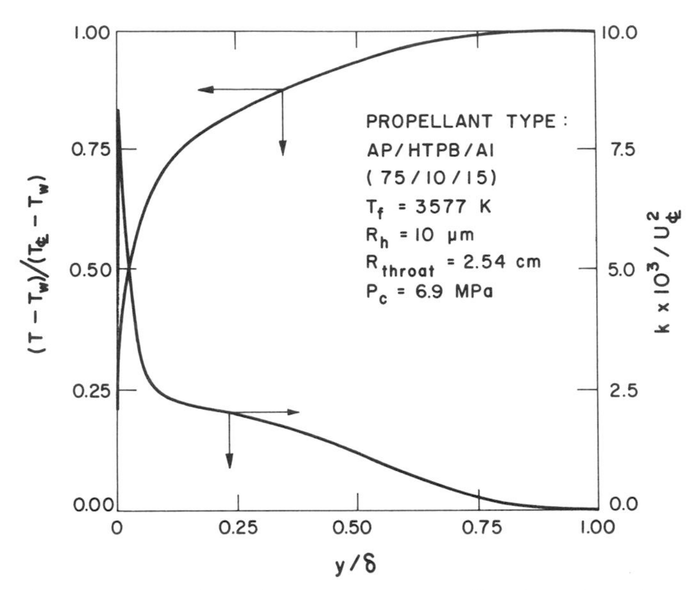

Figure 9.28 Temperatures and turbulent-kinetic-energy profiles in the boundary layer at the throat location.

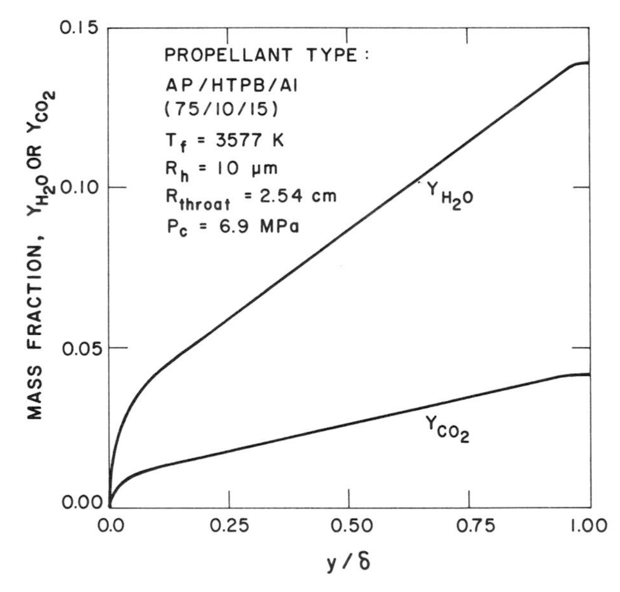

Figure 9.29 Concentration profiles of H_2O and CO_2 in the boundary layer at the throat location.

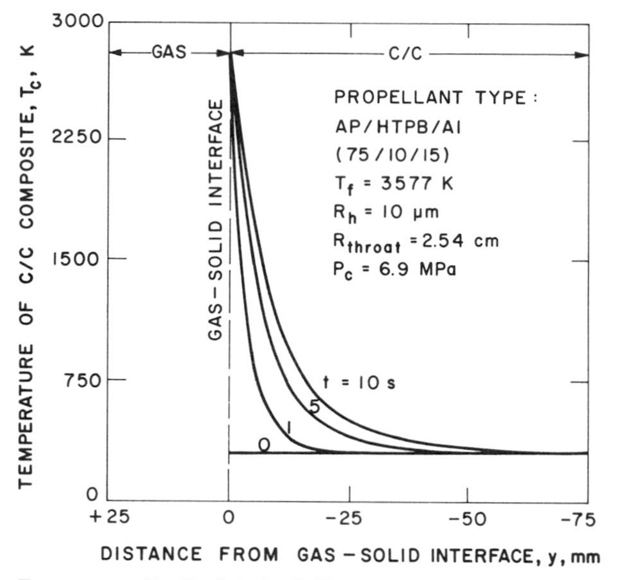

Figure 9.30 Temperature distributions in C/C composite at the throat for various times.

throat are plotted in Fig. 9.28. As clearly seen from the plot, the temperature gradient is highest near the nozzle surface. The predicted turbulent kinetic energy rises sharply from the surface and reaches a peak value very close to the wall, where the shear stress is the highest, as governed by the velocity profile shown in Fig. 9.27. The turbulent kinetic energy decreases abruptly as the velocity gradient decreases, and goes to zero near the edge of the boundary layer where the velocity gradient is negligible. The concentration profiles of H_2O and CO_2 at the nozzle throat are shown in Fig. 9.29. The mass fractions of these two species decrease from the edge of the boundary layer to the nozzle surface, due to their consumption in the heterogeneous reactions at the surface. The gradients of the concentrations are largest near the wall, since the turbulent mass diffusivity drops to zero as the viscous sublayer is approached near the wall, resulting in much lower effective mass diffusivity there.

Figure 9.30 shows the temperature profile in the C/C nozzle at 0, 1, 5, and 10 sec. Near the gas–solid interface, there exist large gradients in the temperature profiles which may lead to severe thermal stresses in the nozzle material. There is considerable temperature rise in the material below the surface; for example, at a distance of 1 cm, the temperature reaches 1000 K at $t = 10$ sec.

Figure 9.31 shows the nozzle surface temperature as a function of time for varying aluminum content of the propellant. The plot shows that the nozzle surface temperature for all cases rises very rapidly, reaching a temperature of 2400 K in less than 0.1 sec. Once the surface has reached a sufficiently high temperature, heterogeneous reactions which are endothermic become significant. This results in a lower rate of increase of the surface temperature. The

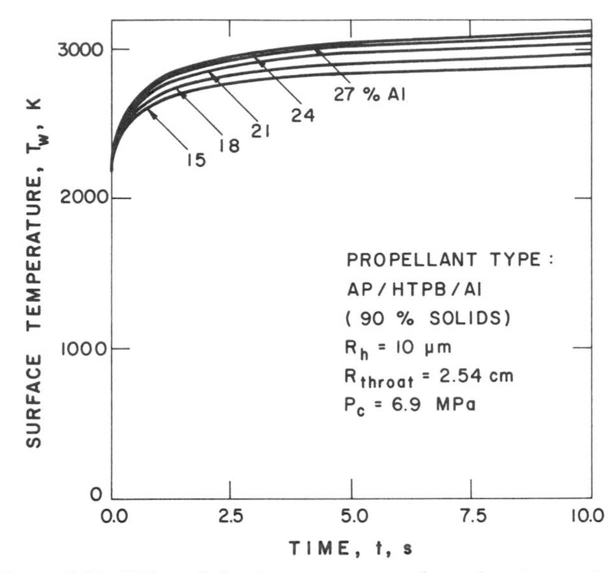

Figure 9.31 Effect of aluminum content on the surface temperature.

calculated surface temperature is higher for propellants with higher aluminum content. The greater the aluminum content of a propellant, the lower the concentration of oxidizing species, such as H_2O and CO_2. And since the recession of the nozzle surface occurs due to the oxidation of carbon to CO, the propellants which have higher aluminum content should result in lower endothermic surface reactions, giving higher surface temperatures.

Figure 9.32 shows the recession rate at the throat as a function of time for propellants with varying aluminum content. It can be seen that the recession rate decreases significantly as the aluminum content of the propellant is increased. This decrease is due to the lower concentration of H_2O and CO_2 with increasing aluminum content.

It can also be seen from Fig. 9.32 that the recession rate rises quickly and then increases very slowly with time. The region of rapid rise in the recession rate corresponds to the time period of chemical-kinetics control of the recession process. However, as the recession event progresses and the nozzle surface temperature increases, the chemical-kinetic rates increase exponentially and the recession rate is controlled by the diffusion process, which is relatively independent of the surface temperature. Therefore, as time increases, the recession rate approaches the diffusion-controlled recession rate and hence remains relatively constant. Figure 9.32 also reveals that the recession rate for a propellant with a higher aluminum content approaches the diffusion-controlled recession rate faster. This is explained by the faster surface-temperature rise for propellants with higher aluminum content as shown in Fig. 9.31. This reduces the time during which the chemical kinetics dominate, so that the recession process approaches the diffusion-controlled stage faster.

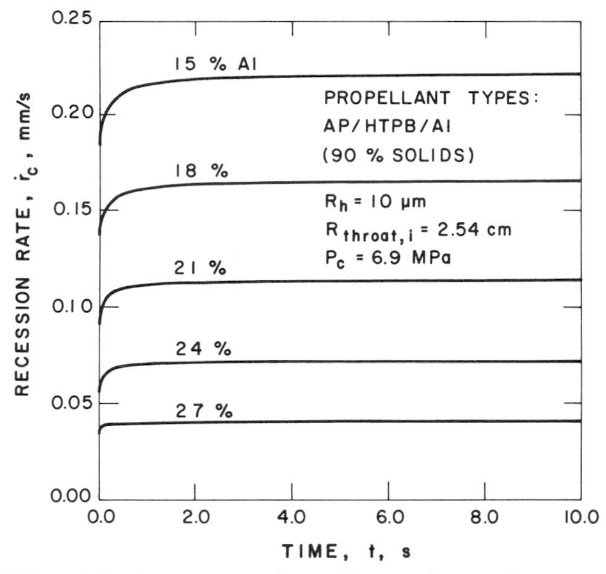

Figure 9.32 Effect of aluminum content of propellant on the recession rate at the throat.

The total recession at the throat as a function of time and aluminum content of the propellant is shown in Fig. 9.33. The data obtained by Geisler for similar operating conditions are also plotted on the figure. There is good agreement between the measured data and the predicted recession except for the propellant with an aluminum content of 21%. The difference between predicted results and the experimental data is within the range of repeatability of recession experiments.

Figure 9.34 shows the effect of surface roughness on the recession rate as a function of time at the nozzle throat. Numerical solutions have been plotted for equivalent sand-grain roughnesses (not the physical roughness of the surface) of 10, 20, and 30 μm. These equivalent roughness heights are quite typical, as revealed by the postfiring examination of C/C composites reported by Frankle and Lebiedzik.[77] An increase in the nozzle-surface equivalent sand-grain roughness from 10 to 20 μm causes a 14% increase in the recession rate at time 10 sec. The surface roughness enhances the mass transfer rate of oxidizing species to the nozzle surface, resulting in an increase in the recession rate.

Figure 9.35 shows the influence of chamber pressure on the instantaneous recession rate at the nozzle throat. The plotted results show that chamber pressure has a very strong influence on the recession rate. A change in the chamber pressure from 6.90 MPa (1000 psi) to 2.76 MPa (400 psi) causes a decrease in the recession rate of about 60% at times greater than 2 sec. A reduction in pressure causes a reduction in the density of the gas phase. The mass transfer rate across the turbulent boundary layer is directly proportional

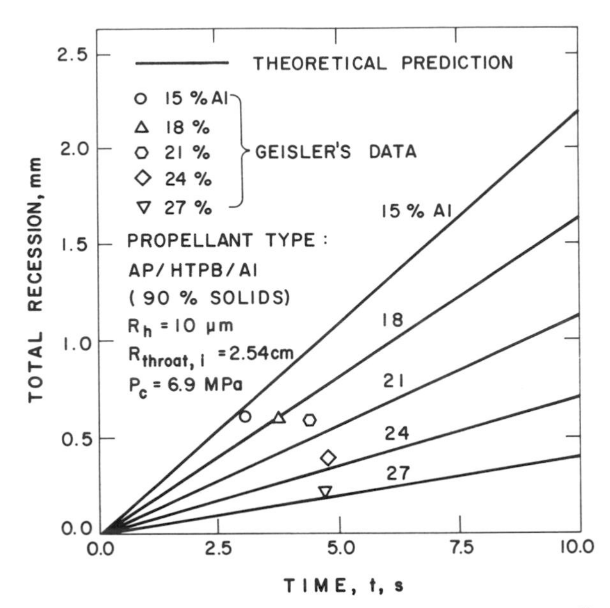

Figure 9.33 Total recession at the throat as a function of time for propellants with different aluminum content.

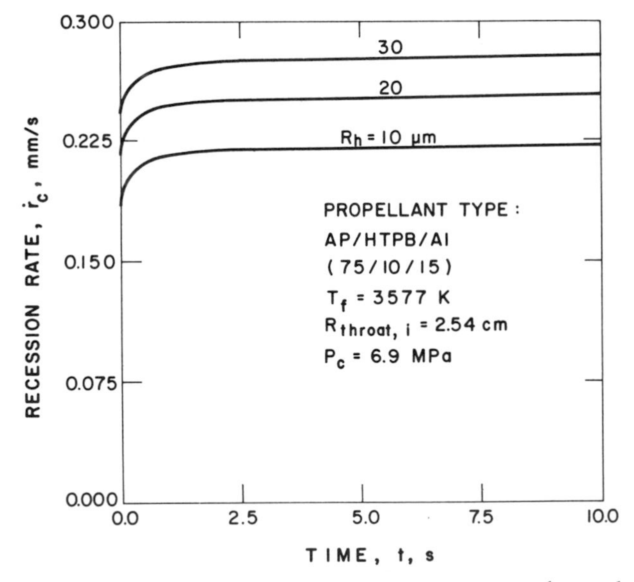

Figure 9.34 Effect of surface roughness on recession rate at the nozzle throat.

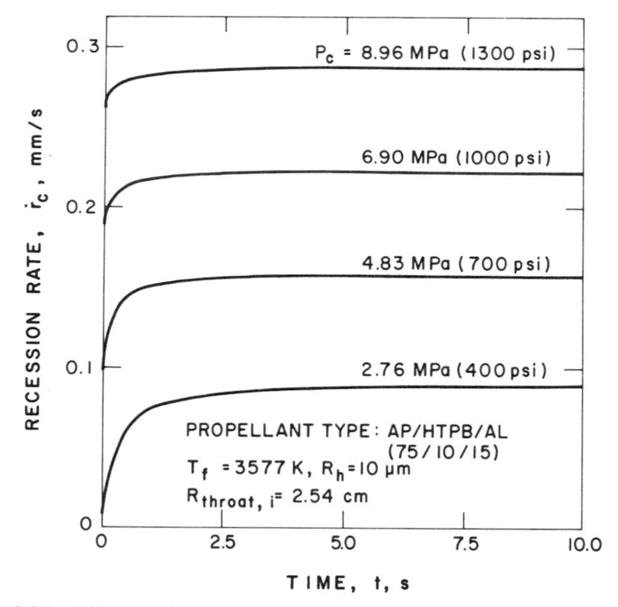

Figure 9.35 Effect of chamber pressure on recession rate at the nozzle throat.

to the density. Hence, a decrease in pressure results in a decreased rate of supply of oxidizing species to the nozzle surface. Hence, pressure has a strong influence on the recession rate. One other interesting point to note is the slower rate of increase in recession rates at lower pressures. This is caused by a slower rate of surface temperature rise due to reduced heat-transfer coefficients at lower pressures.

Many investigators in the past have correlated the recession rate with the density of the C/C composite material or bulk graphite showing higher r_C with lower ρ_C. The model developed by Kuo and Keswani[14] predicts that the recession rate will be inversely proportional to the density of the nozzle material. This is evident from examining the expressions for $\dot{r}_{C,d}$ and $\dot{r}_{C,ch}$ [see Eqs. (9-259), (9-262), and (9-263)], both of which have the density of the nozzle material in the denominator. This explains the experimentally determined inverse relationship between the nozzle-material density and the recession rate.

The calculations show that chemical kinetics influence the recession rate only during a short time interval at the beginning of the recession event. After the initial phase of surface temperature increase, the recession rate is controlled mainly by diffusion. Since the diffusion process is not influenced by the thermophysical properties of the solid, C/C composite materials and various bulk graphites should have nearly equal recession rates after the short chemical-kinetics-controlled time interval, if their densities are equal. This finding is in agreement with Geisler's experimental observation that C/C and bulk graphite have similar thermochemical recession rates at the same densities.

7.6 Turbulent Wall Fires

An important topic in fire-safety research is the investigation of fire plumes. A *fire plume* is the region of hot gases above a fuel source at a fixed position. Fire plumes are classified into two categories:

A. Free Plumes. Hot gases originating at the source of the fire move upward unimpeded. There are no boundaries to influence the flow pattern and temperature distribution. Examples are a pan of oil burning in the open air, or a forest fire.

B. Confined plumes. There is an interaction between plume gases and contact surfaces. The flow field is affected by the presence of solid boundaries. Examples are a wall plume resulting from a fire at the base of a vertical wall, or a ceiling plume resulting from a fire located under a large horizontal ceiling.

This section addresses solely the problem of confined plumes near a vertical wall. On this topic, Ahmad and Faeth[78,79] have made a significant contribution in considering a turbulent wall fire consisting of chemically reacting flow above a fire located at the base of a vertical wall. The wall plume is of fundamental importance in understanding confined plumes, since it combines plume behavior and wall boundary-layer effects. The wall plume also has practical significance for the problems of upward fire spread on surfaces and preheating of unburned fuel elements in a fire environment.

Four major regions can be distinguished in a fire on a vertical surface, as shown in Fig. 9.36. The characteristics of each region are as follows:

Region 1: Pyrolysis Zone. In this region, the surface material is decomposing to provide combustible gases. The fire extends over the pyrolyzing surface, oxidizing a portion of the fuel gases. Heat feedback from the fire to the surface sustains the decomposition process by providing energy of gasification of the fuel, as well as energy to compensate for heat losses from the pyrolysis zone, so that temperatures can be maintained high enough for significant pyrolysis to occur.

Region 2: Combusting Plume. Oxidation of all the fuel produced in the pyrolysis zone is completed in the combusting plume; however, pyrolysis does not occur at the wall surface and no additional fuel is produced. Heat transfer from the flame to the surface completes preheating of the wall material to the point at which significant pyrolysis occurs.

Region 3: Strongly Buoyant Plume. Although no additional combustion occurs in this region, gas temperatures are high. Therefore, variable-property effects are important in this region. Heat transfer from the plume to the wall continues the preheating process.

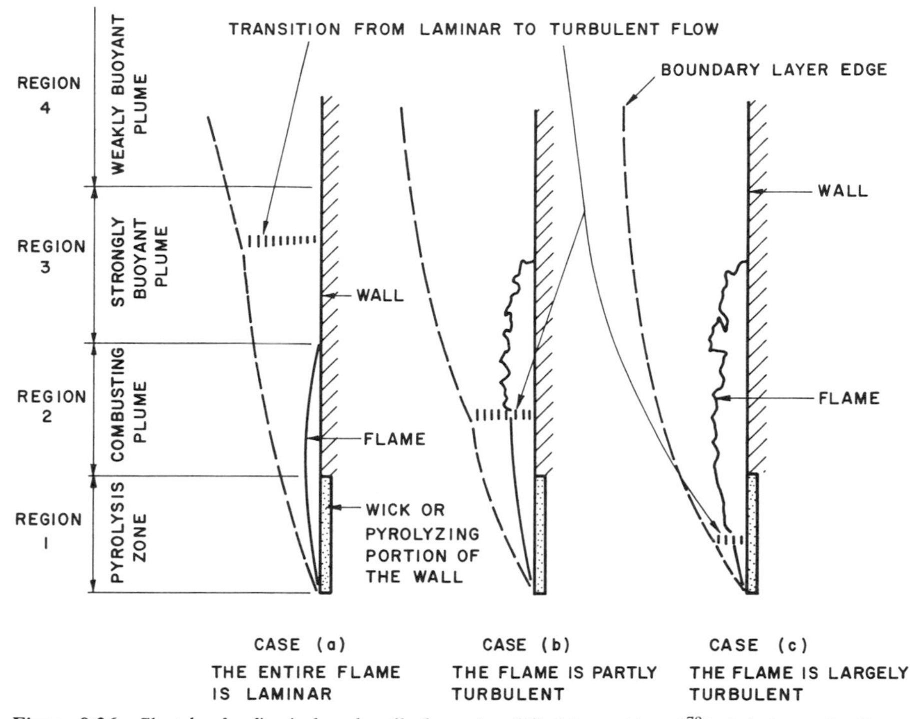

Figure 9.36 Sketch of a fire-induced wall plume (modified from Ahmad[79]): (*a*) the entire flame is laminar; (*b*) the flame is partly turbulent; (*c*) the flame is largely turbulent.

Region 4: Weakly Buoyant Region. This is the uppermost region of the flow, where entrainment of ambient gases has cooled the plume. Variable properties and radiation are not significant, except for the treatment of buoyant forces.

In Fig. 9.36, there are three cases. In case (a), the transition from laminar to turbulent flow takes place in the noncombusting region and the entire flame is laminar. In case (b), the transition occurs in the combusting portion of the plume, and the flame is partly turbulent. In case (c), the transition takes place near the leading edge of the pyrolysis zone, and the flame is largely turbulent. In their attempt to understand the behavior of the wall plumes, Ahmad and Faeth[78,79] considered the two limiting cases (a) and (c).

Prior to the work of Ahmad and Faeth,[78,79] the laminar pyrolysis zone of a wall fire had been investigated by Kosdon, Williams, and Buman.[80] They developed a similarity theory for the laminar natural-convection boundary layer along a vertical wall which considered property variations, mass addition in the boundary layer at the wall, and boundary-layer combustion. The study assumed an infinitely thin diffusion flame and no gas radiation. Experimental measurements of the standoff distance of the flame enveloping burning α-cellulose cylinders agreed reasonably well with the theoretical calculations.

Kim, de Ris, and Kroesser[81] completed an analysis of the laminar pyrolysis zone of a free-burning fuel surface, similar to that of Kosdon et al.[80] They solved the conservation equations numerically for a wide variety of fuels. Theoretical predictions of the burning rate agreed reasonably well with measurements, particularly for fuels with low molecular weights, whose properties most closely correspond to the assumptions of the variable-property model. Their results confirm that the only important dimensionless parameter governing the fluid-mechanical behavior is the Grashof number, while the chemical effects are primarily controlled by the mass-transfer driving potential B and a stoichiometry parameter composed of the ambient oxygen concentration, stoichiometric coefficients, and molecular weight of fuel and oxidizer. Blackshear and Murty[82] studied rates of heat and mass transfer to, from, and within freely burning cellulosic solids. They discovered a strong dependence of the burning rate on size for cellulosic cylinders of various diameters.

The turbulent pyrolysis zone of wall fires has been extensively studied by Markstein and de Ris[83] and Orloff et al.[84,85] This group of researchers studied experimentally the upward spread of turbulent wall fires. They found that the fire spread rate accelerates to an asymptotic value, the acceleration being characterized by a power-law relationship between the pyrolysis spread rate and the length of the pyrolyzing zone. Upward spread of turbulent fires over a fuel surface was studied in greater detail by Orloff et al.,[84] and their results indicate that the flame radiation accounts for 75–80% of the total heat flux to the solid-fuel (polymethylmethacrylate) surface in some cases. de Ris and Orloff[85] performed experiments to study the role of buoyancy direction and radiation in turbulent diffusion flames over surfaces. They report that for a free-burning vertical surface, the radiant heat flux to the surface is smaller than the outward flux to the surroundings, possibly because of absorption by combustion products and intermediates of reaction near the fuel surface. They also found that the ratio of radiant to total heat transfer to the fuel surface increases with increasing mass-transfer driving potential B.

In the study of strongly buoyant plumes, Pagni and Shih[86] solved the problem of laminar free- or forced-convection burning of synthetic polymer slabs. Using initial conditions for the pyrolysis-zone results and adopting an integral scheme proposed by Yang,[87] they formulated integral solutions for the overfire region. The results of their analysis were limited to predictions of flame length, although the model is capable of calculating other quantities as well.

Liburdy and Faeth[88] theoretically investigated a steady, laminar thermal plume resulting from a line heat source located along the base of an adiabatic vertical wall. They formulated a variable-property solution for velocity and temperature profiles over a wide range of Prandtl numbers. They compared the results with unconfined laminar line plumes having the same energy flow per unit length as the wall plume, and found that in laminar flow, wall thermal plumes above a line source and free line plumes are governed by the same scaling laws. The results also indicated that the wall plume has a lower maximum velocity and a higher maximum temperature than the corresponding free plume.

Turbulent buoyant diffusion flames along walls have been theoretically investigated by Plumb and Kennedy.[89] They modified a k–ε–g model originally proposed by Spalding[90] to include the effects of buoyancy on the generation of turbulent kinetic energy and turbulent dissipation. Because of lack of experimental flame data, they only compared the results with nonreactive flow data; they found good agreement in this case.

Lockwood and Ong[91] have presented results on the combustion of a vertical fuel jet along a vertical insulated wall. They employed the finite-difference method of Patankar and Spalding to analyze the combined free- and forced-convection boundary layer, taking into account the chemical reaction and variations in fluid properties. They also measured the flame length and the concentration of an inert contaminant along the wall in the overfire region and found the theory in reasonable agreement with the measurements.

Recently, Pagni and Shih[92] analyzed a wake turbulent flame downstream of a burning fuel slab using the k–ε–g model of turbulence and a flame-brush profile equation. They obtained a rounded temperature peak and overlapping reactant profiles, which are typical of turbulent combustion processes. Their results also indicate that due to stronger mixing, the fuel is consumed faster than in the laminar case, causing the turbulent flame to be shorter than the laminar flame.

In a study of weakly buoyant plumes, Grella and Faeth[93] measured the mean velocities and temperatures in a turbulent plume above a line heat source at the base of a vertical adiabatic wall. They obtained correlating expressions for the data by assuming streamwise similarity of mean flow quantities and a fixed entrainment constant and local skin-friction coefficient. Their results indicate that the rate of entrainment of the wall plume is substantially lower than that of unconfined line plumes, resulting in higher velocities and greater density defects (or temperature defects in a thermal plume) in the wall plume than in a free plume under comparable conditions. Local flow properties were found to depend only weakly upon the value of the skin-friction coefficient.

Faeth and his coworkers[94,95] also conducted an experimental investigation of turbulent thermal plumes along an isothermal wall. They measured profiles of mean velocity and the turbulent quantities and correlations such as u', v', w', T', $\overline{u'v'}$, $\overline{u'T'}$, and $\overline{v'T'}$ for different values of the thermal energy flux in the plume and distances above the source. They found that the isothermal wall plume is thicker than the natural-convection boundary layer and the adiabatic wall plume. The results also indicate that the isothermal plume has greater velocities and temperatures at a given condition than the free-line plume, because of its reduced entrainment, in spite of the losses to the wall. A local-similarity integral model was found to provide a reasonably good correlation of the mean flow properties. Further analysis by numerical solution of the turbulent boundary-layer equations, using a mixing-length model employed by Cebeci and Khattab[96] for turbulent natural-convection flows, provided good agreement with their measurements.[95]

The objective of Ahmad and Faeth's investigation[78] was to study the transport characteristics and structure of turbulent fires on vertical surfaces under natural convection and conditions. Both the pyrolysis zone and the wall-plume region above the pyrolysis zone were studied. Their aim was to see how well the numerical solutions of the boundary-layer equations and the integral models with their underlying assumptions compare with measurements.

7.6.1 Experimental Methods Used by Ahmad and Faeth

New measurements of wall fires were made. These measurements include burning rates in the pyrolysis zone; convective and radiative heat fluxes to the wall, and radiative heat fluxes to the surroundings, in the plume; and profiles of mean velocity, temperature, and concentrations in both regions of the flow. The measurements were then compared with integral models. The results in the noncombusting plume were also compared with a solution of the turbulent-boundary-layer equations, employing a mixing-length model.

The test arrangement of Ahmad and Faeth[78] was a vertical surface, 660 mm wide, with two side walls 178 mm wide extending the full length of the apparatus. A 1-mm-wide strip, extending 2 mm from the wall, tripped the flow at the base of the pyrolysis zone. The pyrolysis zone was simulated with fuel-soaked wicks. Burning rates were determined by measuring the weight loss of the wick after burning for a timed interval. The details of their experimental setup are given in Ref. 79.

Wall heat fluxes were measured by the transient heating of a copper wall, segmented into a number of blocks by slots milled into the rear surface. Measuring the rate of temperature rise of each block after ignition yielded the total wall heat flux as a function of position. The convection heat flux was determined by the transient heating of gold-plated plugs, which absorb little radiation. The radiant heat flux from the flame to the surroundings was measured with a gas-purged, water-cooled sensor mounted 20–40 mm from the surface, depending on the flame standoff distance.

Measurements of profiles of mean quantities required stable test conditions for long time periods. In this case, the wicks were continuously fed with fuel through a manifold and the wall was cooled with water, which circulated through coils mounted on the rear surface. Mean temperatures were measured with a 7.6-μm-diameter, butt-welded and silicone-coated Pt–Pt : 10% Rh thermocouple. Mean velocities were measured with a 1-mm-ID Pitot-static probe. Pressure differences were measured with an electronic manometer. Gas samples were obtained isokinetically through a 1-mm-ID water-cooled stainless-steel probe and analyzed with a gas chromatograph.

7.6.2 Development of the Ahmad–Faeth Correlation

The model developed by Ahmad and Faeth[78] considers a fire at the base of a flat wall which is inclined at an angle ϕ from the vertical. A pyrolysis zone of length x_0 extends along the surface from the lower edge of the wall.

The major assumptions of the analysis are as follows: the flow is a two-dimensional, steady, turbulent boundary layer; ambient conditions are constant; the molecular physical properties approximate a Howarth–Dorodnitzyn gas (i.e., $\rho\mu$, $\rho\lambda$, and Pr are constant); surface conditions and the energy of gasification are constant in the pyrolysis zone, and the wall heat flux from the flame provides the energy of gasification of the fuel (i.e., there is no external source of heat flux to the wall, such as an external radiant flux); the wall temperature in the plume region is constant and equal to the ambient temperature; radiation is neglected; and a one-step reaction is assumed to occur within an infinitely thin flame zone. The last two assumptions are particularly questionable due to the known importance of radiation in turbulent wall fires[97,98] and the unmixedness of turbulent combustion processes. Nevertheless this approach has been successful for similar problems.[98,99] The remaining assumptions have been used in many previous studies of laminar wall fires.

An integral model was developed to provide algebraic expressions for data correlation. In the noncombusting plume, an earlier solution of the turbulent-boundary-layer equations, which employs a mixing-length model of the turbulence, was compared with the measurements.

The integral equations for the turbulent compressible boundary layer were obtained following Morton.[100] This implies low turbulence intensities, so that products of fluctuating quantities are small in comparison with products of mean quantities, when integrated across the flow. The analysis was simplified by employing the variable-property transformation suggested by Marxman.[99]

A one-step reaction is assumed as follows:

$$\nu_F' F + \nu_O' O \rightarrow \nu_P'' P \tag{9-264}$$

For integral analysis of turbulent flow in the pyrolysis zone, empirical information is required to represent friction and heat and mass transfer at the wall. Mass transfer at the wall causes a blowing effect, which reduces wall friction from the values found for a nonevaporating surface. This is treated using the blowing correction for the skin-friction coefficients with and without mass transfer, suggested by Marxman[99] and others:

$$C_f = C_{f0} \frac{\ln(1 + B)}{B} \tag{9-265}$$

The transport of heat and mass at the wall was related to the shear stress through the Reynolds–Colburn analogy. The form suggested by Rotta[101] for cases where the turbulent Prandtl number is not unity was employed:

$$\frac{\mathrm{St}}{C_{f0}/2} = \xi_R \tag{9-266}$$

where ξ_R is Rotta's Reynolds analogy factor and St is the Stanton number. Existing measurements indicate $\xi_R < 1$ for wall flows. Here, ξ_R was left as a parameter used to fit the data, since the earlier results did not involve blowing, combustion, or the like, and different values were expected.

The details of the remaining portions of the analysis, profile-shape assumptions, integration, and so on, are reported in Ahmad's Ph.D. thesis.[79] Assuming completely turbulent flow from the lower edge of the wall, the average burning rate for a smooth wall was found to have the following relationship with other governing parameters:

$$\bar{\dot{m}}''x\,\mathrm{Pr}^{2/3}\mathscr{S}/\mu_\infty = 0.0430\xi_R^{0.8}\psi\,\mathrm{Ra}_x^{0.4} \tag{9-267}$$

where the parameters $\mathscr{S}$, η_f, and r are defined as

$$\mathscr{S} \equiv \left[\frac{1+B}{B\ln(1+B)}\right]^{1/2}\left[\frac{1+0.5\,\mathrm{Pr}/(1+B)}{3(B+\tau_0)\eta_f+\tau_0}\right]^{1/4} \tag{9-268}$$

$$\eta_f \equiv 1 - \left[\frac{r(B+1)}{B(r+1)}\right]^{1/3} \tag{9-269}$$

$$r \equiv \frac{Y_{O\infty}W_F\nu_F'}{Y_{FT}W_O\nu_O'} \tag{9-270}$$

where Y_{FT} represents the fuel mass fraction of pyrolyzing material, and B is the mass-transfer driving potential defined as

$$B \equiv \frac{Y_{O\infty}Q/(W_O\nu_O') - h_{w0}}{L} \tag{9-271}$$

In the above equation, Q represents the heat of reaction per ν_F' moles of fuel, L is the effective heat of vaporization, and h_{w0} is enthalpy of the wall in the pyrolysis zone.

The parameter ψ in Eq. (9-267) is defined as

$$\psi \equiv \mathrm{Pr}^{4/15}\left[\frac{(1+B)\ln(1+B)}{B}\right]^{0.3}\left[\frac{3(B+\tau_0)\eta_f+\tau_0}{1+0.5/\xi_R(1+B)}\right]^{0.4} \times \left[\frac{1+0.5\,\mathrm{Pr}/(1+B)}{3(B+\tau_0)\eta_f+\tau_0}\right]^{1/4} \tag{9-272}$$

and the Rayleigh number Ra_x in Eq. (9-267) is defined as

$$\mathrm{Ra}_x \equiv \mathrm{Gr}_x \mathrm{Pr} \tag{9-273}$$

The parameter ψ is approximately equal to 1.15 for a wide range of fuels and ambient conditions, and this constant value was employed in the following correlations.

The average burning rate for laminar flow was also found from an integral model, as follows:[79]

$$\frac{\bar{\dot{m}}'' x \,\mathrm{Pr}^{2/3} \mathscr{S}}{\mu_\infty} = 0.66\,\mathrm{Pr}^{-1/12} \mathrm{Ra}_x^{1/4} \tag{9-274}$$

The smooth-wall analysis was found by Ahmad and Faeth[78] to give the best results in the pyrolysis zone and was extended into the combusting plume region. No similarity can be applied to this region; hence the solution must be obtained by numerically integrating a system of ordinary differential equations for the velocity, the thickness of the boundary layer, and the species concentrations. The results of these calculations yield the wall heat flux, the flame length, etc. The detailed analysis is described in Ref. 79.

At large distances above the fire, the flow approximates a weakly buoyant plume on an isothermal surface. Local similarity analysis of this region has been completed.[95] The turbulence was represented by a mixing length model proposed by Cebeci et al.[102] for forced convection, which has successfully modeled natural-convection boundary layers.[96] The theoretical results suggest that the parameters

$$\frac{g\beta x^3 \,\Delta T}{\nu^3} \mathrm{Gr}_x^{*\,-2/3}, \qquad \frac{ux}{\nu} \mathrm{Gr}_x^{*\,-1/3} \tag{9-275}$$

are primarily functions of y/x, where x is the distance along the wall and Q_x is the local thermal energy flux in the plume. The modified Grashof number, Gr_x^*, is defined as

$$\mathrm{Gr}_x^* \equiv \frac{g\beta x^3 Q_x}{\rho c_p \nu^3} \tag{9-276}$$

where β is the coefficient of volumetric expansion. The analysis also yields the following expression for the wall heat flux:

$$\frac{x\dot{q}'' \mathrm{Pr}^{2/3}}{Q_x} = 1.9\,\mathrm{Ra}_x^{*\,-0.095} \tag{9-277}$$

The modified Rayleigh number Ra_x^* is defined as

$$Ra_x^* \equiv Gr_x^* Pr \tag{9-278}$$

Equation (9-277) and the two parameters in Eq. (9-275) were compared with the measurements of Ahmad and Faeth,[78] evaluating physical properties at the arithmetic-mean temperature of the flow for each value of x.

7.6.3 Results on Wall Fires

Figure 9.37 is an illustration of mean velocity, temperature, and concentration profiles at various points within an ethanol-fueled wall fire burning in air. The profiles are plotted as a function of y/x, which is an appropriate similarity variable for the weakly buoyant region of the plume,[93-95] but is not a similarity variable for other portions of the flow.[79]

In the combusting portions of the flow ($x/x_0 = 1$ and 2) there is significant overlap of the fuel and oxygen profiles, which is typical of turbulent combustion processes. Wall heat-flux profiles indicated (cf. Fig. 9.39 below) that combustion was complete in the range $x/x_0 = 3$–4 for ethanol wall fires under the present test conditions. Measurements of concentrations at $x/x_0 = 3.9$ showed only trace concentrations of ethanol. After combustion is complete, air diffuses toward the wall, and the concentration of combustion products continues to decrease: see the plot for $x/x_0 = 5.9$ in Fig. 9.37. Carbon monoxide is still present at the highest wall position, suggesting that wall quenching has increased the production of this toxic substance by the fire.

Gas temperatures are highest in the pyrolysis zone and monotonically decrease in both the combusting and noncombusting portions of the plume. This happens because the wall plume is not adiabatic. The temperature profile is inflected between the wall and the maximum-temperature position in the pyrolysis zone. This could be an effect of blowing by the transfer of mass at the wall, since it is not observed in other portions of the flow. Decomposition between the wall and the flame may also be a factor, since temperatures are highest in the pyrolysis zone.

Maximum gas velocities were found to increase with distance into the pyrolysis zone, which agrees with theory.[79] Velocities vary more slowly in the plume region and tend to decrease in the noncombusting portion of the flow. The latter behavior agrees with the characteristics of a weakly buoyant wall plume, where the maximum velocity decreases as the plume thermal energy flux is reduced by heat loss to the wall.

The mass burning rates measured by Ahmad and Faeth[78] for methanol, ethanol, and 1-propanol, along with data from a number of other sources,[81,97,84,85,98,79] are illustrated in Fig. 9.38. The measurements are summarized according to fuel type, pressure (taken to be atmospheric unless indicated), and the ratio of the convective to the total heat flux from the flame (in those instances where an estimation of this quantity is available). In each

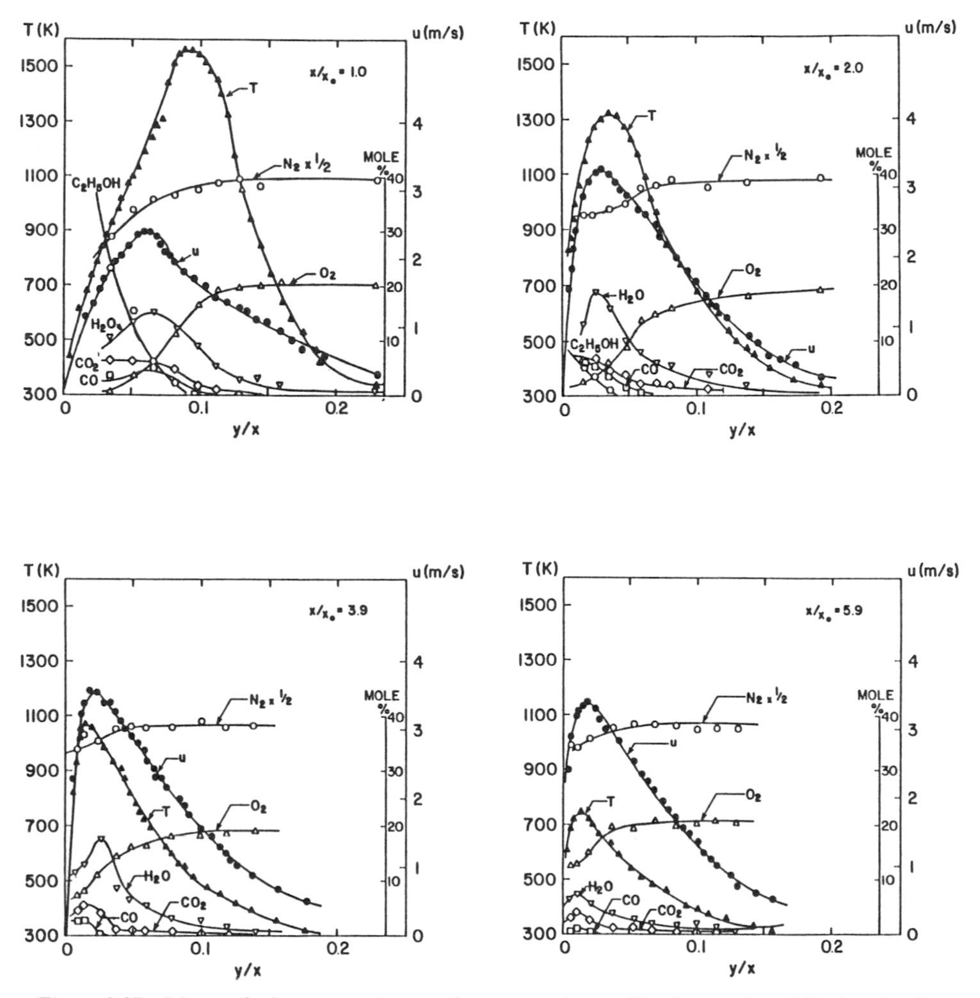

Figure 9.37 Mean velocity, temperature, and concentration profiles for an ethanol fire burning in air on a vertical wall. Pyrolysis zone length 101 mm. (After Ahmad and Faeth.[78])

case, the above ratio was determined at the upper end of the pyrolysis region over which the average mass burning rate had been measured.

As shown in this figure, transition from laminar to turbulent flow does not strongly influence the burning-rate relation. The data in Fig. 9.38 show a slight change in slope at $10^8 < \mathrm{Ra}_x < 10^9$. The laminar theoretical prediction (9-275) provides a good correlation of the data for $\mathrm{Ra}_x < 10^8$–10^9, with maximum errors less than 40%. The laminar data generally involve low levels of radiation from the flame to the burning surface, less than 10% of the total heat flux measured by Ahmad and Faeth.[78] The laminar integral model was compared with results of an exact solution of the boundary-layer equations;[79] errors in

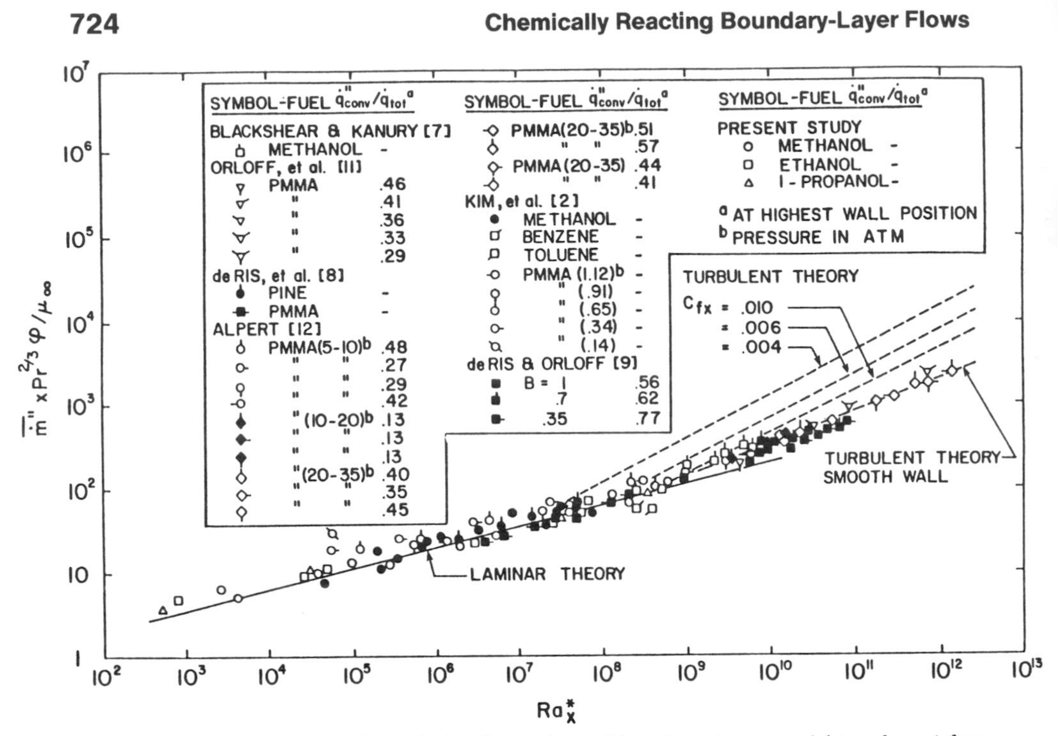

Figure 9.38 Comparison of theoretical and experimental burning rates on upright surfaces (after Ahmad and Faeth[78]).

the integral model were on the order of 10% for values of B and r in the range of most practical fuels ($2 < B < 20$, $0.05 < r < 0.5$). Thus, a portion of the apparent scatter of the data around the predictions, shown in Fig. 9.38, represents systematic errors due to changes in B and r as the fuel is varied.

The turbulent-burning-rate prediction for a smooth wall, Eq. (9-267), provides a good correlation of the data for $Ra_x > 10^8$–10^9. This agreement was achieved by selecting $\xi_R = 0.5$. Measurements by Smith[103] in a turbulent natural-convection boundary layer, and by Liburdy et al.[95] in a turbulent isothermal wall plume, suggest $\xi_R = 0.4$. Therefore, the results of Ahmed and Faeth[78] appear to be reasonable in view of the complications of mass transfer and combustion in the flow. The theoretical results obtained using constant friction factors of various values and $\xi_R = 0.5$ exhibit a steeper slope with respect to Ra_x than the data, and generally overestimate the burning rate.

The radiation flux to the surface does not appreciably influence the correlation illustrated in Fig. 9.38 for ratios of convective to total heat flux in the range 0.13–0.77. This was not expected for a relation which was developed from models which neglect radiation. This effect has also been noted by Kanury;[104] a satisfactory theoretical explanation of it has not been established. According to Ahmad and Faeth,[78] it can be concluded that the quantitative accuracy of the models is partly fortuitous—possibly due to overestimating

TABLE 9.6 Radiative Heat Fluxes in the Plume[a]

	Heat Flux (kW/m^2)			
x/x_0	$\dot{q}''_{w\,tot}$	$\dot{q}''_{w\,rad}$	$\dot{q}''_{\infty\,thin}$	$\dot{q}''_{\infty\,thick}$
1.12[b]	26.92	4.41	4.98	3.96
2.37[b]	22.96	3.38	4.46	3.55
3.61[b]	17.34	3.03	4.00	3.18
4.85	13.21	1.33	2.45	1.95
6.10	7.13	0.78	1.05	0.83

[a] Ethanol burning in air, 101-mm pyrolysis zone length. After Ahmad and Faeth.[78]

[b] Combusting-plume region.

convection heat transfer rates by neglecting unmixedness, radiative heat losses, and dissociation in the flame, while neglecting the direct radiation component to the burning surface. Another factor improving the relation is that the lower regions of the burning surface, which also contribute to the average, are less influenced by radiation than the point where the radiation flux was measured.

Table 9.6 is a summary of a portion of the heat-flux data in the plume, corresponding to the conditions pictured in Fig. 9.38. Radiation to the wall is largest near the end of the pyrolysis zone, but is never more than 10–20% of the total wall heat flux for the present tests. Following Orloff et al.,[84] two estimates are provided for the outward radiative heat flux, assuming thin and thick flames respectively. The true value lies between these two limits but is probably closer to the thin-flame approximation for the test conditions of Ahmad and Faeth.[78] The outward radiation flux based on the thin-flame approximation is larger than the radiant flux to the wall. This effect was also observed by Orloff et al.,[84] who attributed it to radiation absorption by the fuel- and product-rich region near the wall; cf. Fig. 9.37.

The wall heat-flux measurements in the plume are illustrated in Fig. 9.39. The heat flux remains relatively constant in the combusting portion of the plume. The region of relatively constant wall heat flux ends as the average position of the end of the visible portion of the flame is approached.

The heat flux prediction of the integral model overestimates the wall heat-flux measurements in the combusting plume by 10–29% except near the tip of the flame, where the errors are somewhat larger. The data suggest an additional influence of x_0 and Gr_{x_0} on the wall heat flux, not represented by the model of Ahmad and Faeth.[78]

In general, existing laminar and turbulent burning-rate measurements on upright surfaces were correlated by Ahmad and Faeth[78] with maximum errors of 40% using expressions suggested by integral models. A remarkable feature of this result is that much of the turbulent data involved substantial fractions of radiant heat flux to the burning surface (up to 86% of the total heat flux).

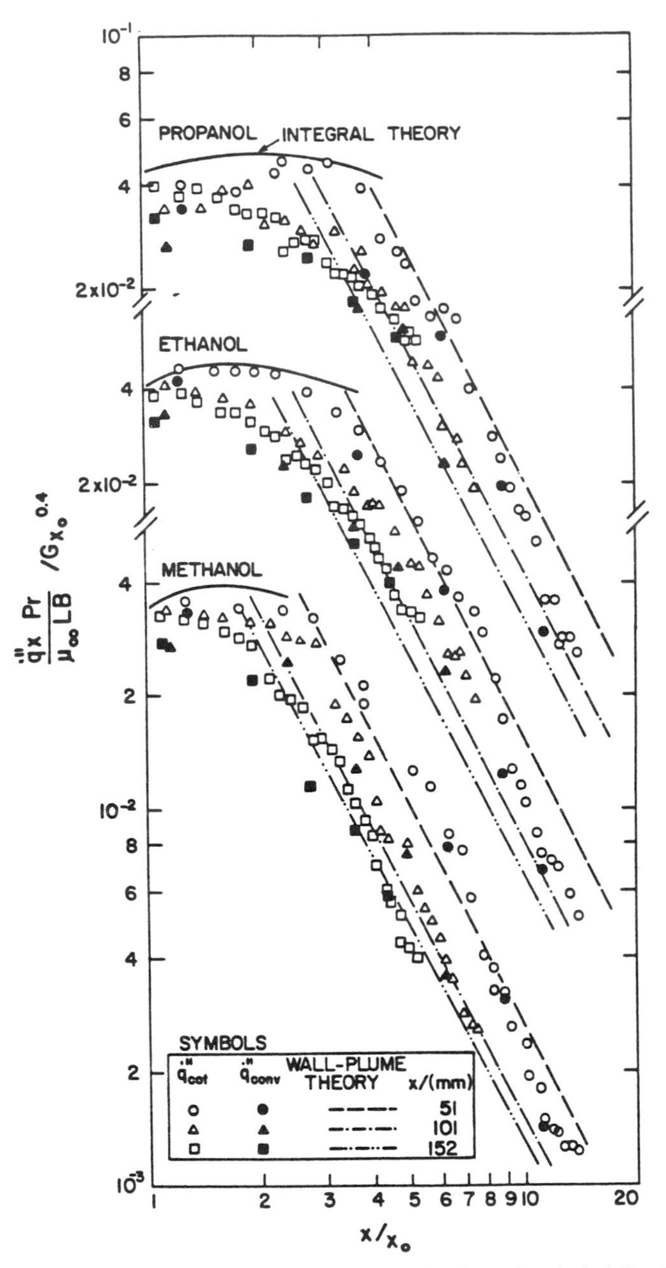

Figure 9.39 Predicted and measured wall heat fluxes in the plume for alcohol fires burning in air on a vertical wall (after Ahmad and Faeth[78]).

Thus, while the theory is useful for data correlation, since radiation was neglected in the model, the overall agreement between theory and experiment is fortuitous for high radiation levels. A satisfactory explanation of this helpful behavior should receive further attention. An accurate determination of the burning characteristics of large-scale vertical surfaces is needed, but this has to be preceded by development of reliable models of turbulent radiative combustion processes. Clearly more work is needed in developing analytical models which can predict all necessary parameters of the plume and wall fires before they can replace extensive, time-consuming experimentation.

REFERENCES

1. Th. von Karman and G. Millan, *Fourth International Symposium on Combustion*, pp. 173–177, Williams and Wilkins Co., Baltimore, 1953.
2. F. E. Marble and T. C. Adamson, Jr., *Jet Propulsion*, Vol. 24, No. 85, 1954.
3. H. W. Emmons, *Z. Angew. Math. Mech.*, Vol. 36, No. 60, 1956.
4. L. Lees, "Convective Heat Transfer with Mass Addition and Chemical Reactions," *Third Combustion and Propulsion Colloquium*, pp. 451–498, AGARD, Pergamon Press, New York, 1958.
5. W. H. Dorrance, *Viscous Hypersonic Flow*, McGraw-Hill Book Company, New York, 1962.
6. P. M. Chung, "Chemically Reacting Nonequilibrium Boundary Layers," *Advances in Heat Transfer*, (J. P. Hartnett and T. F. Irvine, Jr., Eds.), pp. 109–270, Academic Press, New York, 1965.
7. S. V. Patankar and D. B. Spalding, *Heat and Mass Transfer in Boundary Layers*, Intertext Books, London, 1970.
8. E. C. Anderson and C. H. Lewis, *Laminar or Turbulent Boundary-Layer Flows of Perfect Gases or Reacting Gas Mixtures in Chemical Equilibrium*, Contractor Report NASA CR-1893, October 1971.
9. R. M. Evans, *Boundary Layer Integral Matrix Procedures*, Aerotherm Report UM-75-64, Acurex Corporation, Mountain View, Calif., July 1975.
10. M. K. Razdan and K. K. Kuo, "Erosive Burning Study of Composite Solid Propellants by Turbulent Boundary-Layer Approach," *AIAA J.*, Vol. 17, No. 11, pp. 1225–1233, 1979.
11. D. B. Spalding, "Mixing and Chemical Reaction in Steady Confined Turbulent Flames," *Thirteenth Symposium (International) on Combustion*, pp. 649–657, 1971.
12. X. Wu, R. Arora, M. Kumar, and K. K. Kuo, "An Aerothermochemical Model for Erosive Burning of Double-Base Propellants in Turbulent Shear Flow," *Proceedings of 13th Aerospace Technology and Science Meeting*, Tokyo, Japan, June 26–July 3, 1982.
13. R. J. Burick and J. R. Osborn, "Erosive Combustion of Double-Base Solid Rocket Propellants," *CPIA Publication 162*, Vol. II, pp. 57–69, 1967.
14. K. K. Kuo and S. T. Keswani, "A Comprehensive Theoretical Model for Carbon–Carbon Composite Nozzle Recession," *Combustion Sci. Tech.*, Vol. 42, pp. 145–164, 1985.
15. H. Schlichting, *Boundary-Layer Theory*, McGraw-Hill Book Company, New York, 1968.
16. F. A. Williams, *Combustion Theory*, Chapter 12, pp. 288–323, Addison-Wesley Publishing Company, Reading, Mass., 1965.
17. S. C. Lin and J. D. Teare, AVCO Res. Lab. Res. Rept. 115, 1962.
18. S. M. Scala, IAS Paper No. 62-154, 1962.

19. P. L. Walker, F. Rusinko, and L. G. Austin, *Adv. Catalysis*, Vol. 11, p. 1333, Academic Press, New York, 1959.
20. L. Lees, *Jet Propulsion*, Vol. 26, p. 259, 1956.
21. J. A. Fay and F. R. Riddell, *J. Aeron. Sci.*, Vol. 25, p. 73, 1958.
22. W. E. Welsh, Jr. and P. M. Chung, *Proc. Heat Transfer Fluid Mech. Inst.*, p. 146, Stanford Univ. Press, Stanford, Calif., 1963.
23. W. E. Welsh, Jr., Aerospace Corp., SSD-TDR-63-193, 1963.
24. N. S. Diaconis, P. D. Garsuch, and R. A. Sheridan, IAS Paper No. 62-155, 1962.
25. F. G. Blottner, M. Johnson, and M. Ellis, "Chemically Reacting Viscous Flow Program for Multi-component Gas Mixtures," Sandia Laboratories Research Report SC-RR-70-754, Dec. 1971, 316 pages.
26. J. O. Hirschfelder, C. F. Curtiss, and R. B. Bird, *Molecular Theory of Gases and Liquids*, John Wiley and Sons, New York, 1964.
27. R. M. Kendall, *An Analysis of the Coupled Chemically Reacting Boundary Layer and Charring Ablator, Part V: A General Approach to the Thermochemical Solution of Mixed Equilibrium–Nonequilibrium, Homogeneous or Heterogeneous Systems*, NASA CR-1064, June 1968; Aerotherm Report 66-7, Part V.
28. R. B. Bird, "Diffusion in Multicomponent Gas Mixtures," presented at 25th Anniversary Congress Society of Chemical Engineers (Japan), November 6–14, 1961, published in abbreviated form in *Kagaku Kohaku*, Vol. 26, pp. 718–721, 1962.
29. E. P. Bartlett, R. M. Kendall, and R. A. Rindal, *A Unified Approximation for Mixture Transport Properties for Multicomponent Boundary Layer Applications*, Aerotherm Corporation, Final Report 66-7, Part IV, March 14, 1967; NASA CR-1063.
30. J. O. Hirschfelder, "Heat Conductivity in Polyatomic, Electronically Excited, or Chemically Reacting Mixtures," *Sixth Symposium (International) on Combustion*, pp. 351–366, Reinhold Publishing Corporation, New York, 1957.
31. R. M. Kendall, M. W. Rubesin, T. J. Dahn, and M. R. Mendenhall, *Mass, Momentum, and Heat Transfer Within a Turbulent Boundary Layer with Foreign Gas Mass Transfer at the Surface, Part I–Constant Fluid Properties*, Vidya Division, Itek Corporation, Final Report 111, 1964.
32. T. Cebeci and A. M. O. Smith, "A Finite-Difference Method for Calculating Compressible Laminar and Turbulent Boundary Layers," *J. Basic Engrg.*, Paper No. 70-FE-A, 1970.
33. T. Cebeci and A. M. O. Smith, *Analysis of Turbulent Boundary Layers*, Academic Press, 1974.
34. T. Cebeci, *A Model for Eddy-Conductivity and Turbulent Prandtl Number*, Report MDC-J0747/01, McDonnell-Douglas Corporation, May 1970.
35. I. E. Beckwith and D. M. Bushnell, *Calculation by a Finite-Difference Method of Supersonic Turbulent Boundary Layers with Tangential Slot Injection*, NASA TN-D-6221, April 1971.
36. F. M. White, *Viscous Fluid Flow*, McGraw-Hill, New York, 1974.
37. E. G. Van Driest, "On Turbulent Flow Near a Wall," *J. Aeron. Sci.*, Vol. 23, p. 1007, 1956.
38. R. L. Simpson, W. M. Kays, and R. J. Moffat, *The Turbulent Boundary Layer on a Porous Plate: An Experimental Study of Fluid Dynamics with Injection and Suction*, Report HMT-2, Stanford University, Stanford, Calif., December 1967.
39. D. G. Whitten, W. M. Kays, and R. J. Moffat, *The Turbulent Boundary Layer on a Porous Plate: Experimental Heat Transfer with Variable Suction, Blowing, and Surface Temperature*, Report HMT-3, Stanford University, Stanford, Calif., December 1967.
40. D. S. Johnson, "Velocity and Temperature Fluctuations in a Turbulent Boundary Layer Downstream of a Stepwise Discontinuity in Wall Temperature," *J. Appl. Mech.*, Vol. 26, p. 325, 1959.

41. H. Ludweig, "Bestimmung des Verhaltnisses der Austauschkoeffizienten für Wärme und Impuls bei Turbulenten Grenzschichten," *Z. Flugwiss.*, Vol. 5, p. 73, 1956.

42. P. S. Klebanoff, *Characteristics of Turbulence in a Boundary Layer with Zero Pressure Gradient*, NACA Report 1247, 1955.

43. J. B. Peterson, Jr., et al., *Further Investigation of Effect of Air Injection through Slots and Porous Surfaces on Flat Plate Turbulent Skin Friction at Mach* 3, NASA TN-D-331, March 1966.

44. L. O. F. Jeromin, *An Experimental Investigation of the Compressible Turbulent Boundary Layer with Air Injection*, A.R.C. Vol. 28, p. 549, London, November 1966.

45. D. M. Bushnell and I. E. Beckwith, "Calculation of Nonequilibrium Hypersonic Turbulent Boundary Layers and Comparisons with Experimental Data," *AIAA J.*, Vol. 8, No. 8, p. 1462, 1970.

46. E. P. Bartlett and R. M. Kendall, *Nonsimilar Solution of the Multicomponent Laminar Boundary Layer by an Integral Matrix Method*, Aerotherm Corporation, Final Report 66-7, Part III, March 14, 1967; NASA CR-1062.

47. A. Favre, "Equations des Gaz Turbulents Compressibles," *J. Mechanique*, Vol. 4, No. 3, pp. 361–390, 1965.

48. B. E. Launder and D. B. Spalding, *Mathematical Models of Turbulence*, p. 9, Academic Press, New York, 1972.

49. A. D. Gosman, F. C. Lockwood, and S. A. Syed, "Prediction of a Horizontal Free Turbulent Diffusion Flame," *Sixteenth Symposium (International) on Combustion*, pp. 1543–1555, Combustion Institute, Pittsburgh, 1976.

50. F. C. Lockwood and A. S. Naguib, "The Prediction of the Fluctuations in the Properties of Free, Round-Jet, Turbulent, Diffusion Flames," *Combustion and Flame*, Vol. 24, No. 1, pp. 109–124, 1975.

51. S. E. Elghobashi and W. M. Pun, "A Theoretical and Experimental Study of Turbulent Diffusion Flames in Cylindrical Furnaces," *Fifteenth Symposium (International) on Combustion*, pp. 1353–1365, Combustion Institute, Pittsburgh, 1974.

52. H. B. Mason and D. B. Spalding, "Prediction of Reaction Rates in Turbulent Pre-mixed Boundary Layer Flows," *Combustion Institute, First European Symposium*, pp. 601–606, 1973.

53. F. C. Lockwood, "The Modeling of Turbulent Premixed and Diffusion Combustion in the Computation of Engineering Flows," *Combustion and Flame*, Vol. 29, No. 2, pp. 111–122, 1977.

54. R. Arora, K. K. Kuo, and M. K. Razdan, "Turbulent Boundary-Layer Flow Computations with Special Emphasis on the Near-Wall Region," AIAA Paper No. 81-1001-CP, AIAA 5th Computational Fluid Dynamics Conference, Palo Alto, Calif., June 22–23, 1981; *AIAA J.*, Vol. 20, No. 11, pp. 1481–1482, 1982.

55. R. J. Baker and B. E. Launder, "The Turbulent Boundary Layer with Foreign Gas Injection —II. Predictions and Measurements in Severe Stream-wise Pressure Gradients," *Int. J. Heat Mass Transfer*, Vol. 17, pp. 293–306, 1974.

56. T. Cebeci and K. C. Chang, "Calculation of Incompressible Rough-Wall Boundary-Layer Flows," *AIAA J.*, Vol. 16, No. 7, pp. 730–735, July 1978.

57. T. L. Chambers and D. C. Wilcox, "Critical Examination of Two-Equation Turbulence Closure Models for Boundary Layers," *AIAA J.*, Vol. 15, No. 6, pp. 821–828, June 1977.

58. E. M. Sparrow and M. S. Yu, "Local Non-similarity Thermal Boundary Layer Solutions," *Trans. ASME, J. Heat Transfer*, pp. 328–334, 1971.

59. R. K. Lobb, E. M. Winkler, and J. Persh, *Experimental Investigation of Turbulent Boundary Layers in Hypersonic Flow*, NOVORD Rept. 3880, 1955.

60. R. H. Korkegi, "Transition Studies and Skin-Friction Measurements on an Insulated Flat Plate at a Mach Number of 5.8," *J. Aeron. Sci.*, Vol. 23, pp. 97–107, February 1956.
61. A. J. Laderman, "Effects of Wall Temperature on a Supersonic Turbulent Boundary Layer," *AIAA J.*, Vol. 16, No. 7, pp. 723–729, July 1978.
62. A. J. Laderman and A. Demetriades, "Turbulent Shear Stresses in Compressible Boundary Layers," *AIAA J.*, Vol. 17, No. 7, pp. 736–744, 1979.
63. V. A. Sandborn, *A Review of Turbulence Measurements in Compressible Flow*, NASA TMX-62, 337, March 1974.
64. W. J. Yanta and R. E. Lee, "Measurements of Mach 3 Turbulence Transport Properties on a Nozzle Wall," *AIAA J.*, Vol. 14, No. 6, pp. 725–734, June 1976.
65. M. K. Razdan and K. K. Kuo, "Measurements and Model Validation for Composite Propellants Burning under Cross Flow of Gases," *AIAA J.*, Vol. 18, No. 6, pp. 669–677, June 1980.
66. T. Marklund and A. Lake, "Experimental Investigation of Propellant Erosion," *Am. Rocket Soc. J.*, Vol. 3, No. 2, pp. 173–178, 1960.
67. J. Laufer, *The Structure of Turbulence in Fully Developed Pipe Flow*, NACA TR 1174, 1954.
68. J. O. Hinze, *Turbulence*, 2nd ed., p. 730, McGraw-Hill Book Company, New York, 1975.
69. A. Gany, L. H. Caveny, and M. Summerfield, "Aerothermochemistry of Metal Erosion by Hot Reactive Gases," *ASME J. Heat Transfer*, Vol. 100, No. 3, pp. 531–536, August 1978.
70. A. Gany and L. H. Caveny, "Mechanism of Chemical and Physical Gas–Metal Interactions in Very High Shearing Regimes," *Nineteenth Symposium (International) on Combustion*, pp. 731–740, The Combustion Institute, Pittsburgh, 1982.
71. A. C. Alkidas, E. G. Plett, and M. Summerfield, "A Performance Study of Ballistic Compressor," *AIAA J.*, Vol. 14, No. 12, pp. 1752–1758, 1976.
72. A. P. Colburn, "A Method of Correlating Forced Convection Heat Transfer Data and a Comparison with Fluid Friction," *Trans. AIChE*, Vol. 29, p. 174, 1933.
73. E. G. Plett, A. C. Alkidas, R. E. Shrader, and M. Summerfield, "Erosion of Metals by High Pressure Combustion Gases: Inert and Reactive Erosion," *ASME J. Heat Transfer*, Vol. 97, No. 1, pp. 110–115, 1975.
74. R. L. Geisler, "The Relationship between Solid Propellant Formulation Variables and Nozzle Recession Rates," presented at JANNAF Rocket Nozzle Technology Subcommittee Workshops, Lancaster, Calif., July 12–13, 1978.
75. K. Klager, "The Interaction of the Efflux of Solid Propellants with Nozzle Materials," *Propellants and Explosives*, Vol. 2, pp. 55–63, 1977.
76. P. A. Libby and T. R. Blake, "Burning Carbon Particles in the Presence of Water Vapor," *Combustion and Flame*, Vol. 41, pp. 123–147, 1981.
77. R. S. Frankle and J. Lebiedzik, "A New Technique for Characterizing Surface Roughness," presented at JANNAF Rocket Nozzle Technology Subcommittee Meeting, Monterey, Calif., 1980; CPIA Publication 328, pp. 415–431, November 1980.
78. T. Ahmad and G. M. Faeth, "Turbulent Wall Fires," *17th Symposium (International) on Combustion*, pp. 1149–1160, 1979.
79. T. Ahmad, *Investigation of the Combustion Region of Fire-Induced Plumes along Upright Surfaces*, Ph.D. thesis, The Pennsylvania State University, August 1978.
80. F. J. Kosdon, F. A. Williams, and C. Buman, "Combustion of Vertical Cellulosic Cylinders in Air," *Twelfth Symposium (International) on Combustion*, pp. 253–264, The Combustion Institute, Pittsburgh, 1969.
81. J. S. Kim, J. de Ris, and F. W. Kroesser, "Laminar Free Convective Burning of Fuel Surfaces," *Thirteenth Symposium (International) on Combustion*, pp. 949–961, The Combustion Institute, Pittsburgh, 1971.

82. P. L. Blackshear, Jr. and K. A. Murty, "Heat and Mass Transfer to, from, and within Cellulosic Solids Burning in Air," *Tenth Symposium (International) on Combustion*, pp. 911–923, The Combustion Institute, Pittsburgh, 1965.
83. G. H. Markstein and J. de Ris, "Upward Fire Spread over Textiles," *Fourteenth Symposium (International) on Combustion*, pp. 1085–1097, The Combustion Institute, Pittsburgh, 1973.
84. L. Orloff, J. de Ris, and G. H. Markstein, "Upward Turbulent Fire Spread and Burning of Fuel Surface," *Fifteenth Symposium (International) on Combustion*, pp. 183–192, The Combustion Institute, Pittsburgh, 1974.
85. J. de Ris and L. Orloff, "The Role of Buoyancy, Direction and Radiation in Turbulent Diffusion Flames on Surface," *Fifteenth Symposium (International) on Combustion*, pp. 175–182, The Combustion Institute, Pittsburgh, 1974.
86. P. J. Pagni and T. M. Shih, "Excess Pyrolyzate," *Sixteenth Symposium (International) on Combustion*, pp. 1329–1343, The Combustion Institute, Pittsburgh, 1976.
87. K. T. Yang, "Laminar Free-Convection Wake above a Heated Vertical Plate," *ASME Trans. Ser. E, J. Appl. Mech.*, Vol. 86, pp. 131–138, 1964.
88. J. A. Liburdy and G. M. Faeth, "Theory of a Steady Laminar Thermal Plume along a Vertical Adiabatic Wall," *Lett. Heat and Mass Transfer*, Vol. 2, pp. 407–418, 1975.
89. O. A. Plumb and L. A. Kennedy, "Prediction of Buoyancy Controlled Turbulent Wall Diffusion Flames," *Sixteenth Symposium (International) on Combustion*, pp. 1699–1707, The Combustion Institute, Pittsburgh, 1976.
90. D. B. Spalding, "Concentration Fluctuations in a Round Turbulent Free Jet," *Chem. Engrg. Sci.*, Vol. 26, pp. 95–107, 1971.
91. F. C. Lockwood and P. H. Ong, "Study of Combined Free and Forced Convection Turbulent Boundary Layer with Variable Fluid Properties and Chemical Reaction," *Heat and Mass Transfer in Boundary Layers* (N. Afgan, Z. Zaric, and P. Anastasijevic, Eds.), Vol. 1, pp. 339–340, Pergamon Press, New York, 1972.
92. P. J. Pagni and T. M. Shih, "Wake Turbulent Flames," ASME Paper 77-HT-97, presented at ASME/AIChE 17th National Heat Transfer Conference, Salt Lake City, Utah, August 1977.
93. J. J. Grella and G. M. Faeth, "Measurements in a Two-Dimensional Thermal Plume along a Vertical Adiabatic Wall," *J. Fluid Mech.*, Vol. 71, pp. 701–710, 1975.
94. J. A. Liburdy and G. M. Faeth, "Heat Transfer and Mean Structure of a Turbulent Thermal Plume along Vertical Isothermal Wall," *Trans., ASME, Ser. C, J. Heat Transfer*, Vol. 100, No. 2, pp. 177–183, 1978.
95. J. A. Liburdy, E. G. Groff, and G. M. Faeth, "Structure of a Turbulent Thermal Plume Rising along an Isothermal Wall," *Trans. ASME, Series C, J. Heat Transfer*, Vol. 101, No. 2, pp. 249–255, 1979.
96. T. Cebeci and A. Khattab, "Prediction of Turbulent Free Convection from Vertical Surfaces," *ASME Trans., Ser. C, J. Heat Transfer*, Vol. 97, pp. 469–471, 1975.
97. J. de Ris, A. M. Kanury, and M. C. Yuen, "Pressure Modeling of Fires," *Fourteenth Symposium (International) on Combustion*, pp. 1033–1044, The Combustion Institute, Pittsburgh, 1973.
98. L. Orloff, A. T. Modak, and R. L. Alpert, "Burning of Large-Scale Vertical Surfaces," *Sixteenth Symposium (International) on Combustion*, pp. 1345–1354, The Combustion Institute, Pittsburgh, 1977.
99. G. A. Marxman, "Combustion in the Turbulent Boundary Layer on a Vaporizing Surface," *Tenth Symposium (International) on Combustion*, pp. 1337–1349, The Combustion Institute, Pittsburgh, 1965.
100. B. R. Morton, "Modeling Fire Plumes," *Tenth Symposium (International) on Combustion*, pp. 973–982, The Combustion Institute, Pittsburgh, 1965.

101. J. C. Rotta, "Temperaturverteilungen in der Turbulenten Grenzschicht an der Ebenen Platte," *Int. J. Heat and Mass Transfer*, Vol. 7, pp. 215–228, 1964.

102. T. Cebeci, A. M. O. Smith, and G. J. Mosinskis, *J. Heat Transfer*, Vol. 92, p. 499, 1970.

103. R. R. Smith, *Characteristics of Turbulence in Free Convection Flow Past a Vertical Plate*, Ph.D. Thesis, Queen Mary College, University of London, 1972.

104. A. M. Kanury, "The Science and Engineering of Hostile Fires," *NAS/NRC Fire Research Abstracts and Reviews*, 1978.

HOMEWORK

1. Show that by differentiating $h = \sum_{i=1}^{N} Y_i h_i$, one can obtain

$$dh = C_p\, dT + \sum_{i=1}^{N} h_i\, dY_i$$

where $C_p \equiv \sum_{i=1}^{N} Y_i C_{pi}$.

2. Show that the conservation of energy for a two-dimensional boundary layer can be written in the following total-enthalpy form:

$$\rho\left(u\frac{\partial h_t}{\partial x} + v\frac{\partial h_t}{\partial y}\right) = \frac{\partial}{\partial y}\left\{\frac{\mu}{\Pr}\left[\frac{\partial h_t}{\partial y} + (\Pr - 1)\frac{\partial(u^2/2)}{\partial y} + \sum_{i=1}^{N}(\mathrm{Le}_i^{-1} - 1)h_i\frac{\partial Y_i}{\partial y}\right]\right\}$$

where

$$\mathrm{Le}_i \equiv \frac{\lambda}{\rho \mathscr{D}_i C_p} \quad \text{and} \quad \Pr \equiv \frac{C_p \mu}{\lambda}$$

3. What is the relationship between the molar concentration C_i and the mass fraction Y_i of species i in a multicomponent system? Is it the same as that between C_i and X_i (the mole fraction of species i)?

4. Show that by following Levy–Lees's transformation

$$\xi \equiv \int_0^x \rho_e \mu_e U_e\, dx$$

$$\eta \equiv \frac{U_e}{\sqrt{2\xi}}\int_0^y \rho\, dy$$

$$\rho u = \frac{\partial \psi}{\partial y}, \qquad \rho v = -\frac{\partial \psi}{\partial x}$$

$$F(\xi, \eta) \equiv \frac{\psi}{\sqrt{2\xi}}$$

the axial and transverse velocity components of two-dimensional boundary-layer flow can be expressed as

$$u = U_e F_\eta$$

$$v = -\frac{1}{\rho}\left[\left(\sqrt{2\xi}\, F_\xi + \frac{1}{\sqrt{2\xi}} F\right)\frac{\partial \xi}{\partial x} + \sqrt{2\xi}\, F_\eta \frac{\partial \eta}{\partial x}\right]$$

5. Show that the momentum equation

$$\rho\left(u\frac{\partial u}{\partial x} + v\frac{\partial u}{\partial y}\right) = -\frac{dp}{dx} + \frac{\partial}{\partial y}\left(\mu\frac{\partial u}{\partial y}\right)$$

can be transformed to

$$\left(lF_{\eta\eta}\right)_\eta + FF_{\eta\eta} + 2\left(\frac{\xi}{U_e}\frac{dU_e}{d\xi}\right)\left[\frac{\rho_e}{\rho} - F_\eta^2\right] = 2\xi\left(F_\eta F_{\eta\xi} - F_\xi F_{\eta\eta}\right)$$

according to the transformation described above. Also, l is defined by $l \equiv \rho\mu/\rho_e\mu_e$.

CHAPTER 10 Ignition

1 INTRODUCTION

Ignition is a transition from a nonreactive to a reactive state in which external stimuli lead to thermochemical runaway followed by a rapid transition to self-sustained combustion. The general reason for studying ignition is to achieve basic understanding of the detailed physicochemical processes involved in the above transition. Some specific objectives for ignition studies include:

1. To prevent fire hazards
2. To develop igniters and reproducible energy sources for ignition and combustion
3. To investigate the ignitability of a material under given initial conditions with a fixed energy input
4. To determine the minimum energy required for attainment of ignition
5. To study the effect of various physical and chemical parameters on ignition delay

Ignition processes are usually very complex and involve many intricate physical and chemical steps. Ignition is inherently transient, usually triggered by some transient heating processes. In order to predict or interpret certain ignition phenomena, detailed chemical kinetics must be known; however, the measurement of species concentrations in a short time interval is very difficult. As a result, the major mechanism of chemical kinetics during the ignition process is not known in advance. Scientists and engineers have to postulate certain basic mechanisms for chemical kinetics under specific operating conditions. These mechanisms can also vary significantly with ambient conditions and external stimuli, and the variations are usually unknown.

In addition to the complications in chemical kinetics, there are important problems in the mixing of fuel and oxidizer species. Both turbulent flow structures and the interactions of turbulence with chemical kinetics are required for comprehensive modeling of the ignition process. Realistic mathematical simulation of an ignition process will usually result in a three-dimensional and time-dependent model, which is difficult but not impossible to solve.

In the theoretical model formulation, a definition of the onset of ignition must be given. There are many ignition criteria one can select, and it is hard to be sure which is the most appropriate one. Also, there can be a compatibility problem between the ignition criterion used in the theoretical model and that adopted in the experimental observation.

The usual conditions for ignition are given by a "3T" rule of thumb. The three T's stand for:

Temperature. Must be high enough to cause significant chemical reactions and/or pyrolysis.

Time. Must be long enough to allow the heat input to be absorbed by the reactants so that a runaway thermochemical process can occur.

Turbulence. Must be high enough so that there is good mixing between fuel and oxidizer and heat can be transferred from the reacted media to the unreacted media.

In general, there are many parameters which can affect ignition: mixture composition, pressure, pressurization rate, duration of heating, total energy deposited in the system, initial radius of energy deposition, ambient oxidizer concentration, velocity of the convective stream, turbulence scale and intensity, thermal and transport properties of the material being heated, catalysts, inhibitors, and so on. The ignition process also depends on the geometry and material of the environment and the operating condition. Therefore, one must be cautious in interpreting the results of any ignition experiment and in generalizing ignition data.

2 IGNITION STIMULI AND DEVICES

There are various means of achieving ignition. External stimuli can be classified into the following three categories:

1. *Thermal Energy Stimuli.* Transfer of thermal energy to the reactants by conduction, convection, radiation, or any combination of these basic modes of heat transfer
2. *Chemical Stimuli.* Introduction of hypergolic reactive agents
3. *Mechanical Stimuli.* Mechanical impact, friction, or shock waves

There exist many different types of ignition devices in both industrial applications and basic research. These devices include spark plugs, hot wires, electric squibs, pyrotechnic igniters, pyrogen igniters, hypergolic ignition devices injecting gaseous fluorine or liquid chlorine trifluoride, percussion primers, and so on. There is also a variety of equipment which can be used for ignition studies. The commonly used include shock tubes, arc-image furnaces, high-powered CO_2 lasers, solar-energy concentrators, and impact testers. The use of some of these devices is described in the subsequent sections.

3 SPONTANEOUS IGNITION

Apart from the ignition processes introduced by external sources such as sparks, pilot flames, and hot wires, there is a process called spontaneous ignition, (or sometimes referred to as self-ignition, autoignition, or homogeneous ignition) without an ignition source other than hot vessel walls. The definition given by Spalding[1] is as follows: "when a reactive mixture is formed, raised to a definite temperature and pressure, and then left alone, it may burst into flame after a certain time." At the onset of spontaneous ignition, there is usually a rapid rise in temperature, emission of visible radiation, and rapid chemical reactions.

Even though the definition of spontaneous ignition requires that the mixture be left alone until self-ignition is reached, there are many applications of the theory of spontaneous ignition to practical systems in which the ignition is deliberate. For example, when liquid fuel is injected into the combustion chambers of a diesel engine and compressed to elevated pressure and temperature levels, the fuel vapor rapidly mixes with air and achieves self-ignition after a delay of a few milliseconds. Strictly speaking, the process in this case is not spontaneous ignition, since there is no sharp distinction between the mixture preparation and ignition phases; the mixture is not truly "left alone" during the ignition delay. Nevertheless, the essential feature of spontaneous ignition is present in this example, since there is no input of hot gases from an external source, and the gaseous mixture inflames as a result of internal effects after being compressed.

Spontaneous-ignition analysis can also be applied to the part of the gaseous fuel–air mixture that is far from the spark in the cylinder of a gasoline engine. Although the main ignition of the mixture is initiated by the spark and subsequently spreads over the unburned mixture, the flame-spreading process may require several milliseconds to complete. In remote corners of the combustion chamber, the gaseous mixture, being compressed by both the piston and the burned gases, may already have reached spontaneous ignition condition before the flame has spread to them. This phenomenon is the origin of gasoline-engine knock, and it falls into the category of spontaneous ignition. In general, when the time required for the formation of a certain gaseous mixture is short in comparison with chemical reaction time and also the gaseous mixture is not affected by external sources except hot vessel walls, one can apply spontaneous-ignition analysis to the ignition of the mixture.

3.1 Thermal Ignition

In the study of self-ignition, a concept of "ignition temperature" has been introduced by many researchers.[2] It must be noted that the application of this concept, discussed below, as a criterion may be dangerous and highly inappropriate. One reason is that the chemical-reaction rate is in fact nonzero for all temperatures according to the following Arrhenius equation:

$$k_f = A \exp\left[-\frac{E_a}{R_u T}\right] \tag{10-1}$$

Nevertheless, in the context of spontaneous ignition, the concept of ignition temperature is still useful.

The ignition temperature has been defined by van't Hoff as the temperature at which the rate of heat loss due to conduction is equal to the rate of heat production by chemical reactions. It is useful to consider an idealized Le Châtelier pyrometer (see Fig. 10.1a) in the study of spontaneous ignition. This apparatus consists essentially of a container which is heated to a desired temperature and evacuated prior to the admission of reactant gases. The heat capacity of the latter is too small to alter the temperature of the container appreciably on their admission. The reactant mixture in the container is heated or cooled by conduction from or to the container walls. The initial temperature of the gaseous mixture is T_0 and the chamber pressure is p_0 after the injection of the reactant gases. The time variation of the bulk temperature of the gaseous

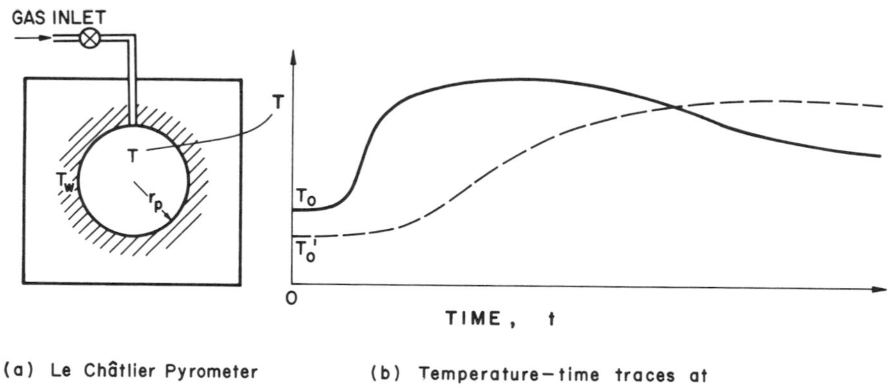

Figure 10.1 (a) Le Châtlier pyrometer. (b) Time variation of the bulk temperature of the gaseous mixture in the pyrometer.

mixture in the vessel is shown in Fig. 10.1*b*. In this plot there are two temperature–time traces: the solid line corresponding to a higher initial temperature T_0, and the dashed line corresponding to a lower initial temperature T_0'. The rise in temperature is due to heat transfer and chemical reaction. The time interval between the initial temperature and the peak temperature depends upon the rate of chemical reaction, which can be on the scale of microseconds, milliseconds, seconds, hours, or days.

The rate of heat evolution, $\dot{q}_R$, due to chemical reaction in a chamber volume V can be expressed as

$$\dot{q}_R = \Delta H_r V\, \mathrm{RR} = \Delta H_r V\left[A \exp\left(-\frac{E_a}{R_u T}\right)\prod_{i=1}^{n} C_i^{\nu_i'}\right] \tag{10-2}$$

where ΔH_r is the heat of reaction in J/kg, and RR is the rate of reaction in kg/sec m^3. For a second-order reaction, the above equation reduces to

$$\dot{q}_R = C_1 C_2\, \Delta H_r V A \exp\left(-\frac{E_a}{R_u T}\right) \tag{10-3}$$

The rate of heat loss to the walls of a vessel having a surface area of S, radius of r_p, and surface temperature of T_w can be written as

$$\dot{q}_L = -S\lambda \left.\frac{\partial T}{\partial r}\right|_{r_p} \approx S\lambda \frac{T - T_w}{L} \tag{10-4}$$

where λ is the thermal conductivity of the gaseous mixture at the wall temperature and L is the characteristic thickness of the thermal layer near the wall. It is quite evident from Eqs. (10-3) and (10-4) that both $\dot{q}_R$ and $\dot{q}_L$ are functions of vessel geometry, and therefore the ignition temperature to be derived from these equations will depend upon the geometry of the apparatus. In the ignition process, the pyrometer may first work as a heat source. It heats up the cold reactant mixture to a higher temperature at which chemical reaction and the associated energy release process occur. After the gas temperature has increased beyond T_w, heat is transferred by conduction to the pyrometer wall, which then behaves as a heat sink.

If one plots $\dot{q}_R$ and $\dot{q}_L$ versus the bulk temperature of the reactant gaseous mixture, the former is nonlinear and the latter is linear, as shown in Eqs. (10-3) and (10-4), respectively. Depending upon the wall temperature, there are three different cases as shown in Fig. 10.2.

In case (a), the wall temperature T_{w1} of the vessel is high enough to cause immediate chemical reactions of the gaseous mixture, and the resulting heat generation causes the reaction to accelerate progressively until completion. In this case, $\dot{q}_R(T)$ is always greater than $\dot{q}_L(T)$.

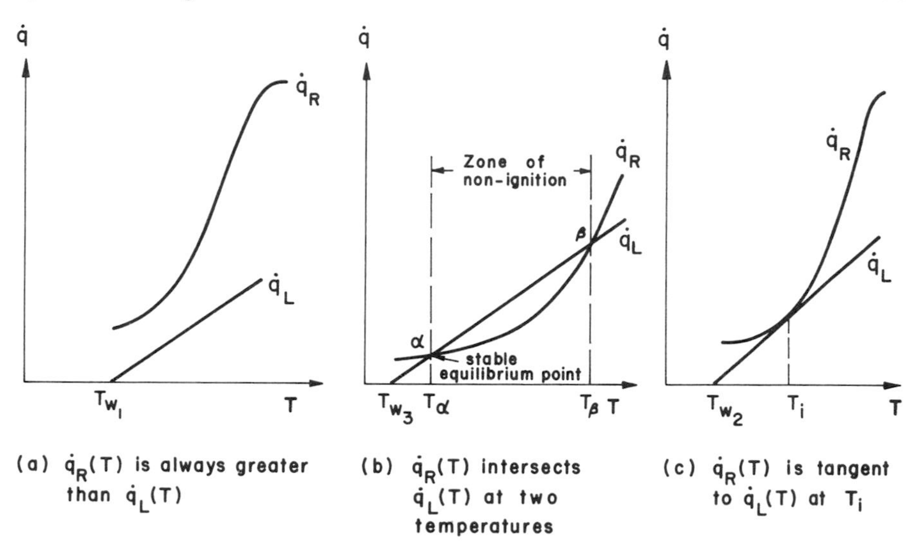

Figure 10.2 Three possible relationships between the rate of heat generation and the rate of heat loss of a reacting gaseous mixture in a vessel with controlled wall temperature: (*a*) $\dot{q}_R(T)$ is always greater than $\dot{q}_L(T)$; (*b*) $\dot{q}_R(T)$ intersects $\dot{q}_L(T)$ at two temperatures; (*c*) $\dot{q}_R(T)$ is tangent to $\dot{q}_L(T)$ at T_i.

In case (*b*), the wall temperature T_{w3} of the vessel is low enough to allow the heat generation by chemical reaction to be significant, and the profile of $\dot{q}_R(T)$ near T_{w3} is very flat. In this case, $\dot{q}_R(T)$ intersects $\dot{q}_L(T)$ at two temperatures, T_α and T_β. The lower point α represents a stable equilibrium; the reactant mixture will self-heat to T_α, but no further, because $\dot{q}_L$ is greater than $\dot{q}_R$ for any small perturbation of temperature above T_α. In order to have ignition, one has to heat the mixture by some other means to point β and thereby create a runaway situation. The critical wall temperature for ignition is T_{w2}.

In case (*c*), the $\dot{q}_R(T)$ curve is tangent to $\dot{q}_L(T)$ at the ignition temperature T_i. This implies that after the reactant gases have been introduced into the vessel, they will self-heat to T_i and then run away. The ignition temperature T_i can be defined mathematically by equating the magnitude and slopes of $\dot{q}_R$ and $\dot{q}_L$ at the tangent point:

$$\dot{q}_L(T_i) = \dot{q}_R(T_i) \tag{10-5}$$

$$\left.\frac{d\dot{q}_L}{dT}\right|_{T_i} = \left.\frac{d\dot{q}_R}{dT}\right|_{T_i} \tag{10-6}$$

Substituting Eqs. (10-3) and (10-4) into the above two equations and expressing the geometry parameters in terms of radius of the pyrometer, we have

$$k_1 \lambda r_p (T_i - T_w) = k_2 C_1 C_2 \, \Delta H_r \, r_p^3 A \exp\left(-\frac{E_a}{R_u T_i} \right) \tag{10-7}$$

and

$$k_1 \lambda r_p = \frac{E_a}{R_u T_i^2} k_2 C_1 C_2 \, \Delta H_r \, r_p^3 A \exp\left(-\frac{E_a}{R_u T_i} \right) \tag{10-8}$$

Dividing Eq. (10-7) by (10-8) and rearranging the terms, we have

$$\frac{R_u T_i^2}{E_a} - T_i + T_w = 0 \tag{10-9}$$

The lower root gives the minimum ignition temperature,

$$T_i = \frac{1 - \sqrt{1 - 4R_u T_w / E_a}}{2R_u / E_a} = T_w + \frac{R_u T_w^2}{E_a} + \frac{2R_u^2 T_w^3}{E_a^2} + \cdots \tag{10-10}$$

By neglecting higher-order terms in the series, we have

$$T_i = T_w + \frac{R_u T_w^2}{E_a} \tag{10-11}$$

For a reasonable value of the activation energy, one can verify that the difference between T_i and T_w is in the order of 10–25 K. In view of the close proximity of these two temperatures, the critical wall temperature T_w can also be regarded as the ignition temperature.

3.2 Effect of Various Parameters on the Self-Ignition Temperature

For a fixed chamber volume, the slopes of $\dot{q}_L$ versus T in Fig. 10.2*a*, *b*, and *c* are nearly constant and are given by $S\lambda/L$ in Eq. (10-4). The effect of chamber volume can be seen from Fig. 10.3. The solid lines in this figure are traced from Fig. 10.2*c*. The dashed lines, corresponding to a case of smaller volume, show no spontaneous ignition when wall temperature is at T_{w2}. This figure clearly shows that by decreasing the chamber volume, an ignitable system becomes nonignitable. The physical reason for the vessel size to have an effect on the self-heated ignition process is the increase of heat loss and reduction of heat generation as the vessel size becomes smaller. Also, the pressure in the container can influence the self-ignition process: the higher the pressure, the higher the density, or else the larger the chamber volume. The critical size for compounds such as ammonium nitrate as a function of the vessel temperature has been studied by Hainer.[3] In general, the higher the vessel temperature, the smaller the mass has to be in order to prevent disaster.

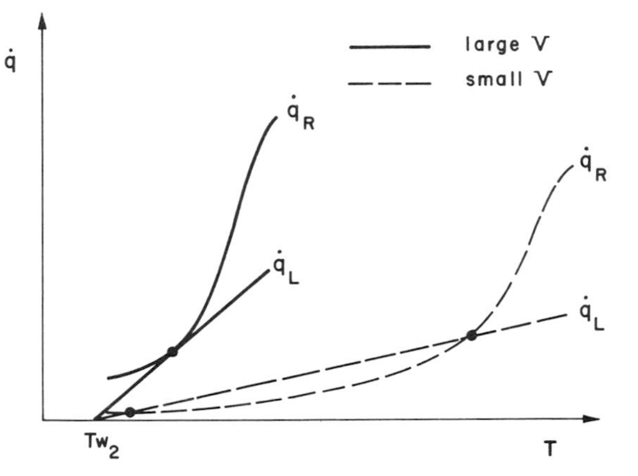

Figure 10.3 Effect of chamber volume on a self-heated ignition process.

To study the spontaneous ignition process in an infinitely large system, one does not have to construct a very large container. An infinite system can be simulated by controlling the wall temperature to be always equal to the gas temperature inside the vessel.

The reactant composition also has an effect on the ignition temperature, since $\dot{q}_R$ is a function of the reaction rate, which depends upon the composition. A typical relationship between T_i and the fuel/oxidant ratio, F/O, is shown in Fig. 10.4. The minimum T_i does not necessarily correspond to the stoichiometric mixture composition.

The ignition temperature is also a function of time. From Arrhenius's law, one can write

$$\frac{1}{\mathrm{RR}} \propto \text{time} \propto \exp\left(\frac{E_a}{R_u T_i}\right) \tag{10-12}$$

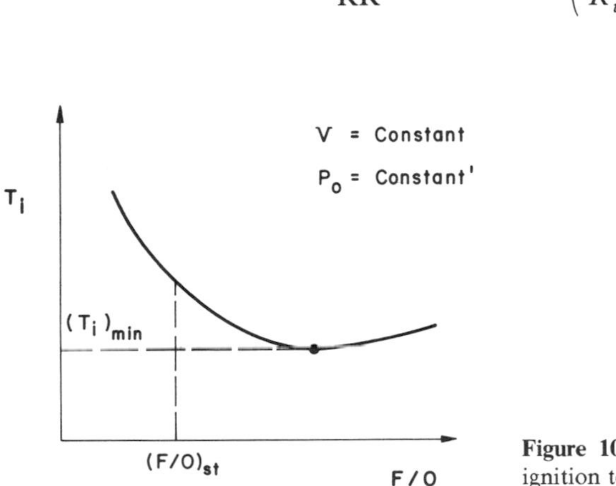

Figure 10.4 Effect of fuel/oxidant ratio on ignition temperature.

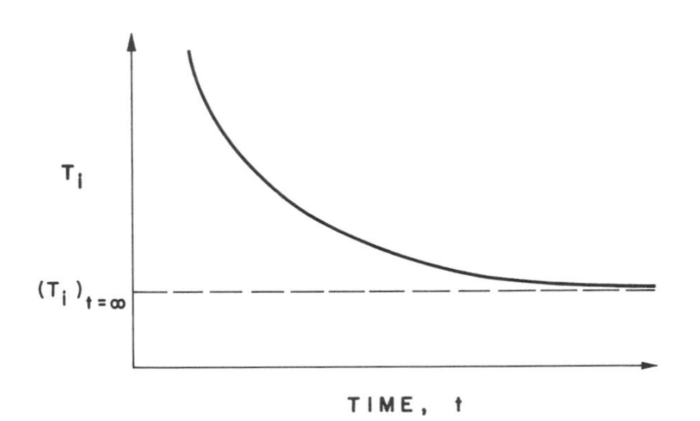

Figure 10.5 Effect of storage time or flow residence time on ignition temperature.

The dependence of T_i on time is shown in Fig. 10.5. The fact that T_i decreases with time implies that in order to achieve self-heated spontaneous ignition at a specified level of T_i, a sufficient time must be given for runaway to occur. In a flow system, the induction time required for self-heated ignition to occur corresponds to the minimum flow residence time of a given gaseous mixture under specified flow conditions.

In general, the ignition temperature is a function of the size, shape, and material of the apparatus, the initial temperature of the mixture, the reactant composition, the activation energy of the governing chemical reaction of the reactants in the mixture, the time, the pressure, the velocity of the fluid element, and the turbulence scale and intensity of the flow. It is meaningful only for a well-defined system. One must be extremely cautious in interpreting and extrapolating measured experimental data.

4 IGNITION OF SOLID PROPELLANTS

The study of the ignition of solid propellants is important for many combustion and propulsion applications. An extensive review of research work performed in this area was conducted 16 years ago by Price et al.[4] Many ignition studies have been conducted since then; a detailed survey of subsequent literature is given by Kulkarni, Kumar, and Kuo in Ref. 5, in which over 100 publications are cited. This paper brings together the developments to date and the difficulties encountered in order to establish the state of the art of solid-propellant ignition. An excellent summary of solid-propellant ignition theories and experiments is also given by Hermance.[6]

In general, ignition of a solid propellant is a complex phenomenon which involves many physicochemical processes, as depicted in Fig. 10.6. The ignition

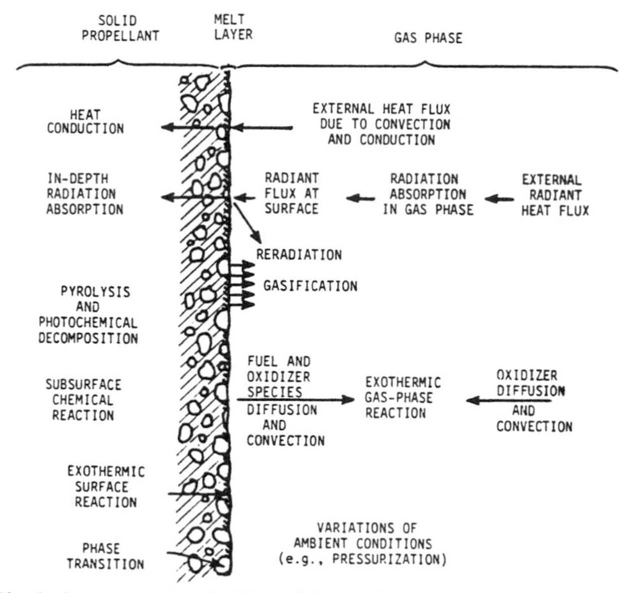

Figure 10.6 Physical processes involved in solid-propellant ignition (from Kulkarni, Kumar, and Kuo[5]). Copyright American Institute of Aeronautics and Astronautics and reprinted with permission of the AIAA.

consists of the following sequence of events:

- Transfer of energy by conduction, convection, and/or radiation
- Inert heating: in-depth radiation absorption, conduction inside the solid propellant
- Phase transition of oxidizer particles
- Decomposition of solid phase
- Pyrolysis of fuel binder
- Photochemical decomposition of the condensed phase
- Subsurface chemical reactions
- Diffusion of pyrolyzed species from the surface and counterdiffusion of oxidizer species from surrounding to reaction site
- Heterogeneous reaction between gaseous species and condensed phase
- Gas-phasc rcaction
- Abrupt increase of temperature and/or highly reactive radicals
- Emission of light from reaction zone
- Development of thermal wave and gaseous flame.

When the net heat evolved from chemical reactions overcomes heat losses, sustained ignition is achieved. It is generally understood that ignition is incomplete if steady-state combustion does not follow the ignition event after the removal of the external energy stimulus.

The time period from the start of the external stimulus to the instant of sustained ignition, called the *ignition delay*, is one of the most important parameters in the study of ignition. Generally, it comprises the inert heating time, mixing (diffusion plus convection) time, and reaction time. The ignition delay, however, is not simply the sum of these three characteristic time intervals, since there is no clear demarcation between the mixing process and the chemical reactions; they may overlap. Also, it is very difficult to identify precisely the instant of sustained ignition.

Following the review of Price et al.,[4] ignition theories are classified into three major groups: (1) gas-phase, (2) heterogeneous, and (3) solid-phase. The gas-phase ignition theory considers the ignition process to be controlled by the chemical reaction between vaporized fuel-rich and oxidizer-rich mixtures and ambient oxidizer gases. In the heterogeneous ignition theory, the reaction between the solid-phase fuel and ambient oxidizer at the interface is viewed as the controlling mechanism. The solid-phase ignition theory does not consider heat release and mass diffusion in the gas phase; rather, the temperature rise inside a solid propellant is supposed to be achieved by the heat release caused by subsurface chemical reaction and/or external heating from the surroundings.

Within each of the three major theories described above, several models have been proposed. These models differ in the governing equations considered, assumptions made, interfacial conditions imposed, ignition criteria used, and types of propellants studied. A complete set of governing equations and boundary conditions, together with the physical meaning of each term in the formulation, is given below.

The solid-phase energy equation is

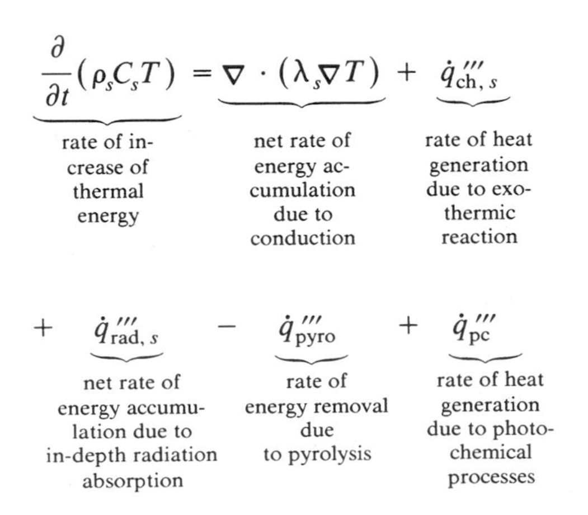

$$\underbrace{\frac{\partial}{\partial t}(\rho_s C_s T)}_{\text{rate of increase of thermal energy}} = \underbrace{\nabla \cdot (\lambda_s \nabla T)}_{\text{net rate of energy accumulation due to conduction}} + \underbrace{\dot{q}'''_{ch,s}}_{\text{rate of heat generation due to exothermic reaction}}$$

$$+ \underbrace{\dot{q}'''_{rad,s}}_{\text{net rate of energy accumulation due to in-depth radiation absorption}} - \underbrace{\dot{q}'''_{pyro}}_{\text{rate of energy removal due to pyrolysis}} + \underbrace{\dot{q}'''_{pc}}_{\text{rate of heat generation due to photochemical processes}} \qquad (10\text{-}13)$$

For spatially uniform pressure, the gas-phase energy equation is

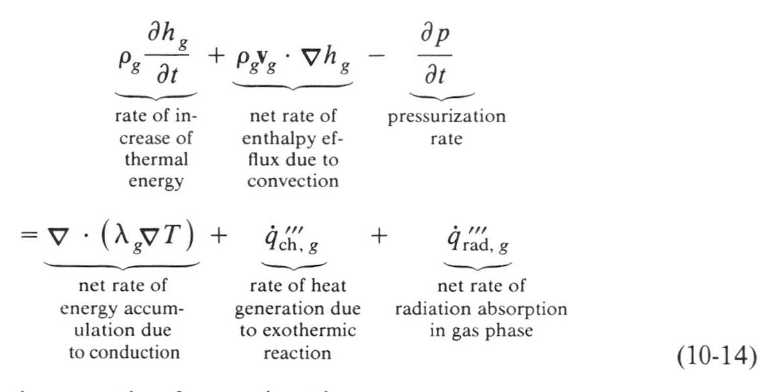

$$\underbrace{\rho_g \frac{\partial h_g}{\partial t}}_{\substack{\text{rate of in-}\\\text{crease of}\\\text{thermal}\\\text{energy}}} + \underbrace{\rho_g \mathbf{v}_g \cdot \nabla h_g}_{\substack{\text{net rate of}\\\text{enthalpy ef-}\\\text{flux due to}\\\text{convection}}} - \underbrace{\frac{\partial p}{\partial t}}_{\substack{\text{pressurization}\\\text{rate}}}$$
$$= \underbrace{\nabla \cdot (\lambda_g \nabla T)}_{\substack{\text{net rate of}\\\text{energy accum-}\\\text{ulation due}\\\text{to conduction}}} + \underbrace{\dot{q}'''_{\text{ch},g}}_{\substack{\text{rate of heat}\\\text{generation due}\\\text{to exothermic}\\\text{reaction}}} + \underbrace{\dot{q}'''_{\text{rad},g}}_{\substack{\text{net rate of}\\\text{radiation absorption}\\\text{in gas phase}}} \tag{10-14}$$

The conservation equation for species i is

$$\underbrace{\rho_g \frac{\partial Y_i}{\partial t}}_{\substack{\text{rate of in-}\\\text{crease of}\\\text{species } i}} + \underbrace{\rho_g \mathbf{v}_g \cdot \nabla Y_i}_{\substack{\text{mass efflux}\\\text{of species } i\\\text{due to con-}\\\text{vection}}} = \underbrace{\nabla \cdot (\rho \mathscr{D} \nabla Y_i)}_{\substack{\text{mass influx of}\\\text{species } i \text{ due}\\\text{to diffusion}}} + \underbrace{\dot{\omega}_i'''}_{\substack{\text{rate of gener-}\\\text{ation of species}\\ i \text{ due to chemi-}\\\text{cal reaction}}} \tag{10-15}$$

The overall continuity equation is

$$\underbrace{\frac{\partial \rho_g}{\partial t}}_{\substack{\text{rate of mass}\\\text{accumulation}}} + \underbrace{\nabla \cdot (\rho_g \mathbf{v}_g)}_{\substack{\text{mass efflux}\\\text{rate}}} = 0 \tag{10-16}$$

The energy flux balance at the gas–condensed-phase interface is

$$\underbrace{\lambda_s \nabla T|_-}_{\substack{\text{conductive}\\\text{heat flux}\\\text{into solid}}} = \underbrace{\lambda_g \nabla T|_+}_{\substack{\text{conductive}\\\text{heat flux}\\\text{from gas}\\\text{phase}}} + \underbrace{a\dot{\mathbf{q}}'''_{\text{rad}}}_{\substack{\text{net radiation}\\\text{flux absorbed}\\\text{at surface}}}$$
$$- \underbrace{\varepsilon \mathbf{E}_b}_{\substack{\text{radiation}\\\text{emitted}\\\text{from}\\\text{surface}}} + \underbrace{\mathbf{r}_b \rho_s (h_{t,s} - h_{t,g})}_{\substack{\text{net flux of}\\\text{total enthalpy}\\\text{convected into}\\\text{interface}}} + \underbrace{\dot{\mathbf{q}}''_{\text{ch}}}_{\substack{\text{rate of heat}\\\text{generation at}\\\text{interface due to}\\\text{chemical reaction}}}$$
$$- \underbrace{\dot{\mathbf{q}}''_{\text{pyro}}}_{\substack{\text{rate of energy}\\\text{consumed due to}\\\text{pyrolysis of}\\\text{solid phase}}} - \underbrace{\rho_g \sum_{i=1}^{n} h_i Y_i \mathbf{V}_i \Big|_+}_{\substack{\text{rate of energy loss}\\\text{at interface due to}\\\text{mass diffusion in}\\\text{gas phase}}} \tag{10-17}$$

The mass-flux balance at the solid–gas interface is

$$\underbrace{\rho_s r_b Y_i|_-}_{\substack{\text{rate of}\\ \text{mass flux}\\ \text{of species}\\ i \text{ from}\\ \text{subsurface}\\ \text{due to}\\ \text{regression}\\ \text{of solid}\\ \text{phase}}} = \underbrace{\rho_g v_g Y_i|_+}_{\substack{\text{rate of mass}\\ \text{flux of}\\ \text{species } i\\ \text{convected}\\ \text{away from}\\ \text{interface}}} - \underbrace{\rho_g \mathscr{D} \mathbf{n} \cdot \nabla Y_i|_+}_{\substack{\text{rate of mass}\\ \text{flux of}\\ \text{species } i\\ \text{diffused}\\ \text{away from}\\ \text{interface}}} - \underbrace{\dot{\omega}_i''}_{\substack{\text{rate of mass}\\ \text{depletion}\\ \text{of species}\\ i \text{ due to}\\ \text{chemical}\\ \text{reaction}}} \tag{10-18}$$

where $\mathbf{n}$ is the outward unit vector at the interface.

Selection of a proper ignition criterion is probably the most controversial issue in ignition study. There has been no universally acceptable definition. In the literature, many different ignition criteria have been proposed in theoretical models, and many different criteria used in experimental studies.[5] This is probably due to the fact that ignition depends not only on the mode of energy deposition onto the solid propellant, but also on propellant characteristics and ambient conditions. Since calculated as well as observed values of the ignition delay depend on the choice of an ignition criterion, conclusions drawn from comparisons of theory and experiments are also affected by this somewhat arbitrary choice.

Commonly used ignition criteria in theoretical studies are as follows:

- ☐ When the surface temperature exceeds a critical value
- ☐ When the rate of rise of the surface temperature exceeds a critical value
- ☐ When there is a point of inflection in the surface-temperature–time trace
- ☐ When the subsurface temperature at a given distance exceeds a critical value
- ☐ When the temperature at any point in the gas phase exceeds a critical value
- ☐ When the gas temperature rises after the removal of external stimuli
- ☐ When the rate of rise of the maximum gas temperature approaches a low value, signifying the attainment of a near-steady-state condition
- ☐ When the light emitted by the hot combustion gases exceeds a critical intensity
- ☐ When the maximum reaction rate in the gas phase exceeds a critical value
- ☐ When the spatially integrated gas-phase reaction rate exceeds a critical value
- ☐ When the rate of rise of the spatially integrated gas-phase reaction rate exceeds a critical value
- ☐ When the gas-phase heat generation is balanced with the heat loss to the solid phase

Ignition criteria used in experimental studies are as follows:

- □ First appearance of flame recorded in high-speed motion pictures
- □ Onset of light emission detected by a photocell
- □ Attainment of a certain light intensity detected by a photodiode
- □ Entry into the ignition zone by passing the go–no-go ignition boundary
- □ Abrupt rise in a thermocouple output
- □ Onset of mass loss of the propellants
- □ Abrupt change in voltage–current characteristic of an electrically heated wire
- □ Onset of signal from an ionization pin which is submerged immediately below the propellant surface

In gas-phase models, the ignition criterion is generally based on variations in gas-phase temperature or reaction-rate distributions, whereas in heterogeneous and solid-phase models the ignition criterion is usually based on the attainment of a critical temperature or a critical rate of temperature increase at the propellant surface. The most commonly used ignition criteria in experimental studies are the first detection of flame and the go–no-go test. The desired characteristics of a definition are (1) compatibility between experimental and theoretical studies, and (2) insensitivity to the values of the constants employed to characterize the runaway condition.

General ignition-behavior boundaries for an arbitrary propellant have been suggested by De Luca et al.[7] Figure 10.7 is a general map derived from their radiative ignition of double-base propellants and results of many previous investigators. This figure can easily be understood if one considers the time sequence of events occurring when a propellant is subjected to a fixed incident radiant flux at constant chamber pressure. Corresponding to these boundaries, De Luca et al.[8] have recorded some interesting high-speed shadowgraph movies. Figure 10.8*a* shows a number of photographs illustrating flame development on noncatalyzed double-base propellants. Figure 10.8*b* shows the formation of a carbonaceous layer adjacent to the surface of catalyzed double-base propellants. In general, the ignition delay versus heat flux is a function of pressure, as shown in Fig. 10.9, reported by Beyer et al.[9] The higher the pressure, the closer the curve of self-sustaining ignition to the boundary of incipient gas evolution (represented by the line of slope -2.0).

Most of the important experimental studies and gas-phase, solid-phase, and heterogeneous models are summarized in separate tables in Ref. 5. For each investigation, the tables include the source, basic assumptions, theoretical formulation, ignition criterion, results, and review comments. Test setup and measurement techniques used in the experimental studies are also included.

In the gas-phase ignition theory, most models (except that proposed by Kumar and Hermance[10] and Kumar, Wills, Kulkarni, and Kuo[11]) are one-

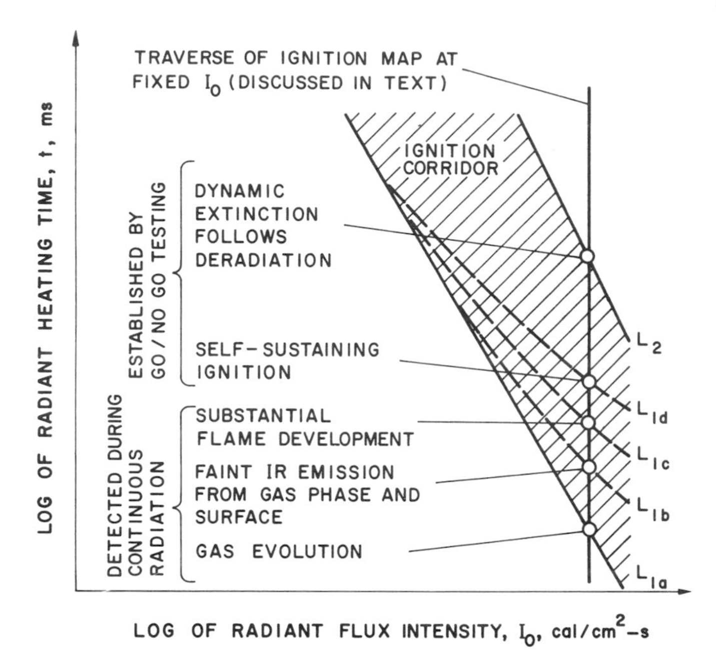

Figure 10.7 Generalized ignition map showing event limits or signals that occur during radiant heating of solid propellants (from De Luca et al., 1976).[7]

dimensional. The gas-phase theory is usually more complicated than the solid-phase or heterogeneous ignition theory, because gas-phase species and energy equations must be included in the formulation. The major advantage of the gas-phase theory is its ability to predict the dependence of ignition delay on such ambient conditions as oxidizer concentration and pressure. In composite propellants whose fuel-binder ablation temperature is considerably lower than that of the oxidizer particles, the fuel may react with the ambient oxidizer before any significant decomposition of the solid-phase oxidizer occurs. Under such circumstances, the gas-phase ignition theory is more realistic.

Compared with gas- and solid-phase theories, little attention has been given to heterogeneous ignition theory, probably because the ignition process in the usual rocket motor does not depend chiefly on a rate-controlling heterogeneous reaction. A majority of the models proposed in this theory use asymptotic methods to obtain solutions. The heterogeneous ignition theory is also called *hypergolic* when the surface reaction begins immediately following the introduction of reactive fluids at room temperature. The hypergolic ignition theory also differs from the gas- and solid-phase theories with regard to the external heat flux, which can be absent in the hypergolic situation but is required in the other two.

Mathematical formulation of the solid-phase theory is greatly simplified in that gas-phase equations are not considered. Thus, the solid-phase theory

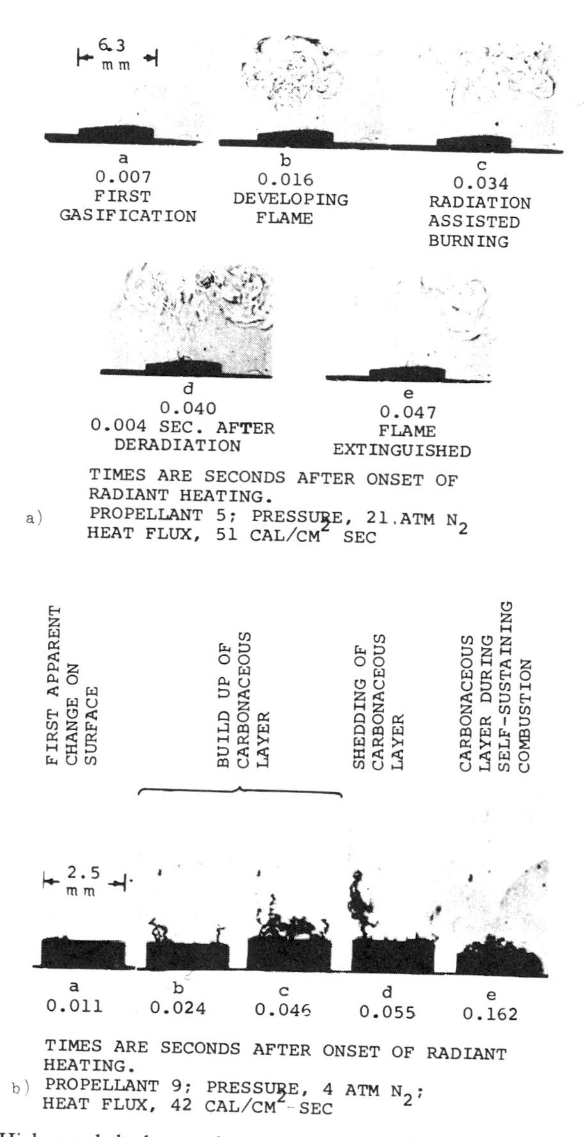

Figure 10.8 (*a*) High-speed shadowgraph movie illustrating flame development on noncatalyzed DB propellant; note the visible flame in (*c*) and (*d*) and the large standoff distance. Sample diameter is larger than the 2.5-mm size used elsewhere in this study. (*b*) High-speed shadowgraph movie showing carbonaceous layer formation on the surface of catalyzed DB propellant. (From De Luca et al.[8])

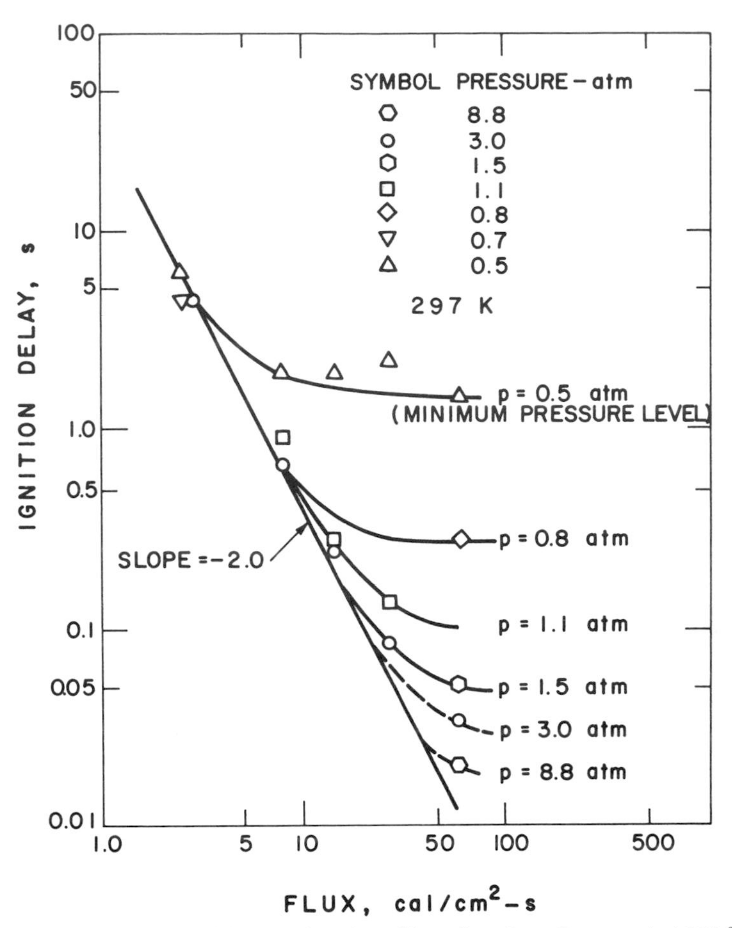

Figure 10.9 Ignition delay as a function of heat flux (from Beyer et al., 1965).[9,4]

generally cannot predict the effect of conditions in the surrounding atmosphere. However, under conditions of low external heat flux, high ambient oxidizer concentration, and high pressure, the inert heating time is much longer than the diffusion or chemical-reaction time; the ignition delay predicted by the solid-phase theory may then agree closely with experimental observations.

The actual ignition process of solid propellants is, in general, too complex to be described by any one gas-phase, solid-phase, or heterogeneous theory under all operating conditions. It is desirable, therefore, to develop a theory which allows simultaneous reactions in the gas phase and solid phase and at the interface, and which also employs an ignition criterion that is flexible enough to allow a runaway condition at any site. Such an effort has been made by Bradley,[12] Kulkarni et al.,[13] and Kumar et al.[11] A number of useful results have been obtained.

The major difficulties encountered in a theoretical analysis of solid-propellant ignition include: selection of an appropriate reaction mechanism for the propellant under specific operating conditions; detailed specification of chemical kinetics of the precombustion reactions; unavailability of adequate thermal-, chemical-, and transport-property data; multidimensionality of the ignition process resulting from heterogeneity of the propellant; and the selection of the ignition criterion. In addition, the numerical solution of the model is intricate and cumbersome.

The experimental study of ignition is just as complex as the theory. Adequate simulation of the ignition process in a rocket motor in a laboratory is one of the major difficulties encountered in solid-propellant ignition studies. The actual igniter generates hot gases and particles flowing over or impinging upon a solid propellant surface, whereas most of the laboratory experiments use idealized methods of energy transfer which cannot fully simulate the actual process. In addition, the time period of the entire ignition event is very short (several hundreds of microseconds to a few milliseconds for many commonly used solid propellants under normal conditions), and the region of major activity is extremely small (usually on the order of several hundred micrometers). Thus, because it is difficult to probe and observe the ignition region, no broad experimental data base is as yet available.

It is usually observed that the ignition delay decreases[5] with increasing rate of external energy transfer to the propellant, ambient pressure, ambient oxidizer concentration, gas-to-solid density or thermal-conductivity ratio, ambient pressurization rate, exothermic chemical reaction rates, and initial temperature of the propellant or surrounding gas. An increase in the oxidizer particle size decreases the ignition delay. In general, the types of fuel binders, oxidizers, catalysts, opacifiers, and burning-rate modifiers used (together with their relative mass fractions) have some effect on ignition delay.

A significant amount of work has been done in the area of solid-propellant ignition; qualitative (and in some cases quantitative) predictions for the ignition delay and its dependence on several parameters can be made by adjusting various constants and by making *a priori* assumptions about the controlling mechanism. However, these predictions are semiempirical, and a significant amount of research in this area is needed to obtain accurate, systematical, and extensive data for comparison with theory. Overall, although the foundation for solid propellant ignition research is laid, ignition studies are still far from complete.

5 IGNITION OF LIQUID-FUEL SPRAYS IN A FLOWING AIR STREAM

The ignition of liquid-fuel sprays, such as kerosene spray, in a flowing air stream is a process of considerable importance in many practical combustion devices. It is of special significance for aircraft gas-turbine engines in view of the need for a rapid reignition in the event of flame extinction at high altitudes. Since the ignition process in most combustion chambers is complicated by wide variations in flow velocity, turbulence, and air–fuel ratio in the region adjacent to the igniter, many basic studies of ignition have been largely concerned with quiescent mixtures or with flowing air streams for a single fuel droplet. For a more general and realistic application, this section describes the ignition of multidroplet mists in a flowing air stream. The experiment, based on the ignition and flame-quenching theory of Ballal and Lefebvre,[14–16] shows strong dependence of the minimum ignition energy on fuel drop size, air velocity, and air–fuel ratio, as discussed in the following. Some criteria for the design of combustion chambers will also be explained.

5.1 Test Apparatus of Subba Rao and Lefebvre and the Concept of Minimum Ignition Energy

A diagram of the test facility used by Subba Rao and Lefebvre[17] is shown schematically in Fig. 10.10. Basically, the test section comprises a circular stainless-steel tube, about 7 cm in diameter and 36 cm long, through which air is supplied at atmospheric pressure by means of a suction pump. The tube is fitted with glass windows to allow the onset of the spark and subsequent growth of the flame kernel to be visually observed and photographed. At its upstream end are two tungsten spark electrodes of 3-mm diameter, mounted in insulating bushes. The upper electrode has a micrometer adjustment to provide an accurate control of the spark-gap width. Downstream from the test section is an exhaust duct which is fitted with four drain points to facilitate the removal of liquid fuel precipitated on the duct walls.

Fuel injection was accomplished using simplex swirl atomizers located at the center of the convergent entry duct. In order to achieve a wide variation in air–fuel ratio and mean drop size, five atomizers were employed, having flow

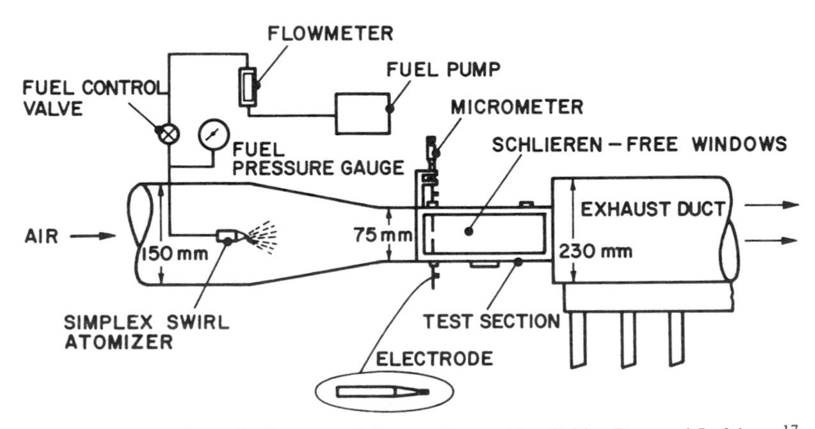

Figure 10.10 Schematic diagram of the test rig used by Subba Rao and Lefebvre.[17]

numbers of 0.3, 0.4, 0.5, 0.6, and 0.7. [The flow number is defined as (fuel flow rate in U.S. gallons per hour)/(fuel injection pressure in psi)$^{1/2}$.]

Mean drop sizes were measured with each atomizer, using a diffractive light-scattering technique[18] between the electrodes. The Sauter mean diameter (SMD) is defined as the diameter of a drop whose volume-to-surface ratio is equal to that of sprays as a whole (see Chapter 8 for its mathematical definition). Air flow rates were measured on a venturi meter fitted at the entrance of the tube. Fuel flow rates were obtained by dividing the total amount of fuel used at any fixed test condition by the duration of the fuel flow. The air–fuel ratio was then obtained as the ratio of air flow rate to fuel flow rate.

The ignition unit of the test rig was designed to supply damped capacitance sparks whose energy and duration could be varied independently. Spark energies were measured by mounting two probes on the electrodes and feeding their output into a two-channel oscilloscope. Traces were recorded on a Polaroid camera, and the spark energies were obtained as the integral of the product of the current and voltage readings. Based on the work of Swett[19] and Ballal,[20] the optimum spark duration and the optimum spark gap width correspond to minimum ignition energy; 100-μsec duration and 0.3-cm gap width were chosen. The main specifications of the unit were:

Breakdown voltage = 0–15 kV

Maximum current in spark = 22 A

Available spark energies = 19.3, 26, 38, 50.7, 65, and 130 mJ

Spark duration = 100 μsec

The process of ignition is envisaged to occur in the following manner. Passage of the spark creates a small, roughly spherical volume of a mixture of gas and liquid drops (referred to henceforth as the spark kernel), whose temperature is sufficiently high to initiate rapid evaporation of the fuel drops within it. Mixing times are extremely short in the spark kernel, and reaction rates can be regarded as infinitely fast, so that any fuel vapor created within the spark kernel is instantly transformed into combustion products at the stoichiometric flame temperature. If the rate of heat release by combustion of evaporated fuel with air exceeds the rate of heat loss by thermal conduction at the surface of the inflamed volume, then the spark kernel will grow in size to fill the entire combustion volume. If, however, the rate of heat release is less than the rate of heat loss, the temperature within the spark kernel will fall steadily until fuel evaporation ceases altogether. Thus, of crucial importance to ignition is the size of the spark kernel at which the rate of heat loss at its surface is just balanced by the rate of heat release, due to the instantaneous combustion of fuel vapor, throughout its volume. This concept leads to the definition of *quenching distance* as the critical size which the inflamed volume must attain in order to propagate unaided, while the amount of energy required from an external source to attain this critical size is termed the *minimum ignition energy*.

All the tests were carried out at atmospheric pressure with air temperature between 290 and 295 K. The minimum fuel flow rate for ignition was determined for various values of the atomizer flow number and spark energy. At any fixed values of atomizers flow number, flow velocity, and spark energy, the fuel flow rate was gradually increased until after each spark a flame was visible in the window. This procedure was carried out for all atomizers, using all six levels of spark energy, at air velocities of 20, 30, 37.5, and 46 m/sec. The results discussed in the following subsections correspond to weak ignition limits.

5.2 Ignition Results for Kerosene Sprays in a Flowing Air Stream

The results obtained by Subba Rao and Lefebvre[17] with an air velocity of 37.5 m/sec are shown in Fig. 10.11, in which the minimum ignition energy is plotted against air–fuel ratio for four different values of the mean drop size. For each curve the left-hand side represents a region of no ignition and the right-hand side a region of ignition. This figure clearly shows how the weak-ignition limit is considerably improved by a reduction in mean drop size. The effect is especially pronounced at the smallest drop size. Another interesting observation from Fig. 10.11 is that the curves of minimum ignition energy versus air–fuel ratio become steeper with increasing mean drop size. This indicates that when the mean drop size is large, a much greater increase in spark energy is needed to extend the weak-ignition limit than for small mean drop size. The practical significance of these results is that atomizers of low flow number are far more effective from an ignition viewpoint than atomizers of high flow number.

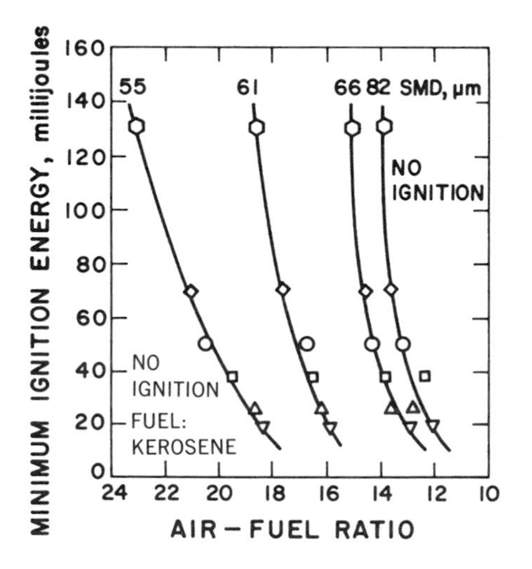

Figure 10.11 Influence of air–fuel ratio on minimum ignition energy. Air velocity = 37.5 m/sec (after Subba Rao and Lefebvre[17]).

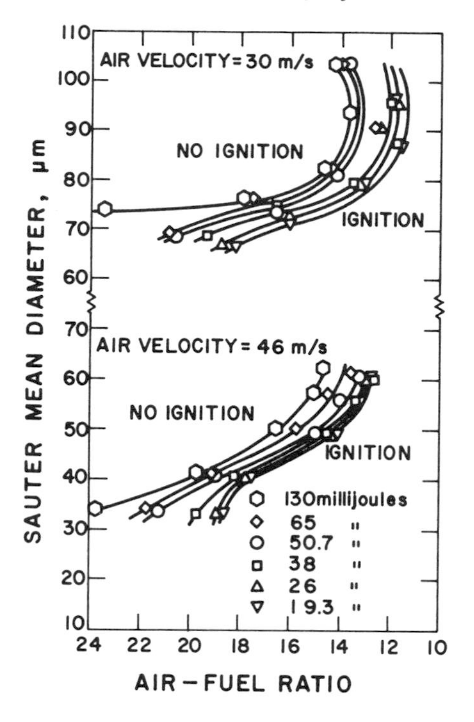

Figure 10.12 Weak-ignition limits obtained at air velocities of 30 and 46 m/sec for different levels of spark energy (after Subba Rao and Lefebvre[17]).

In Fig. 10.12 the ignition limits obtained at air velocities of 30 and 46 m/sec for all six levels of spark energy are plotted in terms of SMD versus air–fuel ratio. For both velocities the weak-ignition limits are widened by an increase in spark energy, presumably because of the resulting increase in the temperature and the size of the spark kernel. An increase in temperature accelerates chemical-reaction rates, while an increase in the kernel size makes the ignition less susceptible to the quenching effects of the surrounding fresh mixture. As the air–fuel ratio is reduced towards the stoichiometric value of 14.8, the more rapid chemical reaction within the spark kernel liberates more heat for further fuel evaporation and allows the ignition limits to extend into a region of larger drop sizes. At mixture strengths richer than stoichiometric the lines of constant spark energy become increasingly steep, until a region is encountered in which the weak-ignition limits become sensibly independent of fuel drop size. This may be explained on the grounds that for fuel-rich mixtures a partial evaporation of the total fuel contained within the spark kernel can produce local mixture strengths that are close to stoichiometric, and therefore more conducive to ignition. Thus under these conditions, an increase in drop size, with consequent reduction in drop surface area and in fuel evaporation rate, will actually improve ignition.

Figure 10.13 shows the variation of minimum ignition energy with the SMD of the spray for stoichiometric mixtures at four velocities. The most striking feature of this figure is that at all values of velocity, large increases in spark

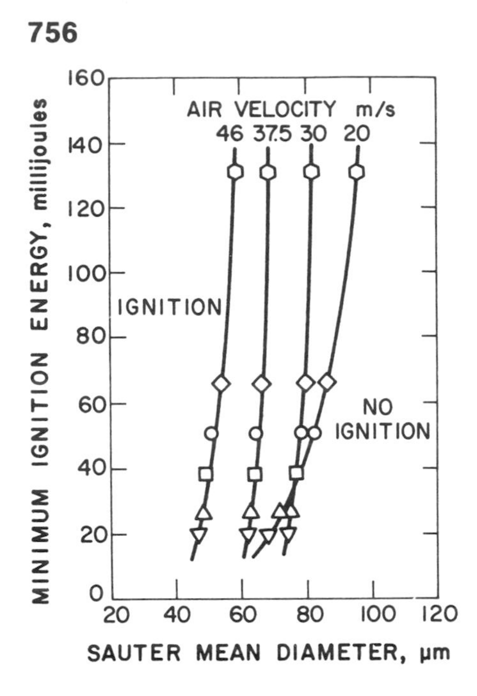

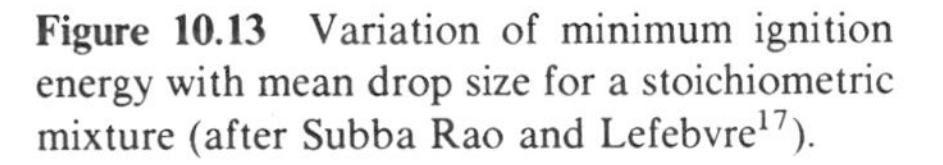

Figure 10.13 Variation of minimum ignition energy with mean drop size for a stoichiometric mixture (after Subba Rao and Lefebvre[17]).

energy are needed to compensate for quite modest increases in mean drop size. From a practical viewpoint, this figure illustrates the crucial role of the fuel injector in the attainment of good ignition performance.

The data obtained with stoichiometric mixtures are presented in Fig. 10.14. The conclusion drawn from this figure is that flow velocity has a strong effect on ignition performance and that relatively small increases in velocity entail fairly substantial increases in ignition energy.

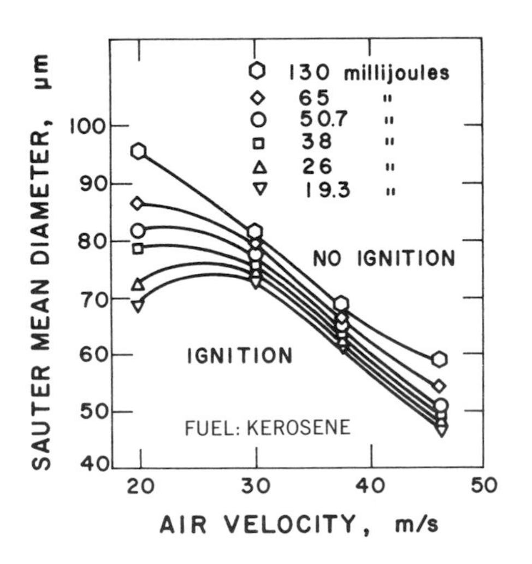

Figure 10.14 Influence of air velocity on ignition limits for a stoichiometric mixture (after Subba Rao and Lefebvre[17]).

A study of the growth of the spark kernel was also made by direct flame photography at a speed of several thousand frames per second. A typical pattern of flame initiation and development in a kerosene–air mixture flowing at a specific velocity was recorded by Subba Rao and Lefebvre.[17] Their observation supports the view that passage of the spark creates a kernel in which high gas temperatures are attained, partly from the energy supplied in the spark, and partly from the heat liberated by rapid combustion of the fuel drops. This high initial temperature of the spark kernel then falls as heat is lost by diffusion to the fresh mixture in contact with the outside surface of the kernel, and to the remaining fuel drops undergoing evaporation within the kernel.

In general, the results obtained clearly demonstrate that for kerosene sprays injected into a flowing air stream the minimum spark energy for ignition is markedly raised by increases in mean fuel drop size and air velocity. Improvements in fuel atomization during the ignition sequence and minimization of the flow velocity in the ignition zone can significantly enhance the ignition of liquid-fuel sprays in flowing air streams.

5.3 Lefebvre's Evaporation Model for Quenching Distance and Minimum Ignition Energy in Liquid-Fuel Sprays

Lefebvre[21] presented a theoretical model for the prediction of quenching distance and minimum energy in the ignition of liquid-fuel sprays. This model was based on the evaporation of the liquid-fuel droplet and assumed infinitely fast reaction rates and mixing times. The sole criterion for successful ignition was considered to be an adequate supply of fuel vapor in the ignition zone, since the major factor limiting ignition performance was a deficiency of fuel vapor in the ignition zone.

In this model, the passage of a spark into the mixed gases was assumed to create a spark kernel, as described in the previous section. The spark kernel is a small, roughly spherical volume of gas whose temperature is sufficiently high to initiate evaporation of the fuel drops within the volume. Since the reaction rates and mixing times were assumed infinitely fast, the fuel vapor within the spark kernel was assumed instantly transformed into combustion products at the stoichiometric flame temperature.

The essential condition for propagation of the spark kernel (i.e., for the combustion to continue and ignition to be successful) was that the rate of heat generation by the fuel evaporation exceeds the rate of heat loss. If this condition was not met, the temperature in the spark kernel would decrease until fuel evaporation ceased. If, however, the condition was met, the spark kernel would grow and eventually fill the entire combustion volume.

The *quenching distance* d_q was defined as the critical size at which the inflamed spark kernel would grow unaided. The least amount of energy required from an external source to create such a spark kernel whose size is equal to the quenching distance was defined as the *minimum ignition energy*

E_{min}. Thus, E_{min} is the amount of energy needed to raise a spherical volume of air–fuel mixture of diameter d_q up to the stoichiometric flame temperature.

From the condition that the rate of heat generated by the fuel evaporation exceeds the rate of heat loss, equations for E_{min} were developed for both stagnant and flowing mixtures:

$$E_{min} = C_{p\,air}\rho_{air}\Delta T_{stoich}\frac{\pi}{6}d_q^3 \tag{10-19}$$

where, for stagnant mixtures,

$$d_q = D\left[\frac{\rho_{fuel}}{\rho_{air}\phi \ln(1+B)}\right]^{0.5} \tag{10-20}$$

and for flowing mixtures,

$$d_q = \frac{0.30\rho_{fuel}(\mathrm{Tu}/100)U^{0.4}D^{1.4}}{\phi\rho_{air}^{0.6}\mu_{air}^{0.4}\ln(1+B)} \tag{10-21}$$

where

D = drop diameter

ϕ = equivalence ratio

B = Spalding transfer number

Tu = percentage turbulence intensity = $100u'/U$

u' = RMS value of fluctuating velocity

U = air velocity

These equations were compared with experimental results of Ballal and Lefebvre[22] and Rao and Lefebvre.[23] The first set of comparisons was found to be only qualitatively reasonable. The experimental methods used in Ref. 22 were a severe test for the evaporation model, since these experiments were carried out at low pressures, where the assumption of infinitely fast chemical reaction rates is least valid. The data from Rao and Lefebvre[23] are shown in Fig. 10.15, which cover wide ranges of velocity, equivalence ratio, and mean drop size (obtained at relatively high pressures). Closer agreement between the evaporation model and the data was obtained. The experiments of Rao and Lefebvre[23] were conducted in the same type of apparatus as shown in Fig. 10.10 for measurements of E_{min} of kerosene sprays injected into flowing air streams under atmospheric pressure and temperature.

The spark duration and spark gap were also optimized. Earlier studies on premixed gaseous mixtures detailed the need to place the spark electrodes at a

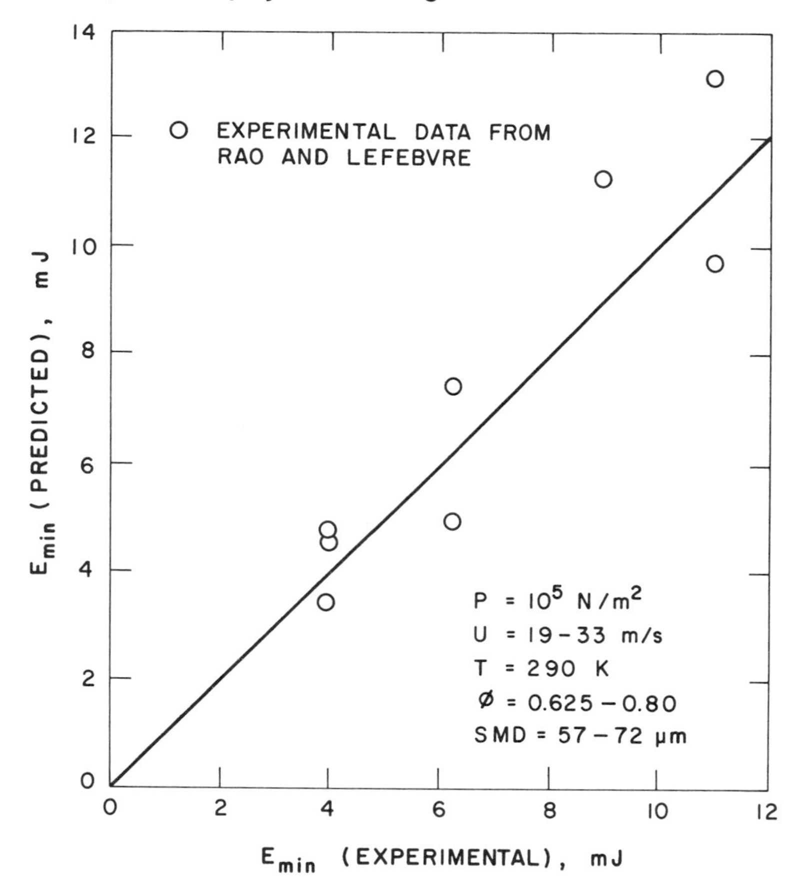

Figure 10.15 Comparison of predicted minimum ignition energy from evaporation model with experimental data of Rao and Lefebvre.[23]

gap distance equal to or slightly larger than the quenching distance. The quenching distance was determined over a wide range of conditions to be always less than 3 mm; thus a constant spark-gap width of 3 mm was used in the experiments. For any given mixture strength and flow condition, there was one particular value of spark duration for which the rate of energy release was minimized for ignition. Thus, E_{min} was defined as the total electrical energy dissipated from the weakest spark that would ignite the reactant mixture when the spark gap was 3 mm and the spark duration had the optimum value. Since this optimum spark duration varied with the flow conditions, all experiments were preceded by tests to determine the optimum duration.

The results of these experiments indicated that the most important parameters affecting E_{min} were drop size, air velocity, and fuel–air ratio. Since all these parameters were interrelated, much care was necessary in the experimental procedure to distinguish the separate effects. The ignition point was defined

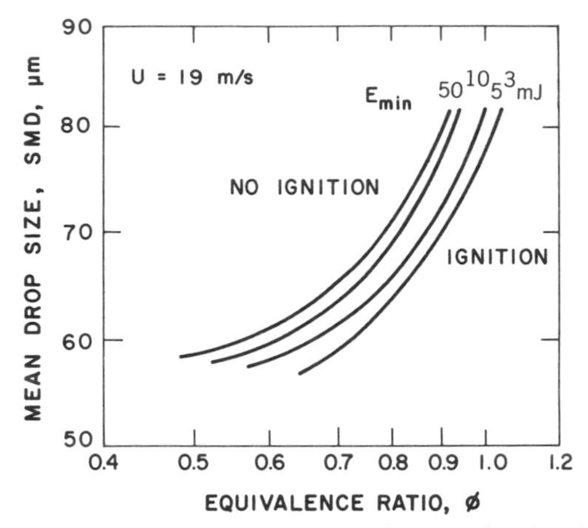

Figure 10.16 Ignition limits obtained at constant flow velocity and various levels of spark energy (after Rao and Lefebvre[23]).

as the condition at which a flame was visible in the mixture. At high velocities the flame visibility was poor; however, ignition was signaled by an audible sound accompanied by a tongue of flame at the exhaust pipe.

The parameter with the most pronounced effect on E_{min} was the fuel drop size. This effect is shown in Figs. 10.16 and 10.17. Even a slight improvement in atomization quality for any given air velocity or equivalence ratio would

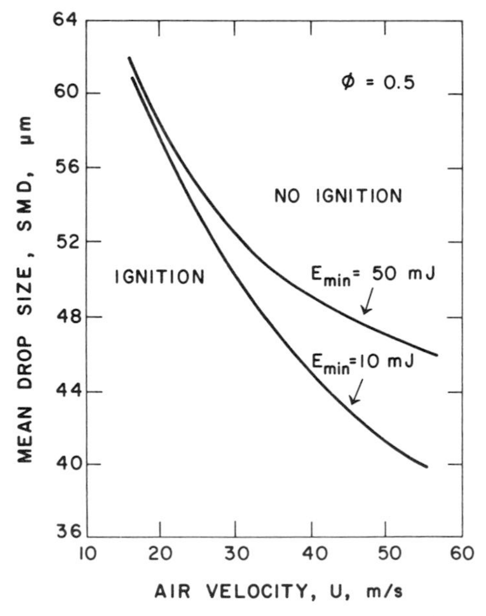

Figure 10.17 Influence of flow velocity on ignition limits for two levels of spark energy (after Rao and Lefebvre[23]).

greatly decrease E_{min} and thus be very beneficial to ignition. E_{min} was also appreciably reduced by an increase in equivalence ratio, as shown in Fig. 10.16. Also, for any given value of the drop size, an increase in air velocity increased E_{min}, as shown in Fig. 10.17. However, these effects were small compared to the effect of drop size on E_{min}.

Chin[24] studied the effect of oxygen addition on E_{min}. He gathered experimental data and developed empirical correlations to show that oxygen addition can reduce E_{min} significantly, especially at low pressures and flowing condition. A series of ignition test results was used to determine the effect of oxygen addition on the high-altitude ignition performance of a combustion chamber taken from an existing engine. Oxygen was injected into the air stream at the inlet of the test rig so that the oxygen was uniformly supplied to the igniter. The results showed that the ignition-velocity limits were increased significantly with oxygen addition, especially at low pressures. In another test, the altitude blowout and relight performance of an experimental combustor segment was found to be greatly improved with a small amount of oxygen addition.

Chin presented data that showed the optimum E_{min} for stagnant mixtures could be correlated by

$$(E_{min})_{opt} = Ap^{-n}[O_2]^{-4} \tag{10-22}$$

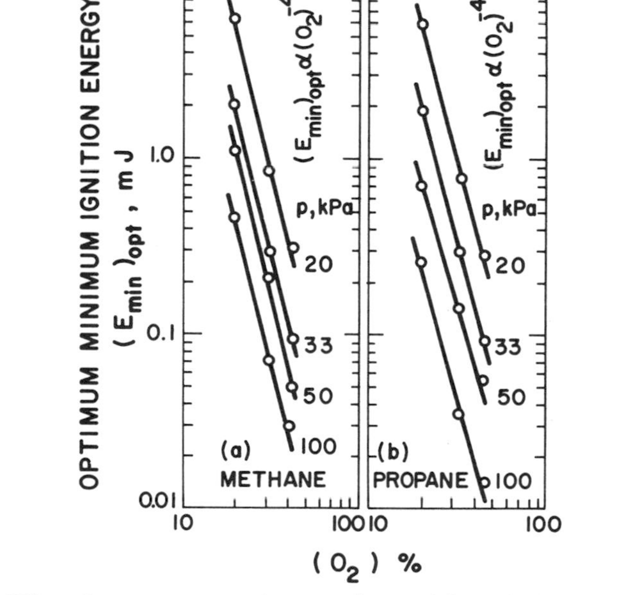

Figure 10.18 Effect of oxygen concentration on optimum minimum ignition energy for quiescent mixtures (after Chin[24]).

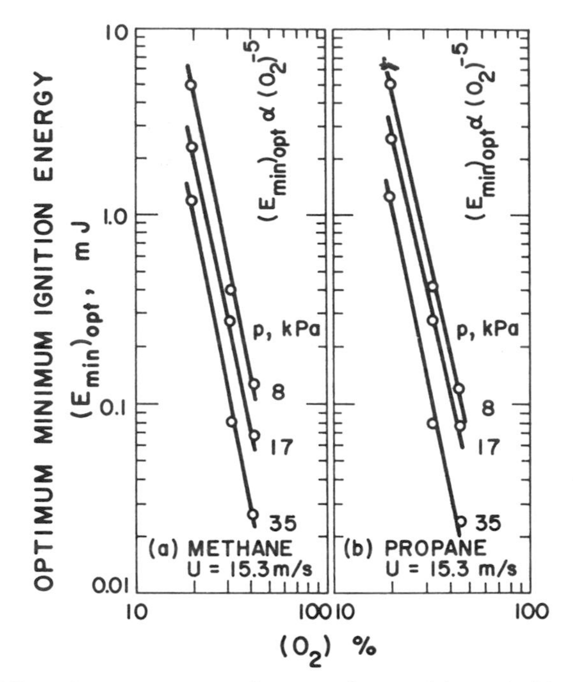

Figure 10.19 Effect of oxygen concentration on optimum minimum ignition energy for flowing mixtures (after Chin[24]).

and for flowing mixtures,

$$(E_{\min})_{\text{opt}} = Bp^{-m}[\mathrm{O_2}]^{-5} \tag{10-23}$$

where the exponents n and m and coefficients A and B are empirical constants. Typical results are shown in Figs. 10.18 and 10.19.

Chin found that for a specific fuel with a given intial percentage of 0_2 (before oxygen addition), there exists an optimum oxygen addition which gives the best ignition performance. Oxygen addition beyond this optimum value would not improve the performance. In an earlier work, Ballal and Lefebvre[22] had obtained minimum ignition energy data for flowing mixtures of methane (or propane), oxygen, and nitrogen at several subatmospheric pressures. One set of their data is given in Fig. 10.20, in which the dashed lines show the optimum minimum ignition energy $(E_{\min})_{\text{opt}}$.

To determine the effect of oxygen addition on $E_{\min}$ for stagnant mixtures, classical thermal-ignition theory was used. This, as in Lefebvre's analysis, assumed that successful ignition occurred when the electric spark supplied sufficient energy to form a high-temperature spark kernel of sufficient size to reach self-sustained flame propagation (i.e., a rate of heat release by chemical reaction exceeding the rate of heat loss). With this postulation, and assuming

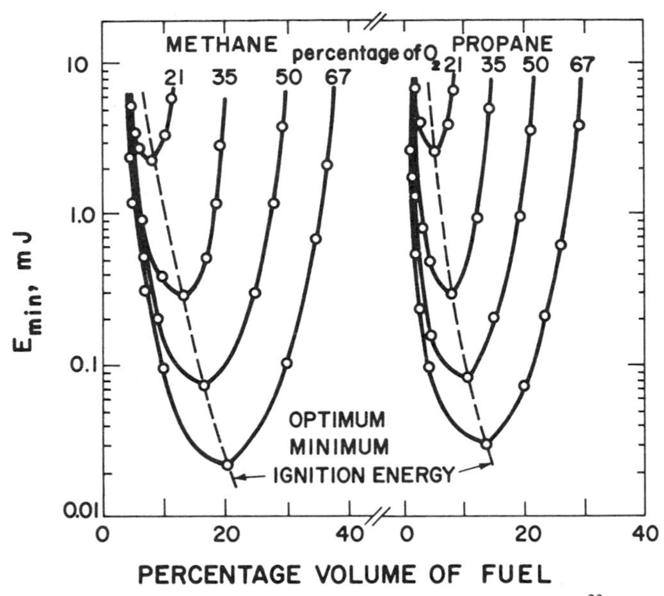

Figure 10.20 Influence of oxygen concentration on minimum ignition energy[22] ($P = 0.17$ atm; $U = 15.3$ m/sec).

(1) the exothermic chemical reaction is second order, (2) diffusion is negligible, (3) all spark energy is converted into thermal energy, and (4) the flame reaction zone thickness is proportional to the quenching distance, Chin developed an equation for the quenching radius of the spark kernel:

$$r = \left[\frac{3\lambda(T_f - T_0)\exp(E_a/R_u T_f)}{CAQ_f X_f X_{O_2} \rho^2} \right]^{1/2} \tag{10-24}$$

where

λ = thermal conductivity, kcal/m sec K

C = scale factor defined in such a manner that $(T_f - T_0)/(Cr)$ represents the effective temperature gradient for heat loss

A = Arrhenius coefficient

Q_f = heat of reaction of the fuel, kcal/mole

ρ = mixture density, $kg_f\ sec^2/m^4$

E_a = activation energy, kcal/mole

X_f = mole fraction of fuel

X_{O_2} = mole fraction of oxygen

Comparing with experimental data, Chin found good agreement, except at very high oxygen concentration. The effect of oxygen addition on E_{min} for a stagnant mixture was considered to be due mainly to the change of oxygen concentration (from X_{O_2}) and the effect of change of flame temperature on the chemical reaction rate.

For flowing conditions, Chin used an ignition theory proposed by Ballal and Lefebvre,[22] where for low-turbulence conditions ($u' < 2S_L$)

$$E_{\min} = C_p \rho \Delta T \frac{\pi}{6} \left[\frac{10\lambda}{C_p \rho (S_L - 0.16u')} \right]^3 \tag{10-25}$$

ΔT in Eq. (10-25) represents the temperature rise due to combustion.

Chin compared this equation with experimental data and found good agreement. The main effect of oxygen addition on the $E_{\min}$ of a flowing mixture is believed to be caused by the increase in the flame speed and decrease in the quenching distance.

In summary, fuel drop size was found to have the most significant effect on $E_{\min}$, with comparatively small effects due to air velocity and fuel–air ratio. The evaporation model developed by Lefebvre[21] was found to be quite satisfactory when compared with experimental data. This model would even be more useful if applied to some of the alternative fuels now being developed for gas-turbine engines, since the evaporation rates of these fuels will be low and thus likely to be a controlling factor in the combustion performance.

6 IGNITION OF BORON PARTICLES

In the past several decades, an increasing interest in high-energy-density fuels for use in combustion and propulsion systems has resulted in a growing emphasis on the advancement of slurry-fuel technology. Slurry fuels, which combine the advantages of a liquid fuel (pumpability, spray injectability) with those of a solid fuel (high energy content per unit volume), usually consist of a finely milled solid particulate fuel mixed into a conventional liquid-fuel carrier. A large variety of different solid fuels have been proposed as additives to liquid hydrocarbon carriers in order to meet the demands for a higher-volumetric-energy fuel; one of the most interesting at this time is particulate boron. The considerable research efforts of the past several decades which have been directed particularly at furthering the understanding of boron ignition are presented and discussed in this section. Although a great deal of research has focused on full-fledged combustion of boron as well, the viability of boron fuels depends heavily on the residence times required for particle ignition. Therefore, this section will concentrate primarily on boron ignition.

Solid particulate boron has received considerable attention in recent years due to its high gravimetric and volumetric heating values. As indicated by King,[25] the heating values of boron are considerably higher than those of carbon, aluminum, and magnesium (see Table 10.1). This shows that boron is an ideal candidate for the development of a high-energy fuel. However, while

TABLE 10.1 Heating Values of Various Fuels[a]

Fuel	Gravimetric Heating Value (cal/g)	Volumetric Heating Value (cal/cm^3)
JP5	10,150	8,230
Shelldyne-H	9,860	10,000
Carbon	7,830	17,700
Aluminum	7,420	20,000
$(CH_2)_n$	10,400	9,600
Boron	13,800	32,200
Magnesium	5,910	10,300

[a]Assuming H_2O (gas), CO_2 (gas), and metal oxide (liquid) as products. After King.[25]

the combustion processes of most typical hydrocarbon fuels are well understood, present knowledge of the mechanism of ignition of solid particulate boron is incomplete. The necessity for the understanding of both boron ignition and combustion should be evident from Table 10.1. The potential for savings in weight and/or volume for a given propulsion system by switching to boron fuels is enormous.

Researchers have taken a variety of approaches to the problem of boron ignition, and have applied a number of different assumptions and boundary conditions to their particular set of experiments. A generalized model of the boron ignition processes is illustrated in Fig. 10.21. As shown in this figure, the boron ignition process is complicated by the presence of a liquid layer of boron oxide (B_2O_3) at temperatures between the melting and boiling points of this oxide (723–2316 K). The general model shown accounts for the convective and radiative heat flux to the particle, the reaction of the boron at the B–B_2O_3

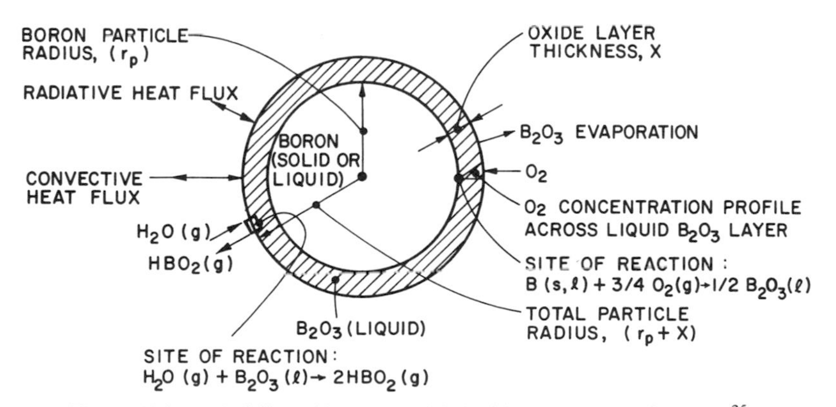

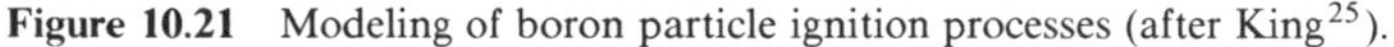
Figure 10.21 Modeling of boron particle ignition processes (after King[25]).

TABLE 10.2 Ignition-Time Data[a]

Gas temperature (K)	Y_{O_2}	Y_{CO_2}	Y_{H_2O}	Ignition Time (msec) $d_p = 35\ \mu m$	Ignition Time (msec) $d_p = 44\ \mu m$
2280	0.23	0.30	0	4.4	5.5
2430	0.20	0.33	0	4.8	5.7
2870	0.23	0.34	0	3.4	5.0
2450	0.37	0.34	0	2.1	3.3
2490	0.08	0.34	0	3.6	7.4
2240	0.19	0.11	0.16	4.0	7.2
2330	0.21	0.12	0.16	3.5	5.8
2430	0.19	0.13	0.19	3.8	6.1
2640	0.20	0.15	0.21	2.6	5.6

[a] Macek flat-flame-burner data at $p = 1$ atm. After King.[25]

interface the evaporation of the B_2O_3 at the gas–liquid interface, the diffusion of oxygen through the oxide layer to the B–B_2O_3 interface, and the reaction of $B_2O_{3(l)}$ with water vapor to form HBO_2. Because many researchers have studied the ignition of crystalline as well as amorphous boron particles, the idealized geometry shown in Fig. 10.21 may not be entirely accurate for all modeling studies, but it gives a good general understanding of the complexities of the problem.

Macek and Semple[26] presented a comprehensive study of the ignition of crystalline boron in the postflame zone of a flat-flame burner. They studied two powder samples with average diameters of 34.5 and 44.2 μm, respectively, at gas temperatures which varied from 1800 to 2900 K. The oxygen and water-vapor mole fractions in these flames ranged from 0.08 to 0.37 and from 0 to 0.21, respectively. A summary of all flame conditions used is given in Table 10.2.

In this experiment, boron particles were injected into the hot gas streams through a hypodermic needle using a minute flow of helium carrier gas at a velocity of 1.5–2.5 m/sec. Particles were photographed using a stroboscopic disk at a frequency of 2860 Hz. This technique allowed for the determination of the particle velocity at ignition, the ignition delay, and the total burning times of the boron particles.

Macek and Semple discovered several prominent features of boron particle ignition. The first is that ignition temperature is always well defined, as will be explained later. The second is that the flame structure of a combusting boron particle consists of three zones—a bright central core roughly the diameter of the particle, a wider, symmetrical, somewhat less luminous zone 300–400 μm wide, and a green envelope zone up to 1 cm wide. The third, and perhaps most interesting, characteristic of boron ignition is that it is a two-stage process. In the first stage, the particle ignites and burns brightly for a short period of time, and then seems to extinguish. In the second stage, the particle reignites and burns more brightly than before in a full-fledged combustion process.

As previously mentioned, ignition temperatures were always well defined in this study. These temperatures were determined by simultaneous integration of three differential equations,

$$\frac{dT_p}{dt} = \frac{6}{\rho_p C_p d_p}\left[\frac{\lambda_g \mathrm{Nu}}{d_p}(T_g - T_p) - \sigma\epsilon T_p^4\right] \tag{10-26}$$

$$\frac{du_p}{dt} = \frac{3C_D\rho_g}{4d_p\rho_p}(u_g - u_p)^2 \tag{10-27}$$

$$\frac{dx}{dt} = u_p \tag{10-28}$$

In these equations,

T_p = particle temperature

T_g = gas temperature

u_g = gas velocity

u_p = particle velocity

x = distance from ignition point to particle location

ρ_p = particle density

ρ_g = gas density

C_p = particle specific heat

d_p = particle diameter

$$C_D = \text{drag coefficient} = \begin{cases} 24/\mathrm{Re} & \text{for} \quad \mathrm{Re} < 1 \\ (100/\mathrm{Re})^{2/3} & \text{for} \quad \mathrm{Re} > 1 \end{cases}$$

λ_g = gas thermal conductivity

σ = Boltzmann's constant

ϵ = particle emissivity

TABLE 10.3 Ignition Temperatures from Heatup Data[a]

Gas Mix No.	d_p = 34.5 μm: $\bar{x}_i$ (cm)	d_p = 34.5 μm: $\bar{T}_i$ (K)	d_p = 44.2 μm: $\bar{x}_i$ (cm)	d_p = 44.2 μm: $\bar{T}_i$ (K)
1	2.37 ± 0.18	1980	3.97 ± 0.63	1970
2	2.50 ± 0.20	2000	3.31 ± 0.20	1970
3	1.88 ± 0.28	1990	3.14 ± 0.41	1990
5	4.83 ± 0.64	1840	10.75 ± 1.14	1930
6	4.98 ± 0.56	1850	9.66 ± 0.72	1880
7	4.08 ± 0.40	1810	7.45 ± 1.12	1800
8	7.09 ± 1.42	1960	8.11 ± 0.98	1810
10	2.50 ± 0.37	2030	3.42 ± 0.17	2010
11	2.26 ± 0.25	1990	—	—

Overall averages: 1992 ± 16 K (X_{H_2O} = 0)
1860 ± 24 K (X_{H_2O} = 0.16–0.21)

[a] After Macek and Semple.[26]

Using the assumptions that particles behave as spheres of stated average diameters and that particle self-heating during ignition is negligible, the experimentally determined values of u_p, u_g, and T_g were used in Eqs. (10-26) to (10-28), which were then solved to yield complete particle temperature histories. Ignition temperatures calculated in this manner are given in Table 10.3. Statistical analysis of these data indicates that there is no significant trend of ignition temperatures either with particle size or with gas temperature. There is, however, a notable trend toward decreasing ignition temperature with increasing ambient water-vapor concentration, which suggests that water vapor promotes the ignition of boron to some extent.

In work published by Gurevich et al.,[27] crystalline and amorphous boron particle ignition in hot gas streams of a plasma generator was studied. The experimental setup consisted of a particle feeding apparatus where the particles were sifted through a screen, entrained by a carrier gas, and carried into the hot-gas zone. The hot-gas zone consisted of a tungsten cathode, a copper anode, and an argon plasma which mixed with the specified oxidizer and vented into the test chamber. The test chamber was fitted with a mica window so that the burning particles could be photographed with a 35-mm motion-picture camera.

Five particle sample sizes ranging from 50 to 260 μm were separated using microsieves. These samples were run through the test chamber continuously while the gas temperatures were varied until the ignition-limiting gas temperature was determined. This technique was used for 15, 30, and 45% oxidizer

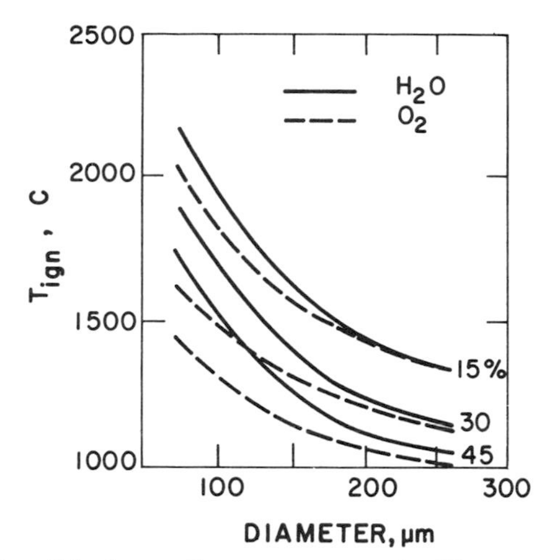

Figure 10.22 Effect of particle size, oxidizer percentage, and oxidizer type on ignition temperature (after Gurevich et al.[27]).

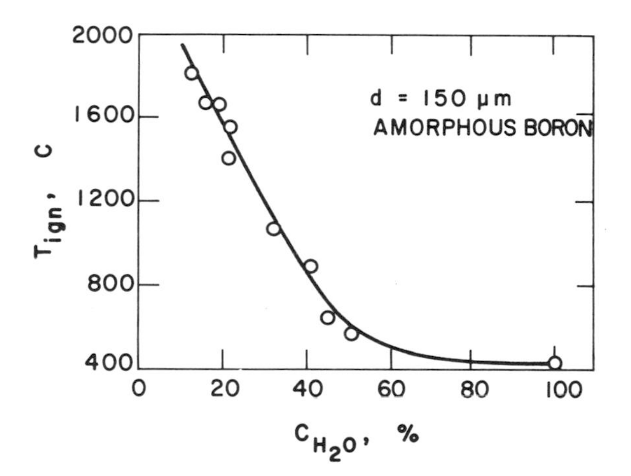

Figure 10.23 Effect of water-vapor percentage on the ignition temperature of boron particles (after Gurevich et al.[27]).

mixtures, where both water vapor and oxygen were tested as oxidizers. Typical experimental results are shown in Figs. 10.22 and 10.23.

Figure 10.22 shows two interesting trends: (1) ignition temperature decreases with increasing water-vapor content, in agreement with Macek's observations, and (2) ignition temperature decreases with increasing particle size —a trend not evident in Macek's data. (This discrepancy may be attributed to the fact that Macek and Semple studied only two sample sizes, for which the size difference was only about 20%.)

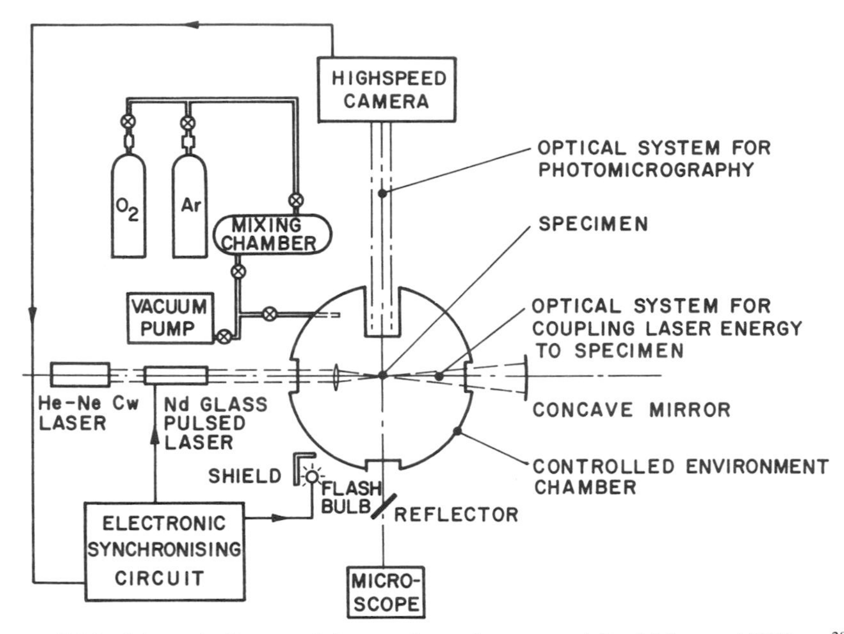

Figure 10.24 Schematic diagram of the experimental apparatus (after Mohan and Williams[28]).

Figure 10.23 shows the decrease in ignition temperature with increasing ambient water-vapor content for a 150-μm particle of amorphous boron. Gurevich et al.[27] noted that in all cases, amorphous boron was more easily ignited than crystalline boron. Note also that all data presented so far indicate that the presence of water vapor may enhance the removal of the oxide layer in the particle ignition process, thereby reducing ignition temperature and ignition delay.

Mohan and Williams[28] also conducted an investigation of ignition of crystalline and amorphous boron in the 100-μm range. The particles were ignited by a pulsed laser in a chamber filled with known mixtures of oxygen and nitrogen. These particles were then photographed with a high-speed motion-picture camera. The experimental setup used in these tests is shown in Fig. 10.24.

Note that in this work, unlike the previous studies, the boron particles were supported at the tip of a 10-μm glass fiber. These particles were then ignited by the radiative flux of a neodymium-doped glass laser (beam diameter 3 mm) rated to deliver 1 J at a wavelength of 1.06 μm in a 0.6-msec pulse. This emission was focused onto the surface of the boron particle by a convex lens of 6.5-cm focal length and a concave mirror of 13-cm radius of curvature to insure that both sides of the particle received approximately equal radiative fluxes. During ignition, the particles separated from the glass fiber and remained in the view of the camera long enough to determine ignition times and burning times. Because the particle was suspended in a closed chamber,

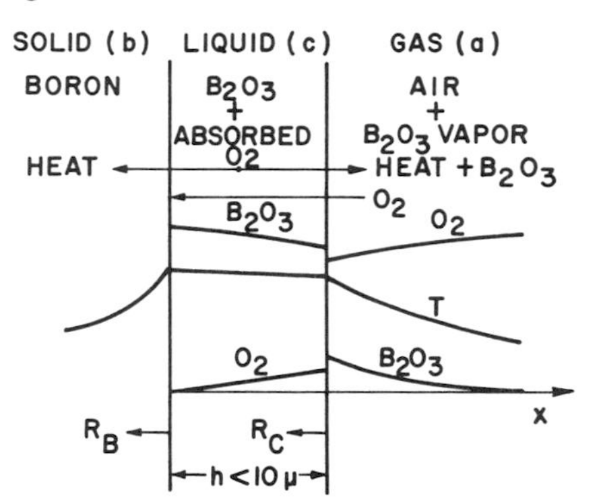

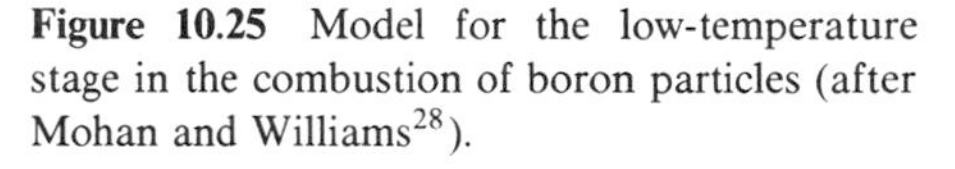

Figure 10.25 Model for the low-temperature stage in the combustion of boron particles (after Mohan and Williams[28]).

the pressure and composition of the oxidizing environment could be controlled.

The effect of increasing the ignition stimulus from the laser on the ignition of crystalline boron was observed. Mohan and Williams[28] noted, in agreement with Gurevich et al.,[27] that crystalline boron was significantly more difficult to ignite than amorphous boron. Mohan and Williams also verified that particle self-heating is indeed negligible during ignition (as Macek and Semple[26] had assumed), which implies that first-stage ignition will lead to second-stage ignition only if temperature conditions are adequate. Violent ignition of an amorphous boron particle, originally spherical and 150 μm in diameter, was also recorded from the setup shown in Fig. 10.24. These particles were found to fragment into very fine particles, which then burned according to the two-stage ignition process noted by other investigators. This explosive behavior is attributed to the fact that laser penetration is much more effective on amorphous boron because particles of it are actually agglomerates of submicrometer particles, and thus have a much greater specific surface area than crystalline samples.

A model for the ignition of boron particles presented in Ref. 28 is shown in Fig. 10.25. This model is quite similar to the general model shown in Fig. 10.21, except that in Fig. 10.25 we assume (1) planar geometry (valid only for a very thin oxide layer) and (2) that removal of the oxide from the particle is gas-diffusion-controlled (evaporation kinetics is negligible). In this model,

$$R_B = \text{regression rate of surface } B$$

$$R_C = \text{regression rate of surface } C$$

$$h = \text{oxide-layer thickness}$$

and the change in oxide-layer thickness with time obeys the equation

$$\frac{dh}{dt} = R_B - R_C \tag{10-29}$$

Assuming also that the thickness and thermal resistivity of the oxide layer are small enough for the layer to be approximately isothermal, the energy balance for this layer can be written as

$$h\rho_c C_c \frac{dT_c}{dt} = q_B \rho_B R_B - q_C \rho_c R_C - L \tag{10-30}$$

where

T_c = oxide-layer temperature

C_c = heat of capacity of liquid oxide

ρ_B = density of solid boron

ρ_c = density of liquid boron oxide

q_B = heated liberated per unit mass of boron consumed

q_C = heat absorbed per unit mass of B_2O_3 vaporized at C

L = sum of heat losses by radiation and conduction

Assuming absorptive equilibrium for O_2 at surface C and one-dimensional steady-state diffusion of O_2 through the oxide layer to a diffusion-controlled reaction at surface B, the rate of regression of surface B is given by

$$R_B = \frac{Xk\mathscr{D}}{h} \tag{10-31}$$

where

X = the mole fraction of O_2 in the ambient atmosphere.
k = effective distribution coefficient of O_2 (related to the ratio of the mole fraction of O_2 absorbed in $B_2O_{3(l)}$ to the gas-phase mole fraction).
$\mathscr{D}$ = diffusion coefficient for absorbed O_2 in $B_2O_{3(l)}$.

The distribution coefficient k is given by

$$k = \frac{\frac{4}{3}k' p W_B}{\rho_b R_u T_c} \tag{10-32}$$

where

k' = equilibrium ratio of liquid phase to gas phase O_2 concentration.
W_B = atomic weight of boron.

Equation (10-31) assumes an ideal-gas mixture at C. The product of $k\mathscr{D}$ is proportional to p, and both k and $\mathscr{D}$ must be evaluated at T_c.

It was determined that for a wide range of conditions, R_C is controlled by outward gas-phase diffusion of $B_2O_{3(g)}$, where

$$R_C = \frac{2X_a \mathscr{D}_a \rho_a}{\rho_c d_p} \tag{10-33}$$

in which

X_a = ratio of equilibrium vapor pressure of B_2O_3 at T_c to the total pressure.
d_p = diameter of the particle.
$\mathscr{D}_a$ = binary diffusion coefficient of $B_2O_{3(g)}$ in air at T_c.
ρ_a = density of a pure ideal gas of B_2O_3 at ambient pressure and temperature T_c.

The product $\rho_a \mathscr{D}_a$, which is independent of p, has been presumed independent of temperature as well, and X_a has been assumed to be small compared to unity for T_c less than 2100 K. The term L in Eq. (10-30) takes account of heat losses by inward conduction, outward conduction, and outward radiation, which are all roughly comparable in magnitude.

Equations (10-29) and (10-30) yield a steady-state solution such that

$$R_B = R_C = \frac{q_C \rho_c R_C + L}{q_B \rho_b} \tag{10-34}$$

The stability of this solution can be determined by linearization about the steady state according to the following simplifications:

$$\frac{\partial R_B}{\partial h} = -\frac{R_B}{h}, \qquad \frac{\partial R_B}{\partial T_c} = \frac{A}{R_u T_c}\left(\frac{R_B}{T_c}\right) \tag{10-35}$$

$$\frac{\partial R_C}{\partial h} = 0, \qquad \frac{\partial R_C}{\partial T_c} = \frac{H}{R_u T_c}\left(\frac{R_C}{T_c}\right) \tag{10-36}$$

$$\frac{\partial L}{\partial h} = 0, \qquad \frac{\partial L}{\partial T_c} = \frac{dL}{dT_c} \tag{10-37}$$

where $\rho_c C_c$ = constant, $q_B \rho_b$ = constant, and $q_C \rho_c$ = constant. Equation (10-35) follows from Eq. (10-31) if we assume that $k\mathscr{D}$ is expressed in Arrhenius form as

$$k\mathscr{D} \propto \exp\left(-\frac{A}{R_u T_c}\right) \tag{10-38}$$

where A is the difference between the activation energy for diffusion and the heat of absorption. Equation (10-36) follows from Eq. (10-32) if we assume that $\mathscr{D}_a \rho_a/(\rho_c d)$ changes negligibly and that

$$X_a \propto \exp\left(-\frac{H}{R_u T_c}\right) \tag{10-39}$$

where H is the molar heat of evaporation of B_2O_3 at temperature T_c. Equation (10-37) results from the nature of the heat-loss processes.

The stability analysis reveals that the steady-state solution is stable only if both

$$T_c \frac{dL}{dT_c} > \frac{LH}{R_u T_c} \tag{10-40}$$

and

$$\frac{H}{R_u T_c} + \frac{\rho_c C_c T_c}{q_B \rho_b} + \left(T_c \frac{dL}{dT_c} - \frac{LH}{R_u T_c}\right)(q_B \rho_b R_B)^{-1} > \frac{A}{R_u T_c} \tag{10-41}$$

Activation energies for absorption or liquid-phase diffusion seldom exceed 20 kcal/mole, while H in this case is 87.6 kcal/mole, so that Eq. (10-41) generally holds true as long as Eq. (10-40) holds true. In the laser ignition experiments of Mohan and Williams,[28] however, the losses L are quite large because the particle remains in a 300-K environment after the initial energy pulse. This, coupled with the fact that T_c is quite high after the laser pulse stops, results in a condition in which Eq. (10-40) is not satisfied, so that the first-stage ignition process in this experiment is unstable. This explains the inability of these tests to produce a second-stage ignition process.

The ignition temperature can be determined from this analysis by combining Eqs. (10-33) and (10-34) to yield the steady-state equality

$$p_e = \frac{R_u T_c \rho_c}{2\mathscr{D}_a W}\left(\frac{L d_p}{q_B \rho_b - q_C \rho_c}\right) \tag{10-42}$$

where p_e is the equilibrium vapor pressure of B_2O_3 at temperature T_c and W denotes the molecular weight of B_2O_3. For laser ignition experiments, the solution to Eq. (10-42) which corresponds to an unstable condition should define the particle ignition temperature T_c. This solution is presented graphically in Fig. 10.26.

The dashed line in Fig. 10.26 represents the condition in which the two sides of Eq. (10-40) are equal, corresponding to the boundary between the stable (below the dashed line) and the unstable (above the dashed line) solutions. The dashed line therefore corresponds roughly to the experimental data of Macek

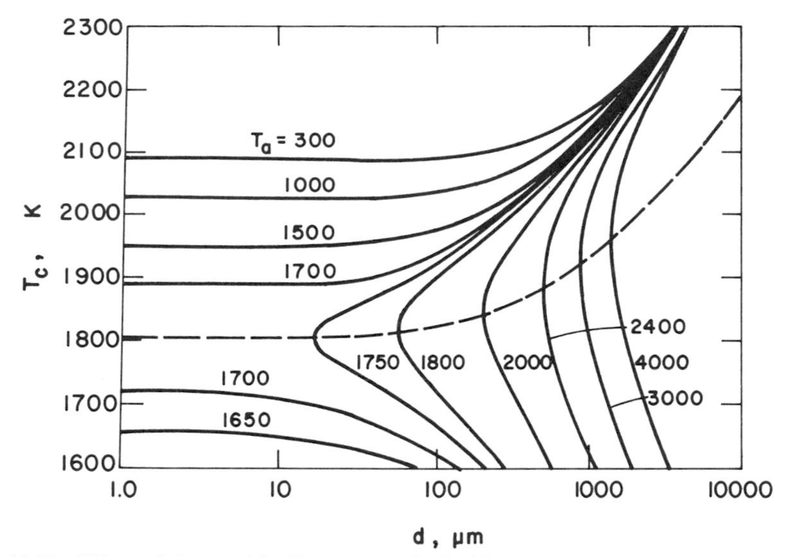

Figure 10.26 Effect of the particle diameter on the ignition temperature of boron at different ambient temperatures and at atmospheric pressure (after Mohan and Williams[28]).

and Semple.[26] For a particle size of 40 μm, their observed ambient gas ignition temperature was 1925 K, while this data would predict 1780 K. This discrepancy could be attributed to the fact that Mohan and Williams were using the concentrated energy of a pulsed laser to ignite their particles, while Macek and Semple used the hot gas stream of a flat-flame burner.

Boron particle ignition is clearly a complex and intriguing phenomenon, and continues to elude complete understanding, despite the considerable attention the subject has received. Several very significant advances have been made on this problem, however, which should serve well to guide the direction of further research. Some of the key pieces to the puzzle, as evident in the research work outlined in this section, will become the materials of current and future work in this area.

It is clear, for example, from the results of Macek and Semple,[26] Gurevich et al.,[27] and Mohan and Williams[28] that the dynamics of the boron oxide layer which forms upon heatup of the boron particle plays a crucial role in the overall ignition process. The efficient removal of this oxide layer is the most important step in insuring rapid ignition and complete combustion of the boron particle.

It was shown in Refs. 26 and 27 that the removal of the oxide layer can be enhanced by the presence of water vapor in the ambient gases. This enhanced removal led to both shorter ignition delay and lower ignition temperatures—two very important considerations in the design of a propulsion-system combustor. The mechanism for this enhanced removal, however, is not entirely understood at this time. It is generally agreed that the probable cause of the presence of a significant amount of HBO_2 as shown in Fig. 10.21 is the reaction

$$B_2O_{3(l)} + H_2O(g) \rightarrow 2HBO_{2(g)} \tag{10-43}$$

Clearly, further research is necessary to determine the magnitude of the importance of this reaction, as well as to determine the feasibility of boron ignition in relatively dry environments.

Glassman et al.[29] have presented a preliminary study of the effects of the boron oxide layer on the ignition and combustion of boron, and have hypothesized that oxidation of boron in the ignition stage is governed not by the oxide diffusion to the boron surface, but by boron diffusion through the oxide layer to the oxide surface. Verification or negation of this hypothesis is another area of further research that is currently in need of completion.

7 IGNITION AND FLAME SPREADING OVER A SOLID FUEL IN A HOT OXIDIZING GAS STREAM

The study of ignition and flame spreading in solid fuels heated by high-temperature oxidizing gases is important for both fire prevention and the development of high-performance propulsive devices. The ignitability of a solid polymeric hydrocarbon fuel (PBAA) by a hot oxidizing gas stream was studied experimentally by Kashiwagi et al.[30] in a shock-tube wind tunnel with a flat-plate specimen placed parallel to the flow. Their main objective was to identify the physical mechanism involved in the ignition process. Two alternative mechanisms were postulated as responsible for the development of ignition: (1) an exothermic gas-phase reaction in the boundary layer, and (2) a heterogeneous reaction at the solid surface. Each of the two theories takes into account the changing profiles within the boundary layer during the induction period prior to ignition, the simultaneously changing thermal profile below the surface of the fuel due to convective heating, and the gradually rising rate of reaction. Ignition is said to occur at that time and at that location at which the reaction rate reaches a suitably defined runaway condition.

In their theoretical studies, Kashiwagi et al.[30] found that the surface-reaction model (heterogeneous-reaction theory) predicts that ignition will always occur near the leading edge of the plate of fuel; although the ignition time is sensitive to the oxygen concentration, gas temperature, and other physical parameters, the location of ignition is not. The gas-phase model, however, predicts that ignition will occur at some downstream position, the distance increasing with increasing flow velocity and with decreasing oxygen concentration. They concluded that ignition of PBAA solid fuel in hot oxygen–nitrogen mixtures occurs mainly by means of a gas-phase reaction in the boundary layer, the reactants being the oxygen of the hot gas stream and the fuel-like gases evolved from the pyrolytic heating of the solid fuel.

7.1 Experimental Investigation

The schematic drawing of the test apparatus used by Kashiwagi et al.[30-32] is shown in Fig. 10.27. After the shock wave reflects off the end wall of the shock tube, the doubly compressed high-temperature and high-pressure test gas flows through the area-contracted test section into the low-pressure exhaust tank at a rate controlled by a choked nozzle. The flow in the test section reaches a steady condition in a time that is a small fraction of the ignition delay. The free-stream velocities can be adjusted by using different sizes of nozzles. The flat-plate specimen of PBAA with a rounded leading edge was placed 2° downward to avoid the separation of boundary layer at the leading edge.

In their data acquisition system, six RCA 1P28 photomultipliers with Corning 7-54 filters connected with fiber optics were used to detect the onset of flame for successive regions along the sample surface; these signals were simultaneously recorded on a Honeywell 7600 magnetic tape recorder. High-speed photographs were also used to record the onset of flames, and these results agreed very well with the observations of the photomultipliers.

Effects of the free-stream oxygen content (20–100%), free-stream temperature (1270–2100 K), free-stream velocity (77–275 m/sec), and diluent inert gases (N_2, Ar) on ignition and flame-spread behavior were studied. It was

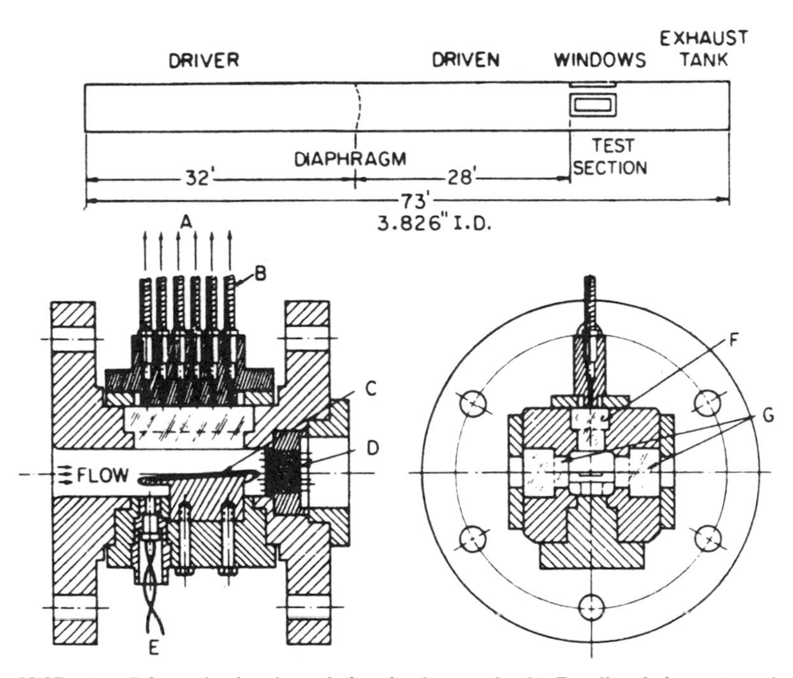

Figure 10.27 (*a*) Schematic drawing of the shock tunnel. (*b*) Details of the test section: A, photomultipliers; B, fiber optics to photomultipliers; C, fuel specimen; D, multinozzle plate to exhaust tank; E, pressure transducer; F, top window; G, side windows. (After Kashiwagi et al.[31])

observed[31] that if the external heating rate is high or if the sample tends to pyrolyze at low temperatures, the heat feedback from the exothermic gas-phase reaction is not important and the effects of free-stream oxygen content are small for both ignition and flame spread. However, if the external heating rate is small or the sample tends to pyrolyze only at high temperatures, the heat feedback from the exothermic gas-phase reaction is important and effects of free-stream oxygen content are significant for ignition and flame spread.

Figure 10.28 shows the effect of free-stream oxygen content on the first ignition position and subsequent flame spread for PBAA (polybutadiene acrylic acid) fuel. A parabolic curve means that there are two noteworthy events at any particular position, the lower (earlier) one being the onset of ignition at that location, and the upper (later) one being the disappearance of flame at that same location (blowaway phenomenon). Therefore, for times inside the parabola, flame is sustained at the given position. The location of the first ignition is described by the minimum of the lower branch of the parabola. The slope of the parabola indicates the rate of flame spread. It can be seen from Fig. 10.28 that the point of first ignition and the subsequent flame spreading are sensitive to the free-stream oxygen content. At a high free-stream oxygen content (more than 60% O_2), ignition occurred in the forward section

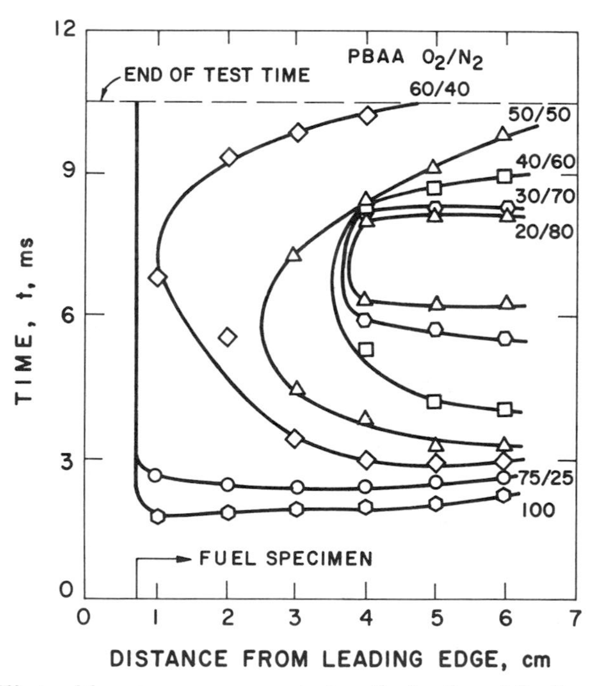

Figure 10.28 Effects of free-stream oxygen content on the location of the flame versus time for PBAA: $T_e = 1620$ K, $U_e = 275$ m/sec, $P = 260$ psi (after Kashiwagi et al.[31]).

and flame spread downstream with the flow. With decreasing free-stream oxygen content, the point of first ignition shifted downstream and the flame spread both downstream and upstream against the flow. This behavior was also observed with PIB (polyisobutylene) fuel.

It is evident in Fig. 10.28 that the flame does not spread upstream all the way to the leading edge in the presence of a low free-stream oxygen content. There exists some standoff distance between the leading edge and the position where the flame spreads furthest upstream. The amount of standoff distance depends strongly on the free-stream oxygen content. This behavior can be explained in terms of the Damköhler number of the first kind (ratio of chemical time to flow time). Increasing the free-stream oxygen content decreases the gas-phase chemical-reaction time and reduces the Damköhler number if the flow time (defined as the sample length divided by the free-stream velocity) is kept constant. Hence, a shorter standoff distance was observed in this study.

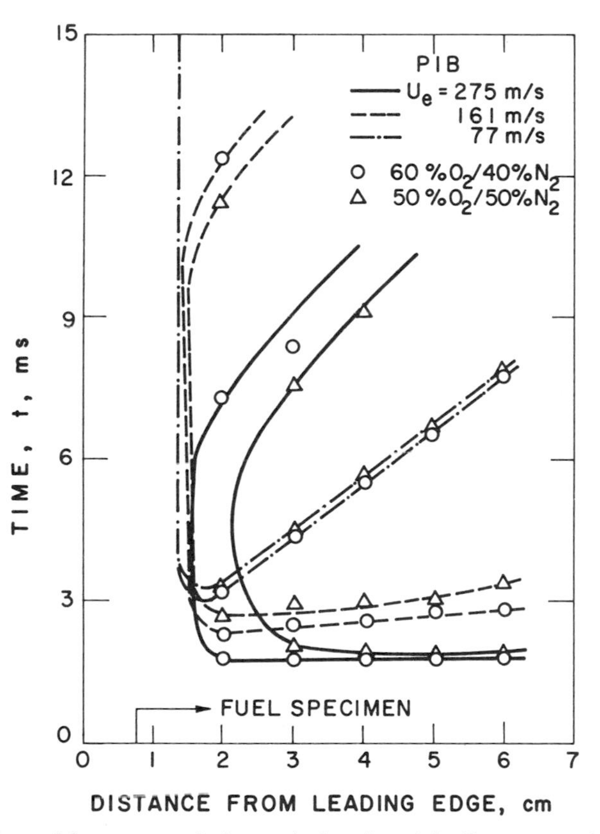

Figure 10.29 Effects of free-stream velocity on the location of the flame versus time: $T_e = 1620$ K, $P = 260$ psi (after Kashiwagi et al.[31]).

Figure 10.29 shows the effects of free-stream velocity. Reduction in free-stream velocity increases the ignition delay time because it lowers the external convective heating rate. It increases the flow time and thereby shifts the location of the first ignition further upstream, equivalent to reducing the Damköhler number. It is interesting to note that the order of the flame shift distance is about 0.01 m, and the difference in the free-stream velocity in Fig. 10.29 is roughly 100 m/sec. Then a rough estimate of the gas-phase chemical-reaction time is about (0.01 m) ÷ (100 m/sec) = 0.1 msec. The rate of flame spread becomes slower with decreasing free-stream velocity because such a shift requires more heat feedback from the flame (due to a lower external heating rate). Effects of other important physical parameters such as inert diluent gas and free-stream temperature on ignition and flame spread were also studied. These results are reported in Refs. 31 and 32.

7.2 Theoretical Formulation

To predict quantitatively the ignition delay time, the first ignition position, and the flame-spreading behavior after the ignition, Kashiwagi and Summerfield[33] formulated a comprehensive model which is useful for studying both fuel-rich solid-fuel rocket-motor combustion and general fires. In their theoretical work, the model was solved numerically in order to predict the ignition of a solid polymeric fuel in a hot oxidizing flow field and also to compare results with their earlier work[30] based upon the local-similarity approximation and the integral approximation. In Ref. 33, the complete nonsimilarity calculation was carried out to replace the previously used approximations in solving the condensed-phase energy equation.

7.2.1 Basic Assumptions

1. A quasi-steady gas-phase boundary layer is assumed, because the characteristic time for transient heat conduction within the solid phase is typically much longer than the time for various processes in the gas phase.
2. The solid-phase energy equation takes the one-dimensional form in y-direction perpendicular to the gas flow direction.
3. The boundary layer is laminar along the flat plate.
4. The surface regression rate V_s is expressed by an Arrhenius pyrolysis law, and depends only upon the surface temperature.
5. Within the solid phase, chemical reactions are negligible, and the specific heat, thermal conductivity, and density are constant.
6. The mixture of gases behaves like a perfect gas.
7. The specific heats and diffusion coefficients of all gaseous species are equal and constant.

8. The molecular weights of all species are equal.
9. The Prandtl, Schmidt, and Lewis numbers are equal to unity.
10. The gas-phase reaction is assumed to be a one-step irreversible chemical reaction, subject to second-order Arrhenius kinetics:

$$[F] + n[O_2] \rightarrow (n+1)[P] \tag{10-44}$$

7.2.2 Governing Equations

The governing equations are formulated as follows:

Continuity Equation:

$$\frac{\partial(\rho u)}{\partial x} + \frac{\partial(\rho v)}{\partial y} = 0 \tag{10-45}$$

Momentum Equation:

$$\rho u \frac{\partial u}{\partial x} + \rho v \frac{\partial u}{\partial y} = \frac{\partial}{\partial y}\left(\mu \frac{\partial u}{\partial y}\right) \tag{10-46}$$

Energy Equation:

$$\rho u C_p \frac{\partial T}{\partial x} + \rho v C_p \frac{\partial T}{\partial y} = \frac{\partial}{\partial y}\left(\frac{\mu C_p}{\Pr}\frac{\partial T}{\partial y}\right) + Q\rho^2 Y_O Y_F A_R \exp\left(-\frac{E_R}{R_u T}\right) \tag{10-47}$$

where E_R is the activation energy for reaction and A_R is the corresponding Arrhenius factor.

Oxidizer Species:

$$\rho u \frac{\partial Y_O}{\partial x} + \rho v \frac{\partial Y_O}{\partial y} = \frac{\partial}{\partial y}\left(\rho \mathscr{D} \frac{\partial T}{\partial y}\right) - n\rho^2 Y_O Y_F A_R \exp(-E_R/R_u T) \tag{10-48}$$

Fuel Species:

$$\rho u \frac{\partial Y_F}{\partial x} + \rho v \frac{\partial Y_F}{\partial y} = \frac{\partial}{\partial y}\left(\rho \mathscr{D} \frac{\partial Y_F}{\partial y}\right) - \rho^2 Y_O Y_F A_R \exp(-E_R/R_u T) \tag{10-49}$$

State:

$$\rho T = \rho_e T_e \tag{10-50}$$

Solid-Phase Energy:

$$C_s\rho_s\frac{\partial T_s}{\partial t} + C_s\rho_s v_s\frac{\partial T_s}{\partial y} = \lambda_s\frac{\partial^2 T_s}{\partial y^2} \tag{10-51}$$

7.2.3 Boundary and Initial Conditions

At $y = 0$,

$$u = 0 \tag{10-52}$$

$$v = v_w = \frac{\rho_s v_s}{\rho_w} = \frac{\rho_s}{\rho_w}A_p\exp\left(-\frac{E_p}{R_u T}\right) \tag{10-53}$$

$$-\lambda\frac{\partial T}{\partial y} = -\lambda_s\frac{\partial T_s}{\partial y} - \rho_s v_s L_v \tag{10-54}$$

$$T = T_s \tag{10-55}$$

$$-\rho\mathscr{D}\frac{\partial Y_O}{\partial y} + \rho v Y_O = 0 \tag{10-56}$$

$$-\rho\mathscr{D}\frac{\partial Y_F}{\partial y} + \rho v Y_F = \rho_s v_s \tag{10-57}$$

where E_p and A_p are the activation energy and Arrhenius factor of the surface pyrolysis.

At $y = \infty$,

$$u = u_e, \quad T = T_e, \quad Y_O = Y_{O,e} \quad Y_F = Y_{F,e} \tag{10-58}$$

At $y = -\infty$,

$$T_s = T_0 \tag{10-59}$$

Initial condition for solid-phase energy equation at $t \leq 0$:

$$T_s = T_0 \tag{10-60}$$

or

$$T_s = T_s(y) \tag{10-61}$$

7.2.4 Transformation into Similarity Coordinates

All the equations above are transformed into similarity coordinates with the assumption of $\rho\mu = \rho_e\mu_e =$ constant throughout the flow field and with the Illingworth–Levy transformation, namely,

$$s = \int_0^x \rho_e\mu_e u_e\, dx \tag{10-62}$$

and

$$\eta = \frac{\rho_e u_e}{(2s)^{1/2}} \int_0^y \frac{\rho}{\rho_e}\, dy \tag{10-63}$$

Next the stream function ψ is defined by the following equations:

$$\rho u = \frac{\partial \psi}{\partial y} \tag{10-64}$$

$$\rho v = -\frac{\partial \psi}{\partial x} \tag{10-65}$$

Also, introduce the following transformations:

$$\psi = (2s)^{1/2} f(s, \eta) \tag{10-66}$$

$$\theta = \frac{T}{T_e}, \qquad \theta_s = \frac{T_s}{T_e} \tag{10-67}$$

$$\eta_s = \left(\frac{u_e}{L\alpha_s}\right)^{1/2} y, \qquad \tau = \frac{u_e t}{L} \tag{10-68}$$

The transformed governing equations can be written in the following form: Equation (10-46) becomes

$$\frac{\partial^3 f}{\partial \eta^3} + f\frac{\partial^2 f}{\partial \eta^2} = 2s\left(\frac{\partial^2 f}{\partial s\,\partial \eta}\frac{\partial f}{\partial \eta} - \frac{\partial f}{\partial s}\frac{\partial^2 f}{\partial \eta^2}\right) \tag{10-69}$$

Equations (10-47) to (10-51) become

$$\mathrm{Pr}^{-1}\frac{\partial^2 \theta}{\partial \eta^2} + f\frac{\partial \theta}{\partial \eta} = 2s\left(\frac{\partial f}{\partial \eta}\frac{\partial \theta}{\partial s} - \frac{\partial f}{\partial s}\frac{\partial \theta}{\partial \eta}\right) - C_s\frac{Y_O Y_F}{\theta}\exp\left(-\frac{\theta_R}{\theta}\right) \tag{10-70}$$

$$\mathrm{Sc}^{-1}\frac{\partial^2 Y_O}{\partial \eta^2} + f\frac{\partial Y_O}{\partial \eta} = 2s\left(\frac{\partial f}{\partial \eta}\frac{\partial Y_O}{\partial s} - \frac{\partial f}{\partial s}\frac{\partial Y_O}{\partial \eta}\right) + Bns\frac{Y_O Y_F}{\theta}\exp\left(-\frac{\theta_R}{\theta}\right) \tag{10-71}$$

$$\mathrm{Sc}^{-1}\frac{\partial^2 Y_F}{\partial \eta^2} + f\frac{\partial Y_F}{\partial \eta} = 2s\left(\frac{\partial f}{\partial \eta}\frac{\partial Y_F}{\partial s} - \frac{\partial f}{\partial s}\frac{\partial Y_F}{\partial \eta}\right) + Bs\frac{Y_O Y_F}{\theta}\exp\left(-\frac{\theta_F}{\theta}\right) \tag{10-72}$$

$$\frac{\partial \theta_s}{\partial \tau} - \frac{\beta}{(2s)^{1/2}}\left[f + 2s\frac{\partial f}{\partial s}\right]_{\eta=0}\frac{\partial \theta_s}{\partial \eta_s} = \frac{\partial^2 \theta_s}{\partial \eta_s^2} \tag{10-73}$$

where

$$C = \frac{2QA_R}{\mu_e u_e^2 C_p T_e}, \qquad \theta_R = \frac{E_R}{R_u T_e}$$

$$\theta_p = \frac{E_p}{R_u T_e}, \qquad B = \frac{2A_R}{\mu_e u_e^2}$$

$$\beta = \left(\frac{\Pr}{\chi}\right)^{1/2} \frac{C_s}{C_p} (\rho_e \mu_e u_e L)^{1/2}$$

$$\chi = \frac{C_s \rho_s \lambda_s}{C_p \rho_e \lambda_e} \tag{10-74}$$

The transformed boundary conditions are as follows: At $\eta = 0$,

$$f + 2s\frac{\partial f}{\partial s} = -\frac{(2s)^{1/2}}{\rho_e \mu_e u_e}(\rho v)_w \tag{10-75}$$

$$\frac{\partial f}{\partial \eta} = 0 \tag{10-76}$$

$$\frac{\partial \theta}{\partial \eta} = \left(\frac{2\Pr \chi s}{\rho_e \mu_e u_e L}\right)^{1/2} \frac{\partial \theta_s}{\partial \eta_s} - \frac{L_v \Pr}{C_p T_e}\left(f + 2s\frac{\partial f}{\partial s}\right) \tag{10-77}$$

$$\frac{\partial Y_F}{\partial \eta} = \mathrm{Sc}\left(f + 2s\frac{\partial f}{\partial s}\right)(1 - Y_F) \tag{10-78}$$

$$\frac{\partial Y_O}{\partial \eta} = -\mathrm{Sc}\left(f + 2s\frac{\partial f}{\partial s}\right)Y_O \tag{10-79}$$

At $\eta = \infty$,

$$\frac{\partial f}{\partial \eta} = 1, \quad \theta = 1, \quad Y_O = Y_{O,e}, \quad Y_F = 0 \tag{10-80}$$

At $\eta = -\infty$,

$$\theta_s = 1 \quad \text{or} \quad \theta_s = \theta_s(\eta_s) \tag{10-81}$$

The initial condition along the s-coordinate is specified by the local-similarity solution[30] at some short distance from the leading edge.

7.2.5 Solution Method

STEP 1. Guess a surface temperature; iterate it until Eq. (10-77) is satisfied.

STEP 2. Replace the nonlinear terms in Eqs. (10-69) to (10-72) by first-order Taylor series expansion terms obtained from the previous solution.

STEP 3. Put Eq. (10-77) into an ordinary-differential-equation form in η by substituting a three-point formula for the first derivatives with respect to s, and solve for f subject to the boundary conditions (10-75), (10-76) and (10-80).

STEP 4. Cast the PDEs (10-70), (10-71), and (10-72) into finite-difference forms, based on the Crank–Nicholson scheme.

STEP 5. Invert the tridiagonal matrix form step 4 and solve for θ, Y_F, and Y_O.

The above procedure was repeated until a prescribed convergence was achieved for certain values of successive iterations. After all stations along s were calculated, the same procedure was repeated for another step in time, until ignition was reached.

For the definition of ignition, an ignition criterion must be selected. As discussed in previous sections, in general, a theoretical ignition criterion may not directly relate to an experimental one. In most experiments, the measurements of the ignition delay time are based on the detection of light emission. However, this criterion cannot easily be applied as the theoretical ignition criterion, unless detailed values of the physical parameters of the chemical-

TABLE 10.4 Values of Physical and Chemical Properties Used in the Calculations[a]

E_R (kcal/mole)	14.1, 16, 18
Order of reaction	2
n	6, 9, 12
$Q \times A_R$ (cal/cm^3 sec)	1.2×10^{11}
T_e (K)	1620
E_p (kcal/mole)	9.3, 10, 6.12
A_p (cm/sec)	100
L_v (cal/g)	−120
p (psi)	270, 400, 500
$\int_0^\infty$ RR dy (g/cm^2 sec) (ignition criterion)	3.1×10^{-3}
u_e (m/sec)	161, 275

[a]After Kashiwagi and Summerfield.[33]

reaction kinetics are well known. Since in fact they are only poorly known, Kashiwagi and Summerfield[33] selected the following ignition criterion depending upon the chemical-reaction rate:

$$\int_0^{\infty} (\text{reaction rate})\, dy \geq c^* \tag{10-82}$$

This criterion is close to the experimental criterion of a minimum detectable light intensity, seen by photomultipliers viewing the boundary layer. Values of physical and chemical properties used in the calculations are shown in Table 10.4.

7.3 Ignition and Flame-Spreading Results

Figure 10.30 shows that the effects of activation energies from theoretical calculations of both gas-phase reaction and pyrolysis are significant. With about 20% reduction of the gas-phase reaction activation energy, the ignition delay time is shortened by 40%, the first ignition position shifts upstream, and the flame-spreading rate decreases. With about 20% reduction of the pyrolysis reaction activation energy, the ignition delay time is shortened by 70%, the first ignition position shifts downstream, and the flame-spreading rate increases significantly. These results can be explained physically by postulating that the rate-controlling step in ignition and flame spreading is the fuel-supply process

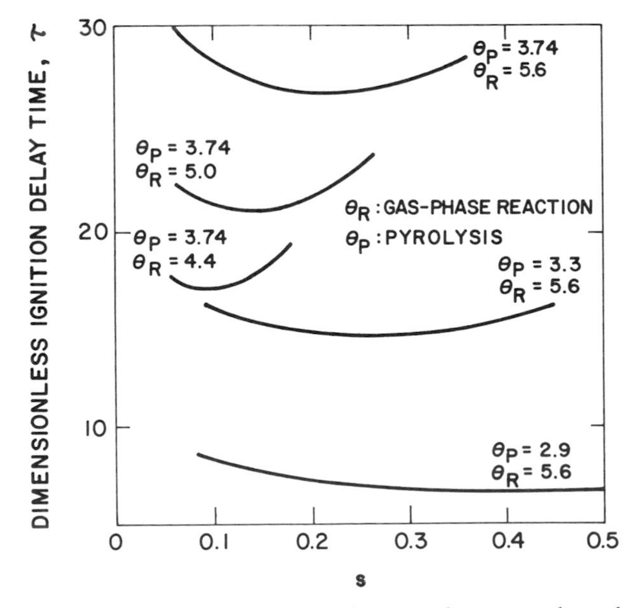

Figure 10.30 Effect of activation energies of gas-phase reaction rate and pyrolysis (U_e = 275 m/sec, $Y_{O,e}$ = 0.5) (after Kashiwagi et al.[33]).

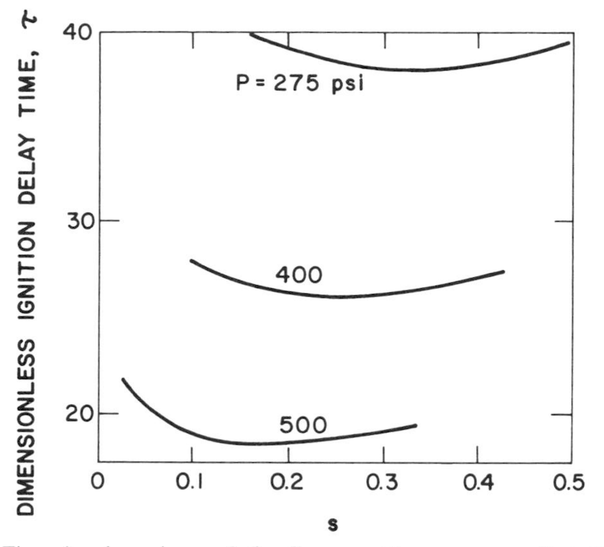

Figure 10.31 Timewise dependence of the flame position; pressure effect ($U_e = 275$ m/sec, $Y_{O,e} = 0.5$) (after Kashiwagi et al.[33]).

(i.e., pyrolysis process). No experimental data are available for comparison, since extreme complexity is encountered in chemical kinetics.

Figure 10.31 shows the calculated effect of the free-stream pressure on the ignition delay time. As the pressure increases, the ignition delay time decreases, the first ignition position shifts upstream, and the flame-spreading rate increases. These effects are probably due to the enhancement in the gas-phase reaction rate and the increase in heat transfer to the surface of the solid fuel. No experimental data were obtained for comparison.

Figure 10.32 shows the effect of inert diluents (He, N_2, Ar) according to theoretical calculations. The ignition delay times are in decreasing order for Ar–O_2, N_2–O_2, and He–O_2. The flame-spreading rates are in decreasing order for He–O_2, N_2–O_2, and Ar–O_2; the physical explanation is that the thermal diffusivities and thermal conductivities are in decreasing order for He–O_2, N_2–O_2, and Ar–O_2; therefore, the heat transfer from the free stream to solid phase is fastest in the He–O_2 stream.

In general, Kashiwagi et al.[30–33] found that the local-similarity model is just able to predict the trend of first ignition position. It fails to predict the ignition delay time and the flame-spreading rate. However, results from the nonsimilarity solution are fairly good in predicting the trends of ignition delay time, first ignition position, and flame-spreading rate. The ignition time is controlled by the fuel-supply process. In other words, low pyrolysis activation energy will make ignition easier. The thermal properties of the free-stream gas are found to be important for the rate of heat transfer to the solid phase during the preheating time. Neither the local-similarity model nor the nonsimilarity model

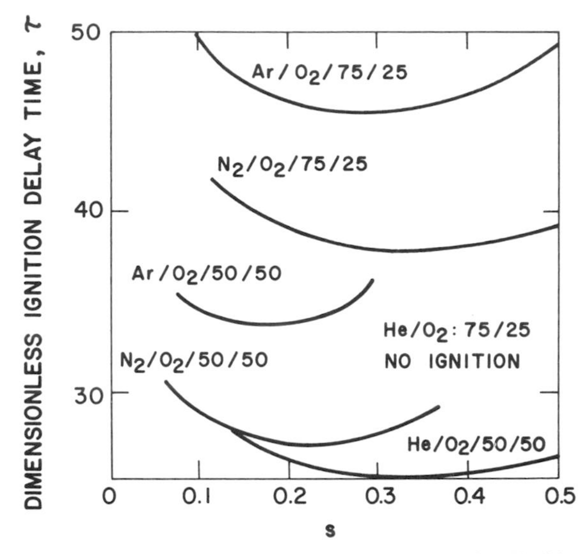

Figure 10.32 Effect of diluent inert gas (U_e = 275 m/sec) (after Kashiwagi et al.[33]).

can accurately predict the ignition delay time, the first ignition position, and the flame-spreading rate.

In order to improve the predictability of the theoretical model, some accurate kinetic data for solid-fuel pyrolysis and gas-phase reaction must be obtained and incorporated into the governing equations so that quantitative values for the ignition delay time, first ignition position, and flame-spreading behavior (rate and direction) can be determined. Also, it would be useful to measure the surface temperature of the solid fuel as a means of clarifying the degree of surface heat generation. Further refinement of the nonsimilarity model by inclusion of non-steady-state governing equations—particularly the energy, oxidizer-species, and fuel-species equations—may help to predict and explain the blowoff phenomenon observed experimentally.

REFERENCES

1. D. B. Spalding, *Combustion and Mass Transfer*, Chapter 15, pp. 272–293, Pergamon Press, New York, 1979.
2. F. J. Weinberg, *Optics of Flames*, Chapter 3, pp. 40–115, Butterworth & Co., London, 1963.
3. R. M. Hainer, "The Application of Kinetics to the Hazardous Behavior of Ammonium Nitrate," *Fifth Symposium (International) on Combustion*, pp. 224–230, Reinhold Publishing Corp., New York, 1955.
4. E. W. Price, H. H. Bradley, G. L. Dehority, and M. M. Ibiricu, "Theory of Ignition of Solid Propellants," *AIAA J.*, Vol. 4, pp. 1153–1181, September 1966.
5. A. K. Kulkarni, M. Kumar, and K. K. Kuo, "Review of Solid Propellant Ignition Studies," AIAA Paper 80-1210, July 1980; *AIAA J.*, Vol. 20, No. 2, pp. 243–244, 1982.

6. C. E. Hermance, "Solid Propellant Ignition Theories and Experiments," *Fundamentals of Solid Propellant Combustion*, AIAA Progress Series, Vol. 90 (K. K. Kuo and M. Summerfield, Eds.), October 1984.
7. L. De Luca, L. H. Caveny, T. J. Ohlemiller, and M. Summerfield, "Radiative Ignition of Double-Base Propellants: I. Some Formulation Effects," *AIAA J.*, Vol. 14, No. 7, pp. 940–946, July 1976.
8. L. De Luca, T. J. Ohlemiller, L. H. Caveny, and M. Summerfield, "Radiative Ignition of Double-Base Propellants: II. Pre-ignition Events and Source Effects," *AIAA J.*, Vol. 14, No. 8, pp. 1111–1117, August 1976.
9. R. S. Beyer, R. Anderson, R. O. MacLaren, and W. J. Corcoran, "Ignition of Solid-Propellant Motors under Vacuum," United Technology Center, Sunnyvale, Calif., Final Report UTC-2079-FR, Contract AF 04 (611) 9701, April 8, 1965.
10. R. K. Kumar and C. E. Hermance, "Gas Phase Ignition Theory of a Heterogeneous Solid Propellant," *Combustion Sci. and Technol.*, Vol. 4, pp. 191–196, 1972.
11. M. Kumar, J. E. Wills, A. K. Kulkarni, and K. K. Kuo, "A Comprehensive Model for AP-Based Composite Propellant Ignition," *AIAA J.*, Vol. 22, No. 4, pp. 526–534, 1984.
12. H. H. Bradley, Jr., *A Unified Theory of Solid Propellant Ignition*, NWC TP 5618, Pts. 1–3, Naval Weapons Center, China Lake, Calif., December 1975.
13. A. K. Kulkarni, M. Kumar, and K. K. Kuo, "A Comprehensive Ignition Model for Composite Solid Propellants," *18th JANNAF Combustion Meeting*, Pasadena, Calif., October 1981, CPIA Publication 347, Vol. III, pp. 215–223.
14. D. R. Ballal and A. H. Lefebvre, "The Influence of Flow Parameters on Minimum Ignition Energy and Quenching Distance," *15th Symposium (International) on Combustion*, pp. 1473–1481, The Combustion Institute, Pittsburgh, 1974.
15. D. R. Ballal and A. H. Lefebvre, "The Influence of Spark Discharge Characteristics on Minimum Ignition Energy in Flowing Gases," *Combustion and Flame*, Vol. 24, pp. 99–108, 1975.
16. D. R. Ballal and A. H. Lefebvre, "Ignition and Flame Quenching of Quiescent Fuel Mists," *Proc. R. Soc. London Ser. A*, Vol. 364, pp. 277–294, 1978.
17. H. N. Subba Rao and A. H. Lefebvre, "Ignition of Kerosene Fuel Sprays in a Flowing Air Stream," *Combustion Sci. and Technol.*, Vol. 8, pp. 95–100, 1973.
18. C. Godfrey, *An Assessment of a Light-Scattering Technique for Drop-Size Measurement*, M. Sc. thesis, Cranfield Institute of Technology, Cranfield, U.K., 1969.
19. C. C. Swett, "Spark Ignition of Flowing Gases Using Long Duration Discharges," *Sixth Symposium (International) on Combustion*, p. 523, Reinhold Publishing Corp., New York, 1957.
20. D. R. Ballal, *Ignition of a Turbulent Gaseous Mixture*, Ph.D. thesis, Cranfield Institute of Technology, Cranfield, U.K., 1972.
21. A. H. Lefebvre, "An Evaporation Model for Quenching Distance and Minimum Ignition Energy in Liquid Fuel Sprays," presented at Fall Meeting of the Combustion Institute (Eastern Section), 1977.
22. D. R. Ballal and A. H. Lefebvre, "Ignition and Flame Quenching in Flowing Gaseous Mixtures," *Proc. Roy. Soc. Ser. A*, Vol. 357, pp. 163–181, 1977.
23. K. V. L. Rao and A. H. Lefebvre, "Minimum Ignition Energies in Flowing Kerosene-Air Mixtures," *Combustion and Flame*, Vol. 27, pp. 1–20, 1976.
24. J. S. Chin, *The Analysis of the Effect of Oxygen Addition on Minimum Ignition Energy*, AIAA paper 82-1160, 1982.
25. M. K. King, "Ignition of Boron Particles and Clouds," *J. Spacecraft*, Vol. 19, No. 4, pp. 294–306, July–August 1982.

26. A. Macek and J. M. Semple, "Combustion of Boron Particles at Atmospheric Pressure," *Combustion Sci. and Technol.*, Vol. 1, pp. 181–191, 1969.
27. M. A. Gurevich, I. M. Kir'yanov, and E. S. Oserov, "Combustion of Individual Boron Particles," *Combustion, Explosion, and Shock Waves*, Vol. 5, No. 2, pp. 150–153, 1969.
28. G. Mohan and F. A. Williams, "Ignition and Combustion of Boron in O_2/Inert Atmospheres," *AIAA J.*, Vol. 10, No. 6, pp. 776–783, 1972.
29. I. Glassman, F. A. Williams, and P. Antaki, "A Physical and Chemical Interpretation of Boron Particle Combustion," *19th JANNAF Combustion Meeting*, CPIA Pub. 366, Vol. 1, pp. 63–72, October 1982.
30. T. Kashiwagi, B. W. MacDonald, H. Isoda, and M. Summerfield, "Ignition of a Solid Polymeric Fuel in a Hot Oxidizing Gas Stream," *Thirteenth Symposium (International) on Combustion*, pp. 1073–1085, The Combustion Institute, Pittsburgh, 1971.
31. T. Kashiwagi, G. G. Kotia, and M. Summerfield, "Experimental Study of Ignition and Subsequent Flame Spread of a Solid Fuel in a Hot Oxidizing Gas Stream," *Combustion and Flame*, Vol. 24, pp. 357–364 (1975).
32. T. Kashiwagi, *Ignition of Solid Polymeric Fuels by Hot Oxidizing Gases*, Ph.D. thesis, Princeton University, 1970.
33. T. Kashiwagi and M. Summerfield, "Ignition and Flame Spreading over a Solid Fuel: Non-similar Theory for Hot Oxidizing Boundary Layer," *Fourteenth Symposium (International) on Combustion*, pp. 1235–1247, The Combustion Institute, Pittsburgh, 1973.

PROJECT

The potential of boron as solid fuel in propulsion systems has long been recognized, because it has extremely high heat of combustion (57.6 MJ/kg). Solid boron particles could be introduced into a combustor either as components of slurries or as ingredients of solid propellants. For a simplified case, let us consider the ignition and combustion of a single spherical boron particle in a gaseous oxidizing environment as shown in the accompanying figure.

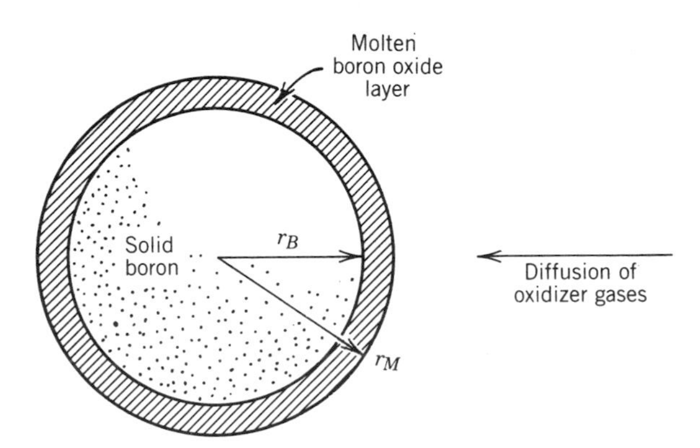

To help your theoretical formulation of this problem, the following phenomena based upon experimental observations are given:

1. Boron particles generally enter the hot combustor environment as low-temperature solids with an extremely thin solid boron oxide coating on the order of 10 Å.
2. Heat transfer from the gas causes the particle temperature to rise, with the oxide coating melting at roughly 720 K.
3. Boron and/or oxidizer gas diffuse across the oxide layer and tend to react more rapidly as particle temperature increases.
4. As reaction rates increase following the first-stage ignition, the luminosity increases.
5. As the oxide layer thickens, the rates of reactant diffusion reduce, and reaction rate slows down.
6. Further increase of particle temperature causes rates of evaporation of the relatively volatile oxide to increase. This in turn reduces the thickness of the oxide layer.
7. Eventually a condition is reached (~ 1900 K) where rates of oxide evaporation are sufficiently large to remove most of the oxide layer. At this point, a second-stage ignition is observed and relatively rapid boron oxidation follows.
8. If ambient temperatures or rates of reaction are sufficiently high, the boron particle melts (~ 2450 K).
9. It has been found that both liquid-phase B_2O_3 and gas-phase HOBO are formed from the heterogeneous reactions in wet flames according to

$$2B + 2O_2 + H_2 \rightarrow 2HOBO_{(g)} \qquad \Delta H_r = -1151 \text{ kJ/mol}$$

$$\rightarrow B_2O_{3(l)} \qquad \Delta H_r = -1468 \text{ kJ/mol}$$

Approximately one-third of the energy release is involved in the generation of $HOBO_{(g)}$. The rest is in the generation of $B_2O_{3(l)}$.

Assignment

1. Use the information given above to formulate a transient one-dimensional theoretical model for boron particle ignition and combustion. Write the governing equations, boundary and initial conditions or the condensed-phase boron, molten boron oxide layer, and the gas surrounding the particle. List your assumptions.

2. Sketch a set of your anticipated profiles of concentrations and temperature as functions of radius.

APPENDIX

Constants and Conversion Factors Often Used in Combustion Problems

Universal gas constant:

$$R_u = 1554 \frac{\text{ft lb}_f}{\text{lb}_m\text{-mole °R}} = 1.9872 \frac{\text{Btu}}{\text{lb}_m\text{-mole °R}}$$

$$= 1.9872 \frac{\text{cal}}{\text{g-mole K}} = 0.08206 \frac{\text{atm liter}}{\text{g-mole K}}$$

$$= 8.3166 \times 10^7 \frac{\text{erg}}{\text{g-mole K}}$$

$$= 82.06 \frac{\text{cm}^3\text{ atm}}{\text{g-mole K}} = 8314.4 \frac{\text{J}}{\text{kg-mole K}} = 84{,}786.85 \frac{\text{g}_f\text{-cm}}{\text{g}_m\text{-mole K}}$$

$$= 0.729 \frac{\text{ft}^3\text{ atm}}{\text{lb}_m\text{-mole °R}} = 10.71 \frac{\text{psi ft}^3}{\text{lb}_m\text{-mole °R}}$$

$$= 8.313 \frac{\text{J}}{\text{g-mole K}}$$

Standard acceleration of gravity:

$$g_c = 32.174 \frac{\text{ft lb}_m}{\text{lb}_f\text{ sec}^2} = 1 \frac{\text{g cm}}{\text{dyne sec}^2} = 1 \frac{\text{kg}_m\text{ m}}{\text{N sec}^2}$$

$$= 1 \frac{\text{slug ft}}{\text{lb}_f\text{ sec}^2} = 980.665 \frac{\text{g cm}}{\text{g}_f\text{ sec}^2}$$

Avagadro's number:

$$\tilde{N} = 6.0254 \times 10^{23} \frac{\text{molecules}}{\text{g-mole}}$$

Stefan–Boltzmann constant:

$$\sigma = 5.6699 \times 10^{-5} \frac{\text{erg}}{\text{cm}^2 \text{ sec K}^4}$$

$$= 1.35514 \times 10^{-12} \frac{\text{cal}}{\text{cm}^2 \text{ sec K}^4}$$

Work/energy conversion factor:

$$J = 778 \frac{\text{ft lb}_f}{\text{Btu}} = 42664.9 \frac{\text{g}_f \text{ cm}}{\text{cal}}$$

Energy units:

$$1 \text{ cal} = 4.18585 \text{ J} = 0.003974 \text{ Btu}$$

Author Index

Subject Index

About The Author

Professor Kenneth K. Kuo joined the Mechanical Engineering Department of The Pennsylvania State University in September of 1972. He received his B.S. from National Taiwan University, his M.S. from the University of California-Berkeley, and his Ph.D. from Princeton University. His special interest lies in the field of combustion, propulsion, and ballistics, involving solid propellant combustion and two-phase flows. Besides this book, he was the co-editor of a recent AIAA Progress Series Volume on "Fundamentals of Solid Propellant Combustion." He has been the principal investigator of many large research contracts, and has received various honors for the quality of his research work. He has published over 60 papers in the combustion field, and has presented numerous invited lectures around the world. Dr. Kuo has developed two major laboratories, a High Pressure Combustion Laboratory and a CO_2 Laser Laboratory. These labs contain many state-of-the-art equipment and test facilities. These labs represent one of the best-equipped University facilities for combustion and propulsion studies in the U.S.A. Dr. Kuo has also been very active in developing several new graduate level combustion courses. He has recently been awarded the title of "Distinguished Alumni Professor of Mechanical Engineering." His family includes his wife, Olivia, and two daughters.